英汉塑料技术词汇

AN ENGLISH – CHINESE DICTIONARY OF PLASTICS TECHNOLOGY

钱德基 刘好菊 编

中国石化出版社

内 容 提 要

本词典内容涉及塑料原料、生产工艺、高分子材料、塑料加工和机械设备、塑料助剂、塑料焊接和技术、增强及复合材料等，内容丰富，涉及面广，实用价值强，吸收了国外先进技术、专利文献，是塑料领域的科技人员阅读英文资料时的必备工具书。

图书在版编目(CIP)数据

英汉塑料技术词汇：英汉对照/钱德基，刘好菊编
—北京：中国石化出版社，2015.5
ISBN 978-7-5114-3156-1

Ⅰ.①英… Ⅱ.①钱… ②刘… Ⅲ.①塑料-词汇-英、汉 Ⅳ.①TQ320.6-61

中国版本图书馆 CIP 数据核字(2015)第 088437 号

未经本社书面授权，本书任何部分不得被复制、抄袭，或者以任何形式或任何方式传播。版权所有，侵权必究。

中国石化出版社出版发行
地址：北京市东城区安定门外大街58号
邮编：100011　电话：(010)84271850
读者服务部电话：(010)84289974
http://www.sinopec-press.com
E-mail:press@sinopec.com
北京柏力行彩印有限公司印刷
全国各地新华书店经销

*

850×1168 毫米 32 开本 18.25 印张 917 千字
2015 年 7 月第 1 版　2015 年 7 月第 1 次印刷
定价：98.00 元

前　言

为适应塑料工业的飞速发展，吸收国外先进技术，我们着手汇编《英汉塑料技术词汇》，供从事塑料行业的技术人员阅读英文文献资料时参考。

本词汇搜集塑料专业及相关专业术语近 47000 条，内容涉及塑料原料、生产工艺、高分子材料、塑料加工和机械设备、塑料助剂、塑料焊接和技术、增强及复合材料等。

由于水平有限，错误在所难免，恳切欢迎读者批评指正。

编者

使 用 说 明

一、同一英文词,有数种释义时,词义相同或相近者用","号分开,词义不同者用";"号分开。

二、英文条目均按字母顺序编排。为了查找方便,复合词组均归纳在一起,组成一系列词组,与上列重复的单词用"~"号代替,如:

 plastic   塑料
 ~ film   塑料薄膜
 ~ welding   塑料焊接

三、本词汇中,可省略的释义用圆括号( )表示,可替换的释义用方括号[ ]表示,注释性的词用尖括号〈 〉表示。

目 录

字典正文 ……………………………………………… 1~566
附录一 聚合物常用缩写词 ……………………………… 567-570
附录二 塑料专业术语 …………………………………… 571-572
附录三 聚合物的容度参数表 …………………………… 573-574
附录四 塑料专业常用法定计量单位换算表 …………… 575-576

A

A - polymer [addition polymer]　加聚物
A - polymerization [addition polymerization]　加成聚合
A - stage　甲阶段,A - 阶段,可溶可熔阶段〈酚醛树脂〉
A - stage resin　甲阶酚醛树脂,可溶可熔酚醛树脂
A - type calender　A 型压延机,三角形三辊压延机
AAS [acrylonitrile acrylate styrene copolymer]　丙烯腈 - 丙烯酸酯 - 苯乙烯三元共聚物
abandon　放弃,抛弃
abate　减少,降低;消除;废除
abatement　减少,降低;消除;废除;废料
abbe number　色散系数
abbr. [abbreviation]　缩写
abbreviation　缩写
ABFA [azobisformamide]　偶氮二甲酰胺,发泡剂 AC
abherent　阻黏剂〈表面〉,防黏剂〈塑料薄膜〉,脱模剂〈塑模〉
abhesion　失黏,脱黏〈表面〉
abhesive　防黏剂;防黏材料
～ layer　防黏层,防黏面
ability　能力,性能
ablation　烧蚀;脱离;切除
～ polymer　烧蚀聚合物
～ rate　烧蚀速度
～ resistance　耐烧蚀性
～ velocity　烧蚀速度
ablative　烧蚀材料;烧蚀的
～ material　烧蚀材料
～ performance　烧蚀性能
～ plastics　烧蚀塑料
～ polymer　烧蚀聚合物
～ property　烧蚀性能
ablator　烧蚀材料,烧蚀剂
abnormal　反常的,不规则的

～ density　反常密度
～ dispersion　反常色散
～ fibres　异状纤维
～ temperature　异常温度
above　高于,在…之上;在上面;上述的
～ ground installation　在地面(上)铺设〈管〉
ABR [acrylate - butadiene rubber]　丙烯酸酯 - 丁二烯橡胶
abradability　磨耗性,磨损度
abradant　研磨剂,磨料
abrade　磨(损,耗),研磨
abrader　磨损试验机;研磨机,磨光机
abrading　磨耗;磨光
～ device　磨光机,研磨机
abrasion　磨耗,磨损,磨蚀
～ chamber　研磨室
～ coefficient　摩擦系数
～ cycle　磨耗周期
～ disc method　盘状研磨
～ hardness　磨耗硬度
～ index　磨损指数
～ loss　磨耗量,磨耗损失,磨耗减量
～ loss of gloss　光泽磨损
～ machine　磨损机
～ mark　擦伤痕
～ pattern　磨纹
～ proof　耐磨(的)
～ quality　耐磨性
～ resistance　耐磨性
～ resistance index　耐磨指数
～ resistant surface　耐磨面
～ resisting phenolic moulding powder　耐磨酚醛模塑粉
～ service test　磨损应用试验
～ strength　耐磨强度
～ test　磨耗试验,磨损试验
～ tester　磨耗试验机
～ - testing machine　磨损试验机

abrasive
- ~ value 磨耗值
- ~ wear 磨耗(量),耐磨性

abrasive 磨料;研磨剂;研磨的
- ~ and cutting-off machine 研切机
- ~ belt 砂带
- ~ binder 磨料粘结剂
- ~ cloth 砂布,金刚砂布
- ~ coated paper 磨涂纸
- ~ composition 磨料,磨蚀剂
- ~ disc 研磨盘
- ~ fabric 砂布
- ~ filler 磨料
- ~ finishing 打磨修整,磨光
- ~ finishing machine 抛光机,磨光抛光机
- ~ grain 研磨粉,研磨粒
- ~ grinding wheel 磨轮
- ~ hardness 研磨硬度,耐磨硬度
- ~ machining 磨削加工
- ~ material 磨料
- ~ paper 砂纸,金刚砂纸
- ~ particle 磨粒
- ~ powder 磨粉
- ~ resistance 耐磨性;耐磨牢度
- ~ substance 研磨材料
- ~ test 磨损试验
- ~ wear 磨损,磨耗〈美〉
- ~ wheel 砂轮
- ~ wheel binder 砂轮黏合剂

abrasiveness 磨耗性
abrasor 磨损试验机
abrupt failure 猝裂
abruption 断裂,破裂
- ~ test 断裂试验

Abs. [absolute] 绝对的;纯的
ABS [acrylonitrile-butadiene-styrene]
ABS共聚物,丙烯腈-丁二烯-苯乙烯三元共聚物

abscissa 横坐标
absolute 绝对的;纯粹的
- ~ atmosphere 绝对大气压
- ~ brightness threshold 绝对亮度阀
- ~ compliance 绝对柔量
- ~ density 绝对密度
- ~ dielectric constant 绝对介电常数
- ~ dry weight 绝对干重
- ~ dynamic modulus 绝对动态模量
- ~ elongation 绝对伸长
- ~ error 绝对误差
- ~ humidity 绝对湿度
- ~ intensity 绝对强度
- ~ judgement 绝对判断,无可比物
- ~ modulus 绝对模量
- ~ peel strength 绝对剥离强度〈粘接〉
- ~ pressure 绝对压力
- ~ scaling 绝对定标
- ~ specific gratively 真密度,绝对相对密度
- ~ temperature 绝对温度
- ~ value 绝对值
- ~ velocity 绝对速度
- ~ viscosity 绝对[动态,动力]黏度
- ~ zero 绝对零度

absorb 吸收
- ~ fillers 添加填充剂,添加填料

absorbability 吸收性
absorbance 收吸度;吸光度
absorbed 吸收的
- ~ dose 吸收剂量
- ~ dose rate 吸收剂量速率
- ~ energy 吸收能

absorbency 吸收能力
absorbent 吸收剂;吸音材料;能吸收的
- ~ pack 吸收包装
- ~ power 吸收能力
- ~ quality 吸收性;吸湿性
- ~ resin 吸附树脂

absorber 吸收剂;吸收器;过滤器;缓冲器,减震器
absorbing 吸收的
- ~ glass-fibre material 浸渍(树脂)的玻璃纤维材料
- ~ medium 减震材料,减震介质
- ~ power 吸收本领

absorptance 吸收性
absorption 吸收
- ~ band 吸收光谱带
- ~ behaviour 吸收性能

~ capacity 吸收能力
~ cell 吸收池
~ cell with charcoal filling 带有木炭填料的吸收池
~ chamber 吸收室
~ coefficient 吸收系数
~ column 吸收柱
~ curve 吸收曲线
~ dynamometer 吸收功率计
~ fabric 吸收性织物
~ factor 吸收率
~ peak 吸收峰值
~ power 吸收能力
~ rate 吸收率
~ ray 吸收射线
~ spectrometer 吸收分光计
~ spectrometry 吸收光谱测定
~ spectroscopy 吸收光谱法
~ spectrum 吸收光谱
~ test 吸收试验
~ value 吸收值
~ factor 吸收能力［率，系数］
absorptivity 吸光率
~ coefficient 消光系数
abstr. ［abstract］ 文摘，摘ași
abstraction 分离；除去
abt. ［about］ 大约；关于
abund. ［abundant］ 丰富的，大量的
abundant 丰富的，大量的
abutting 互相结合［连接］
abv. ［above］ 上文，上述
Ac ［acetyl radical］ 乙酰基
ac. ［acid］ 酸〈的〉
Ac ［Actinium］ 锕〈元素符号〉
Ac ［alternating current］ 交流电(流)
Ac current 交流电(流)
AC permeability 透交流电性，交流导电率
acaroid resin 禾木树脂
acc. ［according］ 依照，相应，根据
ACCCE ［Association of Consulting Chemists and Chemical Engineers］ 顾问化学家和化学工程师协会
accelerant 促进剂

accelerated 加速的
~ ageing 加速老化
~ ageing enviroment 加速老化的环境
~ ageing test 加速老化试验
~ ageing test by oxygen pressure method 加压氧气加速老化试验
~ creep 加速蠕变
~ indoor method 室内加速试验法
~ light ageing test 加速光老化试验
~ outdoor exposure test 户外加速暴露试验
~ oxidation test 加速氧化试验
~ ozone ageing test 加速臭氧老化试验
~ resin （加有促进剂的）速固化树脂
~ test 加速试验
~ weathering resistance 加速耐气候性
~ weathering test 加速耐候试验
accelerating 加速的
~ agent 促进剂
~ plastic flow 加速塑性流动
accelerative thickening 加速稠化
accelerator 促进剂；加速器
~ activator 促进活化剂
acceptable 容许的；合格的；验收的；令人满意的
~ daily intake ［ADI］ 允许日剂量
~ life 有效使用寿命
~ quality level 验收质量标准
acceptance 验收；认可
~ certificate 验收单
~ check 验收，合格验收
~ condition 合格条件；验收条件
~ level 验收标准
~ sampling 验收取样
~ test 验收试验
accepted 验收的；合格的
~ product 合格产品
~ stock 精炼物料，优质物料；合格浆料
accessible 相容的，可混的
accessory 辅助的，从属的，副的；次要的
~ material 辅助材料
~ pigment 辅助色料
~ substance 副产物

accommodation 适应,调节
　~ coefficient 调节系数
accommodator 调节器
accompanying substance 伴生物质
accumulate blow 储料吹塑,积聚吹塑
accumulater type blow mo(u)lding machine 储料器型吹塑成型机
accumulation chamber 储料室〈吹塑〉
accumulator 储料器;储料缸;储压器;蓄压器;蓄电池
　~ battery 蓄电池
　~ box 蓄电池箱,蓄电池槽
　~ case 蓄电池外壳
　~ chamber 储料罐内腔
　~ cylinder 储料罐〈吹塑机〉
　~ head 储料罐式机头〈吹塑机〉
　~ jar 储料容器
　~ plunger 储料罐〈推〉料杆
　~ process 储料罐法;熔体注塑发泡法〈发泡〉
　~ separator 蓄电池隔板
　~ system 储料罐系统
　~ tank 储[蓄]料槽;蓄电池槽
　~ vessel 蓄电池容器
accuracy 准确(度);精确度
　~ in gauge 测量准确度
　~ in measurement 测量准确度
　~ in measuring 测量准确性
　~ in size 尺寸准确度
　~ of the mean 平均值准确度
accurate to dimensions 尺寸精确的,符合加工尺寸
acetal 聚甲醛;乙缩醛;缩醛
　~ copolymer 聚甲醛共聚物
　~ fibre 缩醛纤维
　~ homopolymer 聚甲醛均聚物
　~ plastics 缩醛塑料;聚甲醛塑料
　~ resin 缩醛树脂;聚甲醛树脂
acetaldehyde 乙醛
　~ resin 乙醛树脂;聚甲醛树脂
acetate 乙酸盐[酯],醋酸盐[酯]
　~ butylate 乙[醋]酸丁酯
　~ fibre 乙[醋]酸纤维
　~ film 乙[醋]酸纤维(素)薄膜
　~ rayon 乙[醋]酸人造丝
　~ staple fibre 乙[醋]酸短纤维
acetic 乙(酸)的,醋酸的
　~ acid 乙酸,醋酸
　~ acid yield 乙酸含量
　~ aldehyde 乙醛
　~ anhydride 乙酸酐,醋(酸)酐
　~ peroxide 乙酰化过氧,过氧化乙酰
acetoguanamine 乙酰胍胺
　~ resin 乙酰胍胺树脂
acetone 丙酮
　~ extraction 丙酮萃取,丙酮抽提
　~ formaldehyde resin 丙酮甲醛树脂
　~ immersion test 丙酮浸泡试验
　~ resin 丙酮树脂
　~ soluble matter 丙酮可溶物
acetonitrile 乙腈〈溶剂〉
acetyl 乙酰〈基〉
　~ butyl cellulose 乙酰丁纤维素
　~ cellulose 乙酰纤维素,乙[醋]酸纤维素
　~ cellulose sheet 乙酰纤维素薄片
　~ fibre 乙酰化纤维
　~ radical [Ac] 乙酰基
　~ tri-n-butyl citrate 乙酰柠檬酸三正丁酯〈增塑剂〉
　~ triethyl citrate 乙酰柠檬酸三乙酯〈增塑剂〉
　~ tri-(2-ethylhexyl) citrate 乙酰柠檬酸三(2-乙基己酯)〈增塑剂〉
　~ tri-n-octyl(n-decyl citrate) 乙酰柠檬酸三(正辛正癸酯)〈增塑剂〉
　~ value 乙酰值
acetylacetone 乙酰丙酮
acetylating agent 乙酰化剂
acetylation 乙酰化
N-acetylethanolamine N-乙酰乙醇胺
acetylene 乙炔,电石气
　~ black 乙炔〈烟〉黑,乙炔炭黑〈填料〉
　~ derivative 乙炔衍生物
　~ polymer 乙炔聚合物
　~ tetrabromide 四溴化乙炔〈即均四溴乙烷,催化剂〉

acid. [to acidify] 酸化;使发酵
acid 酸(的)
 ~ acceptor 酸性中和剂
 ~ amide 酰胺
 ~ anhydride 酸酐
 ~ azo dyes 酸性偶氮染料
 ~ boiking test 酸煮沸试验
 ~ catalysis 酸催化
 ~ catalyst 酸催化剂
 ~ catalyst polycondensation 酸催化缩聚作用
 ~ catalyzed lacquer 酸催化漆
 ~ catalyzed resin 酸催化树脂
 ~ chrome dyes 酸性铬〈媒〉染料
 ~ content 含酸量
 ~ cure resin 酸固化树脂
 ~ dyes 酸性染料
 ~ egg 酸蛋,蛋形升酸器
 ~ etched 酸蚀的
 ~ fast 耐酸的
 ~ fastness 耐酸牢度,耐酸性
 ~ forming 成酸的
 ~ fume scrubber 酸烟洗涤器
 ~ hardening 酸硬化的
 ~ imide 酰亚胺
 ~ index 酸度指数
 ~ number 酸值
 ~ pigments 酸性颜料
 ~ proof 耐酸性
 ~ -proof adhesive 耐酸黏合剂
 ~ -proof cement 耐酸粘接剂
 ~ -proof coating 防酸保护,防酸面层
 ~ purification [acid washing] 酸洗
 ~ radical 酸基;酰基
 ~ resistance 耐酸性
 ~ resistant coating 耐酸涂层
 ~ resisting 耐酸(性)的
 ~ resisting paint 耐酸涂料
 ~ separator 酸分离器
 ~ test 酸性试验
 ~ value 酸值
acidic ion exchange resin 酸性离子交换树脂
acidification 酸化
 ~ tower 酸塔
acidify 酸化;使发酵
acidimeter [acidometer] 酸(液)比重计,酸度计
acidity 酸性;酸度
 ~ index 酸度指数
acidn. [acidification] 酸化
ACM elastomer 聚丙烯酸酯弹性体
acme thread 梯形螺纹
acoustic(al) 传声音的
 ~ absorption 消声,吸音
 ~ board 吸音板
 ~ coating 消音涂层,消音涂料
 ~ flooring insulation 吸音地板,隔音地板
 ~ foam 吸音泡沫(塑料)
 ~ insulating material 隔音材料
 ~ insulation 消声,隔音(层、材料)
 ~ insulation board 消声板,隔音板
 ~ material 消声材料,隔音材料
 ~ panel 吸音板
 ~ spectrametry 声谱测定法
 ~ spectrum 声谱
acridine dyes 吖啶染料
acrolein 丙烯醛
 ~ resin 丙烯醛树脂
acronym 缩写词
across 跨越,横过,横断;交叉
 ~ cutting 横切,横向切割
 ~ direction 横向
 ~ machine 横向机(器)
acrylate 丙烯酸盐[酯]
 ~ acrylonitrile copolymer 丙烯酸酯-丙烯腈共聚物
 ~ butadiene rubber [ABR] 丙烯酸酯-丁二烯橡胶
 ~ resin 丙烯酸酯树脂
 ~ resin adhesive 丙烯酸酯树脂黏合剂
acrylic 丙烯酸(类);丙烯酸系衍生物;聚丙烯腈系纤维
 ~ acid 丙烯酸
 ~ copolymer 丙烯酸系共聚物
 ~ elastomer 丙烯酸系弹性体
 ~ ester 丙烯酸酯

acrylics

~ fibre 丙烯酸系纤维
~ foam 丙烯酸系泡沫塑料
~ plastics 丙烯酸系塑料
~ resin 丙烯酸系树脂
~ resin coating 丙烯酸系树脂涂料
~ rubber [AR] 丙烯酸系橡胶
~ - styrene - acrylonitrile copolymer [ASA] 丙烯酸酯-苯乙烯-丙烯腈共聚物
~ - styrene - copolymer 丙烯酸脂-苯乙烯共聚物
~ system strongly basic anion exchange resin 丙烯酸系强碱性阴离子交换树脂

acrylics 丙烯酸系树脂衍生物
acrylonitrile [AN] 丙烯腈

~ acrylate copolymer 丙烯腈-丙烯酸酯共聚物
~ acrylate styrene copolymer [AAS] 丙烯腈-丙烯酸酯-苯乙烯共聚物
~ - butadiene copolymer 丙烯腈-丁二烯共聚物
~ - butadiene rubber [NBR] 丙烯腈-丁二烯橡胶-丁腈橡胶
~ - butadiene - styrene [ABS] 丙烯腈-丁二烯-苯乙烯(共聚物), ABS(共聚物)
~ - chlorinated polyethylene - styrene [ACS] 丙烯腈-氯化聚乙烯-苯乙烯(共聚物), ACS(共聚物)
~ - chlorinated polyethylene - styrene copolymer [ACS] 丙烯腈-氯化聚乙烯-苯乙烯共聚物, ACS 共聚物
~ isoprene rubber [NIR] 丙烯腈-异戊二烯橡胶
~ - itaconic acid ester copolymer fibre 丙烯腈-衣康酸酯共聚物纤维
~ methyl methacrylate [AMMA] 丙烯腈-甲基丙烯酸甲酯(共聚物)
~ methyl methacrylate plastics 丙烯腈-甲基丙烯酸甲酯塑料
~ starch copolymer 丙烯腈-淀粉共聚物
~ - styrene copolymer [AS] 丙烯腈-苯乙烯共聚物
~ - vinylcarbazole - styrene copolymer [AVCS] 丙烯腈-乙烯咔唑-苯乙烯共聚物
~ - vinyl chloride copolymer 丙烯腈-氯乙烯共聚物

ACS [acrylonitrile - chlorinated polyethylene - styrene] 丙烯腈-氯化聚乙烯-苯乙烯(共聚物)
ACS [acrylonitrile - chlorinated polyethylene - styrene copolymer] 丙烯腈-氯化聚乙烯-苯乙烯共聚物
ACS [thermoplastic blend of a copolymer from acrylonitrile and styrene with chlorinated polyethylene] 丙烯腈-苯乙烯共聚物与氯化聚乙烯的热塑性共混物
act. [active] 活性的;有效的
act 法规,条例;作用,行动;对…起作用
Actinium, [Ac] 锕〈元素符号〉
action 活(性)化
activated 活化的;活性的;有效的

~ anionic polymerization 活性阴离子聚合(作用),催化阴离子聚合
~ carbon 活性炭
~ carbon method 活性炭法
~ lacquer 活性漆
~ pigments 活性颜料
~ polymerization 活化聚合(作用)
~ resin 活性化树脂

activation 活化,激活

~ energy 活化能
~ temperature 活化温度
~ time 活化时间
~ zone 活化区(段)

activator 活化剂,促进剂;硬化剂;催化剂〈发泡〉

~ mixture 活化剂混合物

active 活性的;有效的,实际的

~ anvil 活动砧
~ area 工作面积,有效面积
~ carbon 活性炭
~ carbon for the mercuric chloride 氯汞用活性炭〈催化剂〉

~ carbon for the zinc acetate 乙酸锌用活性炭〈催化剂〉
~ centers 活性中心
~ current 有效电流
~ filler 活性填料
~ force 作用力,有效力
~ group 活性基团
~ ingredient 活性组分
~ length 有效长度
~ life 使用寿命,使用期限
~ plasticizer 活性增塑剂
~ power 有效功率
~ solvent 有效溶剂;活性溶剂
~ volume 有效体积
activity 活动;活性;作用;功率
~ of a catalyst 催化剂活性
actual 实际的,有效的
~ cavity fill time 模腔填充(料)实际时间,横腔(实际)充模时间
~ crystal 实际晶体
~ draft 实际牵伸
~ efficiency 实际效率
~ fill time of the mo(u)ld 模具填充(料)实际时间,模具(实际)充模时间
~ joint 接缝
~ parameter 实际参数
~ stress 有效应力
~ value 实值
~ weight [AW] 实际重量
actuate 操纵,控制
actuated ball valve 自动球阀
actuator 调节器;传动装置
~ rod 加料杆;自动动作装置杆
acuity 敏锐(度)
acylation 酰化作用
acyl group 酰基
ad. [advertisement] 广告,公告
AD [air-dry] 空气干燥〈漆〉
Adams' chromatic value diagram 阿达姆色值图
adapt 适应,适合
adaptability 适应性
adaptable 可适应的

adaptation 适应
adapted 适合的,恰当的
adapter [adaptor] 接头,接套(挤塑机)
~ bearing 带卡紧接套的轴承
~ block 模头连接器
~ bush 接套〈管〉
~ coextrusion 接套式共挤塑
~ insert 插入接套
~ plate 载模板;合模板
~ plate area 模板面积
~ ring 模头连接圈,模头接套,口模接套
~ sleeve 卡紧套,固定套
adaptor [adapter] 连接,连接件;配件;连接管
~ disc 连接垫片
~ nipple 对接头
~ socket 异径接头
~ union 螺丝紧旋连接
ADC [allyl diglycol carbonate] 烯丙基二甘醇碳酸酯
add [addendum] 附加物
add 加成〈化学〉;辅助挤塑〈挤塑机〉;加〈数学〉;添加,加入〈塑料〉
addendum 附加物;附录
adding 加成〈化学〉
addition 加成,加添
~ agent 添加剂,助剂
~ compound 加(成化)合物
~ condensation 加成缩合
~ energy 结合能
~ polycondensation 加成缩合作用
~ polymer 加聚物
~ polymerization 加聚作用
~ power 加合能力,加成本领
additional 附加的,补充的
~ equipment 附加设备,辅助设备
~ heating 补偿加热
~ heating system 补偿加热系统〈流道〉
~ surface treatment 表面补偿处理,表面后处理〈粘接面〉
~ thermal treatment 加热后处理,加热补偿处理

additive

~ treatment 补偿处理,后处理
additive 添加剂;加成的
~ blooming 添加剂渗出
~ compound 加(成化)合物
~ copolymerization 加成共聚合
~ extrusion 添加剂渗出
~ for adhesive 黏合剂用添加剂
~ mixture of colours 加色混合
~ plasticizer 外加增塑剂
~ resin 加成树脂
addn. [addendum] 附加物
addn. [addition] 加成,加入,附加
addnl. [additional] 附加的,额外的
adequate stimulus 适当刺激
adh. [adhesive] 黏合剂;附着的
adhere 黏附,附着
adherence 黏附;粘接性;黏着(力)
adherend 被黏物,黏附体,黏着物
~ failure [fracture] 黏附体破损[破裂]
~ preparation 黏附体配制
~ surface 黏附体表面
adherent 黏着;黏合
~ surface 黏附表面(积),粘接表面(积)
adhering 黏附,黏着
~ joint 粘接缝
adherometer 黏附计,黏附力计
adhesion 黏合(力),黏附(力);黏附性
~ agent 黏合剂
~ [adhesive] capacity 黏合(能)力
~ factor 黏合系数
~ failure 黏合破坏,粘结失效
~ force 粘结力
~ loss 失黏(性)
~ [adhesive] power 黏合(能)力
~ promoter [promotor] 增黏剂,胶黏剂
~ properties 粘接性能
~ ratio 黏力比
~ strength 黏合强度,粘接强度
~ test 黏合试验,粘接试验
adhesional 黏附,黏合,粘接
~ energy 黏合能

~ wetting 黏合润湿,黏润作用
adhesive 胶黏剂,胶黏剂;黏合的
~ agent 黏合剂,胶黏剂,粘结剂
~ agglutination 黏附(作用)
~ assembly 黏合组装
~ -backed tape 胶黏带
~ bar 棒状黏合剂
~ base 黏附(基)体
~ bond 粘接
~ -bonded part 粘接件
~ bonding 粘接,胶接,胶黏
~ bonding of metals 金属粘接,金属黏合
~ capacity 黏合能力
~ coating 黏合层
~ component 胶黏剂组分,胶黏剂成分
~ container 黏合剂容器
~ dispersion 胶黏剂分散(作用),黏合剂扩散(现象)
~ emulsion 黏合乳胶
~ fabric 黏合布
~ failure 黏合破坏,粘结失效
~ fillet 填角黏合
~ film 黏合膜,膜状黏合剂,胶黏薄膜,胶膜
~ force 黏合力
~ for plastics 塑料用黏合剂
~ formula [tion] 黏合剂组分[配方]
~ interlayer 黏合凝固层
~ [glue] joint 粘接接头,粘接点
~ laminating 粘接[胶黏]层合
~ lamination 粘接[胶黏]层合
~ lamination of foam 泡沫材料粘接层合
~ layer 黏合层,胶黏层
~ -line thickness 黏合层厚(度)
~ material 结合材料,黏合剂,胶黏剂
~ mixer 调胶机
~ mixture ready for use 备用混合黏合剂
~ peel joint 剥离黏合接头
~ pellets 粒料黏合
~ plaster 黏着膏剂,黏附膏剂

~ power 黏附本领,黏合(能)力
~ primer 底胶[漆]
~ promoter 增黏剂
~ receptivity 黏合吸收能力
~ resin 黏合用树脂
~ soluble in organic solvent 可溶于有机溶剂的黏合剂
~ solution 黏合溶液
~ spread consumption 黏合剂涂敷消耗量,胶黏剂涂布消耗量
~ spread on adherend 涂布黏合
~ spreader 黏合剂涂布机,涂胶机
~ spreading machine 黏合剂涂布机,涂胶机
~ stick 黏附
~ strength 黏合强度,黏附强度
~ strength under shear 剪切黏合强度
~ stress 黏合应力
~ surface 黏合表面
~ tape 黏胶带,胶黏带
~ tape dispenser 黏胶带供给器
~ testing 黏合试验
~ thickness 黏合(层)厚度
~ varnish 黏合清漆
adhesiveness 黏合(能)力,粘接性
adhesivity 黏着性;黏合力
adhesivobonding 胶接
adhint 粘接头
ADI [acceptable daily intake] 允许日剂量
ADI number [acceptable daily intake number] 允许日剂量
adiabatic 绝热的,不传热的
~ change 绝热变化
~ compression 绝热压缩
~ contraction 绝热收缩
~ cooling curve 绝热冷却曲线
~ elasticity 绝热弹性
~ expansion 绝热膨胀
~ extruder 绝热挤塑机
~ extrusion 绝热挤塑
~ operation 绝热运转
~ process 绝热过程,绝热作用〈在挤出机中〉

adiabator 保温材料;绝热材料
adipamide 己二酰二胺
adipate 己二酸盐[酯]
~ plasticizers 己二酸酯类增塑剂
adipic 己二酸(的)
~ acid 己二酸
~ ester 己二酸酯
~ hexamethylene diamine 己二酸己二胺盐,AH盐
~ nitrile carbonate 碳酸己二腈
adiponitrile 己二腈
adj. [adjacent] 相邻的,邻近的
adjacent 相邻的,邻近的
~ circuits 缠绕
adjust 调节,调整
~ tension 调节松紧
adjustable 可调节的
~ angle sound head 可调角度(声波)探头
~ barrier 可调节流元件〈挤出机口模〉
~ bleeder holes 可调渗料孔
~ curvature (辊)可调曲率,可调曲[弧]度
~ gib 可调导板;可调节流栓
~ in height 高度可调
~ lip 可调模唇〈挤塑片材模头〉
~ load 可调负载
~ nozzle 可调喷嘴
~ opening die 调口式模头
~ ports 可调节孔
~ restrictor bar 可调节流元件〈挤出机口模〉
~ ring 调节环
~ roll 可调节辊
~ screw 调节螺丝,调节螺钉
~ speed 调速度
~ stop 可调挡板;可调制动器
adjusting 调节的,调整的
~ bolt 调整螺栓
~ device 调节装置
~ gear 调节齿轮
~ screw 调节螺丝,调节螺钉
~ screw nut 调距螺母

adjustment 调整,调节
~ of lip 模唇调节
~ of tension 张力调节
~ range 调节范围
~ thread 调节螺纹
adjuvant 掺ług料,混合料;填料
admixtion 掺混物,混合物;掺混,混合
admixture 掺混物,混合物;掺混,混合
adsorbability 吸附能力;吸附性
adsorbate 吸附物,吸附体
adsorbed 吸附的
~ material 吸附物,吸附体
~ substance 吸附物,吸附体
adsorbent 吸附剂,吸附体
adsorption 吸附
~ chromatography 吸附色谱〈测定高分子量〉
~ isotherm 吸附等温线
~ layer 吸附层
adsorptive capacity 吸收能力,吸附能力
adulterant 掺混剂,掺杂剂
adulterating agent 掺混剂,掺杂剂
advanced 先进的,高级的
~ composite 高级复合材料
~ fibre 高级纤维
~ plastics 高级塑料;新型塑料
~ technique 先进技术
dvertisement 广告,公告
aeolotropy 各向异性
aerated plastics 泡沫塑料,多孔塑料
aerating agent 发泡剂
aeration 充气,吹气;通风
aerial mast 天线杆〈电视〉
aero lubrification 空气润滑,气垫
aero-shaft 充气膨胀轴心,气动轴
aerogel 气凝胶
aerosol 气溶胶;烟雾剂
~ can packing 密封外壳包装
~ package 密封包装,加压包装
~ propellant 气溶胶型喷雾剂
aerospace application 航空应用
AES [ethylene-propylene-styrene-acrylonitrile copolymer] 乙烯-丙烯-苯乙烯-丙烯腈共聚物

AES [thermoplastic quaterpolymer from acrylonitrile, ethylene, propylene, styrene] 丙烯腈-乙烯-丙烯-苯乙烯的热塑性四元共聚物
affect 时效;效应,效果
~ oven 老化烘箱
~ property 老化性质
~ resistance 耐老化性
~ stability 老化稳定性
~ temperature-time curve 老化温度-时间曲线
~ test 老化试验
~ time 老化时间
affine deformation 仿射形变
affinity 亲和力,亲和力
affinity for 和…有亲合力
after 在…之后
~ annealing 后退火〈脱模〉
~ -bake 后固化,后烤,后熟化
~ baking 后烘烤
~ -chloride 后氯化
~ -chlorination 后氯化
~ chlorinated PVC resin 后氯化聚氯乙烯树脂
~ contraction 后收缩,残余收缩
~ crystallization 后结晶
~ cure 后固化
~ -drawing 后拉伸
~ -effect 后效
~ -exposure 曝光后处理
~ -heating 后加热
~ mo(u)ld stress 成型后应力
~ -polymerization 后聚合(作用)
~ -pressure 继续压力
~ -pressure phase 续压阶段〈注塑时〉
~ -pressure time 续压时间
~ shrinkage 后收缩,后收缩率〈塑料成型件〉
~ -stove 后烘烤
~ -stretch 后拉伸
~ -stretching 后拉伸
~ tack 回黏
~ treatment 后处理

~ vulcanization 后硫化
~ working 后效应
~ - yellowing 变黄
after effect 后效应;弹性后效
afterflame 余(火)焰
~ time 余焰时间
afterglow 余辉
aftermade size 制成尺寸
afterpressure 续压
~ phase 续压阶段〈注塑时〉
~ time 续压时间
aftertreatment 后处理
AG [Aktiengesellschaft]〈德〉股份公司
Ag [silver] 银
against 反对,防备
agar 琼脂
agar - agar 琼脂
AGE [allyl glycidyl ether] 烯丙基缩水甘油醚〈环氧树脂的反应稀释剂〉
age 时代;时期;寿命;老化
~ hardening time 最终硬化时间
~ hardness 时效硬度
~ - inhibiting additive 防老化添加剂
~ - protecting agent 防老(化)剂
~ resister 防老剂
~ resisting 耐老化(的)
ageing [aging] 老化,熟化
~ behaviour 老化性能
~ coefficient 老化系数
~ time 老化时间
agent 剂
~ ageprotecting 防老化剂
agg. [aggregate] 总数;共计
agglomerate 附聚物;料团,料疙瘩
agglomerated grain 团粒,结块颗粒
agglomerating agent 附聚剂;凝结剂;烧结剂
agglomeration 附聚作用;烧结
agglutinant 凝集剂;烧结剂;粘结剂
agglutinating property 烧结性
agglutination 凝集;烧结
~ by adhesion 黏附着
~ by diffusion 渗浸黏合
aggregate 集料;聚集体

~ condition 聚集状态
aggregation of nodules 链球
aggregational structure 聚集态结构
aggregative fluidization 聚集液态化〈在沸腾层内浓和稀悬气共存,与粒子不同〉
aggressive tack 干黏性
aging [ageing] 老化,熟化
agitate 搅拌
agitated autoclave 搅拌高压釜
agitating 搅拌
~ blade 搅拌浆叶
~ pan 搅拌锅,搅拌容器
agitation 搅拌
agitator 搅拌器
~ ball mill 搅拌式球磨机
~ bath 搅拌槽
~ blade 搅拌浆叶
~ dryer 搅拌干燥器
~ mixer 搅拌混合器
~ tank 搅拌槽
AGM extruder 锥型双螺杆挤塑机
agmt [agreement] 一致
agn [again] 又,再
agricultural 农业的
~ applications 农业中的应用
~ film 农用薄膜
agriplast 农用塑料
AI [amid - imide polymer] 聚酰胺 - 酰亚胺
AIBN [azobisisobutyronitrile] 偶氮二异丁腈〈引发剂,发泡剂,硫化剂〉
aid 助剂
air 空气
~ accumulator 蓄气器
~ agitator 充气搅拌器
~ - assist forming 气胀真空成型,气胀成型
~ - assist vacuum thermoforming 气胀真空热成型
~ - assisting vacuum forming 气胀真空成型
~ bag under pressure 加压气袋
~ blade coater 气刀涂布机

air

- ~ blast 吹起〈吹塑成型机〉
- ~ – blast mixing machine 吹[鼓]风混合器,空气搅拌器
- ~ bleeding valve 排气阀,通风阀
- ~ blow 鼓风,吹气
- ~ blowing 吹气,压气成型〈吹塑成型机〉
- ~ blowing into female mould 气胀阴模成型〈热成型〉
- ~ blowing with plug assist 气胀塞助成型〈热成型〉
- ~ blown foam plastic 空气发泡塑料
- ~ bomb test 空气弹状瓶试验
- ~ borne 气拱
- ~ borne dryer 悬浮气体干燥器
- ~ borne dust 烟雾尘
- ~ borne noise 空气噪声
- ~ borne web dryer 浮带干燥器
- ~ bottle 压缩空气罐,压缩空气钢瓶
- ~ – box 鼓风箱
- ~ – box dryer 鼓风箱式干燥器
- ~ brush 气刷
- ~ brush coating 气刷刮涂
- ~ bubble 气泡;气垫
- ~ bubble viscometer 气泡黏度计
- ~ cap 干燥罩;气帽
- ~ circulation 空气循环
- ~ classifier 吹气分离器
- ~ cleaning 空气净化
- ~ conditioned cabinet 通风柜
- ~ conditioning 空气调节,空调
- ~ – conditioning cabinet 通风柜
- ~ consumption 空气消耗(量)
- ~ content 空气含量,含空气量
- ~ conveyer 压缩空气输送器
- ~ cooling 空气冷却
- ~ cooling ring 空气冷却环,风冷环〈吹膜机〉
- ~ current 空气流
- ~ cushion 气垫;充气垫
- ~ cushion dryer 气垫干燥器,悬浮干燥器
- ~ cushion forming 气垫成型〈热成型〉
- ~ diffuser 空气扩散器
- ~ – diffusion chamber 空气扩散室
- ~ – doctor 气刮刀;气刷
- ~ – dried 风干的
- ~ – dry 空气干燥
- ~ – dryer 空气干燥器
- ~ – drying 空气干燥〈涂层〉,风干
- ~ – drying coating 空气干燥涂层;自然干燥涂料
- ~ – drying lacquer 空气干燥漆
- ~ – drying primer 空气干燥底漆,空气干燥底涂料
- ~ duct (通)风道,空气管道
- ~ ejection 气力脱膜,压力脱模,气压脱模,喷气脱模
- ~ ejector 气力脱模销,气力顶出销〈注塑机〉
- ~ escape 排气,放气
- ~ – flow oven 气流烘箱
- ~ gap (空)气隙
- ~ gun 喷气枪
- ~ hammer 气锤
- ~ heater 空气加热器
- ~ hole 气孔
- ~ humidifier 空气增湿器
- ~ humidity 空气湿度
- ~ hydraulic accumulator 蓄气器
- ~ impermeability 气密性
- ~ inclusion 气泡
- ~ – inflated 充气的
- ~ – inflated cushion [air – cushion] 气垫
- ~ – inflated double – skin tent 气动吹胀双层帐篷
- ~ – inflated double – walled structures 气动吹胀双层结构
- ~ – inflated double – walled tent 气动吹胀双层帐篷
- ~ – inflated tube 充气软管
- ~ inlet 供气(口)
- ~ – intake 进气口
- ~ jet 喷气(刷),气动刷;空气喷嘴
- ~ – jet lift 气动升降机
- ~ – jet mill 喷气研磨机
- ~ joint 气密连接

~ knife 气刀,喷气刀;气刷
~ knife coater 气刀涂布机,气刷涂布机,喷气刀涂刮机
~ knife coating 气刀涂布,气刷涂布,喷气涂刮
~ leakage test 气密试验
~ less spraying 无气喷涂
~ lift 气动提升机
~ – lift agitator 气拌器
~ – lock 气沟
~ locks 气窝
~ loss 风干失重
~ lubrication 空气润滑
~ nozzle 喷气嘴
~ occlusion 气泡〈制品〉
~ – operated clamp 气动夹紧器
~ outlet 排气口〈挤出模具〉
~ oven 空气烘箱,热空气箱
~ oven ageing test 热空气老化试验
~ – oven orientation 热空气取向法
~ permeability 透气性;透气度
~ permeable plastics 透气性塑料
~ piston 气动活塞
~ pocket 气穴,气泡〈制品〉
~ pollution 大气污染
~ porosity 透气性;透气度
~ pressure [compressed air] 压缩空气
~ – pressure centrifugal casting 旋转铸塑
~ pressure heat aging test 压缩空气热老化试验
~ pressure injection mo(u)lding 气压注塑
~ – pressure injection process 气压注塑法
~ proof 不透气的,气密的
~ receiver 蓄气器,空气蓄压器,气罐
~ ring 气环,风圈,空冷环
~ separator 空气分离器
~ silk 空心丝,气泡丝
~ slip forming 气胀包模成型
~ – slip method 气胀包成型法
~ – slip vacuum thermoforming 气胀包模真空热成型

~ sluice 气沟〈气泡中的〉
~ spray coating 气压喷涂
~ spraying 气压喷涂〈漆〉
~ – stream floatation 气流浮选〈材料层〉
~ – supported 气拱
~ – supported dome [hall] 气承帐篷〈由塑料涂层织物制成〉
~ tight 不透气的,气密的
~ vent 排气孔,排气口
~ vessel 蓄气器,气罐
~ wiper 空气除水器
airborne dust 烟(道)尘,烟(囱)灰
airing 通风;空气循环
~ plant 通风装置
~ tube 通风管(道)
airless 无空气的
~ blast deflashing 无气冲击除边
~ spraying 无气喷涂
airproof 气密(封)的
airtight 气密(封)的
~ barrel 密封桶〈塑料〉
ajusta bow 拉幅杆,拉幅辊〈美〉
AK [alkyd resin] 醇酸树脂
Al., Alum [aluminium] 铝
~ foil 铝箔
alcohol 醇;乙醇
alcoholate 醇化;醇化物
alcoholization 醇化(作用)
Ald. [aldehyde] 醛
aldehyde 醛;乙醛
~ resin 聚醛树脂,醛类树脂
aldol 羟基丁醛
~ – α – naphthylamine 2 – 羟基丁醛 – α – 萘胺〈防老剂〉
~ α – naphthylamine(resin) 丁间醇醛 – α – 萘胺〈高相对分子质量,树脂状〉,防老剂 AH
~ resin 醇醛树脂
alembic 蒸馏器[釜]
alfin 醇烯
~ catalyst 醇烯催化剂
~ polymer 醇烯聚合物
~ rubber 醇烯橡胶

alginate [alginic] fibre 藻酸(酯)纤维，海藻纤维
alginic acid 藻酸
alginic fibres 藻酸纤维
alicyclic 脂环〈族〉的
- ~ epoxyresin 脂环〈族〉环氧树脂
- ~ hydrocarbon 脂环烃

align 校正；调整
aligned fibre composite 纤维排列复合材料
aligner roll 自动调节辊
alignment 对准〈模具〉；定位〈分子〉
aliphatic 脂〈肪〉族的
- ~ hydrocarbon 脂肪烃
- ~ plasticizer 脂(肪)族增塑剂
- ~ polyamine 脂(肪)族聚胺
- ~ polyester 脂(肪)族聚酯

aliquation 层化；起层
alk – [alkaline] 碱的
alkali 碱〈指强碱、苛性碱〉
- ~ cellulose 碱纤维素
- ~ cleaner 碱性洗涤剂
- ~ glass 碱玻璃〈增强材料〉
- ~ fastness 耐碱性
- ~ -proof paper 耐碱纸
- ~ resistance 耐碱性
- ~ resistant cellulose 耐碱纤维素
- ~ resisting cellulose 耐碱纤维素

alkalify 碱化；加碱
alkaline 碱(性)的〈指强碱、苛性碱〉
- ~ cleaning 碱性清洗
- ~ cleaning bath 碱性洗涤浴〈表面处理〉
- ~ degreasing 碱性脱脂〈金属胶接部件处理〉
- ~ earths 碱土
- ~ medium 碱性介质
- ~ soaker 碱性浴
- ~ strength 碱含量

alkalinity 碱性，碱度
alkaliproof 耐碱的
alkine 炔烃
alky. [alkalinity] 碱度；碱性
alkyd 醇酸(树脂)

- ~ moulding compounds 醇酸树脂模塑料
- ~ plastic 醇酸塑料
- ~ resin 醇酸树脂
- ~ resin adhesive 醇酸树脂黏合剂
- ~ resin mo(u)lding compound 醇酸树脂模塑料
- ~ resin varnish 醇酸树脂清漆

alkyl 烷基
- ~ cyanoacrylate adhesive 腈基丙烯酸烷基酯黏合剂
- ~ phenol-formaldehyde resin 烷基(苯)酚(甲)醛树脂〈增黏剂〉
- ~ phenol resin 烷基酚醛树脂
- ~ phosphate diethanolamine salt 磷酸烷基酯二乙醇胺盐，防静电剂 P

C_7 ~ C_{10} **tetrahydrophthalate** 四氢邻苯二甲酸 C_7 ~ C_{10} 烷基酯〈增塑剂〉

alkylate 烷基化；烃化
alkylcellulose 烷基纤维素
alkylchlorosilane 烷基氯(甲)硅烷
alkylene oxide polymer 环氧烷(类)聚合物，烯烃化氧聚合物
alkyne 炔烃
all 全(部)；所有的
- ~ cotton 全棉的
- ~ -cotton cloth 全棉织物
- ~ -cotton stretch yarn 全棉弹力纱
- ~ -hydraulic injection machine 全液压注塑机〈美〉
- ~ -plastic 全塑料
- ~ -plastic article 全塑制品
- ~ -plastics barrel vault 全塑桶顶
- ~ -purpose adhesive 通用黏合剂，万能胶
- ~ -skin rayon fibre 全皮层黏胶纤维

all-purpose 通用的
- ~ adhesive 通用黏合剂
- ~ gum 通用胶
- ~ rubber 通用橡胶

alleviate stresses 降低应力
Allied Chemical Corporation foam mo(u)lding 联合化学公司泡沫成型，

气体反压泡沫塑料注塑法
Allied ihjection process 联合化学公司注塑法
allomer 异质同晶体
allomorph 同质异晶
allotropism 同素异形
allotropy 同素异形
allowable (可)容许的,(可)允许的
~ error 允许误差
~ load 允许负荷
~ pressure 允许压力
~ slip angle 允许偏离角〈玻璃纤维缠绕〉
~ stress 允许应力
~ value 允许值
~ variation 允许偏差
allowance 允许,容许
~ error 容许误差
~ for shrinkage 容许收缩量
allowed value 容许值
alloy 合金;掺混聚合物
allyl 烯丙基
~ alcohol 烯丙醇
~ alcohol polymer 烯丙醇聚合物
~ cellulose 烯丙基纤维素
~ chloride low polymer 烯丙基氯低聚物
~ diglycol carbonate,[ADC] 烯丙基二甘醇碳酸酯
~ dilycol carbonate resin 烯丙基二甘醇碳酸酯树脂
~ ester 烯丙酯
~ ester polymer 烯丙酯聚合物
~ glycidyl ether(AGE) 烯丙基缩水甘油醚〈环氧树脂的反应稀释剂〉
~ glycol carbonate resin 烯丙基甘醇碳酸酯树脂
~ plastics 烯丙基塑料
~ polymer 烯丙基聚合物
~ resin 烯丙基树脂
~ allylic resin 烯丙基树脂
alm -[almost] 几乎,差不多
alpha α〈希腊字母〉
~ cellulose 甲种纤维素,α-纤维素

~ cyanoacrylate α-氰基丙烯酸酯
~ fibre α-纤维
~ helical conformation α-螺旋构象
~ loss peak α-损耗峰
~ methylstyrene α-甲基苯乙烯
~ ray α-射线
alterarion 改变,变化
alternate 交替,交错,交变
~ bending test 交变弯曲试验
~ copolymer 交替共聚物
~ crosslayers 交向层,交错层〈胶合板〉
~ elastomer 交替型弹性体
~ flexural stress 交变弯曲应力
~ folding 交错折叠
~ material 替换材料
~ stress 交变应力
~ wet and dry test 干湿交替试验
alternting 交替的,交错的,交变的
~ bending strength 交变弯曲强度
~ climate 交变气候
~ copolymer 交替共聚物
~ copolymerization 交替共聚(作用)
~ current 交流电
~ damp heat atmosphere 交替湿热气候
~ deformation 交替形变
~ die gap adjustment 交替模隙调节
~ humidity atmosphere 交替湿度气候
~ strain amplitude 交变应变振幅
~ stress 交变应力
~ stress amplitude 交变应力振幅
~ stresses load 交替应力负载
~ structure 交错结构
alum 矾
alumina 矾土,氧化铝〈填料〉
~ trihydrate 三水合氧化铝〈阻燃剂〉
~ white 矾土白
aluminium [Al] 铝
~ distearate 双硬脂酸铝〈热稳定剂〉
~ foil 铝箔
~ hydrate[hydroxide] 氢氧化铝〈阻燃剂〉
~ oxide 矾土,氧化铝〈填料〉

aluminizied polyester film

~ oxide trihydrate 氧化铝三水合物〈阻燃剂〉
~ powder 铝粉
~ silicates 硅酸铝
~ stearate 硬脂酸铝〈热稳定剂〉
~ triethyl [ATE] 三乙基铝
~ trihydroxide 氢氧化铝〈填料〉
aluminizied polyester film 镀铝聚酯薄膜
aluminum - [aluminium] 铝
~ triethyl [ATE, triethyl aluminum] 三乙基铝〈聚合催化剂〉
alveolar 牙槽的；气泡的
Am - [ammonium] 铵
Amaranth 苋菜红〈染料〉
amber 琥珀；琥珀色(的)，淡黄色的
amberwood 酚醛树脂胶合板
ambient 周围的,环境的
~ air 环境空气
~ conditions 周围条件,环境条件
~ dried 环境干燥的
~ gas 环境气氛
~ moisture 环境湿度
~ pressure 环境压力
~ temperature 环境温度,室温
American National Standards Institute [ANSI] 美国国家标准研究所
American Society for Testing and Material [ASTM] 美国材料试验学会
amide 酰胺
~ epoxy resin 酰胺-环氧树脂
~ hardener 酰胺固化剂,酰胺硬化剂
~ - imide polymers [AI] 聚酰胺-酰亚胺聚合物
~ - imide resin 聚酰胺-酰亚胺树脂
γ - amidinothiopropyltrihydroxysilane γ-脒基硫代丙基三羟基(甲)硅烷〈偶联剂〉
amido colo(u)r 酰胺染料
amidoamine 氨基酰胺〈环氧树脂固化剂〉
amine 胺
~ aldehyde resin 胺醛树脂
~ antioxidant 胺类抗氧剂

~ cellulose 胺类纤维素
~ - cured epoxy 胺硬化环氧树脂
~ curing agent 胺硬化剂
~ hardener 胺硬化剂
~ resin 胺类树脂
amino 氨基
~ - acids 氨基酸
~ - aldehyde plastic 氨醛塑料
~ - aldehyde resin 氨醛树脂
~ alkyd resin coating 氨基醇酸树脂涂料
~ - formaldehyde resin 氨醛树脂
~ plast 氨基塑料
~ resin 氨基树脂
aminoethyl 氨乙基
N - β ~ - γ - aminopropylmethyl dimetho - xy silane N - β - (氨乙基) - γ - 氨丙基甲基二甲氧基(甲)硅烷〈偶联剂〉
N - β ~ - γ - aminopropyltrimethoxy silane N - β - (氨乙基) - γ 氨丙基三甲氧基(甲)硅烷〈偶联剂〉
N - ~ piperazine N - 氨乙基哌嗪〈固化剂〉
aminoketone dyes 氨基酮染料
3 - aminomethyl - 3,5,5 - trimethylcyclohexylamine 3 - 氨甲基 - 3,5,5 - 三甲基环己胺〈固化剂〉
aminoplastic 氨基塑料
~ moulding. composition 氨基塑料模塑料〈美〉
~ moulding compound [material] 氨基塑料模塑料
~ resin 氨基树脂
γ - aminopropyltriethoxysilane γ-氨丙基三乙氧基(甲)硅烷〈偶联剂〉
aminos [aminoplastics] 氨基塑料
AMMA [acrylonitrile methyl methacrylate] 丙烯腈-甲基丙烯酸甲酯(共聚物)
ammeter - [ampere meter] 安培计；电流表
ammon. - [ammoniacal] 氨的
ammonia 氨

ammoniacal 氨的
ammonium [NH_4^+, NH_3R^+, $NH_2R_2^+$, NHR_3^+, NR_4^+] 铵
~ bicarbonate 碳酸氢铵,酸式碳酸铵,重碳酸铵〈发泡剂〉
carbonate 氨基甲酸铵
~ persulfate 过硫酸铵〈引发剂,氧化剂〉
amorphism 无定形性,非晶性,不结晶性
amorph. [amorphous] 无定形的,非晶形的
amorphous 无定形的,非晶的
~ cellulose 无定形纤维素,非晶态纤维素
~ filament 非晶态丝
~ phase 无定形相,非晶相
~ plastic 无定形塑料
~ polymer 无定形聚合物,非晶态聚合物
~ region 无定形区,非晶区
~ state 无定形态,非晶态〈塑料〉
~ structure 无定形结构,非晶态结构
~ substance 无定形物
~ thermoplastic 无定形热塑性塑料,非晶热塑性塑料
amount 总计;量
amphoteric 两性的
~ ion exchange membrane 两性离子交换膜
amplifier 放大器
amplitude 振幅
Amsler type tester 阿姆斯勒型万能试验机
amt. [amotint] 共计;数量
amylaceous 含淀粉的
amylum 淀粉
AN [acrylonitrile] 丙烯腈
anaerobic 厌氧的
~ adhesive 厌氧黏合剂,厌氧胶
~ polymerization 无氧聚合(作用)
anal. [analysis, analyses] 分析
analogue 相似;模拟
~ result 模拟(试验)结果

analogy model 模拟模型
analyse 分析,研究,分解
analysis 分析;分解
~ by fractional distillation 分馏分析
~ sample 化验样品,分析试样
analytical 分析的
~ balance 分析天平
~ chemistry 分析化学
~ chromatograph 分析色谱仪
~ column 分析柱
~ data 分析数据
~ electron microscope [AEM] 分析电子显微镜
analyzer 分析器;测定器;检偏振器
anchor 锚定,固定,定位〈模具〉;冷料钩〈注塑模〉
~ agitator 锚式搅拌器,U形搅拌器
~ bolt 锚栓;地脚螺栓
~ coating 固定涂层
~ mixer 锚式混合器
~ nut 固定螺帽
~ pin 定位销〈模具〉
~ screw 锚栓,地脚螺栓
~ -type stirrer with baffles 备有隔板的锚式搅拌器
anchorage 嵌embed
anchoring 嵌定;黏固
~ agent 胶黏剂
~ strength 粘结强度
ancillary 辅助的
~ equipment 辅助设备
~ material 辅助材料
anelasticity 滞弹性
angle 角,角度;斜的
~ branch 弯管,肘管
~ check valve 直角型止逆阀
~ head 角向机头,斜角机头,倾斜挤出机头,弯机头
~ joint 角接接头
~ mo(u)ld 角型模具
~ mo(u)lding press 压角机
~ of bending 弯曲角
~ of contact 接触角
~ of feed 供料角

~ of preparation　斜角,破口角度
~ of refraction　折射角
~ of repose　休止角,倾斜角,坡度角
~ of tear　撕裂角
~ of torsion　扭转角
~ of twist　扭曲角
~ of vee　坡口角〈焊接〉
~ peeling strength　直角剥离强度〈粘接〉
~ peeling test　直角剥离试验〈粘接〉
~ pipe　弯管,肘管
~ ply laminate　角铺敷层合板,角铺设层合板
~ press　角式〈液〉压机,直交双活塞式压机
~ -seat check valve　斜座止回阀
~ -seat valve　斜座阀
~ section　角型材
~ tear test　直角撕裂试验
~ tee　直角枝管,三通管
~ valve　角阀
angular　角的,角度的;倾斜
~ cam　斜导柱〈模具〉
~ extrusion head　倾斜挤出机头
~ head　斜向机头,弯机头〈挤塑机〉
~ pin　斜位销,斜向合模销
~ speed　角速度
~ strain　角应变
~ surface　斜面
~ test　弯曲试验
~ velocity　角速度
anhyd.　[anhydrous]　无水的
anhydride　脱水物;(酸)酐
~ hardener　干硬化剂,无水硬化剂
anhydrous　无水的
~ aluminium chloride　无水三氯化铝〈催化剂〉
aniline　苯胺
~ dyes　苯胺染料
~ finishing　苯胺涂饰
~ formaldehyde resin　苯胺甲醛树脂
~ modified phenolic mo(u)lding powder　苯胺改性酚醛模塑粉
~ resin　苯胺〈甲醛〉树脂

anilinomethyl triethoxysilane　苯胺甲基三乙氧基(甲)硅烷〈偶联剂〉
anilinoplast　苯胺塑料
anilox roll　涂刷辊
animal　动物;动物的
~ base adhesive　动物〈性〉黏合剂
~ fillers　动物性填充剂
~ gelatin　动物明胶
~ gelatine adhesive　(动物)明胶黏合剂
~ glue　动物胶〈黏合剂〉
~ glue size　动物胶黏剂
~ lubricant　动物润滑剂
~ resin　动物树脂
anion　阴离子
~ -cation composite ion exchange membrane　双极离子交换膜
~ exchange　阴离子交换(的)
~ exchange resin　阴离子交换树脂
~ exchanger　阴离子交换剂
~ ion exchange membrane　阴离子交换膜
~ surface active agent　阴离子型表面活性剂
anionic　阴离子的
~ catalyst　阴离子催化剂
~ coorfinate polymerization　阴离子配位聚合
~ copolymerization　阴离子共聚
~ dye　阴离子染料
~ grafting　阴离子接枝
~ initiator　阴离子引发剂
~ isomerization polymerization　阴离子异构化聚合
~ polymerization　阴离子聚合(作用)
~ surface agent　阴离子型表面活性剂
anisotropic　各向异性的
anisotropy　各向异性
~ of laminates　层压材料的各向异性
~ ratio　各向异性比
annealed　经退火的
~ in vacuum　真空退火
annealing　退火,热处理
~ chamber　退火室,热处理室

~ cdo(u)r 烤色;退火色
~ godet 牵引辊固定装置〈薄膜取向装置〉
~ furnace 退火炉
annual production [output] 年产量
annular 环(形)的
~ blank 环状坯料
~ buffing section 环形抛光段
~ clearance 环形间隙
~ die 环形口模
~ die gap 环形口模间隙,环形模隙
~ die plate 环形口模板,环形模板
~ extrusion 环形挤出,环形口模挤出
~ gap 环形模缝
~ gate 环形浇口
~ groove 环状槽
~ parison 环形毛坯〈吹塑〉
~ piston 环状活塞
~ piston accumlator 环状活塞储料器〈吹塑〉
~ plunger 环状阳模;环状柱塞
~ slot 环状间隙〈挤出机口模〉
~ - slot element 环状间隙元件〈螺杆〉
anode 阳极
anodic 阳极的
~ oxidation 阳极氧化
~ treatment 阳极处理
anodizing 阳极处理[氧化]
anomalous viscosity 反常黏度,可变黏度
ANSI [American National Standards Institute] 美国国家标准研究所
antechamber nozzle 长颈注嘴
antiadhesive 防黏剂
~ lining 防黏衬里
antiager 防老剂〈美〉
antiageing agent 防老(化)剂
antiblocking agent 抗黏连剂,开口剂;防粘结剂,防结块剂
antiblowing agent 消泡剂
antiblushing agent 防变色剂〈漆〉
anticaking agent 防结块剂
antichecking agent 防龟裂剂
anti - clockwise 反顺时针方向〈旋转〉

anticoagolant 阻凝剂,抗凝剂
anticoking 防焦添加剂
anticorrosive 防(腐)蚀的
~ agent 防腐蚀剂
~ coating 防腐蚀涂料,防腐蚀涂层
~ paint 防腐蚀涂料
~ primer 防腐蚀底漆
anticrackle agent 防裂纹剂
anticreaming agent 防(膏)冻剂,防结皮剂(涂层)
anticrease finish 防皱整理
antidecomposition additive 防分解添加剂
antidegradant 防降解剂
antifatigue agent 防疲劳剂
antiflaming 阻燃的
antiflammability agent 阻燃剂
antiflooding agent 防泛色剂,防变色剂〈漆〉
antifoam(er) 消泡剂
antifoaming 消泡的
~ agent 消泡剂
~ emulsion 消泡沫乳剂
anti - fog 防雾剂
antifogging agent 防雾剂
antifouling coating 防污涂料
antifreeze pack 防冻包装
anti - freezing 防冻的
~ agent 防冻剂
~ property 防冻性,耐寒性
antifriction resin 耐磨(性)树脂
antifroth [defoaming]agent 消泡剂
antigelling agent 防胶凝剂
antiimpact bumper 耐冲保险杠
antilivering agent 防硬化剂,防凝结剂
antimicrobial stability 防微生物稳定剂
antimist agent 防雾剂〈薄膜覆盖层〉
antimony [Sb] 锑
~ oxide 氧化锑;三氧化二锑,锑白〈阻燃剂〉
~ silicooxide 硅合氧化锑〈阻燃剂〉
antinoise 防噪声的,消噪声的
~ coating 消声涂层
antioxidant [AO] 抗氧剂,防老剂

antioxidant agent 抗氧剂
antioxidizing agent 抗氧剂
antiozonant 抗臭氧剂
antiperoxide additive 抗过氧化物添加剂
antiphotosensitive additive 蔽光添加剂
antiplasticization 反增塑(作用)
antiplasticizer 反增塑剂,反软化剂
antisag agent 防流淌剂,增稠剂
antiseizing property 防黏性
anti-setting agent 防沉降剂
anti-skid 防滑性〈地板〉
antiskinning agent 防结皮剂(涂层)
antislip agent 防滑剂
anti-snag 防流淌
antisofter 防软剂
antisoiling 防污的
 ~ agent 防污剂
 ~ finish 防污涂层
antisplash guard 挡泥板
antistatic 防静电的;防静电剂
 ~ additive 防静电添加剂
 ~ agent 防静电剂
 ~ coating 防静电涂料,防静电涂层;防静电涂布
 ~ fibre 防静电纤维
 ~ finishing 防静电修饰,防静电处理
 ~ treatment 防静电处理
antisticking agent 防黏剂
antistripping agent 防剥离剂
antisun material 耐光材料
antiswelling 防溶胀性
antitack agent 防黏剂
anti-toxic 防毒性的;解毒剂
antiwear additive 耐磨添加剂
anti-wrinkle 防皱(的)
 ~ slat expander 防皱板条展幅机
AO [antioxidant] 抗氧剂,防老剂
aperture 孔;孔隙;裂缝
 ~ of lips 唇开度
app- [appendix] 附录
APP [atactic polypropylene] 无规(立构)聚丙烯
apparatus 仪器,仪表;设备

apparent 表观的
 ~ area 表观面积
 ~ colour 表观颜色
 ~ constant 表观常数
 ~ density 表观密度
 ~ elastic modulus 表观弹性模量
 ~ flow 表观流动
 ~ flow curve 表观流动曲线
 ~ hardening curve 表观硬化曲线
 ~ melt viscosity 表观熔体黏度
 ~ modulus 表观模量
 ~ modulus of elasticity 表观弹性模量
 ~ molar mass 表观摩尔质量
 ~ plasticity index 表观塑性指数
 ~ porosity 表观孔隙率
 ~ powder density 粉末表观密度
 ~ ralative molecular mass 表观相对分子质量
 ~ shear rate 表观剪切速率
 ~ shear viscosity 表观剪切黏度
 ~ specific gravity 表观相对密度
 ~ viscosity 表观黏度
 ~ volume 表观体积
appearance 外观
 ~ defect 外表缺陷
 ~ failure 外表损伤
 ~ fracture test 断裂外观试验
 ~ inspection 外观检查
 ~ of flame 火焰(的)外观
 ~ quality 外观质量
apple sauce 条痕,条纹〈吹塑〉
appliance 设备,装置,仪器;附件,零件
application 应用,使用;施加(力);涂敷
 ~ of adhesive 黏合剂涂刷
 ~ of force 作用力,施加力
 ~ roll 涂布辊
 ~ test 应用试验
applicational property 应用性能,使用性能
applicator 涂布器
 ~ and dip roll 涂布辊和浸渍辊〈涂敷机〉
 ~ [coating] roll 涂布辊
applied 应用的

~ chemistry 应用化学
~ chemistry research 应用化学研究
~ research 应用研究
~ stress 外加应力
applique welding 镶嵌焊接
apply 涂敷〈涂料〉
appr. [approximately] 大约,近似地
apprec. [appreciable] 可估计的
approach 近似;方法;料道
~ direction （熔体）通道流向,熔料流向
approaching passage 料道〈挤出机头〉
apron 挡板
~ nip 辊距;辊隙
~ plate 围护板,挡板
~ roll 接收辊
aq [aqua, aqueous] 水;(含)水的
aquaglass preform method 玻璃水预模塑法
aqueous （含）水的
~ extract 水提物
~ phase 水相
~ polymerization 水溶液聚合
~ AR [acrylic rubber] 丙烯酸系橡胶
Ar [Argon] 氩
aramid fiber 芳(香)族聚酰胺纤维
~ reinforced plastics (AFRP) 芳纶增强塑料
aramide 聚芳基酰胺,芳(香)族聚酰胺;芳酰胺纤维〈增强剂〉
arbor press 芯棒压力机
arc （电）弧
~ cutting 电弧切割
~ extinguishing plastics 灭弧性塑料
~ discharge 电弧放电
~ lamp 弧光灯
~ resistance 耐电弧性
~ resistance test 耐电弧试验
~ suppressor 消弧器
~ tester 电火花检验装置
~ suppressor （弧）击穿〈电〉
~ tracking 电弧迹,径迹
~ welding 电弧焊
arch 拱

arching of a roll 辊外凸,辊中高
arcing 成拱;(电)弧
area 面积;表面;区域
~ burning rate 面积燃烧速率,表面燃烧速率
~ factor 面积系数
~ of fracture 断裂面,断裂面积
~ of heating 加热表面,加热面积
~ of indentation 压痕面积
~ of injection plunger 注射柱(活)塞截面积
~ of the effective surface 有效表面积
~ of the surface exposed to a catalyst 一种催化剂的有效表面积
~ of stress concentration 应力集中区
arenaceous quartz 石英砂
Argon [Ar] 氩
arid 干燥〈气候〉
arisings 旧材料;废材料
arm 臂;杆;桨叶
~ mixer 桨叶式混合器
~ stirrer 桨叶式搅拌器
armature coil 电枢线圈
armed pipe 复合管
armour 铠装〈电缆〉
armoured 防护,防具
~ barrel 带套机筒〈注射机或挤出机〉
~ hose 钢丝软管,铠装软管
~ screw 挡板螺杆〈注塑机或挤出机〉
armouring 加强,增强〈塑料材料〉
arom. [aromatic] 芳族的;芳香的
aromatic 芳(香)族
~ amine 芳(香)族胺〈硬化剂〉
~ azo polymer 芳(香)族偶氮聚合物
~ compounds 芳(香)族化合物
~ copolyamide microfibre 芳(香)族共聚酰胺超细纤维
~ hydrocarbon 芳(香)烃
~ ladder polymer 芳(香)族梯形聚合物
~ polyamide 芳(香)族聚酰胺
~ polyamide fiber 芳族聚酰胺纤维
~ polyamide-imide 芳族聚酰胺-酰亚胺

~ polyamine 芳族多胺;芳族聚胺
~ polyester 芳族聚酯
~ polymer 芳族聚合物
~ polymerization 芳(香)族聚合(作用)
~ polysulfon 芳族聚砜
~ side chain 芳族侧链
arrangement 排列,布置;装置,设备
~ of fibres 纤维排列
~ of the rolls 滚筒排列
array of fibres 纤维排列
arrest 制动(装置)
arresting 制动(的)
~ spring 制动簧片
arsenic 砷
article 制品,制件,物件
artificial 人工的,人造的;模拟的,仿真的
~ ageing 人工老化
~ climate laboratory 人工气候试验室
~ cooling 人工冷却,强制冷却
~ cotton 人造棉
~ crystallization nucleus 人造晶核
~ environment 人造环境,人工环境
~ fibre 人造纤维
~ filament 人造长丝
~ gum 人造橡胶
~ horn 人造角〈塑胶〉,酪素塑料
~ leather 人造革
~ limb 人工补装的假手足
~ resin 人造树脂
~ silk 人造丝
~ tooth 人造牙齿
~ ventilation 人工通风
~ weathering 人工气候老化
arylation 芳基化(作用)
As [arsenic] 砷〈元素符号〉
AS [acrylonitrile - styrene copolymer] 丙烯腈-苯乙烯共聚物
as - delivered condition (按)供给(时的)原来)状态
as - drawn fibre 初拉伸纤维
as - formed fibre 初纺纤维
ASA [acrylonitrile - styrene - acrylate copo - lymer] 丙烯腈-苯乙烯-丙烯酸酯共聚物
ASA rubber [acrylonitrile - styrene - acrylic rubber] 丙烯腈-苯乙烯-丙烯酸酯橡胶
asbestic 石棉的
asbestos 石棉〈增强剂〉
~ board 石棉板
~ cord mo(u)lding compound 石棉绳模塑料
~ fabric 石棉织物
~ fabric base laminate 石棉织物层合板
~ fibre 石棉纤维
~ fibre - board 石棉纤维板,石棉板
~ fibre reinforced plastics 石棉纤维增强塑料
~ fibre sheet 石棉纤维板
~ floats 石棉绒
~ joint 石棉填衬,石棉密封;石棉接头
~ laminate 石棉层合制品〈石棉纸或石棉织物〉
~ paper base laminate 石棉纸基层合板
~ phenol resin laminate 石棉酚醛树脂层合板
~ reinforced plastic 石棉增强塑料
~ reinforced thermoplastic 石棉增强热塑性塑料
~ reinforcement 石棉增强材料
~ roving 石棉粗纱
~ tape 石棉带
~ yam 石棉绳
asbestosis 石棉沉着病〈工业疾病〉
ascending gas 上升酸气〈流化床〉
ash 灰分
~ analysis 灰分分析
~ composition 灰成分
~ content 灰分(含量),含灰量
~ determination 灰分测量
~ test 灰分试验
ashen 灰(色)的
ashing 湿法抛光,灰(砂)抛光〈塑料表

~ sample 灰化试样
aspect ratio 纵横比;长阔比〈纤维〉;长径比〈螺杆〉
asperity 粗糙度
assembling 装配
　~ department 装配间
　~ drawing 装配图
　~ jig 装配夹具
assembly 组装,装配;装配件;黏合结构;(塑料)叠合
　~ adhesive 装配黏合剂
　~ diagram 装配图
　~ dimension 装配尺寸;粘接[叠合]面积
　~ drying ambient temperature 黏合干燥环境温度
　~ fixture 装配夹具;黏合夹紧装置,黏合夹具
　~ glue 装配胶黏剂
　~ gluing 组装胶合,装配黏合,二次黏合
　~ instruction 装配规程
　~ line 流水线,装配线
　~ line production 流水生产,流水作业,装配线生产
　~ parts 组合件
　~ program 装配程序
　~ time 装配时间;黏合持续时间
assistant 辅助的
assistants 助剂
assisting plug 辅助模塞
association 缔合,结合
　~ complexes 缔合络合物
　~ degree 缔合度
　~ heat 缔合热
　~ polymer 缔合聚合物
associative ability 缔合能力
assorted 各种各样的,各色俱备的,混合的
　~ colo(u)rs 杂色
　~ fibre 混杂纤维
assortment 种类,花色品种,分类
astatic 不稳定的,无定向的

~ coating 无定向涂层
ASTM [American Society for Testing and Material] 美国材料试验学会
　~ mothed 美国材料试验学会方法
　~ specification 美国材料试验学会技术规范
　~ standards 美国材料试验学会标准
astrol 鲜红
asymmetric 不对称的
　~ carbon atom 不对称碳原子
　~ induction polymerization 不对称诱导聚合
　~ selective polymerization 不对称选择聚合
asymmetrical monomer 不对称单体
asymmetry 不对称性
at-mo(u)ld combination 现场配料
atactic 无规的
　~ block 无规(立构)嵌段
　~ polymer 无规(立构)聚合物
　~ polymerization 无规聚合(作用)
atm. atmosphere 大气;大气压
atmosphere 大气,大气压;环境
atmospheric 大气的,空气的
　~ ageing 大气老化,自然老化
　~ conditions 大气条件
　~ cracking 大气龟裂,暴露龟裂,环境龟裂〈环境影响〉
　~ degradation resistance 耐气候降解性
　~ exposure test 大气暴露试验,耐气候性试验,风蚀试验,耐气候老化性试验
　~ fading 大气褪色
　~ humidity 大气湿度,空气湿度
　~ pollution 大气污染
　~ pressure 大气压
　~ temperature 大气温度,常温
　~ valve 空气阀,放空阀
at. no. [atomic number] 原子序数
ATO [antimony trioxide] 三氧化(二)锑〈阻燃剂〉
atom 原子
　~ nucleus 原子核
atomic 原子的

atomization

~ bond　原子键
~ number　原子序数
~ weight　相对原子质量
atomization　雾化,喷雾
atomizer　喷雾器,雾化器;细粉磨机
atomizing　雾化的
~ nozzle　喷雾嘴
ATR［attenuated total reflectance］衰减全反射
~ spectroscopy　衰减全反射光谱学
~ technique　衰减全反射技术
attachment　附件,配件;装置;夹具;连接,固定
~ area　夹持面
~ point　固定点
~ screw　紧固螺钉,连接螺钉
attack　侵蚀,腐蚀
attendance　维护,保养,维修〈机器〉
attenuated total reflectance［ATR technique］衰减全反射
attenuation　拉细;减薄,衰减;稀释
~ coefficient　衰减系数〈流变学〉
~ constant　衰减常数
attrition　摩擦,磨损,磨耗
~ mill　研磨机
~ rate　磨耗率,磨损程度
~ resistance　耐磨耗性
~ resistant　耐磨的
~ test　磨损试验
attritor　磨碎机
at. wt. = atomic weight　相对原子质量
AU［a polyurethane rubber type］一种聚氨酯橡胶类
Au［gold］（黄）金;金黄色
audio frequency　声频
auto［automotive］自动(车)的,汽车的
autoacceleration　自动加速(作用)
~ effect　自动加速效应
autoadhesion　自黏(力)
auto body　汽车车身
autocatalytic reaction　自动催化反应
autoclave　热压釜,热压锅,热压罐,蒸缸,高压锅〈俗称〉
~ bag moulding　气袋层压成型

~ blanket　热压锅隔膜,热压釜压力传递膜
~ bonding　热压粘结(黏合)
~ membrane　热压锅隔膜
~ moulding　热压釜成型〈层压塑料〉
~ press　热压锅,热压釜
~ vacuum bag method　热压釜真空袋法
autocontrol　自动控制
autocrosslinking　自交联
autodecomposition　自动分解
autodemo(u)lding polymer　自动脱模聚合物
autoejection　自动顶出,自动脱模〈注塑〉
autoelectronic emission　自动电子放射〈黏合理论〉
autofeed　自动加［装,供］料,自动供给
autofixed nozzle　自动固定喷嘴
autoflying transfer　自动快速换辊〈薄膜卷取〉
autogenous　自焊的;自热的
~ control　自动熔接控制〈挤塑机〉
~ extrusion　自热挤塑
~ ignition　自燃
~ ignition temperature　自燃温度
~ operation　自热运转,绝热运转
autohesion　自黏〈同类物质黏合〉
autoignition　自燃
~ temperature　自燃温度
autoinjection　自动注塑
~ control　自动注塑控制
automated　自动(化)的
automatic　自动化的
~ adhesive binder　自动黏合器
~ bag making machine　自动制袋机
~ block system　自动联锁装置
~ blow moulding machine　自动吹塑成型机
~ circuit breaker　自动断路器
~ compression moulding　自动压塑成型
~ control　自动控制
~ control device　自动控制装置
~ control equipment　自动控制设备

~ cut - off nozzle　自动关闭式注嘴
~ cut - out　自动关闭
~ cycle　自动循环
~ drier [dryer]　自动干燥机
~ ejection　自动顶出,自动脱横〈注塑〉
~ feed　自动装 [加,供] 料,自动供给
~ feeder　自动装料器
~ filling machine　自动灌装机,自动装填机
~ flow control　流量自动调节
~ frequency control　自动频率调节
~ injection control　自动注塑控制
~ injection device　自动注塑装置
~ injection mo(u)lding　自动注塑成型
~ injection mo(u)lding machine　自动注塑机
~ lubrication　自动润滑
~ machine set - up　自动调整机
~ material loader　自动装料机〈注塑机用〉
~ mo(u)ld　自动模具
~ mo(u)lding　自动模塑
~ mo(u)lding machine　自动成型机,自动模塑机
~ packaging unit　自动包装机
~ parision controller　型坯自动控制器
~ pneumatic control　自动气动控制
~ preforming machine　自动预成型机
~ preset　自动程序调整
~ press　自动模压机
~ production　自动化生产
~ programming　程序自动化
~ reel change　自动更换卷轴〈薄膜卷取〉
~ regulation　自动调节,自动控制
~ scrap recovery　边角料自动回收
~ screen pack changer　自动换(过滤)网器
~ spray coating　自动喷涂
~ spray equipment　自动喷涂设备
~ spray gun　自动喷枪
~ spraying　自动喷涂
~ switch　自动开关
~ system　自动化系统
~ temperature controller　自动温度控制器
~ thin - layer spreader　自动薄层涂布器
~ time switch　自动定时开关
~ timer　自动定时器
~ transfer press　自动压铸机,自动传递(成型)压机
~ traverse　自动横向进刀
~ vacuum feeding　自动真空给料
~ vacuum forming　自动真空成型
~ weigher　自动秤
~ weighing　自动称量
~ weight feeder　自动称量喂料器
~ welder　自动焊接机
~ welding　自动焊接
automatical laser inspection　自动激光检验〈薄膜厚度〉
automaticity　自动化程度
automation　自动化
automotive　自动(车)的;汽车的
~ engineer　汽车工程师
~ engineering　汽车工程
~ industry　汽车制造业
~ mouldings　汽车用模塑件
autooxidation　自动氧化(作用)
autopolymer　自聚物
autopolymerization　自聚合
autoscreen system　自动滤网系统
autoshut nozzle　自闭喷嘴
autothermal　自(加)热
~ extrusion　自热挤塑
auto - transfer [automatic reel changing]　自动更换卷轴〈薄膜卷取〉
autoxidation　自动氧化
aux. [auxiliary]　辅助的,附属的
auxiliary　辅助的
~ agent　助剂
~ apparatus　辅助设备
~ attachment　附件
~ cylinder system　辅助液压缸系统
~ drive　辅助传动
~ equipment　辅助设备

auxochromic

~ extruder 辅助挤塑机
~ material 辅助材料,辅料
~ operations 辅助操作程序
~ ram 辅助活塞
~ ram transfer nonlidug 辅助活塞式压铸,辅助柱塞式传递成型
~ sprue 辅助浇口
auxochromic 助色的
av. [average] 平均的
available 有效的
~ capacity 有效能力,有效容量
~ energe 有效能量
~ life 有效寿命
~ load 有效负荷
~ power 有效功率
~ surface 有效表面
~ time 有效时间
AVCS [acrylonitrile – vinylcarbazole – styre – ne copolymer] 丙烯腈–乙烯咔唑–苯乙烯共聚物
average 平均(的);平均值
~ degree of polymerization 平均聚合度
~ density 平均密度
~ depth 平均深度
~ deviation 平均偏差
~ effciency 平均效率
~ error 平均误差
~ fibre length 平均纤维长度
~ flow 平均流量
~ life [period] 平均寿命
~ molecular weight 平均相对分子质量
~ orientation 平均取向度
~ output 平均产量
~ particle diameter 平均粒径
~ particle size 平均粒径,平均粒度
~ piston speed 平均活塞速度
~ pressure 平均压力
~ quality level 平均质量水平
~ rate 平均速率;平均挤出能力
~ room condition 平均室内条件,正常室内气候(25℃时40%相对湿度)
~ sample 平均试样
~ sample number 平均取样数
~ size 平均尺寸

~ size of crystallite 晶粒的平均粒度
~ stiffness 平均劲度
~ stress 平均应力
~ temperature difference 平均温差
~ value 平均值
~ velocity 平均速度
~ viscosity 平均黏度
AW [actual weight] 实际重量
away 离开;去掉
axial 轴(向)的
~ angle 光轴角
~ area of screw channel 轴向螺槽面积
~ clearance 轴向间隙
~ compression 轴向压缩,轴向压力
~ direction 轴向
~ expansion 轴向膨胀,轴向伸长
~ extruder head 轴向挤塑机头,直通式挤塑机头
~ flight land width 轴向螺棱面宽度
~ flow 轴向流
~ force 轴向力
~ groove 轴向槽纹
~ head 轴向机头,直机头〈挤塑机〉
~ length 轴长
~ mixing 轴向混合
~ movement of core 模芯轴向运动〈吹塑〉
~ orientation 沿轴取向
~ Seal 轴向密封
~ serew channel width 轴向螺槽宽度
~ skewing 轴交叉
~ strain extensometer 轴向应变(拉伸)延伸仪〈美〉
~ strain sensor 轴向应变(拉伸)传感器
~ strength 轴向强度〈管〉
~ stress 轴向应力
~ symmetry 轴对称
~ tension 轴向拉伸
~ tensionmeter 轴向拉伸仪
~ thrust 轴向推力〈螺杆〉
~ thrust bearing 轴向止推轴承
~ velocity 轴向速度
~ winding 轴向缠绕

axially symmetric injection mo(u)lding 轴对称注塑;轴对称注塑件
axis 轴(线,心)
~ crossing 辊－轴交叉
~ of abscissae 横坐标轴
~ of ordinate 纵坐标轴
axle 轴
~ bearing 轴承
~ box liner 轴箱内衬
~ box slide 轴箱滑板
~ guard sliding block 轴护挡滑块
azelaic acid 壬二酸
azelate 壬二酸酯
Azlon fibres 人造蛋白质纤维,再生蛋白质纤维
azeotrope 共沸[恒沸]混合物
azeotropic 共沸的,恒沸的
~ copolymer 恒分共聚物,共沸共聚物
~ copolymerization 恒比共聚合(作用),恒分共聚合(作用),共沸共聚(作用)
~ drying 共沸干燥
~ graft polymerization 共沸[恒沸]接枝共聚
~ mixture 共沸[恒沸]混合物
~ point 共沸点
~ temperature 共沸温度
azide SB 苯磺酰叠氮〈发泡剂〉
azidine dyes 叠氮染料
azine dyes 吖嗪染料,对氮(杂)蒽型染料

azo 偶氮
~ α,α' ~ －bis－2,4－dimethyl－valeronit－rile α,α'－偶氮双－2,4－二甲基戊腈〈引发剂,发泡剂〉
~ colo(u)r 偶氮染料
~ compound 偶氮化合物
~ coupling 偶氮偶合
~ dye 偶氮染料
~ forming agent 偶氮类发泡剂
~ initiator 偶氮引发剂
~ lake 偶氮色淀
~ pigment 偶氮颜料
azobenzide 偶氮苯〈发泡剂,引发剂,促进剂〉
azobisformamide [ABFA]偶氮二甲酰胺[发泡剂AC]
azobisisobutyronitrile [AIBN] 偶氮二异丁腈〈引发剂,发泡剂,硫化剂〉
azodicarbonamide 偶氮二甲酰胺〈发泡剂〉
azodicarboxylic acid 偶氮二羧酸,偶氮二甲酸
azoform E 偶氮二甲酸乙二酯〈发泡剂〉
azoic 偶氮
~ colour 偶氮染料
azoic dye 偶氮染料
2,2'－azoisobutyronitrile 2,2'－偶氮(二)异丁腈〈引发剂,发泡剂,硫化剂〉
azure 天青(色)

B

B [boron] 硼〈元素符号〉
B－stage B阶段,乙阶段,半熔阶段,不溶可熔阶段〈酚醛树脂〉
~ resin 乙阶酚醛树脂,B阶(酚醛)树脂,熔酚醛树脂,不溶可熔酚醛树脂
Ba [barium] 钡

back 背面,反面;底板〈胶合板〉
~ axle 后轴
~ bearing 后轴承
~ cloth [backing cloth] 背衬布;底布
~ coat 背面涂层,底面涂层
~ coating 背面涂层,背面涂布

backbone

- ~ crossing 底部交叉阻隔层〈胶合板〉
- ~ draft 反斜度,反锥度;后牵伸;逆流通风
- ~ eddies 可逆涡旋〈挤塑机头〉
- ~ filling 回填(土);背面上浆,单面上浆
- ~ flow 背压流;回流;返流
- ~ flow barrier 止回流器
- ~ flow condenser 回流冷凝器
- ~ flow cut-off 止回流装置
- ~ flow facter 返流系数
- ~ flow index 返流指数
- ~ flow stop [valve] 止逆阀
- ~ hand weld 反手焊接
- ~ lash 轴隙〈压延机〉
- ~ layer (层合板)底板
- ~ lining 背衬
- ~ matter 附件〈资料〉
- ~ mixing 回混,逆向混合
- ~ number 过期杂志;过期刊物
- ~ nut 支承螺母;锁紧螺母,后螺母
- ~ offset roll 后侧辊〈L型压延机〉
- ~ out 旋出,旋下
- ~ plate [backing plate] 垫模板
- ~ pressure 背压,返压,反压
- ~ pressure control 返压控制〈注塑机〉
- ~ pressure relief port 减返压口,背压泄料机
- ~ pressure spindle 返压螺杆
- ~ pressure valve 背压阀,反压阀,止压阀〈注塑机〉
- ~ roll 后滚筒,后辊;支承辊,轧辊
- ~ shaft 后轴
- ~ shop 辅助车间,修理车间
- ~ sizing 背面涂布,单面上浆
- ~ stress 反应力;背应力
- ~ stroke 回程
- ~ swing 回摆
- ~ taper 反斜度;倒拔梢,倒锥度
- ~ -up 支承;后备,备用
- ~ -up and mo(u)ld mounting plate 模板和装模板〈注塑模具〉
- ~ -up coat 背衬涂层,外涂层
- ~ -up material 填隙料
- ~ -up plate 底板,垫模板
- ~ -up ring 保护圈,垫圈
- ~ -up roll(er) 垫辊,托辊;支承辊〈压延机〉
- ~ -up unit 连续装置,随动装置
- ~ view 后视图
- ~ water 逆水

backbone 主链
- ~ chain 主链,主链结构
- ~ structure 主链结构

backed fabric 涂胶浆织物,衬里织物
backer 衬里
backface 反面,背面
background 本底
- ~ colo(u)r 底色
- ~ material 底材,基材

backhand [backward] welding 右旋转焊接,右向焊接,反手焊接
backing 衬料;涂胶浆〈薄膜与织物背衬〉背面,反面
- ~ bead 底焊;背面焊道
- ~ board 底托板
- ~ cloth 背衬布;底布
- ~ coat 底涂层,底漆
- ~ fabric 衬料织物
- ~ layer 衬垫层
- ~ material 衬料,衬垫材料
- ~ plate 垫模板
- ~ plate on ejection side 顶出侧上的模板
- ~ plate on the feed side 喂料侧模板〈注塑模具〉
- ~ press 贴合机
- ~ pressure 背压,反压力
- ~ ring 垫环
- ~ roll 反压辊;垫辊,支承辊
- ~ run 封底焊道
- ~ strip 垫板
- ~ weld 底焊,底焊焊缝

backlining 衬板,背衬
backplate 后挡板,护板;背面板
backspring 反向弹簧,回动弹簧
backward 向后的,反向的
- ~ running 反转

bactericide 杀菌剂
bacteriostatic 抑菌剂
~ fibre 抑菌性纤维
~ plasticizer 抑菌性增塑剂
baffle 导流塞;限流片;导流板;挡板
~ board 隔板,挡板;折流板,反射板
~ member 截流板
~ plate 折流板,限流板,挡板
~ - plate impact mill [impeller breaker] 叶轮式轧碎机
~ position 节流阀位置
baffler 隔板,挡板,折流板
bag 袋
~ closing machine 封袋机
~ closure 封袋
~ filling and closing (or sealing) machine 装袋和封袋机
~ header 封袋机带
~ - in - bag 双层袋
~ in can 罐衬袋
~ liner 袋衬里
~ - making machine (膜)袋焊封机,制袋机
~ no(u)ld 气胎模,(膜)袋模
~ no(u)lding 袋压成型,气袋成型
~ of cement 水泥袋
~ of the bottom - fold type 底部折叠式袋〈由板材制〉
~ of the flat type 扁平型袋〈由板材制〉
~ of the side - fold type 侧面折叠式袋〈由板材制〉
~ packing machine 装袋机
~ seal 封袋
~ sealer [welder] 封袋机,(膜)袋焊封机
~ sealing (膜)袋焊封
~ sealing machine (膜)袋焊封机
~ stitching machine 缝袋机
~ template 袋(焊封)型板〈袋焊封机〉
~ with tear tape 用撕裂膜条制成的袋,撕裂膜袋
bagger 装袋机;制袋机

~ shrink tunnel 袋收缩巷道
bagging machine 装袋机
bake 烘(烤)干,烧硬(固)
~ cycle 烘烤周期
~ oven 烘箱
~ after 后烘焙
Bakelisator [baking oven] 烘烤炉
bakelite 酚醛树脂,电木,胶木〈俗〉
~ A 甲阶酚醛树脂,可溶可熔酚醛树脂
~ B 乙阶酚醛树脂,半熔酚醛树脂,不溶可熔酚醛树脂
~ C 丙阶酚醛树脂,(不溶)不熔酚醛树脂
Baker - Perkin's kneader 贝克尔 - 珀金捏和机,乙型捏和机
~ mixer 贝克尔 - 珀金捏和混合机;西格马捏和机
baking 烘烤,烘干;硬化〈塑料表层〉
~ coating 烘烤涂层
~ enamel 烘干瓷漆
~ finish 烘(干面)漆
~ oven 烘箱,烤炉;硬化炉〈树脂〉
~ press [block press] 压块机,模压机
~ resin 烘干树脂
~ temperature 烘干温度〈65.6℃以上〉
~ varnish 烘干清漆,烤漆
balance 平衡,均衡
~ of assembly 装配程序
~ runner 平衡流道,对称流道
~ weight 平衡锤;砝码
balanced 平衡的
~ construction 平衡构造,对称叠层
~ design 平衡设计,等应力缠纹
~ film 双向均衡拉伸薄膜
~ gating 对称浇口
~ laminate 均衡层合板
~ mo(u)ld 对称模槽式模具
~ orientation 平衡定[取]向结构〈塑料〉
~ oriented film 均衡定向薄膜
~ plywood construction 平衡结构胶合板,各向同性胶合板

bale
　~ runner　平衡流道,对称流道〈注塑模〉
　~ state　平衡状态
　~ -type isotension head contour　平衡型等张力封头曲面〈玻璃纤维绕缠〉
bale　叠合板坯;打包
　~ breaker　拆包机
　~ press　包装机,打包机
baler　打包机
　~ sack　装袋机
　~ twine plant　捆扎绳装置
baling　打包;包装
　~ machine　打包机
　~ press　包装机,打包机
ball　球;滚珠
　~ and ring softening point　环球法软化点
　~ and ring test　环球法试验,环球试验
　~ and-socket joint　万向节,球窝关节
　~ bearing　滚珠轴承
　~ check nozzle　球阀注嘴,球式止逆喷嘴
　~ check valve　球式止逆阀
　~ cock　球状旋塞小型球阀
　~ crusher　球磨机
　~ impact test　落球冲击试验
　~ indentation hardness　球压痕硬度
　~ indentation test　球压痕试验
　~ joint　万向节,球窝关节
　~ mill　球磨机
　~ milling　球磨
　~ molecule　球形分子
　~ rebound test　落球回弹试验
　~ seal　球形密封
　~ test　球压试验
　~ tester　球压(法)硬度试验器
　~ thrush bearing　滚珠止推轴承
　~ -up　自聚
　~ valve　球(形)阀
　~ viscosimeter　落球黏度计
banana　香蕉
　~ liquid　香蕉水
　~ oil　香蕉油,乙酸戊酯
　~ roller　香蕉形辊

Banbury mixer　斑伯里混炼机,斑伯里密炼机
banburying　密炼
band　带,条
　~ conveyor　带式运输器,传送带
　~ dryer　带式干燥机
　~ guide　带的导向装置
　~ heater　带式加热器,电热圈,电热套
　~ label　条带状标签
　~ leveller　抛光辊组
　~ polymer　带状聚合物
　~ saw　带锯
　~ seal　密封带
　~ sealing　密封带,带式焊封
　~ -separation　谱带离析〈光谱学〉
　~ spectrum　带光谱
　~ spinning　带式(熔融)纺丝
　~ spinning machine　带式(熔融)纺丝机
　~ -type film casting　带状薄膜流延
　~ viscometer　带式黏度计
　~ welding [band sealing]　带式焊封
　~ width　带宽
bandage　轮箍;绷带
bank　辊隙存料,料垄
　~ mark　料垄纹
bar　棒,条;巴〈非法定压力单位,等于10^5Pa〉
　~ agitator　棒型搅拌器
　~ coater　刮条涂布机
　~ cutter　切条机
　~ drawing　棒材拉伸
　~ electrode　棒状电极
　~ gate　棒状浇口
　~ mark　条痕
　~ material　棒料
　~ mould　条模,镶条式模具
　~ pressure　大气压
　~ sealing　热封带式封合
　~ section　型材;棒形断面
　~ sprue　条形浇道〈注道〉
　~ stock　棒材,棒料
　~ tension device　棒状张力器
barbed needle　刺针

Barcol 巴科尔,巴氏
~ hardness　巴氏硬度,巴科尔硬度
~ hardness test　巴科尔硬度试验
~ impressor　巴科尔硬度计
bare‑glass 未梳理的玻璃(纱线、纱束、织物)
barite 重晶石〈填料〉
barium [Ba] 钡
~ azodicarbonate　偶氮二碳酸钡,偶氮二甲碳酸钡盐〈发泡剂〉
~ azodicarboxylate　偶氮联羧酸钡〈发泡剂〉
~ cadmium stabilizer　钡镉稳定剂
~ laurate　月桂酸钡〈热稳定剂〉
~ metaborate　偏硼酸钡〈阻燃剂〉
~ peroxide　过氧化钡
~ ricinoleate　蓖麻醇酸钡〈热稳定剂〉
~ stearate　硬脂酸钡〈热稳定剂,润滑剂〉
~ sulfate　硫酸钡〈填料〉
barrel 机筒,料筒〈注、挤塑机〉;压延辊〈压延机〉;滚筒
~ cooler　机筒冷却器
~ colo(u)ring　滚筒着色
~ control thermocouple　料筒控温热电偶
~ cooling　机筒冷却〈挤塑机〉
~ finish　滚磨光
~ heater　机筒加热器
~ heating　机筒加热〈挤塑机〉
~ heating zone　机筒加热区
~ jacket　机筒外套〈挤塑机〉
~ liner　料筒衬套〈挤塑机〉
~ lining　机筒衬里〈挤、注塑〉
~ mill　辊磨机
~ mixer　鼓式混合机,滚筒式混合机
~ polishing [tumbling]　滚筒抛光〈压模件〉
~ ‑shaped conical roller bearing　圆锥形滚珠轴承
~ sleeve　机筒衬套〈挤、注塑〉
~ support　机筒支架
~ temperature　机筒温度
~ venting　机筒排气
~ wear　机筒磨损
~ with internal armouring　内套机筒〈注射机或挤出机〉
~ zone temperature　机筒段温度
barreling 鼓形卷筒
~ machine　滚光机
barrier 隔板,阻挡层
~ adhesive　防渗黏合剂
~ board　绝缘(纸)板
~ coat　隔离涂层,防渗涂层;防渗涂料
~ coating　防渗涂层
~ component　防渗透材料组分
~ film　防渗透薄膜,隔气性薄膜
~ layer　阻挡层
~ material　防渗透材料
~ plastic　防渗性塑料,隔离性塑料
~ property　防渗性,隔离性,防护性
~ resin　防渗性树脂
~ screw　屏障螺杆
~ sheet　隔离板,防渗板
~ type mixing screws　屏障型混炼螺杆
~ wrap　防渗覆盖层
Barus effect 巴勒斯效应,离模膨胀效应
baryte 重晶石〈填料〉
base 基材;底座;碱
~ cloth　基布
~ coat　底涂层
~ colo(u)r　基本色;底色
~ elastomer　基本弹性体
~ fabric　底布
~ face　基准面
~ flame　底座,支架
~ lacquer　底漆
~ level　基准水平,基准面,地平面
~ line　基线
~ material　原(材)料;基材;接合底材
~ mix　原始混合物
~ part　接合件
~ ‑part material　接合底材
~ plane　基准平面
~ plate　底座,基板
~ point　基点
~ polymer　基元聚合物,原料聚合物

basebox

- ~ pressure　基准压力
- ~ product　基础产物
- ~ runner　主流道
- ~ stock　基本原料;基本组分
- ~ strength　基本强度
- ~ unit　基本单元;链节
- **basebox**　基箱;标准箱
- **basic**　基本的;碱的
  - ~ carbonate white lead　碱式碳酸铅;白铅粉〈热稳定剂〉
  - ~ compound　基础混合料,塑料基料,母料
  - ~ colo(u)r　碱性颜料
  - ~ die blank　模套;〈挤塑机〉模坯料
  - ~ dyes　碱性染料
  - ~ dyeing　碱性染色
  - ~ fibre　未经处理的(玻璃)纤维
  - ~ ion　正离子,阳离子;碱离子
  - ~ ion exchange resin　碱性离子交换树脂
  - ~ lead carbonate　碱式碳酸铅,白铅粉〈热稳定剂〉
  - ~ lead chlorosilicate-sulfate complex　碱式氯硅酸铅-硫酸铅络合物〈热稳定剂〉
  - ~ lead sulfate　碱式硫酸铅〈热稳定剂〉
  - ~ lead sulfate-phthalate　碱式硫酸铅-邻苯二甲酸铅〈热稳定剂〉
  - ~ lead sulfite　碱式亚硫酸铅〈热稳定剂〉
  - ~ lead sulfo phosphite complex　碱式磺基亚磷酸铅络合物〈热稳定剂〉
  - ~ level　基级
  - ~ machine　主机
  - ~ magnesium carbonate　碱式碳酸镁
  - ~ material　基(础)材(料),(基本)原料
  - ~ parameter　基本参数
  - ~ polymer　基元聚合物
  - ~ polymer chain　聚合物基链
  - ~ property　碱性
  - ~ research　基础研究
  - ~ resin　基础树脂

- ~ shaft　主轴
- ~ size　基准尺寸
- ~ solvent　碱性溶剂
- ~ specification　基本参数
- ~ stress　基本应力
- ~ structure　主要构架,骨架
- ~ symbol　基本符号,主要符号
- ~ tolerance　基本公差
- ~ white lead carbonate　碱式碳酸铅,铅白
- **basket weave**　筐形编织,席纹编织,交织
- **bassetite**　铁铀云母
- **bast fibre**　韧皮纤维
- **batch**　批(料),分批;颜料浓缩物
  - ~ agitator　间歇式搅拌器
  - ~ charging　间歇装料
  - ~ distillation　分批蒸馏,间歇蒸馏
  - ~ dryer　间歇式干燥器
  - ~ dyeing　分批染色
  - ~ hopper　定量加料斗
  - ~ impregnating plant　分批[间歇]浸渍装置
  - ~ kneader　分批捏和机,间歇式捏和机
  - ~ mixer　分批混合机,间歇式混合机
  - ~ mixing　分批混合,间歇混合
  - ~ mixing time　批混时间
  - ~ number　批号
  - ~ operation　分批运转;间歇操作[运转]
  - ~ packing　分批包装
  - ~ polymerization　间歇聚合
  - ~ process　成批生产,分批生产,分批加工
  - ~ processing　间歇加工,分批加工
  - ~ production　间歇生产
  - ~ reactor　间歇反应器
  - ~ sampling　分批取样
  - ~ test　间歇试验;分批试验
  - ~ treating　分批处理
  - ~ -up [wind-up]　收卷、卷取
  - ~ weighing　每批称重
  - ~ weight　每批重量
- **batcher**　定量器;配料器,计量给料斗

batching 分批;分批加料
batchwise 分批地;间歇的
~ polymerization 分批聚合;间断聚合
~ production 分批生产,间断生产
bath 浴,槽
~ [suspension] of fluidized powder 流体化粉末浴
~ solution 槽液
~ stretch 浴内拉伸
battery 蓄电池,电池
~ acid 蓄电池酸液
~ age 蓄电池寿命
~ box 蓄电池箱,蓄电池外壳
~ case 蓄电池壳体
~ separator 蓄电池隔板
~ solution 蓄电池溶液,电解液
~ support 蓄电池支架
Baum'e(B'e) 波美度
Bayer method 拜耳公司法〈泡沫塑料加工〉
bayonet 卡口
~ base 卡口插座
~ catch 卡口式连接,插销节
~ catch lid 卡口式连接盖
~ cover 卡口式(连接)盖
~ joint 卡口式连接,插销节
~ lock 卡口式连接
~ socket 卡口套接
BBP [benzyl butyl phthalate] 邻苯二甲酸苄丁酯〈增塑剂〉
BCF [bulked continue filament] 膨化长丝
BCM [bakelite copolymer] 酚醛(树脂)共聚物,酚醛(树脂)-苯乙烯-丙烯腈共聚物
BCT [box compression test] 箱式压缩试验
BD [butanediol] 丁二醇
BDMA [benzyl dimethylamine] 苄基二甲胺
Be [beryllium] 铍〈元素符号〉
bead 珠粒料;焊蚕,焊珠,凸缘
~ board [popcorn plastics] 米花状塑料

~ catalyst 颗粒催化剂
~ cutter 切边机
~ forming 弯边,折边,卷边
~ glass 玻璃珠
~ parting line 圆脊合模缝
~ polymer 珠状聚合物,悬浮聚合物
~ polymerization 珠状聚合,悬浮聚合
~ weld 珠焊,堆焊
~ welding 堆焊焊道
beaded 珠粒状的
header 卷边工具;弯管机
beading 卷边,折边,涨边
~ die 卷边模
~ machine 卷边机
~ mandrel 卷边筒〈美〉
~ press 压弯机,卷边机
~ profile 卷边型材
~ ring 卷边卡环〈美〉
~ roll 波纹轧辊
beads 微珠;空心颗粒
beaker 烧杯
beam 梁;(搅拌)梁叶
~ agitator 桨叶式搅拌机
~ with undulating web 波纹状加固梁
beamed yarn 经轴纱,束状纱
bearing 支承,承载;轴承
~ accuracy 定位精度
~ area 支承面
~ box 轴承箱
~ bush 轴承衬套,轴瓦
~ cage 轴承罩
~ capacity 承载能力
~ plate 支承板,垫板,底板
~ race (滚珠轴承)坐圈
~ sleeve 轴承套筒
~ strength 承载强度,负载[载重]能力;载荷量
~ stress 承载应力
~ struction 承载结构
~ surface 支承表面
~ with plastics lining 塑料里的滑动轴承
beat 研磨,磨碎;打浆
beater 打浆机;捣碎机〈美〉

beating

- ~ addition　打浆加料
- ~ roller　打浆辊

beating　打浆,搅拌
- ~ engine　打浆机；捣碎机

bed　床,底,垫,层,台面
- ~ die　阴模；固定压模
- ~ knife　底刀,固定刀〈切粒机〉
- ~ piece　垫板,底板
- ~ plate　台板；底板；支承板

behaved wood　改性木材,(树脂)浸渍木材

behaviour　情况,状态,行为；性能,特性
- ~ in fires　燃烧行为,着火性

behenamide　山萮酸酰胺〈防粘连剂,爽滑剂〉

Beilstein　拜尔斯坦
- ~ reaction　拜尔斯坦反应
- ~ test　拜尔斯坦试验

Beken　贝肯〈人名〉
- ~ duplex kneader　贝肯双联式捏和机
- ~ mixer　贝肯混合机

bell　插管,套管〈美〉
- ~ and spigot joint　窝接,套接〈美〉

belled pipe end　套接管端

belling equioment　承插口制作装置

bellows　波纹管

belt　带
- ~ assembly line　流水装配线
- ~ conveyor　带式输送机,传送带
- ~ drive　带式传动
- ~ (conveyor) dryer　带式干燥机
- ~ dynamometer　传动式测力计,带状功率计
- ~ grinding machine　带式研磨机

guide　带式导向装置
- ~ haul-off　带式牵引装置
- ~ leather　带皮,带革
- ~ pipe puller　带式管材牵引装置
- ~ press　压带机
- ~ roller　压带轮
- ~ sander　磨带,环形带式磨光机
- ~ -type dryer　带式干燥机
- ~ -type take-off　带式牵引
- ~ weigher　带式称量器,带式计量器

- ~ weigher with feeder meter　带有计量加料器的带式计量器

belting press　压带机

bench　工作台,试验台
- ~ mark　标线,基准标记
- ~ scale research　扩大试验
- ~ test　小型试验

bend　弯曲,挠曲
- ~ bar　弯曲模板；卷边夹板
- ~ brittle point　弯曲脆点,低温断裂温度
- ~ pipe　弯管
- ~ radius　弯曲半径
- ~ stress　弯曲应力,挠曲应力
- ~ strip　弯曲试条
- ~ strength　弯曲强度
- ~ test　弯曲试验

bending　弯曲(的),挠曲(的)
- ~ angle　弯曲角(度)
- ~ brittle point [bend brittle point]　弯曲脆点
- ~ crack　弯曲开裂
- ~ creep　弯曲蠕变
- ~ die　弯曲模
- ~ elasticity　弯曲弹性
- ~ endurance test　弯曲疲劳试验,耐弯曲试验
- ~ experiment　弯曲试验
- ~ failure　弯曲破坏
- ~ fatigue　弯曲疲劳
- ~ fatigue life　弯曲疲劳寿命
- ~ fatigue resistance　耐弯曲疲劳性
- ~ flexure　弯曲挠度
- ~ force　弯曲力
- ~ jig　弯管机〈美〉,弯管夹具
- ~ load　弯曲负荷
- ~ machine　弯曲机
- ~ moment　弯矩,挠矩
- ~ -moment factor　弯(曲力)矩率
- ~ peel test　弯曲剥离试验
- ~ press　压弯机；卷边机
- ~ pulsating test　脉动弯曲试验〈玻纤〉
- ~ radius　弯曲半径
- ~ rigidity　弯曲刚度

~ rolls 辊子弯卷机
~ shear test 弯曲剪切试验,弯曲折断试验
~ stiffness 弯曲劲度,弯曲刚度
~ strain 弯曲应变,弯曲变形
~ strength 弯曲强度
~ stress 弯曲应力,挠曲应力
~ stress tester 弯曲应力试验机
~ support 弯折支架
~ test 弯曲试验
~ test for rigid cellular plastics 硬质微孔塑料弯曲试验
~ vibration test 振动弯曲试验
bent 弯曲
~ lever 曲杆,曲臂,弯杆,弯臂
~ pipe 弯管,肘管
~ plate 曲板,弯板
~ strip specimen 弯曲试样,弯曲试条
~ strip test 弯条试验
~ tube 弯管,曲管
~ wrench 弯头扳手
bentonite 膨润土〔岩〕,皂土,浆土〈填料〉
benzaldehyde 苯甲醛
benzanil colo(u)rs 亚苯基苯胺染料
benzene 苯
~ nucleus 苯核
~ ring 苯环
benzenesulfon butylamide 苯磺酰丁胺〈增塑剂〉
benzidine 联苯胺
~ dyes 联苯胺染料
benzimidazole 苯并咪唑
benzine 汽油;挥发油
benzo – colo(u)rs 苯并染料
benzoate fibre 苯甲酸酯纤维
benzofuran 苯并呋喃
~ resin 苯并呋喃树脂,古马隆树脂
benzoguanamine 苯并胍胺
~ alkyd resin coating 苯并胍胺醇酸树脂涂料
~ resin 苯并胍胺树脂
benzoic acid 苯甲酸
benzol(e) 苯;粗苯

benzophenone 苯(甲)酰苯二苯(甲)酮,苯酮〈俗称〉
~ 3,3',4,4' – ~ tetracarboxylic dianhydride [BTDA] 苯酮四羧酸二酐〈固化剂〉
benzothiazyl – 2 – sulphene morpholide 苯并噻唑 – 2 – 次磺酰吗啉〈促进剂〉
benzoxazole 苯并噁唑
benzoyl [Bz.] 苯(甲)酰基
~ benzene 苯(甲)酰苯,二苯甲酮,苯酮〈俗〉
~ chloride 苯(甲)酰氯
~ peroxide [BPO] 过氧化苯甲酰,苯酰化过氧〈引发剂,硫化剂,漂白剂,氧化剂,交联剂〉
~ peroxide paste 苯酰化过氧浆糊
4 – benzoyloxy – 2,2,6,6 – tetramethyl piperidine 4 – 苯甲酰氧基 – 2,2,6,6 – 四甲基哌啶〈光稳定剂744〉
benzyl 苄基;苯甲基
~ acetate 乙酸苄酯,醋酸苄酯
~ benzoate 苯(甲)酸苄酯
~ – n – butyl adipate 己二酸正丁·苄酯〈增塑剂〉
~ cellulose 苄基纤维素
~ cyanoethyl cellulose 苄基氰乙基纤维素
~ formate 甲酸苄酯
~ 2 – methyl imidazole 1 – 苄基 – 2 – 甲基咪唑〈固化剂〉
~ octyl adipate 己二酸苄基辛酯
~ trimethylammonium chloride 氯化苄基三甲基铵
Berl saddle 马鞍形填料〈蒸馏〉
beryllium [Be] 铍〈元素符号〉
best setting 最佳调整
beta (〈希腊字母〉β
~ cellulose β – 纤维素
~ loss peak β – 损耗峰
~ particle β – 粒子
~ – ray β – 射线
~ – ray gauge β – 射线测厚仪
~ – ray thickness gauge β – 射线测厚仪
between product 中间产品

bevel 斜切;倾斜;斜切的
～ angle 坡口角度,斜切角
～ face 斜面
～ - fit valve 斜座阀
～ gear 斜齿轮,伞[锥]齿轮
～ - gear teeth 斜齿轮啮合
～ joint 斜接
～ tooth 斜齿轮啮合
bevelled 斜切的;倾斜的
～ edge 斜边,斜面
～ lap joint 斜搭接(头)
bevelling 斜削坡口;斜切;倾斜
Bewoid size 局部皂化树脂胶,贝沃德胶
bezel 尖棱
BF₃ 三氟化硼
～ - monoethylamine [BF₃ - MEA] 三氟化硼 - 单乙胺
～ - monoethylamine complex 三氟化硼 - 单乙胺络合物〈固化剂〉
～ - piperidine complex 三氟化硼 - 哌啶络合物〈固化剂〉
～ - triethanolamine complex 三氟化硼 - 三乙醇胺络合物〈固化剂〉
BHET [bis(2 - hydroxyethyl) terephthalate] 双(2 - 羟基乙基)对苯二甲酸酯
Bi [bismuth] 铋〈元素符号〉
bias 偏离;倾斜;偏压
～ cut 斜切
～ cutter 倾斜切割机
～ mop 卷曲抛光[磨光]轮,布轮
～ seam 斜接缝
biased roll 鼓形辊,偏向辊
biaxial 双轴的
～ creep 双轴(各)蠕变
～ crystals 双轴晶(体)
～ deformation 双轴向变形
～ drawing 双轴(向)拉伸
～ extension 双轴向延伸
～ fabric 双轴向织物
～ load 双轴向负荷
～ normal force stress 双轴受正交力,双轴标准应力
～ orientation 双轴取向,双轴拉伸;圆周拉伸〈瓶类〉
～ stress 双轴应力
～ stress - strain measurement 双轴向应力 - 应变测量
～ stretching 双轴拉伸,双轴取向拉伸
～ wind 双轴缠绕
～ winding 双轴缠绕
biaxially 双轴
～ oriented film 双轴取向薄膜
～ oriented polyester film 双轴取向聚酯薄膜
～ oriented polypropylene film [BOPP] 双轴取向聚丙烯薄膜
～ stretched 双轴向拉伸
～ stretched [oriented] tape 双轴向拉伸[取向]带
bibl [bibliography] 书目(提要),文献目录
bicolourimeter 双色比色计
bicolourimetric 双色比色的
bicomponent 双组分
～ fibre 双组分纤维
～ film 双层薄膜,复合薄膜;双组分薄膜
～ film yarn 双组分薄膜纱
～ spinning device 双组分纤维纺丝装置
～ structure 双组分结构
～ tape 双层薄膜条[丝],复合膜胶带;双组分膜胶带
bicone 双锥形筒子〈纺纱锭〉
bidirectional 双向的
～ fabric 双向同性织物
～ chromatography 双向色谱(法)
～ laminate 双向直交层合板,直交层合板
Bierbaum scratch handness test 比尔鲍姆刮痕硬度试验
Bierer - Davis ageing test apparatus 比勒 - 戴维斯老化试验仪
bifunctional 双官能的
～ catalyst 双官能催化剂
～ molecule 双官能分子
～ polycondensation 双官能团缩聚

作用
　~ structural unit　双官能结构单元〈聚合物化学〉
big　大的
　~ inch line　大直径管线
　~ inch pipe　大直径管
　~ injection　大件[大尺寸]注塑
　~ mouth barrel　大口桶
　~ repair　大修(理),大(检)修
Bigelow mat　绗缝毡(垫),比奇洛毡(垫)
biisopropylxanthogenate　二异丙基黄原酸酯,调节剂J
bilateral　双边的;双面的;双向的
　~ drive　双向驱动
　~ fibre　并列型纤维
　~ structure　双面结构,并列型结构
　~ tolerance　双向公差
　~ vulcanization　双面硫化
bilayer　双层
billet　坯料,毛坯;锭料;塑坯预塑
　~ forging　锭料锻塑
billow　气胀
　~ plug forming　塞助气胀成型〈热成型〉
　~ snap-back forming　气胀快速回收成型〈热成型〉
　~ forming　气胀成型〈热成型〉
　~ forming with reversed draw　气胀回成型〈热成型〉
bimolecular　双分子的
　~ disproportionation　双分子歧化,歧化终止
　~ termination　双分子终止
bin　仓,库
　~ activator　料仓防堵器
　~ stability　储存稳定性
　~ stock　储料
binary　二元的
　~ alloy　二元合金
　~ catalyst　二元催化剂
　~ compound　二元化合物
　~ copolymerization　二元共聚合
　~ mixture　二元混合物

binder　粘结剂,黏合剂,黏料
　~ content　粘结剂含量
　~ fibre　黏合织物,无纺布
　~ granules　粒状粘结剂
　~ mat [bonded mat]　粘结纤维毡片
　~ resin　粘结性树脂;树脂粘结剂
　~ spraying　黏料喷涂〈玻纤〉
　~ twine　捆扎绳〈由膜状丝制〉
binding　粘结,黏合
　~ agent　粘结剂
　~ effect　黏合效应
　~ energy　结合能
　~ film　黏合薄膜
　~ force　黏合力
　~ layer　黏合层
　~ material　粘结剂
　~ medium　粘结剂
　~ power　黏合力
Bingham　宾汉
　~ body　宾汉体
　~ flow　宾汉流动
　~ model　宾汉模型
　~ plastic fluid　宾汉塑性流体,宾汉体
　~ plasticity　宾汉塑性
　~ viscometer　宾汉黏度计
　~ yield value　宾汉屈服值
biocide　杀生物剂,抗微生物剂
biocopolymer　生物高分子,生物共聚物
biocyclic thermoplastic　生物区热塑性塑料,可生物循环的热塑性塑料
biodegradability　生物降解能力
biodegradable　可生物降解的〈塑料〉
　~ mulch　可生物降解的地膜
　~ plastic　可生物降解的塑料
　~ polymer　可生物降解的聚合物
biodegradadon　生物降解
　~ pathway　生物降解途径
biological　生物(学)的
　~ agent　生物剂,生物降解抑制剂〈塑料助剂〉
　~ decomposition　生物分解
　~ degradability　生物降解性
　~ degradation　生物降解
　~ oxidation　生物氧化

biomacromolecule

~ polymer 生物高分子,生物聚合物
~ treatment 生物处理
biomacromolecule 生物高分子;生物大分子
biomedical 生物医学的
~ polymer 生物医用聚合物
biopolymer 生物高分子,生物聚合物
biopolymerization 生物聚合
bi‑orienting[biaxial orientation] 双轴取向,双轴拉伸
biostabilizer 微生物稳定剂
biotite 黑云母
bipolar 双极的
~ ion exchange membrane 双极离子交换膜
bipolymer 二元聚合物,二元共聚物
bird's eye 鸟眼〈黏合木材的黏痕〉
~ fabric 鸟眼花纹织物
~ view 鸟瞰图,俯视图
birefringence 双折射
~ relaxation 双折射松弛
bis‑ 〈词头〉双
1,3‑ ~ ‑(aminomethyl) cyclohexane 1,3‑双(氨甲基)环己烷〈固化剂〉
3,9‑ ~ ‑(3‑aminopropyl)‑2,4,8,10‑tetraoxaspiro [5.5] undecane adduct 3,9‑双(3‑氨基丙基)‑2,4,8,10‑四氧杂螺[5.5]十一碳烷加成物〈固化剂〉
~ [3,3‑bis(3'‑tert‑butyl‑4'‑hydroxy phenyl)‑butyric acid] glycol ester] 双[3,3‑双(3'‑叔丁基‑4'‑羟苯基)丁酸]乙二醇酯〈抗氧剂〉
~ (4‑tert‑butylcyclohexyl) peroxy di‑carbonate 过氧化二碳酸双(4‑叔丁基环己酯)〈交联剂,引发剂〉
2,2‑ ~ ‑(tert‑butyl peroxy) butane 2,2‑双(过氧化叔丁基)丁烷〈交联剂〉
1,1‑ ~ ‑(tert‑butyl peroxy) cyclohexane 1,1‑双(过氧化叔丁基)环己烷〈交联剂〉
α,α‑ ~ ‑(tert‑butyl peroxy) diisopropyl benzene α,α‑双(过氧化叔丁基)二异丙苯〈交联剂〉
1,3‑ ~ (2‑tert‑butyl peroxy isopropyl) benzene 1,3‑双(2‑过氧化叔丁基异丙基)苯〈交联剂〉
1,1‑ ~ (tert‑butylperoxy)‑3,3,5‑trimethyl cyclohexane 1,1双(过氧化叔丁基)‑3,3,5‑三甲基环己烷〈交联剂〉
2,2‑ ~ [4‑(2,3‑dibromopropoxy)‑3,5‑dibromophenyl] propane 2,2‑双[4‑(2,3‑二溴丙氧基)‑3,5‑二溴苯基]丙烷〈阻燃剂〉
~ (2,3‑dibromopropyl)‑dichloropropyl phosphate 磷酸双(2,3‑二溴丙酯)二氯丙酯〈阻燃剂〉
~ (2,3‑dibromopropyl)‑trans‑malate 双(2,3‑二溴丙基)反丁烯二酸酯〈阻燃剂 FR‑2〉
~ (3,5‑di‑tert‑butyl‑4‑hydroxy behzyl) sulfide 双(3,5‑二叔丁基‑4‑羟基苄基)硫醚,甲叉4426‑S〈抗氧剂〉
N,N'‑ ~ [3,‑(3,5‑di‑tert‑butyl‑4‑hydroxyphenyl)‑propionyl]‑hex‑amethylenediamine N,N'‑双[[3‑(3,5‑二叔丁基‑4羟基)丙酰基]己二胺[抗氧剂1098]
2,2‑ ~ (4,4‑di‑tert‑butyl peroxy cyclohexyl) propane 2,2‑双(4,4‑二叔丁基过氧环己基)丙烷〈交联剂〉
~ (dibutyl thio‑carbamyl) disulfide 双(二丁基硫代氨基甲酰)二硫化物〈促进剂,硫化剂〉
~ (2,4‑dichlorobenzoyl peroxide) 双(2,4‑二氯过氧苯甲酰)〈硫化剂〉
~ (diethyl thio‑carbamyl)‑disulfide 双(二乙基硫代氨基甲酰)二硫化物〈促进剂,硫化剂〉
N,N'‑ ~ (1,4‑dimethyl pentyl)‑p‑phenylene diamine N,N'‑双(1,4‑二甲基戊基)对苯二胺〈抗氧剂4030〉

~ (dimethyl thiocarbamyl) disulfide 双
(二甲基硫代氨基甲酰)二硫化物〈促
进剂,硫化剂〉
~ (2,3-epoxy cyclopentyl) ether 双
(2,3-环氧环戊基)醚
~ (ethoxy ethoxyethyl) phthalate 邻苯
二甲酸双(乙氧乙氧基乙)酯〈增塑
剂〉
N,N'- ~ (1-ethyl-3-methyl pentyl)-p-phenylene diamine N,N'-
双(1-乙基-3-甲基戊基)对苯二
胺〈抗臭氧剂〉
N,N'- ~ furfurylidene-1,6-hexylene diimine N,N'双呋喃亚甲基-1,6-
亚己基二亚胺〈硫化剂〉
1,2- ~ (2-hydroxy benzoyl) hydrazide 1,2-双(2-羟基苯甲酰)肼
〈稳定剂,铜抑制剂〉
4,6- ~ (4-hydroxy-3,5-di-tert-butyl phenoxy)-2-n-octyl thio-1,
3,5-trizine 4,6-双(4-羟基-3,
5-二叔丁基苯氧基)-2-正辛基硫
代-1,3,5-三嗪〈抗氧剂〉
1,1- ~ (4-hydroxy phenyl) cyclohex
-ane 1,1-双(4-羟基苯)环己烷
〈抗氧剂〉
N,N',- ~ (2-hydroxyethyl) alkyl amine N,N'-双(2-羟基乙基)烷
基胺〈防静电剂〉
~ (β-hydroxyethyl)-γ'-aminpropyl
-triethoxy silane 双(β-羟经乙基)
-γ氨丙基三乙氧基(甲)硅烷
N,N'- ~ (2-hydroxyethyl)-N-(3'
-dodecyloxy-2'-hydroxy propyl)
methyl ammonium methosulfate N,N'
-双(2-羟基乙基)-N-(3'-十二
烷氧基-2'-羟基丙基)甲铵硫酸甲
酯盐〈防静电剂〉
2,2'- ~ (p-hydroxyphenyl) propane
2,2'-二对羟苯基丙烷,二酚
丙烷,双酚 A〈热稳定剂,抗氧剂,增
塑剂〉
1,4- ~ (methoxymethyl) benzene phenol 对苯二(甲氧甲基)苯酚树脂

~ N,N'-methyl-butylmethylene di-ethylene triamine 双-N,N'-甲基
-丁基甲叉二乙撑三胺〈固化剂〉
N,N'- ~ (1-methyl heptyl)-p-phenylene diamine N,N'-双(1-甲
基庚基)对苯二胺〈抗氧剂 288〉
~ (nonyl phenyl) phenyl phosphite 亚
磷酸苯·双(壬基苯)酯〈抗氧剂〉
~ (2-phenyl ethoxy) peroxy dicarbonat
[BPPD] 双(2-苯基乙氧基)过氧
化二碳酸酯〈引发剂〉
~ (phenyl isopropvlidene)-4,4'-di-phenyl amine 双(苯基异丙叉)-4,
4'-二苯胺〈抗氧剂〉
~ (8-quinolinato) copper 双(8-羟
基喹啉基)铜〈防霉剂〉
~ (2,2',6,6'tetramethyl-4-piperi-dine) sebacate 双(2,2',6,6'-四甲
基-4-哌啶癸二酸酯〈光稳定剂
770〉
1,2- ~ (2,4,6-tribromophenoxy) ethane
1,2-双(2,4,6-三溴苯氧基)乙
烷[阻燃剂 FR-3B]
~ (tri-n-butyltin) oxide 双(三正基
锡)氧化物〈防霉剂〉
~ (tributyltin) sulfide 双(三丁基锡)
硫化物〈防霉剂〉
~ (tri-n-butyltin) sulfosalicylate 双
(三正丁基锡)磺基水杨酸盐
~ tridecyl phosphite 亚磷酸双十三烷
酯〈抗氧剂〉
~ (3,5,5-trimethyl hexanoyl peroxide)
过氧化双(3,5,5-三甲基己酰)
〈引发剂 K〉
~ (2,2,4-trimethyl-1,3-pentane-diol monoisobutyrate) adipate 双(2,
2,4-三甲基-1,3-戊二醇单异丁
酸酯)己二酸酯
biscuit 坯〈压塑制品〉;饼干色〈浅灰棕
色〉;模压;粒料
bishydroxyethylsulfone 双羟乙基砜〈交
联剂〉
bismuth [Bi] 铋〈元素符号〉
bisphenol 双酚

~ A 双酚 A
~ A based epoxy resin 双酚 A 型环氧树脂
~ A containing pendant group epoxy resins 双酚 A 侧链型环氧树脂
~ A diglycidyl ether 双酚 A 二缩水甘油醚
~ A disalicylate 双水杨酸双酚 A 酯〈紫外线吸收剂〉
~ A epoxide 双酚 A 环氧化物
~ A epoxy resin 双酚 A 型环氧树脂
~ A polycarbonate 双酚 A 型聚碳酸酯
~ F diglycidyl ether 双酚 F 型二缩水甘油醚
~ F epoxy resin 双酚 F 型环氧树脂
~ PA diglycidyl ether 双酚 PA 二缩水甘油醚
2,2′ - ~ propane 2,2′ - 双酚基丙烷,2,2′ - 双(4 - 羟基苯基)丙烷
~ S diglycidyl ether 双酚 S 二缩水甘油醚
~ S epoxy resin 双酚 S 环氧树脂
bisphenols 双酚类
p,p′ - bisphenyl phenol A 4,4′ - 双苯基酚 A;双酚 A
bistre [bister] 深褐色
bite 渗入力,渗胶能力;侵蚀〈溶剂〉
Bitruder 双螺杆挤塑机
bitumen 沥青
~ - bonded asbestos felt face 沥青粘接的石棉 - 毡覆盖层
~ mould material 沥青模塑材料
~ plastics 沥青塑料
~ resin 沥青树脂
bituminous 沥青的
~ adhesive 沥青黏合剂
~ bound 沥青粘结的
~ paint 沥青涂料
~ plastic 沥青塑料
~ resin 沥青树脂
~ substance 沥青状物质
~ varnish 沥青清漆
bivinyl 1,3 - 丁二烯
~ rubber 丁二烯(基)橡胶

bkn. [broken color] 复色;配合色
black 黑色(的);炭黑;黑斑,污点;无镀层的
~ box 集合腔〈不同物料进行共挤塑时〉,混合室
~ dot 黑斑
~ fibre 黑(色)纤维
~ - filled 炭黑填充的
~ nut 防松螺母
~ patches 黑斑点
~ - plate 无镀层板〈包装〉
~ red 暗红
~ shot 黑迹〈镀敷时的变黑现象〉
~ spot 斑痕,黑斑
~ spun 黑色着色纺丝
~ streak 黑条纹
blackboard core 轻质薄木芯〈夹层结构件的〉
blackening 变黑,发黑;涂黑;黑度
blacking 黑色涂料
blackness 黑(色),黑度,毛面
blade 桨叶〈搅拌〉,搅棒;刮刀,刮板
~ coater 刮板涂布机
~ ejector 扁脱模销,平面脱模销
~ mixer 桨式混合机
~ speed 叶片圆周速率[度]
~ thickness 叶片厚度
~ velocity coefficient 叶片速度系数
blanching 发白
blank 毛坯,坯料,型坯,预制品〈吹塑〉;冲切;空白;无色的
~ experiment 空白试验
~ flange 盲板
~ holder 型坯夹具
~ moulding part 模塑件型坯
~ run 空转
~ stock 无色物料
~ test 空白试验
blanket 涂布单板;压垫,垫板
~ coater 垫带式刮涂机,刀带涂布机
~ coating 垫带式刮涂
~ heating 覆毡加热法
~ spreader [blanket coater] 垫带式刮涂机

~ spreading 垫带式刮涂
blanking 模压件,冲压件;冲裁,模切,冲切
~ and piercing 冲切
~ die 冲裁模,切料模,冲压模
~ press 冲压机,冲切机
two outlines [blanking and piercing] 冲切
blast 鼓风,吹风;喷砂;吹气
~ deflashing 冲击除边
~ drawing 鼓[吹]风牵伸法〈短纤维〉
~ ejection 喷气脱模
~ engine 鼓风机
~ finishing 冲击除边,冲击退光;喷磨修饰
~ grit 喷砂
blasting treatment 喷砂处理
bleached pulp 漂白浆
bleaching 漂白
~ agent 漂白剂
~ fastness (耐)漂白坚牢度
~ of colo(u)r 颜色漂白
~ solution 漂白液
bleb 发泡,起泡
bleed 渗开,渗色〈染色〉;渗料
~ air 放出空气
~ hole 渗料口
~ out 泅出
~ outlet 渗料口
~ plug 渗料口旋塞
~ testing 渗色试验
~ through 渗胶〈胶黏剂〉;渗透
~ valve 放泄阀,放气阀
bleeder 渗料口
~ screw 渗料口旋塞
bleeding 渗色,泅色;渗化,渗开;色料扩散;渗料
~ of additive 添加剂的渗出现象
~ of colourants 颜料扩散
~ resistance 防渗性〈增塑性〉
~ valve 放泄阀,放气阀
bleedout 泛流,树脂渗出
blemish 瑕疵
blend 掺混,混配,混合,共混,混合物,

共混料,混配料
~ composition 掺合组分,混合成分
~ feeder 混合加料器〈注塑机〉
~ from ethylene copolymers with bitumen 乙烯共聚物和沥青的共混物
~ of tetrahydrophthalic anhydride and hexahydrophthalic anhydride 四氢(邻)苯(二甲酸)酐和六氢(邻)苯(二甲酸)酐的共熔混合物〈固化剂〉
~ polymer 共混聚合物
~ ratio 掺合比,调合比
blendable 可掺合
blended 混配的,混合的
~ dyes 混合染料,混配染料
~ product 掺合产品
blender 混合机,掺混机
~ loader 掺混机装料器,旋风料斗
blending 混配,共混,掺混,混合
~ agent 共混剂,掺混剂
~ compound 掺合组分
~ constituent 掺合成分
~ method 调合方法
~ ratio 混配比,混合比
~ resin 掺混用树脂
~ stock 调和原料
~ tank 掺合槽
blind 挡板;盲的,封闭的
~ blocking 模糊压纹;无间隙粘连〈薄膜〉
~ cavity 盲腔
~ end[blind opening] 盲孔;泡罩〈包装〉
~ flange 盖板,盲板
~ hole 盲孔〈塑件〉
~ -pin 盲销
~ opening [blind end, blind orifice] 盲孔;泡罩〈包装〉
~ orifice [blind opening] 盲孔;泡罩〈包装〉
blister 起泡;气泡,空泡;型罩,鼓胀
~ package 泡罩型包装(法)
~ packaging 型罩包装
~ wrapper 泡形罩包装
blistered casting 多孔铸件

blistering 起泡
block 单元,部件;区段;坯料;块料;嵌段;滑车
~ bearing 止推轴承
~ construction 部件结构
~ copolymer 嵌段共聚物,镶嵌共聚物
~ copolymer of ethylene oxide and propylene oxide 氧化乙烯和氧化丙烯的嵌段共聚〈润滑剂,脱模剂〉
~ copolymerization 嵌段共聚(作用)
~ diagram 方块图;结构图
~ flow diagram 方块流程图
~ number 批号
~ polycondensation 嵌段缩聚
~ polymer 嵌段聚合物
~ polymerization 嵌段聚合
~ press 压块机,模压机,压板机
~ program 程序框图
~ resistance 防黏着性
~ rubber 块状橡胶,橡胶块
~ shear test 块切片试验
~ skiving 块料刨片
~ slicing 块料刨片
~ slicing machine 块料刨片机
~ valve 断流阀;节流阀
blocked curing agent 阻塞凝固剂,封闭型固化剂,堵漏固化剂;块状固化剂
blocking 粘连,压黏,黏附,粘住;阻碍,阻塞
~ fibre 嵌段纤维
~ foil 压黏膜,热烫合膜
~ force 粘连力
~ layer 阻挡层
~ point 热烫合温度〈塑料膜〉
~ property 粘结性能
~ resistance 防粘连性,滑溜性〈薄膜〉
~ test 粘连试验
blood red 血红色;氧化铁红
bloom 起霜,渗霜;喷霜
~ - free plasticizer 不起霜的增塑剂
~ out 起霜〈表面〉
bloomed film 霜状薄膜
blooming 起霜〈塑料件表面〉
~ action （丝纱）增艳作用

~ agent 喷霜剂
~ of additive 添加剂起霜
blot 污点,疵点
blow 吹风,吹气;喷;冲击;爆发
~ air 吹塑空气
~ air introduction 吹塑空气输入
~ bending test 弯曲冲击试验
~ bubble forming 气胀成型〈热成型〉
~ case 吹气（扬气）箱,蛋形升液（酸）器
~ core pin [blow mandrel] 吹塑芯棒
~ die 吹塑成型机头
~ dieing 缩口式制品的气胀成型〈美〉
~ extruded film 挤吹塑薄膜
~ extrusion 挤吹〈薄膜〉
~ extrusion plant （薄膜）挤吹装置
~ extrusion process 挤吹塑法
~ extrusion of film 吹塑薄膜挤塑法
~ forming 吹模成型,压气成型
~ gun 清density气枪,气压枪
~ head 吹塑机头
~ hole 气孔,气泡
~ indentor 冲击压痕试验器
~ - in pipe 进气管
~ [inflation] mandrel 吹气[进气]芯棒
~ mo(u)ld 吹塑模具
~ mo(u)ld parting line 吹塑模具合模线
~ mo(u)lded article 吹塑制品
~ mo(u)lded container 吹塑成型容器
~ moulded part 吹塑制件
~ mo(u)lder （挤）吹塑机
~ mo(u)lding 吹塑（成型）;吹塑制品
~ mo(u)lding cycle 吹塑成型周期
~ mo(u)lding die 吹塑成型机头
~ mo(u)lding equipment 吹塑成型装置
~ moulding in the first heat 第一次加热吹塑成型
~ mo(u)lding machine 吹塑成型机
~ mo(u)lding of plastics 塑料挤吹
~ mo(u)lding process 吹塑成型法

- ~ mo(u)lding tool 吹塑模具
- ~ nozzle 吹塑嘴
- ~ -off 吹出,放气
- ~ -off pipe 排气管
- ~ -off pressure 停吹气压
- ~ -off valve 放泄阀,放空阀
- ~ out 吹出,爆裂〈刑坯〉
- ~ -out switch 放气开关
- ~ pin 进气杆,吹气芯棒〈美〉
- ~ pipe 吹管
- ~ pressure 吹塑压力,吹胀压力
- ~ rate 吹胀速率[度]
- ~ ratio 吹胀比
- ~ speed 吹胀速率[度]
- ~ stick 进气杆,吹气芯棒
- ~ tank 受料槽,出料槽
- ~ test 冲击试验
- ~ torch 喷灯,焊接灯
- ~ torch test 喷焰试验
- ~ undercuts 缩口式制品的气胀成型
- ~ -up ratio 吹胀比
- ~ -up test 吹胀试验
- ~ valve 送风阀
- ~ vent 排气口,通气口
- ~ wash 吹洗
- **blower** 吹[送、鼓]风机
- ~ -cooled 吹风冷却的
- ~ cooling 吹风冷却
- **blowing** 吹气,吹塑;起泡,发泡;气胀成型,压气成型
- ~ agent 发泡剂,起泡剂
- ~ agent bubble 发气泡剂
- ~ cavity 吹塑模腔
- ~ -cavity mould 吹塑模
- ~ engine 鼓风机
- ~ head [film blowing head] 吹塑机头
- ~ hollow article die 吹塑成型机头
- ~ in a mo(u)ld 模具吹塑成型
- ~ in a single-piece mould 整体模具吹塑成型
- ~ in a solid mo(u)ld 固体模具吹塑成型
- ~ in a split mo(u)ld 对开模具吹塑成型
- ~ in free space 无模压气成型,无模吹塑成型
- ~ into a mould 有模具吹塑成型
- ~ into the open 无模具吹塑成型
- ~ mandrel 吹气芯棒,进气杆
- ~ mo(u)ld 吹塑模(具)
- ~ of a hot-injected preform 热注塑型坯吹塑
- ~ of an extruded preform 挤塑型坯吹塑
- ~ of an extruded tube 挤塑膜管吹塑
- ~ oven 发泡烘箱,塑化发泡箱
- ~ pipe 吹塑管
- ~ pressure 吹塑压力
- ~ speed 吹胀速率[度]
- ~ station [blowing unit] 吹塑工位
- ~ unit [blowing station] 吹塑单元
- ~ -up zone 吹胀区
- ~ with a rubber blanket 用橡胶垫气胀成型
- ~ with plug 模塞吹塑
- **blown** 吹气,吹塑
- ~ article 吹塑制品
- ~ bottle 吹塑瓶
- ~ container 吹塑成型容器
- ~ -extrusion method 挤吹塑成型法
- ~ film 吹塑薄膜
- ~ film chiller 吹塑薄膜冷却器
- ~ film die 吹塑薄膜口膜
- ~ film die head 吹塑薄膜模头
- ~ film extrusion 吹塑薄膜挤塑法
- ~ -film line 吹塑薄膜机组
- ~ film tower 吹塑薄膜机架
- ~ film tubing 吹塑薄膜软管
- ~ oil 吹塑用油
- ~ piece 吹塑制件
- ~ to shape 吹塑成型的
- ~ tubing 吹塑膜管
- ~ tubular film 吹塑管状薄膜
- ~ -up ratio 吹胀比
- **blue** 蓝,青
- ~ paste 蓝糊;蓝色涂料
- ~ pigments 蓝色颜料
- ~ scale 蓝色标

blued　变蓝的
blueing　发蓝,泛蓝
bluish [blueish]　带蓝色的,浅蓝色的
blunt　钝的,无锋的
　~ edge knife heater bar　钝型焊接加热棒〈用于焊接长缝〉
　~ nose rasp　粗锉〈用于锉塑料、橡胶〉
blush　发白,泛白;温晕
blushing　发白,泛白,温晕
　~ test　发白试验
BMC [bulk mo(u)lding compound]　整体成型料,团状模塑料
BOA [benzyl octyl adipate]　己二酸苄基辛基酯〈增塑剂〉
board　(厚)板
　~ machine　纸板机,圆网造纸机
boat varnish　船用清漆
bob　布轮,擦光毡;浮子〈黏度计压力元件〉
　~ unit　布轮,擦光毡;浮子〈黏度计压力元件〉
bobbin　绕线轴,线轴,筒管
　~ spinning　筒管(式)纺丝
　~ spinning machine　筒管(式)纺丝机
body　身体,机体,壳体;增黏;体质颜料,稠厚性〈漆〉;型坯;挤干〈美〉;干含量〈黏合剂〉;罩,外包封〈包装〉
　~ case　壳体,外壳
　~ coat　体质涂料
　~ colo(u)r　体色,不透明色;车厢面漆
　~ ferce　体力,质量力,惯性力〈物理〉
　~ of equipment　设备壳体
　~ of paint　色漆稠度
　~ of the roll　辊体
　~ of varnish　清漆调度
　~ putty　车身腻子
　~ seat　外表密封
　~ stock　坯料
　~ up　增黏,增稠〈漆,涂料〉
bodying　稠化,变稠
　~ agent　增稠剂
　~ speed　增稠速度,热炼聚合速度
　~ -up　增黏

boil　煮;暗泡
　~ in-pouch　蒸煮袋
　~ -in-the-bag ⒠　蒸煮袋
　~ in-the-bag film　耐煮(沸)包装膜
　~ -in-the-bag package　耐煮袋包装
boiled　煮沸过的
　~ oil　熟油,清油,干性油
　~ water cure　煮沸固化
boilg. [boiling]
boiling　沸腾,煮沸
　~ bed　沸腾床;沸腾层
　~ flash　烧瓶
　~ fastness　耐煮性
　~ heat transfer coefficient　沸腾传热系数
　~ point [b. p.]　沸点
　~ range　沸腾范围
　~ resistance [strength]　耐蒸煮性
　~ temperature　沸点温度
　~ test　煮沸试验〈硬度测定〉
　~ water absorption　耐煮沸性
　~ water test　煮沸试验
boilproof　耐煮的
package　耐煮(沸)包装
bolster　模箍,模套,模圈,模框,支模板
　~ plate　压板
bolt　螺栓;栓接
　~ cap　螺栓帽
　~ die　板牙,螺丝板
　~ -hole　螺栓孔
　~ on　上紧螺栓
　~ on nozzle　上紧螺栓喷嘴
　~ up　用螺栓紧固
bolted joint　螺栓连接
bomb-type calorimeter　弹式量热器
bond　粘接(缝料),胶接;键
　~ area　粘接面,黏合面
　~ clay　黏土
　~ coat　黏合层
　~ -coat weight　涂料用量
　~ energy　键能,结合能,黏合能
　~ failure　粘结失效,粘结不良
　~ fracture　粘结损坏

~ improvement 粘接改善
~ length 键长
~ line 粘接缝
~ properties 粘接性能
~ resistance 耐粘接性
~ shear strength 粘接抗切［抗剪］强度,粘接剪切强度
~ strength 粘接［黏合、粘结］强度
~ stress 黏合应力
~ tester 黏合试验机
~ value 黏合(力)值
bondability 黏连(能)力
~ index 黏合指数
bondant 黏合剂
bonded 结合的,黏合的
~ adhesive 黏合剂,胶黏剂
~ area 粘接面,黏合面
~ assembly 粘接装配
~ composite‐to‐metal joint 复合材料与金属粘连(接)
~ honey comb sandwich panel 粘接的蜂窝夹层板
~ (joint) 胶接接头
~ laminate 黏合层合制件
~ mat 粘结纤维毡片
~ material 胶黏材料
bonder 粘接剂
bonding 粘接,胶接
~ adhesive 粘结剂,胶黏剂
~ agent 粘接剂,胶黏剂
~ area 粘连面,黏合面
~ autoclave 粘接固化用蒸缸
~ cement 黏合胶,填缝黏合剂
~ clay 黏合黏土
~ condition 黏合条件,黏合状况
~ effect 黏合效应
~ energy 结合能;黏合能,键能
~ explosive forming 粘接爆炸复合成型
~ fixture [jig] 粘接夹具,黏合夹紧装置
~ force 黏合力
~ joint 粘接接头,粘接点,粘接部
~ layer 黏合层

~ machine 黏合机
~ material 黏合材料
~ medium 粘接［合］剂,黏合［接］介质
~ method 粘接方法
~ of broken bones 骨折粘接
~ of concrete 混凝土粘结
~ of metals 金属黏合,金属粘接
~ of plastics 塑料粘结,塑料黏合
~ of rubber to metal 橡胶与金属粘接
~ of sheets 薄片［片材］粘接,片材黏合
~ of wood 木材粘接,木材黏合
~ of wood to metal 木材与金属粘接
~ permanency 粘接稳定性,耐粘接性
~ points 黏合点
~ position 黏合位置;键合位置
~ power 粘连(能)力
~ press 粘接压力机
~ pressure 黏合压力
~ process 粘接工艺［技术］,粘接法
~ property 粘接性,黏合性能
~ resins 粘结树脂
~ state 黏合状态
~ strength 粘接强度,黏合强度
~ surface 粘连面,粘合面
~ test 粘接试验
~ tool 粘接［黏合］夹具
~ wire 接合线
~ zone 粘接区段
bone 骨(头)
~ binder 骨胶〈美〉
~ dry 干透(的)
~ dry fiber 绝干纤维
~ dry strength 绝干强度〈纱线〉
~ dry weight 绝干重量,无水干重
~ flour 骨粉
~ glue 骨胶
~ meal 骨粉
~ test 刚性试验
booster 增压器;增效剂
~ box 增效箱〈还原染料着色〉
~ compressor 增压压缩机
~ hose 耐压管

~ ram 增压活塞
~ ram system 增压活塞装置
BOP [benzyl octyl phthalate] （邻）苯二甲酸苄基辛基酯
BOPP [biaxially oriented polypropylene] 双轴定向聚乙烯,双轴取向聚丙烯
BOPP [biaxially oriented polypropylene film] 双轴定向聚丙烯薄膜
bordeaux 枣红
bore 孔;孔径;钻孔
~ core 钻芯
~ hole 镗孔;钻出的孔
~ size 内径
bored roll 钻孔式辊筒,空心滚筒〈压延机辊〉
boric acid 硼酸
boron [B] 硼〈元素符号〉
~ carbide fibre 碳化硼纤维
~ -containing polymer 含硼聚合物
~ fibre 硼纤维
~ fibre reiforced plastics (BFRP) 硼纤维增强塑料
~ resin 硼树脂
~ trilluoride-amine [BF_3-amine] 三氟化硼胺〈硬化剂类〉
~ trifluoride ether complex 三氟化硼乙醚络合物〈催化剂,固化剂〉
boss 毂;凸起部,凸轮
botany bay resin [acaroids resin] 禾木树脂
bottle 瓶
~ blowing machine 吹（塑）瓶机
~ bobbin 瓶型筒子
~ bottom 瓶底
~ -box tool 瓶箱[筐]模具,瓶用周转箱模具
~ cap 瓶盖
~ capping machine 瓶盖机
~ closing machine 封瓶机
~ closure 瓶塞
~ hood 瓶盖
~ jackct 瓶(外)壳
~ muff [bottle jacket] 瓶(外)壳
~ neck 瓶颈

~ pack system 瓶的吹塑-充液系统〈由吹塑成瓶直到充入液体的系统〉
~ packing machine 瓶包装机
~ sleeve [bottle jacket] 瓶(外)壳
~ stopper 瓶塞
~ wall 瓶壁
~ bottling 装瓶,灌注
~ plant 装瓶机
bottom 底,下部
~ backing plate 下垫板,底背板,冲压件垫
~ belt roller 底部导向辊
~ blow 下吹〈吹塑〉
~ blowing 底吹(法),下吹〈吹塑〉
~ board 底板
~ cap strip 底盖材
~ clamp 底夹
~ clamp plate 下夹模板,底夹模板〈注塑模具〉
~ clearance 径向间隙
~ coat 底涂层
~ colo(u)r 底色
~ core 下模芯
~ diameter 底径;(螺纹)内径
~ die 底模
~ discharge 底部卸料
~ dyeing 底层染色
~ ejection 下脱模,底脱模〈注塑〉
~ fastening bolt 底螺栓
~ feed 下部进料
~ filling 底部装料〈包装〉
~ flash 底飞边
~ force 下模塞〈压塑〉
~ force press 下压式压机
~ gate 底浇口
~ grid 下槽架,下槽板
~ insert 底部嵌件
~ knockout 下脱模
~ knockout bar 下脱模架
~ knockout mould 下脱模式模具
~ knockout pin 下脱模销
~ mounting plate 下装模板
~ of screw channel (螺杆)螺槽底
~ of thread 螺纹的根部

~ plate 下模托板〈模具〉,底板,垫板
~ platen 下压台
~ plug 模腔嵌件,下模塞〈压塑模具〉
~ press platen 下压台
~ ply 下部轨道
~ ram 下活塞〈压机〉;阴模
~ ram mo(u)ld 下活塞式模具
~ ram press 下压式压机
~ ram transfer mo(u)lding 下活塞式传递模塑,下活塞式注压成型
~ redius 〈螺纹〉底部圆角半径
~ roll 下辊,底辊〈压延机〉
~ seam weld 底部缝焊〈塑料袋〉
~ stack roll 下托棍,支承棍
~ up 颠倒,底朝上
~ view 底视图,下视图
~ weld 底部焊接〈袋〉
~ width 底宽
~ zone 底区;底层
bottomer 底层机〈袋生产〉
bottoms 下脚料,残料
bounce 回弹,弹跳;反冲
~ -back 反弹,反冲
~ impact force 回弹冲击力
~ impact resilience hardness 回弹冲击硬度
bouncing putty 黏弹性材料
bound energy 结合能
boundary 界面,边界
~ condition 边界条件
~ curve 边界曲线
~ dimensions 边界尺寸,轮廓尺寸
~ effect 界面效应
~ film 边界膜
~ friction 边界摩擦
~ layer 边界层,临界层,界面层
~ layer cooling 边界层冷却
~ material 界面物质
~ roll 限定辊,界面辊
~ strength 界面强度
~ surface 界面
~ tension 界面张力
~ value 边界值
~ zone 界面区

bounding 边界的;黏附的
~ surface 边界曲面
bow 弯曲;扭曲;凹陷
bowl 压延辊〈压延机的〉;轧辊;辊筒
~ calender 辊筒压延机
~ mill 球磨机
~ of a calendar 压延机压辊
~ temperature 滚筒温度
~ width〔face width〕 辊宽度
box 箱,匣,盒;框;外壳,套,罩
~ blank 箱毛坯〈包装〉
~ chamber 箱式模腔
~ for spares 备件箱
~ frame 箱框
~ insert〔box tray〕 箱填充物〈包装〉
~ -shaped injection-moulded part 箱〔盒〕形注塑件
~ spanner 套筒板手
~ spinning 罐式纺丝
~ tray〔box insert〕 箱式料盘
~ tread 内螺纹,阴螺纹
~ wrench 套筒板手
b.p〔boiling point〕 沸点
B.P.〔British Patent〕 英国专利
BPBG〔butyl phthalyl butyl glycollate〕 丁基邻苯二甲酰乙醇酸丁酯〈增塑剂〉
BPO〔benzoyl peroxide〕过氧化苯甲酰〈引发剂,硫化剂,漂白剂,氧化剂,交联剂〉
Br〔bromine〕 溴〈元素符号〉
BR〔cis-1,3-butadiene rubber〕 顺(式)-1,3-丁二烯橡胶,顺丁橡胶
Brabender 布拉本法
~ plastograph 布拉本法塑性仪
~ plastometer 布拉本法塑性仪
~ torque rheometer 布拉本法扭矩流变仪
braced structure〔framework structure〕 构架结构
bracket 托架,支架
brackish water 微〔半〕咸水
Bragg's angle 布喇格角
braid 编织;编织带,编织物
~ reinforcement 编织增强材料

braided hose　内衬编织物软管
brainpower　科学工作者,科学家
brake　制动;制动器,刹车
　~ action　制动作用
　~ actuating lever　制动操纵杆
　~ adjusting　制动调整
　~ apparatus　制动装置
　~ bar　制动拉杆
　~ drum　制动鼓
　~ force　制动力
　~ frictional torque　制动器摩擦力矩
　~ hand lever　制动器手把
　~ hose　制动液软管
　~ lining　刹车垫
　~ pedal　制动踏板
Bramley – Beken kneader mixer　勃拉姆莱－贝肯揉和混合机
Bramley – Beken mixer [Beken – Duplex kneader]　勃姆莱－贝肯混合机
branch　分支
　~ circuit　支管线路
　~ high polymer　支化高聚物
　~ pipe　分流管,支管
　~ point　支化点
　~ tube　支管
branched　分支的,分流的,支化的
　~ chain　支链
　~ chain compound　支链化合物
　~ chain polymer　支链聚合物
　~ chain reaction　支链反应
　~ isomer　支化异构件
　~ junction　分支交联点〈聚合物化学〉
　~ line　支线
　~ molecule　支链分子
　~ monomer　支化单体
　~ polyethylene　支化聚乙烯
　~ polymer　支化聚合物
　~ structure　支化结构
branching　支化,分支;分流
　~ agent　支化剂
　~ coefficient　支化系数
　~ degree　支化度
　~ denstity　支化(密)度
　~ point　支化点

　~ ratio　分支比
　~ reaction　支链反应
　~ unit　支化单元
brand　烘烤〈皮革〉;商标,标记,牌号
　~ mark　商标
brass　黄铜;黄铜色
　~ master　黄铜母模
　~ master mo(u)ld　黄铜母模
　~ slicker　黄铜刮刀
brazing　硬钎焊,铜焊
breadth　宽度,幅度
break　断裂,破坏;破裂面;开模
　~ - away　开模;断开,破裂,裂开
　~ - away speed　开模速度,开启速度〈注塑机或压机〉
　~ in　试车,试运转;嵌入
　~ length　断裂长度
　~ line　破裂线
　~ - off method　断开法,扯开法;撕分法;破坏法
　~ - out material　易磨损材料
　~ - resistant　耐破碎的,耐断裂的,不易碎的,不易破坏的
　~ strength　断裂强度
　~ surface　断裂表面
　~ test　破坏试验
　~ the mo(u)ld　开模
　~ through　穿透,击穿
　~ - up　断裂〈聚合物链〉
　~ - up model　拆散模型
breakable　易破碎的,脆的
breakage　破坏,破裂,断裂
　~ by separation [cleavage breakage]　破碎;脆性断裂
breakdown　破坏,断裂;(介)电击穿,介电强度
　~ filed strength　击穿电场强度
　~ of boundary film　界面膜破坏
　~ of cell structure　泡沫结构破裂
　~ point　击穿点;屈服点
　~ potential　击穿电压
　~ strength　击穿强度
　~ test　断裂试验;介电击穿试验
　~ voltage　击穿电压

breaker [crusher]　破碎器;多孔板
　~ baffle　挡板,折流板,缓冲板;筛板〈挤塑机〉
　~ beater [broke better]　废物磨碎机
　~ plate　多孔板,分流板;筛板,滤料板〈挤塑机〉
　~ strip　垫层
breaking　破坏,破裂,断开,断裂;破乳
　~ behaviour　破碎性,断裂性状;破乳〈乳液〉
　~ down test　耐(电)压试验,击穿试验;断裂试验
　~ elongation　断裂[致断]伸长率,总伸长率
　~ energy　断裂能
　~ extension　断裂伸长
　~ force　断裂力
　~ length of yarn　纱线断裂长度
　~ load　断裂负载
　~ moment　断裂力矩
　~ point　破坏点,断裂点
　~ rolling-mill　滚筒破碎机
　~ strain　断裂应变
　~ stress　断裂应力,破裂应力
　~ strength　断裂强度
　~ strength of yarn　纱线断裂强度
　~ test　破坏试验,断裂试验
breast roll　胸辊〈带气刷〉
breathable　呼吸的,透气的
　~ film　透气性薄膜
　~ leather　透气性皮革
breathe　放气,通气〈模塑材料加工〉;透气
breathing　放气,排气,脱气〈模具〉;透气
　~ a mo(u)ld　开模排气,开模放气
　~ fabric　透气人造革
　~ film　透气薄膜
　~ material　透气材料,多(微)孔材料
　~ package　透气包装
　~ property　透气性
breather　透气材料
bridge　桥;桥接,跨接
　~ bond　桥键

　~ pavement　桥(路)面〈塑料〉;桥面(涂层)
bridging　架桥,搭桥〈分子间〉;架桥现象,结拱〈料斗〉
　~ fibre　搭桥纤维
bridle　张力惰轮[紧带滑轮]机构;缰绳
bright　光亮的,明亮的;嫩的〈颜色〉
　~ coloured　淡色的
　~ crystalline fracture　光亮的晶体断裂面
　~ film　干净薄膜
　~ staple fibre　有光短纤维
　~ yam　有光纱线,有光丝
brighten　使光泽
brightener　增亮剂;抛光剂,增艳剂
brightening　增亮,增白;澄清
　~ agent　增白剂;增亮剂,增艳剂
brigtness　亮度;光泽;透明度
　~ contrast　亮度对比
bril. [brilliant]　明亮的;鲜艳的
brilliance　亮度,光泽;鲜艳〈颜色〉
brilliant　明亮的,鲜明的;艳的〈染料〉
　~ yam　亮光纱线
brimful　充满的,满到边的〈包装〉
Brinell　布里涅耳,布氏
　~ ball test　布氏球压试验
　~ figure　布氏硬度值
　~ hardness　布氏硬度
　~ hardness number　布氏硬度值
　~ hardness test　布氏硬度试验
　~ hardness tester　布氏硬度试验机
brisk　活性的,活泼的
bristle　鬃,刚毛
British Standard [BS]　英国(工业)标准
British Standard Specification [BSS]　英国标准规范
brittle　脆(性)的
　~ -ductile transition temperature　脆性-延性转变温度
　~ failure　脆性断裂
　~ fibre　脆纤维
　~ -flexible transition　脆性彻性转化
　~ fracture　脆性破裂
　~ lacquer　脆性喷漆,挥发性(磁)漆;裂纹漆

brittleness

~ lacquer analysis 裂纹漆法
~ lacquer coating 脆性漆涂层
~ point 脆化点〈塑料材料〉
~ resistance 耐脆性
~ temperature 脆化温度
~ temperature test 脆化温度试验
~ transition temperature 脆性转化温度
~ varnish 脆裂清漆,脆性清漆
brittleness 脆性;脆度
~ point [brittle temperature] 脆化点
~ temperature 脆化温度〈塑料材料〉
~ temperature test 脆化温度试验
broach 绞孔,扩孔;绞刀;拉刀,削刀
broaching 绞孔,扩孔;拉削
~ cutter 绞切;拉(削)刀
~ machine 绞孔机,拉床
broad 宽(阔)的
~ - line NMR 宽谱核磁共振
~ fabrics 宽幅织物
broadloom fabric 宽幅织物
broke beater [broker beater] 废物磨碎机
broken colour 复色,配合色
bromine [Br] 溴〈元素符号〉
~ index 溴指数
~ number 溴值
~ test 溴化试验
~ value 溴值
3 - bromo -1, 2 - epoxy propane 3 - 溴-1,2 - 环氧丙烷〈阻燃剂〉
bromocresol 溴甲酚
bronze 青铜;青铜色
bronzine 青铜制的;青铜色
bronzing 青铜斑;镀青铜
~ liquid 镀青铜液
brown striation 棕条纹,黑条纹
Brownian 布朗的
~ motion 布朗运动〈分子运动〉
~ motion of chain segments 链节布朗运动
~ movement 布朗运动〈分子运动〉
Brunauer - Emmett and Teller method
BET 法〈借助惰性气体物理吸附测定比表面的方法〉

brush 刷;刷涂;刷子
~ cleaning device 刷洗器
~ coater [brush spreader] 刷涂机
~ coating 刷涂
~ fibre 制刷纤维
~ holder 电刷夹持器
~ mark 刷痕
~ method 涂刷法
~ polish 刷光剂;上光剂
~ roll 刷辊
~ roller 涂刷滚棍
~ scraper 刷式刮刀[刮具,刮板]
~ spread coater 刷涂机
~ spreader 刷涂机
~ spreading 刷涂
~ spreading machine 刷涂机
brushability 刷涂性
brushing 刷涂
~ compound 刷涂料
~ consistency 刷涂稠度
~ lacquer 涂(刷)漆
~ property 涂刷性
BS [British Standard] 英国(工业)标准
BS [butadiene - styrene copolymer] 丁二烯 - 苯乙烯共聚物
BSH [benzene - sulfonyl hydrazide] 苯磺酰肼〈发泡剂〉
BSI [British Standard Institute] 英国标准学会
BSS [British Standard Specification] 英国标准规范
BT [thermoplastic poly (1 - butene)] 聚(1 - 丁烯)热塑性塑料
bubble 气泡,暗泡;膜泡〈吹塑薄膜〉
~ cap 泡罩〈塔盘〉
~ collapsing board [guide] 夹膜泡板〈薄膜吹塑〉
~ collapsing rolls 夹膜泡辊〈薄膜吹塑〉
~ expansion zone 膜泡膨胀区〈膜泡吹塑〉
~ films 泡膜
~ formation 起泡,气泡形成
~ forming 气胀成型,膜泡成型

~ forming zone 膜泡成型[膨胀]区〈膜泡吹塑〉
~ instability 膜泡不稳定性
~ open 碎泡
~ mold cooling 用冷却液冷却模具〈注塑〉
~ nucleation 气泡成核作用〈泡沫〉
~ nucleator 气泡成核剂〈泡沫〉
~ pack 泡罩型包装
~ package 泡罩型包装法
~ plate [bubble tray] 泡罩(塔)板
~ rubber 海绵胶
~ tray 泡罩塔板
~ tray column 泡罩塔
~ viscometer 气泡黏度计
bubbler mo(u)ld cooling 喷水式塑模冷却
bubbling 冒泡;沸腾
~ bed 沸腾床
~ point 起泡点
~ polymerization 气泡聚合
buckle 纵向弯曲;皱曲;蜷缩
buckling 蜷缩,皱曲,卷缩;纵向弯曲,翘曲
~ deformation (纵向)弯曲变形
~ depth 弯曲深度
~ force (纵向)弯曲力
~ load (纵向)弯曲负载
~ resistance 耐(纵向)弯曲性
~ strain 弯曲应变
~ strength (纵向)弯曲强度
~ stress 弯曲应力
~ test (纵向)弯曲试验
buff [buffing wheel] 抛(擦)光轮;缓冲;浅黄色;米色;黄色厚革
~ wheel 磨光轮
buffed 磨了面的
~ leather 磨面革
buffer 缓冲器[垫],缓冲装置,阻尼器,保险杠〈由塑料或弹性体制成〉;缓冲剂
~ arm 保险杠
~ bar 缓冲杠
~ coat 缓冲涂层,过渡涂层

~ layer 缓冲层
~ value 缓冲值
buffered 缓冲的
~ silica gel 缓冲硅胶
buffing 抛[磨]光,软皮(布)抛光
~ aggregate 抛[磨]光机
~ compound 抛光剂
~ fabric 抛光(用)织物
~ machine 抛[磨]光机
~ plate 抛光板
~ ring 抛光轮[环]〈压制件去除倒刺〉
~ wheel 抛光轮
build 建立,制造;组合;构造,造型,结构
build-in stress 内应力〈加工形成〉,加工应力
build-up 装配,安装;组成,构造;形成;增长,增加
~ cavity 拼合模槽
~ height 模板间距〈模具〉
~ member 装配部件,拼装构件
~ mo(u)ld 拼合模具
~ of frictional heat 摩擦热形成
~ of pressure 压力构成,压力增长
~ of stresses 应力构成,应力集中
~ tolerance 装配公差
~ type 组合式,装配式
builder 组分
building 建筑,建造;组合,组装
~ board 组装板
~ element 结构元件
~ facade 建筑正面
~ material 建筑材料
~ plastics 建筑用塑料
~ room 装配间
~ site 建筑场所
~ structure 建筑结构
~ technology 建造工艺学
~ trade 建筑行业
built-in 内装的,埋设的;嵌入的;内含的
~ accessories 内装附件
~ contaminat 内含的杂质
~ ejector 模内装顶出器

bulit-up

~ heating 内装加热
~ impeller 内装旋转混合器
~ lubrication 自润滑,无油润滑
~ strain 内应变
~ stress 内应力〈加工形成〉,加工应力
~ thermal balance plant 内装热均衡装置
~ type 内琴式;固定式

bulit-up 组合,组成;装配,增长
~ construction mould 组合[拼合]结构模具
~ foam slab 层状[片状]泡沫塑料
~ mo(u)ld 组合模(具)
~ of interchangeable elements 可更换组合元件
~ pattern 组合模(具)
~ shaft 组合轴
~ stabilizer 复合稳定剂

bulb 球管〈化学〉;灯泡〈电〉

bulge 鼓胀,膨胀;凸出
~ exatrusion 鼓泡挤塑法,吹塑薄膜挤塑法
~ forming (液压)膨胀成型
~ package [crown pack, squeeze packing] 鼓胀包装,拱形包装

bulging (表面)鼓胀〈塑料缺陷〉

bulk 体积;物质;大块(的);散装的
~ article 大型制件;松散[粒状,颗粒,散粒]材料
~ articles 大宗商品
~ compliance 体积柔化
~ compressibility 体积压缩性
~ compression 体积压缩
~ conductivity 体积电导率,导电性
~ creep compliance 体积蠕变柔量
~ [powder] density 表观密度,体积密度;松密度,堆密度
~ elasticity 体积弹性
~ factor 体积系数
~ fibre 散纤维
~ filler 增量填料
~ goods 松散[粒状,颗粒,散粒]材料
~ index 体积系数

~ injection 整体注塑
~ material 松散物料,散装物料;块状材料
~ modulus 体积模量
~ modulus of elasticity 体积弹性模量
~ moulding compound [BMC] 整体成型料,团状模塑料
~ plastics 日常塑料,大宗塑料
~ polymer 本体聚合物
~ polymerization 本体聚合(作用),整体聚合(作用)
~ production 批量生产
~ -properties 本征特性,内在性质,态变特性;堆积性能
~ resin 日常用树脂,大宗塑料
~ sample 大样
~ sepeific gravity 相对表观密度,体积密度,堆密度,松密度〈粒状物〉
~ shipment 整装运
~ storage 大批存储;散装存储
~ strain 体积应变
~ stress 体积应力
~ stretch yarn 膨体(弹力)纱
~ temperature 模塑材料温度
~ viscoelasticity 体积黏弹性
~ viscosity 体积黏度,本体黏度
~ volume 松散体积;总体积
~ weight 体积重量;整重
~ specific gravity 比重

bulked 膨化的
~ continue filament [BCF] 膨化长丝
~ continuous filament yarn 膨化长丝纱
~ yarns 膨体纱

bulkhead 隔壁,隔板

bulking 膨胀
~ density 表观密度,假密度,松密度
~ property 膨胀性;松散性
~ value 单重体积,重量体积;捣实[固]体积

bulky 容积(大)的;体积大的
~ good 大型物品
~ material [voluminous material] 大型材料

bull ring 大型夹模环

bump　碰撞；冲击；隆起
　~ test　冲击试验
bumper　保险杠，挡板，缓冲器［垫］,阻尼器〈由塑料或弹性体制成〉
　~ bar　保险杠杠〈汽车〉
　~ beam　缓冲梁，保险杆杠
　~ collision　保险杠碰撞
bumping　撞击，冲击；放气
bun foom　块状泡沫〈未修边的聚氨酯〉
buna ［buNa］　布纳（橡胶），丁钠橡胶〈德国生产的丁二烯橡胶〉
　~ N　丁腈橡胶，丁二烯-丙烯腈橡胶
　~ rubber　布纳橡胶
　~ S　丁苯橡胶，丁二烯-苯乙烯橡胶
bundle ［bale］　（一）束，捆，包；（胶合板）叠合板坯
bung　栓,塞
buoyancy　漂浮性［能力］,浮力
buoyant　易浮的,有浮力的；漂浮的
　~ force　浮漂（能）力,浮力〈测定塑料密度〉
burial　埋入,埋藏
　~ layer　埋层
　~ trials　土埋试验
buried　埋入的；浸没的
　~ cable　埋置电缆
　~ layer　埋层
　~ pipe　埋入管道,地下敷设管道
　~ pipeline　地下敷设管线
burl　斑点，粒结〈玻璃丝织物上〉
hurling machine　修补机〈玻纤〉
burn　燃烧，焦烧；烧痕
　~ mark　焦斑〈制品缺陷〉
　~ -off　烧损〈粘接〉
　~ -out　烧尽，烧完
　~ -out temperature　完全燃烧温度
　~ spots　焦斑〈注塑缺陷〉
burned　焦烧；焦斑
　~ area　燃烧面积,烧毁面积
　~ mark　煳斑,焦斑
　~ spot　燃烧斑点,烧痕,焦斑〈制品缺陷〉
burner　燃烧器
burning　燃烧

　~ area　燃烧面积
　~ behaviour　燃烧行为,燃烧性能
　~ capacity　燃烧量
　~ -off　烧除［去］〈涂层〉
　~ point　燃烧点
　~ rate　燃烧速率
　~ resistance　耐燃性
　~ speed　燃烧速率［度］
　~ temperature　燃烧温度
　~ test　燃烧试验
　~ time　燃烧时间
　~ velocity　燃烧速度
burnish　整平；磨光
burnisher　磨光辊，抛光辊；磨光器,抛光器
burnishing　磨光；抛光；光泽
　~ brush　抛光刷
　~ roller　抛光棍
burnt　燃烧过的
　~ orange　橙色
　~ sienna　深褐色
　~ weld　焊焦
burr　毛口，毛刺；模压接缝
　~ free　无毛刺
burring　除毛刺,去毛口
burrs　卷边；毛刺；飞边
burst　破裂；爆裂，爆破
　~ pressure　爆破压力〈美〉
　~ speed　破裂速率［度］
　~ strength　脆裂强度；爆裂［破］强度；耐压强度〈美〉
　~ strength tester　爆破强度试验器
　~ stress　爆破应力
　~ test　爆破试验
bursted　填料缺陷〈层合〉
bursting　破裂,开裂,裂纹
　~ disc　爆破用垫圈〈防护〉
　~ layer　爆破层
　~ pressure　爆裂压力,爆破压力
　~ reinforcement　防爆增强材料
　~ strength　爆破强度,破裂强度；开裂强度；脆裂强度；耐压强度
　~ stress　爆裂应力
　~ surface　爆裂表面

bush
~ test 爆破试验,开裂强度试验
bush 衬套;套筒
bushing 衬套;套筒;橡胶活接
~ bearing 衬套轴承
~ retainer ring 衬套夹圈
Buss kneader 布斯捏和机,蜗杆捏合机〈带有动蜗杆和固定捏和齿的捏和机〉
butadiene 丁二烯〈通常指1,3-丁二烯〉
~ - acrylonitrile elastomer 丁二烯-丙烯腈弹性体
~ copolymer 聚丁二烯共聚物
~ fibre 聚丁二烯纤维
~ nitrile rubber 丁腈橡胶
~ resin 聚丁二烯树脂
cis-1,3-~ rubber [BR] 顺(式)-1,3-丁二烯橡胶,顺丁橡胶
rubber [BR] 聚丁二烯橡胶
sodium polymer 丁纳聚合物
sodium rubber 丁纳橡胶
~ - styrene copolymer [BS copolymer] 丁二烯-苯乙烯共聚物
~ viny copolymer 丁二烯-乙烯共聚物
butane - butene fraction 丁烷-丁烯馏分,C_4馏分
butanol [hutyl alcohol] 丁醇
butene 丁烯
~ plastic 聚丁烯塑料
~ rubber 聚丁烯橡胶
cis - butenedioic acid diallyl ester 顺丁烯二酸二烯丙酯,马来酸二烯丙酯
cis - butenedioic anhydride 顺丁烯二酐,马来酐
butane 丁炔
butoxyethyl 丁氧基乙基
~ laurate 月桂酸丁氧乙酯〈增塑剂〉
2 - ~ oleate 油酸2-丁氧乙酯〈增塑剂〉
~ pelargonate 壬酸-2-丁氧基乙酯
2 - ~ stearate 硬脂酸-2-丁氧乙酯〈增塑剂〉
butt 平接,对接

~ end 接合端
~ fusion 对头焊接,对接熔焊
~ fusion joint 对头焊接头
~ fusion jointing [butt fusion] 对头焊接,对头熔接
~ joint 对接;对接接头
~ - join tensile test 对接拉伸试验
~ - pin 对头成孔销
~ seal 对头封接
~ seam 对接缝
~ seam welding 滚对焊
~ strap 对接搭板
~ weld 对头焊缝,对接焊缝
~ welding 对头焊接
~ wrap 对接卷绕带
butterfly valve 蝶(形)阀
button 钮(子);按钮〈开关〉
~ blank 纽扣毛坯
~ measurements 纽扣尺寸
~ mo(u)ld 纽扣模具
~ switch 按钮开关
buttress thread 锅齿状螺纹
butyl 丁基
~ acetate 乙酸丁酯;醋酸丁酯
n ~ acetate 乙酸正丁酯〈溶剂,萃取剂〉
sec ~ acetate 乙酸仲丁酯
~ acetoxystearate 乙酰氧基硬脂酸丁酯〈增塑剂〉
~ acetylricinoleate 乙酰蓖麻醇酸丁酯〈增塑剂〉
n ~ - acrylate 丙烯酸正丁酯
~ alcohol 丁醇
n ~ ~ benzoate 苯甲酸正丁酯
N - tert - benzothiazole - 2 - sulfenamide N-叔丁基-2-苯并噻唑次磺酰胺〈促进剂 NS〉
~ benezyl cellulose 丁苄纤维素
~ benzyl sebacate 癸二酸丁苄酯〈增塑剂〉
n ~ ~ - 4,4 - bis (tert - butylperoxy) valerate 4,4-双(过氧化叔丁基)戊酸正丁酯〈交联剂〉
~ cellulose 丁基纤维素

n - ~ cyclohexyl phthalate 邻苯二甲酸正丁环己酯〈增塑剂〉
~ decyl phthalate [BDP] 邻苯二甲酸丁·癸酯〈增塑剂〉
n - ~ n - decyl phthalate 邻苯二甲酸正丁·正癸酯〈增塑剂〉
~ epoxyoleate 环氧油酸丁酯〈增塑剂〉
~ epoxystearate 环氧硬脂酸丁酯〈增塑剂〉
~ ester of epoxyfatty acids 环氧脂肪酸丁酯〈增塑剂,稳定剂〉
~ 2 - ethylhexyl phthalate 邻苯二甲酸丁酯·2 - 乙基己酯
tert - ~ hydroperoxide 氢过氧化叔丁基〈交联剂〉
2 - (3′ - tert - ~ 2′ - hydroxy - 5′ - methylp - henyl - 5 - chlorbenzotriazole 2 - (3′ - 叔丁基 - 2′ - 羟基 - 5′ - 甲基苯基) - 5 - 氯苯并唑〈紫外线吸收剂 UV - 326〉
~ isodecyl phthalate [BIDP] 邻苯二甲酸丁·异癸酯〈增塑剂〉
~ lactate 乳酸丁酯
~ lithium 丁基锂〈催化剂〉
n - ~ myristate 肉豆蔻酸正丁酯〈增塑剂〉
~ octadeanoate 十八酸丁酯,硬脂酸丁酯
~ octyl phthalates [BOP] 邻苯二甲酸丁辛酯〈增塑剂〉
~ oleate 油酸丁酯〈增塑剂〉
n - ~ palmitate 棕榈酸正丁酯,软脂酸正丁酯〈增塑剂〉
tert - ~ peroxyacetate 过氧(化)乙酸叔丁酯〈交联剂〉
tert - ~ peroxybenzoate 过氧(化)苯甲酸叔丁酯〈交联剂,引发剂〉
tert - ~ peroxy(2 - ethyl hexanoate) 过氧(化)(2 - 乙基己酸)叔丁酯〈交联剂〉
tert - ~ peroxyisopropylcarbonate 过氧(化)叔丁基碳酸异丙酯〈交联剂〉
tert - ~ peroxylaurate 过氧(化)月桂

酸叔丁酯〈交联剂〉
n - ~ phosphoic acid 磷酸正丁酯
tert - ~ peroxy pivalate 叔丁基过氧化新戊酸酯〈引发剂〉
~ phthalyl butyl glycollate [BPBG] 丁基邻苯二甲酰基乙醇酸丁酯〈增塑剂,溶剂〉
~ ricinoleate 蓖麻醇酸丁酯〈增塑剂〉
~ rubber [BR] 丁基橡胶〈粘接材料〉
~ stearate 硬脂酸丁酯
n - ~ stearate [BS] 硬脂酸正丁酯〈增塑剂,润滑剂,添加剂〉
~ tetradecyl phthalate 丁基邻苯二甲酸十四烷基酯〈增塑剂〉
~ titanate [tetrabutyl titanate] 钛酸四丁酯
butylamine cellulose 丁胺纤维素
butylated 丁基化的
~ melamine - formaldehyde resin 丁醇改性三聚氰胺甲醛树脂
~ melamine resin 丁醇改性三聚氰胺树脂
~ octylated phenol 丁基化辛基苯酚〈抗氧剂〉
~ resin 丁基化树脂
~ styrenated cresol 丁基化苯乙烯化甲酚〈抗氧剂〉
butylene 丁烯
~ glycol 丁二醇
1,4 - ~ glycol bis(β - aminocrotonate) 1,4 - 丁二醇双(β - 氨基丁烯酸酯)〈热稳定性〉
~ plastics 聚丁烯塑料
butylidene 亚丁基,丁叉〈俗〉
4,4′ - ~ bis(6 - tert - butyl - m - cresol) 4,4′ - 丁叉双(6 - 叔丁基间甲酚)〈抗氧剂〉
4,4′ - ~ bis(3 - methyl - 6 - tert - butyl phenol) 4,4′ - 丁叉双(3 - 甲基 - 6 - 叔丁基苯酚)〈抗氧剂〉
4,4′ - ~ bis[(3 - methyl - 6 - tert - butyl phentl) - di - tridecyl phospite]
4,4′ - 丁叉双[亚磷酸(3 - 甲基 - 6 - 叔丁基苯基)二(十三烷基)酯]〈抗

氧剂〉
p – tert – butylphenol　对叔丁基苯酚
tert – butylidene – acetylene resin　叔丁基苯酚乙炔树脂〈增黏剂〉
p – tert – formaldehyde resin　对叔丁基苯酚甲醛树脂〈硫化剂〉
p – tert – butylphenyl salicylate[TBS]　水杨酸对叔丁基苯酚〈紫外线吸收剂〉
butyraldehyde aniline condensate　正丁醛与苯胺缩合物[促进剂 808,防老剂 BA]
butyric acid　丁酸
BVU[British Viscosity Unit]　英国黏度单位
by – pass　旁路,支路
～ extruder　排气式螺杆挤塑机,旁路挤塑机
～ line　旁路管线
by – product　副产品
Bz.[benzoyl]　苯(甲)酰

C

C = [carbon]　碳
C = [centigrade]　百分度
℃[degree centigrade]　摄氏度
C – polymer[condensation polymer]　缩聚物
C – polymerization[condensation polymerization]　缩聚(作用)
C – stage　C – 阶段,丙阶段,不溶不熔阶段〈酚醛树脂〉
～ resin　丙阶酚醛树脂,不溶不熔酚醛树脂
C – type press　C 型压机,颚式压机
Ca[calcium]　钙
CA[cellulose acetate]　乙酸纤维素,醋酸纤维素
CA[Chemical Abstracts]　化学文摘
cab.[cabriolet]　篷式汽车
CAB[cellulose acetate butylate]　乙酸丁酸纤维素
cabinet　橱;箱
～ drier　干燥橱,橱式干燥机
～ dryer　干燥橱,橱式干燥机
cable　电缆;多股线
～ accessories　电缆附件[配件]
～ clave moulding　电缆热压成型
～ clip[cable clamp]　电缆夹(子)
～ coating　电缆护套
～ compound　电缆料
～ core　电缆芯
～ cover　电缆包皮
～ covering　电缆包覆
～ covering die　电缆包覆机头
～ extrusion head　电缆(护套)挤塑机头
～ hose　电缆软管
～ insulant　电缆绝缘,电缆(绝缘)料
～ insulation　电缆绝缘
～ jacket　电缆护套
～ lacquer　电缆(绝缘)漆
～ protective sleeve　电缆保护套
～ sealing compound　电缆密封[绝缘]料〈塑料〉
～ sheath(ing)　电缆护套
～ sheathing compound　电缆护套料
～ sheathing plant　电缆护套挤出装置
～ socket　电缆套管
～ wrapping　电缆缠绕
cabled　加捻的
～ glass filament yarn　玻璃纤维加捻长丝纱
～ glass staple fibre yarn　玻璃纤维加捻短丝纱
～ yarn　加捻多股纱;缆线
cabling　扭绞,绞合〈电缆〉;电缆敷设
cabtyre cable　橡胶护套电缆
cadmium[Cd]　镉

~ ethylhexoate 乙基己酸镉〈热稳定剂〉
~ laurate 月桂酸镉〈热稳定剂〉
~ pigment 镉颜料
~ ricinoleate 蓖麻醇酸镉〈热稳定剂〉
~ stearate 硬脂酸镉〈热稳定剂〉
caesium [Cs] 铯
cafeteria tray 自助菜托盘
Cage 笼(状的);壳体;机架,骨架构造
~ blade 笼式搅拌桨叶〈用于糊状黏合剂〉
~ mill 笼式磨机
~ mixer 笼式搅拌机
~ structure 笼式结构
cake 滤饼;丝饼;试块;团块
~ dyeing 丝饼染色
~ sizing 丝饼上浆
~ spinning 丝饼式纺丝
caking 结块;粘结
~ capacity 块结性
~ power 粘结力
~ property 粘结性
cal. [calorie] 卡〈热量非SI单位〉
calcd. [calculated] 计算的;有计划的;适合的
calcimine 胶黏涂料
calcination 煅烧
calcine 煅烧
calcite 方解石〈一种碳酸钙矿石,填料〉
calcium [Ca] 钙
~ carbonate 碳酸钙〈填料〉
~ Laurate 月桂酸钙〈热稳定剂〉
~ oxide 氧化钙〈填料〉
~ phosphate 磷酸钙
~ phosphate teriary 三代磷酸钙,正磷酸(三)钙〈稳定剂〉
~ ricinoleate 蓖麻醇酸钙〈稳定剂〉
~ silicate 硅酸钙,硅灰石〈填料〉
~ stearate 硬脂酸钙〈稳定剂,润滑剂,脱模剂〉
~ sulfate 硫酸钙,石膏〈填料〉
~ thiocyanate 硫氰酸钙〈溶剂〉
~ -zinc phosphate complex 亚磷酸钙锌合物〈热稳定剂〉

~ -zine stabilizer 钙-锌稳定剂
calculate 计算,预测;打算,计划
calculated 被计算出来的;设计负载
~ area 计算面积
~ average life 平均计算寿命
~ capacity 计算容量
~ load 计算荷载,设计负载
~ stress 计算应力
~ value 计算值
calendar 压延;砑光;压延机;砑[轧,压]光机
~ bowl 压延机辊;砑[轧,压]光机辊
~ coater 涂[镀]压贴机,压延涂布机
~ coating 压延涂布,辊压贴合
~ colo(u)ring 碾光上色
~ configuration 压延机构型
~ effect 压延效应,压延纹理,纹理效应
~ equipment 压延机组[设备]
~ feeding 压延机供料
~ finish 整理压延
~ follow-up unit 压延机跟踪装置
~ grade resin 压延级树脂
~ grain 压延纹理
~ line 压延机生产线
~ plant 压延机组[装置]
~ plate-out 压延结垢,压延沉析贴层
~ roll 压延辊
~ stack 压[轧]光机组
~ train 压延机组,压延生产线,压延成型系统,压延成型工艺流程
~ types 压延机类型
calenderability 压延性能
calendered 压延(的)
~ cellular vinyl 压延法微孔聚氯乙烯
~ film 压延薄膜
~ goods 压延制品
~ sheet 压延片材
~ sheeting 压延片材
~ web 压延片幅
calendering 压延;压[砑]光;压制成型
~ composition 压延混合料
~ compound 压延混合料

calibrate
- ~ condition 压管条件
- ~ direction 压延方向
- ~ equipment 压延装置
- ~ fabric/foam/fabric 泡沫双层压延织物
- ~ force 压延力
- ~ glazing 砑光[压光]
- ~ line 压延生产线
- ~ roll 压延辊;扎光辊
- ~ roll counter bending 压延辊逆向弯曲
- ~ shrinkage 压延收缩
- ~ spots 压延损伤
- ~ temperature 压延温度
- ~ train [calender line] 压延生产线

calibrate 校准;刻度;测量[定]口径;标定,标准化

calbrated shot 定量注射

calibrating 校准;定径;定型〈型材挤塑〉
- ~ and cutting bushing 定型和切割套
- ~ die 定(口)径模(头)
- ~ section 定径部件
- ~ tube 定径套;管式定径器

calibration 校准;标准化;定(口)径〈管材挤塑〉,定型〈型材挤塑〉
- ~ blow mandrel 定型吹塑芯棒
- ~ by mould closing 闭模定型
- ~ curve 标准曲线,校准曲线
- ~ mandrel 定型芯棒
- ~ mark 校准标记
- ~ sample 校准试样
- ~ test 校准试验,标准试验

calibrator 标准器;定型模,定径套〈管材挤塑〉;定径芯杆〈吹塑机〉;定型套〈型材挤塑〉

caliper 卡尺,卡规
- ~ variation 厚度误差

calium titanite fibre 钛酸钾酯纤维

calk 生石灰

caloric 热量(的)
- ~ capacity 热容量
- ~ effect 热效应
- ~ power 热值;发热量
- ~ receptivity 吸热量;热容量

- ~ unit 热量单位〈旧用卡现规定用焦耳,1卡≈4.185焦耳〉
- ~ value 热值,卡值〈旧〉

calorific 热的,热量的,发热的
- ~ balance 热平衡
- ~ efficiency 热效率
- ~ loss 热量损失
- ~ potential 热能发热潜能,潜热能
- ~ power [C.P.] 发热量,热值,卡值
- ~ value 发热量,热值,卡值

calorimeter 量热器[计],卡计〈旧称〉,热量计

calorimetry 量热法

cam 凸轮,偏心轮
- ~ control 凸轮[偏心轮]齿轮;凸轮控制(机构)
- ~ element (螺杆)凸轮混合元件
- ~ press 凸轮式压机
- ~ shaft 凸轮轴
- ~ type tabletting machine 凸轮式压片机

camber 滚筒[辊]中高度;膨胀;起泡;弯曲;突出部〈焊缝〉
- ~ (curving) of cylinder 滚筒[辊]中高度[凸度,鼓度]
- ~ of sheet 板材的翘曲
- ~ test 翘曲试验〈板材〉

cambered bowl 中高辊筒

camel 驼色

camouflage colo(u)ring 伪装染色

camphor 樟脑

CAN = cellulose acetate nitrate 乙酸硝酸纤维素

can 罐〈头〉;圆筒
- ~ dryer 圆筒(式)干燥器
- ~ lid 罐头盖
- ~ package 罐头包装
- ~ packing 罐头包装
- ~ stability 耐储存性,储存稳定性

canister 罐〈包装〉

canter 芯板

cantilever 悬臂梁
- ~ beam impact strength 悬臂梁(式)冲击强度

~ softening point 悬臂梁法软化点
caoutchouc 天然橡胶,生橡胶
~ adhesive 天然橡胶黏合剂
CAP [cellulose acetate phthalate] 乙酸(邻)苯二甲酸纤维素
CAP [cellulose acetate propionate] 乙酸丙酸纤维素
cap 帽;盖;罩;塞
~ bar 皮辊架
~ bottom 底盖
~ closure 盖封〈包装〉
~ joint 梯形连接,锥形连接〈美〉
~ making machine 制盖机
~ nut 螺母;外套螺帽;锁紧螺帽
~ positioning 配盖
Capability 能力,本领,可能性;性能,功率,生产率
capacitance 电容(量)
~ of a capacitor 电容器容量
capacity 容量;电容(量);能力,本领;生产能力;功率
~ current heating 介电加热损耗;高频加热功率〈焊接〉
~ facior 功率;利用率〈材料〉;容量因素
~ for deformation 变形能力;模塑性
~ load 满负荷,额定载荷
~ of heat transmission 热传导能力,传热量
~ of the cooling system 冷却(系)水容量
~ operation 满载操作
~ production 生产能力
~ rating 额定生产率,额定功率
capillarity 毛细(管);毛细(现象)
capillary 毛细管的,毛细的
~ action 毛细(管)作用〈黏合剂表面〉
~ capacity 毛细吸湿量
~ crack 毛细裂纹,细裂缝
~ die 毛细管模头
~ effect 毛细管效应
~ extrusion 毛细管挤塑
~ extrusion rheometer 毛细管挤塑流变仪
~ flow 毛细管流动,细流
~ flow measuring method 毛细管黏度测定法
~ force 毛细作用力
~ leak 毛细管泄漏
~ nozzle 毛细管喷嘴
~ rheometer 毛细管流变仪
~ surface area 毛细管表面积
~ tension 毛细张力
~ tube 毛细管
~ viscometers 毛细管黏度计
~ viscosimeter 毛细管黏度计
capillator 毛细管比色计
capillometer 毛细试验仪
capital 基本的,主要的
~ expenditure 基建费用
~ repair 大修
capping cement 封端胶
caprolactam 己内酰胺
Capron 〈商〉卡普纶〈耐纶6纤维〉
capryl 辛基
~ alcohol 辛醇〈消泡剂〉
~ octyl phthalate 邻苯二甲酸仲·异辛酯〈增塑剂〉
capstan (拉丝)卷筒,牵引盘[辊],绞盘装置〈挤出物用〉
capsular 胶囊的
~ adhesive (微)胶囊黏合剂
capsule 盒,容器;封壳;胶囊
caption 标题,题目;在绘图上说明
car (小)汽车
~ body 汽车车身
~ bottom coating 车底涂层
~ bumper 汽车保险杠
carajura 秋海棠红
caramel 酱色
caramelization 热脱[退]〈玻璃纤维〉
caramelize 热脱[退]浆〈玻璃纤维〉
carbamide 脲;尿素
~ resin 脲醛树脂,碳酰胺树脂
carbanion 碳阴离子;碳负离子
~ polymerization 碳负离子聚合
Carbanthrene dyes 〈商〉卡朋士林染料

carbazol(e) 〈蒽醌还原染料〉
carbazol(e) 咔唑
carbide 碳化物;碳化钙
carbinol 甲醇
carbitol 卡必醇,二甘醇单乙醚〈溶剂,乳化剂,柔软剂〉
carbocyanine dyes 羰花青染料
carbocyclic ladder polymer 碳环梯形聚合物
carbohydrate 碳水化合物,糖类
carbolic acid 苯酚;石碳酸
carbon [C] 碳
~ arc 碳弧
~ arc weathering equipment 碳弧灯耐候试验机
~ bisulphide 三硫化碳
~ black 炭黑,烟黑〈光稳定剂、填料、着色剂〉
~ black dispersion 炭黑分散体
~ black filler 炭黑填充剂
~ chain 碳链
~ chain fibre 碳链纤维
~ chain polymer 碳链聚合物
~ content 碳含量
~ dioxide 二氧化碳
~ dioxide laser(beam) 二氧化碳激光(束)
~ microsphere 碳微球〈填料〉
~ tetrachloride 四氯化碳
~ fibre 碳纤维〈增强剂〉
~ fibre composite 碳纤维复合材料
~ fibre plastic composite 碳纤维塑料复合材料
~ fibre prepreg material 碳纤维预浸渍料
~ fibre reinforced plastics [CFRP] 碳纤维增强塑料
~ fibre reinforced plastic [CRP] 碳纤维增强塑料
~ fibre reinforced polymer composite 碳纤维增强聚合物复合材料
~ fibre reinforced thermoplastic 碳纤维增强热塑性塑料
~ oxide 氧化碳
~ reinforced plastics, CRP 碳增强塑料
~ spheres 碳球
carbonate 碳酸盐[酯]
carbonated 含碳酸的
carbonic acid 碳酸
carbonitrile 腈
carbonium 碳鎓,阳碳
~ colours 碳鎓染料
~ ion 碳鎓离子;阳碳离子
~ ion polymerization 碳鎓离子聚合
carbonization 碳化[焦化](作用)〈塑料燃烧试验〉
~ process 碳化过程
carbonized 碳化的
~ fibre 碳化纤维
~ polyacrylonitrile fibre 聚丙烯腈碳纤维
carbonyl 羰基;碳酰
~ group 羰基
~ resin 羰基树脂
carborane 碳硼烷
~ polymer 碳硼聚合物,聚碳硼烷
~ -siloxane polymer 碳硼烷-硅氧烷聚合物
carborundum 碳化硅,碳硅砂,(人造)金刚砂
Carbowax 〈商〉聚乙二醇〈水溶性有机润滑剂〉
carboxyl nitrile rubber 羧基丁腈橡胶
carboxylated styrenebutadiene 羧基丁苯橡胶
carboxylic 羧基
~ acid anhydride 羧酸酐
~ polymer 羧基聚合物
~ rubber 羧基橡胶
carboxymethyl 羧甲基
~ cellulose [CMC] 羧甲基纤维素
~ hydroxyethyl cellulose [CMHEC] 羧甲基羟乙基纤维素
carburizing 渗碳;碳化
carcase [carcase] 支架,框架,车架,胎体
carcinogenic 致癌的

~ properties 致癌性
~ substance 致癌物
card 卡片,插件;梳理机;花板〈纺丝〉
~ sealing and protecting film 护卡膜〈用于相册等〉
cardan 万向节
cardboard 卡片纸板,硬厚纸板
carded 粗梳的
~ spinning 粗纺法
~ wool 粗纺毛
~ wool yarn 粗纺毛纱
carding 梳理
~ action 梳理作用
carking 颜料沉积〈在涂料中〉
carmine 胭脂红,洋红(色)
carneous 肉色的
carpet 地毯
~ back 地毯背面
~ backing 地毯背衬
~ fiber 地毯纤维
~ underlay 地毯衬垫
~ yarn 地毯纱线
carriage 支架,托架,滑架
carrier 载体,担体,吸收剂;支承物;衬垫;模芯支柱
~ bag 承载袋〈包装〉
~ dyeing 载体染色
~ effect 载色效应
~ fibre 伴侣纤维,载体纤维
~ gas 载气
~ material 载体;基材,基片,衬垫
~ paper 承载纸〈夹层〉
~ pin 嵌件定位销〈模塑品嵌件〉
carrousel 转盘模塑,转盘注射成型
~ lin 模具旋转传送系统〈制备泡沫聚氨酯〉
carrying 装载的;承载的
~ capacity 载质量,装载〈支承〉能力
~ wall 承载壁
cartridge 夹头;管壳;套筒
~ heater 筒式加热器,加热筒
~ shell 筒壳
~ -type filter 筒式过滤器
cascade 分级;级联,串联

~ agitator 串联搅拌机
~ condenser 串联冷凝器
~ connection 级联,串联
~ control (system) 串联控制(系统)〈用于注塑机和挤塑机〉
~ dryer 阶式干燥器
~ dyeing 冲流染色
~ extruder 串联挤塑机
~ flow 阶式流动
~ heat exchanger 阶式热交换器
~ reaction 阶式反应器
~ system 阶式系统;串联系统
~ temperature control 级联温度控制
~ type couler 阶式冷却器
case 外壳,壳体;容器,盒;箱;表面
~ -chilled 表面冷激的
~ dispenser 附计量器的容器
~ harden 表面硬化
~ hardening 表面硬化
~ hardness 表面硬度
Cased [Casinged] **plastic** 凯辛法极化塑料
casein [CS] 酪素,酪蛋白,酪朊〈旧称〉
~ adhesive 酪蛋白黏合剂
~ cement 酪蛋白黏合剂
~ fiber 酪蛋白纤维
~ formaldehyde resin 酪蛋白甲醛树脂
~ glue 酪蛋白胶
~ plastic 酪蛋白塑料
cashew 槚如(树),腰果(树)
~ modified resin 槚如改性树脂
~ nut resin 槚如坚果仁树脂
~ nut shell liquid resin 槚如坚果壳液树脂
~ resin 槚如树脂,腰果树脂
Casing 凯辛法〈氟塑料和聚乙烯的极化法〉
~ crosslinking 凯辛交联法〈惰性气体激活离子交联〉
~ method [process, technique] 凯辛法〈氟塑料和聚乙烯的极化法〉
casing 浇铸,铸塑;外壳,机壳,罩套,外皮
~ hardening 表面硬化,表面淬火〈铸

cast
塑制品热处理〉
~ seal 套管密封
cast 铸塑；流延〈薄膜〉；色调
~ adhesive 浇铸黏合剂
~ casting 流延；热辊层压法
~ coater 流延涂布机
~ cold 低温浇注
~ embossing 流延刻花
~ enblock 整体铸塑
~ film 流延薄膜；浇铸平膜
~ film die 缝口模头，流延薄膜模头
~ film die head 流延薄膜机头
~ film extrusion 流延薄膜挤塑，平膜挤塑
~ foil 流延片材
~ form 铸塑成型；铸型
~ gate 浇口，流道
~ heater 铸封式电热器
~ – in aluminum heater 铝铸封式加热器
~ inblock 整体铸塑
~ – in heater 铸封式电热器
~ – in – place 现场浇铸
~ integral 整体铸塑
~ mo(u)lding 铸塑，铸塑成型品
~ nylon 铸塑尼龙[聚酰胺]
~ plastic 铸塑塑料
~ polymer 浇铸用聚合物
~ polymerization 浇铸聚合
~ product 铸塑制品
~ profile 铸塑型材
~ resin 浇铸树脂；流延树脂
~ resin embedment 浇铸树脂埋[嵌]入
~ resin mo(u)lded material 浇铸树脂成型料
~ sheet 流延片材
~ speed 浇铸速度
~ temperature 浇铸温度
~ welding 铸焊
castability 可铸塑，可流延性
castable 可铸塑的
~ resin 铸塑树脂
caster 铸塑辊；流延机；骨轮；滑轮

casting 浇铸，铸塑；铸塑制品；流延
~ adhesive 浇铸黏合剂，热熔黏合剂
~ area 浇铸面积；流延面积
~ belt 流延环带
~ belt speed 流延环带速度
~ box 铸塑箱[框]，成型箱，模塑箱
~ chamber 铸塑腔
~ compound 流延用混合物；铸塑用混合物
~ department 铸塑车间
~ die 薄膜挤塑模头〈吹塑装置〉
~ drum 流延转鼓
~ film 流延薄膜；浇铸平膜
~ head 涂布头
~ into open moulds [open casting] 敞口铸塑
~ line 铸塑痕；流延痕
~ machine 流延机；铸塑机
~ material 浇铸(材)料
~ matrix 浇铸基体
~ method 浇铸方法
~ mo(u)ld 浇铸模具，铸塑模(具)
~ mo(u)lding 浇铸成型品
~ of film 薄膜流延
~ of resin 树脂浇铸；树脂流延
~ phenolic plastic 浇铸酚醛塑料
~ polymerization 铸塑聚合
~ process 铸塑法；流延法
~ resin 铸塑树脂，浇铸树脂；流延树脂
~ resin into hollow mould 敞口模铸塑树脂
~ roll 流延辊〈可逆辊涂敷〉；定型辊〈挤板机〉
~ skin 铸塑件表面硬层
~ stress 铸塑件内应力
~ support 流延支持体
~ syrup 浇铸浆；铸塑胶液
~ temperature 浇铸温度
~ wheel 流延转鼓轮
castor 蓖麻；铸塑辊；滑轮
~ oil 蓖麻油
Castor – Severs 卡斯托 – 西弗斯
~ extrusion rheometer 卡斯托 – 西弗

斯挤出式流变仪
~ viscosity 卡斯托-西弗斯黏度
catal · [catalogue] 目录
catalin 铸塑酚醛塑料
Catalure 〈商〉脲醛树脂类黏合剂
catalysis 催化(作用)
catalyst 催化剂;促进剂〈固化〉
~ carrier 催化剂载[担]体
~ efficiency 催化剂效率
~ particle 催化剂粒子
~ powder 催化剂(粉)
~ pretreatment 催化剂预处理
~ promoter 助催化剂
~ temperature 催化剂温度
catalytic 催化的;固化的
~ action 催化作用
~ activity 催化活性
~ agent 催化剂
~ hardener 固化促进剂
~ nonselective polymerization 催化非选择性聚合
~ plasticizer 催化增塑剂
~ polymer 催化聚合物
~ polymerization 催化聚合
catalyze 催化
catalyzed 催化了的
~ anionic polymerization 催化阴离子聚合
~ lacquer 催化固化型漆
~ resin 催化树脂,固化型树脂
catalytically hardening paint system 催化固化涂漆系统
catch 捕捉;制动
~ bolt 止动螺栓,防松螺栓
~ frame 挡泥板
~ pan 收集盘,存屑盘;集料器〈涂布机〉
~ plate 收集盘;挡板
catchment tray 集水槽
caterpillar 履带(传动);履带式的
~ band 履带
~ drive 履带传动
~ traction 履带牵引
cathode 阴极,负极

~ breakdown 阴极击穿
~ drive 阴极激励
~ electrochemical polymerization 阴极电化学聚合
~ ray 阴极射线
~ sputtering 阴极溅射〈镀金属〉
~ sputtering process 阴极溅射法
cation 阳离子,正离子
~ exchange 阳离子交换
~ exchange redox polymer 阳离子交换氧化还原聚合物
~ exchange resin 阳离子交换树脂
~ exchanger 阳离子交换剂
~ ion exchange membrane 阳离子交换膜
cationic 阳离子的
~ additive 阳离子掺合剂
~ coordinate polymerization 阳离子配位聚合
~ dye 阳离子染料
~ grafting 阳离子接枝
~ initiator 阳离子聚合引发剂
~ isomerization polymerization 阳离子异构化聚合
~ polyelectrolyte 阳离子型聚电解质
~ polymerization 阳离子聚合
~ softener 阳离子型柔软剂
~ urea resin 阳离子性脲醛树脂
cationite 阳离子交换剂
car's eye 猫眼,鱼眼〈缺陷〉
caul 衬板,模板;垫板〈美〉
cauliflower polymer 卷心菜式聚合物
caulk 衬板;模板;填隙,密封
~ welding 填缝焊接
caulking 密封;填塞,填缝
~ cement 填隙粘接剂
~ compound 填缝料
~ gun 填缝焊枪
~ material 填缝料
~ seam 填缝;焊缝
cauprene 聚戊烯橡胶
caustic 苛性地,腐蚀性的
~ liquor 苛性(碱)液
~ soda 苛性苏打,苛性钠

caustics 焦散线
caution label 警告标签〈包装〉
cavitation 空洞现象;空化
　~ damage 空蚀耗损
cavity 模腔,型腔〈模具的〉;空洞〈泡沫塑料缺陷〉;阴模
　~ back-up plate 夹板,垫板
　~ block 阴模,模腔
　~ cluster 模腔,阴模
　~ depth 模腔深度
　~ die 阴模,凹模,型腔模
　~ filament 气泡丝
　~ flashing 制品飞边,模塑料溢出〈从模具空腔〉
　~ forming 模腔成型
　~ insert 模槽嵌件
　~ mo(u)lding 浇铸成型品
　~ plate 阴模托板
　~ plug 阴模嵌件
　~ pressure control 模腔压力控制
　~ retainer 阴模固定板
　~ retainer plate 阴模托板,模腔套板
　~ side 阴模侧,模腔边
　~ side part 模腔侧部分
　~ space 模槽空间
　~ wall 模槽壁
Cb [columbium] 铌
CBA [chemical blowing agent] 化学发泡剂
cc. [cubic centimetre] 立方厘米
Cd [cadmium] 镉
　~ -pigment [cadmium pigment] 镉颜料
CDP [cresyl diphenyl phosphate] 磷酸二甲苯酯
Ce [cerium] 铈
CEC [cyanoethyl cellulose] 氰乙基纤维素
ceiling 天花板
celadon 雾青(色),灰绿色
celanese colo(u)rs 纤烷丝染料
celanol colo(u)rs 纤烷醇染料
celestol 青(色)
cell 泡孔,微孔;(蓄)电池;容器;槽

~ box 电池箱
~ collapse 微孔萎塌,瘪泡
~ formation 微孔成型〈泡沫塑料成型时〉
~ gas 发泡[起泡]气体,气体发泡剂
~ membrane 泡孔隔膜〈泡沫塑料〉
~ partition 槽穴分布〈模具〉
~ size 泡孔直径,孔眼大小〈泡沫塑料〉
~ srructure 泡孔结构
~ urethane foam 蜂窝状聚氨酯泡沫塑料
~ wall 孔壁〈泡沫塑料〉
cellophane 赛璐玢,玻璃纸〈俗称〉,胶膜
cellular 蜂窝状的,微孔的,泡沫(状)的,多孔(状)的
~ adhesive 微孔黏合剂,泡孔黏合剂
~ body 蜂窝状体
~ cellulose acetate 微孔乙酸纤维素(塑料)
~ construction 蜂窝状结构
~ core 微孔芯核
~ core layer 微孔芯核层
~ leather cloth 微孔人造革
~ material 微孔材料
~ plastics 微孔塑料,泡沫塑料,多孔塑料
~ plastic with open and closed cells 开孔和闭孔微孔塑料
~ polyolefin film 泡沫聚烯烃薄膜
~ polyvinyl chloride 泡沫聚氯乙烯
~ profile 泡沫塑料型材
~ stration 泡孔条纹
~ structure 微孔结构,蜂窝状结构,多孔性结构〈泡沫塑料〉
cellule 微孔,泡孔〈泡沫塑料〉
celluloid 赛璐珞,硝化纤维素塑料,纤维素硝酸酯塑料
~ sheet 赛璐珞板
cellulose 纤维素
~ acetate [CA] 乙酸纤维素
~ acetate butyrate [CAB] 乙酸-丁酸纤维素
~ acetate butyrate plastic 乙酸丁酸纤

维素塑料
- ~ acetate film 乙酸纤维素薄膜
- ~ acetate flakes 乙酸纤维素碎片
- ~ acetate foam 乙酸纤维素泡沫塑料
- ~ acetate mo(u)lding composition 乙酸纤维素成型料〈美〉
- ~ acetate mo(u)lding compound 乙酸纤维素成型料
- ~ acetate nitrate [CAN] 乙酸硝酸纤维素
- ~ acetate phthalate [CAP] 乙酸苯二甲酸纤维素
- ~ acetate plastics 乙酸纤维素塑料
- ~ acetate propionate [CAP] 乙酸丙酸纤维素
- ~ acetate succinate 乙酸琥珀酸纤维素
- ~ aceteto-butyrate 乙酸-丁酸纤维素
- ~ acetobutyrate [CAB] 乙酸丁酸纤维素
- ~ adhesive 纤维素黏合剂
- ~ base fiber 纤维素纤维
- ~ blends 纤维素纤维混配制品
- ~ butyrate 丁酸纤维素
- ~ butyrate fiber 丁酸纤维素纤维
- ~ -coate plate 涂纤维素板
- ~ derivative 纤维素衍生物
- ~ diacetate 二乙酸纤维素
- ~ ester 纤维素酯
- ~ ether 纤维素醚
- ~ fibre 纤维素纤维
- ~ film 纤维素薄膜,(黏)胶(薄)膜,赛璐玢〈俗称玻璃纸〉
- ~ filler 纤维素填料
- ~ filler best general [CFG] 改性用的纤维素填料
- ~ filler best impact [CFI] 改善冲击韧性的纤维素填料
- ~ filler high eleclric [CFE] 改善电性能的纤维素填料
- ~ glycolate 羧甲基纤维素
- ~ ion exchange powder 离子交换纤维素粉

- ~ ion-exchanger 离子交换纤维素
- ~ lacquer 纤维素漆
- ~ membrane 纤维素膜
- ~ nitrate [CN] 硝酸纤维素
- ~ nitrate coating 硝酸纤维素涂料
- ~ nutrate plastics 硝酸纤维素塑料
- ~ plastics 纤维素塑料
- ~ plate 纤维素板
- ~ powder 纤维素粉〈填料〉
- ~ propionate [CP] 丙酸纤维素
- ~ propionate fibre 丙酸纤维素纤维
- ~ resins 纤维素树脂
- ~ triacetate [CT,CTA] 三乙酸纤维素,醋酸纤维素
- ~ varnish 纤维素清漆
- ~ xanthate 黄原酸纤维素
- ~ xanthogenate 黄原酸纤维素

cellulosic 纤维素衍生物;纤维素性的
- ~ fibre 纤维素纤维
- ~ plastics 纤维素塑料

cellulosics 纤维素制品;纤维素塑料
Celuka structural-foam process 塞卢卡结构泡沫塑料成型法
cement 胶浆,胶水
- ~ -lasting 粘接(鞋帮)
- ~ properties 黏合性能

cementation 胶接作用,硬化
cemented 胶黏的,胶合的
- ~ bell and spigot joint 胶黏套接〈美〉
- ~ collar 胶黏管卡圈
- ~ sleeve [socket] joint 胶黏窝接
- ~ sole 胶黏底〈鞋〉
- ~ spigot and socket joint 胶黏窝接
- ~ tube 胶合管

cementing 胶黏;胶合;黏合
- ~ agent 粘接剂
- ~ power 粘接能力
- ~ property 粘接性
- ~ technique 粘接技术,粘接法

cent. [centimeter] 厘米
centage 百分率
center [centre〈英〉] 中心;中间;中心板,中间板〈层压板〉
centigrade 百分度的;摄氏温度的

centigram

~ temperature 摄氏温度
~ thermometer 摄氏温度计
centigram [c.g] 厘克
centimeter [cm] 厘米
centipoise [cP] 厘泊〈动态黏度,非 SI 单位,等于 10^{-3} Pa·s〉
centistokes [cSt] 厘沲〈运动黏度,非 SI 单位,等于 $10^{-6} m^2/s$〉
central 中央的,中心的
~ control 集中控制,中央控制
~ drive 中心传动
~ feed head 集中喂料机头
~ gate 中心铸口〈浇道〉,主注口
~ lubrication 集中润滑
~ mix 集中拌和
~ plate 中间板,分层板〈多层压板〉
~ spindle (中)心轴
~ sprue 中心铸口〈浇道〉,主注口
~ water treatment 集中水处理
centralized control 集中控制
centralized lubrication 集中润滑
centrally-mounted 集中装配的
centre [center〈美〉] 中心;中间;芯板;中间板〈层压板〉
~ bore 定中心孔
~ bowl 中辊
~ cut-out 中间废料〈塑料冲压时〉
~ ejection 中心脱模
~ ejector 中心脱模销
~ fed 中心喂料〈吹塑机头〉
~ -fe(e)d die head 中心喂料模头〈吹塑〉
~ fed head 中心给料机头〈注塑〉
~ flow 溢边;溢边,飞边
~ folded sheeting 中间折叠薄膜
~ galed mo(u)ld 中心浇口模具
~ gate 中心浇口
~ -gated injection mould 中心浇口注塑模具
~ -gated mould 中心浇口模具〈注塑〉
~ guid pin 中心合模销
~ hole 中心孔
~ layer 中间层

~ panel 中央面板
~ roll 中辊〈L型压延机第二辊〉
~ section 中间压板
~ shot pin gate feed 中心针孔浇口进料
~ sprue 中心浇口
~ stepped scarfed joint 中心多级斜口粘接
~ stretch 带拱形〈涂层时〉
~ waste 中间废料〈塑料冲压时〉
~ winder 中心卷取(机)
centreless gringding 无(中)心研磨
centrifugal 离心机;离心的
~ blender 离心搅拌机
~ casting 离心铸塑,离心浇铸,离心成型
~ casting machine 离心铸塑机
~ casting process [technique] 离心铸塑工艺[技术],离心铸塑法
~ dryer 离心干燥机
~ effect 离心效应
~ filtration 离心过滤
~ force 离心力
~ forming 离心成型
~ grinder 离心研磨机
~ impact mill 离心冲击式研磨机
~ impact mixer 离心冲击式混合机
~ impeller mixer 离心叶轮式混合器
~ machine 离心机
~ mixing 离心混合
~ mo(u)lding 离心模塑,离心熔塑
~ polymerization 离心聚合
~ pressure casting 离心压铸(塑)
~ stirrer 离心搅拌机
centrifugally cast laminate 离心铸塑层压板
centrifuge 离心;离心机
~ stock 离心处理原料
centring roller 定心辊
ceramic 陶瓷的
~ adhesive 陶瓷黏合剂
~ fibers 陶瓷纤维〈增强剂〉
~ fibers reinforcement 陶瓷纤维增强
~ fiber reingorcement plastics 陶瓷纤

维增强塑料
ceramoplastics 陶瓷纤维增强塑料
ceraplast 陶瓷塑料
cerise 樱红
certificate of quality 质量证(明)书
certification 证明,确认,鉴定
　~ of quality mark　检验标志〈塑料模制品上〉
　~ of quality of fitness　合格证(书)
　~ of quality of proof　检验证书
　~ of quality test　检定试验,考核试验
cerulean 天蓝色的
cesspipe 污水管
cesspit 污水池
cestnut 栗色
CEtPr [ethylene-propylene-copolymer] 乙烯-丙烯共聚物〈法〉
cetyl 鲸蜡基;十六(烷)基
　~ alcohol　鲸蜡醇〈润滑剂〉
　~ cellulose　十六烷基纤维素
　N,N ~ morpholinium ethosulfate　乙基硫酸 N,N-十六烷基吗啉盐
　~ palmitate　棕榈酸十六(烷)醇酯,软脂酸鲸蜡醇酯〈润滑剂〉
cf [confer;compare] 参考,参see
CF [cresol-formaldehyde resin] 甲酚-甲醛树脂
CFE [cellulose filler high electric] 改善电性能的纤维素填料
CFG [cellulose filler best general] 改性用的纤维素填料
CFI [cellulose filler best impact] 改善冲击韧性的纤维素填料
c.f.m. [cubic feet per minute] 每分钟立方英尺
CFRP [carbon fiber reinforced plastics] 碳纤维增强塑料
c.g. [centigram] 厘克
C-glass [glasfs reinforcement] 玻璃纤维增强
c.g.s [centimeter-gram-second] 厘米·克·秒
chain 链
　~ alignment　链排列
　~ bind　链束
　~ branching　链支化
　~ break　链的断裂
　~ breaking reaction　断链反应
　~ bridging　链桥
　~ carrier　链(增长)载体;链反应活性中心
　~ cessation　链终止
　~ cleavage　链开裂
　~ cleavage additive　链开裂剂〈聚合〉
　~ compound　链化合物
　~ configuration　链构型
　~ conformation　链构象
　~ conveyor　链条输送器
　~ coupling agent　链偶合剂,链偶联剂
　~ crawler　链型履带牵引装置
　~ degradation　链降解
　~ drive　链条传动
　~ element　链节;链单元
　~ end　链终止
　~ extender　增链剂〈聚氨酯添加剂〉
　~ flexibility　链的柔顺性
　~ folded lamellar　折叠链片晶
　~ folded structure　折叠链结构
　~ folding　链折叠
　~ folding structure　链折叠结构〈热塑性塑料〉
　~ formation　链的形成
　~ growth　链增长
　~ growth period　链增长期
　~ -growth polymerization　链增长聚合(作用)
　~ initiation　链引发
　~ length　链长(度)
　~ length distribution　链长分布
　~ molecule　链型分子;线型分子
　~ ordering　链序列
　~ orientation　链取向
　~ polycondensation　链型缩聚
　~ polymer　链型聚合物
　~ polymerizaton　链型聚合
　~ propagation　链增长(作用)
　~ propagation reaction　链增长反应
　~ reaction　连锁反应

chainlike molecule

~ reaction polymerization 连锁(反应)聚合
~ reducing gear 减速链式传动〈挤塑机〉
~ regularly 链结构规则性〈热塑性塑料〉
~ rigidity 链刚性
~ rupture 断链
~ scisson 断链(作用)
~ segment 链段
~ shiftiness 链刚性
~ -stitched 链状缝纫
~ structure 链结构
~ termination 链终止
~ terminator 链终止剂
~ track 履带
~ transfer 链转移;链传递,链条传动
~ transfer agent 链转移剂
~ transfer polymerization 链转移聚合
~ transfer reaction 链转移反应
~ uncoiling 链伸展
~ unit (分子)链单元

chainlike molecule 链状分子
chalk 白垩〈一种碳酸钙〉
~ mark 粉斑,白垩斑
chalking 起垩;起霜
chamber 室,腔
~ area 料腔面积
~ area of transfer mo(u)ld 压铸模料腔面积
~ body of (Banbury) mixer (斑伯里)密炼机密炼室
~ bored roll (加热)内腔辊
~ drier 箱式烘燥机
chambered roll 空腔式辊筒
chamfer 斜切;倒角;倒棱;焊接坡口
chamfering machine 斜切机〈塑料半制品〉
change 变化;更换
~ -can 更换罐〈行星式搅拌器时〉
~ can mixer 换罐式混合器
~ drive 变速传动
~ in [of] state 物态变化
~ in weight 重量变化

~ of lap 换卷,调卷
~ -over 转变;转换;倒转〈机器〉
~ -over time 更换(品种)时间
~ parts 可更换零件[部件]
~ rate 变速率〈组织结构〉
~ speed 变速
~ -tank 更换罐
changer 变换器
changing 更[变,转]换;变化
~ in full operation [changing under full load] 全负载变[转]换
~ of temperature 温度变化
channel U形截面;螺槽;料道;(模具内)冷却水道
~ black 槽黑,槽法炭黑〈补强剂,颜料,填料〉
~ depth 螺槽深度〈螺纹〉
~ depth ratio 螺槽深度比
~ iron 槽钢,U形钢
~ mould U形型材模具〈玻璃纤维增强硬塑料〉
~ section U形型材
~ shaped 槽状的
~ volume ratio 螺槽容积比
channelling 槽沟
~ fluidized bed 槽状流化床
char 烧焦,炭化
~ layer 炭化绝缘层
characteristic 特性;特性的
~ absorption band 特性吸收谱带
~ component 特性组分
~ curve 特性曲线
~ data 特性数据;参数
~ property 特有性能
~ value 特征值〈塑料材料〉
~ viscoelastic parameter 黏弹性特征参数
~ X-ray 特征X射线
characterization of polymer 聚合物的表征
charge 装料,装模;负荷;电荷
~ cavity [charge pot] 装料模腔
~ chute 装料槽
~ mixture 配料

~ of an electron 电子电荷
~ of surety 安全荷载,容许荷载
~ ratio 投料配比
~ sheet 配料单
~ transfer polymerization 电荷转移聚合
charged membranes 荷电膜
charging 给料,装料,加料
~ area 装料场
~ capacity 装载量
~ deck 加料台
~ device 加料装置
~ door 装料口
~ funnel 装料斗
~ hopper 装料斗
~ machine 装料机;加料机
~ tray 装料盘〈压塑模具装料〉
~ trough 加料槽
Charpy 沙尔贝
~ impact strength 沙尔贝冲击强度,简支梁冲击强度,摆锤式冲击强度
~ impact test 沙尔贝冲击试验,简支梁冲击试验,摆锤式冲击试验
~ impact test machine 沙尔贝冲击试验机,简支梁冲击试验机,摆锤式冲击试验机
~ impact test specimen 沙尔贝冲击试样,简支梁冲击试样,摆锤式冲击试样
~ impact tester 沙尔贝冲击试验仪,简支梁冲击试验仪,摆锤式冲击试验仪
~ method 沙尔贝法〈耐冲击试验〉
~ value 沙尔贝冲击值
charred weld 烧焦焊接
charring 烧焦,炭[焦]化〈塑料燃烧试验〉
chart magnification range 放大倍数自动记录仪
chase 模箍,模套,模圈,模框
~ ring 防涨圈
chassis 车底架〈汽车〉
chatter 振动
~ mark 振纹,振痕〈塑料半成品〉
Ch. E [chemical engineer] 化学工程师

check 校验,检验;开裂,细裂缝[纹],龟裂
~ analysis 校验分析;控制分析
~ bolt 防松螺栓
~ crack 细裂缝,收缩裂纹,网裂
~ nozzle 防漏喷嘴
~ sample 对照试样;核对试样
~ surface 表面龟裂
~ test 对照试验;核对试验
~ valve 止回阀;操纵阀;检验阀
~ weigher 复核杆
checkered sheet 花纹板,网纹板
checking 校验,检验;裂纹,微裂〈薄膜〉
cheese 筒子纱
chelate 螯合;螯合物
~ complex 螯合物
~ compound 螯型化合物
~ fibre 螯合纤维
~ initiator 螯合引发剂
~ polymer 螯型聚合物
chelating 螯合的
~ agent 螯合剂
~ cation-exchange resin 螯合型阳离子交换树脂
~ resin 螯合型树脂
~ resin containig aminoacetic group 胺羧基螯合树脂
~ type ion exchange membrane 螯合型离子交换膜
chelation 螯合作用
chem. [chemical] 化学的
chemical 化学(制)品,化学药品,化合物;化学的
~ addition agent 化学添加剂
~ additive 化学添加剂
~ analysis 化学分析
~ attack 起化学反应,化学侵蚀
~ blowing 化学发泡
~ blowing agent [CBA] 化学发泡剂
~ bond 化学键
~ bonding 化学链接;化学黏合(工艺)
~ catalyst 化学催化剂
~ cellulose 化学纯纤维素

~ change 化学变化
~ composition 化学成分;化学组成
~ compound 化合物
~ constitution 化学结构;化学组成
~ corrosion 化学腐蚀
~ cotton 化学纤维素,化学棉
~ coupling 化学偶合
~ cracking 化学裂解
~ creep 化学蠕变
~ crosslinking 化学交联
~ decomposition 化学分解
~ degradation 化学降解〈塑料材料〉
~ desizing 化学脱浆
~ destaticizer 化学防静电剂
~ embossing 化学刻花
~ energy 化学能
~ engineer 化学工程师
~ engineering 化学工程
~ engraving 化学蚀刻
~ equilibrium 化学平衡
~ etching 化学溶蚀,化学侵蚀〈塑料加工模具〉
~ etching treatment 化学侵蚀处理
~ fibre 化学纤维,化纤
~ foaming 化学发泡
~ foaming agent 化学发泡剂
~ hardware 化学仪器;化工设备
~ industry 化学工业
~ inerness 化学惰性
~ inhibitor 化学抑制剂
~ intermediate 化学中间体
~ isomeril form 化学异构形成
~ leather 化学用革
~ literature 化学文献
~ mechanism 化学反应历程;化学机理
~ milling 化学刻花,化学刻蚀〈美〉
~ modification 化学改[变]性,化学改变
~ oxygen demand 化学耗氧量
~ plasticizer 化学增塑剂
~ plasticizing 化学增塑
~ plating 化学镀膜
~ pretreatment 化学(表面)预处理
〈表面粘接〉
~ processing 化学加工,化学处理
~ properties 化学性质
~ reaction 化学反应
~ resistance 耐化学(药品)性
~ screw locking 螺栓的化学锁紧法〈用单组分胶黏剂、厌氧剂防止螺栓松动〉
~ shift 化学位移
~ silvering 化学镀银
~ stability 化学稳定性
~ stitch process 用聚氨酯泡沫材料或胶黏剂粘接法〈织物〉
~ stress 化学应力
~ stress relaxation 化学应力松弛
~ synthesis 化学合成(法)
~ technology 化学工艺
~ tendering 化学脆化
~ festing 化学(性质)试验
~ treatment 化学处理
~ woodpulp [CWP] 化学木浆
chemically 化学(性质)上
~ blocked adhesives 化学稳定型黏合剂
~ foamed plastics 化学泡沫塑料,化学发泡塑料
~ initiated polymerization 化学引发聚合
~ modified fibre 化学变性纤维
~ modified natural polymers 化学改性天然聚合物
~ modified rubber 化学改性橡胶
~ pure[c. p.] 化学纯
~ stabilized wood 化学改性木材
chemilumiescence 化学发光,化学荧光
chemisorption 化学吸附
chemistry 化学;化学性质
~ laboratory 实验室;试验室
chequered 花纹;网纹
chicken-skin 桔皮纹;鸡皮〈缺陷〉
chill 激冷,冷激,骤冷;冷却
~ cast film 骤冷平膜
~ casting 冷激铸塑件
~ harden 激冷裂纹

~ harden 冷硬化
~ mark 冷斑
~ mo(u)ld 冷铸模
~ point 冰冷点,冻结点,凝固点
~ spinning 骤冷纺丝
~ time 激冷时间
~ roll 冷却辊,骤冷辊
~ roll cast film 冷却辊浇铸薄膜
~ roll casting 冷却辊流延法
~ roll coextrusion 冷却辊式共挤塑
~ roll extrusion 冷却辊式挤塑
~ roll method 冷却辊法
~ roll take-off machine 冷却辊牵引机〈挤塑薄膜〉
~ roll unit 冷却辊单元〈挤型薄膜〉
chilled 冷却了的
~ iron roll 冷铸铁辊〈压延机〉
~ roll 冷却辊,骤冷辊
~ spinning 骤冷纺丝
chilling 激冷;冷却
~ machine 冷却器;冷却辊
~ point 冷冻点
~ roll 骤冷辊,冷却辊
~ system 骤冷系统,冷却系统
~ temperature 冷冻温度
~ unit 冷冻装置
chimney 纺丝仓[室];烟囱
~ -type instrument 烟囱式仪器〈测定塑料燃烧性〉
china clay 高岭土,白土,白陶土,瓷土〈填料〉
chip 碎片,木屑
~ board 粗纸板;刨花板
~ material (木)屑料
~ off 削掉,切掉;成片剥落
~ pan 承屑盘
~ proof 防鳞片状;防剥离
~ room 排屑槽
~ wood 刨木屑
chipped area 模压缺陷,碎破面〈缺陷〉
chipping 铲除表面缺陷,修整;剥落
~ off 脱层;剥离
chloracetic acid 氯乙酸
chloramine 氯胺

chlorazol colo(u)rs 氯唑染料
chlorendic 氯茵的
~ acid 氯茵酸〈增塑剂,阻燃剂,防霉剂〉
~ anhydride 氯茵酸酐〈阻燃剂,固化剂〉
chlorester 氯(代)酯
chloride 氯化物
~ of lime 漂白粉
~ of lime bleaching 漂白粉漂泊
chlorinate 氯化;氯化物
chlorinated 氯化了的
~ additive 氯化添加剂
~ biphenyl 氯化联苯〈增塑剂〉
~ butyl rubber 氯化丁基橡胶
~ paraffin 氯化石蜡〈增塑剂、润滑剂〉
~ polyether 氯化聚醚,聚氯醚
~ polyethylene [CPE,PEC] 氯化聚乙烯〈耐冲改性剂〉
~ polyolefine 氯化聚烯烃
~ polypropylene [PPC] 氯化聚丙烯
~ polyvinyl chloride [CPVC] 氯化聚乙烯
~ polyvinyl chlodride fibre 氯化聚乙烯纤维
~ resin 氯化树脂
~ rubber 氯化橡胶
~ rubber paint 氯化橡胶漆〈防水防锈〉
chlorine [Cl] 氯
~ gas 氯气
~ hydrate 八水合氯
~ -resist resin 耐氯树脂
chlorisatide 氯靛红
chloro- 〈词头〉氯代;氯基
1-~ -2,3,4,5,6-pentabromocy-clohexane
1-氯-2,3,4,5,6-五溴环己烷〈阻燃剂〉
4-~ -2-phenyl phenol 4-氯-2-苯基苯酚〈防霉剂、防蚁剂〉
chloroacetylamino dyes 氯乙酰胺(型)染料

chlorobenzothiazole dyes 氯苯并噻唑（型）染料
chlorobutadine rubber 氯丁橡胶
chlorobutyl rubber 氯丁基橡胶
clorethylene 氯乙烯
chloroflucarbon 氯氟碳
~ plastic 氯氟碳塑料
~ resin 氯氟碳树脂
chlorofluorohydrocarbon plastic 氯氟烃塑料
chlorohydrine 氯化醇
chlorhydrin rubber 氯乙醇橡胶
choroparaffin 氯化石蜡〈阻燃剂〉
chloroprene 氯丁二烯
~ polymer 氯丁二烯聚合物
~ rubber [CR] 氯丁二烯橡胶,氯丁橡胶
~ rubber adlhesive 氯丁胶黏合剂
γ - **chlorpropyltrimethoxysilane** γ - 丙基三甲氧基(甲)硅烷〈偶联剂〉
chloropyrimidine dyes 氯嘧啶(型)染料
chlorosulfonated polyethylene [CSPE] 氯磺化聚乙烯
chlorotrifluoroethylene [CTFE] 三氟氯乙烯
~ polymers 三氟氯乙烯聚合物
~ resins 三氟氯乙烯树脂
- **vinylidenefluorid copolymer** 偏氯乙烯－三氟氯乙烯共聚物
CHM [cyclohexylmethacrylate] 甲基丙烯酸环己酯
chocolate brown 巧克力棕色,棕褐色
choked neck 节流口颈,窄颈
choker 节流门,调节闸阀
~ bar 节流栓;调节排〈挤板模头〉
choking 堵塞;堵模;未塑化颗粒〈模塑件〉
chop 切断;碎块
~ - out die 冲模;冲切
chopped 切断的,短切的,碎的
~ cotton cloth 棉碎片〈填料〉
~ cotton cloth filled plastic 棉碎片填充的塑料
~ cotton fabric 棉碎布片〈填料〉

~ fiber 短切纤维
~ fiber pellets 短纤维粒料
~ filled plastic 碎片填充塑料
~ glass fiber 短切玻璃纤维
~ glass strands 短切玻璃丝束,切断玻璃丝束〈增强材料〉
~ strand 短切原丝(短切纤维)
~ strand glass mat 短切玻璃丝毡
~ strand mat 短切原丝毡
~ strand mat reinforced laminate 短玻璃丝毡增强层合板
~ strand mat reinforcement 短玻璃毡增强材料
chopper 切碎机
chopping 切断,短切,切碎〈纤维〉
chord modulus 弦线模量
chroma 色(泽)度;彩度
chromatic 色的
~ aberration 色(象)差
~ difference 色差
~ dispersion 色散(现象)
~ treatment 色度处理
chromaticities 色值
chromaticity 色度
~ coordinate 色度坐标
~ diagram 色度图,色调图(用于颜色选择)
~ index 色度指数
chromatogram 色谱
chromatograph 色谱仪
chromatographic 色谱的
~ adsorption 色谱吸附
~ analysis 色谱分析
~ column 色谱柱
~ fractionation 色谱分离
chromatography 色谱法
~ of ions 离子色谱法
chromazol violet 铬唑紫
chrome 铬
~ colo(u)rs 铬(处理)染料
~ - plated 镀铬的
~ - plated mould 镀铬模具
~ plating 镀铬
~ - plation 镀铬

~ treatment 镀铬,铬处理
~ vermillion 铬朱红
chrome fast cyanine 铬坚牢花青
chromic 铬的
   ~ oxide coating sheet 镀铬(氧化物)薄板
   ~ sulphuric acid pickling process 硫酸铬浸渍法
chromium [Cr] 铬
~ plating 镀铬
chromogen 发色团;铬精(染料)
~ colours 铬精染料
chromogen azurine 铬精天青
chrommometer 比色计
chromphore 发色团
chromophoric group 发色基团
chromosol colours 铬溶染料
chrysotil [hydrated magnesium silicate] 水合硅酸,烷滑石粉
chucking device 夹具,卡紧装置
chunk 开面模具,敞口模具
churning 涡流;搅拌
chute 溜槽,斜槽,滑槽
~ calibration 斜槽计量,出料槽计量
~ feed 溜槽加料
~ lining 料槽衬里
C. I. [Colour Index] 染料索引
C. I. [Colour Index No.] 色指数
cigarette 卷烟,香烟
~ - filter tow 卷烟过滤嘴用丝束
~ - proof sheet 耐烟卷烧层合板
~ - proof veneer 耐烟卷烧胶合板
~ - resistance 耐烟卷灼烧性
CIL - viscosimeter CIL 型黏度计,常压毛细管黏度计
cinereous 灰色的
cinnabar 朱红的
cinnamon 红棕色的,肉桂色的
cinnamonitrile 肉桂酰腈
circ. [circumferential, circular] 圆周的,环形的
cire - winding [circumferential winding] 环向[圆周]平面缠绕
circle 圆,圆周;循环

~ test 循环试验
circuit 电路;循环;绕圈
~ [printed] board 印刷线路板
~ breaker 断电器
circuits per layer [Cl] 每层绕圈
circuits per pattern [Cp] 每种花纹循环
circular 圆的;循环的
~ blade 圆盘刀,圆盘锯片
~ Couelte [Couette] flow 凯脱型圆周流动
~ die 环形模头,圆形冲模
~ disk 圆盘
~ filler rod 圆形焊条
~ folded plate structure 圆形板折叠设备
~ [rotor] knife 圆盘刀
~ path 圆形路径
~ pitch [G. P.] 圆周齿距
~ saw 圆盘锯
~ shelf 圆架,回转架[盘]
~ winding 圆形巷绕
~ woven fabric 圆形机织物
circulating 循环
~ air 循环空气
~ hot air 热空气循环
~ (- air) oven 循环(热气)炉
~ purep 循环泵
~ tank 循环容器,环流槽
~ water 循环水
~ - water cooling unit 循环水冷却器(压型模具)
circulation 循环
~ cell 循环池,环流槽
~ flow 循环流动
rcumferential 圆周的,环形的
~ groove 环形槽沟
~ joint 环形连接
~ speed 圆周速度
~ stress 圆周应力
~ wind 环形缠绕
~ winding 环形[圆周]平面缠绕
~ wrapping 环形缠绕
~ cis - butadiene rubber 顺(式)丁(二烯)橡胶

cissing 收缩;塌陷
cistactic polymer 顺式有规聚合物
citrate plasticizers 柠檬酸酯增塑剂
citric acid 柠檬酸
citrine 浅橄榄色
citrus 柑橘黄
civil engineering 土木工程
Cl [chlorine] 氯
Cl [circuits per layer] 每层绕圈(长丝缠绕)
clad 涂敷;镀金属;贴面
~ fibre 涂层纤维,包层纤维
~ plate 包装板,装饰板
cladding 镀敷涂层;表面处理;贴面;外罩
~ material 镀盖层,覆盖材料
clam shell 蛤壳
~ mo(u)lding process 蛤壳式模铸造法
~ vacuum forming 蛤壳式真空成型
clamp 夹,夹具,夹板;合模
~ bolt 夹紧螺栓,拉紧螺栓(塑料加工机器)
~ capacity 锁模力,合模力
~ coupling 合模联轴节
~ cylinder 合模油缸
~ daylight 合模开挡
~ device 夹紧装置;合模装置
~ end 合模侧〈注塑〉
~ fastener 持夹器
~ force 夹紧力;合模力
~ holder 持夹器
~ iron 保压铁夹板,旋螺夹钳
~ mo(u)ld 夹固型模具
~ plate 载模板,装模板;夹板
~ ram speed 合模活塞速度,锁模活塞速度
~ roll 夹辊,拉紧辊
~ screw 制动[压紧,紧固]螺钉,夹紧螺栓,扣紧螺栓
~ stroke 合模冲程,锁模冲程
~ time 模压时间;合模时间
~ with dish – shaped spring – washers 盘式弹簧垫圈夹具,带圆盘弹簧垫的

夹紧装置(用于胶黏)
clamped 夹持的,固定的
~ at bottom 下部夹持的
~ joint 固定连接(塑料管)
~ sheet 夹持腹片
~ test specimen 夹持试样
clamper 夹持器
clamping 夹紧,夹住,紧固;合模
~ apparatus 夹具,卡具
~ belt 张力带
~ block 夹板〈胶合板〉
~ capacicy 合模力〈注塑〉
~ cylinder 合模液压缸〈注塑〉
~ device 合模装置,夹紧机构
~ flange 固定座盘,固定凸缘
~ force 锁模力,合模力〈注塑机〉
~ frame 夹持框架〈热成型机〉
~ grid [draw grid] 拉伸框格
~ hole 固定孔〈在夹板中〉
~ jaw 夹片器,夹持器,夹具〈断裂试验机〉
~ lever 合模拉杆
~ mechanism 合模机构,液压锁模机构
~ of test specimen 试样夹持
~ plate 合模板,注塑模载模板,压机压板,载模板,装模板;夹板
~ plunger 合模活塞
~ pressure 合模压力,锁模压力
~ ring 锁模圈,夹环[圈]
~ side 顶出侧(注塑)
~ side mould part 顶出侧模具部件
~ speed 合模速度
~ stroke 合模冲程
~ time 压粘时间;合模时间
~ tonnage 合模(总)吨数(注塑)
~ unit 合模[锁模]装置;夹持装置(注塑机)
clarifier 透明剂,澄清剂
clarity 亮度;透明度,透明性
Clash – Berg 克拉什 – 伯格
~ flexure temperature 克拉什 – 伯格柔韧性温度
~ point 克拉什 – 伯格点

~ temperature 克拉什-伯格温度
~ test 克拉什-伯格试验,塑料扭转刚度试验
class 种类,等级
~ of accuracy 精度等级
~ of precision 精度等级
~ of resins 树脂种类
classification 分类;分级;分粒
classified moulding material 标准模塑料
classifier 分选机,分类机;分粒机
classify 分级,分类,分选;分粒
claw clutch 爪形离合器(注塑机)
clay 陶土,黏土(填料)
~ coating 黏土涂层
clean 清洁(的),干净(的)
~ -out 清除,清理
cleaning 清洗,净化(连接件表面)
~ blower 吹洗器[机](压模)
~ roll 清洁辊,抛光辊(除去压模件棱角)
~ treatment 净化处理
cleanliness 清洁(度),净度
cleansing agent 清洁剂,净化剂
clear 透明的;纯的;净的
~ area 有效截面(面积);净面积
~ coating 透明涂料
~ distance 净距离,净空
~ headroom 净高
~ height 净高,间隙高度
~ lacquer 透明漆,清漆;硝基清漆;罩漆(大漆)
~ opening 净宽;净孔
~ point 透明点
~ sheet 透明片材;透明膜
~ span 净间距
~ varnish 透明漆
~ width 净宽,内径
clearance 间隙;净空;间距
~ adjustment 间隙调整
~ between rolls 轧辊间隙,轧辊开口度
~ flow 隙流
~ for expansion 膨胀间隙

~ gap 隙距
~ height 净空高度;外廓高度
~ hole 孔隙
~ length 外形长度
~ of a gear 齿距
~ of a press 模压机开程[距]
~ of a screw 螺纹间距
~ ratio 隙距比
~ width 外形宽度;外廓宽度
clearness [clarity] 透明性,透明度
cleavage 劈裂,裂开;层间剥离
~ breakage 破碎
~ crack 劈裂
~ failure 断裂,裂缝
~ fracture 裂碎
~ in a test specimen [wedge-type kerf] 楔形劈裂,试样劈裂
~ of hydrogen chloride 氯化氢裂解
~ of the bond 粘接分裂
~ product 裂解产物;开裂产品
~ strength 劈裂强度
~ [cleaving]test 耐裂试验,裂开试验(粘接)
clevis plate U形夹具[夹板](粘接)
clicker 冲切机
~ die 冲坯模
~ die cutting press 冲坯模切压机
~ press 冲床,冲坯机
~ clicking 冲切,冲坯;切料
~ climate 气候
~ investigation 耐气候性试验
~ protection test 防大气试验,耐气候性试验
~ test chamber 气候试验室[箱]
climatic 气候的
~ cabinet 耐气候性试验机
~ conditions 气候条件
climatizer 气候实验室
climbing drum peel test 爬高滚筒剥离试验
cling 黏着,黏住
~ compound 防[抗]滑(料)
~ property 防[抗]滑性能
clip 夹,夹住,夹紧;切除;

clipper

- ~ buffer 管夹垫
- ~ -on ga(u)ge 夹具测量仪

clipper 剪板机
clipsing 卡住,扣住(按钮系统)
clockwise 顺时针(方向)的(旋转)
clog 障碍(物),阻塞(物)
clogging 堵塞
close 密闭的;紧密的

- ~ contact glue 触压密封胶
- ~ packed structure 密封装结构(塑料材料)
- ~ packing 密封装(塑料结构)
- ~ tolerance 精密公差

closed 闭紧的;封闭的;紧密的

- ~ assembly 紧密装配,叠装(粘接)
- ~ assembly time 叠合时间,叠装时间(粘接)
- ~ assembly time in glueling 粘接时的叠合时间
- ~ cell 闭孔(泡沫塑料)
- ~ -cell cellular material 闭孔微孔材料
- ~ -cell cellular plastic 闭孔微孔塑料
- ~ -cell foam 闭孔泡沫塑料
- ~ -cell foamed plastics 闭孔泡沫塑料
- ~ -cell porosity 闭孔多孔性,闭孔孔隙率
- ~ -cell urethane foam 闭孔聚氨酯泡沫塑料
- ~ clrcuit 闭路
- ~ coat 紧密涂层
- ~ contact glue 密实接触粘结剂
- ~ cup flash point test 闭杯闪点试验
- ~ end 闭合端
- ~ -end preform 闭合端型杯
- ~ height 闭合高度(上下压模)
- ~ hole 盲孔,不通孔
- ~ joint 无间隙焊缝(热气焊接),无间隙接头
- ~ -loop system 闭路循环系统
- ~ -loop water cooling 闭合式水冷
- ~ mo(u)ld 闭合模具
- ~ mo(u)ld moulding 闭合模模塑

- ~ preform process 密闭预成型法(增强塑料)
- ~ surface 密实面层
- ~ weld 无间隙焊缝(热气焊接)

closing 闭合

- ~ by gravity 重力闭合
- ~ cylinder 合模筒,闭模液压缸(注塑机或压机)
- ~ force 锁模刀,合模力(注塑机)
- ~ joint 合模接缝(模具)
- ~ machine 闭合机,封口机(包装)
- ~ motion 闭合动作
- ~ pin (喷嘴)针阀;合模销
- ~ plate 密封板
- ~ pressure 闭合压力
- ~ ram 闭合活塞(液压挤塑机)
- ~ sleeve 夹紧套筒
- ~ slide 闭锁阀
- ~ speed 闭合速度;闭模速度
- ~ stroke 闭合行程;闭模行程
- ~ time 闭合时间,闭模时间
- ~ time of thermosetting moulding materials 热固性模塑料闭模时间
- ~ travel 闭合行程;闭模行程
- ~ unit 合模[锁模]装置;夹持装置(注塑机)
- ~ wall 板桩墙,槽舌接合壁

closure 密闭,封闭

- ~ cap 密封盖
- ~ of mo(u)ld 合模,闭模

plug 密封塞
cloth 织物;布

- ~ buff 布(抛光)轮,抛光布轮
- ~ disk 布(抛光)轮
- ~ laminate 布基层合塑料,层合塑料;层压布
- ~ wrapper 布包辊(薄板盘卷)

cloud 混浊(模制品面;涂层)

- ~ point 浊点,浊化点
- ~ test 浊点试验

cloudiness 浊度;浑浊性;云斑(疵点)
clouding 发白;无光泽;云状花纹;云斑
cloudiness 混浊,不透明性
cloudy 混浊的,不透明的;云状的;云

斑的
- ~ dyeing 染色云斑

CLPVC [crosslinked polyvinyl chloride] 交联聚氯乙烯

cluster 熔融体中分子聚集体

clustering of water vapour 蒸气聚集体（聚合物中）

clutch 离合器
- ~ brake 离合制动器
- ~ coupling 离合联轴节

cm [centimeter] 厘米

CM [composite materials] 复合材料

CMC [carboxymethyl cellulose] 羧甲基纤维素

CMHEC [carboxymethyl hydroxyethyl cellulose] 羧甲基羟乙基纤维素

CMR [contact microradiography] 触压X射线显微照相检验法

CN [cellulose nitrate] 硝酸纤维素

Co [cobalt] 钴

Co [compary] 公司

CO [elastomeric polychloromethyl oxiranel; polyepichlorohydrin] 聚氯甲基环氧乙烷弹性体；聚表氯醇

Co – Blow [compression – blowing] 压缩(型杯)吹塑

co – condensation 共缩合
- ~ polymer 共缩聚物
- ~ polymerization 共缩聚

co – curing agent 辅助固化剂,共固化剂

co – injection 共注塑
- ~ blow mo(u)lding 共注吹塑(成型)
- ~ mo(u)lding 共注塑(成型)

co – plasticizer 辅助增塑剂

co – rotating twim screw extruder 同向旋转双螺挤塑机

co – solvency 共溶解性

co – stabilizer 辅助稳定剂

coacervation 凝聚作用

coadhesive 共黏剂

coagel 凝聚胶

coagulant 凝结剂

coagulated filament [precipitated filament] 凝结长丝

coagulating 凝结
- ~ agent 凝结剂
- ~ bath 凝固浴,纺丝浴
- ~ point 凝固点,凝结点
- ~ power 凝固力,凝结力

coagulation 凝聚作用,凝结作用
- ~ bath [precipitation bath] 凝固浴
- ~ phenomena 凝固现象
- ~ point 凝固点,凝结点
- ~ preventing drying 防凝结干燥
- ~ treatment 凝结处理
- ~ value 凝固值

coagulum 凝块;凝结物

coal 煤
- ~ tar 煤焦油(软化剂)
- ~ tar resin 煤焦油树脂,库马龙[古马隆]树脂

coarse 粗的,粗糙的
- ~ adjust 粗调(整)
- ~ adjustment 粗调
- ~ breaking 粗压碎
- ~ cloth 粗织物
- ~ control 粗控,粗调(节)
- ~ crusher 粗碎机
- ~ denier 粗的细度,粗旦
- ~ fibre 粗纤维
- ~ fibre structure 粗纤维结构
- ~ filament 粗长丝
- ~ grain 粗粒(料),粗晶粒,粗纹
- ~ grained 粗粒的;粗纹理的
- ~ grinding 粗磨
- ~ knurl insert 粗滚花嵌件
- ~ meshed 粗网格的
- ~ particles 粗粒
- ~ plate 厚板
- ~ powder 粗粉
- ~ product 粗产品
- ~ reduction (尺寸)粗略缩小
- ~ staple fibre 粗短纤维

coarseness 粗度;粒度

coat 涂层;镀层;涂料
- ~ content 涂膜百分率
- ~ formation 涂层结构,涂层形成
- ~ hanger die 衣架式模头

coated
~ hanger manifold　衣架式歧管料道（挤塑）
~ of plastic　塑料涂层
~ of synthetic resin　合成树脂涂层
~ with paint　涂漆;涂色
~ with plastic　包塑料,覆盖塑料层
coated　涂敷的,涂覆的
~ fabric　涂覆织物,涂敷织物,涂层织物
~ fibre　涂层纤维
~ paper　涂覆纸,涂料纸
~ pigments　涂敷颜料;包核颜料
~ sand　涂胶砂,上胶砂
~ web　涂覆织物
coater　涂布机;涂层机,涂料器
coating　涂层(产品);涂布(方法)
~ agent　涂布剂,涂饰剂,涂层剂
~ applicator　实验涂布装置
~ blank　涂布坯件
~ blanket　涂布垫带
~ calender　涂层压延机
~ composition　涂覆材料;涂料组分
~ compound　涂覆材料,涂料
~ dam　涂布挡板
~ damage　涂层损坏
~ device　涂敷装置(涂层)
~ die　涂布模头
~ equipment　涂布设备
~ extruder　涂布挤塑机
~ film　涂层,涂膜(层)
~ head　涂布头,涂布机头
~ knife　涂布刮刀
~ layer　涂布层(共挤塑)
~ line　涂布生产线
~ machine　涂布机;喷漆机
~ material　涂覆材料,涂料
~ medium　涂料介质
~ method　涂布方法
~ nip　涂布辊隙(挤塑)
~ of many hairs　起毛涂ähr
~ of wires and cables　钢丝和电缆涂层
~ pan　涂料盘[槽];涂漆滚筒(美)
~ process　涂布方法
~ property　涂敷性能

~ resin　涂料用树脂,涂层树脂
~ roll(er)　涂布辊
~ speed　涂布速度,涂敷速度
~ strength　涂层强度,镀层强度
~ substrate　涂布基材
~ tank　浸渍槽
~ test　涂覆试验
~ thickness　涂层厚度
~ thickness tester　涂层厚度测定仪
~ varnish　涂层清漆;罩光清漆
~ web　涂布基材,浸胶基布
~ weight　涂层量;涂层重量
~ with plastics　涂塑料,塑料涂布
~ with powder　涂粉(末),粉料涂布
coatings　涂料
coaxial　同轴的,共轴的
~ cable　同轴电缆
~ cylinder viscometer　同轴圆筒式旋转黏度计
~ feeder　同轴馈(电)线
~ screw preplasticizing injection moulding machine　同轴螺杆预塑化式注塑成型机
cobalt [Co] 钴
~ complex dyes　含钴络合染料
~ naphtenate　环烷酸钴(催化剂,固化促进剂)
~ stearate　硬脂酸钴(热稳定剂)
cobaltous acetate　乙酸钴(催化剂)
cobwebbing　蛛网状(涂层时)
cocatalyst　助催化剂
cock　旋塞,阀门
coconut　椰子
~ fatty acid diethanolamide　椰子脂酸二乙醇胺缩合物(泡沫稳定剂,润湿剂,净洗剂,乳化剂)
~ shell flour　椰子壳粉(填料)
cocoon　蚕茧;蚕茧状防护喷涂层(塑料)
~ packing　蚕茧状喷涂包装
cocooning　蚕茧状喷涂包装,蚕茧状防护喷涂层
cocoonization　蚕茧状喷涂包装;蚕茧状防护喷涂层
cocrystallization　共结晶

cocure 共固化
coefficient 系数;程度
~ of acoustic absorption 吸声(程)度
~ of charge 装料系数,装载系数
~ of compressibility 压缩系数
~ of conductivity 传导系数
~ of contraction 收缩系数
~ of corrosion 腐蚀系数
~ of cross viscosity 横向黏性系数
~ of cubic expansion 体膨胀系数
~ of cubic thermal expansion 体热膨胀系数
~ of cubical elasticity 体积弹性系数
~ of cubical expansion 体膨胀系数
~ of damping 阻尼系数
~ of dilatation 膨胀系数
~ of discharge 卸料系数
~ of dispersion 分散系数
~ of dynamic viscosity 动力黏度系数,动态黏性系数
~ of eddy viscosity 涡流黏性系数
~ of efficiency 有效系数,效率
~ of elasticity 弹性系数
~ of elasticity in shear 剪切弹性系数
~ of elongation 伸长系数
~ of expansion 膨胀系数
~ of expansion due to heat 热膨胀系数
~ of extension 延伸系数
~ of fluidity 流动性系数
~ of friction 摩擦系数
~ of hardness 硬度系数
~ of heat conductivity 导热系数
~ of heat transfer 传热系数
~ of heat transmission 传热系数
~ of heat transmittance 传热系数
~ of hybrid effect 混杂效应系数
~ of impact 冲击系数
~ of insulation 绝缘系数
~ of kinematic viscosity 运动黏度系数
~ of kinetic friction 动摩擦系数
~ of linear expansion 线膨胀系数
~ of linear thermal expansion 线性热膨胀系数
~ of opacity 不透明系数,不透明度
~ of performance 有效系数,效率
~ of polymerization 聚合系数
~ of resilience 回弹系数
~ of resistance 阻抗系数
~ of restitution 恢复系数
~ of rigidity 刚性系数
~ of safety 安全系数
~ of shear 剪切系数
~ of shear viscosity 剪切黏度系数
~ of shrinkage 收缩系数
~ of sliding friction 滑动摩擦系数
~ of sound absorption 吸声(程)度
~ of static friction 静摩擦系数
~ of stiffness 劲度系数
~ of tensile viscosity 拉伸黏度系数
~ of thermal conductivity 导热系数
~ of thermal expansion 热膨胀系数
~ of thermal insulation 绝热系数
~ of thermal transmission 传热系数
~ of transverse elasticity 横向弹性系数
~ of twist contraction 加捻收缩系数(应用于玻璃纤维纺织品)
~ of variation 偏离系数,偏差系数
~ of virtual viscosity 有效黏性系数
~ of viscosity 黏性系数,黏度系数
~ of volume expansion 体膨胀系数
~ of volume viscosity 体积黏度系数
~ of waste 废料系数
coercitivity 矫顽(磁)性
coercive force 矫顽(磁)力
coextruded 共挤塑的
~ blow moulded bottle 共挤塑吹塑瓶
~ corrugated packaging material 共挤塑波形包装材料
~ film 共挤塑薄膜
~ parison 共挤塑型坯(吹塑)
~ sheet 共挤塑片材
~ tube 共挤塑(软)管
coextrusion 共挤塑
~ accumulator head 共挤塑储料器式机头(吹塑)
~ blow mo(u)lding 共挤吹塑(成型)
~ casting 共挤塑平膜;共挤塑流延
~ casting of flat sheet 共挤塑平片材

~ casting of flexible film 共挤塑软(质)薄膜
~ coating 共挤塑贴合
~ containers 共挤(多层)容器
~ die 共挤塑机头(平片材)
~ film blowing 共挤塑薄膜吹塑
~ laminatingn 共挤塑层合
~ of multi-layer tubes and parisons 多层软管型坯共挤塑
~ of multi-1ayer tubular film 多层管状薄膜共挤塑
~ tandem extrusion laminating 共挤塑串联挤塑层合

cograft 共接枝
cohere 黏合,粘结
coherence 粘结;内聚力;内聚现象
coherent 粘结的;黏附的
~ structure (分子的)内聚结构
cohesible 可黏合的;可粘结的
cohesion 黏合(力);抱合力;内聚(力,现象,作用)
~ -adhesion failure 黏合破坏
~ between layers 层间黏合(层压)
~ energy 内聚能
~ energy density 内聚能密度
~ failure 内聚破坏;黏合破坏
~ force 黏合力;内聚力
~ moment 黏聚力矩
~ pressure 粘结压力;内聚压力
~ test 抱合力检验
cohesionless 无黏聚性的,不粘结的;无内聚力的
cohesive 黏合的;内聚的
~ -adhesive failure 黏合破坏(黏连)
~ energy density 内聚能密度
~ failure 内聚破坏,黏合破坏(黏连)
~ force 黏合力;内聚力
~ fracture 内聚破裂
~ material 粘结材料
~ property 粘结性
~ strength 黏合强度;内聚强度(粘连)
cohesiveness 黏聚性,内聚性
coil 线圈,线团,盘管;蛇盘;卷材

~ accumulator 蛇管式存储器,滚筒补偿器(用于连续操作涂敷装置)
~ coater 盘旋式涂敷机;卷筒式涂敷机
~ coating 卷筒式涂敷
~ coating line 卷筒式涂敷装置
~ former 绕线模架
~ handling equipment 卷材送料设备
~ insulation 线圈绝缘
~ pipe 蛇管;蛇管形
~ stock 卷材,卷料
~ un-roller 开卷机,展开机,退卷辊
coiled 卷绕,卷曲
~ grid spinning 格栅熔融纺丝
~ macromolecule 线团大分子
~ molecule 卷曲分子
~ plastic 成卷的塑料落板材料,薄塑料板卷
~ strip 成卷带料
coiler 绕管机,盘管机
coiling 卷绕,卷取;卷曲
~ action 卷绕动作
~ machine 卷取机
~ of the molecule 分子缠结
~ type polymer 螺旋型聚合物
coining (后)铸压,压制,挤压,模压;压花纹
~ die 铸压模具;压花纹模
coinjection 共注塑
~ blow moulding 共注塑(型坯)吹塑(中空制品)
~ moulding 共注塑(法);多组分泡沫注塑法(发泡的);多组分紧密注塑法(未发泡的)
cokey 不完全固化的(缺陷)
~ resin 不完全固化反应树脂
cold 冷;冷的,低温的;常温的;失光
~ adhesive 冷黏合剂
~ bend 耐冷弯曲
~ bend strength 耐冷弯曲强度,耐低温弯曲强度
~ bend temperature 低温揉曲温度
~ bend test 低温弯曲试验
~ bending 冷弯(曲)

~ boiler 真空蒸发器
~ brittleness 冷脆性
~ calendering 常温压延
~ casting 冷浇铸
~ charge 低温装料
~ check resistance 耐冷开裂
~ crack 冷裂纹
~ crack resistance temperature 耐低温开裂温度
~ crack temperature 低温开裂温度
~ cracking 冷裂
~ creep 冷蠕变
~ cure 冷硫化;冷固化,冷熟化
~ cure moulding 常温熟化成型
~ cure urethane 常温熟化聚氨酯
~ curing 冷固化,冷熟化
~ – curing flexible foam 冷固化软泡沫塑料
~ – curing foam 冷固化泡沫塑料
~ cut 冷切(造粒)
~ cut varnish 冷配清漆
~ deformation 冷变形
~ dipping 冷醮浸,冷醮涂
~ draw 冷拉伸
~ draw effect 冷拉效应
~ drawability 冷拉伸性
~ drawing 冷拉伸,冷牵伸
~ drawn 冷拉伸
~ endurance 耐寒性
~ exchanger 冷交换器,冷却器
~ feed 冷加料
~ flex 冷挠曲
~ flex temperature 冷挠曲温度
~ flex tester 冷挠曲试验机
~ flexibility 冷挠曲性,低温挠曲性
~ flow 冷流
~ foam [cold – curing foam] 冷发泡泡沫塑料
~ formable laminating 可冷成型层压材料
~ forming 冷成型(塑料)
~ gluing 冷粘接;冷胶合
~ hardening 冷硬化
~ hobbing 冷切压

~ injection 冷注塑(发泡)
~ lamination 冷层压,常温层压
~ mo(u)lding 冷压模塑,冷成型
~ mo(u)lding compound [material] 冷压模塑料
~ orientation 冷取向,冷拉伸
~ parison 冷坯(吹塑)
~ parison blow moulding 冷坯吹塑法
~ parison process 冷坯(吹塑)法
~ performance test 低温性能试验
~ plastic 低温塑料
~ plastication 低温塑炼
~ plasticity 低温塑性
~ point 冷点;冻点
~ polymer 低温聚合物
~ polymerization 低温聚合
~ press 冷压(粘接),冷层压
~ – press moulding 冷压模塑
~ pressing 冷压(合)
~ – pressing adhesive 冷压黏合剂
~ processing 冷加工
~ proof 耐寒的
~ property 低温特性
~ punching 冷冲
~ refinement [refining] 冷精炼,冷精加工
~ resistance 防冻性,耐寒性
~ – resistant 耐寒的,抗冻的
~ – resisting property 耐寒性
~ roll – forming [rolling] 冷辊压(成型)
~ rubber 低温聚合橡胶
~ runner mould 冷流道模具(注塑)
~ saturated 冷饱和的
~ – seal adhesive 冷封闭剂,冷封闭黏合剂
~ seal paper 冷封合纸
~ setting 常温固化,冷固化
~ setting adhesive 冷固化黏合剂,常温固化黏合剂
~ setting binder 冷固化粘结剂,常温固化粘结剂
~ setting foundary mould binder 冷固化型铸模粘结剂

collapse

- ~ setting glue 冷固化胶黏剂,常温固化胶
- ~ setting lacquer 常温固化漆
- ~ setting resin 常温固化树脂
- ~ setting temperature (冷)固化温度
- ~ shaping 冷成型(塑料)
- ~ short 冷脆
- ~ shortness 冷脆性,常温脆性
- ~ shots →cold slugs pl 不均匀进料时模压缺陷
- ~ slug 早冷料,冷料头(注塑时)
- ~ slug well 冷料井,冷料槽
- ~ slugs [defect-cold shots] 不均匀进料时模压缺陷
- ~ spots 冷疤
- ~ spray 常温喷涂
- ~ spreading 常温涂敷(黏合剂)
- ~ stain 冷变形
- ~ stamping 冷冲压,常温成型
- ~ start protection 冷起动保护(用于塑化螺旋推进器)
- ~ state 冷状态(塑料)
- ~ stretch 冷拉伸
- ~ stretching 冷拉伸
- ~ tapping 冷攻螺纹
- ~ test 冷态试验,常温试验
- ~ treatment 低温处理
- ~ vulcanization 冷硫化(橡胶用黏合剂)
- ~ -water-waxing 冷水上蜡,冷水涂蜡
- ~ well 冷料井,冷料槽
- ~ working 冷加工(塑料)

collapse 瘪泡,瘪孔,塌瘪(微孔塑料)
collapser 稳定板,夹模板,人字板
collapsible 可折叠的;可收缩的

- ~ box 可折叠箱
- ~ container 可折叠(装运)容器
- ~ mandrel 折叠模芯
- ~ mould 分片模;分室模
- ~ tube 软管;收缩管

collapsing 压坏,断裂;伸缩,折叠

- ~ board 夹膜泡沫
- ~ frame 夹膜框,人字架,稳定板
- ~ of bubbles 气泡的破坏
- ~ plate 夹膜板,人字板
- ~ pressure 破坏压力
- ~ rolls 夹膜泡辊
- ~ strength 破坏强度

collar 圈,环;轴环,套环

- ~ -type nozzle 环状注嘴(注塑机)
- ~ with eaternal thread 外螺纹套筒

collective package 集中包装
collector 收集器
collimated roving 平行粗纱(玻璃纤维)
collimation 平行性(长丝缠绕)
collodion 胶棉

- ~ cotton 低氮硝酸纤维素,胶棉,硝棉

colloid 胶体,使成胶态

- ~ mill 胶体研磨机;胶态磨

colloidal 胶体的

- ~ dispersion 胶体分散
- ~ dyes 胶态染料
- ~ solution 胶体溶液
- ~ suspension 胶体悬浮液

Colombo-type screw 可罗姆博型螺杆(双螺杆挤塑)
colophony 松香;树脂(天然树脂)
colour [color(美)] 颜色;着色,染色;料;染料

- ~ agglutinant 色料黏合剂
- ~ analyzer 颜色分析器
- ~ batch 色母料
- ~ bleeding 渗色,洇色
- ~ blend 混色;混色料
- ~ blindness 色盲
- ~ change 变色
- ~ chart 色卡
- ~ combination 配色
- ~ comparator 比色仪
- ~ comparimetry 比色法(分析)
- ~ concentrate 色母料
- ~ conditioning 调色,色彩调节
- ~ contrast 颜色对比
- ~ difference 色差
- ~ difference magnitude 色差值
- ~ difference meter 色差仪
- ~ dispersion 色散

~ drift　色变,色调偏差
~ fading　褪色
~ failure　褪色
~ fast　不褪色的
~ fastness　色牢度
~ fastness of plastics to light　塑料耐光色牢度
~ fastness to atmospheric gases　耐大气色牢度
~ fastness to boiling　耐沸煮色牢度
~ fastness to daylight　耐日光色牢度
~ fastness to exposure to light　曝光色牢度
~ fastness to heat　耐热色牢度
~ fastness to hot water　耐热水色牢度
~ fastness to light　耐光色牢度
~ fastness to rubbing　耐摩擦色牢度
~ fastness to sunlight　耐日晒色牢度
~ fastness to washing　耐洗涤色牢度
~ fastness to water　耐水浸色牢度
~ fastness to weathering　耐气候色牢度
~ filter　滤色片;滤色器
~ grinding mill　色料研磨机
~ heterogeneity　颜色不均匀性
~ hue　色调
~ index　比色指数,颜色指数;(大写)《染料索引》
~ index number　染料索引号,C.I.(编)号(染料)
~ matching　配色,调色;比色
~ measurement　比色法;测色
~ migration　色移
~ mill　色料研磨机,颜料研磨机
~ mixing　混色;调色
~ mixing room　调色室
~ number　色数
~ overlapping　搭色
~ pan　色槽,调色浆锅
~ particle　色料粒子
~ paste　色浆;染料糊
~ pigment　着色颜料
~ printing　彩印,色印
~ properties　颜色性质
~ range of indicator　指示剂变色范围

~ ratio　色比
~ register　调色器
~ retention　保色性
~ retention on exposure to hydrosulfide　耐氢硫化物保色性
~ sample　色样
~ scale　色标(度)
~ scheme　色调
~ ship　着色母料;浓色片
~ space　色区(色比较)
~ stability　着色坚牢度,着色稳定性
~ stability index　着色稳定性指数
~ stability test　着色稳定性试验
~ stabilizer　着色稳定剂
~ stabilizing agent　着色稳定剂
~ standard　颜色标准
~ temperature　色温
~ test　显色试验
~ tone　色调,色光(染料的)
~ triangle　原色三相图
~ unit　染料厂;颜料厂
~ value　明度;色值
colo(u)rability　着色法;着色能力
colo(u)rant　着色剂,色料;颜料;染料
~ layer　染色体
colo(u)ration　着色
colo(u)red　有色的,着色的
~ article　着色制品
~ band　色谱带
~ fibre　有色纤维
~ pattern　配色样板
~ pastics　染色塑料
~ spot　色斑
~ stock　着色物料
colo(u)rfastness　色牢度
colo(u)rimeter　比色计,色度计
colo(u)rimetric　比色的
~ analysis　比色分析
~ method　比色法
~ purity　色纯度
~ test　比色试验
~ titration　比色滴定
colourimetry　比色法
colo(u)ring　着色,染色

colo(u)rity
- ~ agent 着色剂
- ~ discrimination 颜色区别
- ~ equipment 着色[染色]设备(用于颗粒)
- ~ material 着色材料,着色剂;色料
- ~ matter 着色剂;色素;色料
- ~ paste 色浆
- ~ pigment 着色颜料
- ~ power 染色本领

colo(u)rity 颜色;色度
colo(u)rless 无色的
- ~ protecting lacquer 防护(清)漆

columbium [Cb] 铌
column 柱;立柱(压力机);塔
- ~ chromatography 柱色谱
- ~ crystals 柱晶
- ~ equipment 柱式设备
- ~ fractionation 柱上分级,淋洗分级
- ~ packing 柱填充(物)
- ~ press 立柱式压力机
- ~ tray 塔板

com. [commercial] 商业的
comb 脱膜叉;梳型;蜂窝
- ~ chain 梳型链
- ~ polymer 梳型聚合物,梳状聚合物
- ~ structure 蜂窝状结构

combination reinforcement 组合增强材料
combined 联结,组合,复合,混合
- ~ fabric 胶合织物
- ~ filament yarn 混合长丝纱
- ~ material 复合材料
- ~ moulding compound [XMC] 复合模塑料
- ~ plastic 复合塑料
- ~ polymerization 共聚合
- ~ strength 复合强度,组合强度
- ~ stress state 综合应力状态

combining chamber 混合室(挤塑)
comburant [comburent] 燃烧(的)
combustibility 燃烧性,可燃性
combustible 可燃的,易燃的
- ~ component 可燃组分
- ~ fibre 可燃纤维
- ~ material 可燃材料
- ~ substance 可燃物
- ~ zone 燃烧层

combustion 燃烧(作用)
- ~ [burning] test 燃烧鉴别试验
- ~ product 燃烧产物

comformability 梳曲性(玻璃纤维)
commanding apparatus 操纵设备
commercial 商业的;商品的;工业的
- ~ application 在工业上的应用
- ~ colo(u)rs 商品染料
- ~ dyes 商品染料
- ~ efficiency 经济效率
- ~ grade 商品(等)级
- ~ grade vulcanised fibre 商业级硬化纸板
- ~ manufacture 工业制造,工业生产
- ~ plant 工业设备
- ~ plastic 商业[商品]塑料
- ~ process 工业化生产过程
- ~ product 商品
- ~ production 工业化生产,大规模生产
- ~ quality 商品质量
- ~ sample 商品试样
- ~ scale 工业规模
- ~ scale process 大规模生产法
- ~ size 工业规模
- ~ specification 商品规格
- ~ standard 工业标准
- ~ unit 工业设备
- ~ use test 实用试验
- ~ value 工业价值

comminute 粉碎
comminuted polymer 粉末聚合物
comminution 粉碎,磨碎
commission 试运行,投产,起动
commissioning test run 投料试生产
commodity 日常品,商品
- ~ foam 日常用泡沫塑料
- ~ plastic 日常用塑料,大宗塑料
- ~ resin 日常用树脂,大宗树脂

common loading chamber 共用进料腔(多腔模具)

comonomer 共聚(用)单体
~ ration 共聚单体配比
compact 紧密(的),压紧(的);粉粒料,压缩粉,压模塑
~ cloth 紧密织物
~ extruder 粉料挤塑机
~ winding 密致卷缠
compacted bulk density 紧密度
compaction 紧实,密实,压实(塑料粉料)
~ degree 压实度
compactness 紧密性,紧密度
company 公司,社团,商社,商号
~ limited 有限公司
~ standard(specification) 企业标准,公司标准,社团标准
comparative 比较的
~ dyeing 对比染色
~ judgement 比较判断
~ test 比较试验,对比试验
~ tracking index 对比电弧径迹指数
comparison stimulus 比较刺激
compartment 间隔,分隔,分室
~ dryer 分室干燥器,分段干燥器
~ mill 分室研磨机
compatibility 相容性;可混性
~ limit 相容极限
~ of plasticizer 增塑剂相容性
~ test 相容性试验
compatibilization 相容(作用)
compatibilizer 相容剂
compatible 相容的;可混的
~ dyes 可混用的染料
~ plasticizer 相容增塑剂
~ polymer 相容性聚合物
~ UV light absorber 相容紫外线吸收剂
compatilizer 相容剂
compd.[compound] 化合物;混合料
compensate 补偿;抵消
compensating 补偿的
~ loop 补偿环路
~ socket joint 补强球窝[套筒]接合
compensation 补偿

~ loop 补偿弯管(塑料管线)
~ method 补偿法
compensator 补偿器(管)
complementary 互补的,补充的;辅助的
~ color 互补色;辅助色
complete 完全的,完整的
~ -closure mould 瞬时全闭模
~ combustion 完全燃烧
~ coolant equipment 全套冷却设备
~ cycle time 全循环时间
~ discharge 完全卸载,卸空
~ equipment 成套设备
~ inspection 全面检品
~ set 成套,全套
complex 络合物;复合的
~ admittance 复数导纳
~ calcium - stannous - zinc modified compound 钙亚锡锌改性(的)复合物(热稳定剂)
~ compliance 复数柔量(塑性)
~ copolymerization 络合共聚
~ dielectric constant 复数介电常数
~ elastic modulus 复数弹性模量
~ film 复合膜
~ flow 复合流动,混合流动
~ modulus 复数模量
~ modulus of elasticity 复数弹性模量
~ moulded article 复合模塑制品
~ plastic flow 复合塑性流,复数塑性流动
~ sample 复杂试样
~ shear modulus 复数剪切模量
~ utilization 综合利用
~ viscosity 复数黏度
~ Young's modulus 复数杨氏模量,复数弹性模量
complexing agent 络合剂
compliance 柔量,柔韧性(塑性)
~ in extension 拉伸柔量
~ in shear 剪切柔量
compn. = composition 组成;成分
compo board 合成纸板
component 成分,组分;元件,构件,部件
~ analysis 组分分析

- ~ assembly 零件装配
- ~ fibre 组分纤维
- ~ of a formulation 配方组分
- ~ of a yarn 纱线组分
- ~ of adhesive 黏合剂组分
- ~ part 构件
- ~ testing 结构件试验

composed state of stress 复合应力状态
composite 复合材料;复合塑料;复合的
- ~ adhesive film 复合黏合膜
- ~ blown film 复合吹塑薄膜
- ~ board 复合板
- ~ building board 复合组合板,复合材料建筑板
- ~ catalyst 复合催化剂
- ~ cellulosic membrane 复合纤维膜
- ~ dye 拼合染料
- ~ fibre 复合纤维
- ~ film 复合薄膜,多层薄膜
- ~ film extrusion 多层薄膜挤塑,薄膜共挤塑,复合薄膜挤塑
- ~ film lamination 复合薄膜层合
- ~ foam 复合泡沫塑料
- ~ laminate 复合层合板,复合层压板,复合层合塑料板
- ~ laminate board 复合层合板,复合层压板,复合层合塑料板
- ~ materials [CM] 复合材料
- ~ mo(u)ld 什锦模具,复合模具,多模腔模具,组合模具
- ~ mo(u)lding 压注联合成型,复合成型
- ~ package 联合包装,混合包装
- ~ panel 复合板
- ~ plastics 复合塑料
- ~ product 复合材料制品
- ~ profile 复合组分型材
- ~ sample 复合试样
- ~ sheet 复合片material
- ~ structure 复合结构
- ~ structural material 复合结构材料
- ~ surface 复合面
- ~ wood 复合木材
- ~ wood panel 复合木板,多层木板

composition 混合物,混合料;组分,成分
- ~ material 组合材料

compound 混合,掺混;混合料,配混料;化合物;复合的
- ~ bushing 复合衬套(塑料)
- ~ cellulose 复合纤维素
- ~ colo(u)re 混合色;调和色
- ~ curvature 三维曲率
- ~ draft system 混合牵伸装置
- ~ dyeing 拼色染色
- ~ fabric 复合织物,多层织物
- ~ failure 复合破坏,复合破裂
- ~ glass 复合玻璃,不碎玻璃,防弹玻璃,安全玻璃
- ~ mixer 配料混合器
- ~ pipe 塑料衬里钢管,复合[组合]管
- ~ processing 配料
- ~ self-heating 混合料自(动加)热
- ~ shade 复色,拼色
- ~ stress 复合应力,综合应力
- ~ structures 复合结构,多层结构
- ~ tool 复式模具,组合模具
- ~ yarn 股线

compounder 配料机,混料机
compounding 复合,混合,混配,配合;配料;混合料,模塑料
- ~ agent 配合剂
- ~ efficiency 混合效率(挤塑机)
- ~ equipment 混合设备,混合器
- ~ extruder 混料挤塑机
- ~ ingredient 配合成分,配料成分
- ~ line 混料生产线
- ~ material 配合[混配]材料
- ~ of plastics 塑料配配料,塑料混合料
- ~ ratio 配料比
- ~ technique 配料技术

compreg 浸渍压缩木材,胶合板
compregnated 浸渍压(缩)的
- ~ laminated wood 浸渍层合木材,胶合层压板
- ~ wood 浸渍压缩木材,胶合层压板

comprehensive 综合(性)的,全面的
- ~ test 综合试验
- ~ utilization 综合利用

compression

compress 压缩；打包机
compressed 压缩的，压紧的
~ air 压缩空气
~ air ejection 压缩空气脱膜
~ air inlet 压缩空气进气口
~ air spraying 压缩空气喷涂
~ air vessel 压缩空气储槽
~ cushioning 压缩缓冲垫层
~ [densified] laminated wood 浸渍层压木材，胶合层压板
~ preform 压缩（泡状）塑坯预塑
~ wood 浸渍压缩木材，胶合层压板
compressetometer 压缩疲劳试验机
compressibility 压缩率；压缩性
~ coefficient 压缩系数
compression 压缩，压紧
~ and recovery test 压缩恢复试验
~ blow mo(u)lding 压缩（型坯）吹塑（成型）
~ - blowing [Co - Blow] 压缩型坯吹塑
~ buckling 压缩（纵）弯曲
~ charateristics 压缩特性
~ coefficient 压缩系数
~ creep 压缩蠕变
~ creep test 压缩蠕变试验，蠕变试验
~ deflection characteristics 压缩致偏特性
~ deflection load 压缩畸变负荷
~ deformation 压缩变形
~ die 压挤模
~ edge 压塑模棱
~ elasticity 压缩弹性
~ factor 压缩系数（膜塑料）
~ failure 压缩破坏
~ fatigue test 压缩疲劳试验
~ fit 压紧配合
~ flow 压缩流动
~ heat sealing 热压合
~ injection moulding 压缩注塑成型
~ - joint pipe fitting 压紧对接管配件
~ load 压缩负荷
~ melting 高压塑化，压缩塑化（树脂；塑料）

~ member 受压构件
~ mo(u)ld 压缩模具，压塑模具，压制模具
~ moulded 压塑的
~ mo(u)lded material 压塑材料，压缩模塑料
~ mo(u)lded part 压塑件
~ mo(u)lded sheet 压制板材，压塑成型片材
~ mo(u)lded specimen 压缩试样
~ mo(u)lding 压塑，压缩模塑
~ mo(u)lding compound 压塑材料，压缩模塑料
~ mo(u)lding diagram 压塑图解
~ mo(u)lding material 压塑材料，压缩模塑料
~ mo(u)lding press 压塑压机
~ mo(u)lding pressure 压塑模塑压力，压塑压力
~ mo(u)lding resin 压塑树脂，压塑成型树脂
~ of a spring 弹簧压紧
~ of bulk material 压缩松散物料
~ permanent set 压缩永久变形
~ plastometer 压缩式塑度计
~ press 压缩压机
~ ratio 压缩比（美）
~ resiliance 压缩回弹性
~ resistant 耐压缩的
~ ring 压环
~ roll 压辊
~ screw [central spindle] 立螺杆，压缩螺杆
~ section 压缩段（螺杆）
~ set 压缩永久变形
~ spring 压缩回弹；压缩弹簧
~ strain 压缩应变
~ strength 压缩强度
~ stress 压缩应力
~ stress relaxation 压缩应力松弛
~ tcst 压缩试验
~ test machine 压缩试验机
~ test at elevated temperatures 高温压缩试验

~ tool 压塑模具
~ -type press 压塑型压机
~ type viscometer 压缩式黏度计
~ work 压缩功
~ yield point 压缩屈服点
~ zone 压缩段(螺杆)
compressive 压缩的
~ cleavage 压缩分裂(粘结)
~ creep 压缩蠕变
~ crimping 压缩卷曲
~ deformation 压缩形变
~ elasticity 压缩弹性(率)
~ failure 压缩破裂
~ force 压缩力
~ modulus 压缩模量
~ modulus of elasticity 压缩弹性模量, 压缩模量
~ offset yield stress 压缩偏置屈服,应力(压缩试验)
~ parallel plate viesometer 压缩式平行板黏度计
~ properties 压塑性能
~ resistance 耐压性,耐压强度
~ shear strenght of dowel joint 套接压剪强度
~ strain 压缩应变
~ strain at compressive yield stress 压缩屈服应力(时)压缩应变(压缩试验)
~ strain at rupture 破裂(时)压缩应变(压缩试验)
~ strength 压缩强度
~ strength at failure 断裂压缩强度
~ strength of cellular plastics 微孔塑料压缩强度
~ stress 压缩应力
~ stress-strain curve 压缩应力-应变曲线
~ stress-strain diagram 压缩应力-应变图
~ testing 压缩试验
~ transient state 压缩瞬变状态
~ ultimate strength 压缩极限强度
~ yield load 压缩屈服负荷

~ yield point 压缩屈服点
~ yield strength 压缩屈服强度
~ yield stress 压缩屈服应力(压缩试验)
compressometer 压缩计,压缩仪
compressor 压缩机
computer 计算机
~ oriented 与计算机有关的
cone. [concentmted] 浓的
concavity factor 凹度系数
concentrated 浓的;浓缩的;集中的
~ load 集中荷载
~ phase 浓相
~ polymer solution 聚合物浓溶液
concentration 浓度;浓缩,集中
~ constant 浓度常数
~ gradient 浓度梯度
~ of hydrogen ions 氢离子浓度
~ technique 浓缩技术
concentric 同心的,同轴的
~ cable 同轴电缆
~ cylinder rotation viscometer 同心圆筒式旋转黏度计
~ cylinder-type rheometer 同心圆筒式(旋转)流变仪
conchoidal 贝壳状的,螺旋线的
~ fracture 贝壳状破裂,贝壳状裂痕
concn. [concentration] 浓度
concourse roof 群众场所棚顶
concrete 混凝土(粘结用连接材料);浸膏(香料)
~ bonding 混凝土粘结
~ -plastic composite 混凝土-塑料复合材料
cond. [conductivity] 导电率
condensate 冷凝(物);冷凝液;缩合物
condensation 缩合(作用);凝聚(作用)
~ agent 缩合剂
~ compound 缩合物
copolymefization 共缩聚(作用)
~ plastic 缩合塑料
~ polymer 缩聚物,缩合聚合物
~ polymerization 缩聚(作用)
~ product 缩合产物

~ reaction 缩合反应
~ resin 缩聚树脂
~ rebber 缩聚橡胶
~ temperature 凝聚温度
condense dye 缩合染料
condensed 缩合的;冷凝了的
~ film 缩合膜
~ polymer 缩聚物
~ water 冷凝水
condenser 冷凝器;电容器;聚光器
~ hood [condenser bonnet] 冷凝器防护罩
condensing 冷凝;凝聚
~ and distilling kettle 冷凝和蒸馏釜
~ kettle 冷凝釜
condition 条件,状态;调节(空气)
~ of service 运行条件
~ of surface 表面状态
conditioned (有)条件的,调节的
~ - air inlet 调节空气进口
~ test atmosphere 气候试验
conditioning 状态调节;(空气)调节(试验物体)
~ atmosphere 状态调节环境
~ cabinet 状态调节箱,气候调节箱
~ of test specimens 试样调节
~ period 状态调节时间,空气调节时间
~ protection test 防大气试验,耐气候试验
~ room 空气调节室,空调室;状态调节室
~ stage 调节阶段(拉吹)
~ time 状态调节时间;空气调节时间
eonductance 电导
conductible finish 导电(末道)漆;导电(末道)涂料
conducting 传导的,导电的
~ cil 感应圈
~ fibre 导电(性)纤维
~ film 导电薄膜
conduction 传导;导热
~ dryer (间接)加热干燥器
~ of heat 传热,导热

conductive 传导的,导电的
~ adhesive 导电黏合剂
~ chanael black 导电槽(法炭)黑(补强剂,颜料)
~ coating 导电涂料
~ fabric 导电织物
~ furnace black 导电炉(法炭)黑(补强剂,颜料)
~ high - polymeric resin 导电高聚物树脂
~ lacquer 导电漆
~ plastics 导电塑料
~ polymer 导电聚合物
~ polyvinyl chloride plastics 导电聚氯乙烯塑料
~ reinforcement 导电补[增]强剂
~ resin 导电树脂
~ rubber 导电橡胶
conductivity 电导率
~ electrical 导电率
conductometric 传导的;电导的
~ analysis 电导分析
~ titration 导电定量滴定(法),电导滴定
conductor 导体
conduit clip 管夹
Condux dicer 圆盘形切粒机
cone 圆锥体,圆锥形
~ and disc viscometer 锥盘式黏度计
~ and plate rheogoniometer 锥板式流变测定仪
~ and plate rheometer 锥板流变仪
~ and plate viscometer 锥板黏度计
~ and plate viscosimeter 锥板黏度计
~ and screw mixer [epicyclic screw mixer] (圆)锥形螺杆式混合器
~ apex 锥形顶端(锥板黏度计)
~ belt 三角皮带
~ crusher 锥形粉碎机
~ - cylinder screw 锥柱形螺杆
~ gate 锥形浇口
~ impeller 锥形旋转混合器
~ impeller mixer 锥形旋转混合器
~ mill 锥形磨

~ plate viscosimeter   锥板黏度计
~ pulley drive   锥形轮传动
~ winding   锥形筒子卷绕
confer   比较;对比;参看
configurate   模型
configuration   构型;造型
cis-configuration   顺式构型
configuralional   构型的
　～ base unit   构型基本单元
　～ elasticity   构型弹性
　～ repeating unit   构型重复单元
　～ sequence   构型序列
　～ unit   构型单元
conformability   一致性,适应性,顺应性
conformal coating   仿形涂层,螯合涂层,共型涂层
conformation   构象
conformational anaysis   构象分析
conforming to requirements   符合需求
conformity   一致,适应,相似
　～ certification mark   检验合格标志
confusion of adhesive components   黏合剂组分混合
congeal   冻凝;凝固
congealing   冻凝(作用)
congo red method   刚果红法(测定热稳定性)
conical   圆锥形的
　～ bell mill   锥形球磨机
　～ blender   锥形混合[混配]器
　～ core   锥形模芯;锥形螺杆芯孔
　～ disc mill   圆锥磨
　～ drum centrifugal   锥鼓离心机
　～ dry blender   锥形干混机
　～ grinder   锥形研磨机
　～ hom   锥形超声焊接仪
　～ insert   锥形嵌件
　～ mill   锥形研磨机,锥形磨
　～ mixer   锥形混合器
　～ plug   锥形塞
　～ roller bearing   锥形滚柱轴承
　～ screw   锥形螺杆
　～ sleeve   锥形套筒
　～ sonotrode   锥形超声焊接仪

~ sprue bushing   锥形浇道套
~ tapered section   锥形渐变段(螺杆)
~ transition section   锥形过渡段(螺杆)
conicylindrical viscometer   锥筒式黏度计
conjugate   共轭的;配合的,组合的
　～ fiber   组合纤维
　～ spinning   共轭纺丝,复合纺丝(法)
conjugated   共轭的
　～ bond   共轭键
　～ chain   共轭链
　～ compound   共轭化合物
　～ diene   共轭二烯
　～ double bond   共轭双键
　～ hydrocarbons   共轭烃
　～ linkage   共轭键
　～ polymer   共轭聚合物
conjugative   共轭的
　～ effect   共轭效应
　～ mechanism   共轭机理
　～ monomer   共轭单体
　～ polymer   共轭聚合物
conjunct   混合的;结合的
　～ polymer   混合聚合物
　～ polymerization   混合聚合
connect   连接,接合,联系
connected in series   串联(连接)
connecting   连接
　～ box   接线盒,连接箱
　～ head   连接头
　～ pipe   连接管
connection   连接
　～ for exhaust   排气连接(管)
　～ hose   连接软管
connector   接合器,接头,接线柱
conoid   圆锥(形的),锥体
consistancy [consistency]   稠度
　～ coefficient   稠度系数
　～ cup   稠度杯
　～ curve   稠度曲线
　～ index   稠度指数
　～ meter   稠度计
　～ property   相容性质

consistometer 稠度计
consolidation 凝固,凝结;填实,压实
consolute temperature 会溶温度
const. = constant 常数,恒量
constant 恒定的
- ~ atmosphere 恒定气氛;恒定大气
- ~ copolymerization 恒比共聚
- ~ depth screw 等深螺杆
- ~ depth variable pitch screw 等深不等距螺杆
- ~ head 恒定水压
- ~ head tank 恒定水压槽
- ~ land screw 等螺距螺杆(挤塑机)
- ~ lead 等导程(螺杆)
- ~ lead screw 等导程螺杆(挤塑机)
- ~ – level regulator 恒位面调节器
- ~ pitch 等螺距,固定螺距(挤塑机螺杆)
- ~ pitch screw 等螺距螺杆(挤塑机)
- ~ pressure 恒压,等压
- ~ proportion 恒比,定比
- ~ rate 等速,恒速
- ~ rate of elongation 恒定伸长速率,等速伸长
- ~ rate creep 等速蠕变率
- ~ rate of extension test 等速伸长试验
- ~ rate of loading 定速负荷,等加负荷
- ~ rate of loading test 等加负荷试验
- ~ rate of traverse 恒速横移;等速往复
- ~ shear viscosimeter 恒定剪切黏度计
- ~ stirring tank reactor 连续搅拌釜式反应器
- ~ stress creep 恒应力蠕变
- ~ stress type tester 恒定应力型试验机
- ~ taper screw 等距变径螺杆(挤塑机)
- ~ temperature 恒温,等温
- ~ temperature and humidity 恒温恒湿
- ~ temperature bath 恒温浴
- ~ temperature cabinet 恒温箱
- ~ temperature compression 等温压缩
- ~ temperature oven 恒温炉,恒温烘箱
- ~ temperature regulator 恒温调节器
- ~ tension windup 定张力卷取装置
- ~ thread decreasing pitch screw 等螺纹减距螺杆,恒定螺槽深减距螺杆
- ~ torque winding 恒扭矩卷取、恒转矩卷绕
- ~ worm pitch 等距螺纹

constellation 构象
constituent 成分;组分;构成
- ~ of fibre 纤维成分
- ~ of plastic 塑料成分

constitution 结构;构造;成分
constitutional 结构的,构成的
- ~ detail 结构零件
- ~ formula 结构式
- ~ repeating unit 重复结构单元
- ~ sequence 结构序列
- ~ unit 结构单元

constitutive property 结构性质
construction 结构,构造;建筑
- ~ cost 建筑费,工程费

constructional 结构的,建筑的
- ~ detail 结构零件
- ~ element 构件,结构元件
- ~ engineering 建筑工程;结构工程
- ~ formula 结构式
- ~ gluing 构件胶合
- ~ materials 结构材料
- ~ plastic 结构塑料,工程塑料
- ~ repeating unit 结构重复单元
- ~ unit 结构单元

consumer 消费者,用户
- ~ goods 消费品,生活必需品
- ~ package 消费(品)包装
- ~ unit 消费品(包装)容器

consumption 消耗量,使用量
- ~ of plastics 塑料消耗(量)

contact 接触
- ~ adhesive 触压黏合剂,接触型黏合剂
- ~ agent 接触剂
- ~ angle 接触角
- ~ area 接触面(积)
- ~ box 配电箱
- ~ condenser 接触冷凝器
- ~ corrosion (表面)接触腐蚀

contacting plate

- ~ curing pressure 固化接触压力(反应黏合剂)
- ~ dryer 接触型干燥器
- ~ failure in the glue line 黏合接触不良
- ~ fatigue 接触疲劳
- ~ force 触压力
- ~ heating and vacuum forming 接触加热真空成型
- ~ insert 插入式嵌件
- ~ laminate 接触层压板
- ~ laminating 接触层压
- ~ layer 接触层
- ~ layup moulding 低压铺叠成型,接触成型
- ~ manometer 触压型压力表
- ~ microradioaraphy [CMR] 触压X射线显微照相检验法
- ~ mo(u)lding 接触成型
- ~ pressure (接)触压(力)
- ~ –pressure laminate 接触层压(板)
- ~ pressure laminating 触压层合,接触层合
- ~ pressure laminating resin 触压层合树脂
- ~ pressure mo(u)lding 触压成型
- ~ pressure resin 触压树脂,低压固化树脂
- ~ pressure roller 触压辊(卷取机)
- ~ resin 低压固化树脂,触压树脂,触压成型树脂,接触成型树脂
- ~ roller 接触辊,触压辊
- ~ spraying 触压[低压]喷雾;触压喷镀
- ~ stress 接触应力
- ~ surface 接触面
- ~ switch 接触开关
- ~ test 接触试验
- ~ time 接触时间
- ~ ultrasonic welding 超声波接触焊,接触超声波焊接

contacting plate 接触层板(蒸馏)
contactor 接触器,触点,开关
- ~ control 接触控制,触点控制

contained plastic deformation 限定塑性形变
container 容器,箱,罐
- ~ board 盒纸板(包装)
- ~ construction 储存器结构
- ~ dispenser 附计量器的容器

contaminant 沾污,沾染,污染;掺杂物(塑料加工)
- ~ particle 污染物颗粒

contaminated 污染(过)的
- ~ area 污染面积
- ~ surface 污染表面
- ~ particle size 污染物粒度

contaminating material 污染物质
contamination 污染,沾染;沾染物;杂质
content 含量
continator 烘干和预热颗粒用料斗(注塑机)
contingent on application 取决于应用
continuous 连续的,不断的
- ~ adjustment 连续调节
- ~ ageing 连续老化,持续老化
- ~ analysis 连续分析
- ~ apparatus 连续式装置
- ~ automatic spinning 连续自动化纺纱
- ~ baling press 连续式打包机
- ~ beater 连续打浆机
- ~ belt 传动皮带
- ~ bend test 持续弯曲试验
- ~ Brookfield viscometer 连续式布鲁克菲尔德黏度计
- ~ casting 连续铸塑
- ~ centrifuge 连续离心机
- ~ circulation 连续循环
- ~ compounder 连续混料机
- ~ control 连续控制
- ~ cure 连续硫化;连续焙烘
- ~ current 恒定电流,直流
- ~ deformation 连续形变
- ~ drafting 连续牵伸
- ~ drier [dryer] 连续式干燥器
- ~ duty 连续运转
- ~ dyeing 连续染色
- ~ extrusion moulding 连续挤塑,连续

挤出模塑
~ feeding 连续进料
~ fiber pellets 长纤维粒料
~ fiber winding 长丝缠绕
~ fibrillation 连续纤化作用
~ filament 长丝
~ filament glass cloth [woven glass filament fabric] 玻璃纤维长丝织物
~ filament glass yam 玻璃纤维长丝
~ filament mat 长丝垫
~ filament rayon yam 黏胶长丝
~ filament staple fibre 长丝短纤维
~ filament staple fibre moven fabric 长丝定长纤维织物
~ filament woven fabric 长丝织物
~ filament yam 连续长纱,长丝纱,复丝
~ film 连续薄膜
~ –flow calorimeter 续流热量计
~ flow chromatography 连续流动色谱(法)
~ flow curtain coating 帘流涂布
~ grain growth 连续晶粒生长
~ grinder 连续研磨机
~ kneader and compounder 连续捏和配料机
~ kneader mixer 连续捏和混合机
~ laminating 连续层合
~ laminator 连续层合机
~ load 连续载荷
~ lubrication 连续润滑(作用)
~ mill 连续研磨机
~ mixer 连续混合机
~ mixing 连续混合
~ network structure 连续网络结构
~ operation 连续运转,连续操作
~ pattern 连续花纹
~ phase 连续相
~ plase of an emulsion 乳液连续相
~ plant 连续操作装置
~ polymerization 连续聚合
~ polymerization unit 连续聚合装置
~ printing 连续印刷
~ process 连续作业,流水作业

~ production 连续生产
~ production of foam slabstock 泡沫(塑料)块料连续生产
~ proportioning plant 连续式配料设备
~ puddles 连续料垄
~ pultrusion method 连续挤拉法
~ reactor 连续式反应器
~ resin casting plant 树脂连续铸塑装置
~ roller welding 连续辊焊[滚焊]
~ running 连续运转
~ sampling 连续取样
~ sealing machine 连续热合机
~ sheeting 连续成片材(法)
~ slab production 连续块料生产(泡沫塑料)
~ spectrum 连续光谱
~ spinning 连续纺丝
~ spinning process 连续纺丝过程
~ stirred tank reactor 连续搅拌釜式反应器
~ strand mat 长丝束毡
~ stress relaxation 连续应力松弛
~ technique 连续成型
~ textile glass filament 连续纺织玻璃长丝
~ tube process 连续挤塑管状型坯法(吹塑成型)
~ tubular dryer [CT–dryer] 连续运转圆筒形干燥器,连续转动管式干燥器
~ tunnel drier 连续热道干燥器
~ turbo–mixer 连续式涡轮混合器
~ vacuum forming 连续真空成型
~ viscosimetry 连续黏度测量法
~ vulcanization 连续硫化
~ vulcanizing plant 连续交联装置(聚乙烯管)
~ weld 连续焊缝
~ welding machine 连续焊接机
~ winding 连续卷绕
~ X–rays 连续X射线
~ X–ray spectrum 连续X射线谱
~ yarn tester 连续纱线试验仪

continuously 连续不断地
continuum theory 连续(介质)理论
contour 轮廓,外形;仿形的,异形的
~ electrode 异形焊条
~ package 仿型包装
contract 收缩;限定;合同,契约
~ of carriage 输送合同,运输契约
~ service 按照协定规定服务
contraction 收缩(率),压缩,填料,密封
~ allowance 收缩余量(塑料加工模具)
~ cavity 收缩孔
~ coefficient 收缩系数,压缩系数
~ of area 面积收缩
~ percentage 收缩率
~ stress 收缩应力
~ value 收缩参数(塑料)
contrarotating 反转(的),逆转(的)
~ screws 反向旋转螺杆(双螺杆挤塑机)
contrast 对比,对照;反差
~ colur 对比色
~ index 反差指数
~ of tone 色调对比
~ ratio 对比率(颜色)
~ test 对照试验
control 控制,操纵,调节(器)
~ board 控制板,控制盘,操纵台
~ button 控制按钮,操纵按钮
~ cabinet 开关箱,控制柜,配电箱
~ centre 控制中心,操作台
~ core 试发芯块
~ core density 试发芯块密度
~ desk 操纵台,控制板
~ divice 控制装置
~ electronics 电子控制仪(注塑机)
~ element [regulating unit] 控制机构,调节机构
~ engineering (自动)控制技术
~ equipment 控制设备
~ experiment 对照实验
~ index setting 控制指数调整
~ knob 控制按钮,操纵按钮
~ laboratory 检验室
~ magnet 控制磁铁
~ mechanism 操纵机构
~ module 控制元件;控制模件;控制模数
~ of melt temperature 熔体温度控制,熔体调温
~ of monofilaments 单丝(卷绕)控制
~ of reflux ratio 回流比控制
~ panel 控制板,控制盘,操纵机,操纵台
~ pivot 控制轴颈
~ plant 控制装置,检查装置
~ plate 控制板
~ reception 验收
~ roll 导辊,调节辊,传动辊
~ room 控制室
~ sample 对照试样
~ signal 控制信号
~ slab 控制板
~ symbol 控制符号
~ test 对照试验,控制(性)试验
~ thermocouple 控制热电偶
~ time 调节时间,调整时间
~ valve 控制阀,调节阀
controllable 可控制的,可调节的
~ lip 可调模唇(挤塑片材模头)
~ load 可调负荷
~ viscometric flow 可控测黏流动
controlled 可控的,控制的
~ break-away 可控开模速度
~ chain growth polymerization 键增长受控聚合
~ chilling 控制冷却
~ degradation 可控降解
~ polymerization 控制聚合,受控聚合
~ roll 可控辊(压延机或涂布机)
~ shrinkage annealing 控制收缩热处理
~ structure polymer 有规结构聚合物,受控结构聚合物
~ temperature 控制温度
~ variable 控制变量
controller 控制器,调节器
controlling 控制的,调节的

~ means 控制设备;控制手段
~ thermocouple 调温热电偶
convection 对流
~ dryer 对流干燥器
convention 习惯,惯例
conventional 惯用的;普通的
~ base unit 带用基本单元;链节(聚合物)
~ draft 普通牵伸;传统牵伸
~ method 常规方法
~ plastic 普通塑料
~ polyethylene 普通聚乙烯
~ rubber 普通橡胶
~ sign 通用符号
~ spinning 常规纺纱,传统纺纱
~ symbol 习用符号
~ test 普通试验;标准试验
~ type 普通型号,常规类型
convergent 合聚(的),收敛(的)
~ die 聚流式模头(挤塑)
~ polarized light microscopy 聚敛偏光镜检法
converging 会聚;锥形(双螺杆)
~ plate 导向板(薄膜牵引装置)
conversion 变换,转变;换算;再制(二次加工)
~ chart 换算表
~ coating 预处理涂层(粘结件)
~ operation 加工方法(塑料)
~ period 转换时期,转化期
~ rate 转化率
~ table 换算表
convert 转换,转化,转变;换算;再制
converter 转[变]换器,转化器;再制工厂
~ and processor 加工者;加工厂
convex 凸面的
~ bowl 鼓形辊,凸面辊,弧面辊(压延机)
~ joint 凸缝
~ roll 鼓形辊,凸面辊(压延机)
~ weld 凸形角焊缝;加固填角焊
convey 输送,传送
conveyance 输送

~ of water under pressure 水加压输送
conveyer 传送机,输送器
~ belt 输送带,传送带
conveying 传送,输送
~ air 输送空气
~ belf 输送带,传送带
~ capacity 输送能力,输料量
~ chute 输送斜槽
~ - effective feed zone 输送有效喂料段(挤塑机)
~ performance 输送效能(挤塑机螺杆)
~ screw 输送螺杆
~ spiral 输送螺旋
~ stock 输送原料
conveyor 传送机,输送器
~ belt 输送带,带式运输机
~ dryer 履带式烘燥机
~ screw 螺旋输送机
~ stock 待运材料
~ table 流水生产操作台
convolute 卷绕(的)
cook 蒸煮
cooker 蒸煮器
cool 冷却;凉的,冷的
~ at low temperature 低温冷却
~ boiler 真空蒸煮器
~ colo(u)r 冷色(调)
~ down 冷却
~ hot - runner 低温热流道
~ off 冷却
coolant 冷却剂,致冷剂
~ drain pipe 冷却剂排泄管
~ feed pipe 冷却剂供给管
~ pump 冷却剂泵
~ system 冷却剂系统
~ thermostat 冷却恒温器
cooler 冷却剂;冷却器
~ drum 冷却转鼓
cooligomer 共聚低聚物
cooligomerization 共聚低聚(作用)
cooling 冷却(的)
~ agent 冷却剂,致冷剂
~ air 冷却空气

coordinate

~ air baffle 冷却空气挡板
~ annulus 冷却环(薄膜吹塑)
~ bath 冷却浴,冷却槽
~ blast 冷却通风
~ block 冷却定型模
~ calender 冷却滚压机
~ can 冷却滚筒
~ channel 冷却流道(模具)
~ chute 冷却槽
~ circuit 冷却环路,冷却系统(塑料加工模具)
~ coil 冷却盘管,冷却蛇管
~ contraction 冷却(时)收缩
~ curve 冷却曲线
~ cylinder 冷却滚筒;冷却料筒
~ device 冷却装置
~ -down 冷却;冷冻
~ drum 冷却转鼓,冷却辊筒
~ effect 冷却反应
~ fixture 冷却胎模
~ fluid 冷却液
~ jacket 冷却夹套
~ jig 冷却定型模
~ liquid 冷却液
~ mandrel 冷却芯型(管材)
~ means 冷却手段
~ medium 冷却介质
~ method 冷却法(塑料加工)
~ mixer 冷却混合机
~ mixture 冷却混合物
~ of injection moulding tools 注塑模具的冷却
~ oil 冷却油
~ pan 冷却槽
~ plate 冷却板
~ power 冷却能力
~ pump 冷却泵
~ rate 冷却速率
~ rate curve 冷却速率曲线
~ ring 冷却环,空冷环,气环(薄膜吹塑机)
~ roll(er) 冷却辊
~ section 冷却段,冷却区域
~ sleeve 冷却套筒
~ speed 冷却速率[度]
~ stress 冷却应力
~ system 冷却系统
~ table 冷却台
~ tank 冷却槽[罐]
~ time 冷却时间
~ tower 冷却塔
~ train 冷却辊系列
~ trough 冷却槽
~ tube 冷却管
~ velocity 冷却速度
~ water 冷却水
~ water circulation 冷却水循环
~ water tube 冷却水供应管
~ worm 冷却旋管
coordinate 配位(的);配价(的)
~ bond 配价键
~ covalent bond 配位共价键
coordinated 配位的,配价的
~ ionic polymerization 配位离子聚合
~ polymer 配位聚合物
coordination 配位(作用)
~ bond 配价键
~ catalysts 配位催化剂,齐格勒催化剂
~ compound 配位化合物
~ polymer 配位聚合物
~ polymerization 配位聚合
~ type initiator 配位型引发剂
copal(resin) 珂巴(树脂)〈天然树脂〉
~ ester 珂巴酯
cope 铸模顶部
~ pattern 模型上半部
copolyamide 共聚酰胺
copolycondensation 共缩聚
copolyester 共聚聚酯
copolymer 共聚物,异分子聚合物
~ fibre 共聚物纤维
~ latex 共聚物(型)胶乳;共聚物(型)分散(体)
~ of acrylate and butadiene 丙烯酸酯与丁二烯共聚物
~ of-silicon and polyalkoxyether 聚硅氧烷-多烷氧基醚共聚物,发泡灵

~ resin 共聚树脂
copolymerizable monomer 共聚单体
copolymerization 共聚(作用)
~ with crosslinking 交联共聚合
copolymerize 共聚合
copper [Cu] 铜,紫铜,红铜
~ chloride catalyst 氯化铜催化剂
~ - clad 铜箔黏贴
~ clad expexy laminate 铜箔贴面环氧层合板
~ clad laminate 铜箔贴面层合板,附铜层合板
~ deactivator 耐铜剂;铜钝化剂
~ inhibitor 耐铜剂
~ plate printing roll 钢板印刷辊,凹版印刷辊
coprecipitated 共沉淀的
~ barium - cadmium laurate 共沉淀月桂酸钡镉(热稳定剂)
~ barium - cadmium myristate 共沉淀肉豆蔻酸钡镉(热稳定剂)
~ basic silicate - sulfate 共沉淀碱式硅酸铅 - 硫酸铅(热稳定剂)
~ lead orthosilicate and silicagel 共沉淀正硅酸铅 - 硅胶(热稳定剂)
coproduct 副产品
copy 副本;复制品,拷贝
copying machine 仿形机,复制机
cor. [corrected] 校正
cord 条斑(制品缺陷);绳;帘布
~ - adhesive applicator 带状熔融黏合剂涂布机
~ fabric 帘子布
~ rayon 帘子线
cordage thread 绳索用线,绳线(聚合物纤维industry)
core 模芯,芯子;中间层,夹层(层压板);阳模塞(压塑模)
~ and separator 芯棒和分流器(制软管挤塑模具)
~ binder 模[型]芯粘结剂(铸塑);芯板粘结剂(层压板)
~ box vent 芯模箱气孔
~ building machine 模[型]芯制造机

(铸塑)
~ density 芯密度(泡沫塑料)
~ diameter (螺杆)芯直径
~ - draw 型芯抽取
~ - drawing equipment 模[型]芯抽出设备
~ fibre 纱芯纤维
~ insert 模芯嵌件(压塑膜)
~ jaw 模[型]芯夹具(模具)
~ loss 模芯损耗
~ pigments 色核颜料
~ pin 芯杆,进气杆,吹气芯棒;成孔销
~ pin plate 穿孔销托板
~ pin retainer plate 成孔销托模板(模具)
~ plate 芯模托板,阳模板
~ plug 进气杆,吹气芯棒
~ pulling equipment 模芯抽出设备
~ pulling mechanism 芯棒顶出机构,模芯顶出机构
~ - retainer plate 模芯托板,阳模托板
~ rod 芯棒
~ sheet 芯板(层合材料)
~ - shell adhesive 微胶囊型黏合剂
~ side 侧位型芯
~ slide 型芯闸板(模具)
~ stroke 型芯冲程
~ temperature 模芯温度
~ traction 型芯抽取
~ veneer 芯板
~ winding 芯轴缠绕
~ withdrawal 型芯抽出(模具)
~ withdrawing cylinder 型芯抽出液压罐
co - reacting 副反应;交联反应
cored (型)芯的,空心的
~ mo(u)ld 芯塑模;钻孔模具
~ - out mo(u)ld plunger 钻孔模塞
~ roll 中空辊
~ screw 空心螺杆
~ steam plate (加)热板
coring pin 成孔销

cork 软木(塞)
- ~ flour 软木粉
- ~ powder 软木粉(填料)

corner 角;弯头,弯管
- ~ guard 防护角
- ~ joint [weld] 弯头速接;角(接)焊缝;角接接头
- ~ joint with fillet weld 角焊缝焊接接头
- ~ pad 角护板
- ~ welding 角焊

corona 电晕(放电)
- ~ discharge 电晕放电
- ~ discharge treatment 电晕放电处理
- ~ effect 电晕效应
- ~ phenomenon 电晕现象
- ~ pretreatment 电晕(放电)预处理(为了非极性塑料表面极化)
- ~ priming 电晕放电打底,电晕放电处理
- ~ resistance 耐电晕性
- ~ treater 电晕放电处理机
- ~ treatment 电晕放电处理
- ~ voltage 电晕电压

correct 正确的;恰当的,标准的
- ~ mixture 正确混合比
- ~ processing 正确加工
- ~ temperature 设定温度,给定温度

correction 校正

correlated colo(u)r temperature 相关色温

correlation test 相关性试验

corrodibility 腐蚀性

corrosion 腐蚀
- ~ cracking 腐蚀开裂
- ~ creep 腐蚀蠕变
- ~ fatigue 腐蚀疲劳
- ~ index 腐蚀指数
- ~ inhibiting adhesive primer 防腐粘接底漆
- ~ - inhibition paint 防腐漆,防腐蚀涂料
- ~ inhibitor 防腐蚀剂
- ~ prevention 防腐蚀
- ~ - preventive ability 防腐性能,防腐能力
- ~ - proof 耐腐蚀的,不锈的
- ~ protection 防腐蚀
- ~ remover 防腐剂,除锈剂
- ~ resistance 耐腐蚀性
- ~ - resistant 耐腐蚀的;不锈的
- ~ - resistant material 耐腐蚀材料
- ~ resisting 耐腐蚀的;不锈的
- ~ stability 耐腐蚀性
- ~ strength 耐腐蚀性,腐蚀强度
- ~ test 腐蚀试验
- ~ testing method 腐蚀性测试法
- ~ value 腐蚀值,腐蚀度

corrosive 腐蚀剂;腐蚀性的
- ~ agent 腐蚀剂
- ~ atmosphere 腐蚀环境,腐蚀氛围
- ~ sample 腐蚀性试样

corrugated 波纹的
- ~ board 瓦楞(纸)板
- ~ buff 波形抛光轮
- ~ film tape 波形膜带,波形裂膜丝[纤维]
- ~ hose 波纹软管
- ~ iron 波纹薄(钢)板
- ~ laminate 波纹层合板
- ~ mixing rolls 带槽混合辊
- ~ packaging 用波纹材料包装
- ~ panel 波纹板,瓦楞板
- ~ paper 波纹纸,瓦楞纸
- ~ pipe 波纹管
- ~ plate 波纹板,瓦楞板
- ~ polyester glass roof sheeting 波纹聚酯玻璃钢屋面板材
- ~ reinforced sheeting 增强塑料波纹板
- ~ roll 槽纹辊,瓦楞辊
- ~ roofing 波纹屋面材料,波纹(屋)瓦
- ~ sheet 波纹板,瓦楞板
- ~ sheet continuous moulding machine 波纹板连续成型机(增强塑料)
- ~ sheet process 波纹板生产(增强塑料)
- ~ stair tread 有沟槽梯形踏板
- ~ tube 波纹(软)管,瓦楞(软)管

corrugating roll 槽纹辊,瓦楞辊
corrugation 波纹,槽纹;起皱纹;成波纹,轧波纹,压槽纹
~ former 波纹模塑件
corundum 金刚砂(填料,磨料)
cost 成本;费用,价格
~ accounting 成本核算
~ estimating 成本估算
~ of construction and equipment 建筑及设备费
~ of maintenance 养护费,维修费
~ of operation 运转费
~ of overhaul 大修费,彻底检修费
~ of production 生产成本
~ of repairs 修理费
~ price 成本价格
cotruder 锥形双螺杆挤塑机
cotter pin 开口销,开尾销
cotton 棉;木棉
~ cloth 棉布
~ fabric 棉织物;棉布
~ fabric base 棉布基
~ fiber 棉纤维
~ flock 棉绒,棉屑
~ flock filler 棉绒填料
~ flock phenolics 棉绒填充酚醛塑料
~ linters 棉籽绒,棉短绒(填料)
~ phenolics 棉絮酚醛塑料
~ rags 棉屑
~ reinforcement 棉增强材料
~ seed hull 棉籽壳
~ seed meal 棉籽粉
~ web base 棉布基
coulomb yield 电流输出,电流效率(静电粉末喷涂)
coumarone 苯并呋喃,古马隆,香豆酮,氧茚
~ - indene resin 苯并呋喃-茚树脂,古马隆-茚树脂,香豆酮-茚树脂(软化剂)
~ resin 苯并呋喃树脂,古马隆树脂,香豆酮树脂
count 计算;统计;线纱支数;织物经纬密度

counter 相反,反面,反对;计算器
~ bending 反向弯曲(压延辊调节压延片材幅宽方向厚薄的一种方法)
~ blade 对刀,反向刀
~ cast 对铸塑件
~ clockwise 反时针方向的
~ clockwise motion 反时针方向转动
~ clockwise rotation 反时针旋转
~ current 逆[反,对]流;逆电流
~ - current heat exchanger 逆流热交换器
~ - current pan mixer 逆流盘状混合器
~ display 计算机显像
~ die 底模
~ draft 反斜度,反锥度,倒拔梢
~ electrode 反电板
~ flow dryer 逆流式干燥机
~ former 反模塑件
~ knife 对刀,反向刀
~ motion 反向运动
~ pressure 反压力
~ rotating peddle 逆转式桨叶
~ switch 计时开关
~ weight 平衡锤,砝码
counterbalance 平衡重,砝码,托盘天平
countercurrent 逆[反,对]流;逆电流
~ centrifugal force separator 逆流离心式分离机,逆流离心机
~ condenser 逆流冷凝器
counterdraft 反斜度,反锥度,倒拔梢
counterflow 逆流
~ mixing 逆流式搅拌
counterpressure 反压(力),平衡压力
~ roll 反压辊,承压辊
counterrotating 反转,相对旋转
~ screws 反转螺杆
~ twin screw extruder 反转双螺杆挤塑机
countershaft 逆转轴,副轴
counterstain 复染色;对比染色
countersunk 埋头孔
~ head screw 埋头螺钉
~ hole 埋头螺孔(超声波埋嵌件)

~ screw 埋头螺钉
country pipeline 地下(敷设)管线
coupled glass – reinforced polypropylene 偶联玻璃纤维增强聚丙烯
coupling 对,双;偶合,偶联;偶合联接;联接器,接箍
~ agent 偶联剂
~ bath 偶合浴,显色浴
~ bush 连接套筒
~ component 偶合组分
~ compound 偶合化合物
~ dyes 偶合染料
~ effect 偶合效应(层合板)
~ finish 偶联处理剂(玻璃纤维织物)
~ polymerization 偶联聚合
~ reaction 偶联反应
~ screw 连接螺钉
~ shaft 连接轴
~ size 偶联浸润剂,偶联上浆剂
~ sleeve 联接套
~ technique 偶联技术
~ viscosity 偶合黏度
course 过程
~ of reaction 反应过程
~ of receiving 验收过程
~ of state 物态过程
covalence 共价
covalent 共价的
~ bond 共价键
~ compound 共价化合物
~ crosslink 共价交联键
~ crystal 共价结晶
~ link(age) 共价键
~ multiple bonding 共价重键
covalescence 网状键
cover 盖;罩,套;面层;盖板,覆面物
~ coat 面层
~ die 套模
~ joint 覆盖焊缝
~ mo(u)ld 盖模,固定模(注塑模具)
~ plate 盖板
~ sheet 覆盖薄片,覆盖面板(层压)
~ slab 盖板
~ strip 盖条;挡板,护板;补强带(焊接)
~ strip welding 搭接焊,补强带焊接
~ weld 覆盖焊缝,补强焊缝
coverage 覆盖(度),遮盖力;可涂刷性
covering 覆盖;覆盖层,涂层;壳,罩
~ a roll with rubber 涂层橡胶辊
~ colo(u)r 覆盖色,涂敷色
~ disc 盖盘
~ extrusion head 包覆(电缆)挤塑机头
~ layer 覆盖层,涂敷层(层压材料)
~ machine 包覆机
~ material 包覆材料
~ power 覆盖(能)力(涂料)
C. P. [calorific power] 热值;卡值
cP [centipoise] 厘泊(动态黏度非SI单位,等于10^{-3}Pa·s)
CP [cellulose propionate] 丙酸纤维素
c. p. [chemically pure] 化学纯
C. P. [circular pitch] 圆周齿距
Cp [circuits per pattern] 每种花纹循环(长丝缠绕)
CPE [chlorinated polyethylene] 氯化聚乙烯(参见 PEC)
C – process [Croning process] 克洛宁法,壳模铸塑法
CPP [chlorinated polypropylene] 氯化聚丙烯
cps. [centipoise] 厘泊
CPVC [chlorinated polyvinyl chloride] 氯化聚氯乙烯
CPVC [critical pigment volume concentration] 临界颜料体积浓度
CR [chloroprene rubber] 氯丁橡胶
Cr [chromium] 铬
CR [polychloroprene rubber] 氯丁橡胶,聚氯丁二烯橡胶
crack 开裂;裂缝,裂纹,龟裂;裂解
~ branching 裂缝分叉,花状开裂
~ correlator 裂缝(长度)测定仪
~ detection 裂纹检验
~ detector 探伤器
~ drawing 裂膜式拉伸
~ failure 龟裂破坏

~ filler 填缝料
~ formation 裂缝形成,龟裂,开裂
~ formation limit 开裂极限
~ formed during heat–treatment 热处理出现的裂纹
~ –free 无裂缝的
~ growth 裂纹增长
~ growth resistance 耐裂纹扩展性
~ initiation 龟裂始发
~ propagation 裂缝扩展
~ resistance 耐开裂性,耐龟裂性
~ sealer 裂缝填料,封缝剂,封缝料
~ test 裂缝试验,耐裂试验
~ toughness 裂缝韧性,龟裂稳定性
cracked surface 表面有裂纹
cracker 破碎机
~ mill 碾磨机,破碎机
~ roll 碾碎辊,破碎辊
cracking 龟裂,开裂;裂纹,裂缝;裂解,裂化
~ caused by light 光致开裂
~ in weld 焊缝开裂
~ of a film 涂膜(层)开裂
~ resistance 耐开裂性,耐龟裂性
~ test 耐裂试验,裂缝试验
~ tester 龟裂试验机
crackfree 无裂缝的
crackled lacquer 裂纹漆
crammer–fed extruder 强制加料式挤塑机
cramp 夹钳;夹紧,固定
crank 曲柄
~ locking gear (mechanism) 曲柄联锁装置
~ press 曲轴式压机
crash wall 防护外壁
crate 板条箱,柳条箱,周转箱
crater 焊口;熔塌穴,陷坑,坑点(塑料制品缺陷)
cratering 起凹穴,塌陷(塑料涂层)
crawling 起凹穴,塌陷
craze 银纹(聚苯乙烯制品等);细裂纹,微裂,碎裂
~ resistance 耐微裂性

crazing 银纹(聚苯乙烯制品等);网状裂纹(涂层),细裂纹(黏合膜)
cream 乳白(聚氨酯起泡沫)
~ of latex 胶乳
~ time 乳白期
creamed latex 乳胶(泡沫塑料)
creaming 乳浊化(分散体的);乳白化(聚氨酯微孔塑料的)
~ time 乳白期
creasability 折皱性
crease 折痕,皱折;折叠
~ abrasion test 折皱耐磨试验
~ acceptance 允许折皱度
~ and shrink resistant finish 防皱防缩整理
~ angle 折皱角度
~ angle method 折皱角度试验法
~ folding 皱折折叠,折皱
~ line 折线
~ mark 折痕
~ proof 防皱的,耐皱的
~ –proof finish 防皱整修
~ proofing 防皱处理
~ recovery 折皱恢复性
~ recovery tester 折皱恢复测定器
~ resist finish 防皱整理
~ resistance 耐皱性,防皱性
~ resistant 耐皱的,防皱的
~ resistant finish 防皱整理
~ retention 折缝耐久性
~ sensitivity 折皱灵敏性,易折皱性能
~ sharpness 褶裥清晰度
creasing 卷边,弯边;折痕,折叠;皱纹(美)
~ angle 卷边角铁模板;弯曲角(美)
~ machine 卷边机,折叠机
~ power 折皱力
~ press 卷边(压)机,折叠机(美)
~ property 折皱性
~ resistance 防皱性,耐皱性,耐折性
~ rule 弯曲模板,卷边夹板(美)
~ strength 耐折叠强度
~ support 弯曲型架,卷边型架(美)
creel 筒子架;粗纱架,绕丝轴

creep 蠕变
- ~ at room temperature 常温蠕变
- ~ behaviour 蠕变性质,蠕变特征
- ~ compliance 蠕变柔量
- ~ contraction 蠕变收缩
- ~ curve 蠕变曲线,蠕变-时间曲线
- ~ deta 蠕变数据
- ~ - depending - on - time compression test under heat 受热条件下蠕变与压缩试验时间相关(塑料)
- ~ effect 蠕变效应
- ~ elongation 蠕变伸长率
- ~ extension 蠕变伸长率
- ~ failure 蠕变破损
- ~ flow 蠕变流动
- ~ fluidity 蠕变流动性
- ~ fracture 蠕变断裂
- ~ function 蠕变函数
- ~ life 蠕变寿命
- ~ limit 蠕变极限
- ~ line 蠕变线
- ~ measurement 蠕变测定
- ~ modulus 蠕变模量
- ~ path 电弧径迹
- ~ path formation 电弧径迹形式
- ~ rate 蠕变速率
- ~ ratio 蠕变比
- ~ recovery 蠕变恢复,蠕变复原
- ~ region 蠕变区域[范围]
- ~ relaxation 蠕变松弛
- ~ resistance 耐蠕变性
- ~ rupture 蠕变断裂
- ~ rupture strength 蠕变断裂强度,持久强度
- ~ rupture test 蠕变断裂试验,持久试验
- ~ speed 蠕变速率[度]
- ~ state 蠕变状态
- ~ strain 蠕变应变
- ~ strength 蠕变强度
- ~ stress 蠕变应力
- ~ test 蠕变试验,持久强度试验
- ~ tester 蠕变试验机
- ~ - time curve 蠕变-时间曲线
- ~ yield time 蠕变屈服期,蠕变-时间曲线

crepe 绉片,绉网
- ~ fabric 绉织物
- ~ rubber 皱纹橡胶;绉片(一种天然橡胶压延成片)
- ~ weave 绉纹组织
- ~ weave fabric 绉纹织物

cresol 甲酚
- ~ - formaldehyde 甲酚-甲醛
- ~ - formaldehyde resin [CF] 甲酚-甲醛树脂
- ~ mo(u)lding compound [material] 甲酚树脂成型料,甲酚树脂模塑料
- ~ novolac epoxy resin 甲酚甲醛环氧树脂
- ~ resin 甲酚树脂

crest 峰顶;峰值;螺齿顶
- ~ clearance 螺齿顶间隙,径向间隙
- ~ of barrel [saddle of barrel] 鞍形滚筒
- ~ of the corrugations 槽纹顶峰[波峰]
- ~ of thread 螺齿齿顶
- ~ radius 螺齿顶半径

cresyl diphenyl phophate [CDP] 磷酸二甲苯酯(增塑剂)

cresylic 甲酚
- ~ resin 甲酚树脂

crevasse 裂缝,裂隙
- ~ crack 裂缝,龟裂

crevice 裂缝,裂隙
- ~ corrosion 裂缝腐蚀

crimp 卷曲,皱纹;皱缩;脆的
- ~ durability 卷曲持久性
- ~ effect 卷曲效应
- ~ elasticity 卷曲弹性
- ~ energy 卷曲能
- ~ extension 卷曲延伸
- ~ factor 卷曲系数
- ~ force 卷曲力
- ~ frequency 卷曲频率
- ~ index 卷曲指数
- ~ modulus 卷曲模量
- ~ - proof 不皱的,防皱的

~ – proof finish 防皱整理
~ ratio 卷曲率
~ recovery 卷曲回复率
~ region 卷曲区
~ retention 卷曲保持性
~ retraction 卷曲收缩率
~ rigidity 卷曲刚性
~ seal 波纹封口
~ stability 卷曲稳定性
crimped 卷曲的
~ fiber 卷曲纤维
~ fibrillated fibre yarn 卷曲原纤化纤维纱
~ fibrillated network yarn 卷曲原纤化网络纱
~ fibrillated tape 卷曲裂膜条
~ glass fibre 卷曲玻璃纤维
~ length 卷曲长度
crimper 卷曲机;折皱器
crimping 卷曲,皱纹;皱缩
~ capacity 卷曲能力
~ device 卷曲装置
~ force 卷曲力
crimpness 卷曲度;绉缩性
crinkle 皱,卷曲
~ resistant 防皱的
~ varnish 皱纹漆
crisp 卷曲;起卷;脆的
crispness 挺爽性
criss – cross 十字形(的),交叉(的)
~ film 网状薄膜
~ sheeting 网状(增强)片材
critical 临界的
~ coefficient 临界系数
~ condition 临界状态
~ constant 临界常数
~ curve 临界曲线
~ defect 严重缺陷
~ density 临界密度
~ difference 临界差
~ elastic energy 临界弹性性能
~ entanglement molecular weight 临界缠结相对分子质量
~ humidity 临界湿度

~ length of filament 纤维临界长度
~ load 临界负荷
~ loading condition 临界负荷状态
~ moisture content 临界含水量
~ molecular weight 临界相对分子质量
~ parameter 基本参数
~ phenomenon 临界现象
~ pigment volume concentration 临界颜料体积浓度[CPVC]
~ point 临界点
~ pressure 临界压力
~ pressur ratio 临界压力比
~ processing temperature 临界操作温度
~ region 临界区域
~ relaxation time 临界松弛时间
~ shear rate 临界剪切速率
~ shear stress 临界剪切应力
~ state 临界状态
~ strain 临界应变
~ stress 临界应力,屈服应力
~ surface tension 临界表面张力(黏合剂)
~ temperature 临界温度
~ temperature of wetting 临界润湿温度
~ time 临界时间
~ value 临界值
~ velocity 临界速度
crocidolite 青石棉
crock 摩擦脱色
~ resistance 耐摩擦性(染色)
~ test 耐摩擦色牢度试验
crocking 泅色,渗色;摩擦脱色
~ fastness 耐摩擦色牢度
Croning 克洛宁
~ method [process] 克洛宁法,型壳铸造法,壳模铸型法
cross 十字形;十字接头,四通管;横(的);交叉(的)
~ air blasting 侧吹风,横向吹风
~ baffle 横隔板
~ – band veneer 直交胶合板
~ – banding 交叉排列(胶合板)

~ bar 横杆;十字头;横色条(织物)
~ beater mill 十字形锤磨机
~ bending test 横向弯曲试验
~ - blade agitator [mixer] 十字形搅拌器
~ bond 交联键
~ bonding 交叉层压(层合材料),交叉粘接
~ break 横向断裂
~ breaking properties 弯曲性能;横向断裂性能
~ - breaking strength 挠曲强度,纵向弯曲强度;断裂强度,弯曲强度
~ breaking test 横向断裂强度试验
~ cut 横切;正交
~ - cut adhesion 划格法粘接法
~ - cut saw 切割锯,横割锯
~ - cut test 横切割试验
~ - cutter 横切刀
~ - cutter type guillotine 横切刀型剪切机
~ direction 横向;侧向
~ elasticity 交叉弹性
~ extruder head 横向挤塑机头,直角机头
~ - feed multi - gating 星形状供料系列浇口
~ fiber 横向纤维,纬线;十字形纤维
~ gate 横浇口,横浇道
~ hatch adhesion 交叉黏合,网状黏合(涂层)
~ hole 十字形孔
~ laminate 交叉层合,交向层压
~ laminated 交叉层合的,交向层压的
~ laminated sheet 交向层压板,直交层合板
~ - laminated wood 交向层压木板
~ - lamination 交向层合
~ - lap fensile test 交叉粘接件拉伸试验
~ - layers 交叉层(胶合板)
~ mark 交叉痕迹,网纹
~ section 横截面
~ - section area 横截面积;断面积

~ slide 横刀架,横向滑板
~ slide valve 横滑(动)阀
~ staff clutch 横杆联轴器
~ staining 交互污染
~ - staple glass mat 交织玻璃纤维毡
~ striation 横条纹
~ twill 破斜纹组织
~ wound cheese 交叉卷绕卷筒
crossed 十字形的;交叉的
~ beams paddle mixer 交叉桨叶式混合机
~ blades 十字形桨叶
~ roll [skewed roll] 斜交叉辊,交错辊
crosshead 直角机头,斜向机头;十字头
~ die 直角模头,直角口模(挤塑机)
~ ejection 十字杆脱模
~ lift 垂直搅拌横杆;上下可移搅拌横杆
~ link 丁丁形连肘
~ lock 横杆锁定
~ of the press 压力机横杆
~ toggle pin 丁字肘节销
~ type pipe die 直角挤管口模
crossing 交叉;交叉层压(层合材料)
~ angle 丁字尺;交叉角
~ effect 交叉效应
crosslayers 交叉层
crosslink 交联;交联链节
~ density 交联(点)密度,交联度
crosslinkable (可)交联的
~ monomer 可交联用单体
~ plasticizer 可交联增塑剂
~ polyethylene [XLPE] 可交联聚乙烯
crosslinkage 交联;交联键
~ agent 交联剂
~ density 交联密度
~ interlacing 交联
crosslinked 交联(的)
~ copolymerized polymethyl methacrylate 交联共聚甲基丙烯酸甲酯
~ functional polymers 交联功能聚合物
~ gel 交联凝胶

~ material 交联材料
~ network 交联网络
~ polyesteramide 变联聚酯酰胺
~ polyethylene [XIPE] 交联聚乙烯
~ polyethylene foam 交联聚乙烯泡沫塑料
~ polymer 交联聚合物
~ polypropylene 交联聚丙烯
~ polystyrene 交联聚苯乙烯
~ polystyrene beads 交联聚苯乙烯珠
~ polyvinyl chloride [CLPVC] 交联聚氯乙烯
~ polyvinyl chloride foam 交联聚氯乙烯泡沫塑料
~ rubber 交联橡胶
~ thermosetting plastics 交联热固性塑料

crosslinker 交联剂
crosslinking 交联
~ agent 交联剂
~ bond 交联键
~ by activated species of inert gases 惰性气体激活离子交联
~ coefficient 交联系数
~ copolymer 交联共聚物
~ density 交联密度
~ efficiency 交联效率
~ index 交联指数,交联度
~ monomer 交联用单体
~ plasticizer 交联型增塑剂
~ polymer 交联聚合物
~ rate 交联速度
~ reaction 交联反应
~ resin 交联树脂
~ stage 交联阶段
~ system 交联体系
~ time 交联时间

crosswise 交向;横向
~ contraction 横向收缩
~ corrugation 横波
~ direction 横向
~ fold 横向折叠
~ laminate 交向层压板,交向层合制品

~ lamination 交向层合
crow-foot weave 四经破缎纹
crown 中高度(压延滚筒的);焊帽(塑料焊接时形成)
~ compensating 压延辊中高度补偿
~ effect 中高效应
~ filler 上991填料
~ gear 冠状齿轮,碟形齿轮
~ pack 鼓胀包装
crowned 中高的,凸形的,拱起的
~ roll 中高压压辊,凸面压延辊
~ roller 中高涂敷辊
Crown's preplasticator 克洛恩斯预塑化器
crow's feet 条纹迹(在滚压涂层上)
crow's foot mark 山形痕
crowsfooting 爪形皱纹
CRP [carbon fiber reinforced plastics] 碳纤维增强塑料
crucible 坩埚;电炉
~ tongs 坩埚钳
crude 粗的,天然的,未经加工的
~ crepe sheet 粗辊压片材
~ cresol 粗甲酚
~ material 原料
~ oil 原油
~ product 粗制品,半制成品
crudity 粗糙度
crumb stage 脆性状态
crumbly 易碎的,脆的
crush 压碎;碎裂;弄皱
~ -proof 防皱的,耐皱的
~ -proofing 防皱处理
~ resistance 耐皱性
~ -resistant 防皱的,耐皱的
~ test 压裂试验
crusher 破碎机
~ roll 破碎机轧辊
crushing 破碎,压碎
~ mill 破碎机
~ resistance 耐压碎性
~ rolls 滚筒式破碎机
~ strength 耐压碎强度(管)
~ stress 压碎应力

cryogen

~ test 耐压碎强度试验
~ tester 耐压碎强度试验器

cryogen 冷却剂;低温粉碎
~ - extrublas 低温挤吹(利用液氮致冷)
~ - rapid process 骤冷法

cryogenic 冷冻的,低温的
~ cooling 低温冷却
~ grinding 低温研磨(在液氮下)
~ temperature 低温温度

cryoscopic method 冰点下降法

crystal 晶体;结晶
~ - axis 晶轴
~ boundary 晶粒间界
~ - clear (像水晶一样)清澈的,晶莹透明的
~ defect 晶体缺陷
~ distortion 结晶扭变
~ elasticity 晶体弹性
~ face 晶面
~ failure 晶体破坏,晶体破裂
~ fines 细粒晶粒
~ forming polymer 成晶聚合物
~ fracture 晶体破碎
~ grain 晶粒
~ grating 晶格
~ growth 晶体增长
~ lattice (结)晶格(子),晶体点阵
~ monochromator 晶体单色器
~ morphology 晶体形态学
~ nucleus 晶核
~ optics 晶体光学
~ orientation 晶体取向
~ shape 晶体形状
~ size 晶体大小
~ size distribution 晶体大小分布
~ structure 晶体结构
~ system 晶系
~ transducer 晶体传感器
~ transitions 晶体转变
~ twin 孪晶

crystalline 结晶的;晶状的
~ - amorphous transition 晶态-非晶态转变

~ complex 晶体复合物
~ grain 晶粒
~ index 结晶指数
~ lamellae 晶态薄层,片晶
~ material 晶体材料
~ matter 结晶物质
~ melting point 晶体熔点
~ modification 结晶变态,晶体变型
~ plastic 结晶性塑料
~ polymer 结晶性聚合物
~ region 结晶区
~ rubber 结晶性橡胶
~ state (结)晶(状)态
~ structure 晶体结构
~ temperature 结晶温度
~ thermoplastics material 结晶型热塑性塑料

crystallinity 结晶度,结晶性

crystallite 晶粒,晶体;微晶
~ orientation 晶粒定向
~ size distribution 晶粒大小分布

crystallizability 可结晶性

crystallizable (可)结晶的
~ polymer 结晶性聚合物
~ structure 晶态结构

crystallization 结晶(作用)
~ heat 结晶热
~ kineties of polymer 高聚物结晶动力学
~ mechanism 结晶机理
~ morphology 结晶形态学
~ point 结晶点
~ rate 结晶速率[度]
~ system 晶系
~ temperature 结晶温度

crystallize 结晶,晶化

crystallizing pan 结晶盘

crystallographic 结晶的
~ - axis 结晶轴
~ dimension 结晶体大小
~ direction 结晶方向
~ modification 结晶变态
~ parameter 结晶参数
~ plane 晶面

~ system 晶系
crytallography 结晶学
Cs [caesium] 铯
CS [casein] 酪素,酪蛋白
cS, cSt, cStk [centistokes] 厘泡(运动黏度非SI单位)
c/s [cycle/second] 周/秒,赫兹(频率单位)
CTD-dryer 连续转动管式干燥器
CTFE [chlorotrifluoroethylene] 三氟氯乙烯
ct/sec [counts per second] 每秒钟计算次数
cts [cents] 分
Cu [copper] 铜
cu. [cubic] 立方(体)的
cub [cubic]
cube 立方(体)
~ cut 方粒切法;切方粒
~ cutter 方粒切粒机
~ dicer 切方粒机
~ making machine 成方粒机,方块成粒机
~ mixer 立方形混合器
cubes 方形颗粒
cub. ft [cubic foot] 立方英尺
cu. ft [cubic foot] 立方英尺
cubic 立方(体)的,体积的
~ centimeter [cm^3] 立方厘米
~ elasticity 体积弹性
~ expansion 体积膨胀
~ inch [in^3] 立方英寸
~ metre [m^3] 立方米
~ structure 立方结构
~ system 立方晶形,等轴晶系
cubical 立方(体)的,体积的
~ dilatation 体积膨胀
~ expansion 体积膨胀
~ material 立方颗粒材料
cubiform 立方形的
~ granulate 方形颗粒
cuff heater band 电热套
cu. ft [cubic foot] 立方英尺
cu. in [cubic inch] 立方英寸

cul-cle-sac area 死区,死角(模具)
cul-cle-sac of mo(u)ld 模具死区
cull 残料,废品
~ pick-up 残料钩
Culus process 库立斯吹塑法
cu. m [cubic meter] 立方米
cumar resin 香豆树脂
cumarin 香豆素
cumarone 苯并呋喃,古马隆,香豆酮
cumene 异丙(基)苯,枯烯
~ hydroperoxide 氢过氧化异丙苯
cu. mm. [cubic millimeter] 立方毫米
cumulative 累积
~ extension 累积伸长
~ molecular weight distribution 累积相对分子质量分布
cup 杯
~ and ball joint 球窝关节
~ and ball viscosimeter 杯球黏度计
~ and ball viscosity 杯球黏度
~ flow figure 杯溢法流动系数,杯模流动指数
~ flow index 杯模流动指数
~ flow test 杯模流动试验
~ mo(u)ld 杯状模具
~ package 杯式包装
~ test 杯形试验,冲压试验
~ viscometer 杯式黏度计
cupped 深拉的
cupping 深拉,深挤压,杯形挤压
~ apparatus 深拉仪
~ die 深拉模
~ machine 深拉压力机
~ test 深拉[压凹]试验,杯形试验
~ test apparatus 深拉试验仪
cuprammonium rayon 铜铵丝,铜铵人造丝
cuprous chloride 氯化亚铜(催化剂)
curative 固化剂,固化的
cure 固[塑,熟,硬,硫]化
~ by the layer 分层固化
~ cycle 熟化周期
~ index 固化指数
~ model 固化模型

~ process 固化工艺
~ rate 固化速率[度]
~ reaction 固化反应
~ schedule 固化条件
~ shrinkage 固化收缩(率)
~ system 固化体系
~ temperature 固[塑,熟,硫]化温度
~ [setting, set-up] time 固化时间
cured resin 固化树脂,硬化树脂
curing 固化,固[塑,熟,硬,硫]化处理
~ agent 固[塑,熟,硬,硫]化剂
~ blister 固化起泡(反应树脂)
~ catalyst 固化促进剂,硫化促进剂
~ condition 固化条件,硫化条件
~ curve 硫化曲线
~ cycle 固化周期;硫化周期
~ degree 固[塑,硫,熟]化度
~ department 固[塑,硫]化车间
~ fixture 固化用夹具
~ formulation 熟化配方
~ in air 空气(中)固化
~ mechanism 固化机理,固化历程
~ oven 固[塑化]烘箱
~ process 固[塑,硫]化过程
~ property 固化性能
~ rate 固化速率
~ reaction 固化反应
~ speed 固化速率[度]
~ step 固化阶段
~ stress 固化应力
~ temperature 固[硫,塑]化温度
~ test 固[塑,熟,硫]化试验
~ time 固[塑,熟,硫]化时间
~ time under temperature 温度-固化时间,高温固化时间
Curie 居里(人名);(小写)居里(放射性强度单位,符号 Ci)
~ point 居里点,居里温度
~ point pyrolysis 居里点热解
~ temperature 居里温度
curl 卷曲,扭[翘]曲
~ effect 卷曲效应
~ plate 弧形板
~ texture 卷曲结构

~ up 卷起,卷曲
curled 卷曲的
~ buff 卷曲抛光[磨光]轮
~ edge 卷边
curling 卷曲
~ die 卷边模
~ punch 卷边冲头,卷边阳模
~ round the roll 缠辊
~ stability 卷曲稳定性
~ stress 翘曲应力
~ whell 卷曲转盘
curliness 卷曲(度)
current 流,流动;电流;现行的,现代的
~ check 例行检查
~ discharge (电流)放电
~ leakage 漏电
~ shade 流行色泽
~ yield 现时产量
currently 普通地,通常地,当前
curtain 幕;帘;挡;流痕(亮漆涂层上)
~ coater 帘流涂布机,幕涂机
~ coater die 幕涂机模头,熔体喷嘴
~ coating 帘流涂布,幕涂
~ door 环状门,帘状门
~ flow coater 流延涂布机
~ flow coating 流延幕涂
~ rail 幕帘轨
~ stretcher 帘伸展器
~ strips 帘带条
~ wall 帘屏障
~ -walled facade 帘屏正面
~ -wall panel 帘屏板
curtaining 垂落(涂层较大面积下垂)
curve 曲线;弯曲
curved 弯曲的,弧形的
~ axis 弯曲轴,弧形轴
~ blade 弯曲桨叶
~ fibre network 弯曲纤维网结构
~ head window (圆)弧形窗
~ intersection 曲线交叉
~ lever 曲杆
~ platen 弯曲板
~ roll 弧形展幅辊
~ stressed-skirt folded structure 弯曲

应力表层褶皱结构
~ stressed–skin structure 弯曲应力表层结构
~ surface 弯曲表面
~ tunnel gate 弧形隧道浇口
cushion 衬垫;缓冲垫;缓冲器
~ coat 垫层
~ breaker 缓冲层
~ effect 减振作用,缓冲效应
~ material 缓冲材料
~ package 防震包装,缓冲包装
cushioned 缓冲的,减振的
~ flooring 弹性地板,缓冲地板
~ vinyls pl [CV flooring] 缓冲聚氯乙烯地板面层
cushioning 缓冲,减振;软垫
~ action 缓冲作用
~ and shockproofing package 防震包装
~ capacity 缓冲能力
~ effect 减震[缓冲]作用,缓冲效应
~ material 缓冲材料,弹性垫料
~ performance 缓冲性能,减振性能
custom 定制的,按客户要求特制的;常规,常例
~ –built machinery 按客户要求特制的机械
~ designed 按客户要求设计的
~ made 按要求制作的,定制的
~ mo(u)ded 按订货条款成型的
~ moulder 定货模压厂
customer 顾客,用户,订货人
~ request test 按定货者要求的试验
cut 切割,切断,裁切;割料;克特支数制
~ away 切掉
~ away view 剖视图
~ down 切断,切下
~ –down buffing 磨光,抛光
~ edge 切割边
~ fibres 切割纤维
~ film 切膜片
~ film strip 切膜条
~ layers 露层〈层合制品缺陷〉
~ lengths 切割长度,切段长度
~ –off 模缝脊,溢料脊;合模脊;切断

~ –off blade 切刀
~ –off device 切断装置
~ –off die 切断模,修边模
~ –off edge 切刀刃
~ –off machine 切断机,切割机,切片机
~ –off mo(u)ld 溢料式模具,溢式塑模
~ –off nozzle 〈自动〉关闭喷嘴
~ –off plate 调幅杆;模口闸板
~ –off push button 断开按钮
~ –off saw 切割锯
~ –off sprue 自断式浇口
~ –off switch 断路开关,切断开关
~ –off test 停车试验
~ –off time 停车时间
~ –open tubular film 管膜切开成平膜
~ –pile ctrpet 割线地毯
~ –pile hoolc 割绒钩〈簇绒法〉
~ reinforcing fibre 短增强纤维,增强用短纤维
~ resistance 耐切割性
~ roving 切断粗纱
~ sheet 切片
~ sheet goods 切片制品
~ staple 切断纤维,短纤维
~ tape 裁切带子
~ to size 定尺寸剪裁
~ –to–she sheet 定尺寸裁剪片材
cutout 切断,断路
~ safety 安全开关,安全断路器
cutter 切刀;切断机
~ bar 切杆
~ knife 切断刀
~ life 刀具寿命
~ roller 切割辊
~ shaft 切刀轴
cutting 切割,切断,切削
~ blade 切割刀片
~ block 冲模;冲头
~ bush [cutting sleeve] 切杯套筒
~ chamber 切割室
~ device 切割装置

~ die 冲〈切〉模
~ disc 圆盘刀
~ down 割料
~ edge 刀口,切削刃,刀刃
~ edge holder 刀夹,刀架
~ electrode 切割电极
~ hardness 切割硬度
~ instrument 切削刀具
~ knife 切〈割,断〉刀
~ machine 裁切机;切〈割,断〉机
~ method 切〈割,断〉法
~ mill [rotary cutter] 旋转式切断机
~ off 切割,切槽
~ operation 裁切操作,切割操作;切削加工
~ out 切断
~ point 切断位置,切削位置
~ resistance 耐切断性
~ ring 切刀
~ roll 切割辊〈涂敷挤塑机〉
~ sleeve 切杯套筒
~ station 切割工位
~ tip 切割刀片
~ tool 切削刀具
~ up 切割
~ up board 裁切板
~ waste 裁割废料
CV flooring [cushioned vinyls] 缓冲聚氯乙烯 地板面屋
CV-line [continuous vulcanizing line] 连续交联生产线〈聚乙烯管〉
CWP [chemical wood pulp] 化学木浆
cyanaloc 氰基树脂〈防水剂〉
cyanine 花青〈染料〉
~ dyes 花青染料
cyanoacrylate adhesive 氰基丙烯酸酯黏合剂
cyanoacrylic adhesive 氰基丙烯酸〈类〉黏合剂
cyanoethyl 氰乙基
~ cellulose [CEC] 氰乙基纤维素
β- ~ methyl polysiloxane fluid β-氰乙基甲基聚硅氧烷液体
cyanoguanidine 氰基胍〈固化剂〉

cyanol 蒽醌青
cycle 周期;循环;环
~ aging conditions 循环老化条件〈塑料温度与湿度变化试验〉
~ counter 循环[周期]计数器,软数计
~ length 循环时间
~ of load stressing 负载应力周期
~ of operation 运行周期
~ operation 循环作业
~ period 循环周期
~ ratio 疲劳应力周期比
~ sequence 循环期满
~ stress 循环应力
~ test 循环试验〈干湿〉
~ time 循环时间,循环期〈注塑成型〉
~ timer 循环计时器
cyclewise drive 循环式驱动
cyclic 循环的;环〈状〉的
~ compound 环状化合物
~ compression 反复压缩
~ condensation 环化缩合
~ dimer 环状二聚体
~ extension 反复伸长
~ ketone resin 环酮树脂
~ oligomer 环状低[齐]聚物
~ polycondensation 环状缩聚〈作用〉
~ polymer 环状聚合物
~ polymerization 环化聚合
~ polyolefine 环状〈结构〉聚烯烃
~ process 循环过程
~ program 循环程序
~ ring structure 环状结构〈分子〉
~ stress 交变应力,反复应力
~ stress-strain 交变应力-应变
~ tension 反复张力
~ trimer 环状三聚体
~ trimerization 环化三聚作用
~ urea triazine formaldehyde resin 环脲三嗪甲醛树脂
cyclics 环扶化合物
cycling 循环〈的〉
~ life test 周期寿命试验
cyclization 环化〈作用〉

~ condensation 环化缩合
cyclized 环化的
~ fibre 环化纤维
~ rubber 环化橡胶
cycloaddition polymerization 成环加成聚合〈作用〉
cycloaliphatic 脂环族的
~ diepoxide 脂环族二环氧化物
~ epoxy resin 脂环族环氧树脂
~ hydrocarbon 脂环烃
cyclocompounds 环状化合物
cyclocondensation 环化缩合〈作用〉
cyclodisilazane-ring polymer 环二硅氮烷型聚合物
cyclohexane 环己烷
cyclohexanol 环己醇
~ acetate 乙[醋]酸环己〈醇〉酯
cyclohexanone 环己酮
~ formaldehyde resin 环己酮-甲醛树脂
~ peroxide 过氧环己酮〈交联剂〉
~ peroxide paste 过氧化环己酮糊〈不饱和聚酯树脂固化剂〉
~ resin 环己酮树脂
cyclohexyl 环己基
~ acetate 乙[醋]酸环己酯
N-~-2-benzotjuazole sulphenamide N-环己基-2-苯并噻唑次磺酰胺〈促进剂CZ〉
~ β-(3,5-di-tert-butyl-4-hydroxyl phenyl) propionate 环己基-β-(3,5-二叔丁基-4-羟基苯基)丙酸酯〈抗氧剂121〉
N-~-p-ethoxyaniline N-环己基对乙氧基苯胺〈防老剂CEA〉
~ methacrylate [CHM] 甲基丙烯酸环己酯
N-~-p-methoxyaniline N-环己基对甲氧基苯胺〈防老剂CMA〉
N-~-p-toluene sulfonamide N-环己基对甲苯磺酰胺〈增塑剂〉
N-~-N'-phenyl-p-phenylene di-amine N-环己基-N'-苯基对苯二胺〈抗氧剂4010〉

N-~-thiophthalinude N-环己基硫代邻苯二甲酰亚胺〈防焦剂CTP〉
cyclone 旋风;旋风分离器,旋风除尘器
~ air classifier 旋风分离器
~ classifying 风力选分
~ dust collector 旋风集尘器
~ hopper 旋风料斗
~ impeller 圆转混合器,旋风搅拌机
~ impeller mixer [rotor cape or squirrel cage impeller mixer] 旋风叶轮式混合器
~ separator 旋风分离器
~ sintering 流化床烧结,流化床涂布
cycloolefin 环烯烃
~ polymers 环烯聚合物
cyclooligomerization 环〈化,状〉低聚合〈作用〉
cyclopentane 环戊烷
~ tetracarboxylic acid dianhydride 环戊四羧酸二酐〈固化剂〉
cyclopolycondensation 环化缩聚
cyclopolpner 环化聚合物
cyclopolymerization 环化聚合〈作用〉
cyclopolyolefin 环状聚烯烃
cylinder 圆柱体;圆筒;料筒〈注塑机〉;机筒〈挤塑机〉;辊;气罐,油罐;印刷滚筒
~ arrangement 滚辊布排
~ bag 筒形袋
~ barrel 筒形料筒
~ beading 圆筒卷边〈美〉
~ bearer 圆筒托架
~ bending 圆筒弯曲
~ blade 圆盘刀片
~ body 料筒体
~ displacement 滚筒位移
~ dowel 圆筒销钉
~ dryer 圆筒形干燥器
~ flange 料筒法兰
~ gap 滚筒间隙
~ head 料筒头部
~ head volume 料筒头部容积
~ heater 料筒热器
~ holding ring 料筒固定圈

cylindroconical

- ~ in staggered arrangement 交错排列的辊
- ~ jacket 料筒外套
- ~ journal 辊轴承颈
- ~ liner 料筒内衬〈注塑机〉
- ~ mill 圆筒碾磨机
- ~ mixer 圆筒形混合器
- ~ piston viscometer 圆筒型柱塞黏度计
- ~ roll journal 辊轴承颈
- ~ spinning 圆筒纺丝法
- ~ support 滚筒支承;机筒支架
- ~ temperature 滚筒温度;料筒温度〈注塑机〉
- ~ temperature proifile 料筒温度分布
- ~ with drilled channels 有内槽〈沟〉的辊,开沟槽的辊
- ~ with inner partition plates 有内隔板的辊
- ~ drier [dryer] 圆筒形〈的〉
- ~ core 圆柱形〈螺杆〉心
- ~ horn 圆柱形焊接仪〈超声波焊接〉
- ~ -notched pin 圆柱开槽销
- ~ plug 圆柱塞
- ~ screw 柱形螺杆
- ~ screw with conical core 锥心柱形螺杆
- ~ sonotrode 圆筒形焊接仪
- ~ test piece 圆柱形试片
- ~ worm gear 圆柱形蜗轮〈装置〉

cylindroconical 圆锥形的
- ~ ball mill 圆锥形球磨机
- ~ core 圆锥形〈螺杆〉心

D

D [density] 密度
d [nominal size] 公称尺寸〈管材〉
DAC [diallyl chlorendate] 二烯丙基氯菌盐酯
Dacron 达克隆,的确良〈聚对苯二甲酸乙二醇酯纤维,即涤纶,商品名称〉
daily 每日的,日常的
- ~ allowance 日消耗量
- ~ crude capacity 每日加工原料能力
- ~ irispection 日常检验,例行测试
- ~ loss 每日损失
- ~ output 日产量
- ~ production 日产量

DAF [diallyl fumarate] 富马酸二烯丙酯
DAIP [diallyl isophthalate resin] 间苯二甲酸二烯丙酯树脂
Daltolex bonding 达尔托弗莱克斯粘接法〈转移粘接〉
dam 挡板,节流栓,调节排〈挤板 模头〉
- ~ board 挡板
- ~ restriction 节流栓

DAM [diallyl maleate] 顺丁烯二酸二烯丙酯
damage 损坏,损伤;损耗
- ~ by friction 擦伤
- ~ mechanism 损伤机械
- ~ tolerance 损伤容限

damageability 易损坏性,易破损性
damageable 易破损的,易损坏的
damaging 损害的。损坏的,破坏的
- ~ deformation 损坏变形〈如:自由落镖试验〉
- ~ rorce 损伤力〈如:自由落镖试验〉

dammar [damar] 珑玛树脂〈一种天然树脂〉
- ~ resin 珑玛树脂

damp 潮湿;减震,缓冲,阻尼;制动
- ~ course 防潮层,防湿层
- ~ heat atmosphere 湿热气候
- ~ proof 耐潮的,耐湿的
- ~ -proof inatallation cable 防潮装配电缆

~ spinning 湿纺
damped 减震的,阻尼的
- elastic response ［delayed elasticity］ 迟延弹性响应
- oscillation 阻尼振动
- vibration 衰减振动

dampen 弄湿,潮湿;减振,缓冲
dampener 潮湿器
damper 减振［缓冲,阻尼］器;调节板,挡［闸］板,节气闸
- brake 制动闸
- coefficient 阻尼系数,衰减系数
- valve 调节阀

damping 减震〈的〉,缓冲〈的〉,阻尼〈的〉,衰减〈的〉;制动〈的〉;回潮的
- capacity 阻尼振荡能力
- factor 阻尼［衰减］因子
- machine 调湿机
- of material 材料减振
- parameter 阻尼系教
- power 衰减能力
- ratio 阻尼比
- time 衰减时间
- value 衰减值

dampness 潮湿,湿度,含水量
dampproof(ing) 防潮〈湿〉的,不透水的
dancer (松紧)调节辊,浮动滚筒
- control 浮动滚筒控制
- roll (松紧)调节辊,浮动滚筒

danger 危险
- light 警告信号灯光
- signal 危险信号

Dnngerous articles 易燃物品
DAP ［diallyl phthalate resin］ 邻苯二甲酸二烯丙酯树脂
dark (黑)暗(的),暗淡(的);深色的;模糊的
- discharge 无光放电
- face 暗面
- shade 暗色,浓色
- speck 黑斑,暗斑

darker 暗色,深色
darkness 暗度

dart 飞镖
- drop impact test 落锤冲击试验

dash 冲撞,碰撞;控制板;挡泥板
- board 控制板,仪表盘[板];挡泥板,隔板〈汽车〉
- panel 仪表板〈汽车〉
- pot 缓冲筒

Dasher 挡泥板;桨式搅拌器
dashpot 减振［缓冲,阻尼］器
data 数据
- handling 数据处理
- logger 数据记录器
- logging 数据记录
- management 数据处理
- plate 铭牌
- processing 数据处理
- sheet 记录表,一览表,技术数据

DATBP ［diallyl tetrabromophthalate］ 四溴邻苯二甲酸二烯丙酯
daub 底漆,底涂;打底,粗涂
daubing 施胶;涂漆
daylight 压板开距,压板间距,模板间距,压板开档;日光
- between the tie–rods 拉杆间距
- fluorescent pigments 日光荧光颜料,荧光颜料
- opening 压板间距,模板间距〈模具〉
- press 多层压机
- test 耐日光牢度试验

DBAPA ［dibutylaminopropylamine］ 二丁胺基丙胺〈固化剂〉
DBDPO ［decabromodiphenyl oxide］ 十溴二苯醚〈阻燃剂 FR–10〉
DBEA ［dibutoxyethyl adipate］ 己二酸二〈丁氧基乙基〉酯〈增塑剂〉
DBEP ［dibuloxyethyl phffialate］ 邻苯二甲酸双丁氧乙基酯〈增塑剂〉
DBF ［di–n–butyl fumarate］ 富马酸二正丁酯〈增塑剂〉
DBI ［dibutyl itaconate］ 衣康酸二丁酯〈增塑剂〉
DBM ［di–n–butyl maleate］ 马来酸二正丁酯,顺丁烯二酸二正丁酯〈增塑

DBO [N,N - di - n - butyl oleamide] N,N - 二正丁基油酰胺〈增塑剂〉

DBP [dibutyl phthatate] 邻苯二甲酸二丁酯〈增塑剂〉

DBPC [di - tert - butyl - para - cresol] 二叔丁基对甲酚

DBR [2,4 - dibenzoyl resorcinol] 2,4 - 苯甲酰基间苯二酚〈紫外线吸收剂〉

DBS [dibutyl sebacate] 癸二酸二丁酯〈增塑剂〉

DBTDL [dibutyltin dilaurate] 二月桂酸二丁基锡〈热稳定剂,润滑剂,催化剂〉

DBTM [dibutyltin maleate] 马来酸二丁基锡〈热稳定剂〉

DC [direct curtem] 直流电

DCHP [dicyclohexyl phthalate] 邻苯二甲酸二环己酯〈增塑剂〉

DCP [dicapryl phthalate] 邻苯二甲酸二辛酯〈增塑剂〉

DCP [dicumyl peroxide] 过氧化二异丙苯,过氧化二枯基〈交联剂〉

DCPD [dicyclohexyl peroxy dicarbonate] 过氧化二碳酸二环己酯〈引发剂〉

DCS [dicapryl sehacate] 癸二酸二辛酯〈增塑剂〉

DD [dicyandiamide - formaldehyde (condensation) resin] 双氰胺甲醛〈缩合〉树脂

DD [differential dyeing] 差异染色

DDA [di - dccyl adipatc] 己二酸二癸酯

DDM [diamino diphenyl methane] 二氧基二苯基甲烷〈固化剂〉

DDP [didecyl phlalate] 邻苯二甲酸二癸酯〈增塑剂〉

DDS [diamino diphenyl sulfone] 二氨基二苯砜〈固化剂〉

DDSA [dodecenyl succinic anhydride] 十二碳烯基丁二酸酐〈固化剂〉

DDTA [derivative differential therrnal analysis] 微分差热分析

DE [diatomaceous earth] 硅藻土

De Mattia flexing tester 德马蒂亚挠曲试验机

Dead 死的,固定的,不活泼的;暗的
~ colo(u)r 暗色,呆色
~ - end polymerization 死端聚合
~ fold 固定折叠,死褶
~ load 静载荷,净重,固定载荷
~ load stress 静载应力
~ rnatter 无机物
~ - milled 过炼的,塑炼过度的
~ polymer 无活性聚合物
~ pulley 惰轮,空转轮
~ ralling 塑炼过度
~ spot 死点,死角,死斑点
~ stop method 零点法〈分析〉
~ time 静止时间,停歇时间,延迟时间,空载[寂静]时间
~ weight 净重;固定负载荷

deadening 消音;消光
~ agent 消光剂
~ board 隔音板

deerate 脱air,脱气〈模塑材料加工〉

deaerating 脱气,脱泡,排气
~ agent 脱气剂,脱泡剂

deaeration 脱泡,脱气,排气

deaerator 脱气器〈树脂配料〉;脱泡机〈糊料〉

dcair 脱泡,脱气

DEAPA [diethyl amino propylamine] 二乙氨基丙胺〈固化剂〉

debond 脱胶

debubbling 脱气,消泡

debug 排除故障

deburr 去毛刺,倒角,修边

deburller 去毛刺机,修边机

deburring 去毛刺,修边

Debye(induction)forces 德拜〈感应〉力

Debye - Scharrer method 德拜 - 谢勒法〈X射线检验〉

dec. [decompose(s)] 分解

decabromodiphenyl oxide [DBDPO] 十溴二苯醚〈阻燃剂 FR - 10〉

decaborane 癸硼烷
~ polymer 癸硼烷聚合物

decal 贴花纸;贴花〈法〉,转移印花法

decalcomania 贴花纸;贴花法,转移印花
~ process 贴花法,转移印花法
decane 癸烷
decay 衰退,衰变
~ coefficient 衰减系数
~ constant 衰变常数
~ curve 衰变曲线
~ period 衰变期
~ stress 衰变应力
~ time 衰减时间
decelerating 减速
~ flow 减速流动
~ plastic flow 减速塑性流动
decibel 分贝
deck 盖〈层,板〉;控制板;〈平〉台
~ plate 盖板
deckle 调幅杆;框(板)
~ edge 毛边
~ frame 框架
~ rod 调幅杆
decoiler 展卷机,退卷机
decolo(u)rizing fficiency 脱色效率
decolo(u)r 脱色
decolo(u)rant 脱色剂,漂白剂
decolo(u)rahon 脱色,退色
decolo(u)ring 脱色,漂白
decolo(u)risahon 脱色(作用)
decolo(u)riser [decolo(u)rizer] 脱色剂,漂白剂
decolo(u)rization 脱色(作用)
decomp. [decompose(s)] 分解
decompd. [decomposed] 分解了的
decompn [decomposition] 分解〈作用〉
decomposable blowing agent 分解型发泡剂分解型
decompose 分解;衰变;
decomposition 分解;分裂;裂解;衰变
~ by radiation 辐射分解
~ product 分解产物
~ reaction 分解反应
~ temperature 分解温度
decompression 降压,减压
~ time 减压[释压]时间
~ valve 泄压阀
~ zone 降压段〈螺杆〉
decontamination 净化,纯化
decorate 装饰,染色
decorating 装饰
~ foil 装璜膜
~ media 彩饰材料,装饰[装璜]材料
decoration 彩饰,装饰,装璜
~ coating 装饰涂层,装饰漆
decorative 装饰的,装璜的
~ board 装饰板
~ composite panel 装饰复合板,层压装饰板
~ fabric 装饰〈用〉织物
~ laminafe 装饰层压板
~ laminated sheet 装饰层合板
~ overlay 彩饰面层
~ paint 装饰漆
~ plastic laminate 装饰塑料层合板
~ plywood 装饰胶合板
~ sheet 装饰板;装饰片材
~ sheeting 装饰片材,装饰板材
~ unsaturated polyester laminate 不饱和聚酯装饰板
decrease 减少,降低,压缩
~ in weight 重量减轻
~ in width 宽度减幅〈片材〉
decreased 减少,递减;单面涂漆的
decreasing 渐减,递减
~ depth of thread 螺槽渐减深度〈螺杆〉
~ lead screw 递减导程螺杆,减距螺杆,螺距渐减型螺杆
~ pitch 螺距渐减〈螺杆〉
~ pitch screw 螺距渐减型螺杆,减距螺杆
decrement 减少,减缩
~ of velocity 减速
decrimping 退卷曲,使不卷曲,卷曲弛解
decyl 癸基
~ epoxyoleate 环氧油酸癸酯〈增塑剂,稳定剂〉
N-~-n-octyl phthalate 邻苯二甲酸正癸·正辛酯

deep

~ tridecyl phthalate　邻苯二甲酸癸基·十三烷基酯〈增塑剂〉
N-~ trimellitate　偏苯三酸正癸基酯〈增塑剂〉

deep　深,深度,深处;深色的
~ channel screw　深槽螺杆,深螺槽螺杆
~ colo(u)r　深色,浓色,饱和色
~ cut screw　深槽螺杆,深螺槽螺杆
~ draw　深度拉伸,深度撑拉
~ draw forming　深拉成型,深度撑拉成型
~ draw mould　深度撑拉模具
~ drawing　深牵成型〈热成型〉
~ drawing part　深拉制件
~ drawing press　深度拉伸成型压机
~ drawing sheet　深拉成型片材,深度成型片材
~ drawing tool　深度撑拉模具
~ drawn　深拉的,深冲压的
~ drawn article　深拉制件,深腔〈拉伸〉成型制件
~ -drawn package　深拉包装
~ drawn vacuum forming　真空深拉成型
~ flight　深螺纹〈螺杆〉
~ -flight intermeshing screws　深螺纹啮合螺杆
~ flight worm　深螺纹螺杆
~ -flighted　深螺纹的
~ flow　高流动性〈熔融〉;高黏度的
~ -freeze package　深冷包装
~ shade　深色,浓色
~ thread　深螺纹〈螺杆〉
~ -thread feed section　深螺纹进料段
~ well therrnocouple　深孔热电偶〈挤塑机筒〉

deeply threaded screw　深螺纹螺杆
defect　缺陷;疵点〈制品〉
~ detecting test　深伤检查,缺陷检查
~ detector　探伤仪
~ level　缺陷程度,质量水平
defective　有缺陷的,损坏的;不合格的,不良的

~ coupling　耦合不良
~ edge　毛边
~ goods　不合格品,次品
~ insulation　绝缘不良
~ mo(u)ldings　模ločos废品
~ percentage　次品率
~ -rate　次品率
defectoscope　探伤仪
defibrate　分解成纤维,分离纤维,纤化〈作用〉
defibration　纤维分离
defibrator　纤维分离机,梳毛机;捣碎机
defibre [defiber]　分离纤维,纤化〈作用〉
deficiellcy　缺乏,不足;缺陷;不足额,亏数
defino joint　榫-槽胶接〈双重连接〉
deflagrable　易爆燃的,易迅速燃烧的
deflag:ration　爆燃作用
deflash　修边
deflasher　修边机,除边机
deflashing　除边,修边,去毛刺
~ mask [nest]　去毛刺成型刀具
~ trimming machine　修边机
deflatable　可放气的;可紧缩的
~ bag method　真空袋压法
~ bag techruque　真空袋压法,真空橡胶袋模法,收缩袋模法
~ flexible bag mo(u)lding　真空袋压法,收缩软袋模成型法
~ flexible bag technique　其空袋压法,收缩软袋模法
deflate　放气,排气
deflation　放气,排气
~ opening　排气孔
~ valve　放气阀
deflect　挠曲,弯曲;偏斜;折射〈光〉
deflecting blade　反射板,挡板
deflection　挠度;挠曲;偏斜,偏差;折射〈光线〉
~ at break　断裂挠度
~ at rupture　破裂挠度;断裂挠度
~ curve　挠度曲线,变形曲线
~ in bending　弯曲挠度

~ surface 挠曲面
~ temperature 挠曲温度
~ temperature of plastics under load 塑料载荷变形温度
~ temperature under load 载荷挠曲温度,载荷变形温度
~ test 挠曲试验
~ under load 载荷挠曲,载荷变形
deflector 导向板〈塑料膜管〉,导向装置,挡板
~ roll 导向辊〈涂布机〉
deflocculation 消絮凝
~ agent 防凝聚剂
Defo 迪福
~ elasticity 迪福弹性
~ hardness 迪福硬度
~ plastometer 迪福式塑度计
~ valve 迪福值
defoam 消泡
defoament 消泡剂,去沫剂
defoamer 消泡剂,去沫剂
defoaming 消泡,去沫
~ agent 消泡剂,去沫剂
deformability 变形性,变形能力,加工性,模塑性
deformable 可变形的,易变形的
~ model 可变形的模型
deformation 变形,形变,扭〔翘〕曲
~ at break 断裂变形
~ at failure 破损变形
~ at yield 屈服形变
~ band 〈滑移〉形变带,形变条纹
~ behavior 变形性,形变性
~ condition 形变条件
~ energy 变形能,形变能
~ gradient 形变梯度
~ in compression 压缩变形
~ limit 形变极限
~ mechanism 形变机理
~ orientation 交形取向
~ range 形变范围,变形区
~ rate 形变速率
~ recovery 变形回复
~ resilience 形变回弹

~ resistance 耐变形性
~ retraction 形变收缩
~ set 永久变形
~ temperature 形变温度
~ tensor 形变张量
~ test 形变试验
~ under heat 受热形变
~ under load 载荷变形
~ under load test 负载变形试验
~ under stress 应力形变
~ under torsion 扭转形变
~ zone 形变区
DEG [diethylene glycol] 一缩二乙二醇,二甘醇
degas 放气,排气,脱气;脱泡
degassing 放气,排气,脱气;脱泡
~ extruder 排气式挤塑机
~ mould 排气式模具
~ screw [devolatilizing screw] 排气式螺杆
~ temperature 排气温度
– zone 排气段〈挤塑机〉
degate 去浇口料
degating 尾料切除,浇料口切除
degradable plastic 可降解塑料,降解性塑料
degradant additive 降解剂
degradation 降解
~ curve 降解曲线
~ mechanism 降解机理
~ peak 降解峰
~ period 降解期,衰退时间
~ product 降解产物
~ rate 降解速率
~ reaction 降解反应
~ theory 降解理论,降解学说
~ with ageing 老化降解
degrade 降解
degrading test 性能衰退试验
degrease 脱脂
degreaser 脱脂剂,脱垢剂
degreasing 脱脂;除去油腻
~ agent 脱脂剂〈黏胶面预处理〉
degree 度;程度,等级

deguming

- ~ Celsius 摄氏度[℃]
- ~ conversion 转化率
- ~ Fahrenheit 华氏度[℉]
- ~ flexibility 柔软度
- ~ of absorption 吸收度
- ~ of accuracy 准确度,精度
- ~ of adhesion 黏合度
- ~ of adsorption 吸附度
- ~ of alternation 交替度
- ~ of blow–up 吹胀度
- ~ of branching 支化度
- ~ of cleanliness 清洁度;纯度
- ~ of compression 压缩度;压缩比
- ~ of crosslinking 交联度
- ~ of crystallinity [cryslallization] 结晶度
- ~ of cure 固[熟,硫]化(程)度
- ~ of degradation 降解度
- ~ of depolymerization 解聚度
- ~ of dispersion 分散度;色散度
- ~ of distortion 变形度,畸变度
- ~ of draw 撑压度,成型深度,牵伸度
- ~ of efficiency 效率程度
- ~ of exhawstion 抽空度
- ~ of expansion 膨胀度;膨胀比
- ~ of fill 充模度;填充度
- ~ of fineness 细度,粉末度,细末度
- ~ of finish 光洁度
- ~ of flammability 可燃度
- ~ of fluidity 流动度
- ~ of foaming 发泡度
- ~ of functionality 官能度
- ~ of grafting 接枝度
- ~ of grind(ing) 研磨(细)度;粉碎(细)度
- ~ of hardness 硬度
- ~ of inspection 检验等级
- ~ of irregularity 不均匀度,不规则度
- ~ of isotacticity 等规度,全同(立构)规整度
- ~ of mixing 混合度,混炼度
- ~ of moisture 湿度
- ~ of (long–range) order 有序度
- ~ of orientation 取[定]向度
- ~ of plasticity 可塑度
- ~ of plasticization 增塑度;塑化度
- ~ of polymerizationo [DP] 聚合度
- ~ of quality 质量等级,质量优劣程度
- ~ of reaction 反应度
- ~ of resistance to glow heat 耐灼热度
- ~ of safety 安全度,安全系数
- ~ of saturation 饱和度
- ~ of shear 剪切度
- ~ of shrinkage 收缩率,收缩度
- ~ of smearing 塑化度(挤塑)
- ~ of stretching 伸展度;拉伸度
- ~ of swelling 溶胀度
- ~ of syndiotacticity 间同(立构)规整度
- ~ of tacticity 构型规整度,立构规整性
- ~ of tension 张力度;拉伸度
- ~ of uniformity 均匀度
- ~ of unsaturation 不饱和(程)度
- ~ of vacuum 真空度
- ~ of viscosity 黏度
- ~ of volatility 挥发度
- ~ of wrap 抱辊度
- ~ of yellowing 黄变度,发黄度

deguming 脱胶
dehumidification 减湿,去湿;干燥;脱水
dehumidifier 减[去]湿剂;干燥[脱水]装置
dehydrate 脱水物;脱水,干燥
dehydrating 脱水的,干燥的
- ~ agent 脱水剂
- ~ machine 脱水机

dehydration 脱水(作用),干燥
dehydrogenation 脱氢(作用)
- ~ polymerization 脱氢聚合

de–inking 脱墨
- ~ solution 脱色溶液,脱墨溶液

deionizing property 除去离子能力
delaminate 脱层,分层,层[剥]离
delaminating 脱层,层离
delamination 脱层,分层,剥离
delay 延迟,滞后
- ~ action 延迟作用

density

~ in delivery 延迟交付,延迟供应
~ in supply 延迟供料[供给]
~ period 浸涂期;延迟期;停留期
~ power 阻燃剂
~ time 延迟时间;滞燃时间;停留时间;浸涂时间
delayed 延迟的,滞后的
~ action accelerator 迟效促进剂
~ cracking 延迟裂纹
~ deformation 延迟形变
~ [retarded] elasticity 延迟弹性,滞后弹性
~ extension 弛缓伸长
~ fracture 迟后断裂
~ recover 延迟回复
~ tack adhesive 延迟黏合剂
delete welding [radiant heat welding] 辐射热焊接
deleterious 有害杂质;有害的,有毒的
~ effect 有害影响
~ substance 有害物(质)
deliquescent (易)潮解的,易吸收湿气的
delivery 交付,分送;输送;供给(量)
~ capacity 生产额;交付量
~ belt 输料带(模塑料)
~ channel 运出通道
~ cock 供给阀
~ end 卸料端
~ head 水头,压力差(供水)
~ of energy 能量供应;动力供应
~ of goods 交货
~ of set unit 成套供应
~ order 交货单,提货单
~ pipe 输送管(道)
~ pressure 出口压力
~ rate 运出量;输料速度
~ sation 卸出工位
~ screw 卸料螺杆
~ speed 排出速度
~ system 供料装置,供料系统
~ term 交货期限
~ valve 放泄阀
delusterant 消光剂
demagnetization 去磁(作用)

demand 要求,需要
~ on the dimensional stability 对尺寸稳定性的要求
demands in moulding technique 模塑技术需求
demi-rigid film 半硬质薄膜
demineralization 脱矿质(作用)
demineralized water 软化水
demixing 分离(混合物);分层(混合液)
demonomerization behaviour 脱单体性
demonstration 实验,示范
~ plant 实验厂
~ test 示范性试验
demo(u)ld 脱模
~ station 脱模工位
demo(u)lding 脱模
~ aid 脱模部件(塑料模具)
~ device 脱模装置
~ force 脱模力
~ time 脱模时间
demountable 可拆卸的
~ cargo container 可装卸运货客器
den [denier] 旦尼尔,旦〈纤度单位〉
den. [density] 密度
denaturation 变性(作用)
dendrite 树枝状晶体
dendritic(al) 树枝状的
~ structure 树枝状结构,分支结构
denier [den] 旦尼尔,旦(纤度单位,纤维长 9000 米、重 1 克为 1 旦)
~ balance 旦尼尔天平
dense 紧密[稠密]的;波厚的
densification 压实(泡沫塑料)
densified 压紧的,压实的
~ impregnated wood (树脂)浸压木材
~ laminated wood (树脂)浸压层合木板,浸渍层压木材
~ wood 压缩木材
densify fluid 黏滞流体
densimeter 密度计;比重计
densitometer (光)密度计;(液体)比重计
density 密度
~ anisotropy 密度异向

~ [specific-gravity]bottle 密度瓶,比重瓶
~ detector 密度检测器
~ distribution 密度分布
~ effect 有效密度,密度效应
~ gradient 密度梯度
~ in raw state 粗密度
~ index 密度指数
~ of cross-linking 交联密度
~ of foam 泡沫塑料密度
~ of packing 填充密度
~ of plastics 塑料密度
~ of textile glass [-linear density] 线密度,纺织玻璃纤维密度
~ temperature coefficient 密度温度系数
dental 牙齿的,齿的;牙科的
~ plastics 牙科用塑料
~ polymer 牙科聚合物
~ resin 牙科用树脂
denture 假牙齿
~ adhesive 补齿黏合剂
~ mould 假牙模(具)
~ plastics 补齿塑料
deoxidation reagent 脱氧剂
DEP [diethyl phthalate] 邻苯二甲酸二乙酯〈增塑剂〉
dependence 依赖性,相关性
dephlegmator 分馏柱,分馏塔,回流冷凝器
depolarization 解偏振作用
~ factor 解偏振因素
depolymerization 解聚(作用)
depolymerizator 解聚(合)器
depolymerize 解聚(合)
depolymering 解聚(合)
~ agent 解聚剂
~ reactor 解聚(合)反应器
deposit 沉积;沉积物;焊道,焊珠;储藏
~ chamber 沉淀槽
depositing 沉积;沉积物,堆积物;喷镀,蒸镀
deposition 沉积;沉积物;喷镀,蒸镀
~ welding 堆焊

Depot method 淡抛脱法(拜耳公司的泡沫塑料加工)
depression 降低,减低;沉降;凹陷
~ chamber 减压室,低压室
~ gate 沉陷式浇口
~ of freezing point 冰点降低
dept. [department] 部门
depth 深度;厚度;色泽浓度
~ of a thread 螺纹深度
~ of conversion 转化深度
~ of draw 撑拉深度,成型深度,拉伸深度
~ of flight 螺纹深度
~ of fusion 熔焊深度
~ of screw channel 螺槽深度
~ of shade 色度,色泽深度
~ of thread 螺槽深度
derby doubler 绕滞机,卷绕机
deresination 脱树脂(作用)
deriv. [derivative] 衍生物
derivative 衍生物;派生的,导(函)数
~ differential thermal analysis [DDTA] 微分差热分析
~ dilatometry 微分膨胀测定法
~ fibre 衍生纤维
~ spectroscopy 微分光谱(塑料分析)
~ thermogravimetric analysis 微分热重分析法
~ thermogravimetry [DTG] 微分热重法
derive 派生,衍生物
~ high polymer 派生高聚物,衍生高聚物
Derlin (商)德尔林(美国杜邦公司产聚甲醛)
dermateen 漆布
dermatitis 皮(肤)炎(接触环氧树脂所引起)
DES [diethyl sebacate] 癸二酸二乙酯(增塑剂)
desalination 脱盐(作用),淡化(水的)
~ plant 脱盐装置
desiccant 干燥剂;干燥的
~ air circulating oven 空气循环干燥烘

箱
~ cartridge 干燥筒
~ cell 干燥池
desiccating agent 干燥剂
desiccator 干燥器,吸湿器
~ cabinet 干燥箱,箱式干燥器
design 设计;图案
~ life 设计寿命,计算寿命
~ of rims 轮廓造型
~ roll or roller 花纹辊(无限印刷)
~ sample 花样样品
~ strength 设计强度
~ test 鉴定试验
designability 可设计性
~ of material properties 材料性能可设计性
designation strip 铭牌,标条
designer 设计者
desired value 要求值,给定值
desize 脱[退]浆
desizing 脱[退]浆
~ agent 退浆剂
desorption 解吸(作用)
dessicant 干燥剂
dessicator 干燥器
~ cabinet 干燥柜
destabilization 使不稳定,破坏稳定
destaticization 消静电作用
destaticize 消除静电
destaticizer 消静电剂;消静电器
destopping machine 取软木塞机(包装)
destruction 破坏,毁坏
~ test 破坏性试验
destructive 破坏的,毁坏的
~ detector 破坏性检测器
~ distillation 破坏蒸馏
~ examination 破坏性检验法
~ inspection 破坏性检验
~ method 破坏性试验法
~ test 破坏性试验,爆破试验
~ testing 破坏试验,断裂试验,击穿试验
desulphurating [desulphurizing] 脱硫
deswelling 消膨胀;退溶胀

detachable joint 可拆式管接头
detackifier 防黏剂;脱黏剂
detail assembly 细节装配;零件装配
detd. determined 测出的,求得的
detect 检测
detectable sample 可检测试样
detection 检测;检定;检查
detector 检测器
detergent cleaner 洗涤剂,去垢剂
deterioration 变质,劣化;损伤
~ by light 光老化
~ prevention 防(产品)变质
determination 判定;测定;确定
detn. determination 测定
detreader 剥皮机(薄片)
detreading machine 剥皮机(薄片)
detrimental impurity 有害杂质
detrition 磨损,磨耗
Deutsche Industrie Norm [DIN] 德国工业标准(德文)
developed 展开的,显现的
~ dyes 显色染料
~ volume of screw channel 螺槽展开容积(螺杆)
development 发展,改善;研制,开发
~ tests 试制品试验,开发性试验
~ time 研制时间
deviation 偏差,偏离,偏移
~ from 与……不符合,与……不一致
~ of form 模型偏差
~ of line in sight through 透视线差异,透明度误差
~ of position 位置偏差
device 装置,设备;部件
~ control 设备控制
~ for coating thermoplastic adhesives 热塑性黏合剂涂布装置,热熔胶涂布装置
devitrification 透明消失(玻纤缺陷)
devoid of strata 无层的
devolatilization 脱挥发分(作用),排气,脱气
devolatilizer 汽化器,脱气器
devolatilizing [removal of volatile constitu-

de – watering

ents] 脱挥发份;排气
- ~ barrel 排气机筒
- ~ dome 排气顶盖
- ~ extruder [vent extruder] 排气式挤塑机
- ~ opening 排气口,排气孔
- ~ screw 排气式螺杆
- ~ screw with by – pass 旁路排气式螺杆
- ~ surface 排气(表)面

de – watering 脱水,排水
dew point 露点
- ~ hygrometer 露点湿度计
- ~ temperature 露点温度

Dewar flask 杜瓦瓶,真空(壁)瓶
dewetting 去(润)湿
dextrin 糊精
- ~ adhesive [glue] 糊精黏合剂

dextrinized starch 糊精浆
dextro – rotator 右旋的,顺时针方向旋转的
DGEBA [diglycidyl ether of bisphenol A] 双酚 A 二缩水甘油醚
DHP [diheptyl phthalate] 邻苯二甲酸二庚酯(增塑剂)
DHXP [dihexyl phthalate] 邻苯二甲酸二己酯
di – C_7 – C_9 alcohols sebacate 癸二酸二 C_7 – C_9 醇酯(增塑剂)
dia. diameter 直径
diacetic acid 双乙酸
diacetyl 二乙酰(基)

N,N' – ~ adipic acid dihydrazide N,N' – 二乙酰基己二酰基二酰肼(金属螯合剂,铜抑制剂)
- ~ cellulose 二乙酰纤维素
- ~ peroxide 过氧化二乙酰

N,N' – ~ thiodipropionyl dihydrazide 硫代二丙酰基 – N,N' – 二乙酰基双酰肼(铜抑制剂)

dlngonal weave 急斜纹组织,贡斜纹
diagram 图(表)
diagrammatic 图解[表,示]的
- ~ curve 图表曲线
- ~ drawing 草[略]图,示意图
- ~ section (横)截面图
- ~ view 简图

dial 刻度盘
- ~ balance 刻度盘天平
- ~ barometer 气压指示针
- ~ feed press 旋转盘供料式压机
- ~ gauge 刻度盘
- ~ ga(u)ge indicator 刻度盘指示器
- ~ ga(u)ge reading 刻度盘读数
- ~ plate 刻度板,刻度盘
- ~ thermometer 刻度温度计
- ~ type feed mechanism 转盘式加料器

diallyl 二烯丙基
- ~ chlorendate [DAC] 氯菌酸二烯丙酯
- ~ fumarate [DAF] 富马酸二烯丙酯
- ~ isophthalate resin [DAIP] 间苯二甲酸二烯丙酯树脂
- ~ malete [DAM] 顺丁烯二酸烯丙酯,马来酸二烯丙酯
- ~ phthalate [DAP] 邻苯二甲酸二烯丙酯
- ~ phthalate moulding material 邻苯二甲酸二烯丙酯模塑料
- ~ phthalate prepolymer 邻苯二甲酸二烯丙酯预聚物
- ~ phthalate resin [DAP] 邻苯二甲酸二烯丙酯树脂
- ~ tetrabromophthalate [DATBP] 四溴邻苯二甲酸二烯丙酯

diallyls 聚二烯丙基
- ~ resln 聚二烯丙基树脂

dialysate [dialyzate] 渗析液
dialysis 渗析,透析
- ~ membrane 渗析膜

dialytical 渗析的
dialyze 渗析
diam [diameter] 直径
diameter 直径;横断面
- ~ of aperture 孔径
- ~ of screw 螺杆直径
- ~ of thread 螺纹直径
- ~ progressive screw 变径螺杆

~ ratio 内外径比
diametral 直径的,径向的,横向的
- ~ extensometer 横向应变传感器(美)
- ~ pitch (齿轮)径节(距)
- ~ screw clearance 径向螺杆间隙

diamine dye 双胺染料
diamino (词头)二氨基
- ~ diphenyl methane [DDM] 二氨基二苯基甲烷(固化剂)
- ~ diphenyl sulfone [DDS] 二氨基二苯砜,磺基二苯胺(固化剂)

diamond 金刚石;菱形的
- ~ cone 金刚石锥体
- ~ cutter 金刚石切割器,玻璃刀
- ~ knurled insert 菱纹滚花嵌件
- ~ mat 菱纹编织毡
- ~ pattem 金刚石花型
- ~ penetrator hardness 维氏金刚石硬度
- ~ pyramid 金刚石棱锥
- ~ pyramid hardness 金刚钻棱锥压陷硬度,维氏硬度
- ~ pyramid hardness test 维氏硬度试验

diamyl phthalate 邻苯二甲酸二戊酯(增塑剂)
2,5 – di - tert – amyl hydroquinone 2,5 - 二叔戊基对苯二酚(抗氧剂)

dian(e) 双酚 A
diaper 尿布;彩色花纹织物
diaphragm 膜片,隔膜(的)
- ~ disc valve 隔片状阀
- ~ fonning 隔膜成型
- ~ gate 隔膜状浇口
- ~ ga(u)ge 薄膜压力计
- ~ metering pump 隔膜计量泵
- ~ pump 隔膜泵
- ~ valve 隔膜阀

diatomaceous 硅藻土的
- ~ ealth [DE] 硅藻土(填料)
- ~ silica 硅藻土

diatomite 硅藻土
diaxial orientation 双轴取向
diazamine colo(u)rs 重氮胺染料
diazide B 二叠氮对苯二酸(发泡剂)
diazo 重氮

~ dyes 重氮染料,偶氮染料
~ plgment 重氮颜料
diazoalrane polymenzation 偶氮烷聚合
diazoaminobenzene 重氮胺基苯(发泡剂)
DIBA [diisobutyl adipate] 己二酸二异丁酯(增塑剂)
dibasic 二碱价的(酸);二盐基的,二碱式的〈盐〉
- ~ lead phosphite 二盐基亚磷酸铅(热稳定剂)
- ~ lead phthalate 二盐基邻苯二甲酸铅(热稳定剂)
- ~ lead stearate 二盐基硬脂酸铅(热稳定剂)

dibenzothiazyl disulfide 二硫化二苯并噻唑[促进剂 DM]
dibenzoyl 二苯甲酰基
p,p' - ~ quinone dioxime 4,4'-二苯甲酰对醌二肟(硫化剂)
2,4. ~ resorcinol [DBR] 2,4 二苯甲酰基间苯二酚(紫外线吸收剂)
dibenzyl 二苄基
- ~ adipate 己二酸二苄酯
- ~ azelate 壬二酸二苄酯(增塑剂、软化剂)
- ~ phthalate 邻苯二甲酸二苄酯(增塑剂)
- ~ sebacate 癸二酸二苄酯(增塑剂)

DIBP [diisobutyl phthalate] 邻苯二甲酸二异丁酯(增塑剂,黏合剂)
dibromo (词头)二溴
- ~ cresyl glycidyl ether 缩水甘油二溴甲苯基醚(阻燃剂)
- ~ neopentyl glycol 二溴新戊二醇(阻燃剂)
- ~ phenol 二溴苯酚(阻燃剂)
2,3 - ~ propanol 2,3.二溴丙醇(阻燃剂)
- ~ propyl acrylate 丙烯酸二溴丙酯(阻燃剂)
2,6 - ~ styrene 2,6.二溴苯乙烯

dibutoxyethoxyethyl 二丁氧(基)乙氧(基)乙基

dibutoxyethyl

~ adipate 己二酸二丁氧乙氧乙酯(增塑剂)
~ glutarate 戊二酸二丁氧乙氧乙酯(增塑剂)
~ sebacate 癸二酸二丁氧乙氧乙酯(增塑剂)

dibutoxyethyl 二丁氧(基)乙基

~ adipate [DBEA] 己二酸二丁氧乙酯(增塑剂)
~ azelate 壬二酸二丁氧乙酯(增塑剂)
~ glutarate 戊二酸二丁氧乙酯(增塑剂)
~ phthalate [DBEP] 邻苯二甲酯二丁氧乙酯(增塑剂)
~ sebacate 癸二酸二丁氧乙酯(增塑剂)

butyl 二丁基

~ adipate 己二酸二丁酯(增塑剂)
~ arrunopropylamine [DBAPA] 二氨基丙胺(固化剂)
~ butylphosphonate 丁基膦酸二丁酯
~ diglycol adipate 己二酸二丁基二甘醇酯
~ itaconate [DBI] 衣康酸二丁酯(增塑剂)
~ phosphite 亚磷酸二丁酯(抗氧剂)
~ phthalate [DBP] 邻苯二甲酸二丁酯(增塑剂)
~ sebacate [DBS] 癸二酸三丁酯(增塑剂)
~ succinate 琥珀酸二丁酯
~ tartrate 酒石酸二丁酯(增塑剂,溶剂,润滑剂)

di–n–butyl 二正丁基

~ adipate 己二酸二正丁酯(增塑剂)
~ -fumarate [DBF] 富马酸二正丁酯,反丁烯二酸二正丁酯(增塑剂)
~ maleate [DBM] 马来酸二正丁酯,顺丁烯二酸二正丁酯(增塑剂)
N,N- ~ oleamide [DBO] N,N-二正丁基油酰胺(增塑剂)
N,N'- ~ thiourea $N,N,$-二正丁基硫脲(促进剂 DBTU)

~ xanthogenate disulfide 二硫化二正丁基黄原酸酯(促进剂 CPB)

di–tert–butyl 二叔丁基

$2,6$- ~ $4-n$-butyl phenol $2,6$-二叔丁基-4-正丁基苯酚〈抗氧剂〉
$2,6$- ~ -p-cresol $2,6$-二叔丁基对甲酚〈抗氧剂〉
$2,6$- ~ -α-dimethylamino-p-cresol $2,6$-二叔丁基-α-二甲氨基对甲酚(抗氧剂)
~ diperoxyphthalate 二过氧化邻苯二甲酸二叔丁酯(交联剂)
$2,6$- ~ -4-ethylphenol $2,6$-二叔丁基-4-乙基苯酚(抗氧剂)
$2,5$- ~ hydroquinone $2,5$-二叔丁基氢醌,$2,5.$二叔丁基对苯二酚(抗氧剂)
$2,6$- ~ -4-methyl phenol $2,6$-二叔丁基-4-甲基苯酚(抗氧剂)
~ peroxide [DTBP] 二叔丁基过氧化物,过氧化二叔丁基,引发剂 A
~ -para-cresol [DBPC] 二叔丁基对甲酚(抗氧剂)
$2,6$- ~ phenol $2,6$-二叔丁基苯酚(抗氧剂)
$2,4$- ~ phenyl $3,5$- ~ -4-hydroxyben-zoate $3,5$-二叔丁基-4-羟基苯甲酸$2,4$-二叔丁基苯酯[光稳定剂120]

dibutyltin 二丁基锡

~ bis(isooctyl thioglycollate) 双(硫代甘醇酸异辛酯)二丁基锡(热稳定剂)
~ bis(monobutyl maleate) 双(马来酸单丁酯)二丁基锡(热稳定剂)
~ di-2-ethyl hexoate 二-2-乙基己酸二丁基锡
~ dilaurate [DBTDL] 二月桂酸二丁基锡(热稳定剂,润滑剂,催化剂)
~ dilauryl mercaptide 二月桂基硫醇二丁基锡(热稳定剂)
~ laurate-maleate 月桂酸马来酸二丁基锡(热稳定剂)
~ maleate [DBTM] 马来酸二丁基锡(热稳定剂)

~ maleate polymer 马来酸二丁基,锡聚合物(热稳定剂)
~ β - mercaptopropionate β-巯基丙酸二丁基锡(热稳定剂)

dicapryl 二辛基
~ phthalate [DCP] 邻苯二甲酸二辛酯(增塑剂)
~ sebacate [DCS] 癸二酸二辛酯(增塑剂)

dice 小方块
~ shaped granulate 方形颗粒

diced 小方块的,小方片的
~ material 方粒状材料
~ moulding board 方片状模塑版
~ moulding compound 方粒状模塑料

dicer [cube dicer] 切(方)粒机
~ Cumberland 条状切粒装置;阶梯型切粒机
~ Condux 圆盘形切粒机
~ rotor 切粒机滚筒刀
~ with circular blades 圆盘形切粒机

dichloro (词头)二氯
~ benzene 二氯(代)苯
5,6 - ~ - benzoxazolinone 5,6-二氯苯并噁唑啉酮(防霉剂 O.)
~ elhylcne 二氯乙烯
3,3, - ~ - 4,4' - diamino diphenyl-metha - ne 3,3'-二氯-4,4'-二氨基二苯基甲烷(硫化剂 MOCA〉(交联剂)
2,4 - ~ - dibenzoyl peroxide 2,4-二氯过氧化二苯甲酰(硫化剂 DCBP〉
1,1 - ~ - ethene 偏二氯乙烯
2,6 - ~ - styrene 2,6-二氯苯乙烯
~ triazine dyes 三氯三嗪染料

dichroic 二色性的
~ ratio 二色性比

dichroism 二向色性,二色性

dicing 切(方)粒(的)
~ cutter 切(方)粒机
~ machine 切(方)粒机

dicumyl peroxide [DCP] 过氧化二异丙苯,过氧化二枯基(交联剂)

dicyandiamide [DICY] 双氰胺,二氰二胺(固化剂)
~ - formaldehyde (condensation) resin [DD]双氰胺甲醛(缩合)树脂

dicyclohexyl 双环己基,二环己基
~ adipate 己二酸二环乙酯
~ azelate 壬二酸二环己酯(增塑剂)
N,N - ~ - 2 - benzo - thiazole sul-phenan - ide N,N-二环己基-2-苯并噻唑次磺酰胺[促进剂 DZ]
~ peroxy dicarbonate [DCPD] 过氧化二碳酸二环己酯(引发剂)
~ phthalate [DCHP] 邻苯二甲酸二环己酯(增塑剂)

dicycjopentadiene 双环戊二烯

DIDA [diisodecyl adipate] 己二酸二异癸酯(增塑剂)

didecyl 二癸基
~ adipate [DDA] 己二酸二癸酯
~ glutarate 戊二酸二癸酯(增塑剂)
~ phosphite 亚磷酸二癸酯〈抗氧剂〉
~ phthalate [DDP] 邻苯二甲酸二癸酯(增塑剂)

di (dioctylphosphato) cthylene titanate 双(二辛基磷酰氧基)钛酸乙二酯(偶联剂)

DIDG [di - isodecyl glutarate] 戊二酸二异癸酯

DIDP [diisodecyl phthalate] 邻苯二甲酸二异癸酯(增塑剂)

die 模头;口模(挤塑);塑模,模具(模塑);冲模(冲切);喷嘴(焊枪)
~ adapter 模头接套(挤塑模具)
~ adaptor 模头接套(挤塑模具)、
~ adjustment 模口调节
~ adjustment bolt 模口调节螺丝
~ annulus 环状模口
~ annulus gap 环状模口隙距
~ aperture 模口
~ approach 模头料道
~ area 模头区,模头范围(挤塑机)
~ assembly 模头总成;模头装配
~ base 模座
~ bed 底模
~ blade 模唇

die

- ~ blank 模套;模坯料
- ~ block 塑模承套,口模套,模体(挤塑);阴模(冲切)
- ~ body 模体,模头接套
- ~ box 模型箱;模具箱
- ~ box mo(u)ld 阴模
- ~ bushing 口模圈
- ~ carrier 模具托架(泡沫塑料模具)
- ~ carrying plane 载模板,装模板
- ~ casting 压铸;模铸
- ~ cavity 阴模,型腔,模槽
- ~ channel 模腔,料道
- ~ character 口模特性,模头特性
- ~ characteristic 口模特性,模头特性
- ~ characteristic curve 口模特性曲线,模头特性曲线
- ~ clamp attachment 模头夹具
- ~ close time 合模时间,锁模时间
- ~ closing force 合模力,锁模力(注塑机)
- ~ coater 口模式涂布机
- ~ coating 挤涂
- ~ coextrusion 模口共挤模(平膜)
- ~ component 模头部件
- ~ cone 分流锥,分流梭,鱼雷形分流芯
- ~ constarit 口模[模头]常数
- ~ cut 模切,冲切,模冲,冲裁
- ~ cutter 冲切刀
- ~ cutting 模切,模冲,冲切,冲裁
- ~ cutting press 模切压机
- ~ discharge channel 模出料道
- ~ entry region 模头进料区
- ~ face cutter 模口切粒刀
- ~ face granulation 模口造粒
- ~ flow corrstait 模具流动常数
- ~ gap 模口隙距
- ~ gap adjustment 模隙调节
- ~ glazjng 模出料压光(片材)
- ~ glazing unit 模出料压光装置(片材)
- ~ grid 模框
- ~ head 机头(挤塑机)
- ~ head temperature 机头温度
- ~ height adjustment screw 模具高度调节螺丝
- ~ hobbing 模压制模法
- ~ holder 模座,口模套
- ~ holder plate 载模板;电极板(HF焊机)
- ~ holder system 模头夹持系统
- ~ hole 模孔
- ~ inlet(space) 模头入口
- ~ insert 模嵌件
- ~ jaw 模唇
- ~ land 口模成型面
- ~ life 模具寿命
- ~ liner 模衬
- ~ lines 口模印,模线,口模痕
- ~ lip 模唇
- ~ lip opening 模唇间距
- ~ locking force 合模力,锁模力
- ~ mandrel 芯模,模芯
- ~ material 压模材料
- ~ moulding 模成型
- ~ of tubing machine 制管机模
- ~ open time 开模时间
- ~ operung 模口
- ~ opening stroke 模口行程
- ~ orifice 模孔,模口
- ~ passage 模口通道
- ~ piece 模具拼块,拼合模块
- ~ plate 载模板,装模板;口模板;冲模底板;电极板(高频焊接);机头板
- ~ press 模冲压机
- ~ pressing 模切,模冲,冲孔,冲压;对模成型
- ~ pressure 模具内压,口模内压
- ~ programming 口模调节
- ~ punching 冲孔模
- ~ relief 模缝
- ~ reservoir 模头料膜(挤塑);模集料道(共挤塑)
- ~ resistance 模头阻力
- ~ restriction 模头限流,模腔收敛
- ~ ring 口模套圈
- ~ rotator 模头旋转器
- ~ section 拼合模块

~ segment   拼合模块
~ set   模座,成套冲模
~ setting press   模具装配压力机
~ shrinkage   模塑收缩;模塑收缩率
~ sinking machine   冲模仿型铣床
~ size   模具尺寸
~ slot   模缝
~ space   型腔,模腔;口模高度
~ space area   装模面积
~ spider   幅架,模芯支架
~ stamping   模压,压花纹;模冲
~ suppot   模具架
~ support car   模具吊车
~ swell   模口膨胀,离膜膨胀,挤出膨胀(挤塑)
~ swell ratio   离模膨胀比(挤塑)
~ swelling   模口膨胀,离模膨胀
~ – to – roll gap   模辊距
~ torpedo   鱼雷形分流芯,分流梭
~ work   模冲压
dies   口模,模头,模具
dielec [dielectric]
dielectric   电介体,绝缘材料;介电的
~ absorption   介电吸收
~ anomalous dispersion   介电异常色散
~ bonding   介电粘接,高频焊接[黏合]
~ breakdown   介电击穿
~ breakdown strength   介电击穿强度
~ breakdown test   介电击穿试验
~ breakdown voltage   介电击穿电压
~ coefficient   介电系数
~ constant   介电常数,电容率,介电系数
~ curing   介电固化,高频固化(如:热固性树脂)
~ disperlon   介电色散
~ dispersion curve   介电谱,介电色散曲线
~ displacement   介电位移
~ dissipation factor   介电损耗因素,介电损耗(角)正切
~ dryer   介电干燥器;高频干燥器(塑料成型料)

~ embossing   介电压花,高频压花
~ expansion   介电加热发泡
~ fatigue   介电疲劳
~ grade   绝缘等级
~ heat sealing   介电热合;高频热合
~ heating   介电加热,高频加热
~ insulation   介电绝缘
~ loss   介电损耗
~ loss angle   介电损耗角
~ loss angle tangent   介电损耗角正切
~ loss characteristics   介电损耗特性
~ loss factor   介电损耗因素
~ loss heating   介电加热损耗
~ loss index   介电损耗指数
~ loss tangent   介电损耗角正切
~ material   介电体;绝缘材料
~ monitoring   介电监控
~ phase angle   介电相位角
~ phase difference   介电相位差
~ phenomenon   介电现象
~ polarization   介电极化(高频焊接)
~ power factor   介电功率因素
~ preheating   介电预热,高频预热
~ property   介电性能
~ relaxation   介电松弛
~ remanence   介电疲劳;介电后效
~ sealing   介电熔合,介电热合;高频热合
~ spectra   介电谱
~ stability   介电热稳定性
~ strength   介电强度
~ strength test   介电强度试验
~ stress   介电应力
~ welding   介电焊接;高频焊接
dielectrical film   电绝缘薄膜
dielectrometer   介电计
diene   二烯(烃)
~ component   二烯组分
~ polymerization   二烯烃聚合
polymers   二烯聚合物
~ rubber   二烯橡胶
dienophile   亲二烯物;亲二烯的
~ compound   亲二烯物
diethoxyethyl   二乙氧基乙基

~ adipate 己二酸二乙氧基乙酯
~ phthalate 邻苯二甲酸二乙氧基乙酯〈增塑剂〉
diethyl 二乙基
~ adipate 己二酸二乙酯
~ alunumum chloride 一氯二乙基铝（催化剂）
~ aluminum hydride 氢化二乙基铝
~ amine 二乙胺
~ - amino propylamine [DEAPA] 二乙氨基丙胺（固化剂）
O,O - ~ - N,N - bis(2 - hydroxyethyl) aminomethyl phosphonate O,O - 二乙基 - N,N - 双（2 — 羟乙基）氨甲膦酸酯（阻燃剂）
~ carbonate 碳酸二乙酯
~ 3,5 - di - *tert* - butyl - 4 - hydroxybenzylphosphate 3,5 - 二叔丁基4羟基苄基磷酸二乙酯（抗氧剂 1222）
~ dichlorosilane 二乙基二氯（甲）硅烷
~ ester of 3,5 - ditert - buty - 4 - hydroxy - benzylphosphonic acid 3,5 - 二叔丁基 - 4 - 羟基苄基膦酸二乙酯（抗氧剂 1222）
~ itaconate 衣康酸二乙酯（增塑剂）
~ maleate 马来酸二乙酯，顺丁烯二酸二乙酯（增塑剂）
~ phthalate [DEP] 邻苯二甲酸二乙酯（增塑剂）
~ sebacate [DES] 癸二酸二乙酯（增塑剂）
~ silicone oil 二乙基硅油
~ succinate 琥珀酸二乙酯，丁二酸二乙酯
di(2 - ethyl butyl) 二（2 - 乙基丁基）
~ azelate 壬二酸二（2 - 乙基丁基）（增塑剂）
~ phthalate 邻苯二甲酸二（2 - 乙基丁酯）（增塑剂、溶剂）
diethylene 二(-1,2-)亚乙基
~ - triamine [DTA] 二亚乙基三胺（环氧树脂固化剂）

~ glycol [DEG] 二甘醇，一缩二乙二醇
~ glycol benzoate and dipropyleneglycol benzoate blend 二甘醇苯甲酸酯和一缩二丙二醇苯甲酸酯混配物（1:1）（增塑剂）
~ glycol bisallyl carbonate 二甘醇双烯丙基碳酸酯
~ glycol dibenzoate 二甘醇二苯甲酸酯（增塑剂）
~ glycol dipelargonate 二甘醇二壬酸酯（增塑剂）
~ glycol dipropionate 二甘醇二丙酸酯
~ glycoldisteaate 二甘醇二硬脂酸酯（增塑剂）
~ glycollaurate 二甘醇月桂酸酯
~ glycol monobutyl ether 二甘醇单丁醚
~ glycol monoethyl ether 二甘醇单乙醚
~ glycol monolaurate 二甘醇单月桂酸酯（增塑剂）
~ glycol monooleate 二甘醇单油酸酯（增塑剂）
~ glycolmonoricinoleate 二甘醇单蓖麻醇酸酯（增塑剂）
di(2 - ethylhexyl) 二(2 - 乙基己基)
~ adipate 己二酸二（2 - 乙基己基）酯（增塑剂）
~ azelate 壬二酸二（2 - 乙基己基）酯〈增塑剂〉
~ 4,5 - epoxy tetrahydrophthalate 4,5 - 环氧四氢邻苯二甲酸二（2 - 乙基己）酯（增塑剂，稳定剂）
~ fumarate 富马酸二（2 - 乙基己基）酯，反丁烯二酸二（2 - 乙基己基）酯（增塑剂）
~ hexahydro phthalate 六氢化邻苯二甲酸二（2 - 乙基己）酯（增塑剂）
~ isophthalate 间苯二甲酸二（2 - 乙基己）酯（增塑剂）
~ itaconate 衣康酸二（2 - 乙基己）酯（增塑剂，交联剂）
~ maleate 马来酸二（2 - 乙基己）酯，

顺丁烯二酸二(2-乙基己)酯(增塑剂)
~ monophenyl phosphate 亚磷酸一苯二(2-乙基己)酯(抗氧剂)
~ peroxydicarbonat 过氧化二碳酸二(2-乙基己)酯(引发剂)
~ phthalate 邻苯二甲酸二(2-乙基己)酯(增塑剂)
~ -m- phthalate 间苯二甲酸二(2-乙基己)酯
~ sebacate 癸二酸二(2-乙基己)酯(增塑剂)
~ terephthalate 对苯二甲酸二(2-乙基己)酯(增塑剂)
~ tetrahydro phthalate 四氢化邻苯二甲酸二(2-乙基己)酯(增塑剂)
~ 4-thioazelate 4-硫代壬二酸二(2-乙基己)酯(增塑剂)

difference 差别,差异
~ of pressure 压力差
~ of temperature 温度差
~ of wall thickness 壁厚差

differential 差别的;差示(动)的;微分(的)
~ calorimetry 差示量热法
~ colorimeter 色差计
~ dilatometry 差示膨胀测定法
~ distributton curve 微分分布曲线
~ distribution of crosslinking 交联差示分布
~ dyeing [DD] 差异染色
~ gear 差动齿轮
~ number distribution curve 微分数分布曲线
~ piston 差动活塞
~ pressure 压差
~ pressure tube sizing 管材差压定径
~ refractometer 差示折射仪
~ scanning calorimetry [DSC] 差示扫描量热法
~ shrinkage 差示收缩量
~ shrinking 差异收缩
~ test 鉴别试验(塑料种类)
~ thermal analysis [DTA] 差热分析

~ thermogravimetric analysis 差热重量分析
~ thennogravimetry 差热重量法
~ thermometer 差示温度计
~ weight distribution 微分重量分布
~ weight distribution curve 微分重量分布曲线

difficult 困难的
~ bonding polymer 难黏聚合物
~ to-process plastic 难加工的塑料

diffraction 衍射
~ of X-rays X射线衍射

diffuse 扩散;漫射的
~ light transmission factori 漫射光透射系数

reflectance [RD] 扩散反射能力(不透光面)
~ reflection 漫反射
~ scattering 漫散射
~ trahsmission 漫透射(光学)
~ transmittance 漫透射率

diffuser 扩散器,喷雾器;漫射体
diffusibility 扩散能力
diffusible resin 扩散性树脂
diffusion 扩散;漫射;渗滤
~ barrier 阻扩散,扩散(屏)障
~ coefficient 扩散系数
~ constant 扩散常数
~ couple 扩散偶合
~ effect 扩散效应
~ mo(u)lding 扩散成型
~ of gas 气体扩散
~ of light 光散射,光漫散
~ theory of adhesion 黏合扩散理论
~ value (光)散射值
~ velocity 扩散速度

diffusivity 扩散系数,扩散性
diffusor 扩散器;孔板(流动床)
digest 消化;浸渍;蒸煮
digester 消化器;蒸煮器;浸渍器
digital 数字的
~ data 数字数据
~ display 数字显示
~ indication 数字显示

~ indicator 数字显示器
diglycidyl ether 二缩水甘油醚(活性稀释剂)
　~ of bisphenol A [DGEBA] 双酚A二缩水甘油醚
diglycol laurate 二甘醇月桂酸酯
dibeptyl phthalate [DHP] 邻苯二甲酸二庚酯(增塑剂)
dihexyl 二己基
　~ phthalate [DHXP] 邻苯二甲酸二己酯
　~ sebacate 癸二酸二己酯(增塑剂)
di-*n*-hexyl 二正己基
　~ adipate 己二酸二正己酯(增塑剂)
　~ azelate [DNHZ] 壬二酸二正己酯(增塑剂)
　~ phthalate 邻苯二甲酸二正己酯(增塑剂)
dihydroxy 二羟基
　2,4-~ benzophenone 2,4-羟基二苯甲酮(紫外线吸收剂)
　2,2'-~ -5,5'-dichlorodiphenyl metha-ne 2,2'-二羟基-5,5'-二氯代二苯基甲烷(防霉剂)
　2,2'-~ -4,4'-dimethoxybenzophenone 2,2'-二羟基-4,4'-二甲氧基二苯甲酮(紫外线吸收剂)
　2,2'-~ -3,3'-di(α-methylcyclohexyl)-5,5'-dimethyl diphenyl methane 2,2'-二羟基-3,3'-二(α-甲基环己烷)-5,5'-二甲基二苯基甲烷(抗氧剂)
　4,4'-~ diphenyl 4,4'-二羟基联苯(抗氧剂)
　2,2'-~ -4-methoxybenzophenone 2,2'-二羟基-4-甲氧基二苯甲酮
　2,2'-~ -4-*n*-octoxybenzophenone [UV-314] 2,2'-二羟基-4-正辛氧基二苯甲酮(紫外线吸收剂)
diisobutyl 二异丁基
　~ adipnte [DIBA] 己二酸二乙丁酯(增塑剂)
　~ azelate 壬二酸二异酯
　~ phthalatel [DIBP] 邻苯二甲酸二异丁酯(增塑剂,黏合剂)
diisocyanate 二异氰酸(酯)
diisodecyl 二异癸基
　~ adipate [DIDA] 己二酸二异癸酯(增塑剂)
　~ 4,5-epoxytetrahydrophthalate 4,5-环氧四氢邻苯二甲酸二异癸酯(增塑剂、稳定剂)
　~ glutarate [DIDG] 戊二酸二异癸酯(增塑剂)
　~ pentaerythritol diphosphite 二亚磷酸季戊四醇(酯.)二异癸酯(抗氧剂)
　~ phthalate [DIDP] 邻苯二甲酸二异癸酯(增塑剂)
　~ tetrahydrophthalate 四氢化邻苯二甲酸二异癸酯(增塑剂)
diisononyl 二异壬基
　~ adipate [DINA] 己二酸二异壬酯(增塑剂)
　~ phthalate [DINP] 邻苯二甲酸二异壬酯
diisooctyl 二异辛基
　~ adipate [DIOA] 己二酸二异辛酯(增塑剂)
　~ azelate [DIOZ] 壬二酸二异辛酯(增塑剂)
　~ funrarate [DIOF] 富马酸二异辛酯,反丁烯二酸二异辛酯(增塑剂)
　~ isophthalate [DIOIP] 间苯二甲酸二异辛酯(增塑剂)
　~ monoisodecyl trimellitate 偏苯三甲酸二异辛单异癸酯(增塑剂)
　~ phthalate [DIOP] 邻苯二甲酸二异辛酯(增塑剂)
　~ sebacate [DIOS] 癸二酸二异辛酯(增塑剂)
diisopentyl phthalate [DIPP] 邻苯二甲酸二异戊酯
diisopropyl 二异丙基
　~ acetone 二异丙基丙酮
　~ azodiformate [DIPA] 偶氮二甲酸二异丙酯(发泡剂)
　~ benzene hydroperoxide 氢过氧化二异丙苯(引发剂)

N'N - ~ 2 - benzothiazole sulphenamide　N'N - 二异丙基 - 2 - 苯并噻唑次磺酰胺(促进剂 DIBS)
~ peroxydicarbonate　过氧化二碳酸二异丙酯(引发剂)
~ sebacate　癸二酸二异丙酯(增塑剂)
diisostearoyl ethylene titanate　二异硬脂酰基钛酸乙二酯(偶联剂)
diisotachc polymer　双全同立构聚合物
diisotridecyl phthalate [DITDP]　邻苯二甲酸二异十三(烷基)酯
dil [dilute]　稀释;冲淡
dilatability　膨胀性
dilatable　可膨胀的
dilatancy　膨胀性;胀流性
dilatant　膨胀(性)的
~ flow　膨胀流
~ liquid　膨胀性流体
~ plastisol　膨胀性增塑糊
dilate　膨胀
dilatation　膨胀
~ coefficient　膨胀系数
~ constant　膨胀常数
dilatational strain　膨胀应变
dilate　膨胀
dilatometer　膨胀计
dilatometric thermometer　膨胀温度计
dilatometry　膨胀测定法
dilauryl　二月桂基
~ phosphite　亚磷酸二月桂酯〈抗氧剂〉
~ thiodipropionate [DLIP]　硫代二丙酸二月桂酯(抗氧剂)
diluent　稀释剂;冲淡剂
dilutability　稀释度
dilute　稀释,冲淡
~ phase　稀相
~ solutiori　稀溶液
~ solution viscometr　稀溶液黏度法
~ solutiori viscosjty　稀溶液黏度
diluting　稀释的,冲淡的
~ agent　稀释剂,冲淡剂
~ ratio　稀释比
dilution　稀释,冲淡

~ viscometer　稀释黏度计
dimensional　尺寸的
~ accuracy　尺寸精度,尺寸准确度
~ allowance　尺寸裕量
~ change　尺寸变化
~ endurance properties　尺寸持久性
~ inaccuracy due to the manufacture　尺寸制造误差
~ preasion　尺寸精度
~ restorability　尺寸[形状]复原性
~ sensitivity to moisture　尺寸对湿度的敏感性
~ sensitivity to temperature　尺寸对温度的敏感性
~ stability　尺寸稳定性;尺寸精度;因次稳定性
~ stability under heat　加热尺寸稳定性
~ tolerance　尺寸公差
~ uniformiity　尺寸一致性
dimensionally stable　尺寸稳定的
dimer　二聚物
~ of toluene 2,4 - diisocyanate　2,4 - 甲苯二异氰酸酯二聚体(硫化剂)
dimenc　二聚的
~ compounds　二聚化合物
~ polymer　二聚物
dimerisation [dimerization]　二聚作用
dimerising dye　二聚染料
dimethoxy　二甲氧基
p,p' - ~ diphenylamine 4,4' - 二甲氧基二苯胺(抗氧剂)
~ methane　二甲氧基甲烷(溶剂)
Di(2 - methoxy ethyl) phthalate [DMEP]　邻苯二甲酸(二甲氧基乙)酯(增塑剂)
dimethyl　二甲基
N,N - ~ - acetamide [DMA]　N,N - 二甲基乙酰胺
N,N - ~ - acrylamide　N,N - 二甲基丙烯酰胺
~ amine　二甲胺(催化剂)
~ - aminoethyl methacrylate　甲基丙烯酸二甲氨基乙酯
2 - ~ aminomethyl phenol　2 - 二甲氨

dimethylol urea

基甲基苯酚(固化剂)
~ - aminopropylamine [DMAPA]二甲氨基丙胺(固化剂)
~ - aniline 二甲基苯胺
~ - benzene 二甲苯
2,5 - ~ - 2,5 - bis(benzoyl peroxy) hex-ane 2,5 - 二甲基 - 2,5 - 双(过氧化苯甲酰)己烷(交联剂)
2,5 - ~ - 2,5 - bis(tert - butyl peroxy) hexane 2,5 - 二甲基 - 2,5 - 双(叔丁过氧基)己烷(交联剂)
2,5 - ~ - 2,5 - bis(tert - butyl peroxy) hex-3-yne 2,5 - 二甲基 - 2,5 - 双(过氧化叔丁基)己 - 3 - 炔(交联剂)
~ butadiene rubber 二甲基丁二烯橡胶,甲基橡胶
2,4 - ~ - 6 - tert - butyl phenol 2,4 - 二甲基 - 6 - 叔丁基苯酚(抗氧剂)
N - (1,3 - ~ butyl) - N - phenyl - p - phenylene diamine N - (1,3 - 甲基丁基) - N - 苯基对苯二胺[抗氧剂4020]
~ chlorendate 氯茵酸二甲酯
~ cyclohexyl adipate 己二酸二(甲基环己)酯(增塑剂)
~ cyclohexyl phthalate [DMCP] 邻苯二甲酸二(甲基环己)酯(增塑剂)
~ dichlorosilane 二甲基二氯(甲)硅烷
N,N' - ~ - N,N' - di(1 - methyl propyl) - p - phenylene diamine N,N'二基 - N,N' - 二(1 - 甲基丙基)对苯二胺(抗氧剂)
N,N' - ~ - N,N' - dirutroso - tere - phtha - mide dimethyl itaconate 衣康酸 N,N',二甲基 - N,N' - 二亚硝基对苯二酰胺盐·二甲酯(发泡剂)
~ formamide [DMF] 二甲基甲酰胺(溶剂)
~ Slutarate [DMG] 二甲基戊二酸酯〈增塑剂〉
2,6 - ~ - 4 - heptanone 2,6 - 二甲基 - 4 - 庚酮

~ maleate [DMM] 马来酸二甲酯,顺丁烯二酸二甲酯(增塑剂)
~ methylol urea resin 二甲羟甲基脲树脂
~ octadecyl hydroxyethyl ammonium-rutrate 硝酸十八烷基二甲羟乙基季铵盐[抗静电剂 SN]
~ phosphite 亚磷酸二甲酯(抗氧剂)
~ phthalate [DMP] 邻苯二甲酸二甲酯(增塑剂,黏合剂,溶剂)
~ polysiloxane 二甲基聚硅氧烷
~ sebacate 癸二酸二甲酯(增塑剂,软化剂,溶剂)
~ silicone oil 二甲基硅油
~ sulfoxide [DMSO] 二甲基亚砜(溶剂)
~ terephthalate [DMTl] 对苯二甲酸二甲酯

dimethylol urea 二羟甲基脲,脲醛树脂
diminish 减少,缩小,削弱
dimmers 小光灯,车前灯(汽车)
dimple 缩窝,凹痕
dimyristyl thiodipropionate 硫代二丙酸二(十四烷基)酯(抗氧剂)
DIN [Deutsche Industrie Norm] 德国工业标准(德)
DINA [diisononyl adipate] 己二酸二异壬酸(增塑剂)
N,N' - di - β - naphthyl - p - phenylene di - amine N,N' - 二 - β 萘基对苯二胺(抗氧剂)
dineric 二液界面的
~ interfac 二液接触面
dinging [mandrel forming] 弯成圆形,花型卷绕成型
dinitroso 二亚硝基
N,4 - ~ - N - methyl - aruline N,4 - 二亚硝基 - V - 甲基苯胺(增进剂,热处理剂)
N, N' - ~ pentamethylene tetramine [DNDT] N,N' - 二亚硝基五亚甲基四胺(发泡剂 H)
~ terephthlamide, DNTA 二亚硝基对苯二甲酰胺(发泡剂)

dinking [die-cutting]　模切,模冲,冲切
- ~ die　冲切模
- ~ punch　模冲头

dinonyl　二壬基
- ~ adipate [DNA]　己二酸二壬酯(增塑剂)
- ~ phthalate [DNP]　邻苯二甲酸二壬酯(增塑剂)
- ~ sebacate [DNS]　癸二酸二壬酯(增塑剂)

DINP [diisononyl phthalate]　邻苯二甲酸二异壬酯

DIOA [diisooctyl adipate]　己二酸二异辛酯(增塑剂)

dioctyl　二辛基
- ~ adipate [DOA]　己二酸二辛酯(增塑剂)
- ~ azelate [DOZ]　壬二酸二辛酯(增塑剂)
- ~ decyl phthalate [DODP]　邻苯二甲酸二辛癸基酯(增塑剂)
- p,p' - ~ diphenyl amine　p,p'-二辛基二苯胺(抗氧剂)
- ~ fumarate [DOF]　富马酸二辛酯,反丁烯二酸二辛酯(增塑剂)
- ~ isophthalate [DOIP] (ISO)　间苯二甲酸二辛酯(增塑剂)
- ~ itaconate [DOI]　衣康酸二辛酯(增塑剂,交联剂)
- ~ maleate [DOM]　顺丁烯二酸二辛酯,马来酸二辛酯(增塑剂)
- ~ phosphite　亚磷酸二辛酯(抗氧剂)
- ~ phthalate [DOP]　邻苯二甲酸二辛酯(增塑剂)
- ~ sebacate [DOS]　癸二酸二辛酯(增塑剂)
- ~ terephthalate [DOTP]　对苯二甲酸二辛酯(增塑剂)

di - n - octyl　二正辛基
- ~ - n - decyl adipate [DNODA]　二(正辛基正癸基)己二酸酯
- ~ tetrahydrophthalate　四氢化邻苯二甲酸二正辛酯(增塑剂)

N,N' - di - sec - octyl - p - phenylene diamine　N,N'-二仲辛基对苯二胺(抗氧剂288)

di - n - octyltin　二正辛基锡
- ~ dilaurate　二月桂酸二正辛基锡(稳定剂)
- ~ bis(2 - ethylhexyl - thioglycollate)　双(硫代甘醇酸2-乙基己酯)二正辛基锡(稳定剂)
- ~ - S,S' - bis (isooctyl - mercapto acetate)　S,S'-双(巯基乙酸异辛酯)二正辛基锡(热稳定剂)
- ~ maleate polymer　马来酸二正辛基锡聚合物(稳定剂)
- ~ bis(monobutyl maleate)　双(马来酸单丁酯)二正辛基锡(热稳定剂)
- ~ bis(monooctylmaleate)　双(马来酸单辛酯)二正辛基锡(热稳定剂)
- ~ dilaurate　二月桂酸二正辛基锡(热稳定剂)
- ~ maleate　马来酸二正辛基锡,顺丁烯二酸二正辛基锡(热稳定剂)

DIOF [diisooctyl fumarate]　富马酸二异辛酯,反丁烯二酸二异辛酯(增塑剂)

DIOIP [diisooctyl isophthalate]　间苯二甲酸二异辛酯(增塑剂)

dioleyl phosphite　亚磷酸二油醇酯(抗氧剂)

DIOP [diisooctyl phthalate]　邻苯二甲酸二异辛酯(增塑剂)

DIOS [di - iso - octyl sebacate]　癸二酸二异辛酯

di (o - xenyl) monopheyl phosphate　磷酸二邻联苯·单苯酯

Diosna　道斯纳
- ~ impactthill [impact reactor]　冲击式研磨机
- ~ mixer　道斯纳混合器,用水和空气冷却的混合器

1,4 - djoxane　1,4-噁烷,(溶剂,助剂)

1,3 - dioxolane　1,3-二氧戊环(助剂)

dioxy diphenyl urethane　二氯二苯基尿烷

DIOZ [diisooctyl azelate]　壬二酸二异辛酯(增塑剂)

dip 浸渍,浸涂
- ~ and scrape 浸刮
- ~ and scrape method 浸刮法
- ~ and squeeze method 浸挤法
- ~ blow 浸渍吹塑
- ~ blow mo(u)lding [dipping mandrel blow moulding] 浸渍吹塑
- ~ bonding 浸渍黏合
- ~ coater 浸渍涂布机
- ~ coating 浸涂,蘸涂
- ~ coating in powder 粉料蘸涂
- ~ compound 浸渍混合物
- ~ forming 浸渍成型,蘸塑
- ~ mo(u)ld 浸渍成型模(具),蘸塑模(具)
- ~ mo(u)lding 蘸塑,浸渍成型
- ~ package 浸涂包装
- ~ polishing 蘸浸上光
- ~ roll 浸渍辊,蘸料辊(涂布机)
- ~ roll coater 辊蘸涂布机
- ~ solution 浸渍溶液
- ~ tank 浸渍槽,浸浆槽
- ~ tank coating 浸(渍)涂
- ~ tray 浸渍盘,浸渍槽
- ~ treating 浸渍处理
- ~ tube [siphon tube, eductor tube] 浸涂管
- ~ tumbler 浸渍转筒
- ~ varniush 浸渍清漆

DIPA [diisopropyl azodiformate] 偶氮二甲酸二异丙酯(发泡剂)

diphenyl 二苯基
- ~ - amine [DPA] 二苯胺(稳定剂)
- ~ cresyl phosphate [DPCP] 磷酸二苯基甲苯酯(增塑剂)
- ~ decyl phosphite 亚磷酸二苯基癸基酯
- ~ dichlorosilane 二苯基二氯(甲)硅烷
- ~ 2 - ethyl hexyl phosphate 磷酸二苯(基)·2 - 乙基己酯(增塑剂)
- N,N' - ~ ethylene diamine N,N' - 二苯基乙二胺(抗氧剂)
- ~ guanidine 二苯胍(促进剂 D)
- ~ isooctyl phosphite 亚磷酸二苯异辛酯(抗氧剂)
- ~ - isodecyl phosphite 亚磷酸二苯异癸酯(抗氧剂)
- ~ ketone 二苯甲酮
- ~ - methane diisocyanate [MDI] 二苯基甲烷二异氰酸酯
- ~ mono - o - xenylphosphate 磷酸单邻联苯(基)二苯酯(增塑剂)
- ~ nonyl phenyl phosphite 亚磷酸二苯(基)壬基苯酯(螯合剂,抗氧剂)
- ~ octyl phosphate [DPOP] 磷酸二苯辛酯(增塑剂)
- N,N', - p - phenylene diamine [DPPD] N,N' - 二苯基对苯二胺(防老剂 H)
- ~ phthalate [DPP] 邻苯二甲酸二苯酯〈增塑剂〉
- N,N' - ~ propylene diamine N,N' - 二苯基丙二胺(抗氧剂)
- ~ - sulfon - 3,3' - disulfonylhydrazide 3,3' - 二磺酰肼二苯砜〈发泡剂〉
- ~ thiourea 二苯基硫脲(热稳定剂,促进剂 CA)
- ~ - xylyl phosphate 磷酸二苯二甲苯酯

diphenylol propane 二酚基丙烷,双酚 A

dipolar 偶极的
- ~ bond 偶极键
- ~ electret 偶极驻极体
- ~ ion 偶极离子

dipole 偶极
- ~ - dipole force 偶极间力,取向力
- ~ disorientation 偶极解取向
- ~ effect 偶极效应
- ~ layer 偶极层
- ~ molecule 偶极分子
- ~ moment 偶极矩
- ~ orientation 偶极取向
- ~ polarization 偶极极化

dipolymer 二聚物

di(polyoxyethylene) hydroxymethyl phos - phonate 二(聚氧乙烯)·羟甲基磷酸酯(阻燃剂)

DIPP [diisopentyl phthalate] 邻苯二甲酸二异戊酯
dipping 蘸浸,蘸涂
- ~ agent 浸渍剂
- ~ barrel 浸渍筒(吹塑成型浸渍型芯)
- ~ barrel piston [recupcratot, piston] 浸渍筒活塞
- ~ bath 浸渍浴
- ~ compound 浸渍料
- ~ compound for strippable 可剥性涂层浸渍料
- ~ form 蘸塑模
- ~ former 蘸塑模
- ~ in water 水中浸渍试验(泡沫塑料)
- ~ lacquer 浸渍清漆
- ~ machine 浸渍机
- ~ mandrel 浸渍芯杆
- ~ mandrel blow moulding 浸渍芯杆吹塑
- ~ mo(u)ld 浸渍模(具)
- ~ process 蘸塑法,浸渍法
- ~ rack 浸渍挂架
- ~ roll 浸渍辊
- ~ solution 浸渍溶液
- ~ tank 浸渍槽
- ~ technique 浸渍技术
- ~ test 浸渍试验〈测泡沫塑料吸水性〉
- ~ time 蘸浸时间,蘸涂时间
- ~ varnish 浸渍清漆

dipropyl 二丙基
- ~ ketone 二丙基甲酮
- ~ phthalate 邻苯二甲酸二丙酯(增塑剂)

dipropylene glycol 一缩二丙二醇
- ~ dibenzoate 一缩二丙二醇二苯甲酸酯(增塑剂)
- ~ monosalicylate 一缩二丙二醇单水杨酸酯

dir. [direct] 直接的
direct 直接的
- ~ action 直接作用
- ~ blow mo(u)lding 直接吹塑
- ~ calendering lamination 直接压光层合
- ~ coating 直接涂布
- ~ current [DC] 直流电
- ~ dye 直接染料
- ~ dyeing 直接染色
- ~ expansion 直接发泡
- ~ feed 直接进料〈注塑〉
- ~ foaming 直接发泡
- ~ gassing [direct injection gas] 直接充气(泡沫塑料)
- ~ gate 直接浇口,棒形浇口
- ~ gated injection mould 直接浇口注塑模具,无流道注塑模具
- ~ gafing [direct feed, sprue gating] 直接浇口
- ~ gravure coater 直接凹(印)辊涂布机
- ~ heated mould 直接加热模具
- ~ injection 直接注塑
- ~ injection moulding 直接注塑(成型),无流道注塑(成型)
- ~ injection of gas [direct gassing] 直接注气,直接充气
- ~ injection of sole 鞋帮底直接注塑
- ~ melt spinning 直接熔体[熔融]纺丝
- ~ melting 直接熔融法
- ~ piston closing system 活塞闭合液压系统(注塑机)
- ~ plastication 直接塑化
- ~ polymer to web-system 聚合物直接成布(法)(非织造织物技术)
- ~ roll coater 直接辊涂机
- ~ roll coating 直接辊涂
- ~ roving 直接无捻粗纱
- ~ spinning 直接纺丝法
- ~ sprue gate 直接注料浇口
- ~ take-up 直接卷取
- ~ tension device 直接式张力装置
- ~ transrhission 直接透射

directed 直接的
- ~ fibre preform 纤维直接预成型
- ~ fibre preform process 纤维直接预成型法(制备层压板)

direction 方向

directional
~ A and B 纵向和横向(层压塑料)
~ action 定向作用
~ control valve 方向控制阀
~ of approach 熔料流向
~ of draw 拉伸方向
~ of feed 进料方向
~ of orientation 取向方向
~ of rolling 辊滚方向
~ of stress 压力方向
~ of twist 加捻方向

directional 有方向性的,定向的
~ pressure 定向压力
~ properties 各向异性
~ property 定向性能
~ reflectance 定向折射
~ valve 定向阀

directive bonding 定向黏合,半极性黏合
directly 直接(地)
~ driven winder 直接传动卷绕机
~ injection moulding 直接注塑成型
~ proportional 正比例的,成正比(例)

dirt 杂质,杂物;污渣,废屑,灰尘
~ adherence 杂物黏合
~ collector 吸尘器
~ pick-up 积灰
~ trap 集渣器

dirty 弄脏,弄污
DIS-screw [dynamic striation screw] 动态细沟状螺杆
disc [disk] 圆盘,圆片
discharge 排出,卸料;放电
~ barrel section (螺杆)卸料机筒段
~ capacity (排)流量,卸载能力
~ chamber 输出通道
~ chute 卸料斜槽(注塑机制品)
~ conveyer 卸料输送带
~ dopr 卸料门
~ door of bandury mixer 密炼机卸料门
~ duct 排料槽
~ end 卸料端
~ eqlupment of banbury [Banbury] mixer 密炼机卸料装置
~ funnel 卸料斗
~ gas 废气,排出气
~ gate 卸料闸板(粒料储器)
~ hole 排出孔,卸料孔
~ hopper 卸料斗
~ losses 出口损耗,卸料损耗
~ opening 卸料孔
~ pipe 卸料管
~ piston 卸料活塞
~ pump 卸料泵
~ rate 卸料速率
~ screw 卸料螺杆,卸料螺旋运输机
~ section (螺杆)卸料段
~ temperature 排出温度,卸料温度
~ throat 卸料喉道
~ tube 卸料管
~ valve 卸料阀
~ velocity 卸料速度
~ zone 放电区(电量);卸料段

discolo(u)r 变色;脱色;褪色
discolo(u)ration 变色;脱色;褪色
~ method 褪色法

discolo(u)red polymer 变色聚合物,泛色聚合物
discolo(u)ring 变色;脱色;褪色
discolo(u)rization 变色;脱色;褪色
disconnect 拆开,分离(偶联)
discontinuity 不连续性;间断性
discontinuous 不连续的,间断的
~ centrifuge 间歇式离心机
~ extrusion 不连续挤塑,间断挤塑
~ fibre (定长)短纤维,不连续纤维
~ mixing 不连续混合,间歇混合,分批混合
~ operation 间歇操作,间歇运转
~ phase 不连续相
~ process 间歇法
~ running 间断运转,周期性生产
~ weldbead 不连续[间断]焊缝
~ welding 间断焊

discrete 个别的;离散的,分散的,不连续的
~ fibre 分离纤维
~ material 松散材料
~ particles 分散粒子
~ phase 分散相

discrimination 辨别,区别(标记)
dished 凹陷;碟状扭曲,凹状扭曲
　~ bottom　蝶形底(反应器)
disintegrate 粉碎
disintegrator 粉碎机;切片机
disk [disc] 圆盘,圆片
　~ and cone agitators　盘锥式搅拌器
　~ attrition mill　盘式磨碎机
　~ crusher　盘式压碎机
　~ cutter　盘形切口,盘形切断机
　~ dryer　圆盘式干燥器
　~ extruder　盘式挤塑机(美)
　~ feeder　盘式加料器
　~ gate　盘式浇口
　~ grinder　盘磨机,盘式磨碎机
　~ impeller　盘式搅拌器
　~ knife　圆片刀
　~ mill　盘式磨碎机,盘磨机
　~ plate　圆盘板
　~ refiner　盘式精研[磨]机
　~ sander　圆盘式砂磨机
　~ saw　圆盘锯
　~ spring assembly　盘簧组件
　~ tribometer　盘式摩擦计
　~ type extruder　盘式挤塑机
　~ type extruder gap　盘式挤塑机间隙
　~ valve　盘形阀,片状阀
Diskpack [Farrel Diskpack plasticator]
(法莱尔)磁盘式塑化器,无螺杆塑化器
　~ screw　磁盘式螺杆
　~ screwless extruder　磁盘式无螺杆挤塑料机
dislocation 错位,移位
dismantle 拆除,拆卸
disordered 无序的
　~ fold surface　不规则折叠面
　~ orientation　不规则取向
　~ structure　无规结构
dispenser 分配器;给料器;混料装置(黏合剂或树脂组分);自动售货机
　~ roll　撕开辊
　~ valve　喷雾式阀
dispensing 分配;配料;分散的

　~ device [instrument]　混料装置,配料装置(黏合剂或树脂组分)
　~ gun　喷射枪〈发泡塑料〉
　~ package　分散包装
　~ unit　喷射发泡装置;给料装置
dispergate 分散
disperge 分散
dispersable vat dyes　分散型还原染料
dispersal 分散,分开
　~ plant　分支装置
　~ ststem　分散系统
Dispersall mixer　碟斯潘沙尔混合器,分散混合器
dispersancy 分散力
dispersant 分散剂
disperse 分散
　~ dye　分散染料
　~ system　分散体系
dispersed 分散的
　~ component　分散组分
　~ phase　分散相
　~ suspension　分散悬浮液
　~ weave　分散组织
dispersobility 分散性
dispersing 分散的
　~ additive　分散添加剂
　~ agent　分散剂
　~ auxiliary　分散助剂
　~ device　分散装置
　~ gun　喷枪
　~ power　分散能力
dispersion 分散体;分散液;分散(作用);色散
　~ – based adhesive　分散黏合剂
　~ coefficient　分散系数
　~ degree　分散度
　~ disc [dispersion disk]　分散盘
　~ effect　分散效应
　~ force　分散力;色散力;
　~ index　分散指数;色散指数
　~ medium　分散介质
　~ mixer　分散混合机
　~ phase　分散相
　~ polymerization　分散聚合(作用)

~ polytetrafluoreethylene［PTFE］ 分散法聚四氟乙烯
~ ratio 分散率
~ resin 分散型树脂
~ type polyvinyl chloride resin 分散型聚氯乙烯树脂
dispersity 分散度,分散性
dispersive 分散的
dispersiveness 分散度
dispersivity 分散性
dispersoid 分散(胶)体,分散质
displace 排代,置换;移动,位移;挤压
displacement 位移;置换,排代,替换
~ angle 位移角〈长丝缠绕〉
~ chromatography 排代色层分离(法),顶替色谱(法)
~ current 置换电流
~ gauge 排代量规,位移量规
~ stroke 位移行程,调节行程
~ transducer conditioner 移动传感调节器,形体测量放大器
display 显示,展览(品);装潢(包装)
~ film 视觉薄膜〈包装〉
~ label 陈列品标签〈包装〉
~ package 视觉包装
~ panel 标志,标签〈包装〉
~ window 窥视孔〈包装〉
disposable 可任意处理的,任意的
~ blade 一次性使用刀片
container 一次性使用容器
~ item 一次性应用制品;废品
~ laundry 一次性使用衣服
~ packaging 一次性使用包装
~ packing 一次性使用包装
disposables 一次性使用件
disposal 处理,处置;清除;废弃物
~ of waste 废物处理
disposability 处理可能性〈废物〉
disposition 安排;布置
~ of cavity 模腔布局
disproportionation 歧化作用
~ reaction 歧化反应
~ termination 歧化终止
disproportionative condensations 歧化缩合作用
disrupt 破裂;碎裂;断裂
disruptive 破坏性的,破裂性的,击穿的
~ discharge 破坏性放电,击穿放电
~ discharge volatge 破坏性放电电压
~ strength 击穿强度
~ test 击穿试验,耐压试验
~ voltage 击穿电压
dissipation 消耗,散逸;分散
~ coefficient 损耗系数
~ factor 损耗因素,介电损耗正切
~ factor loss tangent 介电损耗角正切
~ of energy 能量消耗,消能
~ of heat 热分散
dissociate 分离,分解,离解
dissociation 离解,分解
~ energy 离解能
~ temperature 离解温度
dissolution 溶解
~ heat 溶解热
dissolve 溶解
dissolvent 溶剂
dissolver 溶解器,溶解装置;拌胶器
~ agitator 溶解搅拌器
dissolving 溶解
~ capacity 溶解能力,溶解度
~ drum 溶解鼓
~ power 溶解力
~ pulp 溶解纸浆
~ tank 溶解槽
dissymmetric(al) 不对称的
~ molecule 不对称分子
dissymmetry 不对称,非对称
~ coefficient 非对称系数
distance 距离,间隔;行程
~ between grips 夹子间距〈试样〉
~ between platens of press 压板开距
~ between shafts 轴间距
~ control 遥控,远距离控制
~ piece 垫片,隔片
~ plate 垫板,隔板
~ ring 隔圈,定距环
~ valve 行程阀
~ velocity log 距离速度滞后

distd. [distilled] 蒸馏过的
distearyl 二硬脂酰,双十八烷酰;二硬脂基,双十八烷基
- 3,5 - di - *tert* - butyl - 4 - hydroxybenzyl phosphate 3,5 - 二叔丁基 - 4 - 羟基苄基磷酸二硬脂基酯〈抗氧剂〉
- epoxy hexahydro phthalate 环氧六氢化邻苯二甲酸二硬脂基酯〈润滑剂,抗静电剂,防雾剂〉
- ether 二硬脂醚
- pentaerythritol cyclic diphosphite 双十八烷基季戊四醇环亚磷酸酯〈抗氧剂 618〉
- pentaerythritol diphosphite 二亚磷酸季戊四醇二硬脂醇酯〈抗氧剂〉
- β,β' - thiodibutyrate β,β' - 硫代二丁酸双十八烷基酯〈抗氧剂〉
- thiodipropionate 硫代二丙酸二硬脂醇酯〈抗氧剂〉

distemper 胶黏涂料,水浆涂料,刷墙粉;色胶
distent 膨胀
distillate 蒸馏出液
distillation 蒸馏(作用)
- column [distilling column] 蒸馏柱
- receiver 蒸馏收集器

distn. – distillation 蒸馏(作用)
distort 变形,扭曲
distortion 变形,挠曲,扭变;畸变,失真
- by welding 焊接变形
- point 扭变点
- temperature 变形温度
- tensor 形变张量
- under heat 热扭变

distributing roller 分配辊
distribution 分布,分配
- coefficient 分配系数
- curve 分布曲线
- function 分布函数
- function of orientations 定向分布函数
- of grain sizes 粒度分布,粒径分布
- of molecular weight (相对)(分子)(质量)分布
- of orientation 定向分布
- of polymerization degree 聚合度分布
- of relaxation time 松驰时间分布
- of residence time 滞留时间分布
- of size 粒度分布
- of stress 应力分布
- of twist 捻度分布
- package 配给包装,分类包装
- packaging 现场包装
- parameter 分布参数
- pillar 电缆箱,配电柜
- ratio 分配比率

distributor 分配器,分布器;传墨辊;分电盘
- block 分料区段,进料区段
- plate with tube system 管系分布板
- roll 布料辊,辊式布料器,分布辊〈涂布机〉

disturbance 扰动;干扰;故障
disturbing quantity 干扰量
disyndiotactic polymer 双间同立构聚合物,双间规聚合物
ditactic 构型的双中心规整性
- polymer 双有规立构聚合物

DITDP [diisotridecyl phthalate] 邻苯二甲酸二异十三烷基酯
ditetradecyl phosphite 亚磷酸双十四烷基酯〈抗氧剂〉
4,4' - dithio dimorpholine 4,4' - 二硫代二吗啉〈硫化剂,促进剂〉
dithiobis 〈词头〉二硫代双
- 2,2' - ~ (4, 6 - di - *tert* - butyl - 3 - methyl phenol) 2,2' - 二硫代双(4,6 - 二叔丁基 - 3 - 甲基苯酚)〈再生活化剂 463〉
- 2, 2' - ~ [mono, bis or tris - (α, α - dimethylbenzyl) phenol] 2,2' - 二硫代双[单,双或参(α, α - 甲基苄基)苯酚]〈再生活化剂 MSP - S〉
- N, N', ~ (morpholine) N, N' - 二硫代双吗啉〈硫化剂 DTDM〉
- 2,2' - ~ (*tert* - butylated p - cresol) 2, 2' - 二硫代双(叔丁基化对甲酚)

ditridecyl

〈再生活化剂 420〉
- ~ xylenes 二硫代双二甲苯〈再生活化剂 703〉
- di-o-tolyl 二邻甲苯基
- N,N'- ethylene diamine N,N'-二邻甲苯基乙二胺〈抗氧剂〉
- ~ ~ guanidine 二邻甲苯脒〈促进剂 DOTG〉

ditridecyl 双十三烷基
- ~ phthalate [DTDP] 邻苯二甲酸双十三烷基酯〈增塑剂〉
- ~ thiodipropionate [DTDTP] 硫代二丙酸双十三烷基酯〈抗氧剂〉

diundecyl 双十一烷基
- ~ phthalate [DUP] 邻苯二甲酸双十一烷基酯〈增塑剂〉

divergent die 分流口模
diversion head 转移机头
divert 转向,变换
divided trough kneader 分槽捏和机
- ~ divider 隔板〈容器〉;中间层;分配器,分配器

divinyl benzene [DVB] 二乙烯基苯〈交联剂〉
di(o-xenyl) monophenyl phosphate 磷酸二邻联苯·单苯酯〈增塑剂〉DLTP [dilauryl thiodipropionate] 硫代二丙酸二月桂酯〈抗氧剂〉
DMA [N,N-dimethyl-acetamide] N,N-二甲基乙酰胺
DMA [dynamic mechanical analysis] 动态力学分析
DMAPA [dimethyl aminopropylamine] 二甲氨基丙胺〈固化剂〉
DMC [dough moulding compound] 料团状模塑料,料团成型料
DMCP [dimethyl cyclohexyl phthalate] 邻苯二甲酸二甲环己酯〈增塑剂〉
DMEP [di(methoxyethyl) phthalate] 邻苯二甲酸二(甲氧基乙)酯〈增塑剂〉
DMF [dimethyl formamide] 二甲基甲酰胺〈溶剂〉
DMG [dimethylglutarate] 二甲基戊二酸酯

DMM [dimethyl maleate] 马来酸二甲酯,顺丁烯二酸二甲酯〈增塑剂〉
DMP [dimethyl phthalate] 邻苯二甲酸二甲酯〈增塑剂,黏合剂,溶剂〉
DMS [dimethyl sebacate] 癸二酸-二甲酯
DMS [dynamic mechanical spectroscopy] 动态力学谱
DMSO [dimethyl sulfoxide] 二甲基亚砜〈溶剂〉
DMT [dimethyl terephthalate] 对苯二甲酸二甲酯
DNA [dinonyl adipate] 己二酸二壬酯〈增塑剂〉
DNHZ [di-n-hexyl azelate] 壬二酸二正己酯
DNODA [di-n-octyl-n-decyl adipate] 二(正辛基·正癸基)己二酸酯
DNP [dinonyl phthalate] 邻苯二甲二壬酯〈增塑剂〉
DNPT [dinitrosopentamethylene tetramine] 二亚硝基 五亚甲基四胺
DNS [dinonyl sebacate] 癸二酸二壬酯〈增塑剂〉
DNTA [dinitrosoterephthalamide] 二亚硝基对苯 二甲酰胺
DOA [dioctyl adipate] 己二酸二辛酯〈增塑剂〉
Doc. document 文件
doctor 刮涂;刮刀
- ~ bar 刮条,刮片,刮胶棒
- ~ blade 刮板,刮胶板
- ~ blade applicator 刮板涂布器
- ~ kiss coater 辊舐刮涂机
- ~ kiss coating 辊舐刮涂
- ~ knife 刮刀,刮胶刀
- ~ knife coater [- inverted knife coater] 刮刀涂布机
- ~ roll 涂布辊,涂布量控制辊 辊涂器
- ~ table 刮刀垫衬

dodecenyl succinic anhydride [DDSA] 十二碳烯基丁二酸酐〈固化剂〉
dodecyl 十二烷基
- ~ mercaptan 十二烷基硫醇〈调节剂〉

6 - ~ - 2,2,4 - trimethyl - 1,2 - dihydro - quinoline 6 - 十二烷基 - 2,2,4 - 三甲基 - 1,2 - 二氢化喹啉〈抗氧剂 DD〉

dodecyloxy 十二烷氧基
~ - 2 - hydroxy - benzophenone 十二烷氧基 - 2 - 羟二苯甲酮
N - (3 - ~ - 2 - hydroxy propyl) ethanolam - ine N - (3 - 十二烷氧基 - 2 - 羟丙基)乙醇胺〈抗静电剂〉

dodge 自动固定螺纹环,自动固定螺纹垫圈

DODP [dioctyl decyl phthalate] 邻苯二甲酸二辛基癸基酯

DOF [dioctyl fumarate] 富马酸二辛酯,反丁烯二酸二辛酯〈增塑剂〉

doff roll 牵引辊,引离辊

dog 狗
~ - bone structure 狗骨架状结构
~ - skin 狗皮纹〈缺陷〉

DOI [dioctyl itaconate] 衣康酸二辛酯〈增塑剂,交联剂〉

DOIP [dioctyl isophthalate] 间苯二甲酸二辛酯〈增塑剂〉

dolly (供料给压延辊)热料辊

dolomite 白云石

DOM [dioctyl maleate] 顺丁烯二酸二辛酯,马来酸二辛酯〈增塑剂〉

domain structure 微区结构

dome 圆顶,拱顶
~ - shaped roof 拱形(屋)顶

domed 拱凸,凸状扭曲
~ baffle 拱形防尘板
~ bottom 拱形底

doming 拱起;成拱形
~ mo(u)ld 凸顶模具

donor - acceptor copolymer 给体 - 受体共聚物

door 门,装料口
~ handle 门把手
~ liner 门衬板
~ lining 门衬里
~ opening plate 门框板
~ panel 门面板
~ rim 门框
~ skin 门外板,门面板
~ trim board 门内装饰板
~ window 门窗

DOP [dioctyl phthalate] 邻苯二甲酸二辛酯〈增塑剂〉

Dope 蒙皮漆,涂布漆;掺杂剂,掺杂料;添加剂;膜溶液〈成膜〉;胶料〈皮草和纺织〉;密封料〈包装〉;纺丝溶液;浆糊
~ dyeing 纺液染色

doping 涂布;掺杂;添加
~ agent 掺杂剂

DOS [dioctyl sebacate] 癸二酸二辛酯〈增塑剂〉

dose 剂量;配料

dosimeter 剂量器

dosimetry 剂量测定法

dosing 定量给料,计[配]量
~ and dispensing machine 定量灌装机
~ and proportioning slide 剂量滑阀
~ apparatus 定量器;计量给料器
~ closure 定量封合〈包装〉
~ feeder 计量给料器
~ insert 定量填充物〈包装〉
~ machine 定量机,计量机
~ plant 定量给料装置
~ pump 定量泵,计量泵
~ scales 定量秤,计量秤
~ slide valve 定量给料滑阀〈美〉
~ system 计量系统,加料系统
~ tank 投配器,量斗
~ unit 计量机构
~ valve 计量阀

dot coating machine 点涂机

DOTP [dioctyl terephthalate] 对苯二甲酸二辛酯〈增塑剂〉

double 双倍的,复式的
~ action press 双[复]动压机
~ action ram 双[复]动活塞
~ arm kneader 双壁捏合机
~ back tape 双曲黏合带
~ bevel butt weld 双斜对接焊缝
~ bevel groove 双斜坡口,K 型坡口

double

- ~ bevel groove weld 双斜坡口焊缝
- ~ - blade 双浆叶
- ~ blade kneader 双浆捏和机
- ~ blade mixer 混合器
- ~ block shear test 块试样剪切试验
- ~ bond 双键
- ~ bond polymerization 双键聚合
- ~ bridge polymer 双桥聚合物
- ~ - caterpillar draw-off system 双履带牵引系统
- ~ channel screw 双槽螺杆
- ~ coated fabrics 双面涂布织物
- ~ coated film 双面涂布膜
- ~ coated pressure-sensitive (adhesive) tape 双面涂布压敏胶带
- ~ coated tape 双面涂布带
- ~ compound 复合物
- ~ cone blender 双锥形混合器〈美〉
- ~ cone dosing device 双锥形计量[定量]装置
- ~ - cone impeller 双锥形旋转混合器
- ~ cone impeller mixer 双锥形桨叶混合器
- ~ cone mixer 双锥形混合器
- ~ - cone pipe extrusion head 双锥管挤塑机头
- ~ cone viscometer 双锥式黏度计
- ~ curvature 双弯曲
- ~ curved 双弯曲的
- ~ covered butt joint 双面层对接
- ~ curved shell 双弯曲壳,双曲面件
- ~ cut saw 双割锯
- ~ draw mechanism 双牵拉装置
- ~ drawing 双区拉伸
- ~ drum diyer 双转鼓式干燥机
- ~ edge 卷边,双缘
- ~ ejection 二道脱模,复式脱模
- ~ face coating 双面涂布
- ~ - faced 双面的
- ~ fill protection 重复装料防止器
- ~ - filled fabric 双纬织物
- ~ filler lap joint 对面搭接焊
- ~ fillet weld 双面角焊缝
- ~ filling warning light 重复装料示警灯
- ~ finger buff 双指形折布抛光轮
- ~ - flanged butterfly valve 双凸缘蝶形阀
- ~ - flighted 双螺纹的〈螺杆〉
- ~ folds 双折边
- ~ force mould 双阳模式模具
- ~ force press 双活塞压机
- ~ gate 复合浇口
- ~ groove 双面坡口
- ~ head 双机头〈吹塑〉
- ~ head spray gun 双头喷枪
- ~ - header coat 双道涂布
- ~ helical gear 双螺旋齿轮
- ~ helical mixer 双螺桨混合器
- ~ - jacketed tank 双套箱
- ~ knockout bar mould 双脱模板式模具
- ~ knockout mould 复式脱模模具
- ~ lap joint 双搭接(接头)
- ~ - layer blown film 双层吹塑薄膜
- ~ - layer diagonal space grid 双层交叉空间栅格
- ~ lead screw 双螺纹螺杆
- ~ motion agitator [mixer] 双动搅拌器
- ~ motion paddle mixer 双动浆式混合器
- ~ naben kneader 双毂捏合机
- ~ orientation 双取向
- ~ overlapped adhesive joint 双搭接粘接
- ~ parison technique 双型坯技术
- ~ pass 双路,双道
- ~ - pass coat [double header coat] 十字形涂布 双道涂布
- ~ - pass system 双路系统
- ~ - pickled 两遍酸洗[连接件预处理]
- ~ pipe 套管
- ~ press moulding 二段压塑成型
- ~ pressing technique 二次压缩模塑技术
- ~ protruding with thread insert 双露头螺纹嵌件

~ ram press 双柱塞压机
~ ram system 双柱塞系统
~ refraction 双折射
~ reverse bend 双反向弯曲
~ roller mixer 双辊混合器
~ rotation system powder moulding 双轴回转粉末成型
~ - runner mill 双流道磨
~ saw 双割锯
~ scarf lap joint 双斜搭接
~ screw 双螺杆
~ screw extruder 双螺杆挤塑机
~ screw mixer 双螺杆混合器
~ - screw reactor 双螺旋式反应器
~ shaft kneader 双轴[臂]捏合器
~ seam 双接缝;双折痕
~ shear butt joint 双剪搭接
~ shear lap joint 双剪搭接
~ shot moulding 双色注塑(成型)
~ side coating 双面涂布
~ side gate 双侧面浇口
~ sizer 双面上胶器;双割锯
~ sizing 双面涂胶
~ - skin 双层
~ spiral mixer 双螺旋式混合器
~ spread 双面涂布;双面涂布胶耗量
~ spread coating 双面涂布
~ stacked mo(u)ld 双层模具
~ staining 双重着色
~ - strand polymer 双股聚合物
~ stranded chain 双股链
~ stranded polymer 双股聚合物
~ thread 双头螺纹
~ thread screw 双螺纹螺杆
~ thrust bearing 双向止推轴承
~ toggle locking system 双肘节锁模装置;双肘节锁模方式
~ tone colour 双色调节
~ tube 套管
~ turn 膨胀圈〈管道〉
~ V - butt joint [weld] 双 V 形对接焊缝,X 形对接焊缝
~ V butt weld 双 V 形对接焊缝
~ V - groove 双 V 形坡口

~ valve 双座阀
~ vee - butt joint [weld] 双 V 形对接焊缝,X 形对接焊缝
~ wall pipe 双壁管
~ - walled [double - skin] 双壁的
~ - walled air - inflated tunnel 双壁充气隧道
~ walled section 双壁截面
~ wave screw 双波纹螺杆
~ zone draft 双区牵伸
doubled flat - laid tubular film 双面平折管状膜〈拉伸过程〉
doubler 加缠面;布层贴合机
doubler equipment 贴合设备[装置]
doubling 贴合
~ calender 贴合压延机
~ device 贴合装置
~ machine 贴合机
~ roller 贴合辊
dough 料团
~ - filter 料团滤器
~ mixer 料团混合机
~ mo(u)lding 料团成型
~ mo(u)lding compound [DMC] 团状模塑料,团状成型料
~ spreading machine 料团涂布机,预混料涂布机
dowel 定位销,合缝销
~ bush 定位销套,合缝销套,合模销套
~ bushing 定位销套,合缝销套,合模销套
~ hole 定位销孔,合模销孔
~ joint 套接接头
~ pin 销钉,定位销,合模销
~ pin bushing 导销套
down 降下,落下,降低
~ channel flow 顺螺槽流
~ - cut 切下
~ pipe 落水管
~ pipe clip 落水管夹箍〈塑料管〉
~ pipe connector 落水管接头〈塑料管道〉
~ - pipe spouts 落水管水槽,落水管

流道
- ~ stream 顺流
- ~ stroke (活塞)向下行程
- ~ - stroke press 下压式压机
- ~ - stroke transfer plupger 下压式压铸成型活塞
- ~ time 停机时间〈机器〉;更换时间〈印刷〉

downstream equipment 〈作业线上〉下游设备,下步骤设备,后续设备

downward 向下的,下降的
- ~ extrusion 向下挤塑
- ~ pressure 向下压力

dowtherm sump 联苯储槽,道氏热载体储槽

DOZ [dioctyl azetate] 壬二酸二辛酯〈增塑剂〉

DP [degree of polymerization] 聚合度

DPA [diphenyl - amine] 二苯胺

DPCP [diphenyl cresyl phosphate] 磷酸二苯基甲苯酯

DPOP [diphenyl octyl phosphate] 磷酸二苯辛酯〈增塑剂〉

DPP [diphenyl phthalate] 邻苯二甲酸二苯酯〈增塑剂〉

DPPD [N,N' - diphenyl - p - phenylenedi - amine] N,N'-二苯基对苯二胺〈防老剂 H〉

draft 斜度,脱模斜度〈脱模〉;牵伸,拉伸;草图[案];通风,气流
- ~ angle 斜度角;锥度角
- ~ constant 牵伸常数
- ~ distribution 牵伸分配
- ~ field 牵伸区
- ~ ratio 牵伸比〈美〉
- ~ resistance 牵伸阻力
- ~ test 牵伸试验
- ~ tube 通流管,通风管
- ~ zone 牵伸区

drafting 牵伸;绘图;制图
- ~ film 绘图薄膜
- ~ roller 拉伸棍

drag 牵引;阻力
- ~ bolt 拉紧螺栓
- ~ coefficient 牵引[阻力]系数
- ~ flow 正流,立流,顺流,推进流〈挤塑机〉;黏性流
- ~ flow along channel 沿槽沟推进流〈挤塑机〉
- ~ flow constant 顺流常数,推进流常数
- ~ flow pump 直流泵
- ~ force 曳力
- ~ line 拉伸痕
- ~ mark 拉伸痕
- ~ of film 薄膜的阻力系数
- ~ reducer 减阻剂
- ~ reducing agent 减阻剂
- ~ roll 牵引[拉伸,空转]辊
- ~ strut 阻力支撑
- ~ test 牵伸试验

dragging 咬模,啃模
- ~ force 曳力

drain 排水,排泄;排水管,污水管
- ~ bin 排水库
- ~ board 排水板,滴水板
- ~ box 排水箱
- ~ pipe 排泄管
- ~ valve 排泄阀,卸料阀

drainage 排水,排泄
- ~ capacity 排水功率
- ~ correction 挤塑体积校正

draining 排水,排泄;滴干
- ~ board 干燥盘;滴水板
- ~ trench 排水沟

drape 悬垂性;覆盖
- ~ assist frame 助压型框,悬框辅助法
- ~ assist frame forming 型框助压成型
- ~ forming 包模真空成型,包模成型
- ~ forming with bubble stretching 气胀包模成型
- ~ forming with draw - grid prestretch (multiple mo(u)ld) 撑框预拉伸多模包模成型
- ~ mould 包覆模
- ~ mo(u)lding 包模(真空)成型
- ~ thermoforming 包模热成型
- ~ vacuum forming 包模真空成型
- ~ vacuum thermoforming 包模真空热

成型
draping 包模成型
draught [draft] 斜度〈脱模〉;通风
~ excluder 防通风器
~ -free chamber 敞开通风室
draw 斜度〈脱模〉;拉伸,牵伸;成型深度;回火
~ die 拉伸模,深拉模;拉拔模;拉丝模
~ direction 牵伸方向
~ -down (型坯)垂伸,预拉伸;刮涂;减少,消耗
~ down ratio 牵伸比,垂伸比
~ down speed 牵伸速率
~ forming 拉伸成型,牵伸成型
~ grid 拉伸框架
~ grid method 拉伸框格法
~ -in bolt 拉紧螺栓
~ -in gear 颗粒供给装置,喂料装置
~ mo(u)ld 铸(塑)模;薄壁模子
~ -off roll 牵伸辊,牵引辊,滚压辊
~ -off unit 牵引装置
~ plate 拉模板
~ press 深冲压机
~ ratio 拉伸比,〈美〉牵伸比
~ ring 拉环
~ resonance 拉引共振
~ roll(er) 拉伸辊,牵伸辊;喂料辊
~ -roll take-off 索引辊引出装置
~ speed 牵伸速度
~ spinning 纺丝拉伸
~ stand 拉伸架,牵伸架
~ stand with up-roll 压辊牵伸架
~ take-up unit 牵伸引出装置
~ tube 伸缩管
~ twister 拉伸加捻机〈纺丝〉
~ twisting 拉伸加捻
drawability 拉伸性;塑性;回火
drawback 放出,排出;卸下,取下;缺点,短处
~ ram 回拉活塞
drawing 拉伸,牵伸;拉伸成型;深拉;冲压成型
~ behaviour 拉伸状态
~ degree 拉伸度,伸展度
~ depth 深冲深度
~ die 拉伸模,深拉模;拉拔模;拉丝模
~ disk 牵伸定径板
~ down 拉细,缩小横截面
~ frame 拉伸框架,深拉框架
~ -in speed 引入速度,供给速度
~ machine 拉伸机
~ -off roller 牵拉辊
~ -off speed 牵伸速度
~ press 深拉压力机,引伸压力机
~ quality 深冲性
~ rate [speed] 牵伸率[速度]
~ ratio 牵引比;牵伸比
~ roller 拉伸辊,牵伸辊
~ stress 拉伸压力
~ zone 牵伸段〈薄膜〉
drawn 拉伸的
~ pin type insert 空心嵌销
~ plastic 牵伸塑料
~ plastic film 牵伸塑料膜
~ resonance 拉伸共振,牵伸共振
~ shell type insert 空心壳型嵌件
~ tube 拉制管
~ up 卷取
~ yarn 拉伸纱
dress 修饰,修整
dressing 修整,整修
~ agent 修饰剂;上光剂
~ machine roll 修饰辊;压光辊
dried [dryed] 干燥的
~ stock 待干燥物;被烘干物;干材料
drier [dryer] 干燥剂;干燥器
drift 位移,移动;变化,变动;弹性后效
~ punch 焊接工具延伸杆〈超声波焊接〉
Drill 钻;钻孔;打孔器
~ -chuck spring ring 钻孔机夹紧环
drilled roll 钻孔辊,钻孔式辊筒
drilling 钻孔
~ machine 钻床,钻机
drip pan 收滴器;回收器
dripless 无液滴的

drive 传动装置
- ~ mechanism 传动机构
- ~ roll 传动辊
- ~ screw 传动螺杆
- ~ shaft 传动轴
- ~ side 传动侧

driven 传动的,从动的
- ~ roll 驱动辊,传动辊〈压延机〉;可控辊
- ~ roller conveyer 辊式传动运输机
- ~ spindle 从动转轴

driving 传动(的)
- ~ end 传动侧
- ~ power 驱动力
- ~ shaft 传动轴

drooling 流诞,滴料

drop 滴;降落,下降
- ~ - away checking device 脱模保险装置
- ~ door 下落式卸料门
- ~ forming 模塞助压成型
- ~ - hammer die 落锤模〈加工金属用铸塑模具〉
- ~ hammer die 落锤冲模〈用于塑料半成品〉
- ~ in viscosity 黏度降
- ~ point 滴点
- ~ test 坠落试验
- ~ - weight test 落锤(冲击)试验

dropdown 泄降,降落;(型坯)下垂

droplet (小)滴

dropper 滴管
- ~ bottle 滴瓶
- ~ cap [dropper closure] 滴封,滴停

dropping 降落(的)
- ~ bolt test 降落螺栓试验
- ~ pipe 下落管,下流管
- ~ point 滴点

drum 转鼓,转筒
- ~ cleaning roll 转鼓式清洁辊
- ~ colo(u)ring 转鼓着色
- ~ cooler 转鼓式冷却器,冷影鼓
- ~ cooling device 转鼓式冷却装置
- ~ dryer 转鼓式干燥器
- ~ dryer with dip tank 附有浸渍槽的转鼓式干燥器
- ~ drying 转鼓式干燥
- ~ dumper 转鼓式倾倒器
- ~ dyeing 转鼓染色
- ~ feeder 转鼓加料机
- ~ filter 转鼓式过滤器,滤鼓
- ~ friction test 转鼓摩擦试验
- ~ grinding 转鼓式研磨
- ~ heater 转鼓式加热器
- ~ mill 转鼓研磨机
- ~ mixer 转鼓式混合机
- ~ polisher 转鼓式抛光器
- ~ polishing 转鼓抛光(压模制)
- ~ polishing technigue 转鼓抛光技术
- ~ sander 转鼓式砂磨机
- ~ sealer 转鼓式热合机
- ~ separator 转鼓式分离器
- ~ surface winder 转鼓缠绕机
- ~ take - up 鼓式卷绕
- ~ tumbler 转鼓混合机(美)

drumming 转鼓加工法

drw. [drawing] 图,草图

dry 干燥(的)
- ~ adhesive 干态黏合剂(美)
- ~ adhesive layer 干胶层,干黏层
- ~ - area 干斑
- ~ - bag method 干袋模压法
- ~ binder 干粘结剂
- ~ blend 干混料;干混配,干混合,干掺和
- ~ blend colouring 干混着色
- ~ - blend extrusion 干混料挤塑
- ~ blender 干混机
- ~ blending 干混,干掺和
- ~ bonding 干黏合
- ~ bulb temperature 干球温度
- ~ colo(u)r 干色料
- ~ colo(u)ring 干混着色
- ~ cycle 空循环
- ~ cycle figure 空循环值
- ~ - cycle time 空循环时间(注塑机)
- ~ cylinder 干燥筒
- ~ edge 干边

~ elongation   干态伸长;干态伸长率
~ end   干燥收卷端
~ extract   挤干;干提出物干萃取物
~ extract of aqueous dispersions   水状分散体挤干
~ extrusion   干式挤塑
~ feel   干(燥)感
~ filler   干填充剂
~ film adhesive   干膜黏合剂;胶膜
~ - film laminating   干模层合,干模复合
~ film thickness   干膜厚度
~ flex   干状挠曲,耐干弯曲强度(层压塑料)
~ friction   干摩擦
~ gasifier   干气化器
~ gluing   干态胶黏
~ grinder   干磨机
~ grinding   干磨
~ - hard   指检干硬化,硬干
~ heat curing   干热空气固化
~ hiding   干遮盖力(树脂含量低的涂料)
~ hiding effect   干涂盖效应(美)
~ ice   干冰
~ - jet wet spinning   干喷湿纺法
~ laminate [laminating]   干式层合;欠胶层合(塑料)
~ lamination   干式层合
~ layup   干式叠合
~ layup lamination   干式叠合层压
~ main   干总管
~ mix compound   干混料(塑料成型料)
~ mixer   干混机
~ - out spraying   干粉喷雾
~ patch   干斑(点)
~ pelletizing method   干式造粒法
~ powder blend   干粉混配料,干粉混合料
~ printing   干印,烫印
~ - process   干法成型
~ product   干(材)料,干产品
~ room conditions   干燥室内环境(29.

5℃,相对湿度15%)
~ running behaviour   干运转性能
~ running performance   干运转性能(塑料滑动轴承)
~ running speed   干运转速率[度]
~ sieving   干筛分
~ solid content   干含量部分,干固体含量
~ spinning   干法纺丝,干纺(化纤)
~ spot   干斑;脱胶
~ strength   干强度
~ substance   干燥物质
~ tack   干黏态,干黏性
~ tensile strength   干态拉伸强度
~ test   干燥试验
~ to handle   手感干燥(黏合剂,漆)
~ - transfer laminating   干转移复合(黏合)
~ wax paper   干态蜡纸
~ web impregnation   干卷料浸渍
~ weight   干重
~ winding   干法缠绕
dryer   干燥剂;干燥器
~ bed   干燥床
~ with agitator   搅拌式干燥器
drying   干燥(的)
~ agent   干燥剂
~ box   干燥箱
~ by evaporating of solvents   溶剂挥发干燥
~ cabinet   干燥箱
~ cabinet with recycle of air   循环空气干燥箱
~ carriage   干燥托架
~ chamber   干燥室;烘房
~ compartment   干燥室
~ condition   干燥条件
~ cupboard   干燥柜;橱式干燥机
~ cycle   干燥周期
~ cylinder   干燥筒
~ drum   干燥鼓
~ in circulating air   循环空气干燥
~ - in the open air   户外空气干燥,露天干燥

~ machine 干燥机
~ oil 干性油
~ oven 干燥(烘)箱
~ plant 干燥装置
~ process 干法
~ rack 干燥架,烘干架
~ rate 干燥速率;干燥速度
~ room 干燥室
~ section 干燥区段
~ shrinkage 干燥收缩率
~ speed 干燥速率,干燥速度
~ stock 干(材)料
~ temperature 干燥温度
~ time 干燥时间
~ tower [vault] 干燥塔
~ tumbler 转鼓式烘燥机,烘干转筒
~ tunnel 干燥隧道,烘道
DS [deep screw] 深螺纹螺杆
DSC [differential scanning calorimetry] 示差扫描量热法
DTA [diethylene triamine] 二亚乙基三胺(固化剂)
DTA [differential thermal analysis] 差热分析
DTBP, DBP [di-*tert*-butyl peroxide] 二叔丁基过氧化物,过氧化二叔丁基[引发剂A]
DTDP [ditridecyl phthalate] 邻苯二甲酸双十三烷基酯(增塑剂)
DTDTP [ditridecyl thiodipropionate] 硫代二丙酸双十三烷基酯(抗氧剂)
DTG [derivative thermogravimetry] 微分热重法,导数热重量分析法
dual 双(的),复式的
~ cavity mo(u)ld 双模槽模具
~ channel screen changer 双流道换滤网机
~ compartment can 双室盒(包装)
~ die extrusion system 双模头挤出装置
~ drive 双传动
~ drum haul-off 双筒牵引
~ extrusion 双色挤塑;双机挤塑;串联挤塑机挤塑

~ extrusion die 双料挤塑模头;双色挤塑模头
~ fillet weld 双面角焊缝
~ helical winding 双螺旋形卷绕
~ nozzle spray gun 双嘴喷枪;双组分喷枪
~ orientation [biaxial orientation] 双轴取向(平板)
~ plunger mo(u)ld 双模塞式模具
~ polar wind 双极缠绕;交叉缠绕
~ polar winding 双极缠绕法
~ purpose additive 双效添加剂
~ scarf joint 双嵌接,双斜接头
~ spray gun 双嘴喷枪;双组分喷枪
duct 导管,管道
~ thermostat 温度调节器
~ winding 管材卷缠
ductile 延性的;韧性的
~ adhesive 延性黏合剂
~ -brittle transition 韧脆转变;冲击韧性
~ -crack-growth fracture 延性裂纹增长断裂
~ failure 韧性破坏
~ fracture 韧性破裂
ductileness 延性
ductility 延性;延伸度,韧性
~ test 延展性试验
ducting 管道
dull 无光(泽)的
~ clear varnish 半透明无光漆,无光泽透明漆(美)
~ film 无光膜
~ finish 无光泽修饰,消光修饰;消光剂
~ finish lacquer 无光泽(罩)面漆
~ lacquer 消光漆,暗漆,无光泽漆
~ polish 无光泽的
~ surface 无光面
dulled 无光泽的;缓和的
dulling agent 消光剂
dumb-bell 哑铃
~ model 哑铃状模型
~ sample 哑铃形试样

~ shaped 哑铃形[状]的
~ specimen 哑铃形试样
~ swell test 哑铃形膨胀试验
~ test piece 哑铃形试(验)片
dummy 模造物;虚的,假的,空的;空包装(包装)
~ hole 盲孔
~ rivet 临时铆钉;定位铆钉
dump 倾倒,卸料;堆放
~ bin 废料箱;储料器
~ valve 底阀,放泄阀
~ wagon 翻斗车
dumped 废弃的,弃烛的,堆积的
~ column packing 不规则柱式模料
~ tower packing 不规则塔模块
dumping 储存(废塑料)
Dunlop 邓洛普
~ fatigue tester 邓洛普疲劳试验仪
~ flexometer 邓洛普揉曲仪,邓洛普疲劳试验仪
DUP [diundecyl phthalate] 邻苯二甲酸双十一烷基酯(增塑剂)
duplex 双(的),双联(的),复式(的)
~ beater 复式打浆机,双辊打浆机
~ extrusion line with direct gassing 直排气式双联挤塑生产线
~ fibre 复合纤维
~ film adhesive 复合膜用黏合剂
~ film blowing 复合膜吹塑
~ mo(u)lding 双柱塞压铸成型
duplicate 重复;双联,成对
~ cavity plate 双联阴模板
~ cavity-plate mo(u)lding 双联阴模板成型
~ injection 双联注塑
~ plate 双联模板
~ plate mo(u)lding 双联模板成型
~ stripper-plate 双联脱模板
~ test 重复试验
duplicating milling machine 仿形磨削机,仿形铣床
durability 耐久性,耐用性
~ test 耐久性试验,疲劳试验,寿命试验

durable 耐久的
~ product 耐用品
~ setting 耐久性定形
Duracon (商)杜拉松(聚缩醛)
duration 耐久,耐用;时间,期间
~ of coagulation 凝结时间
~ of runs 运转时间
~ of service 设备使用年限
duromer 热固性塑料
Duromer foam process 道罗漫尔发泡法(热固性结构泡沫塑料)
durometer 硬度计
~ hardness 硬度计测定硬度
dust 灰尘
~ arrestor 吸尘器
~ cap 防尘盖
~ collector 集尘器,吸尘器
~ cover 防尘罩
~ deposit 灰尘沉积
~ elimination 除尘
~ emission 灰尘散发
~ exhausting [extraction] plant 吸尘装置,除尘装置
~ mask 防尘罩,防尘面具
~ pickup test 吸尘试验
~ removal 除尘
~ remover 除尘器,旋风除尘器
~ respirator 防尘面具
~ separator 防尘器
~ shield 防尘罩
~ sucking plant 吸尘装置
~ -tight 绝尘的,防尘的
dusting 除尘;撒粉;涂粉
~ machine 除尘器
dustproof 防尘,绝尘
Dutch 荷兰的
~ liquid (or oil) 荷兰液(或油)(二氯化乙烯)
~ tiles 荷兰砖瓦(硬纤维板制砖瓦)
~ top 荷兰顶盖
duty 功(率);运行,运转(机器)
DVB [divinyl benzene] 二乙烯基苯(交联剂)
dwell 停压;保压;延长闭模时间

dwelling

~ time 停压时间;保压时间;烫压时间;静态
~ angle 静止角(长丝缠绕)
~ pressure [after-pressure] 继续压力(注塑)
~ tack [permanent tackiness] 持久黏性(黏合)
~ time 静止时间(长丝缠绕);继续压力时间(注塑);滞留时间(机器)

dwelling 停压,保压

dye 染料;染色,着色
~ affinity 染料亲和力
~ dispersing agent 染料分散剂
~ levelling 染料的均染性
~ strength 染料强度

dyes and pigments 染料和颜料

dyeing 染色(的)
~ ability 着色力
~ behaviour 染色性,着色性

dyestuff 染料;染色剂,着色剂

Dynacote process 杜那可特派(借助于中子射线处理表面涂层法)

dynamic 动力的,动态的
~ compliance 动态柔量
~ creep 动态蠕变
~ cure 动态特性曲线
~ elastic modulus 动态弹性模量
~ elasticity 动态弹性
~ fatigue 动态疲劳
~ fatigue resistance 耐动态疲劳性
~ fatigue test 动态疲劳试验
~ flexing 动态挠曲
~ friction 动摩擦
~ infrared dichroism 动态红外二向色性
~ load 动态负荷
~ loading 动态负荷
~ long-term load 动态长期负荷
~ loss 动态损耗
~ loss factor 动态损耗因数
~ loss tensile modulus 动态拉伸耗能模量
~ mechanical analysis [DMA] 动态力学分析
~ mechanical frequency spectrum 动态力学频率谱
~ mechanical loss spectrum 动态力学损耗谱
~ mechanical loss tangent 动态力学损耗角正切
~ mechanical spectroscopy [DMS] 动态力学谱
~ mechanical test 动态为学试验
~ modulus 动态模量
~ modulus of elasticity 动态弹性模量
~ pressure 动压力
~ resilience 动态弹性回复
~ rigidity 动态刚度;动态刚性;动态剪切模量
~ shear compliance 动态剪切柔量
~ storage shear compliance 动态剪切储能柔量
~ strength test(ing) 动态强度试验
~ stress 动态应力
~ striation screw 动态细沟状螺杆
~ structural test 动态结构试验
~ test 动态试验
~ testing machine 动态试验机
~ thermomechanical measurment 动态热机测量
~ viscosimeter 动态黏度计,动力黏度计
~ viscosity 动态黏度,动力黏度,绝对黏度

dynamics 动力学

dynamometer 测力计;功率计
~ test 测功试验,测力试验

Dynstat 戴恩斯塔特
~ method 戴恩斯塔特法((缺口)冲击韧性试验)
~ tester 戴恩斯塔特法试验仪
~ type impact tester 戴恩斯塔特型冲击试验仪

E

E - glass E 玻璃[低碱硼硅酸盐玻璃]（玻璃纤维增强塑料用）
E - glass fiber E 玻璃纤维，无碱电绝缘玻璃纤维
E - Pack system E 包装系统(由环氧树脂粉末固化剂制备的片状胶黏剂)
E - PVC [polyvinyl chloride polymerized in emulsion] 乳液法聚氯乙烯
EAA [ethylene - acrylic acid copolymers] 乙烯-丙烯酸共聚物
early colo(u)r 早期着色
earth 地面;土
~ colour [pigment] 矿物颜料,无机颜料
ease 容易;减轻;放松
~ of ignition 易点燃性
~ of processing 易加工(性)
easel 架台
easily ignitible 易燃的
easy 容易的,轻便的
~ flow 易流动
~ - flowing gronules 易流动颗粒(料),松散粒料
~ manual operation 易手动操作
~ processing 易加工
~ processing channel black 易加工槽黑
~ processing resin 易加工树脂
EB [electron beam] 电子束
EBA [ethylene butylacrylate] 乙烯丙烯酸丁酯
ebbulition 煮沸;沸腾;蒸馏
EBC process [electron beam curing] 电子束固化法
EBM [extrusion blow mo(u)lding] 挤出吹塑模塑
ebonite 硬质橡胶,硬质胶,硬橡皮
ebullated bed 沸腾床;流化床
ebullator 沸腾器

ebulliometer 沸点升高仪
ebulliometry 沸点升高测定法
ebullition 沸腾
EC [ethyl cellulose] 乙基纤维素
eccentric 偏心器;偏心的
~ axis of rotation 偏心旋转轴
~ fitting [off - centre fitting] 偏心配件
~ press 偏心式压机
~ shaft 偏心轴
~ tabletting machine 偏心式压片机
~ toggle (lever) press 偏心肘杆压机
~ tumbler 偏心转鼓,转筒式混合器
~ rumbling mixer [offset rotary tumbler] 偏心转鼓混合器
eccentricity [imbalance] 偏心率,偏心度
ecological 生态学的
ecology 生态学
economic 经济的;节俭的
~ material use 节约用料
economical utilization 经济利用(率)
economy in materials 节约用料
ECTFE [ethylene - chlorotrifluoroethylene copolymer] 乙烯-三氟氯乙烯共聚物
ed. editor 编者
ed. edition 版本,出版
eddy 涡流
~ current 涡流;涡电流
~ current heater 涡电流加热器
~ current heating 涡电流加热;感应加热
~ current loss 涡流电损耗
edge 边缘,棱边;刀刃;侧面
~ banding 边缘镶饰(如:胶合板)
~ bowing 边缘弯曲,旁边
~ control 边缘控制
~ crack 边缘裂缝
~ effect 边缘效应

- ~ face gate   边缘浇口
- ~ fan gate   边缘扇形浇口
- ~ feed   边缘进料
- ~ finishing   折叠;卷边
- ~ fold   卷边,折边
- ~ forming   卷边,折边
- ~ forming roll   卷边辊
- ~ gate   边缘浇口,侧向浇口(美)
- ~ gluing   对边胶合,拼接胶合
- ~ guide equipment   导边器
- ~ joint   拼接接头
- ~ joint adhesive   拼接黏合剂
- ~ – jointing adhesive   拼接黏合剂
- ~ lighting   侧缘折光
- ~ mark   边缘缺陷
- ~ mill   轮碾机
- ~ preparation   焊接缝准备(焊接)
- ~ round [top radius]   棱边圆形(齿轮)
- ~ runner [runner muller, muller wheel]   磨轮,轮碾机
- ~ – runner mill   轮碾机
- ~ runner mixer   轮碾混合器
- ~ – runner type pan mixer   碾盘式混合器
- ~ sealing   封边
- ~ trim removal   修边,去边
- ~ trim roll   修毛边辊
- ~ trimmer   修边机
- ~ veneering   镶边
- ~ weld   端面焊接;对边焊
- ~ welding   边焊

edgewater   防黏剂,防黏层
edgewise   侧向(层压塑料的);沿层方向(层合板)
- ~ compression   沿层压缩
- ~ compressive strength   侧压强度
- ~ load   沿层负荷

edging   边缘,窄边,磨(修)边
- ~ profile   镶边型材;封边型材
- ~ strip   边(狭)条,窄边条

eductor   喷射器
- ~ tube [dip rube]   喷射器管

EEA [ethylene – ethyl acrylate]   乙烯 – 丙烯酸乙酯
EEA resins [ethylene acrylate copolymers]   乙烯 – 丙烯酸酯共聚物
effect   作用;效应,影响,结果
- ~ of heat   热效应
- ~ varnish   美饰漆,真空涂漆

effective   有效的
- ~ area   有效面积
- ~ capacity   有效容量;有效功率
- ~ efficiency   效率,有效功能
- ~ L/D ratio [~ l/d ratio]   有效长径比
- ~ output   有效功率,实际生产量
- ~ range   有效范围
- ~ screw length   螺杆有效长度
- ~ surface   有效表面
- ~ temperature   有效温度
- ~ tension   有效张力
- ~ thread   有效螺纹
- ~ thread turns   有效螺纹终端
- ~ viscosity   有效黏度
- ~ working surface   有效工作(表)面

effectiveness   效率,效能
efficiency   效率,利用率
- ~ test   效率试验

efficient heating space   有效加热空间
efflorescence   风化;粉化;起霜
effluent   流出的;流出物;废水;废液,污水
- ~ drainpipe   废水排泄管
- ~ gas analysis   发散气体分析
- ~ gas detection   发散气体检测
- ~ treatment   污水处理,废水处理

efflux   流出(物);废气流
- ~ velocity   流出速度
- ~ viscometer   流动黏度计

e. g. exempli gratia   (拉)例如
EGA [evolved gas analysis]   逸出气体分析
EGD [evolved gas detection]   逸出气体检测
egg   蛋
- ~ flat   蛋纸板箱(包装)

eggs crate   蛋品周转箱
eguoutteur   水印压机

EHMW polyethylene [extra-high molecular weight polyethylene] 超高分子量聚乙烯

EHMWPE [extra high molecular weight polyethylene] 超高分子量聚乙烯

Eickhoff extruder 艾克霍夫型挤塑机,行星滚柱挤塑机

eject 喷射出;顶出(模具)

ejecting 推出的,挤出的
- ~ blanking cutter 推顶式切料器
- ~ device 推顶装置
- ~ out-line blanking die 推顶式外形冲切模
- ~ press 挤压机

ejection 脱模;推顶,顶出;挤出
- ~ bar 脱模板
- ~ connecting bar 推顶连接杆
- ~ control unit 推顶控制系统
- ~ cross bar 推顶横杆
- ~ force 推顶力
- ~ frame 顶出框架
- ~ gride 顶出框
- ~ knock-out 脱模
- ~ mark 脱模销痕
- ~ pad 推顶衬垫
- ~ period 推顶时间
- ~ pin 顶出杆,推顶销,脱模销
- ~ pin assembly 脱模销总成
- ~ plate 推顶板,脱模板
- ~ point 推顶点,脱模点
- ~ ram 推顶活塞,脱模活塞
- ~ rod 脱模板拉杆
- ~ side 脱模侧
- ~ sleeve 推顶套筒
- ~ temperatuer 脱模温度
- ~ tie bar 推顶连接杆,脱模连接杆
- ~ time 推顶时间

ejector 推顶器,顶出器,脱模销;喷射器
- ~ backplate 脱模销座板
- ~ bar 脱模板
- ~ bar stop pin 脱模板限位销
- ~ blade 扁顶杆
- ~ bushing 推顶套(注塑)
- ~ bushing plate 推顶套板(注塑)
- ~ connecting bar 推顶连接杆;脱模连接杆
- ~ die (顶销式)冲(切)模
- ~ frame 推顶框架,脱模框架
- ~ frame guide 顶框导板
- ~ frame guide bar 顶框导杆
- ~ -free position stroke 推顶解锁行程
- ~ gride 顶出框
- ~ housing 脱模空档
- ~ locking plate 脱模销压板
- ~ marks 脱模销痕
- ~ mechanism 脱模机构
- ~ on nozzle side 注料嘴侧脱模销
- ~ operating arm 推顶操纵臂,脱模操纵臂
- ~ operating clearance 脱模销工作间距
- ~ operating mechanism 脱模销驱动机构
- ~ operating rod 脱模销驱动杆
- ~ pad 脱模垫
- ~ pin 推顶销,脱模销
- ~ pin assembly 脱模销总成
- ~ pin bar 脱模销板
- ~ pin mark 脱模销痕
- ~ pin plate 顶出板
- ~ pin retainer plate 脱模销压板
- ~ pin type injection mould 推顶式注塑模
- ~ plate 推顶板,脱模板
- ~ plate return pin 脱模板回程销
- ~ plug 脱模顶栓
- ~ program 推顶程序,推模程序
- ~ rail 推顶杆,脱模杆
- ~ ram 推顶活塞,脱模活塞
- ~ return pin 顶回杆
- ~ return spring 脱模板回程弹簧
- ~ rod 脱模板拉杆
- ~ rod clamp screw 顶出固定螺丝
- ~ rod flange 推顶杆法兰
- ~ side 顶出侧,脱模侧(注塑)
- ~ side mould part 顶出侧模具配件,脱模侧模具配件(注塑模)
- ~ sleeve 套筒脱模销
- ~ spring 推顶出(回程)弹簧

elastic

~ travel 脱模板行程
elastic 弹性的
　~ after effect 弹性后效,滞后弹性
　~ anisotropy 弹性各向异性
　~ body 弹性体
　~ coefficient 弹性系数
　~ compliance 弹性柔量
　~ component 弹性变形部分
　~ constant 弹性常数
　~ cord 弹性绳带
　~ counter bowl [backing roll] 弹性背辊
　~ deformation 弹性变形
　~ deformation limit 弹性变形极限
　~ distortion 弹性畸变
　~ drift 弹性后效
　~ elongation 弹性伸长
　~ energy 弹性能
　~ entrance effect 弹性流入效应
　~ expansion 弹性膨胀
　~ extension 弹性延伸
　~ fabric 弹性织物
　~ failure 弹性破坏
　~ fatigue 弹性疲劳
　~ fiber 弹性纤维
　~ filler 弹性填料
　~ fluid 弹性流体
　~ force 弹力
　~ fore-effect 弹性前效
　~ gel 弹性凝胶
　~ hardness 弹性硬度
　~ hysteresis 弹性滞后
　~ impact 弹性冲击
　~ isotropy 弹性各向同性
　~ limit 弹性极限
　~ melt extruder 弹性熔体挤塑机
　~ melt swelling 熔体弹性膨胀
　~ memory 弹性记忆
　~ modulus 弹性模量
　~ modulus in tension 拉伸弹性模量
　~ number 弹性模量;弹性模数
　~ nylon 弹性尼龙,弹性锦纶
　~ plastics 弹性塑料
　~ potential 弹性势

　~ property 弹性特性
　~ range 弹性范围
　~ recovery 弹性回复,弹性复原
　~ recovery of curl 卷曲弹性回复
　~ recovery percentage of elongation 拉伸弹性回复率
　~ relaxation time 弹性松弛时间
　~ resilience 回弹性
　~ ring 弹性垫圈
　~ scattering 弹性散射
　~ shear 弹性剪切
　~ shear deformation 弹性剪切形变
　~ snap-back 弹性复原
　~ stage 弹性状态
　~ strain 弹性应变
　~ strain energy 弹性应变能
　~ stress 弹性应力
　~ stretching 弹性伸长
　~ torsion 弹性扭力
elasticator 增弹剂;增塑剂;增韧剂;(抗冻)软化剂
elasticity 弹性
　~ of compression 压缩弹性
　~ of elongation 伸长弹性
　~ of flexure 弯曲弹性
　~ of shape 形态弹性
　~ of torsion 扭转弹性,扭曲弹性
　~ test 弹性试验
elasticlzer 增弹剂
elastodurometer 弹性硬度计
elastodynamic 弹性动力的,弹性动态的
　~ extruder 弹性动力式挤塑机
　~ screwless extruder 弹性动态无螺杆挤塑机
elastomer 弹性体
　~ -metal compunod 弹性体-金属复合物
elastomeric 弹性体(的)
　~ adhesive 弹性体型黏合剂
　~ compounds 弹性体混合料
　~ copolymer of ethylene and propylene [EPM] 乙烯-丙烯共聚弹性体(单体),二元乙丙橡胶(单体)
　~ fibre 弹性纤维

~ plastic　弹性塑料
~ polymer　弹性聚合物
~ polyurethane with polyester segments
　聚酯嵌段的聚氨酯弹性体
~ polyurethane with polyether segments
　聚醚嵌段的聚氨酯弹性体
~ – polyvinil chloride　弹性聚氯乙烯
elastometer　弹性计,弹性仪
elastoplastics　弹性塑料
~ system　弹性塑料系列
elastoplasticity　塑弹性
elastopolymer　弹性高聚物
elastoviscosimeter　黏弹仪,弹性黏度计
elbow　肘管,弯头
~ joint　弯(管接)头
elec. electric　电(气)的
elec. electrical　电(气)的
electret　驻极体
electric　电的
~ arc welding　电弧焊
~ band heater　电加热带,带式加热器
~ blanket heating　电热板覆盖加热
~ bonding machine　电焊机;连续热合
　机〈美〉
~ breakdown endurance　耐击穿电压
~ breakdown strength　击穿电压
~ cable manufacture　电缆制造
~ charge　电荷〈静电涂布〉
~ circuit　电路
~ conductance　电导
~ conduction　导电
~ conductivity　电导率;电导性
~ conductor　电导体
~ discharge　放电
~ dissipation　电气耗散
~ drill　电钻
~ eye control　电眼控制,光电池控制
~ field strength　电场强,电场强度〈高
　频焊接时〉
~ furnace [electric crucible] 电炉
~ fusion　电熔焊〈管〉
~ heating element　电热体
~ hot wire cutting　电热丝切割
~ operation　电动操作

~ oven　电烘箱
~ strength　介电强度〈电〉
~ trolley　电瓶车
~ vibration sifter　电振动筛
~ welding　电焊
electrical　电的
~ ageing　电老化
~ breakdown　(介)电击穿
~ conductance　导电性
~ conductivity　电导率,导电性
~ conductivity detector　电导检测器
~ control panel　电控制板,配电箱
~ discharge detector　放电检测器
~ insulation　电绝缘
~ insulation compound　电绝缘料
~ – insulation fibre　电绝缘纤维
~ insulation compound　电绝缘料
~ insulation material　电绝缘材料
~ insulation paint　电绝缘漆
~ insulation varnish　电绝缘清漆
~ jar [Leyden jar]　电瓶
~ radiation　电离辐射
~ resistance　电阻
~ resistance heating　电阻加热
~ resistance wire　电阻丝
~ strength　介电强度
~ stress　介电应力
~ test　电性试验
~ treating　电处理,电晕放电处理
~ treatment　电处理,电晕放电处理
　〈薄膜表面〉
~ volume resistance　体积电阻
~ volume resistivity　体积电阻率
electrically　电
~ actuated　用电操纵的〈阀〉
~ conductive pressure – sensitive adhe –
　sive tape　导电压敏胶带
~ heated hot gas welder　电(加)热(热
　气)焊机
~ heated jacket　电加热套
electricity　电,电学,电气
~ main　电干线
~ meter board　电量计配电盘
electro　电

electrocoating

~ analysis 电解分析
~ coating 电(泳)涂覆
~ oxidation synthesis 电(解)氧化合成
~ electrochemical 电化(学)的
~ corrosion 电化学腐蚀
~ initiation 电化学引发
~ machining 电化学加工,电解加工
~ polishing 电化学抛光,电解研磨
~ polymerization 电化学聚合
~ pretreatment 电化学预处理
electrocoating 电泳涂装
~ jig 焊条夹具
~ screen 电极屏蔽
~ size 电极尺寸〈高频焊接〉
~ tip 电极头
electrodeless 无电极的
~ discharge process 无电极放电法
~ glow discharge 无电极辉光放电
electrodeposit 电解沉积
electrodeposited coating 电镀层
electrodeposition 电沉积
~ coating 电积附涂层
~ process 电沉积法
electroendosmosis 电(内)渗(现象)
electroerosive mould forming 电浸蚀模具成型
eletroformed 电铸的
~ cavity 电铸阴模
~ mo(u)ld 电铸模具
electroforming 电铸
electrohydraulic 电动液压的
electroinitiated cationic polymerization 电引发阳离子聚合
electroinsulating property 绝缘性能
electrokinetics 电动力学,动电学
electroless plating 无电镀,电解质
electrolysis 电解(作用);电分析
electrolyte 电解液,电解质
electrolytic 电解的
~ polishing 电解抛光
~ polymerization 电解聚合
electromagnetic 电磁的
~ adhesion 电磁粘接
~ adhesive 电磁黏合剂

~ induction welding 电磁感应焊接
~ interference [EMI] 电磁干扰
~ radiation polymerization 电磁辐射聚合
~ separator 电磁分离器
~ vibratery feeder 电磁振动加料器
electrometallization 电镀金属〈塑料〉
electromotive force [e. m. f.] 电动势
electron 电子
~ - accepting monomer 电子受体单体
~ acceptor 电子受体
~ affinity 电子亲合(性)
~ atom ratio 电子原子比
~ beam [EB] 电子束,电子射线
~ beam crosslinking 电子束交联
~ beam curing [EBC Process] 电子束固化法
~ beam microanalysis 电子束微量分析
~ bombardment 电子轰击〈静电镀敷〉
~ crosslinking [electron - beam crosslinking] 电子束交联
~ curing [electron - beam curing] 电子束固化法
~ diffraction 电子衍射
~ - donating molecule 给电子型分子
~ donor 电子给体
~ exchange polymer 电子交换聚合物
~ exchange resin 电子交换树脂
~ - lens 电子透镜
~ microscope [EM] 电子显微镜,电镜
~ microscope photography 电子显微相片
~ microscopic examination [study] 电子显微镜检验
~ microscopy 电子显微术
~ optics 电子光学
~ paramagnetic resonance [EPR] 电子顺磁共振
~ - probe [EP] 电子探针
~ - probe microanalyzer [EPMA] 电子探针显微分析仪
~ radiation 电子辐射
~ resonance 电子共振

~ scattering 电子散射
~ spectroscopy for chemical analysis [ESCA] 化学分析用电子能谱学
~ spin resonance [ESR] 电子自旋共振;顺磁共振
~ spin resonance absorption 电子自旋共振吸收
~ transfer polymer 电子转移聚合物
electronic 电子的
~ charge 电子电荷
~ control 电子控制
~ exchange resin 电子交换树脂
~ generator 高频发生器
~ heat sealing [dielectric sealing] 电子热封,电子热合;高频焊接
~ heating 电子加热
~ heating equipment 高频加热设备〈美〉
~ – hydraulic parison programming 电子液压型坯壁程序控制
~ hydraulic wall 型坯壁厚电子-液压调节〈吹塑成型〉
~ instrument controller 电子仪控制器
~ pretreatment 电子(表面)预处理〈聚烯烃〉
~ scanning 电子扫描
~ sealer 电子热合机,脉冲热合机
~ sewing 电子加热熔合,高频加热熔合
~ sewing machine 高频加热熔合机〈美〉
~ spectrum 电子能谱〈光谱〉
~ transition 电子跃迁〈胶黏理论〉
~ treating 电子处理
~ treatment 电子处理〈薄膜表面〉
~ valve 电子管
~ welding 高频焊接〈美〉
electropainting 电涂
electrophilic 亲电子的
~ reaction 亲电子反应
~ reagent 亲电子试剂
electrophoresis 电泳
electrophoretic 电泳的
~ behaviour 电泳性能
~ coating 电泳涂层
~ deposition 电泳沉积
electroplate 电镀
electroplateable 可电镀的
electroplated plastic 电镀塑料
electroplating 电镀
~ equipment 电镀设备
~ on plastics 塑料电镀
electropolishing 电(化学)抛光,电解抛光
electrosmosis 电渗透
electrostatic 静电的
~ capacity 静电容量
~ charge 静电荷
~ charging property 静电充电性
~ coating 静电涂布
~ diffuser 静电喷雾器
~ dipping 静电浸渍
~ flock finishing 静电植绒整理
~ flock printing 静电植绒印刷
~ flocking 静电植绒
~ fluidized bed coating 静电流化床涂布
~ fluidized bed unit 静电流化床装置
~ heat 静电加热
~ lens 静电透镜
~ load 静电荷
~ potential 静电势,静电位
~ powder coating [spraying] 静电粉末喷涂
~ powder spraying unit 静电粉末喷涂装置
~ printing 静电印刷
~ spray 静电喷涂
~ spray coating 静电喷涂
~ spray gun 静电喷枪
~ spray painting 静电喷涂
~ spraying 静电喷涂
~ theory of adhesion 静电黏合理论
electrostrictive 电致伸缩的
~ compliance 电致伸缩柔量
~ (sonic) transducer 电致伸缩(声)换能器
electroviscous effect 电黏带效应

element 元件,部件,构件
elementary charge 单位电荷
elementosiloxane polymer 元素硅氧烷聚合物
elemi(gum) 天然树脂
elesticizer 致弹剂
elev [elevated] (提)高的
elevated temperature 温升;高温
elevation 上升,提高;高程,标高
eliminate 除去,消去;分离
elimination polymerization 脱除聚合
Elmendoft tearing tester 埃尔曼多夫撕裂试验仪
ELO [epoxidized linseed oil] 环氧化亚麻油〈增塑剂,热稳定剂〉
elongation 伸长;伸长率;延伸
~ at break 断裂伸长
~ at rupture 破裂伸长,断裂伸长
~ at yield 屈服伸长
~ between gages 标距伸长
~ of break [rupture] 断裂伸长;致裂伸长;断裂伸长率
~ percentage 延伸率
~ rate 伸长率
~ strain 伸长应变
~ tensor 伸长张量
~ test 延伸率试验
~ test equipment 伸长(试验)仪
elution 洗脱
~ chromatography 洗脱色谱
~ extracting resin 萃淋树脂
~ fractionation 淋洗分级
~ volume 淋洗体积
EMA [ethylene – maleic – anhydride – copo – lymer] 乙烯-顺(丁烯二酸)酐共聚物,乙烯-马来酐共聚物
EMAA [ethylene – (methyl) acrylic acid copolymer] 乙烯-丙烯酸(甲酯)共聚物
EMAR [epoxidized methyl acetorici – noleate] 环氧乙酰蓖麻醇酸甲酯〈增塑剂〉
emanation thermal analysis 发射热分析
embankment angle 坡度角,倾斜角

embedded 埋入的,嵌入的
~ parts 预埋件,嵌入件
~ wire 埋入电线
embedding 嵌铸;埋封,封入成型
~ compound 埋封料,嵌铸料
embedment 埋入,嵌入
emboss 压花,压纹,模压
embossed 压花的
~ film 压花薄膜
~ pattern 压花图案,压花纹
~ sheet 轧花片材,压花片材
~ surface 压花(表)面
embosser 压花机
embossing 轧花,压花,压纹,滚花
~ calender 轧花压光机,辊筒压花机,压纹机
~ die 压花模(具);印模
~ electrode 压花电极
~ foil 压花箔
~ machine 压花机,压纹机
~ mo(u)lding 模塑压花,压花成型
~ plate 压花板
~ press 压花(压)机
~ pressure 压花压力,压纹压力
~ roll 压花辊
embrittlement 变脆,脆化;脆裂;脆性
~ point 脆化点
~ temperature 脆化温度
~ time 脆化时间
emerge 流出,溢出
emergency 紧急(的)
~ lamp 紧急信号灯
~ measures 紧急措施
~ operating order 紧急操作规程
~ repair 紧急修理
~ shut – down 紧急停车
~ situation 紧急情况
~ switch 紧急[事故]开关
emery 金刚砂;刚玉粉
~ cloth 金刚砂布
~ paper 金刚砂纸
~ roller 金刚砂辊
e. m. f. [electromotive force] 电动势
EMI [electromagnetic interference] 电磁

干扰
emission 发射
- ~ spectroscopy 发射光谱学
- ~ spectrum 发射光谱
- ~ wavelength 发射波长

emissivity 发射率,辐射系数
emulsification 乳化
- ~ time 乳化时间
- ~ work 乳化作业

emulsifier 乳化剂;乳化器
emulsify 乳化
emulsifying agent 乳化剂
emulsion 乳(浊)液
- ~ adhesive 乳液黏合剂,黏合乳剂
- ~ binder 乳液黏合剂
- ~ coating 乳液涂料;乳液涂布
- ~ copolymerization 乳液共聚
- ~ dyeing 乳化液染色
- ~ paint 乳化漆
- ~ polymer 乳液聚合物
- ~ polymerizate 乳液聚合物
- ~ polymerization 乳液聚合(作用)
- ~ polystyrene 乳液法聚苯乙烯〈胶黏剂〉
- ~ polyvinyl chloride [EPVC] 乳液法聚氯乙烯
- ~ resin 乳液法树脂
- ~ stabilizer 乳液稳定剂

enamel 磁[瓷]漆
- ~ hold-out (涂层)保持光泽性
- ~ lacquer 瓷漆

enamelled wire 漆包线
enantiomer 对映体
enantiomeric configurational unit 对映异构体的构型单元
enantiomorph 对映体
encapsulant 包封树脂,封铸树脂;密封剂
encapsulated 密封的;封装的
- ~ adhesive 包封黏合剂,胶囊型黏合剂
- ~ manifold 封装歧管

encapsulating 封入,埋封,包封,封铸,嵌铸〈美〉

- ~ compound 包封料,封铸料
- ~ resin 包封树脂,封铸树脂

encapsulation 包封,囊封
- ~ mo(u)lding 包封成型,封铸

encapsulization 包膜,囊封
enclosed 封闭的,封装的
- ~ flighted screw length 全螺纹螺杆长度
- ~ L/D [l/d] ratio 总长径比
- ~ oil bath 封闭油槽
- ~ volume of screw channel 螺槽总容积〈挤塑机〉

enclosure nozzle (自动)闭锁喷嘴
end 末端,终端;竖直,竖立
- ~ assembly 终端组件,端连接件,管接头
- ~ bond 端键
- ~ box 终端盒;终端电缆套管
- ~ connection 端连接;管连接
- ~ count 单丝支数〈粗纱〉
- ~ dome 末端拱起;缠卷筒的凸度
- ~ group 端基
- ~ group analysis 端基分析
- ~ group method 端基法
- ~ group titration 端基滴定
- ~ -initiation 末端引发
- ~ of(the)pipe 管端
- ~ -over-end drum tumbler 竖转式鼓式拌合机
- ~ -over-end mixer 竖转式混合机
- ~ -over-end type mixer 竖转式混合机
- ~ plate 端板
- ~ plug 端塞〈管〉
- ~ product 最终产品,成品
- ~ properties 成品的性能
- ~ sealing 端部热合〈制袋〉;闭封焊接〈包装〉
- ~ -stepped scarfed joint 阶式斜接接头
- ~ -use 产品用途
- ~ -use test 使用试验
- ~ user 用户
- ~ -view 侧视图

endless 无限的;无端的
  ~ printing 无端印刷,循环印刷
sheeting 无限片材;连续成片
endomethylene tetrahydrophthalic anhydride 桥亚甲基四氢邻苯二甲酸酐,纳迪克酸酐,NA酸酐〈固化剂〉
endothermic reaction 吸热反应
endurance 耐久(性),耐用(性);耐疲劳性,稳定性
  ~ bending test 耐弯曲疲劳试验
  ~ crack 疲劳裂缝,疲劳断裂
  ~ expectation 耐久性估计,预计使用期限
  ~ failure 疲劳破坏
  ~ life 耐久寿命
  ~ life test 耐久性试验
  ~ limit 耐久极限,疲劳极限
  ~ of burning 耐燃烧性
  ~ strength 耐久强度
  ~ tension test 拉伸疲劳试验
  ~ test 耐久性试验,疲劳度试验
  ~ torsion test 扭转疲劳试验
enduring 耐久的
energy 能;能量
  ~ – absorbing resin mat 吸收能树脂毡
  ~ absorption 能量吸收
  ~ balance 能量平衡
  ~ conservation 能量守恒
  ~ – converting polymers 能量转换聚合物
  ~ director 聚能器
  ~ dispersive spectrometer [EDS] 能量色散谱仪
  ~ elasticity 能量弹性
  ~ loss 能量损耗
  ~ of activation 活化能
  ~ of adhesion 黏合能
  ~ of attachment 结合能
  ~ of deformation 形变能,变形能
  ~ of retraction 收缩能;回复能
  ~ – saving 节能的
  ~ to fracture 冲击断裂能量
eng. engineering 工程

engage 啮合
Engelit process 恩格利特法,熔体发泡法
engineer 工程师,工程技术人员
engineering 工程(学),(工程)技术
  ~ cost 工程费
  ~ data 技术数据
  ~ drawing 工程图;工程制图
  ~ equipment 技术设备
  ~ evaluation 技术质量鉴定
  ~ foam 工程用泡沫塑料
  ~ instruction 技术说明书
  ~ legislation 工程技术规范
  ~ material 工程材料
  ~ plastic 工程塑料
  ~ project 工程项目
  ~ strength 工程(计算)强度
  ~ resin 工程树脂
Engler 恩格勒,恩氏
  ~ specific viscosity 恩氏比黏度
  ~ viscometer 恩氏黏度计
  ~ viscosimeter 恩氏黏度计
  ~ viscosity 恩氏黏度
engrave 雕刻;蚀刻,纹纹
engraved 雕刻的,刻纹的
  ~ coating 深凹纹涂布,刻花纹辊涂布
  ~ copper roller 刻花铜辊;花筒
  ~ cylinder 刻花辊
  ~ embossing roll 浮雕压花辊
  ~ pattern roll 雕花纹辊
  ~ roll 浮雕辊,花滚筒
  ~ – roll coater 凹印辊涂布机
  ~ – roll coating 凹印辊涂布
engraving 雕刻,蚀刻;刻纹
enhancer 改性剂
enrich 增浓,浓缩
ensilage 储槽〈塑料原料〉
ent entrance 进入,入口
entanglement 扭结,缠结〈纤维〉;缠结〈聚合物〉
entering pin 穿心销
enthalpy 焓
  ~ relaxation 焓松弛
Entoleter 恩土莱特
  ~ centrifugal mill 恩土莱特型离心研

磨机,冲击磨
~ - mixer 恩土莱特混合器
entrained air 掺气,夹杂空气
entrainment 夹带,夹杂;输送
entrance 进入,入口
~ loss 进料损耗
~ velocity 进料速度
entrapment 封入物;夹入物
entrapped air(- bubble) 气泡〈制品〉
entropy 熵
~ - elastic recovery force 熵 - 弹性回复力
~ elasticity 熵弹性
entry 进入,入口
~ accumulator 进料存储器
~ angle 入口角
~ effect 入口效应〈挤塑〉
~ zone 入口段〈挤塑〉
envelope 包封,包皮;外壳
envenomation 表面变质
environment 环境,周围介质
~ tewperature 环境温度
environmental 环境的,周围的
~ agents 环境适应剂
~ contamination 环境污染
~ cracking 环境龟裂〈塑〉
~ degradation 环境下降解〈塑料〉
~ factors 环境因素,外部因素
~ laboratory 环境模拟试验室
~ monitoring 环境监测
~ protection 环境保护
~ quality 环境质量
~ resistance 耐环境性能
~ stress cracking [ESC] 环境应力龟裂
~ stress cracking behaviour 环境应力龟裂性
~ temperature 环境温度
~ test 环境模拟试验
enzymatic change 酶变化
EP [epoxy resin] 环氧树脂
EP [ethylene - propylene copolymer] 乙烯 - 丙烯共聚物
EP polymers [easy - processing polymers)] 易加工聚合物
EPD [ethylene - propylene - diene terpoly - mer] 乙烯 - 丙烯 - 二烯三元共聚物
EPDM [ethylene propylene diene mono - mer] 乙烯丙烯二烯三元共聚物单体
EPDM [poly(ethylene - propylene) ter - polymer monomer] 乙烯 - 丙烯三元共聚物单体
epibolic stress 弛失应力
epibromohydrin 环氧溴丙烷,表溴醇〈阻燃剂〉
epichlorohydrin 环氧氯丙烷,表氯醇
~ rubber 环氧氯丙烷橡胶,表氯醇橡胶
Epicoat 〈商〉爱偏可脱〈一种环氧树脂〉
epicyclic screw mixer 行星式螺杆混合机
EPM [elastomeric copolymer of ethylene and propylene] 乙烯 - 丙烯共聚弹性体
Epon 〈商〉爱朋〈一种环氧树脂〉
epoxidation 环氧化(作用)
epoxide [EP] 环氧化物
~ dermatitis 环氧树脂型皮(肤)炎〈接触环氧树脂引起〉
~ equivalent 环氧当量
~ equivalent weight 环氧当量
~ group 环氧基
~ group concentration 环氧基浓度
~ index 环氧指数
~ moulding compound 环氧树脂模塑料
~ plasticizer 环氧类增塑剂
~ resin [EP] 环氧树脂
~ ring 环氧环
~ stabilizers 环氧类稳定剂
~ value 环氧值
epoxides 环氧化合物
epoxidized 环氧化的
~ fatty oil 环氧脂肪油
~ linseed oil [ELO] 环氧化亚麻仁油〈增塑剂,热稳定剂〉
~ methyl acetoricinoleate [EMAR] 环

氧乙酰蓖麻醇酸甲酯〈增塑剂〉
~ novolak resin 环氧酚醛树脂
~ soybean oil [ESO] 环氧化豆油〈增塑剂,热稳定剂〉
~ triglyceride 环氧化甘油三酸酯〈增塑剂〉

epoxidizing 环氧化
epoxies [epoxy resins] 环氧树脂
epoxy [EP] 环氧;环氧树脂
~ adhesive 环氧树脂黏合剂
~ bonded 环氧树脂粘接的
~ coating 环氧树脂涂层
~ content 环氧基含量
~ foam 环氧树脂泡沫塑料
~ gel coat 环氧树脂凝胶漆
~ glass laminate 环氧树脂玻璃纤维层合塑料
~ group 环氧基
~ laminate 环氧树脂层压板
~ mo(u)ld 环氧树脂模具
~ no(u)lding compound 环氧树脂模塑料
~ no(u)lding material 环氧树脂模塑料
~ mo(u)lding powder 环氧树脂模塑粉
~ novolac adhesive 环氧-酚醛黏合剂
~ novolac resin 环氧酚醛树脂
~ - phenolic resin adhesive 环氧酚醛树脂黏合剂〈酚醛树脂改性环氧树脂黏合剂〉
~ plastic 环氧塑料
~ plasticizers 环氧增塑剂
~ polybutadiene 环氧聚丁二烯
~ polymer 环氧聚合物
~ - polysulfide adhesive 环氧多硫黏合剂
~ resins 环氧树脂
~ ring 环氧环
~ stabilizer 环氧稳定剂
~ - system weakly basic anion exchange resin 环氧系弱碱性阴离子交换树脂
~ trowelling compound 环氧树脂腻子

~ value 环氧值
epoxyresin [EP] 环氧树脂
~ adhesive 环氧树脂黏合剂
~ adhesive pellets 环氧树脂粒料黏合剂
~ based on diphenylol propane 双酚A型环氧树脂
~ coating 环氧树脂涂料
~ cure 环氧树脂固化
~ enamel 环氧树脂磁漆
~ laminated sheet 环氧树脂层合板
~ mortar 环氧树脂混合浆
~ mo(u)lding compound 环氧树脂模塑料
~ - novolak 环氧酚醛树脂
~ of glycidyl ester 缩水甘油酯型环氧树脂
~ varnish 环氧树脂清漆

β - (3,4 - epoxycyclohexyl) - **ethyl trimeth - oxy silane** β - (3,4 - 环氧环己基)乙基三甲氧基(甲)硅烷〈偶联剂〉

EPR [ethylene propylene rubber] 乙丙橡胶
EPR - g - PVC [ethylene - propylene rubber - vinyl chloride graft copolymers] 乙丙橡胶-氯乙烯接枝共聚物
EPS [expandable polystyrene] 可发性聚苯乙烯,泡沫聚苯乙烯
EPSAN [ethylene - propylene - styrene - acry - lonitrile copolymer] 乙烯-丙烯-苯乙烯-丙烯腈共聚物
epsilon caprolactam ε - 己内酰胺
Epsom salt 泻利盐,泻盐,硫苦〈分散剂〉
EPT [ethylene propylene terpolymer] 乙烯-丙烯三元共聚物
EPVC [emulsion polyvinyl chloride] 乳液法聚氯乙烯
eq. equation 方程式,反应式
equal 相等的,等同的,等于
~ pitch [constant pitch] 相等螺距〈螺杆〉
~ socket 相等直径插口
~ tension 均等张力

equalize 平衡,均衡,相等
equalized pressure 平衡压力
equalizer [straightening press] 平衡器；压平机,矫直机
equalizing 平衡,均衡；调整
equation 方程式；等式；反应式；公式
~ of state 状态方程式
equator [tangent line] 切线〈卷绕〉
equiareal 相等面积的
equibinary polymer 平衡二元聚合物
equil. equilibrium 平衡
equilibrium 平衡
~ compliance 平衡柔量
~ copolymerization 平衡共聚合
~ diagram 平衡图
~ mixture 平衡混合物
~ modulus 平衡模量
~ moisture 平衡湿度
~ moisture content 平衡含湿率；平衡湿量
~ osmotic pressure 平衡渗透压
~ polymerization 平衡聚合
~ relative humidity 相对平衡湿度
~ swelling 平衡溶胀
equipment 设备
equispaced 等(间)距的
equiv [equivalent] 等效；相当；当量
equivalence 等效,等价；当量；相当
equivalent 等效；相当；当量
~ luminance 当量发光密度,当量亮度
~ vulcanization 等效硫化
~ weight [EW] 当量
Erdmenger kneader and compounder 埃德曼格捏合混料机,双螺杆连续捏合混料机
erection 结构；安装,装配〈机器〉；建立,建设〈装置〉
Erichsen bubble [depth] **test** 埃里克森鼓泡试验
Erichsen test 埃里克森试验
erosion 侵蚀
~ breakdown 侵蚀破坏
error 误差
~ allowance 容许误差
~ in reading 读数误差
~ of expectation 估计误差
~ of measurement 测量误差
~ of observation 观察误差
~ rate 误差率
erucyl amide 芥酰胺,芥酸酰胺〈爽滑剂〉
erythro－diisotactic 叠同－双全同立构
ESCA [electron spectroscory for chemial analysis] 化学分析用电子能谱学
ESC [environmental stress cracking] 环境应力龟裂
escape 逃逸,泄漏,排泄
~ cock 排放阀
~ hatch 应急出口
~ of volatiles 挥发组分出口
~ opening 排气孔
~ orifice 排泄口
~ pipe 排气管,排水管
escaping mould 溢料或模具
ESO [epoxidized soybean oil] 环氧化豆油〈增塑剂,热稳定剂〉
esp. especially 特别,尤其,格外
ESR [electron spin resonance] 电子自旋共振
est. [estimate] 估计
est. [estimated] 估计的
established procedure 规定操作程序
ester 酯
~ cellulose 纤维素酯
~ exchange 酯交换
~ exchange polycondensation 酯交换缩聚
~ exchange reaction 酯交换反应
~ gum 酯胶,松香甘油酯
~ of β－aminocrotonic acid β－氨基丁烯酸酯〈稳定剂〉
~ plastic 聚酯类塑料
~ plasticizer 酯类增塑剂
~ value 酯值
esterification 酯化(作用)
esterify 酯化
estertins 酯－锡化合物〈聚氯乙烯用稳定剂〉

estimate 估计,预算
estimation 估计,评价
estn. [estimation] 估计
et al. [et alibi; et alii] (拉) 及其他(地方,人),等等
etc. [et cetera] 〈拉〉等,等等
etch 蚀刻,浸蚀
~ - proof 防腐蚀的
~ treatment 腐蚀处理
etching 蚀刻,浸蚀
~ agent 蚀刻剂
~ medium 蚀刻剂,浸蚀剂
~ method 蚀刻法
~ primer 腐蚀性涂料;反应性底漆
~ process 蚀刻法
~ solution 浸蚀液
~ temperature 腐蚀温度
ETFE [ethylene tetrafluoro ethylene] 乙烯 - 四氟乙烯共聚物
ethane 乙烷
ethanol 乙醇,酒精
~ urea 乙醇脲
ethene plastics 乙烯塑料
ethenoid plastics 乙烯类塑料
ether 醚〈类名〉;乙醚
~ resins 醚类树脂
etherizing (用)醚提取
ethoxy 〈词头〉乙氧基 6 - ~ - 2,4 - trimethyl - 1,2 - dihydro - quinoline 6 - 乙氧基 - 2,2,4 - 三甲基 - 1,2 - 二氢化喹啉〈抗氧剂〉
ethoxylene 环氧
~ resin 环氧树脂
~ resin adhesive 环氧树脂黏合剂
ethyl 乙基
~ acetate 乙酸乙酯,醋酸乙酯
~ acrylate 丙烯酸乙酯
~ acrylate polymer 丙烯酸乙酯聚合物
~ alcohol 乙醇,酒精
~ aluminum dichloride 二氯化乙基铝〈助催化剂〉
~ aluminum sesquichloride 倍半氯化乙基铝
~ benzene 乙(基)苯
~ benzoate 苯甲酸乙酯
2 - ~ butyl acetate 乙酸 - 2 - 乙基丁酯
~ butyl ketone 乙基丁基(甲)酮
~ butyrate 丁酸乙酯
~ cellulose [EC] 乙基纤维素
~ cellulose lake 乙基纤维素漆
~ - α - cyanoacrylate α - 氰基丙烯酸乙酯
~ 2 - cyano - 3,3 - diphenyl acrylate 2 - 氰基 - 3,3 - 二苯基丙烯酸乙酯〈光稳定剂〉
~ dichlorosilane 乙基二氯(甲)硅烷
~ formate 甲酸乙酯
~ - α - hydroxyisobutyrate α - 羟基异丁酸乙酯
2 - ~ imidazole 2 - 乙基咪唑〈固化剂〉
~ lactate 乳酸乙酯
~ levulinate γ - 戊酮酸乙酯
~ methacrylate 甲基丙烯酸乙酯 2 -
~ - 4 - methyl - imidazole 2 - 乙基 - 4 - 甲基咪唑〈固化剂〉
~ oleate 油酸乙酯
~ oxalate 草酸乙酯
N - (2 - ~ phenyl) - N' - (2 - ethoxy - 5 - tert - butyl phenyl) oxalic acid di - amine N - (2 - 乙基苯基) - N' - (2 - 乙氧基 - 5 - 叔丁基苯基)草酰二胺〈紫外线吸收剂〉
~ phthalyl ethyl glycolate 乙基邻苯二甲酰乙醇酸乙酯〈增塑剂〉
~ rubber 乙基橡胶
~ silicone oil 乙基硅油
N - ~ - ortho, para - toluene sulfonamide N - 乙基邻(或对)甲苯磺酰胺〈增塑剂〉
~ triazone resin 乙基三嗪酮树脂
~ trichlorosilane 乙基三氯(甲)硅烷
~ triethoxysilane 乙基三乙氧基(甲)硅烷
~ vinyl ether [EVE] 乙基乙烯醚
ethylbutyl [hexyl] 乙基
ethylene 乙烯
~ acrylate 乙烯丙烯酸酯

~ - acrylate copolymer 乙烯-丙烯酸酯共聚物
~ acrylic acid copolymers [EAA] 乙烯-丙烯酸共聚物
~ benzene 苯乙烯
~ - butadiene copolymer 乙烯-丁二烯共聚物
~ butylacrylate [EBA] 乙烯丙烯酸丁酯(共聚物)
~ - butylene copolymer 乙烯-丁烯共聚物
~ - butylene polymer 乙烯-丁烯聚合物
~ carbonate 碳酸亚乙酯
~ chloride 1,2-二氯乙烷
~ chlorotrifluoroethylene copolymer [ECIEE] 乙烯-三氟氯乙烯共聚物
~ copolymer bitumen mixture [ECB] 乙烯共聚物与沥青共混物
~ copolymer foam 发泡乙烯共聚物
~ bis [tris(2-cyanoethyl) phosphonium bromide] 溴化乙撑双[三(2-氰乙基)磷]〈阻燃剂〉
~ dichloride 1,2-二氯乙烷;二氯化乙烯
~ ethyl acrylate [E/EA] 乙烯-丙烯酸乙酯
~ ethyl acrylate copolymer 乙烯-丙烯酸乙酯共聚物
~ N,N'-ethylenebis(12-hydroxys-tearamide) N,N'-乙撑双(12-羟基硬脂酰胺)〈润滑剂,脱模剂〉
~ N,N'-ethylene bis ricinoleamide N,N'-乙撑双蓖麻醇酸酰胺〈润滑剂,脱模剂〉
~ N,N'-ethylene bis stearamide N,N'-乙撑双硬脂酰胺〈润滑剂〉
~ glycol diacetate 乙二醇二乙酸酯〈增塑剂,溶剂〉
~ glycol monobutyl ether laurate 乙二醇单丁醚月桂酸酯〈增塑剂〉
~ glycol monobutyl ether oleate 乙二醇单丁醚油酸酯〈增塑剂〉

~ group 乙烯基1,2-亚乙基
~ imine resin 乙烯亚胺树脂,氮丙啶树脂
~ - maleic anhydride copolymer [EMA] 乙烯-顺丁烯二酸酐共聚物
~ - (methyl) acrylic acid copolymer [EMAA] 乙烯-(甲基)丙烯酸共聚物
~ methyl acrylate copolymer 乙烯-丙烯酸甲酯共聚物
~ [ethene] plastic 乙烯类塑料
~ polymer 乙烯类聚合物
~ propylene [E/P] 乙烯-丙烯
~ propylene copolymer [EPM] 乙烯-丙烯共聚物
~ propylene diene monomer [EPDM] 乙烯-丙烯-二烯三元共聚物单体
~ propylene diene terpolymer [EPD] 乙烯-丙烯-二烯三元共聚物
~ propylene elastomer 乙丙弹性体
~ propylene rubber [EPR] 乙丙橡胶
~ propylene rubber - vinyl chloride graft copolymer [EPR-g-PVC] 乙丙橡胶-氯乙烯接枝共聚物
~ propylene - styrene - acrylonitrile copo-lymer [EPSAN] 乙烯-丙烯-苯乙烯-丙烯腈共聚物
~ propylene terpolymer [EPT] 乙烯-丙烯三元共聚物
~ resin 乙烯类树脂
~ tetrafluoroethylene [ETFE] 乙烯四氟乙烯共聚物
~ trifluorochloroethylene copolymer 乙烯三氟氯乙烯共聚物
~ thiourea 乙撑硫脲[促进剂NA-22]
~ - urea resin 乙烯脲树脂
~ vinyl acetate copolymer [EVA] 乙烯-乙酸乙烯酯共聚物
~ vinyl acetate copolymeric film for agri-culture 乙烯-乙酸乙烯酯共聚物农用薄膜
~ vinyl acetate rubber 乙烯-乙酸乙烯酯橡胶

~ vinyl acetate vinyl chloride copoly-mer 乙烯-乙酸乙烯酯-氯乙烯共聚物
~ vinyl acetate vinyl chloride graft copolymer [EVA-g-PVC] 乙烯-乙酸乙烯酯-氯乙烯接枝共聚物
~ vinylalcohol copolymer [E/VAL] 乙烯-乙烯醇共聚物
~ vinyl chloride maleic anhydride [VCE-MA] 乙烯-氯乙烯-顺丁烯二酸酐共聚物
~ vinyl chloride vinyl acetate 乙烯-氯乙烯-乙酸乙烯酯共聚物
ethylhexyl 乙基己基
2-~-2-cyano-3,3-diphenylacrylate 2-氰基-3,3-二苯基丙烯酸-2-乙基己酯〈光稳定剂〉
2-~ decyl phthalate 邻苯二甲酸2-乙基己癸酯〈增塑剂〉
2-~ epoxy stearate 环氧硬脂酸-2-乙基己酯〈增塑剂,稳定剂〉
2-~ epoxytallate 环氧妥尔酸-2-乙基己酯〈增塑剂〉
2-~ ester of epoxy fatty acids 环氧脂肪酸-2-乙基己酯〈增塑剂,稳定剂〉
2-~ isodecyl phthalate 邻苯二甲酸-2-乙基己基异癸酯〈增塑剂〉
5-**ethylidene-2-norbornene** 5-亚乙基-2-降冰片烯〈助剂〉
EU [polyether based polyurethane rub-ber〈ASTM〉] 聚醚型聚氨酯橡胶
eucolloid 真胶体
eutacticity 理想的构型,规整性
EVA [ethylene vinyl acetate] 乙烯-乙酸乙烯酯共聚物
EURONORM [Europaische Normen] 欧洲塑料标准
EVAC, E/VAC [EVA]
EVA-g-PVC [etylene-vinyl acetate-vinyl chloride graft copolymer] 乙烯-乙酸乙烯-氯乙烯接枝共聚物
evacuate 抽空;排泄
E/VAL [ethylene vinylalchol] 乙烯-乙烯醇

evaluate 估计;评价;测定,鉴定;计算
evaluation 估计;评价;计算
~ by visual examination 目测评价
~ program 评价程序
~ test 评价性试验
evaporate 蒸发
evaporated film 蒸发膜
evaporating 蒸发;浓缩
~ pan 蒸发皿
evaporation 蒸发(作用)
~ humidification 蒸发增湿
~ index 蒸发系数
~ rate analysis 蒸发率分析
evaporative cooling 蒸发冷却
evaporator 蒸发器
EVE [ethyl vinyl ether] 乙基乙烯醚
EVE [vinyl ethyl ether] 乙基乙烯醚
even 平坦的;均匀的
blend 均匀掺混,均匀混合
~ distribution of heat 均匀加热分布
~ dyeing 匀染
~ heating 均匀加热
~ speed of the rolls 辊速同步
~ speed rolls 同步辊
~ surface 平滑表面;均匀表面
~ tension 均匀张力
evener [straightener] 整平机,矫直机
evenness of the coat 涂层均匀性
evolution of heat 热放出
evolve 放出(气体);离析
evolved gas 逸出气体
~ analysis [EGA] 逸出气体分析〈热分析〉
~ detection [EGD] 逸出气体检〈热分析〉
EW [equivalent weight] 当量
ex. [except] 除……之外
examination 检验;试验;测验;研究
exceed capacity 超过范围
excellent 优良的,极好的
excelsior 锯屑,刨花,木丝
excess 过量;余物
~ load 超载

~ material 过量物料,剩余材料
~ pall 溢流皿
~ pressure 超压,过压
~ production 超额生产
excessive 过度的,过量的
~ load 超载
~ weight 超重
exchange 交换
~ of information 信息交流
~ process 交换过程,混合过程
~ resin (离子)交换树脂
exchangeable 可交换的
exchanger 交换器
excited 受激的
~ polymer 激发(态)聚合物
~ seate 激发态
excluded volume 排除体积(已占体积)
exclusion chromatography 排斥色谱
exergy 有效能
exert 施加〈如:压力〉
exfoliate 层状剥离,成片分离
exhaust 排气,排出;废弃;废气;抽空
~ air 排出气
~ air duct 排出气管道
~ blower 抽风机
~ duct 通风管道,排气管道
~ fan [blower] 排风扇,排气机
~ gas 废气
~ gas stack 废气烟囱
~ opening 排气孔
~ orifice 排气口
~ pipe 排气管
~ steam 废蒸气
exhaustor 排气机
exhibition 展览;显示
exit 出口
~ accumulator 出口存储器
~ conveyor 出料输送带
~ pressure 出口压力
~ tube 出口管
exoelectron 外激电子〈胶黏理论〉
~ emission 外激电子发射〈黏胶理论〉
exotherm 放热量
~ curve 放热曲线

exothermic 放热的
~ reaction 放热反应
expand 膨胀;发泡;扩大,扩张
~ reproduction 扩大再生产
expandability 发泡能力,可发泡性
expandable 可发(泡)性的,可膨胀的
~ beads 可发泡珠粒料
~ -beads moulding 可发泡珠粒成型
~ plastic 可发性塑料
~ polyethylene 可发性聚乙烯
~ polystyrene [EPS] 可发性聚苯乙烯
~ polystyrene bead 可发性聚苯乙烯珠粒料
~ PS mo(u)lding 可发性聚苯乙烯(发泡)成型
~ thermoplastic 可发性热塑性塑料
expanded 发泡的
~ glass beads 发泡玻璃珠粒料
~ material 发泡材料
~ phenolic plastic 发泡酚醛塑料
~ plastic 发泡塑料,泡沫塑料,多孔性塑料
~ polyethylene 发泡聚乙烯,泡沫聚乙烯
~ polystyrene [EPS] 发泡聚苯乙烯,泡沫聚苯乙烯
~ polystyrene sheet 聚苯乙烯泡沫板材
~ polyurethane 发泡聚氨酯,泡沫聚氨酯
~ polyvinyl chloride 发泡聚氯乙烯
~ polyvinyl chloride foam 聚氯乙烯泡沫塑料
~ rubber 发泡橡胶,泡沫橡胶,海绵橡胶
~ sheet 发泡片材
~ sponge 发泡海绵
~ vinyl chloride leather cloth 聚氯乙烯泡沫人造革
expander 发泡器;扩张器;拉幅器;骤冷器〈美〉
~ roll 拉幅辊
~ roll with adjustable 可调弧度拉幅辊
~ roll with curved axis 弧形抽拉幅辊

~ roll with helices　螺旋形拉幅辊
~ roll with progressive thread　递增螺纹拉伸辊
expandibility　发泡性;发泡能力;膨胀性
expanding　发泡(的);膨胀(的)
~ agent　发泡剂;膨胀剂
~ mandrel　膨胀式芯模
~ mill　扩径机〈管材〉
~ mill of an extruded preform　挤塑型坯的膨胀
~ mill of a hotinjected blank [preform]　热注塑型坯的膨胀
~ process　发泡过程;发泡法
~ property　发泡性
~ ratio　膨胀率
~ roller　张力辊;拉幅辊
~ test　扩大试验
~ expansibility　可扩张性,可膨胀性,可延伸性
~ expansion　发泡,膨胀;蒸发
~ bend　膨胀弯管;补偿器
~ by heat　热膨胀
~ casting　膨胀铸塑法
~ casting of polystyrene　聚苯乙烯模压发泡
~ coefficient　膨胀系数
~ cracks　膨胀裂缝
~ cylinder　蒸发汽缸
~ fitting　膨胀[伸缩]连接件
~ insert　胀入式嵌件
~ joint　膨胀节,伸缩缝;补偿器
~ lever　伸缩杆
~ mo(u)lding　发泡成型
~ piece　膨胀件
~ pipe　膨胀管;补偿管
~ ratio　发泡率,膨胀率
~ reservoir　溢流皿
~ valve　膨胀阀,安全阀,减压阀
~ vessel　膨胀容器
~ test　发泡试验,膨胀试验
expansivity　膨胀性
expected　预期的
~ life　预期使用寿命
~ service life　预期使用寿命

expendable　可消费的,一次性使用的
~ pallet　一次性使用托盘
experimental　试验的
~ condition　试验条件
~ data　试验数据
~ model　试验模型
~ mould　试验模具
~ plant　试验装置
~ proceduce　试验程序
~ scale　实验规模
~ setup　实验规模
~ study　试验研究
~ technique　实验技术
~ value　试验值
~ verification　实验验证
expert　专家,行家;鉴定人
exploded view　部件分解图
exploitation　开发;投产,开工,起动
exploratory　研究(探索)工作
~ development　探索性研制
~ experiment　探索性试验
~ research　探索性研究
~ test　探索性试验
explosion proof　防爆的
exponential horn [sonotrode]　指数焊接仪
export　出口
~ packing　出口包装
expose　暴露;曝光
~ exposed surface　有效表面〈催化剂〉
exposition　暴露;曝光;陈列
~ exposure　暴露;曝晒;曝光
~ cracking　曝露龟裂,大气龟裂〈环境影响〉
~ test　曝露试验
~ time　曝晒时间
expt. [experiment]　实验
exptl. [experimental]　实验的
ext. [extract]　萃取,抽提;萃取物
extd. [extracted]　萃取的,抽提的
extend　伸长,延长;拉长
extended　伸长的,延长的;拉长的
~ chain　伸直链
~ chain crystals　伸直链晶体

~ linear macromolecule 伸直线型大分子
~ mandrel method 长心轴法
~ nozzle gate 加长注嘴浇口
~ nozzle mould 加长注嘴模具
extender 增量剂,增充剂
~ pigment 增量颜料,体质颜料
~ plasticizer 增量增塑剂
~ resin 充替树脂
extendibility 增充性,增量度
extending filler 增填料
extensibility 延伸性,伸长性,延展性
extension 伸长,延伸
~ at break 断裂伸长
~ at rupture 断裂伸长;断裂伸长率
~ elongation 拉长
~ modulus 伸长模量;张拉模量
~ nozzle 加长喷嘴[注嘴]
~ piece 延伸件
~ ratioprolamin 伸长比
~ speed 伸长速率
~ strain 伸长变形,伸长应变
extensional 延伸,拉伸
~ deformation 拉伸变形
~ flow 拉伸流动;伸长流动
~ motion 拉伸运动
~ stress 拉伸应力
~ viscosity 拉伸黏度
extensive nozzle for spruleless moulding 无流道成型的加长注嘴
extensometer 伸长计,延伸仪
~ for molten plastics 熔体塑料延伸仪
extensometry 伸长测量,延伸率测定
extent 程度;范围
~ of branching 支化度
~ of polymerization 聚合程度
~ of reaction 反应程度
~ of stresses 应力程度
~ of the flame 火焰蔓延程度(耐灼热性试验)
exterior 外部(的);室外的
~ adhesive 室外用黏合剂,耐风蚀黏合剂
~ cladding 外包层,外覆盖层

~ coating 外部涂层;保护层
~ exposure 户外曝晒
~ exposure test 室外曝晒试验
~ glue 室外用黏合剂
~ jaw 外用夹头〈模具〉
~ paint 外用漆
~ plywood 室外用胶合板
~ support 室外用支架
external 外部的,表面的
~ antistatic agent 外用抗静电剂
~ cooling 外部冷却
~ crack 外表裂缝
~ crest 齿顶(螺纹)
~ diameter 外径
~ dimensions 外形尺寸
~ fibre stress 外层纤维应力
~ finish 外用涂料
~ flame retardant 外(用)阻燃剂
~ heat 外部加热
~ hexagon 外六角形
~ ignition 外源点燃
~ impregnation 外部浸渍
~ lubricant 外用润滑剂(助剂)
~ mo(u)ld 外形模具,浸渍模〈外部施胶〉
~ permanent antistatic agent 外用永久性抗静电剂
~ phase 外相;连续相
~ plasticization 外增塑(作用)
~ plasticizer 外增塑剂
~ plasticizing 外增塑
~ sizing 外定径〈管材〉
~ spider 外用架
~ stress 外应力
~ surface 外表面
~ tab washer 外吊环垫圈
~ thread 外螺纹
extinction 消灭;吸[消]光
~ angle 消光角
~ coefficient 消光系数
extn. [extraction]萃取,抽提
extra 额外的;特别的;超过的
~ high molecular weight polymer 超高(相对)分子(质)量聚合物

extract

~ high molecular weight polyethylene [EHMWPE] 超高(相对)分子(质)量聚乙烯
~ thickness 加工余量,切削余量

extract 提取
extractable material 可提取成分
extractant 萃取剂;提取剂
extracting resin 萃取[萃淋]树脂
extraction 萃取,提取;排气;脱模
~ bar 脱模板
~ fractionation 萃取分级分离
~ resin 萃取树脂
~ zone 排气段

extractor 提取器,萃取器,抽提器;脱水器;顶出器,提出器;脱模器,脱模工具
~ plate 脱模板
~ slot 通风道格栅
~ -type centrifuge 离心筛;筛式离心机

extraneous 外来的;无关的
extrudability 可挤塑性
extrudate 挤出物;挤出坯;挤塑制品
~ distortion 挤出物变形
~ roughness 挤出物粗糙度
~ post swell 挤后膨胀
~ swelling 挤出物膨胀
~ temperature 挤出料温度

extrude 挤塑;挤出;挤压
extruded 挤塑的,挤出的
~ bead 挤塑熔焊条
~ bead sealing 挤出熔条封接
~ bead welding 挤出熔条焊接
~ film 挤塑薄膜
~ net 挤出塑料网
~ parison blow mould 挤塑型坯吹塑模具,挤吹模具
~ plastics mesh 挤出塑料网
~ preform method 挤塑型坯法
~ preform method profile 挤塑异型材
~ preform method section [shape] 挤塑型材
~ preform method sheet 挤塑片材
~ preform method sheeting 挤塑片材

~ preform method tube 挤塑管
~ profile 挤塑型材
~ section 挤塑型材
~ strings 挤塑绳带
~ tube 挤塑(软)管

extruder 挤塑机
~ adapter 挤塑机模头接合器
~ approaching passage 挤塑机头通道
~ attitude 挤塑机类型
~ average rate 挤塑机平均产率
~ barrel 挤塑机筒
~ base 挤塑机机座
~ blanks 挤塑坯料
~ bore 挤塑机口径
~ breaker plate 挤塑机分料板
~ calender 挤塑压延机
~ capacity 挤塑机能力
~ constant 挤塑机常数
~ core 挤塑机模芯
~ cylinder 挤塑机机筒
~ degassing 挤塑机排气
~ die 挤塑机口径[口模]
~ drive 挤塑机驱动装置
~ effect 挤塑机效果
~ feeding 挤塑机供料
~ for wire coating 包覆电线用挤塑机,包线挤塑机
~ for wire covering 包覆电线用挤塑机,包线挤塑机
~ grain 挤塑造粒
~ head 挤塑机机头
~ head connecting flange 挤塑机机头连接法兰
~ length to diameter ratio 挤塑机长径比
~ output 挤塑量
~ parts 挤塑机部件
~ -pelletizer 挤塑造粒机
~ rate 挤塑机产率
~ rear end 挤塑机尾端
~ rifled barrel 挤塑机来复线机筒
~ screen 挤塑机过滤网
~ screen pack 挤塑机过滤网组
~ screw 挤塑机螺杆

~ series 挤塑机系列
~ size 挤塑机规格
~ stand 挤塑机机座
~ straight head 挤塑机直机头
~ take-off 挤塑机牵引
~ throughput 挤塑机产量
~ train 挤塑机机组
~ vacuum zone 挤塑机真空段
~ vented 挤气式挤塑机
~ with cutting blade 带有切刀的挤塑机
~ with grooved feed zone 带有槽沟供料段的挤塑机
~ with side delivery head 直角(机头)挤塑机
~ with straight delivery head 直通(机头)挤塑机
~ with takeaway 带有牵引装置的挤塑机
~ with take-off 带有牵引装置的挤塑机
~ with wet bushes 带有冷却水套筒的挤塑机
~ working curve 挤塑机工作曲线

extruding 挤塑;挤出
~ machine 挤塑机
~ press 挤压机
~ profiles 挤塑型材
~ test 挤塑试验
~ tubes 挤塑软管
~ wire coatings 挤塑电线保护层

extrusion 挤塑,挤出
~ aid 挤塑助剂
~ apparatus 挤塑装置
~ blow mo(u)lding [EBM] 挤出吹塑模塑,挤出吹塑成型
~ blowing 挤出吹塑(成型)
~ blowing mo(u)lding 挤出吹塑(成型)
~ -calendering plant 挤塑压延装置
~ capacity 挤出能力
~ casting 挤塑流延;挤塑铸塑
~ characteristics 挤出特性
~ coater 挤出贴合机

~ coating 挤出贴合,挤出涂布
~ -coating resin 挤塑贴合用树脂
~ colo(u)ring 挤塑着色
~ compound 挤塑(用)料
~ cylinder 挤塑机筒
~ die 挤塑模头
~ die for filaments 单丝挤塑模头
~ die for flat sheet 平片挤塑模头
~ die for laces rods bristles 带材、棒材、鬃材挤塑模头
~ die for pipe 管材挤塑模头
~ die for profiles 型材挤塑模头
~ die for rigid pipes 硬管材挤塑模头
~ die for rods and filaments 棒材和单丝挤塑模头
~ die for tubing 软管挤塑模头
~ die with basic die blank and adapter rings 模套和模块接套挤塑模头
~ efficiency 挤出效率
~ equipment 挤塑装置
~ expansion 挤塑发胀
~ -foamed 挤塑发泡的
~ foaming 挤塑发泡
~ forming 挤塑热成型
~ head 挤塑机头
~ head for ram extrusion 柱塞挤塑用挤塑模头
~ head for tubing 软管挤塑机头
~ head for wire coating 电缆保护层用挤塑模头
~ head with screen basket 带有筛网筐的挤塑机头
~ hopper 挤塑料斗
~ laminating 挤塑层合,挤出复合
~ lamination 挤塑层合
~ letdown 挤塑量下降
~ line 挤塑生产线
~ linear speed 挤塑线速率[度]
~ machine 挤塑机
~ mandrel 挤塑模芯
~ material 挤塑(用)料
~ mixing 挤塑混合
~ mo(u)lding 挤塑(成型)
~ mo(u)lding acrylate copolymer 挤

型丙烯酸酯共聚物
~ nozzle 挤出嘴
~ of lay-flat tubing 平折膜管挤出
~ of profiles 型材挤塑
~ of tabular film 管膜挤塑
~ of thermosetting resin 热固性挤塑树脂
~ orifice 挤塑模孔
~ output 挤出产量,挤塑量
~ output surging 挤出量波动
~ plant 挤塑装置
~ plastometer 挤出式塑度计
~ plunger 挤出柱塞,推压杆
~ press （螺杆式)挤塑机
~ pressure 挤出压力
~ process 挤塑法
~ process for reinforced plastics 增强塑料挤塑法
~ rate 挤出速率;挤塑定额
~ rheometry 挤出式流变测定法
~ rheometer 挤塑式流变仪
~ screw 挤塑螺杆
~ speed 挤出速率[度]
~ spinning 挤出纺丝
~ stress 挤压应力

~ stretch blow moulding machine 挤拉吹成型机
~ swelling 挤出膨胀
~ technology 挤塑工艺
~ temperature 挤塑温度
~ type injection machine 挤出式注塑机
~ type plastometer 挤压式塑性计
~ under vacuum 真空挤压成型
~ welding 挤出(熔条)焊接
~ welding equipment 挤出焊接装置
~ with direct gassing 直接充气挤塑
~ with pregassed beads 预充气珠粒料挤塑
~ zone 挤塑段
extrusiograph 挤塑仪
exudate 渗出(物)
exudating 渗出(物)
exudation 渗出(物)
~ mark 渗出痕迹(缺陷)
~ of pigment 颜料析出
exuded material 渗出物
eye hole 孔眼
eyepiece micrometer 目镜测微计,测微目镜

F

F. [Fahrenheit] 华氏温度
F [fluorine] 氟
f [frequency] 频率(物理)
F_3 [polychlorotrifluoroethylene, PCTFE] 聚三氟氯乙烯[氟塑料3]
F_4 [suspension polytetrafluoroethylene, PTFE] 悬浮法聚四氟乙烯[氟塑料4]
F_{23-14} [chlorotrifluoroethylene – vinylidene fluorid copolymer] 偏氟乙烯-三氟氯乙烯共聚物[氟塑料23-14]
F_{23-19} [chlorotrifluoroethylene vinylidene fluoride copolymer] 偏氟乙烯-三氟氯乙烯共聚物[氟塑料23-19]
F_{30} [ethylene – chlorotrifluoroethylene copolymer] 三氟氯乙烯-乙烯共聚物[氟塑料30]
F_{40} [tetrafluoroethylene – ethylene copolymer] 四氟乙烯-乙烯共聚物[氟塑料40]
F_{42} [copolymers of tetrafluoroethylene – vinylidene fluoride] 四氟乙烯-偏氟乙烯共聚物[氟塑料42]
F_{46} [fluorinated ethylene propylene resin, FEP] 氟化乙丙树脂[氟塑料46]

F – type calender　F 型压延机
FA［furfural acetone resin］　糠醛 – 丙酮树脂，糠酮树脂
fabric　织物,织品；布
~ – backed　织物衬里的
~ backed plastic　织物衬里塑料
~ base laminate　布基层压［层合］(塑料)
~ base laminated sheet　布基层压板
~ bearing　层压胶布板制的轴承
~ bonding with foam　织物与泡沫塑料黏合
~ chips［clippings］　布碎片,布屑＜填料＞
~ coating　织物涂布
~ disk　布轮〈抛光〉
~ – filled moulding compound［material］　布填模塑料
~ filled phenolics　布填酚醛塑料
~ filler　布质填料
~ foam laminate　织物 – 泡沫塑料层合制品
~ hybrid composites　混杂织物复合材料
~ insert　布层(层合料)
~ laminate　布基层合［层压］(塑料)
~ lined pressure – tubing　布衬耐压软管
~ supported sheet　布衬片材
~ surface belt　外敷布传动皮带
fabricability　二次加工性
fabricated shapes　加工型材
fabricating　二次加工
~ of plastics　塑料二次成型
fabrication　二次加工
~ device　加工装置
~ line　加工生产线
~ of plastics　塑料二次成型
~ process　加工方法
~ stress　加工应力
fabricator in plastics　塑料加工者；塑料加工厂
fabroid gear　层压布板制的齿轮
facade　正面,门面,表面

~ paint　门面涂料
face　表面层,面板；层(合)面；辊面
~ coat　表面涂层,面漆
~ coating　表面涂布
~ crossing　面板交错层〈胶合板〉
~ cutter　端面切割机〈美〉
~ cutting machine　端面切割机
~ element　(曲)面元素,表面元素
~ fabric　面料
~ finish　表面修饰；表面涂饰剂
~ of flight　螺纹面〈螺杆〉
~ of teeth　齿轮面〈齿轮〉
~ of the weld　焊缝表面
~ plate　面板
~ width　(辊)面宽(度)
~ yarn　面纱
faceroll　面辊
facia　框,架；正面〈建筑〉；仪表板〈汽车〉
facing　面板［层,材,饰］,覆盖面［层］
~ material　面饰材料
factice　油膏〈橡胶代用品〉,亚麻油(或鱼油)橡皮
factis　油膏〈橡胶代用品〉,亚麻油(或鱼油)橡皮
factor　系数,因数
fadeless　不褪色(的)
fadeometer　褪色仪
fading　褪色；不耐光(照)的
Fahrenheit　华氏(温度)
~ degree　华氏温度,华氏度［F］
~ scale　华氏温度,华氏温标［F］
~ temperature　华氏温度
~ thermometer　华氏温度计
failing load　破坏载荷
failure　失效［事］,故障；破坏,断裂
~ criterion　失效准则
~ envelope　失效包络线
~ load　破裂［断裂］负荷
~ mechanics test　断裂力学试验
~ stress　破坏应力
~ in bend　弯曲破坏
~ test　破坏性试验
fairing　涂覆；贴面〈美〉

falling 降落的
- ~ ball(impact)test 落球冲击试验
- ~ ball viscometers 落球黏度计
- ~ bolf test 降落螺栓(冲击)试验
- ~ cylinder viscometer 圆筒下落式黏度计
- ~ dart 落镖
- ~ dart test 落镖(冲击)试验
- ~ film evaporator 降膜(式)蒸发器
- ~ height 降落高度,坠落高度
- ~ sphere viscometer 落球黏度计
- ~ weight 落锤
- ~ -weight energy at 50% failure 落锤(冲击)平均断裂能
- ~ weight impact strength 落锤冲击强度
- ~ weight impact test 落锤冲击试验
- ~ weight test [drop weight test] 落锤(冲击)试验

false 假的,不真实的
- ~ body 假体;假稠性
- ~ bottom 假底
- ~ neck 假颈(吹塑)

family 族,系列,成套
- ~ mo(u)ld 套模,多模腔(具),成套组件成型模具
- ~ package 成套包装,大包装

fan 扇;通风
- ~ gate 扇形浇口
- ~ motor 风扇电动机
- ~ tail die 扇尾形模头[口模]

fancy 鲜艳的,花色的,装饰的,高档的,精制的
- ~ goods 杂货;装饰用品
- ~ package 花色包装,精美包装
- ~ plywood 装饰胶合板
- ~ ribbon 装饰带〈包装〉
- ~ yarn 花色纱

Farrel Diskpack plasticator 法莱尔磁盘式塑化器〈无螺杆塑化器〉

fascia 框,架;正面〈建筑〉;仪表板〈汽车〉

fashioning 精加工,修饰[整]

fast 快速(的),坚牢(的),耐久(的)

- ~ colours 坚牢染料,坚牢色泽
- ~ cure phenolic moulding powder 快速固化酚醛模塑粉
- ~ curing accelerator 快速固化促进剂
- ~ -curing adhesive 快速固化黏合剂
- ~ -curing mo(u)lding compound 快速固化模塑料;快速固化成型料
- ~ -elastic deformation 急弹性变形
- ~ -mo(u)lding additive 快速成型添加剂
- ~ -neutron irradiation 快速中子辐射
- ~ setting adhesive 快速固化黏合剂
- ~ -setting glue 快速固化胶黏剂(美)
- ~ solvent 速干溶剂
- ~ to light 耐光的;不褪色的,不变色的

fastener 坚固件,接线柱

Fasting vibrating mill 法斯厅振动磨

fastness 坚牢(度);不褪色性;耐……性[度]
- ~ of dye 染料坚牢度
- ~ of dyeing 染色坚牢度
- ~ to alkali 耐碱牢度
- ~ to boiling 耐煮性
- ~ to light 耐光牢度,耐光性,耐晒性
- ~ to wear 耐磨耗性

fat 脂肪,(油)脂

fatigue 疲劳
- ~ bending test 弯曲疲劳试验
- ~ breaking 疲劳破裂
- ~ crack 疲劳裂缝
- ~ crack growth 疲劳裂纹生长
- ~ crack propagation 疲劳裂纹生长
- ~ cracking 疲劳龟裂
- ~ curve 疲劳曲线
- ~ endurance limit 疲劳极限
- ~ failure[fracture] 疲劳破裂
- ~ life 疲劳寿命
- ~ limit 疲劳极限
- ~ limit under completely reversed stress 完全交变应力疲劳极限
- ~ notch factor 疲劳缺口因数
- ~ ratio 疲劳比

~ resistance[strength]	耐疲劳性
~ strength	疲劳强度
~ stress	疲劳应力
~ test(ing)	疲劳试验
~ tester	疲劳试验机
~ under flexing	挠曲疲劳
~ value	疲劳极限值

fatty acids 脂肪酸
fault 缺陷,误差;损坏,失效
  ~ of measurement 测量误差
f. e. [for example] 例如
Fe [iron] 铁
feasible 可用的〈材料〉;可实行的〈方法〉
feature 特性,性能;零件,部件
Federal 联邦(制)的;联盟的;联合的
  ~ Specifications 联邦规范〈美国〉
  ~ Standards 联邦标准〈美国〉
Fedl. [Federal] 联邦(制)的;联盟的;联合的
fee 费用
feed 加料,进料,供料,装料
  ~ adjuster 加料调节器
  ~ agitator 加料搅拌器
  ~ arm 装料臂
  ~ bank 装入料
  ~ bin 料斗箱
  ~ block [distributor block] 分料区〈共挤塑〉
  ~ board 装料板
  ~ bush 注口衬套,浇口衬套
  ~ chamber 加料室
  ~ channel 进料道
  ~ chute 进料槽
  ~ cock 进料阀
  ~ control 进料控制,自动进料
  ~ conveyer 进料输送机
  ~ cylinder 料筒
  ~ door 进料口;进料盖
  ~ end 进料端
  ~ extruder 进料挤塑机
  ~ gear(ing) 进料传送
  ~ gun 进料枪
  ~ hopper 加料斗
  ~ hopper door 进料斗门
  ~ hopper venting 加料斗排气
  ~ hopper with slot die 带有缝型模头进料斗
  ~ inlet 进料入口
  ~ line 供给线
  ~ manifold 进料(歧)管
  ~ material 供料
  ~ mill 进料磨
  ~ nip 加料辊隙
  ~ -off roll 卷取辊
  ~ opening [feed throat] 进料口〈挤塑〉
  ~ orifice 浇口,加料口
  ~ pass 加料辊隙
  ~ performance 供料运行
  ~ pipe 供给管,供料管
  ~ pump 供料泵
  ~ ram 供料活塞
  ~ rate 加料速度
  ~ regulator 进料调节器
  ~ roll(er) 供料辊
  ~ screw 供料螺杆
  ~ section 加料段
  ~ section of screw 螺杆加料段
  ~ side 进料侧
  ~ slide 计量给料滑板,定量加料滑阀
  ~ slide dosing device 定量加料滑阀装置
  ~ slot 进料槽
  ~ socket 装料套接管
  ~ spout 供料斜槽
  ~ stock 原料,装入料
  ~ supply 供料
  ~ system 加料系统(在模具中),注料系统;注料道物料〈美〉
  ~ table 装料台
  ~ throat 进料口,进料喉
  ~ tray 供料盘,加料盘
  ~ valve 供给阀
  ~ water 供水
  ~ water preparation [purifying] 供水准备
  ~ weighing 供料称量,称料

~ zone 供料段
~ zone of screw 螺杆供料段
feedback 反馈;回料
~ coextrusion 多层膜共挤塑,多层板共挤塑
~ control system 反馈控制系统
~ selentor 反馈选择器
feeder 进料器,加料器;馈电线〈美〉
~ hopper 加料斗
feeding 加料〈塑料〉
~ belt 加料带
~ capacity 装料量;进料能力〈螺杆〉
~ device 加料装置
~ head 供料机头
~ mechanism 供给机构
~ process 供料过程〈挤塑机〉
~ screw 供料螺杆;计量螺杆
~ tunnel 加料漏斗
~ of hot material 热料供给
~ and take-off device 加料和牵引装置
feel (织物)手感
feeler 探针;厚薄规
feldspar 长石
felt (毛)毡;成毡
~ guide 毡,导辊
~ polishing disc [bob] 毡抛光轮
felting 毡,缩绒
~ screen 毡网
female 阴模
~ die[form,mo(u)ld] 阴模,凹模;下(半)模
~ screw 阴螺纹;螺母
~ thread 阴螺纹,内螺纹
fender 防护板,缓冲器;挡板,挡泥板
FEP [fluorinated ethylene propylene resin, F_{46}] 氟化乙丙树脂[氟塑料46]
ferric oxide 氧化铁;三氧化二铁〈填料〉
ferrous oxide 氧化亚铁,一氧化铁
festoon 花彩
~ dryer 悬挂式干燥机
festooning oven 垂挂式烘箱
ff. [following] 下页;下述的
FF [furan formaldehyde copolymers] 呋喃甲醛共聚物
FFF [phenol furfural copolymers] 苯酮-糠醛共聚物
f. i. [for instance] 例如
fiberize 纤维化,成纤维;分离纤维
fibre [fiber〈美〉]纤维;纤维制品;钢纸
~ accumulation 纤维堆积
~ adhesive 纤维状黏合剂
~ board 纤维板
~ breakage 纤维断裂
~ bunching 玻纤离析
~ clump 纤维束〈玻璃纤维增强热塑性塑料〉
~ coarseness 纤维粗度
~ content 纤维含量
~ content by volume 纤维体积含量(玻璃纤维增强塑料)
~ cross-section 纤维截面
~ cutter 纤维切断器
~ delivery arm 纤维导杆〈长丝缠绕〉
~ distribution 纤维分布
~ dusting plant 植绒装置
~ -filled moulding material 纤维填充模塑料
~ fineness 纤维细度
~ -finish 纤维整理剂,纤维处理剂〈润滑,抗静电〉
~ finishing 纤维表面处理
~ -foam (玻璃)纤维(增强热塑性)泡沫塑料
~ formation[forming] 纤维成型,成纤
~ -forming material 成纤材料
~ -forming polymers 成纤聚合物
~ glass 玻璃纤维
~ glass cord 玻璃纤维帘(子)线
~ glass laminate 玻璃纤维层合(压)塑料
~ glass mat 玻璃纤维毡
~ glass reinforced laminate 玻璃纤维增强层合(压)塑料
~ glass reinforced plastics 玻璃纤维增强塑料
~ glass reinforced thermoplastics 玻璃纤维增强热塑性塑料

fibrillating

- ~ glass reinforced thermoset 玻璃纤维增强热固性塑料
- ~ glass reinforced thermosetting plastics 玻璃纤维增强热固性塑料
- ~ glass reinforcement 玻璃纤维增强材料
- ~ in tension 拉伸纤维
- ~ insulating board 纤维绝缘板
- ~ length 纤维长度
- ~ length distribution[FLD] （玻璃）纤维长度分布〈玻璃纤维增强成型制品〉
- ~ modification 纤维改性(作用)
- ~ number 纤维纤度
- ~ optics [light transmitting fibres] 纤维光学；光导纤维
- ~ orientation 纤维取向(度)
- ~ orientation distribution[EOD] （玻璃）纤维取向分布〈玻璃纤维增强成型制品〉
- ~ paper 纤维纸，硬化纸板，刚纸
- ~ path[filament path] 丝路〈缠绕法时〉
- ~ pattern 纤维纹〈玻璃纤维增强塑料〉；纤维结构
- ~ pay-off 丝溢出
- ~ period 纤维周期
- ~ prominence 露丝
- ~ pull-out test 纤维牵伸试验
- ~ recovery 纤维回收；纤维再生(利用)
- ~ reinforced composites 纤维增强复合材料
- ~ reinforced composite material 纤维增强复合材料
- ~ reinforced plastic[FRP] 纤维增强塑料
- ~ reinforced thermoplastics[FRTP] （玻璃）纤维增强热塑性塑料
- ~ reinforced thermosetting plastics 纤维增强热固性塑料
- ~ reinforcement 纤维增强材料
- ~ resin sprayer 纤维-树脂喷枪
- ~ seal 纤维密封，纤维不透气
- ~ sheet 纤维板
- ~ show 露丝，纤维外露〈增强塑料〉
- ~ slurry 纤维浆(料)〈层合预成型用〉
- ~ spray gun moulding 纤维喷枪喷射成型
- ~ spraying 纤维喷射
- ~ staining 纤维着色
- ~ strain in flexure 弯曲纤维应变
- ~ strand 纤维束〈层合〉
- ~ streak 纤维条纹
- ~ strength 纤维强度
- ~ stress 纤维应力
- ~ structure 纤维结构
- ~ tear 纤维撕裂
- ~ tow 纤维束〈化纤〉
- ~ volume fraction[FVF] 纤维体积含量〈玻璃纤维增强成型制品〉
- ~ wethess 纤维浸润性
- ~ whitening 纤维发白

fibreglass 玻璃纤维
fibres array 纤维排列
fibrefil mo(u)lding 包纤模塑
fibrid 纤条体〈化纤〉
fibril(la) 原纤维
fibrillate 有原纤维的；形成原纤维
fibrillated 原纤化的
- ~ fibre 原纤化纤维
- ~ fibre tow 原纤化纤维束
- ~ film[film fiber] 原纤化薄膜
- ~ film fibre 原纤化薄膜纤维
- ~ film products 原纤化薄膜制品
- ~ film yarn 原纤化薄膜纱
- ~ network yarn 原纤化网状纱
- ~ strand 原纤化纱；裂膜纱
- ~ tape 裂膜带；原纤化膜带；原纤化扁丝
- ~ tow[fibrillated fibre tow] 原纤化纤维束
- ~ yarn 裂膜扁丝；原纤化丝

fibrillating 原纤化
- ~ equipment 原纤化装置
- ~ film 原纤化薄膜
- ~ process 原纤化法

fibrillation

~ roller 原纤化辊
~ technique 原纤化技术
fibrillation 原纤化作用〈塑料膜〉
~ ratio 原纤化率
fibrillator 原纤化器
~ roller 原纤化辊
fibrilled film fibre 裂膜纤维
fibrous 纤维状的
~ composite structure 纤维复合材料结构
~ glass 玻璃纤维
~ glass reinforced plastics 玻璃纤维增强塑料
~ glass reinforcement 玻璃纤维增强材料
~ reinforcement 纤维增强材料
~ slab 纤维板
~ structure 纤维状结构
~ web 纤维带
fictitious strain energy 虚拟应变能
field 现场,范围;场
~ assembly 现场组装
~ connection 现场装配
~ emission microscopy 场致发射显微术
~ ion microscope [FIM] 场离子显微镜
~ of application 应用范围
~ of dimensional tolerances 尺寸公差范围
~ of joining 并接技术
~ of use 应用范围,应用领域
~ strength 场强,电场强度
~ test 现场试验,实地试验
~ weld 现场焊接
FIFO [first in – first out] 先进 – 先出
fig. [figure] 图
figure 形状;数字
filament 长丝;单丝,单纤维
~ adhesive tape 纤维基材胶黏带
~ crystals 丝晶
~ delivery 导纱〈长丝缠绕〉
~ delivery carriage 导纱滑架〈缠绕〉
~ denier 长丝旦数
~ formation 成丝
~ from film 裂膜丝,裂膜纤维
~ guide 导纱器〈缠绕机〉
~ length 单丝长度
~ number 单丝纤度
~ path 长丝卷绕行程
~ pullout test 纤维拔出试验
~ size 单丝支数
~ structure 长丝构架
~ tension compensator 纤维张力补偿器,纤维张力调节器
~ weight ratio 长丝重量比
~ wind – up 长丝收卷
~ winding 长丝卷缠
~ winding machine 长丝缠绕机
~ winding method 长丝缠绕法
~ winding technique 长丝缠绕技术〈层合〉
~ wound glass reinforced plastics 长丝缠绕玻璃增强塑料
~ wound pipe 长丝缠绕管
~ wound vessel 长丝缠绕容器
~ yarn 长丝纱
filamentation 纤化,成丝
filamentous 似长丝的,由长丝组成的
file 提交,申报〈专利〉;锉〈刀〉
fill 填充
~ – and – wipe 填抹〈涂料〉
~ hole 填充口(孔)
~ – in marking 喷口末上漆
~ orifice 填充孔,浇口
~ – out 填充量〈增强塑料〉
~ rate 填充速率[度]
~ time 充模时间
filled 填充的
~ plastics 填充塑料
~ polymer 填充聚合物
~ polypropylene 填充聚丙烯
~ polyvinyl chloride plastics 填充聚氯乙烯塑料
~ short 填充不足〈模具〉,缺料
~ thermoplastics 填充热塑性塑料
~ volume 填充体积
filler 填料,填充剂;腻子;树脂粘结剂;

装灌机;间隔件〈模具内〉
~ characteristic　填料性能
~ content　填料含量
~ fibre　填料纤维〈复合纤维〉
~ gum　填充胶料
~ hole　注入孔
~ layer　填料层〈玻璃纤维〉
~ material　填(充)料;焊条;焊接添加料
~ orientation　填料取向
~ packing fraction　填料充填系数
~ particle　填料颗粒
~ plate　压塑模装料板,装料板
~ plate mould　装料板式模具
~ reinforcement　填充补强
~ reinforcement effect　填料加强效应
~ rod　焊条;填充条,嵌条
~ seam　填密缝
~ sedimentation　填料沉积
~ sheet　垫片
~ speck　填料斑(瑕)
fillet　圆角,圆角嵌条;胶瘤;填角焊缝
~ radius　圆角半径
~ weld　角焊缝
~ welded joint　角焊缝连接
~ welding　填角焊
filling　充模;填充(料);纬纱〈织造〉
~ agent　填充剂,填料
~ and weighing machine　填装称量机
~ compound　填(充)料
~ chamber　填充室
~ cycle　充模周期
~ factor　填充系数,填充程度
~ hole　填充孔
~ layer　填充层;衬层
~ level　填充程度
~ limit　填装极限〈包装〉
~ material　填充材料,填料
~ mo(u)ld speed　充(灌)模速率[度]
~ point　填装点,填装孔
　~ pressure　充模压力〈注塑〉
~ putty　腻子,填料
~ space　装料腔,装槽〈美〉
~ stage　充模阶段,装料过程

~ thread　纬纱,纬丝
~ time　充模时间
~ yarn　纬纱
film　薄膜;软片,(电影,照相)胶片,胶卷;涂膜,涂层
~ adherability　薄膜黏合性
~ adhesive　膜状黏合剂,黏合膜,胶膜
~ and sheet production　薄膜和片材生产
~ blocking　薄膜粘连
~ blowing　薄膜吹塑,管膜挤吹
~ blowing(die)head　薄膜吹塑机头
~ blowing plant　薄膜吹塑装置
~ bobbin　薄膜卷轴
~ bonding　薄膜粘接[黏合]
~ calender　薄膜压延机
~ caliper　薄膜厚度;薄膜厚薄规
~ capacitor　薄膜电容器
~ casting　薄膜流延
~ casting die　薄膜流延机头
~ casting plant　薄膜流延装置
~ coated bag　敷膜袋
~ coating　薄膜涂布;薄膜贴合
~ coefficient　薄膜系数,膜层散热系数
~ conversion　薄膜再加工
~ cooling　薄膜冷却
~ diffusion　薄膜扩散
~ extruder　薄膜挤塑机
~ extrusion　薄膜挤塑
~ extrusion line　薄膜挤塑生产线
~ die head　薄膜(吹塑)机头
~ drier　薄膜干燥机
~ effect　薄膜效应
~ elasticity　薄膜弹性
~ evaporator　薄膜蒸发器
~ extrusion　薄膜挤塑
~ fibre　裂膜纤维,薄膜纤维
~ filament　裂膜丝,裂膜纤维
~ formation　薄膜形成,成膜
~ former　成膜剂;成膜物〈涂料〉
~ forming agent[substance]　成膜剂;成膜物〈涂料〉
~ forming additive　成膜添加剂
~ forming process　成膜过程

~ forming properties 成膜性能
~ forming resin 成膜树脂
~ forming temperature 成膜温度
~ gate 膜状浇口
~ gauge 薄膜厚度;薄膜厚薄规
~ glue 胶膜;膜状胶
~ guide roll 薄膜导辊
~ haul-off unit 薄膜牵引装置
~ haze 薄膜光雾(度)
~ laminate 层合薄膜
~ laminating 薄膜层合
~ laminating adhesive 薄膜层合黏合剂
~ laminator 挤塑敷层装置;薄膜复合机,薄膜层合机
~ lubrication 油膜润滑
~ memory 薄膜存储器;薄膜弹性记忆,薄膜高温收缩
~ negative 底片〈照相〉
~ orientation 薄膜(分子)取向
~ plasticizer 薄膜增塑剂
~ preparation 薄膜制备
~ printing 薄膜印刷,丝网印刷
~ production 薄膜生产
~ rating 薄膜等级,薄膜规格;薄膜评ս
~ ribbon 膜带
~ roll [film bobbin] 薄膜卷轴
~ rupture 膜破裂
~ simultaneous biaxial stretching 薄膜同时双向拉伸
~ slitting 薄膜分切
~ solution (成)薄膜液
~ spicer 胶黏膜涂敷仪
~ sprue 膜状浇口
~ sprue gate 膜状浇口道
~ strength 膜强度
~ strength additive 薄膜增强添加剂
~ stretching in stages 薄膜逐渐拉伸
~ stretching unit 薄膜拉伸装置
~ strip 膜带;薄膜剥离
~ stripping test 薄膜剥离试验
~ swelling 膜层膨胀
~ take-off unit 薄膜牵引装置
~ take-up roller 薄膜卷绕辊
~ tape 膜条,扁丝
~ tape stretching device 薄膜(条)拉伸装置
~ tape yarn 薄膜带纱
~ temperature 薄膜温度
~ test 薄膜试验
~ thickness 薄膜厚度
~ thickness measuring apparatus 薄膜厚度测定仪
~ treatment 薄膜处理,薄膜精制
~ tube 管状薄膜;膜(软)管
~ type evaporator 薄膜式蒸发器
~ wapper 薄膜包装
~ web 膜幅
~ weight 薄膜重量
~ winder 薄膜卷绕机,薄膜收卷机
~ with protective layer 保护膜层,防护层
filter 过滤器
~ board 滤板
~ cartridge 过滤器滤筒
~ cloth [gauze] 滤布,滤网
~ efficiency 过滤效率
~ for impurities 杂质过滤器
~ gauze 滤布,滤网
~ pack 过滤网叠
~ paper 滤纸
~ plate 滤板
~ press 压滤器
~ screen 滤网
filterability 过滤性
filtering 过滤
~ candle [candle filter] 烛状过滤器
~ centrifuge 过滤式离心机
~ disks 滤片
~ flask [vacuum filter] 吸滤瓶,吸滤器
~ medium 过滤介质
~ sieve 滤筛,滤网
~ tub 滤槽
filtration 过滤
filtrate 滤液
fin 模缝脊;毛刺,飞边,溢料

~ formation 溢料构成
~ seal 翅形封口〈包装〉
~ torpedo 翼片式分流梭
final 最终的,最后的
  ~ assembly 总装配
  ~ blow moulding 最终吹塑
  ~ coat 末道漆,罩光漆,面漆
  ~ cross–section 最终断面
  ~ curing 最终固化
  ~ dimension 成品尺寸
  ~ end–product 成品
  ~ expansion 最终发泡〈泡沫塑料〉
  ~ polishing 最后抛光
  ~ product 最终产品,成品
  ~ temperature 终点温度
  ~ treatment 后处理,最终处理
fine 细(粒)的,薄的,精密的,优质的
  ~ adjust 精细调节
  ~ adjustment 细调,微调
  ~ control 细调
  ~ crack 细裂缝
  ~ crush 细碎
  ~ crusher 细碎机
  ~ denier 细纤度,细旦
  ~ fissure 微裂
  ~ grain 微细粒(料),细晶粒
  ~ –grain product 细粒产品
  ~ –grained 细粒的
  ~ grained spherolithic 细粒球结构,微球结构
  ~ grinder 细研磨机
  ~ grinding 精磨,细磨
  ~ mesh 细网眼
  ~ particle 细粒,微粒子
  ~ particle size synthetic silica 细粒合成硅(石)〈特种填料〉
  ~ powder 细粉
  ~ reduction 精细磨碎
  ~ structure 精细结构
  ~ yarn 细支纱
fineness 细度;纤度
  ~ of grain 晶粒细度
  ~ of grinding 研磨度(机械预处理粘接件)

fines 料末;微粒;碎屑;筛屑
finger 手指,指粒
  ~ blade agitator 指型桨叶式搅拌器
  ~ buff 指型抛光轮
  ~ joint 榫槽连接;指形接合
  ~ lever 指状手柄
  ~ paddle 指型桨叶
  ~ paddle[blade] agitator 指型桨叶式搅拌器
  ~ paddle mixer 指型桨叶式混合器
  ~ pin 指形销,斜向销
  ~ tab–out test 指黏试验,指尖试验
fingernail 指甲
  ~ test 划痕(指甲)试验〈塑料硬度试验〉
fingerprint 指纹(印)
  ~ method 指纹(印)法〈测定胶黏剂黏性〉
finish 修整,修饰;涂饰;涂层;表面特性;处理剂
  ~ –grinding 精抛光
  ~ insert 瓶颈成型嵌具
  ~ –milled 细磨的,精抛光的
finished 精加工的,最后加工的
  ~ article 成器(部件)
  ~ dimension 精整后尺寸
  ~ joint 平联接〈管〉
  ~ part 成品部件
  ~ product 成品
  ~ size 最终尺寸,成品尺寸
  ~ surface 精加工表面
  ~ working 精加工
finishing 修饰最后加工;精制;抛光;整理,修整
  ~ agent 涂饰剂,修饰剂;整理剂
  ~ agent for glass fiber 玻璃纤维处理剂
  ~ allowance 精加工留量,加工余量
  ~ coat[paint] 装饰层;面层涂料,末道涂层
  ~ of web–type material 带状织物修饰
  ~ roll 上光辊,精轧辊
finite strain 有限应变
finless 无飞边

finned torpedo 翼片式分流梭,旋转的分流梭
FIP [formed in place] 就地成型,现场成型
fire 火;着火,燃烧
~ behaviour 燃烧行为〈塑料〉
~ crack 热裂
~ crack transfer mark 热裂印痕
~ detector 火灾探测器
~ hazard 易燃性,易着火性
~ hazardous 易着火的
~ intensity 火焰强度
~ performance 燃烧行为
~ point 燃点,着火点
~ proofness 耐火性,防火性
~ protection equipment 防火设备
~ proof 耐火的;防火的
~ -proof fibre 防火纤维,不燃纤维
~ -proof plywood 防火胶合板
~ proofing 阻燃;耐火;防火
~ proofing agent 防火剂
~ resistance 耐火性
~ resistance unsaturatated polyester resin of low toxin and low smokiness 低毒、低发烟性难燃不饱和聚酯树脂
~ resistant 耐火的
~ -resistant fibre 阻燃纤维
~ resistive 耐火的
~ resistive material 耐火材料
~ retardant 阻燃剂
~ retardant and reinforcement nylon 阻燃增强尼龙
~ retardant fibre 阻燃纤维
~ retardant resin 阻燃树脂
~ retarding 阻燃,防火
~ retarding agent 阻燃剂
~ retarding paint 防火漆
~ retarding resin 阻燃树脂
~ wall 防火墙
firing time test 燃烧速度试验
Firestone flexometer 费尔斯通挠曲仪
firmoviscosity [viscoelasticity] 黏弹性
first 第一的,最初的
~ coat 底涂层,底漆(层)
~ -order transiton 一级转变
~ -order transiton point 一级转变点
~ -order transiton temperature 一级转变温度
~ quality 优质,一等品
~ runnings 初馏物〈蒸馏〉
~ stage cure 第一阶段硫化
~ -surface decorating 正面装饰
~ treatment 预处理,初步处理
fish 鱼,鱼尾板
~ -eye 鱼眼,白点
~ eye effect 鱼眼现象
~ glue 鱼胶,鳔胶
~ net 渔网
~ net yarn 渔网线
~ paper 鱼纸,薄硬化纸板
~ scales 鳞斑
~ -tail 鱼尾(型)
~ -tail-blade 鱼尾型搅拌桨
~ -tail die 鱼尾型口模,缝型口模〈挤塑〉
~ -tail manifold 鱼尾型流道
~ -tail sheet die 鱼尾型板材口模
~ -tail type gate 扇型浇口,鱼尾型口
~ -tail type kneader 鱼尾型捏合机
~ -tail-type kneader arm 鱼尾型捏合机桨
fissure 豁裂;裂缝
fit 适合,适应;装配,安装;调准;配准,配ବ
FIT [flash ignition temperature] 骤燃温度
fitment (密封)配件〈液态包装〉
fitting 配件;管件;装配
~ accuracy 装配精度
~ assembling 装配
~ notched pin 装配开槽销
~ slot 菱形键
~ washer 装配垫圈
~ welding 装配焊接
~ work 装配工作
fix 固定,安装,装配
~ stopper 定位销
fixed 固定的

flame

~ bed　固定床
~ bottom plate　固定下模板
~ die blade　定模唇
~ die head　固定模头〈薄膜吹塑〉
~ die platen　定模板,固定压模台板
~ flange coupling　固定法兰连接器
~ humidity　恒湿
~ jaw[anvil jaw]　固定钳;固定颚板
~ knife　固定刀〈切粒机〉
~ lip　定模唇
~ mo(u)ld　固定压模
~ plate　固定模板〈注塑〉,定压板
~ press platen　定压板
~ pressure　夹板压力;固定压力
~ roll　固定辊
~ stripper　固定模板
~ table　固定台〈压机〉
~ temperature　恒温
~ time testing　定时试验
~ vessel　固定容器
fixing bosses　固定环
fixture　固定装置,夹紧装置,卡具〈试验机〉
flake　薄片;绒屑;干坯料
~ off　剥离,剥落;片落
~ board　(夹木屑)胶合板,木屑板
flaking　剥落〈涂层〉
flaky　(薄)片状的
~ grain　片状颗粒
~ material　片状材料
~ resin　片状树脂
flame　火焰;发火焰
~ bonder　火焰粘接器
~ bonding　火焰熔接
~ flash-over test　火焰飞弧试验
~ fusion　火焰熔接
~ gun　火焰喷枪
~ -hardened　火焰硬化的,火焰淬火的〈金属〉
~ inhibitor　阻燃剂
~ laminating　火焰熔接,火焰层合
~ -out time　消焰时间
~ plating　火焰喷涂〈美〉
~ point tester　闪(燃)点试验仪,着火

点试验仪
~ polishing　火焰[热]抛光(热塑性塑料去毛边)
~ -polishing device　火焰抛光装置〈热塑性塑料去毛边和抛光〉
~ proof　耐火的,阻燃的
~ -proof enclosure　(电气设备)防火外壳
~ -proof fibre　阻燃纤维
~ proofing　防火处理,阻燃处理
~ proofing agent　防火剂;耐火剂
~ proofness　耐火性,防火性
~ propagation　火焰蔓延
~ propagation rate　火焰蔓延速率
~ propagation test　火焰蔓延性试验
~ resistance　耐燃性,阻燃性
~ resistance plastic　耐燃塑料
~ resistance polypropylene　阻燃聚丙烯
~ resistance test　耐火性试验,阻燃性试验
~ -resistant　耐火的,防火的,阻燃的
~ resistant fibre　耐燃纤维
~ resistant resin　耐燃树脂
~ resistivity　耐燃性
~ retardance　阻燃性
~ retardant　阻燃剂;阻燃的
~ retardant additive　阻燃添加剂
~ retardant coating　耐燃性涂料
~ retardant plastics　耐燃性塑料
~ retardant reinforced thermoplastics　阻燃性增强热塑性塑料
~ -retardant[retarded]resin　阻燃树脂
~ retarding polyether　阻燃聚醚
~ retarding polymer　阻燃聚合物
~ sealing　焰熔热合
~ smoke suppressant　焰烟抑制剂
~ spray　火焰喷涂
~ spray coating　火焰喷涂
~ spray gun　火焰喷枪
~ spraying　火焰喷涂(法)
~ spraying gun　火焰喷枪
~ spraying method　火焰喷涂法
~ spraying plant　火焰喷涂装置

flaming

- ~ spread　火焰蔓延,延烧
- ~ spread classification　燃烧性分级
- ~ spread rate　火焰蔓延速率,延烧速率
- ~ spread rating　延燃等级
- ~ spread time　火焰蔓延时间,延烧时间
- ~ test　闪燃点试验,燃烧试验〈塑料〉
- ~ treatment　火焰处理
- ~ welding　火焰熔焊

flaming　火焰处理

flammability　易燃性,可燃性
- ~ index　可燃性指数
- ~ point　燃点
- ~ range　可燃性极限,可燃性范围
- ~ resistance　阻燃性
- ~ standards　可燃性标准
- ~ test　可燃性试验,耐燃性试验

flammable　易燃的,可燃的
- ~ fabric　可燃性织物

flange　法兰(盘),凸缘;折边
- ~ - adaptor　法兰连接套
- ~ connection　法兰盘连接
- ~ joint　法兰连接
- ~ - mounted motor　法兰盘式安装发动机
- ~ of the joint　连接凸缘

flanged　凸缘的,带有法兰的,折边的
- ~ edge　法兰边缘;卷边
- ~ joint　法兰连接
- ~ motor　法兰盘式发动机
- ~ pipe　法兰管,凸缘管
- ~ section　凸缘型材〈制品〉

flanging　凸缘;折边
- ~ machine　折边机,卷边机
- ~ press　折边压力机
- ~ roll　折边辊
- ~ tool　卷边工具,折边工具〈塑料成型〉

flank　(齿轮)齿侧面

flannel disk　法兰绒磨光盘

flap　片状物;折板,挡板
- ~ bag　折翼袋
- ~ lid　折翼盖(包装)
- ~ wheel　片状抛光轮

flash　溢料;溢边,飞边;缝脊;闪现
- ~ ared　溢料面;溢料脊
- ~ chamber　溢料腔[槽]
- ~ chearance　溢料缝,溢料间隙
- ~ cooler　真空冷却器
- ~ dryer　急骤干燥器
- ~ drying　急骤干燥
- ~ edge　溢料边缘,飞边
- ~ face [flash ridge, land]　溢料脊,溢料面
- ~ factor　溢料系数
- ~ fin　溢料,飞边
- ~ formation　飞边(构成),溢料(构成)
- ~ - free　无飞边,无溢料
- ~ - free molding　无溢料成型
- ~ gap　溢料缝
- ~ gate　飞边型浇口,溢料浇口
- ~ groove　溢料槽,溢料缝
- ~ heat　加速加热
- ~ heater　急速加热器
- ~ - ignition temperature　骤燃温度
- ~ land [edge]　溢料边缘,飞边
- ~ lathe　去飞边,倒角
- ~ line　合模线,溢料线
- ~ mixer　快速混合器
- ~ mo(u)ld　溢料式模具
- ~ mo(u)lding　溢式成型
- ~ - off　闪蒸出,蒸发
- ~ - off temperature　闪蒸温度,蒸发温度
- ~ - off time　闪蒸时间,蒸发时间
- ~ - off zone　闪蒸区段,蒸发区段
- ~ over　飞弧;放电
- ~ over test　飞弧试验
- ~ overflow　溢出(料)
- ~ overflow mould　溢料式模具
- ~ point　闪点
- ~ point tester　闪(燃)点试验仪,着火点试验仪
- ~ polymerization　瞬间聚合,爆聚
- ~ relief　溢料缝
- ~ removal　切除毛边
- ~ removing lathe　切毛边机

flatting

~ ridge 溢料脊
~ ring 环形溢料缝
~ spinning 瞬时纺丝
~ subcavity mould 溢料式多槽模具
~ supercavity mould 溢料式多槽模具
~ temperature 闪点,着火温度
~ test 闪点试验
~ trimmer 修边机,除边机
~ – type compression cup mould 溢料式压缩敞口模具
~ – type mould 溢料式模具
flashing 溢料(构成)
flask 瓶;烧瓶
flat 暗淡的,无光(泽)的;平的,扁(平)的
~ bed [back] press 平板压烫机
~ blade agitator 平桨叶搅拌机
~ blade ejector 扁脱模销
~ – bottomed tank (大型)平底容器
~ butt weld 平对接焊(缝),无坡口对接焊(缝)
~ butted seam 对头并接缝合
~ compression strength 平压强度
~ crest 平面齿顶〈螺杆螺纹〉
~ crush resistance 耐平压性〈包装〉
~ die 扁平[平口]模头,缝口[缝型]模头
~ die extrusion 扁平模头挤塑
~ engagement nozzle 平列注嘴〈注塑〉
~ fibre 扁丝,撕裂薄膜
~ film 平(挤薄)膜
~ film die 平膜模头
~ – film extrusion 平膜挤出法
~ – film line 平膜生产线[装置]
~ – film method 平膜(生产)法
~ – film producing plant 平膜生产装置
~ finish 无光清漆
~ gasket 平面密封垫
~ head 扁平机头,缝型模头
~ heater 电热板,平板式加热器
~ lacquer 无光清漆
~ laminate 平层合板
~ lubrication head 平面润滑机头

~ material 扁平材料〈片材〉
~ nosed screw 平头螺杆
~ nozzle 平头注嘴
~ paddle agitator 桨叶式搅拌器
~ pan 平盘;平[板]面
~ plate 平板
~ plate heater 电热板,平板式加热器
~ printing [surface printing] 平板印刷
~ root 平顶
~ roll 平压延辊
~ sample 平试样〈测试〉
~ section 平面部件〈试样〉
~ – shaped 平面状的,平坦的
~ sheet 平片材
~ sheet beading 平板卷边
~ – sheet composite 平板复合材料
~ sheet die 平板模头〈挤塑〉
~ – sheet die extrusion 平板模头挤塑
~ – sheet die for co – extrusion 平板共挤塑模头
~ – sheet die head 平板模头
~ sheet extruder 平板挤塑机
~ sheet extrusion 平板挤塑
~ sheeting die 平片[平口,扁平,缝口,缝型]模头
~ side pin ejector 半圆脱模销
~ slot die 平缝口[缝型,扁平]模头
~ tank 平面槽
~ tape 扁带,扁丝
~ tension strength 平拉强度
~ test 压扁试验
~ varnish 消光清漆,无光(泽)清漆
~ wall paint 无光墙漆
~ weld 平焊
~ yarn 扁丝
flatten 弄平,整平;使无光泽
flattened film [lay flat film] 平板薄膜
flattening 整平;平折〈管薄膜〉
~ agent 防粘结剂,防结块剂;润滑剂
~ mill 轧平机
flatness 平坦;平整度
flatting 消光(性),消光效应
~ agent 消光剂,退光剂

flatwise

~ varnish 涂油灰[腻子]漆,无光漆
flatwise 贯层方向〈层压板〉
~ compression 贯层压缩(强度)
~ compressive strength 贯层压缩强度
~ tensile test 贯层拉伸试验〈测定层压板粘接强度〉
flavour lock 闸板阀
flaw 裂缝[纹](缺陷)
flax 亚麻
FLD [fibre length distribution] (玻璃)纤维长度分布
fleece 羊毛(状),绒头织物
~ fabric 起绒织物
fleecy 羊毛状的,绒毛的
flex 挠曲,弯曲
~ cracking 柔软龟裂,挠曲开裂
~ cracking test 挠曲开裂试验
~ fatigue 挠曲疲劳,弯曲疲劳
~ fatigue resistance 耐挠曲疲劳性
~ fatigue test 挠曲疲劳试验,揉曲疲劳试验
~ -foam 软质泡沫塑料
~ life 挠曲寿命〈美〉
~ life test 挠曲寿命试验
~ lip extruder 活模唇挤塑机
~ lip extrusion 活模唇挤塑
~ resistance 耐挠性
~ resistant moulding 耐挠曲成型制品
~ stiffness 挠曲劲度
~ test 柔曲试验
~ tester 弯曲试验器
flexer 疲劳试验机
flexi blade coater 柔性刮式单辊涂胶机
flexibility 柔软性,柔韧性,挠性,揉性
~ loss 柔软性降低
~ factor 挠曲系数
~ resistance 耐折性,柔韧性
~ test 弯曲试验
flexibilizer 柔韧剂,增韧剂,软化剂
flexible 可弯曲的,挠曲的,柔韧的,软的,灵活的,活动的
~ bag laminating 软袋层合,袋式层合,袋压法
~ bag moulding 软袋成型,软袋模塑

~ -bag process 软袋法
~ cellular material 软质泡沫材料
~ cellular plastics 软质泡沫塑料
~ container 软质容器
~ cords 软线(包装)
~ coupling 挠性联轴节
~ fibre 挠性纤维
~ film 软质薄膜
~ film and sheeting 软质薄膜和片材
~ foam 软质泡沫塑料
~ foamed plastics 软质泡沫塑料
~ hose 柔软管;挠性软管
~ limit 挠曲限度
~ liner macromolecule 挠性线型大分子
~ lip [flexlip] 活动模唇〈平片材挤塑〉
~ membrane mo(u)lding 软胎模成型
~ mo(u)ld 挠性模,软模具
~ package 软(质)包装
~ packing film 软质包装薄膜
~ pipe 挠性管
~ plastic 软质塑料
~ plastic foam 软质泡沫塑料
~ plunger 柔性柱塞,软柱塞
~ plunger mo(u)lding 柔性柱塞成型
~ polyester polyurethane foams 软质聚酯型聚氨酯泡沫塑料
~ polyether polyurethane foams 软质聚醚型聚氨酯泡沫塑料
~ polyethylene 软质聚乙烯
~ polyurethane foam 软质聚氨酯泡沫
~ polyvinyl chloride 软质聚氯乙烯,增塑聚氯乙烯
~ pressure 软胎压力,柔软压力;流体压力
~ printed circuit [wiring] 软质印刷线路
~ profile 软质型材
~ resin 软(质)树脂
~ section 软(质)型材
~ sheet 软质片材,软片
~ transparent film 软质透明薄膜
~ tube 挠性管,软管〈塑料〉

~ unsaturated polyester resin 柔性不饱和聚酯树脂
~ urethane foam 软质聚氨酯泡沫(塑料)

flexing 挠曲
    cycle 挠曲周期
~ elasticity 挠曲弹性
~ fatigue 弯曲疲劳
~ fatigue life 挠曲疲劳寿命,揉曲疲劳寿命
~ life 挠曲寿命〈美〉
~ machine 挠曲疲劳试验机
~ resistance 耐挠性
~ tubing 软管

flexion 挠曲,弯曲
flexiplastic 挠性塑料,柔性塑料
flexlip 活动模唇〈平片材挤塑〉
flexometer 挠曲试验仪
flexural 挠曲的,弯曲的
~ creep 挠曲蠕变
~ deflection 挠曲变形
~ elongation 弯曲伸长
~ endurance properties 弯曲耐久性
~ fatigue 弯曲疲劳
~ fatigue resistance 耐弯曲疲劳性
~ fatigue strength 弯曲疲劳强度
~ fatigue test 弯曲疲劳试验
~ fatigue testing machine 弯曲疲劳试验机
~ impact test 弯曲冲击试验
~ modulus 挠曲模量,弯曲弹性模量
~ modulus of elasticity 弯曲模量,挠曲弹性模量
~ moment 弯矩
~ offset yield strength 弯曲偏置屈服强度
~ plastic 软质塑料,易弯曲塑料
~ properties 弯曲性能
~ resilience 弯曲回弹性,挠曲回弹性
~ resistance 耐挠性,耐弯性
~ rigidity 挠曲刚性,弯曲刚度
~ strain 挠曲应变,弯曲应变
~ strength 弯曲强度
~ stress 挠曲应力,弯曲应力

~ stress at conventional deflection 常规挠度下的弯曲应力
~ [flex] temperature 挠曲温度
~ test [bending test] 弯曲试验
~ yield strength 弯曲屈服强度

flexure 弯[挠,屈,扭]曲
~ strength 弯曲强度
~ stress 弯曲应力
~ test 弯曲试验

flight 螺纹[齿](美);螺线刮板;挡板
~ channel 螺槽〈螺杆〉
~ -clearance 螺杆和料筒的间隙〈挤塑机〉
~ depth 螺槽深度
~ land 螺棱面,螺顶
~ land width 螺棱面宽〈螺杆〉
~ lead 螺纹导程,螺齿距
~ pitch 螺距

flighted 螺纹的,螺线形的
~ dryer 螺旋状干燥器
~ length (螺杆)螺纹长度

flipper 挡泥板
float 涂抹;浮标
~ plating 架空添纱〈纺织〉
~ switch 浮动开关

floatation 浮选;悬浮
~ weight loss method 浮力失重法〈测定相对密度〉

floating 浮动;发花;浮色
~ air nozzle dryer 浮动空气喷嘴干燥器
~ bag 浮袋(包装)
~ bag package 浮袋包装
~ chase 活动[活络]模套
~ chase mo(u)ld 活模套模具,双阳模和活套的模具
~ container 浮动容器
~ controller 浮动控制器
~ core 活动模芯,活动阳模
~ film dryer 浮带式干燥器
~ knife 浮动(刮浆)刀,气动(刮浆)刀
~ knife coater 浮刮刀式涂布机,气动刮刀涂布机

flocculation

~ knife coating　浮刀刮涂,气动刀刮涂
~ knife roll coater　浮刮刀辊式涂布机
~ membrane　活动膜
~ nozzle　浮动喷嘴
~ nozzle with needle seal　针阀封闭式浮动喷嘴
~ package　浮液包装
~ particles strainer　浮粒滤器
~ platen　浮动压板;中间压板
~ platen press　浮压板式压机
~ plug　浮动堵头
~ punch　浮动阳模
~ roll　浮动辊筒
~ roll guide　浮动导辊
~ screw　浮动螺杆
~ web dryer　浮带式干燥器
~ weight　上顶栓(密炼机)
~ weight of banbury mixer　密炼机上顶栓

flocculation　絮凝(作用)
flock　短绒,绒屑;短纤维〈填料〉
~ coating　植绒涂层
~ fibre　植绒纤维〈静电植绒〉
~ filler　绒屑填充剂
~ finishing　植绒整理
~ －printing　植绒印花[印刷]
~ spraying　植绒喷射

flocked carpeting　植绒地毯
flocking　植绒
~ equipment　植绒装置

Flomix　夫洛米克斯液体混合器
flood lubrication　溢流润滑,淹没润滑
flooding　泛白,泛色;溢流
~ point　泛液点,液化点

floor　地板;底面,地面
~ caret　地毯;地板铺覆场
~ covering　地板覆盖层
~ foundation　底板,地板
~ foundation troweling material　地板腻子
~ leather　地板革
~ panel　地板
~ plan　平面图
~ plate　地板砖,铺地砖

~ space　占地面积〈机器〉
~ tiles　地板砖,地砖
~ －type　固定结构,固定式样
~ varnish　地板漆

flooring　地板(材料);铺地面
~ compound　地板料

flour　粉(末)
~ filler　细粉填料

flow　流动;流动性;流量,流径〈熔体〉
~ aid　流动性助剂,助流剂
~ behavior index　流动行为指数
~ birefringence　流动双折射
~ capacity　流量;流动能力
~ casting　流延;流动铸塑;中空铸型法
~ channel　流道〈熔体用〉
~ characteristics　流动特征
~ chart　流程图
~ coater　淋涂机
~ coating　淋涂,流涂
~ coefficient　流量系数
~ condition　流动条件〈物理〉
~ control　流量控制
~ control agent　流动控制剂
~ control system　流量控制系统
~ control valve　流量控制阀
~ cup　黏度杯;稠度杯;流杯〈测定液体流动时间〉
~ －cure behaviour　流动固化行为
~ curve　流动曲线〈塑料熔体〉
~ diagram　(工艺)流程图;流动图解
~ dichroism　流动二向色性
~ distance　流道〈模具内〉
~ disturbance　流动干扰,流体扰动〈塑料熔体时〉
~ dividing bracket　分流辐架
~ drawing　流动拉伸
~ elasticity　流动弹性
~ exponent　流动指数
~ field　流场
~ －gun　流动体喷枪,注塑喷枪
~ index　流动指数
~ instability　流动不稳定性〈熔体〉
~ limit　流动屈服值,流动极限

~ line 熔合纹;合流线;接缝线,合流纹,熔接缝〈美〉;流水线〈生产〉
~ line production 连续生产,流水线生产
~ mark 波流痕迹
~ marker line 波流痕迹〈美〉
~ meter 流量表
~ mixer 流动混合机〈液体或气体〉
~ moulded parison 流动模压型坯
~ mo(u)lding 挤料成型;传递模压法
~ of adhesive 黏合剂流动〈固化时〉
~ of air 气流
~ of plasticity 塑性流动
~ orientation 流动取向
~ out 流出;流延性
~ out temperature 流出温度,流延温度
~ path 料流途径(熔体);流道〈模具〉
~ path angle 流道角
~ path length 流径长
~ pattern 流型,流动结构,流动图;活动模
~ plan 流程图
~ process 流水作业;流动过程
~ properties[property] 流动性
~ quantity 流量
~ rate 流量,流率,流速
~ resistance 流阻,阻流
~ scheme (工艺)流程图
~ sheet (工艺)流程图
~ stress 流动应力,屈服应力
~ tab 流度标尺
~ temperature 流动温度
~ temperature range 流动温度范围
~ temperature test 流动温度试验
~ test 流动试验
~ tester 流动试验仪
~ time 流动时间
~ welding 熔流焊接
~ zone 流动区
flowability 流动性
flowed 流动的
~ - in gasket 浇铸衬垫
~ - in - place gasket 现场浇铸衬垫
flower mark 花斑,花纹痕〈制品缺陷〉
flowing 流动(的)
fl. pt. [flash point] 闪点
fluctuate 波动
fluctuating 波动
~ flow 波流流动,波动熔流
~ load 波动负荷
fluctuation 波动
~ of concentration 浓度波动
~ of temperature 温度波动
flue dust 烟囱灰,烟道尘
fluffy 松散的;绒毛的
~ yarn 毛茸纱
fluid 流体;流动的
~ agitation 流动介质搅拌
~ bed 流化床
~ bed catalysis 流化床催化
~ - bed coating 流化床涂布
~ bed dryer 流化床干燥器
~ density 流体密度
~ dynamics 流体动力学
~ - energy jet 喷射流
~ energy mill 喷射流研磨机
~ film 流体薄膜,润滑油膜
~ filter 流体过滤器
~ flow 流体流动
~ forming 流体成型
~ fracture 流体破裂
~ mixer 流体混合器
~ point 流点
~ pressure 流体压力
~ pressure moulding 流体施压成型,液压模塑
~ resin 流动性树脂
~ rubber 流动性橡胶
~ solid system [fluo solid system] 液固体系统
~ system 液体循环,流体系统
~ tight 不透水的,不渗透的
~ welding 熔焊
fluidising mixer 流态化混炼机
fluidity 流动性;流度;流值
~ at low temperature 在低温下的流

fluidizable

动性
- ~ at rest 静止时流动性
- ~ function 流动性函数
- ~ improver 流动性改进剂
- ~ index 流度指数
- ~ ratio 流度比

fluidizable 可流体化的
fluidization 流态化
- ~ chamber 流化室
- ~ coating 流化涂布
- ~ dip coating 流化浸渍涂布
- ~ drying 流化干燥
- ~ gas 流化气体,载体
- ~ point 流化点

fluidized 流动化的
- ~ bed 流化床
- ~ bed coating 流化床涂布
- ~ bed dip coating 流化床浸渍涂布
- ~ bed dryer 流化床干燥器
- ~ bed dyeing 流化床染色
- ~ bed moulding 流化床成型
- ~ bed process 流化床法
- ~ dust 流化床粉末
- ~ dust circulation 流化床粉末循环
- ~ powder bath 流化粉料浴

fluizing 流化的
- ~ cooler 流化式冷却器
- ~ dryer 流化式干燥器
- ~ gas 流化气体;气态载体
- ~ mixer 流化式搅拌器
- ~ velocity 流化速度

fluorescence 荧光
- ~ microscope 荧光显微镜

fluorescent 荧光的
- ~ bleaches 荧光增白剂
- ~ brightening agents 荧光增白剂
- ~ dye 荧光染料
- ~ film technique 荧光薄膜技术
- ~ material 荧光材料
- ~ pigments 荧光颜料
- ~ plastics 荧光塑料
- ~ substance 荧光物质
- ~ tube diffuser 荧光管扩散器
- ~ whitening dye 荧光增白染料

- ~ X-rays 荧光X射线
- ~ -X-ray analysis 荧光X射线分析

fluoride 氟化物
fluorinated 氟化的
- ~ elastomer 氟化弹性体
- ~ ethylene propylene copolymer 氟化乙烯丙烯共聚物
- ~ ethylene propylene resin [FEP, F_{46}] 氟化乙丙树脂[氟塑料$_{46}$]
- ~ polyester 氟化聚酯
- ~ polymer 氟化聚合物
- ~ polyurethane 氟化聚氨酯
- ~ silicone [FSI] 氟硅弹性体
- ~ silicone rubber 氟硅橡胶〈粘接材料〉
- ~ siloxane copolymer 氟硅氧烷共聚物

fluorinating agent 氟化剂
fluorine [F] 氟
- ~ containing elastomer 含氟弹性体
- ~ containing polymer 含氟聚合物
- ~ containing polymerisate fibre 含氟聚合物纤维
- ~ containing rubber 含氟橡胶
- ~ containing polyurethanes 含氟聚氨酯
- ~ oil 氟油〈低聚三氟氯乙烯〉
- ~ plastic 氟塑料
- ~ rubber 氟橡胶

fluorocarbon 氟碳化合物
- ~ blowing agent 氟碳发泡剂
- ~ plastic 氟碳塑料,氟塑料
- ~ polymers 氟碳聚合物,含氟聚合物
- ~ resin 氟碳树脂,氟树脂
- ~ resin fibre 氟碳树脂纤维
- ~ silicone rubber 氟硅橡胶

N-(fluorodichloromethylthio) phthalimide N-(氟二氯甲基硫代)邻苯二甲酰亚胺〈防霉剂〉

fluoroelastomer 氟弹性体,氟橡胶
fluoroethylene 氟乙烯
- ~ resin 氟乙烯树脂

fluorofibre (含)氟纤维
fluorogen 荧光团
fluorohydrocarbon 氟代烃

~ plastic 氟(烃)塑料
~ resin 氟(烃)树脂
fluoroplastic 氟塑料
~ film 氟塑料薄膜
fluoropolymer 氟聚合物
fluorosilicone rubber 氟硅橡胶
fluorspar 荧石,氟石
fluorothene 聚三氟氯乙烯
fluosilicic acid 氟硅酸
flush 冲洗,漂洗,洗涤
~ box 洗涤箱
~ out 清洗;吹洗
~ with 与……齐平
flushability 可洗涤性
flushed colours 底色
flushing 冲洗;挤水
~ pipe 冲洗管
flute 凹槽,沟纹;波纹
fluted 槽纹的;有(沟)槽的
~ core 槽纹芯材
~ cylinder 槽纹辊
~ mixing section 沟槽混合段〈螺杆〉
~ plunger 螺活塞
~ roll 槽纹辊
~ roller 槽纹辊
~ screw tip 槽螺杆顶
flux 助熔剂;流动改进剂
fluxing 熔化;增塑;塑化
~ pressure 熔压
~ temperature 熔融温度,熔化温度
~ zone 塑化段
fly 飞行,飞速
~ dust 飞扬灰尘
~ knife [rotating knife] 飞刀
~ moment 旋转矩
~ press 冲坯压机
~ -wheel press 飞轮式压机
flyer 锭壳,锭翼;衬锭〈捻丝机〉
bobbin 粗纱(筒)管〈纺丝〉
flying 快速的,飞速的
~ cinders 飞扬灰烬
~ shaft 飞轴
~ shears 快速剪切器〈分开薄膜〉
~ splice [flying transfer or non stop transfer] 快速拼接;快速[自动]换辊
~ transfer 快速变换
~ turret unroll 快速旋转退卷
~ wedge mixing - metering machine 快速楔嵌式混合计量器〈制备聚氨酯泡沫塑料〉
flywheel 飞轮
~ momentum 飞轮动量
foam 泡沫塑料;泡沫
~ adhesive 泡沫黏合剂
~ back 泡沫背衬
~ backed fabric 泡沫塑料衬里织物
~ -backed textiles 泡沫塑料衬里织物
~ backing 泡沫塑料衬里
~ base material [matrix] 泡沫塑料基材料
~ bonding 泡沫材料黏合
~ bun 泡沫塑料块
~ casting 泡沫塑料铸塑
~ cell 泡沫(微)孔
~ choking 泡沫阻塞
~ collapse 泡沫崩溃,泡沫结构破坏
~ composite 泡沫复合材料
~ core 泡沫芯
~ -cored laminate 泡沫夹芯层合塑料
~ crown 泡沫顶,泡沫(表)面
~ cushioning sheet 泡沫塑料缓冲板
~ density 泡沫(塑料)密度
~ dispensing machine 泡沫配料装置〈聚氨酯泡沫塑料〉
~ dyeing 泡沫染色
~ extrusion 泡沫塑料挤塑
~ fabrication 泡沫塑料加工
~ filament 发泡纤维,发泡长丝
~ filling of cavities 泡沫模腔填充
~ -filling technique 泡沫填充技术
~ formation 泡沫形成
~ generating nozzle 发泡喷嘴
~ glue 泡沫胶
~ gluing 泡沫粘接剂
~ in mould 模内发泡

foamability

- ~ in place 现场发泡
- ~ in site 就地发泡
- ~ -in-situ 就地发泡
- ~ in situation 就地发泡
- ~ inhibitor 消泡剂,抑泡剂
- ~ initiation rate 起泡速度
- ~ initiation time 泡沫起发时间
- ~ injection molding 泡沫塑料注塑〈菲利普技术〉
- ~ laminate 泡沫塑料层压材料
- ~ laminate fabric 泡沫塑料层压织物
- ~ laminating 泡沫塑料层压黏合
- ~ laminating machine 泡沫塑料层压机
- ~ lamination 泡沫塑料层压黏合
- ~ layer 泡沫塑料层
- ~ life 泡沫塑料寿命
- ~ matrix 泡沫塑料基体
- ~ melt method 熔体发泡法
- ~ -modified concrete 泡沫塑料改性混凝土
- ~ modifier 泡沫塑料改性剂,泡沫塑料调节剂
- ~ mo(u)lding 泡沫塑料成型
- ~ overblow 过度发泡〈泡沫塑料制造〉
- ~ package 泡沫塑料包装
- ~ pad 泡沫塑料(缓冲)垫
- ~ pipe 泡沫塑料管
- ~ plastic 泡沫塑料,多孔塑料
- ~ polystyrene 发泡聚苯乙烯
- ~ pouring 泡沫塑料浇铸法
- ~ promoter 泡沫促进剂
- ~ reservoir 泡沫塑料槽
- ~ reservoir moulding [FRM] 槽式泡沫塑料成型(法)
- ~ rubber 发泡橡胶
- ~ sandwich construction 泡沫塑料夹芯结构
- ~ sandwich mo(u)lding 夹芯泡沫塑料成型
- ~ sandwich panel 泡沫塑料夹芯板
- ~ sheet 泡沫塑料片材
- ~ sheet line 泡沫塑料片材生产线
- ~ sheeting 泡沫塑料片材
- ~ skin 泡沫塑料皮层
- ~ slab 块状泡沫塑料
- ~ slab stock 块状泡沫塑料
- ~ splitting saw 泡沫塑料分割锯
- ~ spraying 泡沫喷镀,泡沫喷涂
- ~ stability 泡沫稳定性
- ~ stabilizer 泡沫稳定剂
- ~ structure 泡沫结构,微孔结构
- ~ suppressor 抑泡剂;消泡剂
- ~ tape 发泡带〈黏合〉
- ~ yarn 发泡纱

foamability 发泡性
foamable melt 可发泡性熔体
foambacks 泡沫塑料衬里织物
foamed 泡沫的
adhesive 泡沫黏合剂
ceiling 泡沫材料天花板

- ~ core 泡沫(材料)芯
- ~ -cored wall 敞开孔泡沫壁
- ~ cushion 泡沫(材料)缓冲垫
- ~ doubled fabric 泡沫复合织物,泡沫塑料衬里织物
- ~ film 发泡(材料)薄膜
- ~ full mo(u)ld model 泡沫全模型
- ~ in mo(u)ld 模内发泡
- ~ in place 现场发泡
- ~ in place process 现场发泡法
- ~ -in-place sandwich 现场发泡夹芯板[结构]
- ~ in place shoe sole 现场发泡鞋底
- ~ in situ shoe sole 现场发泡鞋底
- ~ interlayer 发泡夹层
- ~ material 泡沫材料
- ~ moulding 发泡成型件
- ~ phenolics 发泡酚醛塑料
- ~ plastic 泡沫塑料,多孔塑料
- ~ polyethylene 泡沫聚乙烯,发泡聚乙烯
- ~ polypropylene 泡沫聚丙烯,发泡聚丙烯
- ~ polystyrene 泡沫聚苯乙烯,发泡聚苯乙烯
- ~ polyurethane 泡沫聚氨酯,发泡聚

氨酯
- ~ polyvinyl chloride 泡沫聚氯乙烯,发泡聚氯乙烯
- ~ rubber 发泡橡胶,泡沫橡胶,海棉橡胶
- ~ sheet 发泡(材料)片材
- ~ thermoplastic resin 泡沫热塑性树脂,发泡热塑性树脂

foamer 发泡机,起泡机
foaminess 发泡性
foaming 发泡
- ~ adhesive 发泡黏合剂
- ~ agent 发泡剂
- ~ capacity 发泡能力
- ~ cycle 发泡周期
- ~ expandable beads 可发泡珠粒料
- ~ in closed moulds 闭模内发泡
- ~ in place 现场发泡
- ~ in site 就地发泡
- ~ -in-situ 就地发泡
- ~ -in-situ between laminated skins 层间表皮之间就地发泡
- ~ in the mo(u)ld 模内发泡
- ~ machine 发泡机,起泡机
- ~ mo(u)ld 发泡模具
- ~ of expandable particles 可发性颗粒发泡
- ~ operation 发泡操作,发泡过程
- ~ pressure 发泡压力
- ~ process 发泡法
- ~ property 发泡性
- ~ restraint 发泡抑制
- ~ structure 泡沫结构
- ~ temperature 发泡温度
- ~ time 发泡时间
- ~ tool 发泡模具

foamless foam 无泡沫泡沫材料
foamy plastic 泡沫塑料
FOD [fibre orientation distribution] (玻璃)纤维取向分布
fog 雾;薄层
fogging 起雾,成雾
foil 箔;薄片;薄膜
- ~ adhesive 薄片用黏合剂
- ~ cutting machine 薄片切割机,切薄片机
- ~ decorating 彩箔装饰
- ~ -faced grade 膜面质量
- ~ for marking 彩印箔,压花薄膜

Fokker bond tester 福克粘接试验仪(金属粘接件)
fold 折叠,叠合;折痕,褶痕,皱纹;弯折;合股〈纱线〉
- ~ back 弯扭(注挤制品缺陷)
- ~ crack 折叠裂缝

foldability 可折叠性
foldable [collapsible] 可折叠的〈包装〉
folded 折叠的;合股的
- ~ bottom box 折叠底箱
- ~ buff [pocketed buff] 折叠抛光轮
- ~ chain 折叠链
- ~ chain configuration 折叠链构型
- ~ chain crystal 折叠链晶体
- ~ container 可叠合容器
- ~ fibre 折叠纤维
- ~ glass filament yarn 合股玻璃长丝纱
- ~ glass staple fibre yarn 合股玻璃短纤维纱
- ~ oblong surface 长方形折叠面
- ~ plate structure 折板结构
- ~ triangular surface 三角形折叠面
- ~ [cabled]yarn 合股线,折叠纱

folder 折叠机
folding 折叠,皱纹;折边
- ~ angle 弯边角铁模具
- ~ bar [creasing rule] 折边板
- ~ bellow 折边兽皮
- ~ bend test 折叠弯曲试验
- ~ bottom 折叠底〈包装〉
- ~ box 折叠箱〈包装〉
- ~ box board 折叠箱纸板〈包装〉
- ~ bracket 折叠式支架
- ~ canopy 可折叠顶篷
- ~ chair 折叠式座椅
- ~ door 折叠门
- ~ edge 折叠边缘
- ~ endurance 耐折性
- ~ endurance test 耐折试验

foliate

~ into sheet metal 折叠入金属片中〈塑料制品〉
~ machine[press] 折叠机
~ quality tester 耐折性试验仪
~ rule 折尺;弯曲模板,卷边夹板
~ resistance 耐折叠性
~ seat 折叠式座椅
~ strength 耐折强度
~ support 弯折支架
~ test 折皱试验,折叠试验
~ tester 折叠试验机
~ welding 折叠焊接

foliate 涂箔;分层
follow 跟随,继续;按照;采纳
~ – on device 连续装置
~ – on tool 导向连续冲割模
~ – on unit 按续的装置,连续装置
~ the instructions 按照说明书(做)
~ – up 随动,跟踪;随后的
~ – up control 跟踪控制
~ – up equipment 随后的设备;连续设置
~ – up pressure 恒压;续压;保压〈注塑〉

follower 随动件,从动件
food 食品,食物
~ dye 食用染料
~ – grade 食品级〈适用于接触食品的塑料〉
~ packaging film 食品包装薄膜

fool – proof 防失误装置,安全装置
foot 基脚,底座;支架;英尺
~ press 脚踏压力机

force 模具,阳模,冲头;力;强制
~ block 阳模塞
~ – deformation curve 力–变形曲线
~ dry 强制干燥〈温度低于80~90℃时〉
~ feed lubricated 强制润滑加料
~ feeder 强制加料器
~ of adhesion 黏附力
~ of cohesion 内聚力;黏合力
~ of flocculation 絮凝力
~ piston 模塞,阳模

~ plate 阳模板,芯模板
~ plug 模塞,模芯,阳模
~ plunger 模塞,阳模
~ point 力点〈物理〉
~ retainer plate 芯模板〈注塑模〉
~ side 阳模侧,动模板侧
~ side of the mould 模具阳模侧〈注塑机〉
~ side part 阳模侧

forced 强制的;加强的
~ air 压缩空气
~ air cooling 强制风冷
~ air supply 强制送风
~ airing 强制通风
~ circulation 强制循环
~ circulation evaporator 强制循环蒸发
~ – circulation mixer [positive mixer] 强制循环混合器
~ continuous circulation 强制连续循环〈干燥烘箱〉
~ control 强制控制
~ convection 强制对流
~ cooling 强制冷却
~ drying 强制干燥
~ drying temperature 强制干燥温度
~ feed 强制进料
~ feed system 强制进料系统
~ oscillation 强制振荡

forcement 增强
forcer 阳模
forcing machine 挤压机
Ford 福特
~ measuring cup 福特量杯〈测定胶黏剂或涂料黏度〉
~ viscosity cup 福特黏度杯

foreign 外国的;外来的
~ body 外来物,杂质
~ impurities 杂质
~ material 外来材料,杂质
~ matter 外来物,夹杂物,杂质
~ matter inclusion 夹杂物
~ matters test 杂质检验
~ particle 杂粒

~ substance 杂质,夹杂物,外来物〈填料或成型料〉
forged 锻(造,制)的
~ cutter [forged knife, forger steel die] 锻制切刀
~ knife 锻制刀
~ steel die 锻制钢冲切模
forging 锻塑〈热塑性塑料〉
fork 分叉,叉状;叉形件
~ chain 支链
~ - lift truck 叉式升降装卸车
form. [formula] 公式;化学式;方程式;药方
form 形状;模型;成型
~ - and mold process [Comoform process] 成型模塑法
~ and - spray 成型喷塑〈增强塑料〉
~ birefingence 形状双折射
~ grinding 成型磨削(法)
~ of a thread 螺纹断面形状,螺纹牙形
~ of pellet 颗粒形状
~ of seam [weld] 焊缝形状
~ - work 模板
~ wound 模绕的
formability 易成型性
formable 可成型的
~ film 可成型薄膜
~ sheet 可成型片材,二次成型片材
~ thermoplastic sheet 可成型热塑性塑料片材,二次成型热塑性塑料片材
formal 缩甲醛
formaldehyde 甲醛
formalin 甲醛水溶液,福尔马林
formamidine sulfinic acid 甲脒亚磺酸〈助剂;脱色剂;敏化剂〉
formation 形成,生成
~ of a gel 凝胶形成
~ of bubbles 起泡〈成型制品,涂层〉
~ of condensed moisture 湿气冷凝
~ of cracks 龟裂,开裂
~ of gas 气体形成;汽化,气化
~ of polymer radicals 聚合物游离基的形成

~ reaction 生成反应
formed 成型的
~ fabric 成型织物
~ in place 就地成型
~ piece 成型件〈片材〉
~ plywood 成型胶合板
~ sheet [shaped sheet] 成型片材
former 定型模;定径管嘴
~ block 型模,模块〈弯折用〉
~ bushing 定径套〈管材〉
~ plate 仿形样板
~ sleeve 定型套
formic 甲酸的
~ acid 甲酸
~ aldehyde 甲酸酐
forming 二次成型;定型,成型,模塑
~ and cutting die (热)定型和冲切模
~ behaviour (热)成型性
~ cake 丝束卷;成型坯
~ contour 成型轮廓 [外形]
~ die 定型模,成型模
~ in situ 现场成型
~ into female mould 阴模成型
~ machine (热)成型机
~ material 成型料
~ mo(u)ld 成型模具
~ of a blank 毛坯成型
~ of plastics 塑料(热,二次)成型
~ orifice 成型孔
~ over male mould 阳模成型
~ package (纤维)卷装 [筒]
~ pad 成型衬垫,热成型膜片 [层]
~ pressure (热)成型压力
~ process 成型法
~ property 成型性能
~ range 成型(温度)范围
~ roll 成型辊
~ sheet 成型片材
~ temperature (热)成型温度
~ with helper 模塞助压成型,真空拉深(热)成型
~ with prestrctch 预拉伸成型
~ work 成型加工
formless 无定形的,不成型的

formol 甲醛
formula 配方;公式;化学式,分子式
~ adjustment 配方调整
~ calculation 配方计算
~ change 配方变更
formulated 公式化的;按配方制造的
~ product 按配方制造的产品
~ resin system 按配方制备的树脂系列
formulation 配方;组分,成分;原始混合物
Formvac 真空吸收法
fortifier 增强剂,补强剂;补强材料
forward 向前(的)
~ edge 前向边缘
~ feed 顺流进料
fose [full overlap scal end] 全搭接封合端〈包装〉
fossil 化石的;陈旧的
~ flour 硅藻土
~ fesin 化石树脂〈天然树脂〉
foundation material 地板腻子料
foundry 铸造,铸塑,浇铸
~ core 铸塑模芯
~ phenolic resin 铸塑酚醛树脂
~ resin 铸塑树脂
~ sand core 铸塑(型)砂(模)芯
fountain 供料槽〈涂布机〉;喷布机〈辊式涂布机〉;传墨器
~ feed 涂料供给〈涂布机〉
~ for coating compound 涂布混合料用供料槽
~ roller 供料辊,传墨辊〈印刷〉
four 四个
~ - bowl calender 四辊压延机
~ - parameter body 四参数模型,线型黏弹性流变模型
~ - parameter model 四参数模型,线型黏弹性流变模型
~ - pass bank cooling coil 四道冷却蛇管
~ - ply flat sheet die 四层平板模头
~ - ply laminate 四层层合(共挤塑)
~ point flexural test 四点弯曲试验

~ - roll calender 四辊压延机
~ roll crusher 四辊碾碎机
~ - roll mill 四辊磨
~ - screw extruder 四螺杆挤塑机
~ - screw kneader 四螺杆捏和机
~ - screw kneader mixer 四螺杆捏和(混合)机
~ - way pallet 四路托盘
fourdrinier dryer 多筒式干燥机
fourdrinier paper machine 长网造纸机
f. p. [flash point] 闪点
f. p. [freezing point] 凝固点;冰点
FR [flame retardants] 阻燃剂
fraction 级分;馏分
~ composition 馏分
~ defective 不合格率,次品百分率
fractional 分部的,分级的
~ combustion 部分燃烧
~ crystallization 分级结晶
~ damage 局部破坏
~ dissolution 分段溶解
~ distillation 分馏
~ error 部分误差
~ load 部分载荷
~ melt 分步熔化
~ package 单独包装
~ porosity 部分孔隙率,相对孔隙率
~ precipitation 分级沉淀
~ saturation 部分饱和
~ solution 分级溶解
fractionate 分馏;分级
fractionated packaging 单独包装
fractionating 分馏;分级
~ column 分馏柱
~ tower 精馏塔
fractionation 分级;分馏
~ by adsorption 吸附分离
~ by precipitation 沉淀分级
fracture 断裂,破裂
~ characteristic 断裂特性
~ cross section 断裂横断面
~ edge 断裂边
~ energy 破裂能
~ in shear 剪切断裂

~ initiation tests 开裂试验
~ kinetics 断裂动力学
~ length 裂断长
~ load 断裂载荷,破坏负荷
~ load by bending 弯曲断裂载荷,挠折载荷
~ mechanics 断裂力学
~ mechanics test 断裂力学试验〈美〉
~ mechanism 断裂机理
~ morphology 断裂形态
~ plane 破裂面,断面
~ process 断裂过程
~ propagation 裂缝扩散
~ section 断裂剖面
~ speed (裂纹)破裂速率[度]
~ stress 断裂应力
~ surface 断裂(表)面,断面
~ surface energy 断裂表面能
~ test 破裂试验
~ toughness 断裂韧性;断裂韧度
~ transition of plastic 塑性断裂转变
~ transition temperature 断裂转变温度
fracturing 断裂;开裂,龟裂;压裂
fragile 易碎的,易裂的;脆(性)的
~ article 易碎品
~ fibre 脆性纤维
fragility 脆性,易碎[裂]性
fragment 碎片
frame 模架,模框;结构
~ construction 构架结构
~ cover 护板,护罩
~ drying 框式烘干[干燥]
~ electrode 裸(电)焊条
~ filter 框式过滤器
~ foundation 机架底座
~ head 机头
~ heater 框式加热器(焊接机用)
~ member 构件
~ plate 框板
~ press 框式压机
~ saw 框式锯
framework structure [braced structure] 构架结构
frangibility 脆性,易碎性

frangible 脆的,易碎的
free 自由的;游离的
~ acidity 游离酸度
~ air 大气;自由空间
~ air temperature 大气温度
~ ammonia 游离氨〈在酚醛塑料模塑件内〉
~ area 有效截面,有效面积
~ baking 无模烧结
~ base 显色基
~ beating 游离状打浆
~ bend test 自由弯曲试验
~ blow 无模吹塑
~ blow forming 无模吹塑成型,自由吹塑成型
~ blowing 无模吹塑成型,自由吹塑成型
~ blowing process 无模吹塑成型,自由吹塑成型
~ -blowing vacuum forming 无模吹塑真空成型,自由吹塑真空成型
~ burning 易燃的
~ camber 自由弯曲度
~ convection 自然对流
~ drawing 无模牵伸,自由撑压
~ energy 自由能
~ expansion 常压发泡;自由膨胀
~ fall 自由降落
~ fall velocity 自由落体速度
~ falling dart test 自由落镖试验
~ falling film evaporator 自由降膜(式)蒸发器
~ falling granules 自由下落颗粒(料),易流动颗粒(料)
~ falling mixer 自由下落式混合器,易流动混合器
~ feed 自由进料
~ film extrusion 无衬薄膜挤塑
~ fit 自由配合
~ flow 自由流动
~ flow coupling 直通管接头
~ flow nozzle 低阻力注料嘴
~ flowing 易流动的;松散的(塑料粉料或粒料)

- ~ flowing feed stock　松散料；易流动原料
- ~ flowing granules　易流动颗粒(料)，松散粒料
- ~ flowing mixer　易流动混合器，自由下落式混合器
- ~ flowing material　流动性材料，松散材料
- ~ flowing powder　易流动粉料，松散粉料
- ~ - flowing product　松散产品
- ~ - flowing property　松散性
- ~ - flowing stock　松散材料
- ~ forming　无模成型，自由成型
- ~ - forming process　无模成型法，自由成型法
- ~ from　没有的……
- ~ from contamination　无污染的；无杂质的
- ~ from exudation marks　无渗出痕迹
- ~ from impurities　无杂质
- ~ from smell　无气味
- ~ from surface skin　无表皮
- ~ gas granulate　无气泡颗粒料
- ~ of contamination　无污染
- ~ of losses　无损耗
- ~ of tackiness　不黏着的
- ~ oscillation　自由振荡
- ~ phenol　未反应酚，游离酚
- ~ plasticizer　游离增塑剂
- ~ radical　自由基，游离基
- ~ radical - initiated polymerization　自由基引发聚合；游离基引发聚合
- ~ radical polymerization　自由基(引发)聚合；游离基(引发)聚合
- ~ radical reaction　游离基反应，自由基反应
- ~ radical termination　自由基终止
- ~ radical transfer　自由基转移
- ~ - rise foaming　自由发泡的；自由发泡法
- ~ roll　游辊，(张力)调节辊
- ~ - roller take - off　游辊引出装置
- ~ running　自由行程，空转
- ~ running extrudate　易流动挤塑物
- ~ - running haul - off　空转(辊)牵引装置
- ~ running liquid　易流动液体
- ~ running multivibrator　自激多谐振荡器
- ~ section　可拆部分
- ~ sintering　无模烧结，自由烧结
- ~ - space vacuum forming [free - vacuum forming]　自由空间真空成型
- ~ stripping　溜滑衬里
- ~ state　游离状态
- ~ stream turbulence　自由流湍流
- ~ surface energy　自由表面能
- ~ time　自由时间，不工作时间
- ~ turning roller haul - off [take - off]　活辊牵引
- ~ vacuum forming　无模真空成型
- ~ vibration measurement　自由振动测量
- ~ vibration method　自由振动法
- ~ volume　自由体积
- ~ vulcanization　无模硫化
- ~ zone　自由区

freedom　自由(度)；间隙；剥离〈涂层〉
- ~ from specks　无斑点

freehand blanking　徒手落料

freely　自由地
- ~ falling body　自由落体
- ~ fitted　自由安装的
- ~ flowing　自由流动的
- ~ soluble　易溶的

freeze　冻结，冷冻；凝固
- ~ drying　冷冻干燥
- ~ grinding [cryogenic grinding of thermoplasts]　冷冻研磨
- ~ point　冰点；凝固点
- ~ replica　冷冻复型
- ~ resistance　耐寒性
- ~ - thaw stability　冰 - 融稳定性〈粘接〉

freezing　冻结的，冷却的；凝固的，冰点的
- ~ agent　冷冻剂
- ~ constant　冰点降低常数，摩尔冰点

下降常数
- ~ curve 冷凝曲线
- ~ mixture 冷冻混合物
- ~ of a melt 熔体固化;熔体凝固
- ~ point 冰点,凝固点
- ~ point depression 冰点降低
- ~ polymerization 冷冻聚合(作用)
- ~ process 冻凝[冻结]过程;玻璃化过程
- ~ resistance 耐寒性
- ~ spinning 冻凝纺丝
- ~ temperature 冰点,冻点,冻结点,凝固点,冻结温度
- ~ test 冷冻试验,耐寒性试验
- ~ time 凝结时间

freight 货运,运送;运费;负重
- ~ car 货车
- ~ elevator 载重升降机

French 法国(式)的
- ~ chalk 滑石
- ~ mo(u)ld 法国式模具〈制备形状不规则或极为精密部件的两瓣模〉
- ~ white 法国白,滑石粉

Freon 氟里昂,氟氯烷

freq. frequency 频率

frequency [frequece] 频率
- ~ characteristics 频率特性
- ~ distribution 频率分布
- ~ effect 频率效应
- ~ factor 频率因子
- ~ profile 频率分布图
- ~ relaxometer 频率松弛计
- ~ response 频率响应
- ~ response function 频率响应函数
- ~ spectrum 频谱
- ~ stability 频率稳定度

fresh 新鲜的
- ~ air circulation 新鲜空气循环
- ~ charge 新鲜进料
- ~ feed 新鲜原料,新鲜供料
- ~ feed pump 进料泵
- ~ feed surge drum 新鲜原料收集器
- ~ food packaging 新鲜食品包装
- ~ material 新料

fretting wear 磨损,磨耗

friability 脆性[度];易碎性;酥脆(聚氨酯泡沫塑料)
- ~ test 脆性试验
- ~ time 酥脆时间〈聚氨酯泡沫塑料〉

friable 脆的,易碎的

friction 摩擦(力)
- ~ bearing 滑动轴承
- ~ behaviour 摩擦状态,摩擦行为
- ~ calender 异速压延机;摩擦轧光机
- ~ calendering 摩擦贴合,摩胶压延;摩擦轧光(工艺)
- ~ cap 压紧盖
- ~ closure 压紧塞
- ~ coefficient 摩擦系数
- ~ disk system 摩擦盘系统;圆盘离合器
- ~ disk welding 摩擦盘熔接
- ~ drag 摩擦阻力
- ~ drive 摩擦传动
- ~ factor 摩擦系数[因子]
- ~ finish 摩擦轧光整理
- ~ force 摩擦力
- ~ – free flow 无摩擦流动;无黏性流动
- ~ loss 摩擦损失
- ~ material 耐磨材料
- ~ pressure 摩擦压力
- ~ ratio 辊速比,滚筒速比,摩擦比
- ~ resistance 摩擦阻力;耐摩擦性
- ~ ring 摩擦环
- ~ ring gear system 环形摩擦传动系统
- ~ roll 压紧辊,摩擦辊
- ~ – slide behaviour 摩擦 – 滑动性
- ~ speed 异速,不同速度
- ~ speed calender 异速压延机
- ~ surface 摩擦面
- ~ test 摩擦试验
- ~ wear 磨耗,磨损
- ~ welding 摩擦焊接
- ~ welding disk 摩擦焊接盘
- ~ welding machine 摩擦焊接机
- ~ wheel 摩擦轮;摩擦盘
- ~ work 摩擦功

frictional 摩擦的
~ behaviour 摩擦性能,摩擦行为
~ characterization 摩擦特性
~ connection 摩擦连接
~ contact 摩擦接触
~ effect 摩擦效应
~ energy 摩擦能
~ failure 摩擦损坏
~ heat 摩擦热
~ index 摩擦指数
~ influence 摩擦影响
~ properties 摩擦性能
~ resistance 耐摩擦性
~ stiffness 摩擦劲度
~ testing machines 摩擦试验机
~ work 摩擦功
frictionless 无摩擦的,光滑的
Friedel – Crafts catalyst 弗里德尔–克拉夫茨催化剂,弗–克催化剂
fridge [refrigerator] 冰箱
frigid zone 寒带
frigorimeter 低温计
fringe pattern 条纹图形
fringed – micelle model 缨状微束模型
FRM [foam reservoir moulding] 槽式泡沫塑料成型
front 前面,正面,端部
~ barrel flange 机筒前端法兰〈挤塑机〉
~ bottom radius （螺杆）前底半径
~ cavity plate 前阴模板
~ clamping plate 前夹模板
~ draft zone 前牵伸区
~ – drive extruder 前部传动挤塑机
~ end 前端;顺流端,出料端
~ face of flight 前螺面
~ flange 机头法兰
~ heat zone 前加热段〈挤塑机〉
~ mo(u)ld 定模
~ offset roll 前偏置辊〈L形压延机〉
~ panel 面板;前挡板
~ roll 前辊〈压延机〉
~ shoe 前模板〈注塑模具〉
~ side 前端〈注塑模〉

~ side mould part 前端模具部分〈注塑模〉
~ splash shield 前挡泥板
~ spoiler 前阻流板〈摩托车〉
~ surface mirror 前（置）镜
~ view 正视图,正面图
frost 起晶;起霜;泛白;冻结
~ crack 冻裂
~ fracture 冻裂
~ glass 毛玻璃,磨砂玻璃
~ line 冷却线
~ line distance 冷却线距离
~ line height 冷却线高度
~ proof 防霜冻的
~ resistance 耐寒性
frosted 闷光的,无光泽的
frosting 起霜,结霜,起晶;磨砂面,无光泽面
~ phenomenon 消光现象
froth 泡沫;起沫,发泡
~ beating 搅动起沫
~ dyeing 泡沫染色
~ promotor 成沫促进剂
~ rubber 泡沫橡胶
~ stability 泡沫稳定性
frother 喷沫机
frothing 起沫
~ agent 泡沫剂
~ foam （机械）发泡泡沫塑料〈聚氨酯〉
~ machine 发泡机,起泡机
~ mechanical 机械发泡
~ percentage 发泡率
~ process 发泡法〈聚氨酯〉
frozen 冻结（的）;冷却
~ extrudate 冷却挤塑物
~ orientation 冻结取向〈大分子〉
~ in stresses 冻结应力
~ rubber 冷冻橡胶
~ slug 冷料;凝固料
~ state polymerization 冻结聚合
~ strains 冻结应变
~ stress 冻结应力
FRP [fiber reinforced plastics] （玻璃）纤

维增强塑料
FRPBT［glass fiber reinforced polybuthylene terephthalate］ 玻璃纤维增强聚对苯二甲酸丁二醇酯
FRPET［glass fiber reinforced polyethylene terephthalate］ 玻璃纤维增强聚对苯二甲酸乙二醇酯,增强涤纶
FRPP［glass fiber reinforced polypropylene］ 玻璃纤维增强聚丙烯
FRPVC［fiber reinforced polyvingl chloride］ (玻璃)纤维增强聚氯乙烯
FRTP［fiber reinforced thermoplastics］ 纤维增强热塑性塑料
fruit crate 水果板条箱,周转箱
ft. ［feet,foot］ 英尺〈0.3048 米〉
FTIR［Fourier Transform Infrared Spectrometer］ 傅立叶,变换红外光谱仪
ft/1b foot/pounds 英尺/磅〈0.138255 米/公斤〉
F–type calender F 型压延机
fuel oil 燃料油
fugacity (易)逸性;(易)逸度
fugitive 易褪色的,不坚牢的
~ dyes 易褪色染料
fugitiveness 褪色性;不稳定性
full 满的,全部的,完全的
~ automatic 全自动的
~ automatic control 全自动控制
~ charge 完全装料;全负荷
~ core fibre 全芯纤维
~ cure 完全硫化
~ disk［whole disc］ 全圆盘
~ –disk buff 全圆盘抛光轮
~ flighted length of screw 螺杆螺纹总长度
~ flighted screw 全螺纹螺杆
~ gloss 高光泽
~ hydraulic clamp unit 全液压合模装置
~ load 满负荷
~ load life 全负荷寿命
~ load operation 满负荷运转
~ load test 满负荷试验
~ loaded 满负荷的

~ mo(u)ld casting 全模铸塑,全模铸塑制品〈模具由上下模构成〉
~ mo(u)ld process 泡沫塑料实型铸造法〈聚苯乙烯泡沫塑料作铸型的分解铸造法〉
~ operation 满(负载)运转
~ overlap seal end 全搭接封合端〈包装〉
~ production 满生产
~ roll width 全辊宽度
~ scale 实际尺寸
~ scale model 实尺模型
~ scale operation 全面运转;大规模生产
~ scale plant 成套设备,大规模生产设备
~ scale production 全规模生产
~ scale test 实物试验
~ set 全套
~ shade 饱和色
~ size 实际尺寸
~ thread 全螺纹
~ time 全部时间
~ tinting strength 全着色力〈颜料〉
~ view 全视图
~ weight 总重量,毛重
~ width 全宽,总宽
~ –width runner gate 宽流道浇口
fully 全部,完全
~ automatic 全自动的
~ automatic control 全自动控制
~ –automatic press 全自动压机
~ charged 装满料的
~ drawn yarn 全拉伸丝
~ enclosed 全封闭的,密封的
~ loaded 满负荷的
~ oriented filament 全取向丝
~ plastic 全塑料
~ positive mould 溢出式压模;开adapted式模
fulminating mercury 雷汞
fumaric acid 富马酸
fume 烟(雾)
~ duct 烟道

~ exhaust 排烟
~ hood 排烟通风柜
~ scrubber 洗烟器
fumed silica [pyrogenic silica] 热解法氧化硅,高度分散二氧化硅〈增稠剂〉
fuming 发烟的
  ~ nitric acid 发烟硝酸
  ~ -off 爆燃
  ~ -off temperature 烟化温度;发烟温度;爆燃温度
  ~ sulphuric acid 发烟硫酸
function 函数;作用,功能
  ~ generator 函数发生器
  ~ test bed 功能试验台
  ~ zone 作用段
functional 函数(的);功能(的),官能(的)
  ~ additive 功能性添加剂
  ~ adhesive 功能性黏合剂
  ~ compound 官能化合物
  ~ design 功能性设计
  ~ inspection 作业检查
  ~ joining 构造接合
  ~ membranes 功能膜
  ~ package 功能性包装
  ~ polymer 功能聚合物
  ~ resin 功能树脂
  ~ stage 作业阶段
  ~ wrap 功能性打包〈包装〉
functionality 官能度;官能性
  ~ number 功能数,官能数
fundament 基础;基本原理
fundamental 基本(的),主要的,根本的
  ~ chain 母链
  ~ characteristics 基本性能
  ~ curve 基本曲线
  ~ element 基本元素
  ~ factor 基本因素
  ~ frequency 基本频率
  ~ function 基本功能;特征函数
  ~ law 基本定律
  ~ principle 基本原理
  ~ research 基本理论研究
  ~ test 基础性试验

~ unit 基本单位
fungal resistance [fungus resistance] 耐霉性
fungicide 杀真菌剂,杀霉菌剂,杀菌剂
funginert 耐霉菌的
funginertness 耐霉性,防霉性,抗霉性
fungistats 抑霉剂
fungus 真菌,霉(菌)
  ~ -proof 耐霉菌的,防霉菌的
  ~ resistance 耐霉性,防霉性,抗霉性
  ~ resistant 耐霉菌的,耐真菌的
  ~ test 防霉性试验
funnel 漏斗
  ~ can 漏斗式罐
  ~ kiln 漏斗式干燥器
  ~ spinning 漏斗纺丝
  ~ spinning process 漏斗纺丝法
furan 呋喃
  ~ plastic 呋喃塑料
  ~ resin 呋喃树脂
  ~ resin adhesive 呋喃树脂胶黏剂
furfural 糠醛
  ~ acetone formaldehyde resin 糠醛丙酮-甲醛树脂,糠酮甲醛树脂
  ~ acetone resin[FA] 糠醛丙酮树脂,糠酮树脂
  ~ -phenol resin 糠醛苯酚树脂
  ~ resin 糠醛树脂
furfuran resin 呋喃树脂
furfurol resin 糠醛树脂
furfuryl 糠基
  ~ alcohol 糠醇
  ~ alcohol modified urea-formaldehyde resin 糠醇改性脲醛树脂
  ~ alcohol resin 糠醇树脂
furnace(**carbon**) **black** 炉法炭黑,炉黑
furnish roll 浸渍辊,蘸料辊,涂布辊〈涂布机〉
furnishing roll 上胶辊,施胶辊;涂布辊
furniture 设备,装置,器具;家具
  ~ covering 家具涂层
  ~ fittings 家具装配件
  ~ varnish 家具漆
  ~ webbing 家具(用)织带

fuse 熔化;保险丝;熔焊
fused 熔融的,塑化的
~ area 熔化区域
~ mass 熔体,熔融物,熔融料
~ pellet 塑化粒料
fusibility 可熔性
fusible 易熔的
~ ceramic adhesive 易熔陶瓷黏合剂
fusibleness 熔性
fusing 熔融;熔合;塑化
~ oven 塑化箱〈涂布〉
~ point 熔(化)点
~ [fluxing]pressure 熔化压力
~ [fluxing]temperature 熔化温度
fusion 熔融;熔化;熔焊
~ bond coating 熔黏涂布
~ bond finish 熔黏涂饰
~ bonding process 熔融胶合法
~ coating 熔合涂布
~ face 熔焊面;坡口面〈熔焊〉
~ heat 熔化热

~ jointing (管)熔接
~ jointing temperature 熔接温度
~ lamination 熔接层压(胶合)法
~ mo(u)lding 烧结成型,熔结成型
~ point 熔(化)点
~ pot 熔化罐〈如:酚醛树脂〉
~ pressure 熔化压力
~ seamer 熔融封口机〈塑料〉
~ socket 熔接插口
~ socket fitting 熔接插口配件
~ spigot 熔接插头〈管〉
~ temperature 熔化温度
~ viscosimeter 熔化黏度计
~ welding 熔焊,熔融焊接
~ zone 熔化区
fuzz 微毛,绒毛
~ and fly 绒毛和飞毛〈玻璃纤维〉
fuzzing 起毛
fuzzy 绒毛的;模糊的
FVF[fibre volume fraction] 纤维体积含量(玻璃纤维增强成型制品)

G

g [gaseous] 气体的,气态的
g [gram] 克
G-glass 硼含量高的玻璃
G type [general purpose trpe] 通用型
G-value G值
Ga [gallium] 镓
gadget 小机件,小配件
gage [gauge] 表,计;量规;压力计;计量;测量
gaging roll 校定辊
gain in weight 重量增加
gaiter 防护包封〈机器〉
gal [gallon] 加仑
gallium [Ga] 镓
gallon [gal.] 加仑
galvanize 电镀,镀锌

galvanizing kettle 电镀锅
galvanometer 电流计
gamma γ(希腊字母)
~ -aminopropyl triethoxysilane γ-氨丙基三乙氧基(甲)硅烷
~ loss peak γ-损耗峰
~ -radiation-induced polymerization γ-辐射引发聚合(作用)
~ ray γ-射线
~ value γ值
gang 全套;组合的
~ dies 多头冲模,复式模
~ mo(u)ld 成组模,多槽模,群腔膜,溢料式多槽模具
~ of cavities 多槽模形,多模穴槽;成组模巢;多巢

~ switch 联动开关
gap 间隙;辊隙;模缝
~ adjustment 间隙调整
~ angle 裂缝角度;裂开角度
~ at joint 缝隙
~ between rolls 辊隙
~ – filling 填缝〈黏合剂〉
~ – filling adhesive 填缝黏合剂
~ – filling cement 填缝粘接剂
~ frame C 型框架
~ frame press C 型(框架)压机;开式单臂压机
~ ga(u)ge 间隙规,厚度规
~ increase 间隙增大
~ setting 间隙调节
~ size 间隙尺寸
Gardner 加德纳
~ bubble viscometer 加德纳气泡黏度计
~ mobilometer 加德纳滴度计
~ vertical viscometer 加德纳竖式黏度计
~ viscosity 加德纳黏度
gas 气体
~ adsorption 气体吸收
~ barrier film 阻气薄膜
~ barrier material 阻气材料
~ barrier property 不透气性;气密性
~ black 烟黑〈炭黑的一种〉
~ blow mixing 气发搅拌
~ blowing reaction 发泡反应
~ blown bubble 气发泡孔
~ blown foam 气发泡沫塑料
~ bubbles 气泡
~ burner 燃气喷嘴
~ cable 充气电缆
~ carbon 烟黑〈炭黑的一种〉
~ cavity 气孔
~ chromatogram 气相色谱(图)
~ chromatograph 气相色谱仪
~ chromatography 气相色谱法
~ content 气体含量
~ counterpressure casting process 气体反压铸塑法
~ – counterpressure injection mo(u)lding(process) 气体反压注塑成型(法)〈制备热塑性泡沫塑料〉
~ – developing agent 发泡剂
~ – development agent 发泡剂
~ expanded plastic 微孔塑料,闭孔泡沫塑料
~ expanded rubber 微孔橡胶,闭孔泡沫橡胶
~ – filled cable 充气电缆
~ film 气膜
~ – forming agent 发泡剂
~ gauge 气压计
~ generator 气体发生器
~ – heated hot gas welder 可燃气体加热的热气焊接仪
~ – heated welding gun[torch] 可燃气体加热的焊枪
~ impermeability test 不透气性试验
~ impermeable 不透气的
~ injection extruder 注气挤塑机〈发泡〉
~ injection port 注气孔〈挤塑机〉
~ leakage test 气密性试验
~ liquid chromatography 气液色谱法
~ main 气体总管
~ permeability 透气性
~ permeability test 透气性试验
~ permeability tester 透气性试验仪
~ permeation 透气
~ phase 气相
~ – phase polymerization 气相聚合(作用)
~ phase polyvinyl chloride 气相法聚氯乙烯
~ pipe 气体管道,气管
~ – plasma treatment 气体等离子(表面)处理,等离子流(表面)处理
~ pocket 气窝〈塑料表面〉
~ – proof 不透气的
~ proofing 不透气
~ proofness 气密性;不透气性
~ propellant 气体发泡剂
~ – resistant 不透气的,阻气的

~ -tight 不透气的;气密的
~ tight test 气密性试验
~ tightness 不透气性;气密性
~ transmission [gas germeability] 透气(性)
~ transmission rate 透气速率
~ tube 充气(软)管;气体软管
~ velocity 气体速度
~ weld 气焊
~ welder 气焊机
~ welding 气焊
~ yield 发气量
gaseous 气体的;气态的
~ corrosion 气相腐蚀
~ film 气态膜
~ polymerization 气相聚合法
~ state 气态
gasket 垫圈,垫片,密封垫,填隙料;密封
~ -mounted valve 垫密片阀,密封衬阀
gasoline [gasolene〈英〉] 汽油〈美〉
~ can 汽油桶
~ tank 汽油箱
gassing 放气;充气
~ zone 排气段
gastight 不透气的;气密的
gastightness 不透气性;气密性
gate 浇口〈在注塑和压铸中〉;浇口冷料;栅门
~ agitator 框式搅拌器
~ area 浇口面积
~ balance 浇口平衡
~ blush 浇口白斑
~ cut-off valve 浇口断流阀
~ cutter 浇口切割器,切浇口器
~ depth 浇口深度
~ diameter 浇口直径
~ feed hopper 框式加料斗
~ freeze 浇口冻结,封口
~ land 浇口面
~ location 浇口定位;浇口位置
~ mark 浇口痕
~ mixer 框式混合器
~ paddle agitator 框叶式搅拌器

~ pin 闭锁销(注塑料嘴)
~ resistance 浇口(流动)阻力
~ sealing 浇口封闭
~ splay 注口失形,浇口白斑
~ stirrer 框式搅拌器
~ switch 栅门开关
~ tension device 栅门式张力器
~ tensioner 栅门式张力器
~ tunnel 浇口隧道〈注塑〉
~ type 浇口类型
~ valve 闸阀,滑门阀
~ width 浇口宽度
gating 浇口;浇注系统
~ point 浇注痕;浇注位置,浇口处,浇口点
~ system 浇注系统
gathering 收集,集合
~ channel 集料槽
~ device 堆料装置;储料装置
gauffer 起绉;绉纹,波纹;轧花
~ calender 浮花压光机,印纹轧压机
gauge [gage] 量规,量计;厚度:隔距
~ adjustment 隔距调整
~ board 测量仪表板
~ distortion 厚薄不匀
~ controller 厚度控制器〈压延〉
~ length 标距,计量长度
~ marks (标距)标记
~ of film 薄膜厚度
~ panel 仪表盘,仪表板
~ pressure 表压,计示压力
~ scratch 划痕测量
~ tank 计量槽
gauging [gaging] 测量,计量
~ and blowing mandrel 定型吹塑芯棒
~ pass 定厚辊隙
~ roll 测定辊,校定辊,定厚辊;定型辊
~ tank 计量槽
Gaussian distribution 高斯分布,高斯标准分布
Gauze [gaze] 纱网;纱布
g/d [grams per denier] 克/旦
Ge [germanium] 锗

gear 齿轮;传动装置
~ blank 齿轮毛坯
~ box 齿轮箱
~ case 齿轮箱
~ case cover 齿轮箱盖
~ drive 齿轮传动
~ dynamometer 齿轮功率计
~ face 齿轮端面
~ guard 齿轮罩,齿轮护板
~ mark 齿纹
~ pump 齿轮泵
~ pump extruder 齿轮泵式挤塑机
~ reducer 齿轮减速器〈美〉
~ reducing 齿轮减速
~ reduction 齿轮减速,减速齿轮传动
~ spinning pump 纺丝齿轮泵
~ -type pump 齿轮泵
~ wheel 齿轮
gearing [gears] 齿轮传动,传动装置
~ embossing 齿轮压花
Geer 吉尔
~ oven 吉尔老化试验箱
~ oven method 吉尔烘箱法
~ test 吉尔试验
Gehman torsion test 格曼扭转试验
Geiger counter 盖革计数器
gel 凝胶(体);冻胶
~ chromatography 凝胶色谱(法)
~ coat 亮光涂层;凝胶涂层,胶衣
~ coat(finish) 亮光涂层;凝胶涂层
~ coating 凝胶涂层
~ coating resin 亮光涂层树脂;凝胶涂层树脂
~ coherence 凝胶内聚力
~ content 凝胶含量
~ effect 凝胶效应
~ extraction 凝胶萃取,凝胶抽提
~ fibre 凝胶纤维
~ formation 凝胶形成
~ forming agent 胶凝剂
~ ion exchange resin 凝胶型离子交换树脂
~ lacquers 凝胶漆
~ network 凝胶网络
~ particle 凝胶粒子
~ permeation chromatography[GPC] 凝胶渗透色谱法
~ point 凝胶点
~ polymer 胶凝聚合物
~ rate 胶凝速率
~ rubber 凝胶体橡胶
~ sensitizer 胶凝剂
~ state 凝胶状态
~ strength 胶凝强度
~ swelling 凝胶溶胀度
~ temperature 凝胶温度
~ test 凝胶试验
~ time 凝胶时间
~ type resin 凝胶型树脂
gelatification 胶凝作用;凝胶化作用
gelatin [gelatine] 明胶〈分散剂〉
~ capsule 胶囊
gelatinate [gelatinize] 胶凝
gelatination 胶凝作用
gelatinization 胶凝作用;明胶化(作用)
~ bath 胶化浴
gelatinize 胶凝
gelatinized surface 胶化表面〈硬化纸板〉
gelatinizer 胶凝剂;稠化剂
gelatinizing agent 胶凝剂
gelatinous 胶的;胶凝状的
gelation 胶凝作用;胶凝化(作用)
~ curve 胶凝曲线
~ mixer 凝胶化混合器
~ performance 胶凝特性
~ temperature 凝胶点,凝胶温度
~ test 胶化试验
~ time 胶凝时间
gelationization 胶凝,胶化
gelimat 胶凝机
gellable 可胶凝的
gellant 胶凝剂
gelled fibre 凝胶纤维,胶化纤维
gelling 胶凝作用;胶凝化作用
~ agent 胶凝剂
~ condition 胶凝条件
~ machine 胶凝机

~ point 胶凝点
~ property 胶凝性能
~ screw 胶(凝)化螺杆
~ temperature 胶化温度
~ time 凝胶时间
gemmatin 埃蕈染料
general 普通的,一般的,总的
~ arrangement 总体布置
~ assembly 总装配
~ assembly drawing 总装配图
~ characteristic 一般特性,通性
~ computer 通用计算机
~ drawing 总图
~ extension 均匀伸长,均匀拉伸
~ formula 通式
~ layout 总平面图;总体布置
~ mo(u)lded goods 一般模制品
~ overhaul 大修(理)
~ plan 总图;总规划
~ polystyrene [GPS] 通用聚苯乙烯
~ principle 普通原理;原则
~ program 通用程序
~ use article 一般制品
~ utility 正常利用率
~ view 总图,全视图;概要
~ purpose 通用的,一般用途的
~ purpose calender 通用压延机
~ purpose computer 通用计算机,通用电脑
~ purpose extruder 通用挤塑机
~ purpose grade 通用级,通用型
~ purpose material 通用材料,一般用途材料
~ purpose plastics 通用级塑料
~ purpose polystyrene 通用级聚苯乙烯
~ purpose resin 通用树脂
~ purpose rubber 通用橡胶
~ purpose type [G type] 通用型
~ purpose varnish 通用清漆
~ term 公项,普通项,一般项
generally labeling 一般标记
generalized 推广的;广义的
~ curve 广义曲线

~ solubility parameter 广义溶度参数
~ stress 广义应力
generate 发电;发生(气体)
generator 发生器;发电机;编制程序
generic name 属名
Genpac Coater 根派克(熔融)涂布机
gentle drying 平缓干燥
geodesic 短程
~ line 短程线
~ line winding 短程缠绕
~ path 短程线路
geometric 几何(学)的
~ design 几何形状设计
~ distribution 几何图形分布
~ grey scale 几何分级灰色卡
~ manipulation operation 几何花纹控制
~ metamerism 几何条件配色
~ porosity 多孔性结构
~ stiffness 几何刚度
geometrical 几何(学)的
~ figure 几何图形
~ isomer 几何异构体
~ moment of inertia 几何惯性矩
~ pattern 几何图案
~ pitch 几何螺距
~ symmetry 几何对称
geometry 几何(学,图形,形状,结构)
~ of screw 螺杆几何(学)
Germanium [Ge] 锗
GF [glue failure] 黏合失效
GFP [glass fiber reinforced epoxy resin] 玻璃纤维增强环氧树脂
GFPBTP [glass–fibre reinforced polybutylene terephthalate] 玻璃纤维增强聚对苯二甲酸丁二醇酯
GFPA [glass fibre reinforced polyamide] 玻璃纤维增强聚酰胺
GFPP [glass fibre reinforced polypropylene] 玻璃纤维增强聚丙烯
GFR–nylon [glass fiber reinforced polyamide] 玻璃纤维增强聚酰胺
GFRP [glass fibre reinforced plastic] 玻璃纤维增强塑料

GFRP [graphite-fibre reinforced plastic] 石墨纤维增强塑料
GFRTP [glass fiber reinforced thermoplastic] 玻璃纤维增强热塑性塑料
gge. [gauge] 量规；厚度；隔距
gib 扁栓；夹条
Gibbs indophenol test 吉布斯靛酚试验
gift package 礼品包装
giga 千兆，十亿，10^9
gilsonite 天然沥青
girth 周围；围绕
　~ strength [hoop strength] 圆周强度，环形强度
　~ welding 环缝焊接
given value 给定值
glacial 冰的
　~ acetic acid 冰乙酸
　~ dye 冰(染)染料
gland (密封)压盖
glandless construction 无密封结构
glass 玻璃
　~ adhesive 玻璃胶黏剂
　~ ball 玻璃球
　~ bat [glass wadding] 玻璃(填)絮
　~ bead 玻璃微球〈塑料增强用〉
　~ bonded mica 玻璃纤维胶结云母〈填料〉
　~ bonding 玻璃黏合
　~ capillary viscometers 玻璃毛细管黏度计
　~ chopper 玻璃纤维束切断机，玻丝切碎机
　~ cloth 玻璃(纤维)布，玻璃纤维织物
　~ cloth reinforced laminates 玻璃布增强层合板
　~ container 玻璃容器〈包装〉
　~ content 玻璃纤维含量
　~ count 玻璃纤维支数
　~ E E玻璃〈低碱硼硅酸盐玻璃，玻璃纤维增强塑料用〉
　~ fabric 玻璃纤维织物
　~ fibre 玻璃纤维〈增强剂〉
　~ fibre content 玻璃纤维含量
　~ fibre cordage 玻璃纤维绳(索)
　~ fibre epoxy laminate 玻璃纤维环氧树脂层压板
　~ fibre filled 玻璃纤维填充的
　~ fibre laminate 玻璃纤维层压板
　~ fibre mat 玻璃纤维毡
　~ fibre orientation 玻璃纤维取向
　~ fibre preforming 玻璃纤维预成型法
　~ fibre production 玻璃纤维生产
　~ fibre reinforcement 玻璃纤维增强材料
　~ fibre roving 玻璃纤维粗纱，玻纤粗纱
　~ fibre sleeving 玻璃纤维套管
　~ fibre surface treatment 玻璃纤维表面处理
　~ fibre reinforced 玻璃纤维增强的
　~ fibre reinforced epoxy resin [GFP, GF-EP] 玻璃纤维增强环氧树脂
　~ fibre reinforced nylon [GFR-nylon] 玻璃纤维增强尼龙，玻璃纤维增强聚酰胺
　~ fibre reinforced phenolic resin [GF-PF] 玻璃纤维增强酚醛树脂
　~ fibre reinforced plastic [GFRP, GRP] 玻璃纤维增强塑料
　~ fibre reinforced polyamide [GFPA, GFR-nylon] 玻璃纤维增强聚酰胺，玻璃纤维增强尼龙
　~ fibre reinforced polybutylene terephthalate [GEPBTP, FRPBT] 玻璃纤维增强聚对苯二甲酸丁二醇酯
　~ fibre reinforced polycarbonate 玻璃纤维增强聚碳酸酯
　~ fibre reinforced polyethylene [GRPE] 玻璃纤维增强聚乙烯
　~ fibre reinforced polyethylene terephthalate [FRPET] 玻璃纤维增强聚对苯二甲酸乙二醇酯，增强涤纶，PET增强塑料
　~ fibre reinforced polyolefine 玻璃纤维增强聚烯烃
　~ fibre reinforced polypropylene [GF-PP, FRPP] 玻璃纤维增强聚丙烯

glass

- ~ fibre reinforced polystyrene 玻璃纤维增强聚苯乙烯
- ~ fibre reinforced polyethene 玻璃纤维增强聚乙烯
- ~ fibre reinforced polyvinyl chloride[FR-PVC] 玻璃纤维增强聚氯乙烯
- ~ fibre reinforced thermoplastic [GFTP, GFRTP] 玻璃纤维增强热塑性塑料
- ~ fibre reinforced thermoplastic of the short-fibre type 短纤维型玻璃纤维增强热塑性塑料
- ~ fibre reinforced thermoset 玻璃纤维增强热固性塑料
- ~ fibre reinforced unsaturated polyester 玻璃纤维增强不饱和聚酯
- ~ fibre reinforced unsaturated polyester resin[GUP, GF-UP] 玻璃纤维增强不饱和聚酯树脂
- ~ fibre tube 玻璃纤维软管
- ~ fibre weave 玻璃纤维织物
- ~ filament 玻璃纤维长丝
- ~ filament braid 玻璃纤维长丝编带
- ~ filament cord 玻璃纤维长丝绳
- ~ filament production 玻璃纤维长丝生产
- ~ filament tape 玻璃纤维长丝带
- ~ filament yarn 玻璃纤维长丝纱
- ~ filled 玻璃纤维填充的
- ~ filled nylon 玻璃纤维填充尼龙
- ~ filled plastic 玻璃纤维填充塑料
- ~ filled thermoplastics 玻璃纤维填充热塑性塑料
- ~ flake 玻璃鳞片[薄片]
- ~ flask 玻璃烧瓶
- ~ flock 玻璃纤维屑
- ~ former 玻璃成型器
- ~ furnace 熔玻璃纤维炉
- ~ helix 玻璃螺旋管
- ~ hosiery 玻璃纤维针织品
- ~ laminate 安全玻璃,夹层玻璃
- ~ mat 玻璃纤维毡
- ~ monofilament 玻璃纤维单丝
- ~ reinforced 玻璃纤维增强
- ~ reinforced epoxy 玻璃纤维增强环氧树脂
- ~ reinforced laminate 玻璃纤维增强层压板
- ~ reinforced mo(u)lding material 玻璃纤维增强模塑料
- ~ reinforced panel 玻璃纤维增强板
- ~ reinforced plastic [GRP] 玻璃纤维增强塑料
- ~ reinforced polyester 玻璃纤维增强聚酯
- ~ reinforced polyester laminate 玻璃纤维增强聚酯层压板
- ~ reinforced polypropylene 玻璃纤维增强聚丙烯
- ~ reinforcement 玻璃纤维增强材料
- ~ -resin bond 玻璃纤维-树脂粘接
- ~ resin concentrate 玻璃纤维增强树脂母料
- ~ ribbon 玻璃丝带
- ~ ring 玻璃环〈蒸馏〉
- ~ rod 玻璃棒
- ~ roving 玻璃纤维粗纱
- ~ -rubber transition 玻璃橡胶态转化
- ~ -saddle 玻璃马鞍〈蒸馏〉
- ~ silk 玻璃丝
- ~ sphere 玻璃微球〈填料〉
- ~ spun yarn 玻璃纤维细纱
- ~ staple fibre 玻璃短纤维
- ~ staple fibre yarn 玻璃短纤维纱
- ~ state 玻璃态,玻璃透明状态
- ~ state temperature 玻璃化温度
- ~ strand 纺织玻璃纤维线
- ~ stress 玻璃纤维应力
- ~ surface treatment 玻璃纤维表面处理
- ~ -to-resin ratio 玻璃纤维-树脂比
- ~ transition 玻璃态转化,玻璃化转变
- ~ transition point 玻璃化转变点
- ~ transition temperature[GTT, Tg] 玻璃化转变温度
- ~ veil 玻璃覆面毡料
- ~ wadding 玻璃(填)絮
- ~ weave 玻璃纤维织物

~ wool 玻璃纤维棉,玻璃纤维绒
~ yarn 玻璃纤维纱
~ yarn layer 玻璃纤维纱层
glassed pump 搪瓷泵
glassine 赛璐玢,玻璃纸,透明纸
glassware 玻璃器皿
glassy 玻璃状的;透明的
~ polyester 玻璃态聚酯
~ polymer 玻璃态聚合物
~ stale 玻璃(状)态;透明状态
~ radiation polymerization 玻璃态辐射聚合
glaze 釉;光泽,色泽;轧光,上光
~ coat 釉涂层
glazed 上光的
~ finish 上光处理
~ joint failure 玻璃状粘接破裂
~ paper 釉纸,蜡光纸〈包装〉
glazer 上光轮;抛光轮
glazing 上光;抛光;涂釉
~ calender 砑光机,上光压延机
~ material 玻璃状材料,透明材料〈塑料〉
~ plastic 透明塑料(代替玻璃)
~ rolls 压[轧]光辊,上光辊
glitter 闪光颜料
~ -gem 闪光宝石
globar 碳化硅棒,炽热棒
globe valve 球阀
globular 球状的
~ molecule 球状分子
~ polymer 球型聚合物
~ powder 球状粉粒
~ swelling 球状溶胀
globulite 球皱晶
gloss 光泽;光泽度
~ agent 光泽剂
~ chill roll 光泽骤冷辊
~ cooling roll 光泽冷却辊
~ evaluation 光泽度评价〈塑料表面〉
~ finish 光泽整理,上光处理
~ finishing paint 上光漆
~ measurement 光泽测定
~ oil 光泽油

~ paint 光泽涂料,有光漆
~ retention 光泽保持性,光泽稳定性
~ stability 光泽稳定性〈漆〉
glossimeter 光泽计,光泽仪
glossiness 光泽性,光泽度
glossy 光泽的
~ surface 光泽面
glove 手套
~ box 手套箱
~ former 手套模型
glow 辉光,灼热
~ bar 灼热棒
~ bar test 灼热棒试验
~ discharge 辉光放电
~ discharge heating 辉光放电加热
~ -discharge polymerization 辉光放电聚合
~ discharge treatment 辉光放电处理
~ discharge tube 辉光放电管
~ -proof 耐灼热的
~ resistance 耐灼热性
~ test 灼热试验
~ -wire test 辉光灯丝试验
glowing 辉光
~ combustion 辉光燃烧,灼热燃烧
~ hot-body test 灼热体试验
glue 胶,胶黏剂;胶合,黏合,粘接
~ -block shear testing 粘试样剪切试验
~ bond 胶合层
~ characteristic 黏合剂性能
~ failure [GF] 粘接失效,黏合失效
~ film 胶膜,胶黏膜,干胶膜;胶层
~ gun 涂胶枪
~ in pearls 颗粒胶
~ joint 胶接(合);粘接点,粘接接头
~ layer 黏合层
~ layer thickness 黏合层厚度
~ line 黏合层,粘接层;胶层
~ line (dielectric) heating 粘接层(高频)加热
~ line strength 粘接层强度
~ line thickness 黏合层厚度
~ machine 涂胶机〈制层合材料〉

〈美〉
~ made from waste leather 废皮胶,废皮革制的胶黏剂
~ penetration 透胶,透胶量
~ smearing machine 涂胶机〈美〉
~ spread 涂胶量
~ spreader 涂胶机
~ spreading 涂胶
~ spreading machine 涂胶机
~ stock 胶料
~ water 胶水
glued 胶黏的
~ board 胶合板
~ insert socket 胶黏插接,承插胶接;套筒胶接
~ joint 胶接
~ wood 胶合板
glueing [gluing] 粘接,胶接
gluer 胶黏机
glues and adhesives 胶黏合剂
gluey 粘接的,胶接的
glueyness 胶黏性
gluing 粘接,胶接
~ cylinder 涂胶辊
~ fault 失黏,粘接失效,粘结不良
~ machine 涂胶机
~ practice 粘接技术
gluish 胶黏的
glutaric aldehyde 戊二酸酐
glycerin 甘油,丙三醇
~ phthalic resin 邻苯二甲酸丙三醇树脂
glycerol 甘油,丙三醇
~ diacetate 甘油二乙[醋]酸酯〈增塑剂、溶剂〉
~ ether acetate 甘油醚乙[醋]酸酯〈增塑剂〉
~ monoacetate 甘油单乙[醋]酸酯〈增塑剂〉
~ monolaurate 甘油单月桂酸酯〈增塑剂〉
~ phthalic resin 丙三醇邻苯二甲(酸)酐树脂
~ triacetate 甘油三乙[醋]酸酯

~ tributyrate 甘油三丁酸酯〈增塑剂,润滑剂〉
~ tripropionate 甘油三丙酸酯〈增塑剂〉
glyceryl 甘油基;丙三基
~ diacetate 二乙[醋]酸甘油酯
~ monoricinoleate 单蓖麻醇酸甘油酯〈增塑剂〉
~ monostearate 单硬脂酸甘油酯〈增塑剂〉
~ triacetoxy stearate 甘油三(乙酰氧基硬脂酸酯)
~ tri(acetyl ricinoleate) 甘油三(乙酰蓖麻醇酸酯)
~ tribenzoate 三苯甲酸甘油酯〈增塑剂〉
glycide 缩水甘油;氧丙环基甲醇;2,3-环氧-1-丙醇
glycidol 缩水甘油
γ - **glycidoxypropyltrimethoxysilane** γ-缩水甘油醚丙基三甲氧基硅烷〈偶联剂〉
glycidyl 缩水甘油基
~ ether resin 缩水甘油醚树脂
~ methacrylate 甲基丙烯酸缩水甘油酯
~ polymer 缩水甘油聚合物
glycol 乙二醇;甘醇
~ phthalate 邻苯二甲酸乙二醇酯
glycolysis process 糖醇解法〈废塑料再加工〉
glyptal resin 丙三醇邻苯二甲酸酐树脂,甘酞树脂
gob 液滴状坯料〈吹塑〉
godet 导丝盘〈纺织〉
~ cylinder 导丝滚筒
~ rolls 导丝辊;牵引辊,拉伸辊
~ unit [draw stand] 导丝装置;拉伸装置
~ wheel 导丝轮
goffer 压纹,压花,起皱纹
goffering 压花,压纹,形成皱纹
going barrel 旋转鼓
gold 黄金;金色

goniophotometry 测向光度法
good 良好的,坚固的;有益的,适当的
~ ageing 老化性能良好
~ colo(u)r 色泽良好
~ conductor 良导体
~ cure 适度硫化
~ storage property 良好的储藏性
Goodrich 古德里奇,古氏
~ flexometer 古德里奇挠曲仪
~ plastometer 古德里奇塑度计
goods 货物;商品,物品;成品
GOST [Russian standard] 俄罗斯标准〈ГОСТ 俄文〉
GP [genenal purpose] 通用的,一般用途的
GPC [gel permeation chromatography] 凝胶渗透色谱法
gpm [gallons per minute] 加仑/分钟
gpm [grams per minute] 克/分钟
grade 品级,等级;分级
gradient 梯度
~ curve 梯度曲线
~ mixing device 梯度混合装置
~ of gravity 重力梯度
~ of velocity 速度梯度
~ rube 梯度管
~ tube density determination 梯度管密度测定(法)
~ tube of density 密度梯度管
gradually 逐渐,逐步
~ variable 渐变的
~ variable gear 渐变齿轮传动装置
graduated 分度的,刻度的
~ disk 刻度盘
~ jar 量瓶
~ jug 量杯
~ pipette 刻度吸管
graft 接枝,嫁接
~ copolymer 接枝共聚物
~ copolymerization 接枝共聚(作用)
~ degree 支化度
~ efficiency 接枝效率
~ fibre 接枝纤维
~ modification 接枝改性
~ point 接枝点
~ polymer[GP] 接枝聚合物
~ polymerization 接枝聚合(作用)
~ polypropylene 接枝改性聚丙烯
~ rubber 接枝橡胶
grafted 接枝的,嫁接的
~ copolymer 接枝共聚物
~ monomer 接枝单体
~ side chain 接枝侧链
~ structure 接枝结构
grafting 接枝
~ on surface 表面接枝
~ reactant 接枝反应物
grain 颗粒,粒料;造粒,粒化;粒纹;粒面;纹理;起纹
~ agitator 粒料搅拌器〈注塑机自动着色〉
~ boundary 晶粒间界,晶界
~ boundary cracks 晶界裂纹
~ boundary diffusion 晶粒间界扩散
~ direction 纹理方向
~ effect 纹理效应
~ embossing 粒面压花
~ fineness 颗粒细度
~ fineness number 平均粒度
~ formation 晶粒形成
~ growth 晶粒增长
~ of wood 木材纹理
~ orientation 晶粒取向
~ printing 木纹印刷
~ refinement 细粒化
~ retention 花纹保持性
~ roller 压花辊
~ size 粒度,粒径,粒径度
~ size distribution 粒度分布
~ structure 颗粒结构
~ surface 粒面
~ type cracking 粒型开裂
grainer 压纹器
graininess 粒度;粒性
graining 起纹,压纹
~ board 压纹板

~ calender 颗纹压延机
~ machine 压纹机
grainy 粒状的;颗粒的
gram[g]克
~ atom 克原子〈即摩尔(原子),mol〉
~ atomic weight 克原子量〈即摩尔(原子),mol〉
~ equivalent[val, e-eq, eq, e] 克当量
~ mole 克分子〈即摩尔(分子),mol〉
~ molecular weight 克分子量〈即摩尔(分子),mol〉
~ molecule 克分子〈即摩尔(分子),mol〉
gramophone record press 唱片录音压机
gran. [granular] 粒状
granite pattern 花岗石花纹
granula 颗粒
granular 粒状(的),颗粒(的)
~ boundary 颗粒界面
~ catalyst 粒状催化剂
~ crystalline 粗晶体
~ crystalline structure 粒状结晶结构
~ feed 粒料供给(颗)粒料供给
~ fracture 晶粒状断裂面
~ membrane 筛网过滤器
~ mo(u)lding compound 颗状,模塑料
~ mo(u)lding material 粒状模塑料,成型粒料
~ polyester 粒状聚酯
~ polymer 粒状聚合物
~ polymerization 成粒聚合
~ rubber 粒状橡胶
~ size 粒度,粒径;颗粒大小;晶粒大小
~ structure 粒状结构
~ zone 粒料区
granularity 粒度
granulate 成粒;颗粒料
granulated 成粒的;粒状的
~ catalyst 粒状催化剂
~ compound 粒(状)料
~ gel 粒状凝胶
~ glass sphere 玻璃珠

~ product 粒状产物
~ rubber 粒状橡胶
granulates 粒料
granulating 成粒,造粒
~ die head 成粒机头
~ hood 造粒防护罩
~ machine 成粒机;破碎(成粒)机〈废塑料破碎〉
~ mill 破碎(成粒)机
granulation 成粒,造粒
~ polymerization 成粒聚合
~ size 粒度,粒径,粒径度
granulator 成粒机;破碎(成粒)机
granule 粒子,颗粒;颗粒料
~ manufacture 造粒
granules 粒料
granulometer 粒度计,颗粒测量仪
granulometry 粒度测定(法)
graphite 石墨〈填料〉
~ fibre 石墨纤维
~ fibre reinforced plastic[GFEP] 石墨纤维增强塑料
graphitization 石墨化(作用)
grasshopper 硬丝束;结节
grate separator 筛选分离器,磁栅分离器〈吸除斗中铁质〉
graver 雕(刻)刀
gravimeter 重力计;重差计
gravimetric 重量(分析)的,测定重量的
~ analysis 重量分析
~ batching [weigh feeding] 计量供料,重量供料
~ feed 重力供料,计量供料
~ thickness 重量厚度
gravitational 重力的
~ separation 重力分离
~ separator 重力分离器
gravity 重力
~ bottle 比重瓶
~ casting 重力浇铸〈无压力浇铸〉
~ closing 重力闭合〈活塞依靠自重闭合〉
~ feed 重力供料〈自然下落供料〉
~ loading 重力装料

gravure

- ~ mill 捣碎机
- ~ roller conveyer 重力式滚轴传送装置
- ~ segregation 重力分离
- ~ separator 重力分离器
- ~ tank 自动送料槽
- ~ test 重力试验,相对密度测定

gravure 照相凹板

- ~ coater 槽辊涂布机,凹印辊涂布机
- ~ coating 槽辊涂布
- ~ cylinder 凹版印刷辊
- ~ laminating 局部粘贴
- ~ printing 凹版印刷;着色压印〈塑料薄膜〉

roll 丝网辊

gray 灰色的;(放射)吸收量;坯布,本色布〈纺织品〉

grease 脂膏;润滑脂

- ~ burnishing 脂膏抛光
- ~ forming 润滑脂压力成型〈深拉成型〉
- ~ gun 润滑脂枪
- ~ mark 油迹;油斑
- ~ mo(u)lding 润滑脂压力成型
- ~ proof 防油脂的,耐油脂的
- ~ -proof package 耐油脂包装
- ~ resistance 耐油脂性,防油脂性

greatest 最大的

- ~ limit 最大极限

green 绿(色)的;新鲜的;未加工的,未成熟的;湿潮的;无污染的〈食品等〉

- ~ chromating treatment 轻金属铬酸盐处理〈粘接金属件技术〉
- ~ liquor 绿液〈纸〉
- ~ permeability 湿透气性
- ~ stock 生料;原始混合料,未硫化胶料
- ~ strength 生胶强度,初黏强度;原始强度
- ~ surface 未加工表面
- ~ resin 生树脂,未熟[固]化树脂
- ~ tack 生胶黏性,触模黏性〈黏合剂〉

greening lacquers 发绿漆

Green's viscometer 格林黏度计

greige [gray] 坯布〈纺织品〉

- ~ width 坯布宽度

Grex number 格雷克斯值〈1 万米纤维或单丝重量〉

grey [gray] 灰(色)的

- ~ face 灰色表面
- ~ scale 灰色标度

GR-I [isobutylene rubber] 异丁烯橡胶

grid 栅条,栅板;栅极

- ~ cover 格栅护罩
- ~ electrode 栅(形电)极
- ~ packing 栅格填料
- ~ sheet 栅形板
- ~ spinning 熔栅纺丝
- ~ support 栅格架
- ~ -type sheet 栅型片材

gride 槽型承模板

Griffith's 格里菲斯的

- ~ fracture criterion 格里菲斯断裂判据
- ~ plastometer 格里菲斯塑度计

Grignard 格利雅

- ~ agent 格利雅试剂,格氏试剂
- ~ reaction 格利雅反应
- ~ reagent 格利雅试剂,格氏试剂

grillage 格状结构

grille 格栅

- ~ guard 栅格保护装置,护栅

grind 研磨;研碎

grindability 可磨性

grinder 研磨机

grinding 研磨的,抛光的,粉碎的

- ~ allowance 磨削加工余量
- ~ chamber [milling section] 研磨室
- ~ crack 磨痕,磨削裂纹
- ~ cutt-off 研削切去
- ~ dust 研磨粉尘
- ~ element 研磨单体
- ~ face 磨削面
- ~ head 研磨摆头
- ~ machine 研磨机
- ~ material 磨料;研磨剂
- ~ media 研磨介质,磨料
- ~ mill 研磨机
- ~ pot 研磨罐

~ pressure 研磨压力
~ resin [dispersion resin] 研磨型树脂
~ ring 研磨盘
~ roller 研磨辊
~ stock 研磨料,磨碎料
~ surface 磨削面,磨面,磨碎面
~ test 研磨试验,磨损试验
~ - type resin 研磨型树脂
~ wheel 砂轮
~ wheel binder 砂轮粘结剂
~ - wheel bonding material 砂轮粘结材料
~ wheel resin 砂轮用树脂
~ width 研磨宽度
~ with emery 用金刚砂研磨〈抛光〉
grip 夹;夹具;夹板
~ slide 夹紧滑块
gripper 夹子;夹丝器
gripping device 夹具,固定器;合模装置
grit blasting 喷砂处理〈用金刚砂〉
GR – M [chloro – butadiene rubber] 氯丁橡胶
grommet 金属环;加强环
grommeter 打孔机
groove 槽,槽纹,辊槽纹;溢料缝,溢料槽;坡口〈焊接〉
~ cable 槽纹电缆
~ depth 槽深
~ face 槽纹面;焊接坡口面
~ of thread 螺纹
~ - spew 溢料缝
~ weld 坡口焊缝
~ welding 坡口焊接
~ width 槽宽
grooved 开槽的,有槽的
~ barrel 槽纹机筒
~ bush 槽纹套筒〈挤塑机〉
~ cylinder 槽纹辊,有槽滚筒
~ extruder 槽纹挤塑机
~ feed section [throat] 槽纹供料段〈挤塑机〉
~ feed zone 槽纹供料段〈挤塑机〉
~ liner （机筒）槽纹内衬
~ nip rolls 槽纹压料辊

~ pattern 凹凸花纹,波纹花纹
~ pressure roller 槽纹压辊
~ ring 槽纹环,槽形密封圈
~ roll 槽纹辊筒
~ shaft 槽纹轴
~ tool for mo(u)lding test specimen 模塑试样用槽纹模具
grooving machine 开槽机,切槽机
gross 总的,全部的;粗大的
~ area 总面积
~ area efficiency 总断面效率
~ calorific value 总热值
~ capacity 总容量
~ density 粗密度
~ effect 有效功率
~ efficiency 总效率
~ error 严重误差,过失
~ heating value 总热值
~ load 毛重,总载量
~ output 总产量
~ output value 总产值
~ power 总功率
~ production 总生产量
~ weigth 毛重
ground 地面;底色;磨过的
~ coat 底涂层,底漆(层)
~ coat paint 底漆
~ colo(u)r 底(涂)色
~ fabric 底布
~ fibers [milled fibres] 磨碎纤维,短纤维
~ flour 磨制木粉
~ - glass stopper 磨砂玻璃塞,毛玻璃塞
~ level 地平高度,地面标高,地平面
~ plan 平面图
~ shade 地色,底色
~ temperature 地面温度
~ - type 研磨型
~ - type paste 研磨型糊,分散型糊
~ type resin 分散型树脂,研磨型树脂
~ weft 地纬,底纬〈纺织〉
grounding 接地〈电〉
group 基团;(小)组;类

growing

~ frequency 基团频率
~ - migration polymerization 基团转移聚合
~ of commodities 商品类
~ transfer polymerization 基团转移聚合(作用)
~ transfer reaction 基团转移反应

growing 生长,增长
~ chain 增长链
~ chain end 增长链末端

growth 生长,增长
~ factor 增长系数
~ of crystallite 微晶增长
~ of crystals 晶体增长
~ of defect 缺陷增长
~ of stress 应力增长
~ orientation 增长取向

GRP [glass fiber reinforced plastics] 玻璃纤维增强塑料
~ liner 玻璃纤维增强塑料制的衬里

GRPE [glass fiber reinforced poly ethylene] 玻璃纤维增强聚乙烯

GR-S [butadiene/styrene rubber] 丁二烯-苯乙烯橡胶

GR-S rubber [Government Rubber Styrene] 布纳-S-橡胶,苯乙烯-丁二烯橡胶,丁苯橡胶

GRTP [glass fiber reinforced thermo-plastics] 玻璃纤维增强热塑性塑料

GTT [glass transition temperature, Tg] 玻璃化转变温度

guanidine aldehyde resin 胍醛树脂

guanamine 胍胺;三聚氰二胺
~ resin 胍胺树脂

guarantee 保证;许诺
~ life 保证(使用)寿命
~ period 保证(使用)期
~ test 保用性(鉴定)试验

guaranty 保证;保用

guard 防护;保护
~ bar 护栏,栏杆,扶手
~ board 护板,拦板
~ fence 护栏,护栅,安全栅栏
~ interlock 护栅连锁装置

~ limit switch 护栅限位开关
~ net 保护网
~ plate 防护板
~ ring 防护圈
~ sheet 护板

guarded 屏蔽的,保护的,防护的
~ electrode 屏蔽电极

guayule 银菊胶〈天然橡胶类〉

gudgeon 螺栓;托架;耳轴
~ pin 十字头螺栓;活塞销;耳轴销

guide 控制;导引;定向
~ and knock-out plate 导向和顶出板
~ and stripper plate 导向和脱模板
~ bend test 靠模弯曲试验
~ board 挡板,导板
~ bracket 导架,导座
~ bush 导销衬套
~ device 导向装置
~ eye 导丝头,绕丝头
~ for strip 冲模导向
~ head 导丝头,绕丝头〈玻璃纤维缠绕〉
~ hole 导销孔
~ mark 标记
~ pillar 导向柱
~ pin 导销,定位销
~ -pin bushing 导销套
~ plate 导(向)板,刮料板;支承板
~ pulley 导(向)辊,导轮,惰轮
~ rail scanning 导轨扫描
~ rod 导杆
~ roll 传动辊,导辊;浸渍辊
~ roller 导(向)辊。传动辊
~ screw 导螺杆,传动螺杆
~ shoe 导块
~ slot 导向槽
~ strip 导向板
~ tube 导管,快速焊嘴
~ value 定向值,定位值
~ value of density 密度定位值〈泡沫塑料〉
~ way 导轨;导向装置
~ wheel 导轮

guider 导向器,导辊

guiding 导向,定向
- ~ device 引导装置,导向装置
- ~ groove 导槽
- ~ tube 导向管;快速焊嘴〈热气焊接仪〉

guillotine 剪切机,截切机,切板机;切断器
- ~ knife 剪切机刀
- ~ shear 闸刀式剪切机

gum 树胶;〈有时指〉橡胶[皮];上胶,胶接
- ~ arabic 阿拉伯树胶
- ~ cement 橡胶粘接剂
- ~ content 橡胶含量
- ~ copal 珀巴树脂
- ~ elastic 弹性橡胶
- ~ filler 橡胶填料
- ~ latex 橡胶浆
- ~ level 胶质含量
- ~ mastic 乳香,乳香胶;玛碲树脂
- ~ plastic 树胶塑料,树脂橡胶共混料,塑料合金〈橡胶改性塑料〉
- ~ resin 树胶(树)脂
- ~ rosin 松香
- ~ rubber 胶料
- ~ rubber tubing 软胶管
- ~ spirit [turpentine] 松节油
- ~ stock 胶料
- ~ tolerance 胶质容许量
- ~ waste 废胶
- ~ water 胶水

gummed tape 胶(纸)带

gumminess 胶黏性

gumming 涂胶
- ~ force 胶黏力,胶着力
- ~ machine 涂胶机;胶黏合机
- ~ test 胶黏试验

gummosity 胶黏性

gummous 胶黏的

gummy 胶黏的
- ~ fibre 胶状纤维

gun 喷枪
- ~ -shaped appliance 静电喷〈静电涂敷用粉末喷枪〉

gunjet 喷枪

gunk 料团;预混模塑料〈增强塑料〉
- ~ mo(u)lding 预混料成型,料团成型

gunnies [gunny sacks] 粗麻袋

gusset 横褶;横褶板,人字板;折叠板;补强板,角撑板
- ~ bar 横褶板,人字板
- ~ plate 加强板,角撑板

gutta 古塔波橡胶,杜仲胶
- ~ -percha 古塔波橡胶,杜仲橡胶

gutter 沟槽,排水沟,雨水口
- ~ connector 沟槽连接管
- ~ drain 沟槽排水管
- ~ head [drain or outlet] 沟槽排水管

GW [gross weight] 总重;毛重

gym 体操;体育馆
- ~ outfit 运动衫裤
- ~ shorts 运动短裤
- ~ suit 运动服,体操衣

gypsum 石膏
- ~ mo(u)ld 石膏模

gyrate 旋转(的),回转(的)

gyrating 旋转,回转
- ~ breaker 回转破碎裂机
- ~ crusher 回转压碎机

gyration 旋转,回转

gyrator 旋转器,回转器

gyratory 旋转(的),回转(的)
- ~ breaker 旋转破碎机
- ~ crusher 旋转压碎机
- ~ milling 旋转磨碎
- ~ mixer 旋转混合器
- ~ screw 旋转筛

gyromixer 回转混合器

H

H [hard] 硬的,坚固的
h [hr. hour] 小时
H [hydrogen] 氢
hafnium [Hf] 铪
hair 毛发
　~ [capillary]crack 毛细裂纹,细裂缝
　~ fibre 毛发纤维
hairline 发丝;裂缝,细缝,细纹
　~ crack 微龟裂,毛细裂纹
half 半,一半
　~ bond 半键
　~ crystallization time 半结晶期
　~ die 半模
　~ – finished 半成品的
　~ – finished goods 半成品,半制品
　~ – finished product 半成品,半制品
　~ foam life period 泡沫半衰期
　~ life 半衰期
　~ life period 半衰期
　~ load 半负荷
　~ mould 半模
　~ period 半衰期
　~ reaction 半反应
　~ – round 半圆的
　~ – round gutter 半圆形沟槽
　~ – round runner 半圆形断面流道
　~ saturated 半饱和的
　~ – sun method 升降中点法
　~ time 半衰期
　~ tone 半色调;中间色调
　~ tone effect 半色调效应
　~ transformation point 半转变点
halide 卤化物
　~ – migration polymerization 卤化物迁移聚合
hall – mark 检查标志〈塑料模制品上〉
halocarbon 卤化碳
　~ plastic 卤代烃类塑料
halogenate 卤化

halogenated 卤代的;卤化的
　~ butyl rubber 卤化丁基橡胶
　~ compound 卤代化合物
　~ hydrocarbon 卤代烃
　~ rubber 卤化橡胶
halogenation 卤化(作用)
halogens 卤素
halohydrocarbon 卤代烃
　~ plastics 卤烃类塑料
halopolymer 卤代聚合物
hammer 锤;锤击
　~ crusher 锤式压碎机
　~ effect enamel 锤纹漆
　~ mill 锤磨机
　~ milled filament 锤磨玻璃丝
　~ pendulum [striking pendulum] 锤摆
　~ – scales varnish 锤纹漆
hand 手,手把;手工,工人的;手感〈纤维织物〉
　~ adjustment 手动调整
　~ assembly 手工装配
　~ bag 手提包
　~ banisters 扶手〈楼梯〉
　~ control 手工控制
　~ crank 手摇柄
　~ crusher 手摇压碎机
　~ – driven screw press 手动螺旋式压机
　~ ejection 手工脱模
　~ – fed welding setup 手工焊接装置
　~ feeding 人工加料
　~ finish 手工整修
　~ foggle press 手动肘杆式压机
　~ handle 手柄
　~ – held welding device 手工焊接装置
　~ inspection 用手检查
　~ – laminate 手糊层压(制品)
　~ lay – up 手工敷层;手工铺袋〈增强塑料〉

~ lay‑up method 手糊成型
~ lay‑up mo(u)lding 手糊(铺料)成型〈增强塑料〉
~ lay‑up open‑mo(u)ld method 手铺敞模成型法
~ lever 手柄
~ ‑made 手工制造
~ method 手工方法
~ mo(u)ld 手动模具
~ mo(u)lding 手工模塑
~ mo(u)lding press 手动模压机
~ ‑operated 用手操作的,手动的
~ ‑operated press 手动压机
~ operation 手动操作
~ preform method 手工预成型法
~ press [hand‑operated press] 手动压机
~ properties 手感
~ punch 手动冲模
~ rail 扶手,拦杆
~ sampling 手工取样
~ saver 保护手套
~ screw press 手动螺旋压机
~ sealer 手动热合机
~ selector switch 手动选择开关
~ shears 手动剪截机
~ tool 手动模具
~ transfer mould 手动压铸模
~ traverse 手动横向进给
~ welding 手工焊接
handle 把手;处理;装卸;手感〈纤维〉
~ boss 把手凸台
~ tape 把手带
handling 加工;处理;装卸〈美〉
~ capacity 处理量
~ characteristics 加工性能
~ ease 便于加工
~ equipment 装卸设备〈机械化或自动化〉
~ of benzene vapor 苯蒸气处理
~ property 处理性能
hang 悬挂
~ ‑up 挂料
hanger 吊钩

hanging 悬挂
~ bar 吊杆
~ bar ejection 吊杆传动顶出[脱模]
~ bar ejection system 吊杆传动顶出系统
~ drier[dryer] 悬挂式干燥器
~ ‑up 挂料
hank 绞丝
hard [H] 硬(质)的,坚固的
~ ‑block‑segment polymer 硬嵌段聚合物
~ board 硬质纤维板
~ chrome plated 镀硬铬的〈塑料模具〉
~ chromium plating 镀硬铬;硬铬镀层
~ copy 硬副本,硬拷贝
~ dry 硬干〈漆〉
~ elastic fibre 硬弹性纤维
~ facing 表面硬化
~ fibre 硬质纤维
~ flow 低流动性,难流动性:黏流的,高黏性的
~ foam plastic 硬质泡沫塑料
~ goods 耐用物品
~ handle 粗硬手感
~ hat 安全帽
~ material 硬质材料
~ paper 硬纸;复合纸
~ point 硬化点
~ polymer 硬(弹性)聚合物
~ putty 硬油灰
~ resin 硬质树脂〈天然〉
~ rubber 硬质橡胶
~ segment 硬链段
~ spherical indenter 布氏硬度测试仪,(球)压痕硬度试验仪
~ spot 硬斑点〈透明塑料〉
~ surfacing 表面硬化,硬质面层
~ ‑wearing 耐磨的
~ ‑wood 硬木〈天然〉
~ winter 严冬
hardboard 硬质纤维板
harden 变硬,硬化
hardenable pressure sensitive adhesive 固化型压敏胶黏剂
hardenability 硬化程度,硬化性

hardened 硬化的
  ~ face 硬化面
  ~ flight land 硬化螺纹棱面
hardener 硬化剂
  ~ tank 硬化剂储罐
hardening 硬化
  ~ agent 硬化剂
  ~ furnace 硬化炉
  ~ paste 固化性糊
  ~ – promoting agent 固化促进剂
  ~ resin 硬化性树脂
  ~ with hexa 六亚甲基四胺固化〈间接固化〉,乌洛托品(间接)固化
hardness 硬度；硬性
  ~ ageing 硬化时效
  ~ change 硬度变化
  ~ number 硬度(值)
  ~ penetration 硬化深度
  ~ scale 硬度标(度)
  ~ segment 硬链段
  ~ test 硬度试验
  ~ tester 硬度试验仪
  ~ testing 硬度测试
hardometer 硬度计
hardwood flour 硬木粉
harmonic 谐波；调和的
  ~ analysis 谐波分析
  ~ oscillators 谐振子
  ~ colo(u)r 和谐色
harness 线束〈电缆〉；通丝〈纺织〉；马具〈农耕〉
hat 帽子〈有边的〉
  ~ block 帽模
  ~ body 帽坯
  ~ press 帽形压机
  ~ press mo(u)ld 帽形压机模具
  ~ press mo(u)lding 帽形压机成型
haul – off 牵引,牵引装置
  ~ device 牵引装置
  ~ equipment 牵引装置
  ~ rate 牵引速率
  ~ roll 牵引辊
  ~ speed 牵引速率[度]
  ~ unit 牵引装置

Hauny 汉尼
  ~ injector 汉尼注料器〈增强塑料〉
  ~ process 汉尼法,气压注塑成型法
Hayashi process 林氏(粉末流化成型)法
haze 雾度；光雾
  ~ retarding agent 防光雾剂
  ~ value 雾度值
hazemeter 雾度仪〈薄膜〉
haziness 雾度,(混)浊度
HBCD [1,2,5,6,9,10 – hexabromocy – lodecane] 1,2,5,6,9,10 – 六溴环十二(碳)烷(阻燃剂)
HD – film [high – density film] 高密度薄膜
hdlss. [headless screw] 无头螺钉
HDPE [high density polyethylene] 高密度聚乙烯
HDT [heat deflection temperature] 热挠曲温度
HDT [heat distortion temperature] 热变形温度
He [helium] 氦
head 头(部)；机头；盖,罩；压头；水头,落差
  ~ clamp 机头夹具
  ~ clamp assembly 机头夹紧装置
  ~ core 机头芯
  ~ – die 机头模
  ~ face 端面
  ~ for side extrusion 侧向挤塑机头〈美〉
  ~ for tube 挤管机头
  ~ loss 压头损失
  ~ motor gear 正齿轮发动机
  ~ space 液面上空间〈容器中〉
  ~ straightening machine 矫直机头
  ~ tank 高位槽
  ~ – to – head arrangement 头 – 头排列〈聚合物〉
  ~ – to – head polymer 头 – 头连接聚合物,对头聚合物
  ~ – to – tail arrangement 头 – 尾排列〈聚合物〉

~ - to - tail polymer 头 - 尾连接聚合物,头尾聚合物
~ - to - tail structure 头 - 尾结构
header 头部;集管
~ box 高位箱
~ pipe 总营
headless screw [hdlss.] 无头螺钉
health hazards 对健康有害的事物
heart 心(脏);中心,核心
~ - shaped curve 心形曲线
~ wood 芯材
heat 热(量);加热
~ absorbing action 吸热作用
~ absorbing reaction 吸热反应
~ absorption 吸热,热吸收
~ absorption capacity 吸热量
~ abstraction 散热,排热,除热
~ accumulation 储热,积热
~ accumulation capacity 储(蓄)热量
~ activated adhesive 热活化黏合剂,热反应型胶黏剂
~ age discoloration 热老化变色
~ aging 热老化
~ ageing additive 防[耐]热老化剂
~ ageing inhibitor 防[耐]热老化(抑制)剂
~ ageing test 热老化试验
~ baffle 隔热板
~ balance 热平衡
~ band 加热带;电热丝
~ bending 热弯曲
~ - body 经热增稠〈漆〉
~ bondable fibre 热黏合纤维
~ bonded fabric 热黏合织物
~ bonding 热黏合
~ brander 商标烫印机
~ breakdown 受热击穿
~ build - up 热积聚
~ capacitor 储热器
~ capacity 热容量
~ carrier 热载体
~ carrying element 热载体
~ chamber 加热室
~ change 热交换;热效应

~ check 热裂
~ - cleaned cloth 经热净化处理的玻璃布
~ - cleaned glass cloth 经热净化处理的玻璃布
~ cleaning 热净化,热退浆(玻璃纤维)
~ colo(u)r 耐热染料
~ - conditioning stage 热处理阶段〈拉伸吹塑〉
~ conduction 热传导
~ conduction paste 导热糊
~ conductivity 导热性;热导率
~ conductor 导热体
~ consumption 耗热量
~ content 热含量
~ convection 热对流
~ convertible resin 热硬化树脂
~ crack 热裂纹
~ - cure 热固化
~ deflection temperature 热变形温度;热挠曲温度
~ deflection temperature under load (负荷下)热变形温度
~ deformation 热形变
~ degradation 热降解
~ degree 热度
~ desizing 热脱[退]浆(玻璃纤维)
~ deterioration 热老化;热劣化
~ difference 温差
~ diffusion 热扩散
~ diffusivity 导热性,热导率
~ - dip coating 热浸渍涂敷
~ dispersion 热散射
~ dissipation 热散逸;热消散
~ dissipation rate 散热率
~ distortion 热变形,热扭变
~ distortion point 热变形点;软化点
~ distortion temperature [HDT] 热变形温度
~ distortion test 热变形试验
~ durability 耐热性
~ effect 热效应
~ efficiency 热效率

heat

~ electric couple 热电偶
~ endurance 耐热性
~ energy 热能
~ equivalent 热当量
~ equivalent of fusion 热熔化当量
~ evolution 放热,发热
~ exchange 热交换
~ exchange coefficient 热交换系数
~ exchanger 热交换器,换热器
~ expansion 热膨胀
~ fast 耐热的
~ fastness 耐热牢度
~ flexibility 热柔性
~ flexibility test 热柔性试验
~ flow modulus 热流动模数
~ flux 热(对)流;热通量
~ forming 热成型
~ generation 热产生
~ gradient 热梯度
~ guidance nozzle 导热注料嘴〈热流道注塑模具〉
~ history 累积热
~ hysteresis 热滞后
~ impulse welding 热脉冲焊接
~ input 热量耗费
~ insulated 绝热的,隔热的
~ insulating 绝热的,隔热的
~ insulating ability 保温性,隔热性
~ insulating jacket 绝热夹套
~ insulation 热绝缘;隔热
~ insulation material 绝热材料,保温材料
~ insulator 绝热材料,保温材料
~ interchange 热交换
~ laminating 热层合,加热复合
~ laminating machine 热层合机,加热复合机
~ liberation 放热
~ loss 热损失,热消耗
~ mark 热斑
~ medium 热介质
~ nozzle 导热注料嘴
~ of combustion 燃烧热
~ of compression 压缩热

~ of crystallization 结晶热
~ of decomposition 分解热
~ of formation 生成热
~ of fusion 熔化热,熔融热,熔解热
~ of melting 熔化热,熔融热,熔解热
~ of mixing 混合热
~ of polymerization 聚合热
~ of reaction 反应热
~ of solidification 固化热
~ of swelling 溶胀热
~ of transition 转化热;转变热;潜热
~ of vulcanization 硫化热
~ output 热功率
~ penetration 加热深度
~ plasticization 热塑化(作用)
~ polymer 热聚物
~ polymerization 热聚合
~ - producing capability 发热量,热值
~ production 发热;放热
~ - proof 防热的,隔热的
~ - proof material 防[隔,耐]热材料
~ propagation 热(的)传播,热传导
~ punch 热铆模〈用于热塑性塑料铆接〉
~ quantity 热量
~ radiation 热辐射
~ rate 加热速率
~ ray sealing 热辐射热合
~ - reactivated adhesive 热再活化黏合剂
~ regenerable ion - exchange resin 热再生离子交换树脂
~ release rate 放热速率
~ removal 排热,去热
~ resistance 耐热性;耐热度
~ resistance test 耐热(性)试验,热稳定性试验
~ - resistant 耐热的,热稳定的
~ resistant fabric 耐热织物
~ resistant paint 耐热涂料
~ resistant polymer 耐热聚合物
~ resistant polymethyl methacrylate 耐热聚甲基丙烯酸甲酯,耐热有机玻璃
~ resistant unsaturated polyester resin

耐热不饱和聚酯树脂
- ~ resister 防[耐]热老化剂
- ~ resisting 耐热(的)
- ~ resisting material 耐热材料
- ~ resisting paint 耐热涂料
- ~ resisting plastic 耐热性塑料
- ~ resisting rubber 耐热橡胶
- ~ retaining 保温的,保热的
- ~ retaining power 保温性,保温能力
- ~ screen 隔热屏,绝热罩
- ~ seal 热合,热封
- ~ seal adhesive 热封胶
- ~ seal coating 热封涂层
- ~ seal range 热合温度范围
- ~ sealability 热合性
- ~ sealer 热合机
- ~ sealing 热合,热封(合)
- ~ - sealing dies 热封合模
- ~ - sealing lacquer (or coating) 热封漆
- ~ sealing press 热合压机
- ~ sealing pressure 热合压力
- ~ - sealing tape 热封合带
- ~ sealing temperature 热合温度
- ~ sensitive 热敏的
- ~ sensitive adhesive 热敏性黏合剂
- ~ sensitive material 热敏性材料
- ~ sensitive plastics 热敏性塑料
- ~ sensitive sensor 热敏传感器
- ~ sensitivity 热敏性
- ~ sensitizer 热敏剂
- ~ sensitizing agent 热敏剂
- ~ - set paint 热固性涂料〈在高温下通过溶剂挥发〉
- ~ setter 热定形机
- ~ setting 热定形〈纤维〉;热固化〈塑料〉
- ~ - setting adhesive 热固性黏合剂(温度100℃以上)
- ~ setting bonding 热定形黏合
- ~ setting machine 热定形机
- ~ shield 隔热板
- ~ shocks 热震荡
- ~ shock test 热震荡试验
- ~ shrink ratio 热收缩比
- ~ shrinkability 热收缩性
- ~ shrinkable 热收缩的〈薄膜〉
- ~ shrinkable fibre 热收缩纤维
- ~ shrinkable film 热收缩薄膜
- ~ shrinkable properties 热收缩性
- ~ shrinkable tubing 热收缩管
- ~ - shrinkage 热收缩
- ~ - shrinking unit 热收缩装置
- ~ sink 保热,保温;吸热层;散热片
- ~ - softened material 热软化性材料
- ~ softened plastics 热软化性塑料
- ~ - solvent sealing 加热-溶剂封合
- ~ - solvent tape sealing 加热-溶剂-胶带封合
- ~ stability 热稳定性,耐热性
- ~ stabilizer 热稳定剂
- ~ - stable 热稳定的;耐热的
- ~ - stable antioxidant 防[耐]热老化剂
- ~ - stable material 耐热材料
- ~ storage test 热积聚试验〈模塑制品热稳定性〉
- ~ stretch 热拉伸
- ~ stretch zone 热延伸区
- ~ stretched fibre 热拉伸纤维
- ~ stretching machine 热拉伸机
- ~ test 耐热试验
- ~ tonality 热效应
- ~ tone 热效应
- ~ transfer 传热
- ~ transfer area 传热面积
- ~ transfer capability 传热能力,传热量
- ~ transfer coefficient 传热系数
- ~ transfer factor 传热因素
- ~ - transfer liquid 传热流体
- ~ transfer medium 传热介质
- ~ transfer rate 传热速率
- ~ transfer surface 传热面
- ~ transfer torpedo 导[传]热鱼雷
- ~ transference 传热
- ~ transferring 传热
- ~ transmissbility 传热性
- ~ transmission 传热,热传导

heated

- ~ transmission coefficient 传热系数,热导率
- ~ transmission resistance 传热阻力
- ~ transmission testing apparatus 传热性试验器
- ~ transport 传热
- ~ treatment 热处理
- ~ up 加热
- ~ up time 加热时间
- ~ value 热值,热当量
- ~ welding 热焊接

heated 加热的
- ~ band welding 电热带焊接
- ~ bar 加热焊接棒[带]〈热合机〉
- ~ bar sealing 热触点焊,热带式热合
- ~ chamber 加热室〈纺丝〉
- ~ flat plate 加热平熔接板
- ~ jacket 加热套
- ~ knife 加热刀〈焊接〉
- ~ oven orientation 加热炉定向
- ~ point orientation 加热点定向
- ~ roller 加热滚筒
- ~ rolls orientation 加执滚筒定向
- ~ tool 加热焊具
- ~ tool butt welding 热焊具对焊(接)
- ~ tool (surface) temperature 热焊具表面温度
- ~ tool welding 热焊具焊接
- ~ wedge 热楔形焊具〈焊接〉
- ~ wedge welding 楔形热焊接

heater 加热器;加热焊具〈焊接〉
- ~ adapter (电)加热器连接套〈挤塑机〉
- ~ band 带式加热器,电热圈,电热套;加热带
- ~ band sealing 加热带封焊
- ~ bar 加热熔接棒
- ~ blanket 电热板
- ~ box 加热箱
- ~ head (电)器端,热料筒头〈注塑机〉
- ~ plate (电)加热板
- ~ range 加热段
- ~ strip 电热带

- ~ tunnel (电)加热器槽

heating 加热(的)
- ~ ageing 热老化
- ~ appliance 加热器
- ~ area 加热面积,传热面积
- ~ barred 加热料筒
- ~ bore 热通道〈辊或加热板中的〉;加热腔〈压延辊〉
- ~ box 加热箱
- ~ bush 加热套筒
- ~ cabinet 烘箱
- ~ capacity 热容量
- ~ cartridge 加热筒
- ~ chamber [heating passage] 加热料筒〈注塑机〉;加热腔〈压延辊〉
- ~ channel 加热内腔
- ~ crack 热龟裂
- ~ curve 热曲线〈在热分析中的〉
- ~ curve determination 加热曲线测定
- ~ cycle 加热循环,加热周期
- ~ cylinder 加热料筒(注塑机);加热机筒〈挤塑机〉
- ~ effect 热效应
- ~ efficiency 热效率
- ~ element 电热元件;加热电极;热焊具
- ~ element butt welding 热焊具对焊(接)
- ~ element welding 热焊具对焊(接)
- ~ (feed) hopper 加热料斗
- ~ incubator 烘箱
- ~ jacket 加热套筒
- ~ loss 热损失
- ~ medium 加热介质
- ~ mirror 加热镜面〈对接焊〉
- ~ oven 加热炉,烘箱
- ~ passage 热通道〈辊或加热板中的〉
- ~ pin 加热销〈压塑模具〉
- ~ pipe 加热管,供暖管
- ~ plasticizing period 热塑化周期
- ~ plate 加热板
- ~ pressure 热焊压力
- ~ rate 加热速率
- ~ rate curve 加热速率曲线〈在热分析

中的〉
~ reflector 热反射器〈用于电阻加热元件焊接〉
~ sample 煅烧试样〈鉴别塑料〉
~ spigot 热插窜〈插口熔焊〉
~ strip 热合力,加热片
~ surface 加热面
~ system 加热系统
~ tune 加热时间
~ unit 加热装置
~ – up 加热;升温
~ – up behaviour 加热特性
~ – up period 加热时间;升温期
~ – up pressure 预热压力;升温压力
~ – up process 预热过程;升温过程
~ – up time 加热时间;升温时间
~ zone 加热段;加热区〈挤塑机〉
heatronic mo(u)lding 高频预热成型;高频预热模塑
htavily domed 深度拱形的
heavy 重(型)的,大的
~ addition 大量添加
~ – bodied 很黏的
~ – body 高黏度(的)
~ castings 大型铸件
~ component 高沸点组分
~ denier 粗纤度,粗旦
~ duty 重型
~ – duty coating 耐用涂料,耐磨涂料
~ duty packaging film 重包装薄膜
~ – duty pipe 大口径管
~ – duty plastic 高性能塑料,高强度塑料
~ – duty sack 重型(包装)袋
~ – duty shrink film 重型(包装)收缩薄膜
~ film 厚膜
~ – flash 厚飞边
~ ga(u)ge sheet die 厚片机头[口模],扁机头,窄缝机头
~ leather 厚革
~ moulding 超型模制品
~ oil 重油
~ petrol 重汽油

~ plate 厚板
~ polymer 重聚合物;大分子聚合物
~ pressure 高压
~ repair 大修
~ shade 饱和色
~ spar 重晶石〈填料〉
~ walled 厚壁的
HEC [hydroxyethyl cellulose] 羟乙基纤维素〈表面活性剂,乳化稳定剂,增黏剂,分散剂〉
heel 凸起〈塑料瓶底〉,拱底;尾料,剩余物
~ radius 拱底曲率半径〈中空吹塑容器〉
~ tap 厚薄不匀〈吹塑容器底部〉
height 高度
~ adjustment 高度调节
~ of echo 回声高度[程度]
~ of layer 料层厚度
~ of lower platen (压机)底板高度
~ of thread 螺纹高度
~ of tooth 齿高
~ to width ratio 高宽比
Heisler process 海斯勒法〈旋转粉末成型法之一〉
helical 螺旋形的;斜齿啮合的〈齿轮〉
~ agitator 螺旋式搅拌器
~ angle 螺旋角
~ blade 螺旋叶片
~ blade stirrer 螺旋形片状搅拌器
~ conformation 螺旋构象
~ conveyer 螺旋输送机
~ cooling water duct 螺旋形冷却水通道〈模具〉
~ corrugation 螺旋波纹
~ crimp 螺旋形卷曲
~ crystalline lattice 螺旋形晶格
~ curve 螺旋曲线
~ dryer 螺旋式干燥器
~ gear 螺旋齿轮;斜齿轮
~ molecule 螺旋形分子
~ path 螺旋形(缠绕)程〈长丝缠绕〉
~ projection 螺线型棱口
~ ribbon [screw ribbon or ribbon screw]

螺旋状带
~ ribbon extruder　螺带式挤塑机
~ ribbon mixer　螺带式搅拌器
~ screw channel [screw thread]　螺旋螺槽
~ screw feeder　螺旋式螺杆加料器
~ spindle　螺旋轴
~ - splined milling cutter　带螺旋槽的铣刀
~ spring　螺旋(形)弹簧
~ structrue　螺旋结构
~ teeth　螺旋齿
~ - toothed　螺旋齿形的
~ toothing　螺旋齿
~ wind　螺旋缠绕
~ wire reinforced pipe　螺旋线缠绕增强管
helically　成螺旋形
~ reinforced　螺旋形增强的
~ wound pipe　螺旋形缠绕管
helium [He]　氦
helix　螺旋(形)
~ agitator　螺旋式搅拌器
~ angle　螺旋角
~ conveyor　螺旋输送器
~ distributor　螺旋形分布器〈挤塑软管和硬管机头〉
~ test　螺旋形流动试验〈测定成型料流动性〉
~ type stirrer　螺旋型搅拌器
hells　残余料,下脚料
helmet　安全帽,防护帽;防护罩
helper　助塞;辅助机构
~ drive　辅助传动
hem　边缘,折边
hemicellulose　半纤维素
hemicolloid　半胶体
hemicrystalline　半晶状的;半结晶的
hemispherical bottom　半球形底〈反应器〉
hemming　边缘;卷边
Henschel nixer　亨舍尔混合机〈推进式高速混合机之一〉
　　1,4,5,6,7,8,8 - heptachloro - 3a,4,7,

　　7a - tetrahydro - 4,7 - methanoindene 1,4,5,6,7,8,8 - 七氯 - 3a,4,7,7a - 四氢 - 4,7 - 亚甲基茚(防白蚁剂)
2 - heptadecylimidazole　2 - 十七烷基咪唑〈固化剂〉
herring - bone　人字形(的),鱼刺形(的)
~ gear　人字齿轮
structure　人字形结构
Hertz,HZ　赫(兹)〈周/秒〉
hesion　黏附本领;黏合力
HET - anhydride [hexachloro endometl - iylenetetrahydrophthalic anhydride]　六氯桥亚甲基四氢(邻)苯(二甲酸)酐,氯菌酸酐〈阻燃剂,环氧树脂固化剂〉
heteroblock copolymer　杂嵌段(烯烃)共聚物
heterocatenary polymer　杂链聚合物
heterochain　杂链
~ fibre　杂链纤维
~ polymer　杂链聚合物
heterochromatic　异色的,多色的
heterochromous　异色的,不同色的
heterocyclic　杂环的
~ compound　杂环化合物
~ ladder polymer　杂环梯形聚合物
~ polyamide　杂环聚酰胺
~ polymer　杂环聚合物
heterodisperse　非均相分散
heterofilm　异相薄膜,非均相膜;复合薄膜
heterofunctional　杂官能的
~ condensation　杂官能缩合
heterogeneity　不均匀性;多相性
heterogeneous　不均匀的;多相的;杂的
~ chain compound　杂链化合物
~ chain polymer　杂链聚合物
~ colour　多色性染料
~ grafting　多相接枝
~ ion exchange membrane　异相离子交换膜
~ material　不均匀材料;多相材料
~ nucleation　异相成核

~ polyblend 非均相共混物
~ polymer 非均相聚合物
~ polymerization 非均相聚合
~ reaction 多相反应
~ vulcanization 多相硫化
~ yarn 多相纱
heterohesion 异性材料粘接
heteromorphism 多晶现象
heterophase 多相,非均相
~ polymerization 非均相聚合,多相聚合
heteropolymer 杂聚物
heteropolymerization 杂聚合(作用),异相聚合(作用)
heterotatic 不均匀有规立构
~ polymer 杂同立构聚合物
~ unit 杂同立构单元
hex. [hexagonal] 六方形的
hexa [hexamine] 六胺;乌洛托品
hexabromo – 〈词头〉六溴(代)
~ – benzene 六溴苯〈阻燃剂〉
~ – biphenyl 六溴联苯〈阻燃剂〉
~ – cyclododecane 六溴环十二烷〈阻燃剂〉
hexachloro – 〈词头〉六氯(代)
~ – bicycloheptenedicarboxylic acid 六氯双环庚烯二羧酸〈阻燃剂〉
~ – bicyclo heptenedicarboxylic anhydride 六氯双环庚烯二羧酸酐〈固化剂阻燃剂〉
~ – endomethylenetetrahy drophthalic acid 六氯桥亚甲基四氢邻苯二甲酸〈阻燃剂〉
~ – endomethylenetetrahy drophthalic anhydride [HET anhydride] 六氯桥亚甲基四氢(邻)苯(二甲酸)酐,氯菌酸酐〈反应型阻燃剂,固化剂〉
n – hexadecyl – 3,5 – di – tert – butyl – 4 – hydro – xy benzoate 3,5 – 二叔丁基 – 4 – 羟基苯甲酸正十六烷醇酯〈光稳定剂〉
hexafluoropropylene 六氟丙烯
hexagon nipple 六角形螺纹接头
hexagonal 六方形的

~ barrel mixer 六角桶混合机
~ system 六方晶系
hexahydrophthalic anhydride [HPA] 六氢(邻)苯(二甲酸)酐〈环氧树脂固化剂〉
hexamethyl phosphoric triamide 六甲基磷酰三胺〈光稳定剂〉
hexamethylene 六亚甲基,六甲撑,1,6环己基
N,N′ – ~ bis(3,5 – di – tert – butyl – 4 – hydroxyhydrocinnamamide) N,N′ – 六亚甲基双(3,5 – 二叔丁基 – 4 – 羟基苯丙酰胺)〈热稳定剂〉
~ dianmine 六亚甲基二胺;1,6 – 己二胺
~ diisocyanate 1,6 – 己二异氰酸酯
~ glycol 1,6 – 己二醇
~ glycol bis[β – (3,5 – di – tert – butyl – 4 – hydroxyphenyl) propionate] 己二醇双[β – (3,5 – 二叔丁基 – 4 – 羟基苯基)丙酸酯]〈抗氧剂〉
~ tetramine 六亚甲基四胺,乌洛托品,海克沙〈促进剂H,固化剂〉
hexamine [hexamethylene tetramine or hexa] 六胺;乌洛托品
liexane 己烷
hexyl 己基
~ acetate 乙酸己酯
~ nicthacrylate 甲基丙烯酸己酯
Hf [hafnium] 铪
hf [half] 半,一半
HF [hf;high frequency] 高频(率)
~ – dryer 介电干燥器,高频干燥器
~ – duct [HF tunnel] 高频烟道
Hg [mercury] 汞,水银
HG [mercury pressure] 水银压力
HI high impact 耐高冲击(强度)
bide 片坯,板坯
hiding 遮盖
~ pigment 覆盖颜料,盖底颜料
~ power 遮盖力〈涂料〉
high 高(度,级)的;浓[深]的〈颜色〉
~ abrasion goods 高耐磨制品
~ activity polyether 高活性聚醚

~ – barrier 高隔板,高档板
~ boiler 高沸(点)化合物〈沸点 113~135℃〉
~ boiling 高沸点的
~ boiling component 高沸点组分
~ boiling point 高沸点
~ boiling solvent 高沸点溶剂
~ – build 高黏滞的〈漆〉
~ – build autotnotive finish 高黏滞汽车用喷漆
~ capacity 大容量
~ – carbon steel 高碳钢
~ class 优质的
~ colour channel black 高着色槽法炭黑
~ compression 高压缩
~ – compression ratio 高压缩比〈螺杆〉
~ consistency viscometer 高稠度黏度计
~ current – arc ignition test 大电流电弧着火试验
~ cycle injection moulding 高速注塑成型
~ deflection temperature [HDT] 热挠曲温度
~ density 高密度
~ density assembly 高密度装配
~ – density film [HD – film] 高密度薄膜
~ density foam 高密度泡沫塑料
~ density plywood 高密度胶合板
~ density polyethylene [HDPE] 高密度聚乙烯
~ – density wood 高密度木材,压制木材,硬木
~ draft 大牵伸
~ – duty 重型,大型;大功率的,高生产率的
~ – duty joint 高强度连接;结构连接
~ – duty machine 高生产率机
~ – duty wood 高密度木材,压制木材,硬木
~ efficiency 高效率,高生产率

~ elastic 高弹性的
~ elastic deformation 高弹(性)形变
~ elastic state 高弹(性)态,橡胶态,热弹(性)态
~ elasticity 高弹性
~ – energy irradiation 高能照射
~ – energy radiation 高能辐射
~ – energy rate forming 高效率成型法
~ expansion material 高发泡材料
~ expansion ratio foaming 高倍率发泡
~ extensibility fibre 高延伸度纤维
~ filler loading capacity 高度填料接受能力,填料高含量,高填充量
~ finish 高度光泽
~ flow 易流动性(的)
~ flow rate 高流速
~ frequency 高频(率)
~ frequency bonding 高频粘接
~ frequency dielectric heating 高频介电加热
~ frequency dielectric heating equipment 高频介电加热装置
~ frequency dryer [HF – dryer] 高频干燥器〈塑料成型料〉
~ frequency drying 高频干燥
~ – frequency equipment 高频设备
~ frequency field 高频电场
~ – frequency generator 高频发生器
~ frequency glueing 高频胶合
~ frequency heating [HF – heating] 高频加热
~ frequency heating equipment 高频加热装置
~ frequency method 高频法
~ frequency mo(u)lding [HF – mo(u)lding] 高频(电感)加热成型
~ frequency predrying 高频预烘干〈模塑料〉
~ frequency performance 高频性能
~ frequency preheater 高频预热器
~ frequency preheating [HF – preheating] 高频预热
~ frequency processing 高频加工
~ frequency radiation 高频辐射

high

~ frequency sealing 高频熔合,高频熔接
~ frequency sewing machine 高频缝合机
~ frequency tarpaulin welder[HF - tarpaulin welder] 高频帆布焊接机
~ frequency vibration 高频振动
~ frequency vulcanization 高频硫化
~ frequency welder 高频电焊机
~ frequency welding [HF - welding][dielectric welding] 高频焊接,介电焊接
~ functional polymer 高功能聚合物
~ - gloss finish 高度光泽
~ - gloss plastics leather 高光泽塑料[人造]革
~ - gloss polyester resin varnish 高光泽聚酯树脂涂料
~ grade product 高级产品,高档产品
~ hardness 高硬度
~ heat treatment 高温热处理
~ hiding colo(u)r 高覆盖着色剂
~ impact[HI] 耐高冲击性;耐高冲击的
~ impact material 耐高冲击材料
~ impact nylon 耐高冲击尼龙
~ impact phenolics 耐高冲击酚醛塑料
~ impact plastics 耐高冲击塑料
~ impact polyethylene 耐高冲击聚乙烯
~ impact polypropylene 耐高冲击聚丙烯
~ impact polystyrene[HIPS] 耐高冲击聚苯乙烯
~ impact polyvinyl chloride 耐高冲击聚氯乙烯
~ - impermeability lining system blow moulding [HILS - blow moulding] 高度不渗透性衬里吹塑成型
~ intensity 高强度的
~ - level performance 高度加工性能
~ level production 高水平生产
~ level production 高水平生产
~ limit 最大限度,最大尺寸

~ - load melt index 高负荷熔体指数
~ loading 高填充量
~ - low temperature cycles test 高低温交变试验
~ lubricating polyacetal 高润滑级聚甲醛
~ - luster 高光泽
~ - melting 高熔的
~ - melting plastics 高熔点塑料
~ mica 优质云母,高级云母
~ - mirror polish 高度镜面抛光
~ mo(u)lding compound 高强度模塑料
~ - modulus weave 高模量(玻璃纤维)织物
~ molecular compound 高分子化合物
~ molecular drugs 高分子药物
~ molecular liquid crystal 高分子液晶
~ molecular polymer 高分子聚合物,高聚物
~ molecular solution 高分子溶液
~ molecular weight[HMW] 高相对分子质量
~ molecular weight compound 高分子(量)化合物
~ molecular weight plasticizer 高(相)对分子(质)量增塑剂
~ molecular weight polyethylene [HM-WPE] 高(相对)分子(质)量聚乙烯
~ molecular weight polyvinyl chloride 高(相对)分子(质)量聚氯乙烯
~ molecule 高分子
~ - peel strength adhesive 高剥离强度黏合剂
~ penetration resistance 耐高穿透性
~ - performance adhesive 高性能黏合剂,高强度黏合剂
~ performance antistatic fibre 高性能防静电纤维
~ performance composite 高性能复合材料
~ - performance plastic 高性能塑料,高强度塑料

~ - performance screw 高性能螺杆,高产量螺杆
~ performance thermoplastics 高性能热塑性塑料
~ point 高点〈帐篷〉
~ polish 高度光泽
~ polymer 高聚物
~ polymer molecule 高聚物分子
~ polymeric material 高聚物材料
~ power 大功率
~ pressure 高压
~ pressure capillary rheometer 高压毛细管流变仪
~ pressure capillary viscosimeter 高压毛细管黏度计
~ pressure foaming machine 高压发泡机
~ pressure hose 高压软管
~ pressure laminate 高压层合制品
~ pressure laminating 高压层合
~ pressure metering unit 高压计量装置〈聚氨酯组分〉
~ pressure mo(u)lding 高压成型,高压模塑
~ pressure plasticization 高压增塑
~ pressure plunger moulding 高压柱塞成型
~ pressure polyethylene [HPPE] 高压(法)聚乙烯
~ pressure polymerization 高压聚合
~ pressure press 高压压机
~ pressure processed [produced] polyethylene 高压(法)聚乙烯
~ pressure spining 高压纺丝
~ pressure spot 高压欠胶疵点
~ production rate 高生产率
~ productive capacity 高生产能力
~ purity grade 高纯度级
~ quality 高质量的,优质的
~ rate 高效率,高速
~ resilience 高回弹力
~ - resilient foam [HR foam] 高弹性泡沫塑料
~ resiliecy cold cured polyurethane flexible foams 高回弹冷熟化聚氨酯软质泡沫塑料
~ resolution 高清晰度,高分辨率
~ resolution nuclear magnetic resonance 高分辨核磁共振
~ resolving electron diffraction 高分辨电子衍射
~ revolving 高转速的,高速的
~ - rise 高(速)起泡〈泡沫塑料〉
~ shear mixer 高剪切混合器
~ - shear viscosity 高剪切黏度
~ - shrinkage fibre [HS, high - shrinking fibres] 高收缩纤维
~ silica glass fiber 高硅氧玻璃纤维
~ solids 高固体含量〈漆〉
~ - solids coating [HS coating] 高固体含量涂料
~ - solids solution 高固体含量溶液
~ speed 高速的
~ speed adjustment 高速调整
~ speed agitator 高速搅拌器
~ speed blender 高速掺混机,高速混配机
~ speed centrifuge 高速离心机
~ speed cold forming 高速冷成型
~ speed cure 快速固化
~ speed extruder 高速挤塑机
~ speed extrusion 高速挤塑
~ - speed hammer mill 高速锤式磨
~ speed injection moulding 高速注塑成型
~ speed kneader 高速捏和机
~ speed kneading machine 高速捏和机
~ speed machine 高速机〈注塑机〉
~ speed mixer 高速混合机
~ - speed nozzle 高速焊嘴
~ speed operation 高速运转
~ speed plunger moulding 高速柱塞成型
~ - speed rotating collet 高速旋转夹头,高速卷绕
~ speed spinning machine 高速纺丝机
~ speed stretch 高速拉伸
~ speed tensile strength 高速拉伸强度

~ – speed testing 高速试验,短时试验
~ – speed thermoforming 高速热成型法
~ speed tube extrusion 高速挤管,高速管材挤塑
~ speed turbine stirrer 高速涡轮式搅拌器
~ speed vulcanization 快速硫化
~ speed welding 高速焊接
~ – speed welding nozzle 高速焊嘴
~ – speed winder 高速卷绕
~ speed wind – up 高速卷取,高速卷绕
~ strength 高强度
~ – strength composite 高强度复合材料
~ – strength flame resistant fibre 高强度耐火纤维
~ – strength material 高强度材料
~ – strength moulding compound [HMC] 高强度模塑料
~ – strength plastic 高强度塑料
~ stretched 高倍拉伸的,高度伸张的
~ temperature 高温(的);耐热(的)
~ temperature ageing 高温老化
~ temperature blowing agent [HTBA] 高温发泡剂
~ temperature crushing 高温粉碎
~ temperature decomposition 高温分解
~ temperature dyeing 高温染色
~ temperature dyeing machine 高温染色机
~ – temperature foam [HT foam] 高温泡沫塑料
~ temperature heated tool welding 热焊具高温焊接
~ temperature life 高温使用寿命
~ temperature plasticizers 高温增塑剂
~ – temperature plastics 高温塑料
~ temperature polymerization 高温聚合
~ temperature reaction product of diphenylamine and acetone 二苯胺和丙酮的高温反应产物〈抗氧剂〉
~ temperature resistancefibre 耐高温纤维
~ temperature resistant adhesive 耐高温黏合剂
~ temperature resistant plastic 耐高温塑料
~ temperature resistant resin 耐高温树脂
~ temperature resistant unsaturated polyester resin 耐高温不饱和聚酯树脂
~ temperature setting 高温定形
~ temperature stability 高温稳定性
~ temperature strengh (耐)高温强度
~ – temperature strain 高温应变〈100℃以上时〉
~ temperature test 高温试验
~ temperature test assembly 高温试验装置
~ temperature thermo – setting 高温热定形
~ temperature treatment 高温处理
~ temperature zone 高温区
~ – tenacity 高强度,高强力
~ – tenacity fibre 高强力纤维
~ – tenacity film [HT – film] 高强度薄膜,高韧性薄膜
~ tenacity nylon filament 高强力耐纶丝,高强力尼龙丝
~ tenacity nylon filament 高强力耐纶丝,高强力尼龙丝
~ tenacity rayon 高强度人造丝
~ tensile strength [HTS] 高态拉伸强度
~ – tension cable 高压电缆
~ – tension engineering 高(电)压工程
~ vacuum 高度真空
~ – vacuum metal deposition [depositing] 高真空镀金属〈金属蒸发镀层〉
~ – vacuum metallizing 高真空镀金属
~ – vacuum plating 高真空电镀
~ velocity 高速(度),快速
~ – viscosity 高黏度,高黏性
~ voltage 高电压

~ – voltage – arc track test 高压电弧径迹试验
~ voltage cable 高压电缆
~ voltage tube 耐高电压管
~ volume production 大量生产,大规模生产
~ wet modulus 高湿模量
~ wet – modulus fibre 高湿模量纤维
~ yield 高产率
~ – yield process 高产率法
higher 较高的
~ bulk 高膨松体
~ polymer 高聚物
highly 高,高度地
~ amorphous cellulose structure 高度非晶态纤维素结构
~ branched 多支的
~ branched chain 多支链
~ branched compound 多支链化合物
~ combustible 易燃的
~ cross – linked polymer 高度交联聚合物
~ disperse silicic acid 高度分散的硅酸,超细二氧化硅粉
~ dispersed 高度分散(了)的
~ elastic material 高弹性材料
~ oriented 高度取向的;高度定向的
~ oriented fibre 高度取向纤维,高度定向纤维
~ oriented film 高度取向薄膜,高度定向薄膜
~ viscous 高黏(性)的
HJLS – BM [high – impermeability lining system blow molding] 高度不渗透性衬里吹塑成型
hindrance 阻碍
~ of thermal expansion 阻挠热膨胀(作用)
hinge 铰链,折叶
~ effect 铰链作用
~ pin 铰(链)销
hinged 铰接的
~ bender 铰链式弯曲机
~ bolt 铰接螺栓

~ bolt connection 活节螺栓连接,铰链螺栓连接
~ control cabinet 铰链式控制柜
~ copolymer 铰链型共聚物
~ cover 铰接盖
~ follower mould [hinged retraction mould] 铰链压模
~ joint 铰链接合,铰接
~ lid 铰链盖
~ – pipe clamp 管铰接卡圈
~ retraction mo(u)ld 铰链压模
~ split mo(u)ld 铰链式对开模具
HIPS [high – impact polystyrene] 耐高冲击聚苯乙烯
history 历史;经历记载;历程;随时间变化关系的图解〈曲线,函数〉
HMC [high – strength moulding compound] 高强度模塑料
HMWPE [high molecular weight polyethylene] 高分子量聚乙烯
hoar 灰白
hob [hub] 冲压[切压,冷压]母模
~ sinking 切压制膜
hobbing 冲[切]压制膜
~ blank 冲[切,冷]压钢坯
~ press 冲压机
~ punch 冲压模头
~ ring 冲压环
hobbock 容器,盛料器,运料器〈塑料成型料输送容器〉
hobe 蜂窝材料
~ material [honey – comb before expansion] 蜂窝材料
Hoesch 赫施
~ impeller 赫施搅拌叶轮,旋笼式搅拌叶轮
~ impeller mixer 赫施搅拌器
hog 重型多刀粉碎机,缩粒机
hogging 弯曲,扭曲
hold 握住,固定;保持
~ – down 固定;保持;夹板
~ – down groove 保位槽〈在模具中〉
~ – down mo(u)ld 保位槽式模具
~ – down pin 定位销

~ – down plate 固定板,夹料框〈用于真空或包模成型,高频焊接〉
~ – down plate for slip forming 阳模滑片成型的夹片框
~ – down support 固定支座
~ – out coat 封闭涂层
~ time 持续时间,停留时间
~ – up 死角〈模内〉
~ up time 保持时间;保压时间〈注塑〉
holder 夹(子);容器;(支)架
~ block 模箍;模圈,模套
holderblock 模套,模穴套板
holding 保持,夹持;固定;调整
~ capacity 容积,容量
~ device 夹持装置,夹具
~ fixture 夹具
~ pressure 保压〈注塑时〉
~ pressure pulsation 保压波动[变化]〈注塑时〉
~ ring 调整环
~ tank 接受器,收集槽
~ time [retention time] 保持时间,停留时间〈超声波焊接时〉
hole 孔
~ diameter 孔(直)径
~ – die 孔模
~ forming pin 成孔销,穿孔针
~ punch 冲孔模头
~ – type die 孔型冲模
holidays 漏涂〈涂布缺陷〉
hollander 打浆机;捣碎机
hollow 空的;中空;中空制品
~ article [body] 空心制品;中空制品
~ bead 空心颗粒
~ body 空心体
~ body – forming 空心体成型
~ casting 中空铸塑;搪塑
~ – con screw 中空螺杆
~ container 中空容器
~ cored fibre 空心纤维
~ fibre 空心纤维,中空纤维
~ filament yarn 空心长丝
~ flatten fibre 空心扁平纤维

~ microspheres 中空微球〈填充剂,改性剂〉
~ mo(u)ld 铸模;敞口模〈铸塑〉
~ mo(u)ld casting 空心模铸塑(法)
~ mo(u)lding 中空模塑[成型]
~ needle 空心针〈吹塑〉
~ – plywood 空心胶合板
~ profile 空心型材
~ profiled fibre 空心异形纤维
~ rivet 空心铆钉
~ pipe 空心管
~ shaft 空心轴
~ shapes 空心型材〈挤塑〉
~ shelf 空心隔板
~ shells construction 空心薄壳结构
~ wall pipe 中空壁管;双壁管
holographic 全息的
~ interferometer 全息干扰仪
~ interferometry 全息干扰测量法〈塑料连接时的缺陷分析〉
holography 全息照相术,全息摄影术
holomicroscopy 全显微镜检查法
holotactic 全规整
home 家(的),本国(产的),国内(的)
~ furnishings 家具,家器
~ – made 国产的,自制的
~ plastic 国产塑料
~ – produced 国产的,自制的
~ products 本国产品
~ roll 主辊
~ scrap 返料
homo – 〈词头〉相同,类似,均匀;高〈指多一个甲基的(化合物)〉
~ – chain polymer 均链聚合物
~ – mixer 高速搅拌器
homoaromaticity 同芳香性
homochromatic 同色的
homochromo – isomer 同色异构件
homochromy 同色
homocrosslinking 均相交联〈聚合物〉
homodisperse 均相分散
homofil 单组分纤维
homogeneity 均匀性
~ to the eye 目测均匀性

homogeneization 均化作用
homogeneous 均匀的;均相的
　~ blending 均匀混配,均匀混和
　~ bond 均匀键;均匀黏合
　~ deformation 均匀形变
　~ degree 均匀度
　~ dyes 单一染料
　~ grafting 均相接枝
　~ ion exchange membrane 均相离子交换膜
　~ ion exchange membrane by method of bulk pdymerization 本体聚合法均质离子交换膜
　~ ion exchange membrane by method of casting 流延法均质离子交换膜
　~ ion exchange membrane by method of impregnation 含浸法均质离子交换膜
　~ material 均匀材料
　~ mixture 均匀混合物
　~ nucleation 同质成核,均相成核
　~ phase 均相
　~ polyblend 均质共混物
　~ polymerization 均相聚合
　~ series 同系列
　~ solution polymerization 均相溶液聚合
　~ strain 均匀应变
　~ stress 均匀应力
　~ temperature 同系温度
　~ thermal expansion 均匀热膨胀
homogenization 均化(作用)
　~ efficiency 均化效率〈挤塑机〉
homogenizer 均化器
homogenizing 均化(作用)
　~ performance 均化特性
　~ zone 均化段,计量段〈挤塑机〉
homologous 相应的,相似的
　~ series 同系列
　~ temperature 同相温度
homopolar 无极的
　~ bond 无极键
　~ compound 无极化合物
　~ link 无极键

　~ linkage 无极键
　~ polymer 无极聚合物
homopolyacrylonitrile fibre 均聚丙烯腈纤维
homopolycondensation 均向缩聚
homopolyester fibre 均聚酯纤维
homopolymer 均聚物
homopolymerization 均聚(合)(作用)
honeycomb 蜂窝(结构)
　~ adhesive 蜂窝黏合剂,蜂窝结构胶黏剂
　~ before expansion 膨胀前蜂窝(材料)
　~ construction 蜂窝状结构
　~ core 蜂窝状芯材(蜂窝结构)
　~ cracks 网状裂缝
　~ effect 蜂窝效应
　~ laminate 蜂窝层压材料
　~ material 蜂窝材料
　~ roll 吸(水)辊,蜂窝状辊
　~ sandwith construction 蜂窝状夹心结构
　~ sandwich 蜂窝状夹心材料
　~ structure 蜂窝结构
　~ structure laminate 蜂窝结构层合塑料
hood 盖,套,外壳,罩,通风柜
hook 吊钩
Hookean 虎克的
　~ elasticity 虎克弹性,理想弹性
　~ spring 虎克弹簧〈模拟弹性的弹簧元件〉
Hooke's 虎克的
　~ joint 万向接头
　~ law 虎克定律
hoop 环,箍,圈
　~ direction 周向;圆周方向
　~ drop relay 落环式继电器
　~ layer 环向(缠绕)层
　~ orientation 圆周定向〈拉吹〉
　~ strength 圆周强度
　~ stress 周向应力
　~ winding 环向缠绕,环形线圈
hopper 料斗

- ~ agitator 料斗搅拌器
- ~ box 料箱
- ~ car handling 料斗车装卸
- ~ chute 料斗式斜槽
- ~ dryer 料斗干燥器
- ~ feed 料斗给料
- ~ feeder 料斗给料机
- ~ filler 料斗装料机
- ~ level meter 料斗中料位测量仪
- ~ loader 料斗装料机
- ~ mill 锥形磨机;漏斗式磨碎机
- ~ outlet 漏斗出口

horizontal 水平的;横向的;卧式的
- ~ adjustment 水平调整
- ~ and vertical flash mo(u)ld 横向和纵向溢料模〈压模〉
- ~ centrifugal screen 卧式离心筛
- ~ cylinder dryer 卧(式圆)筒干燥机
- ~ cylinder drying machine 卧筒干燥机
- ~ cylindrical paddle mixer 卧筒桨式混合机
- ~ drying machine 卧式干燥机
- ~ extruder 卧式挤塑机
- ~ extruder head 卧式挤塑机头
- ~ flash mo(u)ld 横向溢料模
- ~ slash ring 横向环形溢料缝
- ~ injection 卧式注塑
- ~ injection machine 卧式注塑机
- ~ lamination 水平层合〈泡沫塑料〉
- ~ multipass dryer 卧式多程干燥器
- ~ plane 水平面
- ~ rotory dryer 卧式回转干燥器
- ~ spinning extruder 卧式挤压纺丝机
- ~ surface 水平面

horizontally-mounted 水平安装的
horn 焊头〈声波焊接具〉
Horrock extruder 霍洛克式挤塑机〈行星滚柱挤塑机〉
horsepower 功率,马力
- ~ rating 额定功率

horseshoe 马蹄形(的)
- ~ agitator 马蹄式搅拌器,U形搅拌器
- ~ mixer 马蹄式混合器,U形混合器
- ~ stirrer 马蹄式搅拌机

hose 软管
- ~ clamp 软管夹
- ~ connection 软管连接
- ~ coupling nipple 软管用接头
- ~ joint 软管连接头
- ~ nipple 软管(螺纹)接头
- ~ pipe 软管
- ~ tube 软管

hot 热(的);加热
- ~ -acting solvent method 热熔剂法〈纤维素衍生物〉
- ~ air 热空气
- ~ air ageing 热空气老化
- ~ air cure 热空气固化
- ~ air drier[dryer] 热风干燥器
- ~ air gun 热空气喷枪(焊接)
- ~ air heated 热空气加热的
- ~ air orienting oven 热气定向拉伸烘箱
- ~ air prefoaming 热空气预发泡〈含发泡剂聚苯乙烯粒料〉
- ~ air sealer 热空气热合机
- ~ air setting 热空气定形
- ~ air shrinkage 热风收缩
- ~ bar 加热(芯)棒〈焊接〉
- ~ bar sealing 热触点焊,热带式热合
- ~ bench test 热台试验
- ~ bend 热弯曲(试验)
- ~ bending 热弯曲
- ~ box process 热箱法
- ~ brittleness 热脆性
- ~ cast 热铸塑
- ~ coining 热后压铸
- ~ compression 热压
- ~ compression roll 热压辊
- ~ crack 热裂(纹)
- ~ cure 热固化,热硫化
- ~ cure foam 热熟[固]化泡沫塑料
- ~ cure mo(u)lding 热熟[固]化成型
- ~ cut 热切〈切粒〉
- ~ cut in air stream 气流中热切
- ~ cut method 热切法
- ~ cut pellet 热切粒
- ~ cylinder setting 热滚筒定形

~ (die) face cutting 热(模)面切断
~ dip coating 热浸涂布
~ dip compound 热浸涂料
~ dip galvanization 热浸镀锌
~ dip mo(u)lding 热浸成型(法)
~ – dip solution 热浸溶液
~ dip stripping compound 热浸渍脱模料
~ dipping 热浸渍(成型法)
~ drawing 热拉伸
~ drawn 热拉的
~ edge mould 热流道模具(注塑)
~ edge tunnel gate 侧隧道形热浇口〈注塑模〉
~ embossing 热压花〈塑料膜片〉
~ extrusion 热挤塑
~ extrusion mo(u)lding 热挤模塑
~ feed 热加料
~ filament sealing [hot wire welding] 热丝封焊
~ filament welding 热丝焊接
~ foam [hot – moulded foam] 热发泡
~ foil stamping 热箔压印,(彩箔)烫印
~ foil transfer 彩箔转印
~ forming 热成型
~ gas jet spinning 热气喷射纺丝
~ gas sealer 热气焊接仪
~ gas weld 热气焊接
~ gas welder 热气焊接仪
~ gas welding equipment 热气焊接仪
~ gate mo(u)lding 热浇口成型,热流道成型〈注塑〉
~ gluing 热粘接
~ grid spinning 熔栅纺丝
~ hardness 高温硬度
~ jacket 热外套
~ jet gun 热喷枪(焊接)
~ jet welding 热风熔接
~ joining 热粘接
~ laminating 热层合,热复合
~ – leaf stamping 彩箔烫印
~ manifold mo(u)ld 热流道模具〈注塑〉
~ manifold system 热流道系统〈注塑〉
~ melt 热熔体;热熔(化)
~ melt adhesive 热熔黏合剂
~ melt application equipment 热熔体涂布设备
~ melt applicator 热熔体涂布仪
~ melt coater 热熔体涂布机
~ melt coating 热熔体涂布
~ melt die 热熔料模头
~ melt extruder 热熔料挤塑机〈热熔胶涂布〉
~ melt granulator 热熔料造粒机
~ melt gun 热熔料喷枪〈热熔胶涂布枪〉
~ melt head 热熔料机头〈涂布机头〉
~ melt plastic 热熔性塑料
~ melt preimpregnating 热熔预浸渍工艺
~ melt strength 热熔体强度
~ melt tank 热熔体槽
~ melt tester 热熔(黏合剂)测试仪
~ mixture 热混合料
~ mo(u)ld 热模
~ – moulded foam 热模发泡
~ mo(u)lding 热模塑
~ parison 热型坯〈吹塑成型用〉
~ parison approach 热型坯料道〈吹塑〉
~ penetration test 热穿透度试验〈电缆包覆料〉
~ plasticity 热塑性
~ plate (电)热板,加热板
~ plate welding 热板焊接
~ polymer 高温聚合物,热聚合物
~ polymerization 热聚合
~ press 热压机
~ press marking 热压标记
~ press moulding 热压成型
~ pressed 热压的
~ pressing 热压
~ punching 热冲压;热冲切
~ resistance 耐热性
~ roll 加热辊〈薄膜拉伸固定装置〉

~ rolled　热轧的;热滚压的
~ rolling　加热辊压,加热滚压
~ rubber　高温橡胶,热橡胶
~ runner　热流道〈注塑模〉
~ runner block　热流道模板〈注塑〉
~ runner bush fechnique　热流道衬套技术〈注塑〉
~ runner gate　热流道浇口
~ runner injection mo(u)ld　热流道注塑模
~ runner injection moulding　热流道注塑成型
~ runner manifold　热流道支管
~ runner mo(u)ld　热流道模具〈注塑〉
~ runner mo(u)lding　热流道成型〈注塑〉
~ runner system　热流道系统〈注塑〉
~ runner valve gate　热流道阀式浇口
~ seal adhesive　热封黏合剂
~ -sealing　热(封)合
~ set　热固化;热定形
~ set plastics　热固性塑料
~ -setting　热固性的〈塑料〉
~ setting adhesive　热固性黏合剂〈温度100℃以上〉
~ setting resin　热固性树脂
~ shock test　受热龟裂试验
~ short　热脆性的;不耐热的
~ shortness　热脆性
~ spot　热点
~ -spray application　热喷涂
~ sprue bushing　热浇口衬套
~ stage　加热台〈熔体台式显微镜〉
~ stamping　热压印,烫印,烫金
~ stamping foil　烫印箔,热印箔
~ stenter setting　热拉幅定形
~ strength　高温强度,热强度
~ stretch　热拉伸
~ tab gate　热柄形浇口
~ tack　热态黏度;热黏性
~ tack barrier coating　热黏合保护涂层〈美〉
~ tack value　热黏性值〈热熔黏合剂〉

~ tear resistance　耐高温撕裂性
~ tear strength　高温撕裂强度
~ tensile strength　高温拉伸强度〈100℃〉
~ tension test　高温拉伸试验
~ test　热态试验
~ tip bushing　热注嘴套
~ tool welding　热板焊接
~ transfer printing　热传导印刷
~ transfer sheet　传热板,热传导板
~ vulcanization　热硫化
~ water bath take-away　热水浴引出装置
~ water mo(u)lding　热水成型
~ water prefoaming　热水预发泡〈含发泡剂聚苯乙烯粒料〉
~ water resistance　耐热水性
~ water take-away　带热水浴的牵引装置〈挤塑机〉
~ welding　热焊接
~ wet strength　高温湿强度
~ wire cutter　热丝切割机〈泡沫塑料〉
~ wire cutting　热丝切割
~ wire heater　炽热丝加热器
~ wire multiple heating tool　炽热丝多处加热器〈弯曲板〉
~ wire side weld　炽热丝侧焊〈塑料袋〉
~ -wire slicer　炽热丝切片机
~ wire welding　炽热丝焊接
~ work　热加工
~ zone　热带,热区
hour　小时;钟点;时间
~ on stream　操作时间
~ to stream　操作时间
hourly　每小时
~ capacity　每小时生产能力
~ efficiency　每小时生产率
~ output　每小时产量
household package　家用包装
housing　外壳,外罩
hp [horse power]　马力
hr. [hour]　小时
HR [Rockwell hardness]　洛氏硬度

HR foam [high-resilient foam] 高弹性泡沫塑料
HS [high speed] 高速率,高速度
HS coating [high-solids coating] 高固体含量涂料
HS fibres [high-shrinkage fibres] 高收缩纤维
ht. [hight] 高;高度
HTE [hydroxyl terminated polyether] 羟基封端聚醚
HT foam [high-temperature foam] 高温泡沫塑料
HTS [high tensile strength] 高态拉伸强度〈玻璃纤维〉
hub 冲压[切压,冷压]母模
hubbing 冲压;切压
～ blank 冲[切,冷]压钢坯
～ press 切压机
hue 色调,色彩;色相;色光〈染料的〉
～ of color 色相;色光
hull 暗斑
humectant 湿润剂
humectation 湿润,增湿
humidification 湿润,增湿
humidifier 增湿器
humidiometer 湿度计
humidity 湿度,湿气
～ ageing 湿气老化
～ application 湿气条件下应用〈材料〉
～ cabinet test 湿润箱试验
～ chamber 增湿室;湿室
～ control 湿度调节,湿度控制
～ controller 湿度控制器
～ indicator 湿度指示器
～ meter 湿度计
～ -proof 防湿的,防潮的
～ ratio 湿度比
～ resistance 防潮性能,耐湿性〈胶黏合性〉
～ sensitive element 湿(度)敏(感)元件
～ sensor 湿(度)敏(感)元件
～ stability 湿气稳定性
～ test (耐)湿度试验〈胶黏合性〉

～ ventilation 润湿通风
humidizer 增湿剂
humidness 湿度,湿气
hybrid 混合(物);杂化;混杂(料)
～ composite 混杂复合材料
～ effect 混杂效应
～ fiber reinforced plastics 混杂纤维增强塑料
～ interphase 混杂界面
～ material 混杂材料
～ structure 混合结构
～ volume ratio 混杂体积比
hyd. [hydrated] 含水的;水合的
hydation epoxy resin 海因环氧脂,内酰脲环氧树脂
hydrate 水合物
hydrated 水合的
～ alumina 水合氧化铝
～ aluminum oxide 水合氧化铝
～ aluminum silicate 水合硅酸铝
～ cellulose 水合纤维素
～ silica 含水二氧化硅〈即硅酸,催化剂〉
hydraulic 液压的
～ accumulator 液压储蓄器
～ actuator 液压传动装置
～ atomization 液压喷雾
～ baling press 液压式打包机
～ bear-down force 液压支撑力
～ booster 液压助力器,增压器〈注塑机〉
～ calender 液压轧光机
～ circuit 液压循环管线
～ clamp injection machine 液压锁模注塑机
～ closing cylinder 合模液压缸
～ contact 液压连接
～ control 液压控制,液压调整
～ cutter 液压裁剪机
～ decimal press 十进(位制)液压机
～ drive 液压传动
～ -driven press 液压机
～ ejection 液压脱模,液压顶出
～ ejector 液压脱模器,液压顶出器

- ~ extruder 液压挤塑机
- ~ fluid 液压流体
- ~ forming 液压成型
- ~ injection cylinder 液压注射油缸
- ~ intensifier 液压增压机
- ~ locking system 液压锁模装置
- ~ medium 传压介质
- ~ mo(u)lding press 液压成型机
- ~ oil valve 液压油阀
- ~ press 液压机
- ~ pressing machine 液压打包机
- ~ pression extruder 液压柱塞式挤塑机
- ~ pressure 液压
- ~ pressure test 液压试验
- ~ ram 液压柱塞〈注塑机〉
- ~ set 液压装置,液压设备
- ~ tabletting machine 液压压片机
- ~ test 液压试验
- ~ -type automatic injection mo(u)lding machine 液压式自动注塑成型机
- ~ valve 液压阀

hydride 氢化物
- ~ -shift polymerization 氢负离子转移聚合
- ~ transfer 氢负离子转移
- ~ hydroabietyl alcohol 氢化松香醇〈增塑剂〉

hydrocarbon 烃,碳氢化合物
- ~ blowing agent 烃类发泡剂
- ~ casting resin 烃类铸塑树脂
- ~ chain 烃链
- ~ elastomer 烃类弹性体
- ~ -formaldehyde resin 烃,甲醛树脂
- ~ oligomer 烃类低[齐]聚物
- ~ plastic 烃类塑料
- ~ polymer 烃类聚合物
- ~ processing 烃加工
- ~ products 烃类产品
- ~ resin 烃类树脂
- ~ synthesis 烃类合成

hydrochloric acid 盐酸;氢氯酸
hydrocyanic acid 氢氰酸;氰化氢
hydrodepolymerization 加氢解裂(作用)

hydrodynamic 流体的,流体动力的
- ~ effect 流体效应
- ~ extruder 流体动力挤塑机
- ~ film 流体动力薄膜
- ~ lubrication 流体润滑
- ~ orientation 流体动力取向
- ~ water-resistance test 水力式防水试验

hydrofoil 防水薄膜,耐水薄膜
hydroforming 液压成型;临氢重整
hydrogels 水凝胶
hydrogen[H] 氢
- ~ bond 氢键
- ~ bridge 氢桥
- ~ cracking 加氢裂化
- ~ gas 氢气
- ~ migration polymerization 氢转移聚合
- ~ peroxide 过氧化氢
- ~ shift polymerization 氢转移聚合
- ~ sulphide 硫化氢
- ~ transfer polymerization 氢转移聚合

hydrogenate [hydrogenize] 氢化;氢化物
hydrogenated 氢化的
- ~ bisphenol A diglycidylether 氢化双酚A二缩水甘油醚
- ~ bisphenol A epoxy resin 氢化双酚A(型)环氧树脂
- ~ diphenylmethane diisocyanate 氢化二苯甲烷二异氰酸酯
- ~ methyl abielate 氢化松香酸甲酯〈增塑剂〉
- ~ polybutene 氢化聚丁烯
- ~ polymer 氢化聚合物
- ~ propylene tetramer 氢化丙烯四聚物
- ~ rubber 氢化橡胶

hydrogenation 氢化(作用);加氢(作用)
hydrogenize 氢化
hydrolysis 水解(作用)
hydrolytic 水解的
- ~ degradation 水解降解
- ~ depolymerization 水解解聚
- ~ polycondensation 水解缩聚

~ polymerization 水解聚合
~ stability 水解稳定性
hydrolytically degradable plastic 水解降解性塑料
hydromechanical 液压-机械的
~ clamp 液压-机械或锁模装置
~ press 液压-机械式压机
hydrometer (液体)比重计
hydroperoxide decomposer 氢过氧化物分解剂
hydrophilic 亲水(的)
~ chain 亲水链
~ fibre 亲水性纤维
~ group 亲水基团
~ nature 亲水性(能,质)
~ polymer 亲水聚合物
~ yam 亲水性纱线
hydrophility 亲水性
hydrophobic 疏水的,憎水的
~ bond 疏水键
~ chain 疏水链
~ dyes 疏水性染料
~ fibre 疏水性纤维
~ nature 疏水性
~ yarn 疏水性纱线
hydrophobicity 疏水性
hydropolymer 氢化聚合物
hydropolymerization 氢化聚合
hydropress 液压机
hydroquinone 对苯二酚,氢醌〈阻燃剂,阻聚剂〉
~ dibenzylether 对苯二酚二苄醚〈防老剂〉
hydroscopic 吸湿的,收湿的
~ property 吸湿性
~ substance 吸湿物
hydroscopicity 吸湿性,吸水性
hydrosol 水溶胶
hydrospenser 连续供料-混合-铸塑装置
hydrostatic 流体静力的
~ burst test 水压破裂试验
~ creep 流体静力蠕变
~ pressure 静压;水压

~ pressure test (静)水压试验
~ stress [isotropic stress] 静水应力
hydrothemal 湿热的
~ effect 湿热效应
~ stability 湿热稳定性
hydrous 水合的
~ aluminium silicate 水合硅酸铝〈补强填料〉
~ magnesium metasilicate 水合(偏)硅酸镁〈补强剂和填料〉
~ silica 水合二氧化硅〈即硅酸,补强填剂〉
hydroxide 氢氧化物
hydroxy 羟基
~ -acid polyester 羟基酸聚酯
~ acrylic resin 羟基丙烯酸树脂
2-~-4-benzyloxybenzophenone 2-羟基-4-苄氧基二苯甲酮〈紫外线吸收剂〉
2-(2'-3-tert-butyl-5'-methylphenyl)5-chlorobenzotriazole 2-(2'-羟基-3'-叔丁基-5'-甲基苯基)-5-氯代苯并三唑,UV-326〈紫外线吸收剂〉
~ carboxylic resins 羟基羧酸型树脂
2-~-5-chlorobenzophenone 2-羟基-5-氯二苯酮〈紫外线吸收剂〉
6-(4-~-3,5-di-tert-butylanili. no)-2,4-bis(cotylthio)-1,3,5-triazine 6-(4-羟基-3,5-二叔丁基苯胺基)2,4-双(辛基硫代)-1,3,5-三嗪〈抗氧剂〉
2-(2'-~-3',5'-di-tert-butylphenyl) benzotriazole 2-(2'羟基-3',5'-二叔丁基苯基)苯并三唑〈紫外线吸收剂〉
2-(2'-~-3',5'-di-tert-butylphenyl)-5-chlorobenzotriazole 2-(2'-羟基-3',5'-二叔丁基苯基)-5-氯代苯并三唑,UV-327〈紫外线吸收剂〉
2-(2'-~-3',5'-dipentylphenyl) benzotriazole 2(2'-羟基3',5'-二戊基苯基)苯并三唑,UV-328〈紫外

hydroxystearic acid

2 - ~ -4 - dodecyloxybenzophenone 2 -羟基-4-十二烷氧基二苯甲酮〈紫外线吸收剂〉

2 - ~ -4 -(2'- hydroxy 3' acryloxy propyloxy) benzophenone 2-羟基-4-(2'-羟基-3'-丙烯酰氧基丙氧基)二苯甲酮〈紫外线吸收剂〉

2 - ~ -4[2'- hydroxy -3'-(methacryloxypropyloxy)] benzophenone 2-羟基-4-[2'-羟基-3'-(甲基丙烯酰氧基丙氧基)]二苯甲酮〈紫外线吸收剂〉

2 - ~ -4 - methoxy - benzophenone 2-羟基-4-甲氧基二苯甲酮. UV-9〈紫外线吸收剂〉

2 - ~ -4 - methoxy -2'- carboxy benzophenone 2-羟基-4-甲氧基-2'-羟基二苯甲酮〈紫外线吸收剂〉

2 - ~ -4 - methoxy -4'- chlorobenzophenone 2-羟基4甲氧基-4'-氯二苯甲酮〈紫外线吸收剂〉

2 - ~ -4 - methoxy -2',4'- dichloroben-zophenone 2-羟基-4-甲氧基-2',4'-二氯二苯甲酮〈紫外线吸收剂〉

2 - ~ -4 - methoxy -5 - sulfobenzophenone trihydrate 2-羟基-4-甲氧基-5-磺基二苯甲酮三水合物〈紫外线吸收剂〉

2 - (2'- ~ -5'- methyl phenyl) benzotria - zole 2-(2'-羟基-5'-甲基苯基)苯并三唑, UV-P〈紫外线吸收剂〉

2 - ~ -4 - octadecyloxybenzophenone 2-羟基-4-十八烷氧基二苯甲酮〈紫外线吸收剂〉

2 - ~ -4 - n - octoxybenzophenone 2-羟基-4-正辛氧基二苯甲酮, UV-531〈紫外线吸收剂〉

2 - (2'- ~ -4'- n - octoxyphenyl) benzo - triazole 2-(2'-羟基-4'-正辛氧基苯基)苯并三唑〈光稳定剂〉

2 - (2'- ~ -5'- tert - octylphenyl) benzo - triazole 2-(2'-羟基-5'-叔辛基苯基)苯并三唑〈光稳定剂〉

p - hydroxybenzoic acid polymer 对羟基苯甲酸酯聚合物

hydroxyethyl 羟乙基

N - (2 - ~)12 - hydrory stearamide N-(2-羟乙基)-12-羟基硬脂酰胺〈润滑剂,脱模剂〉

N - (β - ~) ricinoleamide N-(β-羟乙基)蓖麻醇酸酰胺〈润滑剂,脱模剂〉

hydroxyethylacetamide 羟乙基乙酰胺

hydroxyethylcellulose[HEC] 羟乙基纤维素〈表面活性剂、乳化稳定剂,增黏剂,分散剂〉

2 - hydroxyethyldiethylene triamine 2-羟乙基二-1,2-乙基三胺〈固化剂〉

hydroxyl 羟基
~ group 羟基
~ number 羟基数
~ terminated polyether [HTE] 羟基封端聚醚
~ value 羟基值

hydroxymethyl 羟甲基
~ bisphenol A diglycidyl ether 羟甲基双酚A二缩水甘油醚
~ bisphenol A epoxy resin 羟甲基双酚A环氧树脂

4 - ~ - 2,6 - di - tert - butylphenol 4-羟甲基-2,6-二叔丁基苯酚〈抗氧剂〉

hydroxypropyl 羟丙基
~ acrylate 丙烯酸羟丙酯
~ methacrylate 甲基丙烯酸羟丙酯

hydroxypropylcellulose 羟丙基纤维素

hydroxypropylglycerin 羟丙基三醇

hydroxypropylmethylcellulose 羟丙基甲基纤维素〈分散剂,增稠剂,乳化剂,稳定剂,胶黏剂〉

hydroxystearic acid 羟基硬脂酸〈润滑剂,抗粘连剂〉
~ triglyceride 甘油三羟基硬脂酸酯〈润滑,爽滑剂,脱模剂〉

hygienic test method　卫生试验法
hygral　湿润;湿度
　~ change　湿度变化
　~ expansion　湿膨胀
　~ expansion index　湿膨胀指数
　~ shrinkage　湿润收缩
hygrograph　湿度计
hygrometer　湿度计
hygrometric state　湿态
hygrometry　湿度测定性
hygroscopic　吸水的,吸湿的,收湿的
　~ agent　吸湿剂
　~ capacity　吸湿量
　~ coefficient　收湿系数
　~ dunnage　吸湿衬垫〈包装〉
　~ expansion　吸湿膨胀
　~ material　吸湿性材料
　~ property　吸湿性
　~ sensitivity　吸湿敏感性
hygroscopicity　吸湿性,吸水性
Hypalon［HYP］　〈商〉海波隆橡胶,氯磺酰化聚乙烯橡胶
Hypar［hyperbolic paraboloide］　双曲线抛物面
　~ roofing　双曲线抛物形屋顶
　~ space grid［double-layer diagonal space grid］　双层对角空间格栅
hyperbolic paraboloide［Hypar］　双曲线抛物面
hyperchrome　浓色团
hyperelastic　超弹性
hypermolecular　超分子的
hyperthermal environmental resistance　耐超高温(环境)性
hypochlorite solution　次氯酸盐溶液
hypoelastic body　次弹性体
hypoelasticity　次弹性
hypsochrome　浅色团
hypsometer　沸点测定仪
hysteresis　滞后(现象)
　~ curve　滞后曲线
　~ effect　滞后效应
　~ error　滞后误差
　~ loop　滞后回线
　~ loss　滞后损耗
　~ property　滞后性能
　~ phenomenon　滞后现象
　~ set　滞后变形
　~ tester　滞后试验机
hysterometer　滞后试验机
hyvac［high-vacuum］　高度真空
Hz［Hertz］　赫兹;周/秒〈频率单位〉

I

i［inactive］　不活泼的
i.［insoluble］　不溶的
IBS-process［injection-blow-stretch process］　注拉吹成型法
ice　冰
　~ crystal　冰晶
　~ point　冰点
ICI［Imperial Chemical Industries］　英帝国化学工业公司,卜内门公司
　~ foam moulding［Imperial Chemical Industries foam moulding］　ICI(公司)泡沫塑料成型法,帝化泡沫塑料成型法
I. D.［inner［inside］diameter］　内径
I. D.［inside dimension］　内尺寸
ideal　理想(的)
　~ chain　理想链
　~ conformation　理想构象
　~ copolymerization　理想共聚作用
　~ crystal　理想晶体
　~ dispersion　理想分散
　~ efficiency　理想效率

~ elastic body 理想弹性体
~ elasticity 理想弹性
~ fluid 理想流体
~ network 理想网络
~ operating temperature 理想操作温度
~ orientation 理想取向
~ plasticity 理想塑性
~ temperature 理想温度
~ viscous fluid 理想黏性流体
ideally elastic 理想弹性的
~ behaviour 理想弹性特性
identification 鉴定,鉴别;标志,符号
~ mark 商标,标志
~ test by burning 燃烧鉴别试验
identify 鉴定;鉴别
idle 惰性的,空闲的;空转的
~ motion 空转
~ operation 空转
~ roll(er) 空转辊,从动辊,惰辊,托辊
~ running 空转
~ time 闲空时间,停歇时间
~ timer 停机计时器
~ wheel 惰轮,空转轮
idler 惰轮,张力惰轮;空转辊;导轮
~ pulley 惰轮;导轮
~ roll 惰轮,空转辊,从动辊〈涂布机〉
idling 空转
~ cut-off 空转切断
i.e. [id est]〈拉〉即;就是
IEN [interpenetrating elastomeric network] 互穿弹性网络
IFR [intumescent flam retardent] 发泡阻燃剂
ignitability 可燃性,着火性
ignitable 可燃的
ignite 着火,点燃
ignited gas jets 燃气喷射器〈粘接〉
ignitible 可燃的,可着火的
ignition 着火,点燃
~ loss 灼烧失重
~ point 着火点
~ resistance 耐点燃性
~ temperature 着火温度,点燃温度

~ test 着火点试验
~ time 着火时间,点燃时间
ig. P. [ignition point] 着火点
IHN [interpenetrating homopolymer network] 互穿均聚物网络
IIR [isobutylene-isoprene rubber] 异丁烯-异戊二烯橡胶,异丁橡胶
ILD [indentation load deflection] 压入载荷挠度〈泡沫塑料〉
illuminating colours 照明体着色(用)染料
illumination 照明;照明度
image 图像
~ analyser 显微图像分析仪〈评定塑料结构〉
imbalance [eccentric weight] 不平衡
imbedded 镶铸的,嵌镶的
imcompressibility 不可压缩性
imitation 模仿;仿造的
~ leather 仿革;人造(皮)革
~ parts 仿制零件,仿造品
immediate 直接的,紧接的;立即的
~ elastic deformation 瞬间弹性变形
~ set 瞬间塑性形变
immerse 浸渍;浸没
immersed 浸渍的
~ guide roll 浸渍(涂料)的导向辊,浸没在涂料中的导向辊〈涂布〉
~ roll 浸没辊
immersion 浸没,浸渍
~ bath 浸浴
~ bath test 浸浴试验
~ coating 浸渍涂覆,浸镀
~ cooling 沉浸冷却
~ heater 浸入式加热器
~ roll 浸渍辊
~ test 浸泡试验
immiscibility 不溶混性,难混溶性,不可混合性
immiscible 不溶混的;不混合的
impact 碰撞,冲击;压紧
~ abrasion 冲击磨损
~ absorbing wall 缓冲墙,缓冲板
~ adhesive 压敏黏合剂

impact

- ~ angle 冲击角
- ~ apparatus 冲击试验夹具
- ~ bar 冲击(用)试棒
- ~ bending 冲击弯曲
- ~ bending strength 冲击弯曲强度
- ~ bending test 冲击弯曲试验
- ~ break 冲击破裂
- ~ breaker[crusher] 冲击式破碎机,锤式破碎器
- ~ brittleness 冲击脆性
- ~ compression test 冲击压缩试验
- ~ crusher 冲击式破碎机
- ~ crushing 冲击破碎
- ~ cutting 冲击式切割
- ~ damper 缓冲器,减震器
- ~ disk 冲击轮
- ~ ductility 冲击韧性
- ~ effect 冲击作用
- ~ elasticity 冲击弹性,冲击韧性
- ~ endurance test 冲击疲劳试验
- ~ energy 冲击能量
- ~ extrusion 冲(挤)压
- ~ fatigue 冲击疲劳
- ~ fatigue strength 冲击疲劳强度
- ~ fatigue test 冲击疲劳试验
- ~ flexural test 冲击弯曲试验
- ~ force 冲击力
- ~ fracture 冲击断裂,冲击破裂
- ~ fracture toughness 冲击断裂韧性;耐冲击性
- ~ grinder 冲击型粉碎机
- ~ grinding 冲击粉碎
- ~ heat sealing 脉冲热封
- ~ insulation rating 冲击吸音程度〈地板〉
- ~ load 冲击负荷
- ~ load stress 冲击负荷应力
- ~ machine (摆锤式)冲击试验机
- ~ mill 冲击式研磨机
- ~ mixer 冲击式混合机
- ~ modifier [toughening agent] 冲击(强度)改性剂
- ~ mould 冲击(成型)模具
- ~ mo(u)lding 冲压模塑
- ~ moulding die 冲压成型模
- ~ noise 冲击噪声
- ~ noise attenuation 冲击噪声衰减〈地板〉
- ~ nozzle 高压注料嘴
- ~ opening 冲击孔,冲击口
- ~ pendulum 冲击摆锤
- ~ pendulum test 摆锤冲击试验
- ~ penetration test 冲击穿透试验
- ~ plastic 耐冲击塑料
- ~ polystyrene 耐冲击(性)聚苯乙烯
- ~ pressure 冲击压力
- ~ pulverizer 冲击型粉碎机
- ~ resilience 冲击回弹性(试验)
- ~ resilience test 冲击回弹性试验
- ~ resilience tester 冲击回弹性试验仪
- ~ resiliometer 冲击回弹性试验仪
- ~ resistance 耐冲击性
- ~ risistance polystyrene 耐冲击(性)聚苯乙烯
- ~ resistant 耐冲击的
- ~ resistant plastics 耐冲击塑料
- ~ resistant polystyrene graftpolymer 耐冲击(性)聚苯乙烯接枝聚合物
- ~ resistant polyvinyl chloride 耐冲击(性)聚氯乙烯
- ~ sound reduction 冲击音衰减
- ~ speed 冲击速度
- ~ strength 冲击强度,冲击韧性
- ~ strength index 冲击强度指数
- ~ strength modifying additive 冲击强度改性添加剂,冲击(强度)改性剂
- ~ stress 冲击应力
- ~ stroke 冲击行程
- ~ styrene material 耐冲击(性)聚苯乙烯料
- ~ tensile test 冲击拉断试验
- ~ tension 冲击拉伸
- ~ test 冲击(韧性)试验
- ~ tester 冲击试验仪
- ~ thermoplastics 耐冲击热塑性塑料
- ~ value 冲击值
- ~ velocity 冲击速度
- ~ viscosity 冲击黏性,冲击韧性

~ wear test 冲击磨损试验
~ weight 冲击荷重,冲击锤
~ wheel 冲击轮
~ wheel mixer [Entoleter mixer] 冲击轮式混合机〈美〉
impactor [hammer mill]
impalpable powder 极细微粒粉末
impedance 阻抗电阻〈电〉
impeller 叶轮;高速混合器
~ blade 叶轮片〈快速搅拌机〉
~ breaker [baffle-plate impact milk] 叶轮式破碎机
~ mixer 高速混合机,叶轮式混合机
~ pump 叶轮泵
impending plastic flow 紧急塑性流动
imperfect 不完全的
~ combustion 不完全燃烧
imperfectly elastic behaviour 不完全弹性特性
Imperial Chemical Industries foam mo(u)lding [ICI foam mo(u)lding] （英）帝（国）化（学工业公司）泡沫塑料成型法, ICI（公司）泡沫塑料成型法
impermeability 气密性,不渗透性
~ test 耐渗透试验
~ to gas 不透气性
impermeable 不渗透的,密封的
~ to air 不透气的
~ to gas 不透气的
~ to water 不透水的
impervious 不透的;密封的
~ to moisture 防潮的
~ to water 不透水的,防水的
imperviousness 不渗透性,不透水性
impining gate 接木形浇口
impreg 浸胶木材
impregnability 浸渍性,浸透性
impregnable 可渗浸的,可浸透的
impregnant 浸渍剂
impregnate 浸渍
impregnated 浸渍过的
~ densified wood 浸渍压缩木材
~ fabric 浸渍织物;树脂浸渍织物

~ foil 浸渍箔
~ mat 浸渍玻璃毡
~ paper 浸渍纸
~ sheet （树脂）浸渍片材〈增强塑料〉
~ web （树脂）浸渍基料〈增强塑料〉
~ wood （树脂）浸渍木material
~ yarn 浸渍纱线
impregnating 浸渍
~ agent 浸渍剂
~ bath 浸渍浴
~ equipment 浸胶机
~ liquid 浸渍液
~ machine 浸渍机
~ material 浸渍（用）材料
~ resin 浸渍（用）树脂
~ time 浸渍时间
~ varnish 浸渍清漆
impregnation 浸渍,浸胶
~ compound 浸渍料
~ process 浸渍工艺
~ technique 浸渍技术
~ vessel 浸渍（容）器
impregnator 浸胶机
impression 型腔:阴模;压痕;底色;底漆层
~ cylinder 压花滚筒
~ injection mould 型腔注塑模
~ mo(u)lding 触压成型
~ resin 低压固化树脂
imprinting 压印法
improve 改进,改善
improved wood [indurated wood, modified wood] 改性木材,压缩木材
improvement 改进,改善
improver 改进剂,添加剂
impulse 脉冲;冲击
~ counter 脉冲计数器
~ force 冲击力
~ heat sealing 脉冲热封
~ heated bar 脉冲加热棒
~ rate 脉冲率
~ sealer 脉冲热合机
~ sealing 脉冲焊接,脉冲热封,脉冲封合

~ signal 脉冲信号
~ strength 冲击强度
~ test 冲击试验
~ voltage 脉冲电压
~ wave 冲击波
~ welder 脉冲焊接机
~ welding 脉冲焊接
impure 不纯的,掺杂的
impurity 杂质;混入物
imputrescence 防腐性
imputrescibility 防腐(烂,败)性
in 在…内,在…方面;用…;成…
~ amount 总计,总结
~ block 整体的,做成一体的
~ building constructional plastics 建筑用塑料
~ bulk 大批,大量,成堆;成块;散装
~ - cavity pressure 模腔压力〈注塑〉
~ - draw 内拉伸
~ - teed rolls 供料辊
~ - feed table 供料台
~ field use 就地使用,现场使用
~ - line 轴向的;流水线的;(在)管线内的;串联的;装配程序
~ - line continuous production 流水线(连续)生产
~ - line die 直通式模头
~ - line discharge screw 轴向式卸料螺杆
~ - line injection streth blow 轴向注拉吹(成型)
~ - line mixer 连续流动混合器
~ - line mixing [blending] 在管线内混合
~ - line multiple pin - point gate 系列复式点状孔浇口
~ - line preplasticizer 顺排预塑化器
~ - line production 流水线生产
~ - line reciprocating screw 轴向式往复螺杆
~ - line reciprocating screw injection machine 轴向往复螺杆式注塑机
~ - line screw 轴向螺杆
~ - line screw plasticator 轴向螺杆式塑化器
~ - line screw type injection machine 轴向螺杆式注塑机
~ - line - screw - type injection moulding machine 轴向螺杆式注塑机
~ - mould coating 模内涂层
~ - mo(u)ld decorating 模内装饰
~ - mo(u)ld foiling 模内敷箔
~ - mo(u)lding - labelling system 模内贴签系统
~ parallel 并联
~ - place foaming 现场发泡
~ - place mo(u)lding 现场模塑,就地成型
~ - place process 就地法
~ - place test 现场试验,实地试验
~ - plant colo(u)ring 装置内着色;自着色
~ - process 加工过程中的
~ - process recycling 制造过程中回收利用
~ - series 串联
~ - series continuous production 顺序连续生产
~ - service conditions 使用[运转]条件
~ - service test 实用试验
~ - service testing time 使用[运转]试验时间(机器启动后,检验稳定性)
~ - site polymerization 就地聚合
~ - situ 现场
~ - situ blending 现场混配,现场掺混
~ - situ foam 现场发泡
~ situ foaming 现场发泡(法)
~ - situ polymerization 现场聚合,就地聚合
~ - situ test 现场试验
~ stock 备有;现有,现存
~ store 库存
~ the open 在露天,在户外
~ - use condition 使用条件
~ - use performance 使用效率,使用性能
~ - use testing 实用试验

in. [inch] 英寸,吋
In [Indium] 铟
inaccuracy 不准确性,不正确,误差
～ of dimensions 尺寸不合格
～ to size 尺寸偏差
inaction period 不作用期间,未开动期间,停工期间,故障期间
inactive 不活泼的,钝的
～ filler 惰性填料
inadaptability 不适应性
inadequacy 不相适应
inadequate surface preparation 表面预处理不充分,表面加工有缺陷
inadherent 不粘结的
inadhesion 不黏性,不黏合
inadhesive 不能粘结的
inadvertent 粗心的,无意的
inblock cast 整体铸塑
inc. [inclusive] 包括
incandesce 灼烧,灼热
incandescence 白炽;白热,炽热;灼热
～ bar test 炽热棒试验
～ rasistance 耐炽热性
incandescent 白炽的;白热的
incendiary 燃烧的
～ behaviour 燃烧性,着火性
～ material 燃烧材料
inch [in.] 英寸,吋
inching 降速闭模〈注塑,压塑〉
incidence of light 光入射
incident light 入射光
incipient plastic flow 起始塑性流动
inclination 倾斜;斜度;倾角
～ balance 摆锤式天平
inclined 倾斜的
～ blade 斜桨叶
～ boring 斜孔
～ flow-plate 斜面测流板
～ joint 斜接,斜接缝
～ manometer 倾斜压力计
～ pin 斜销钉
～ plane testing 斜面(摩擦)试验
～ plane viscometer 斜面黏度计
～ roll(er) 斜辊

～ type calender 斜型压延机
～ Z-(type) calender [S-type calender] 斜Z型压延机,S型(四辊)压延机
inclusion 夹杂物;杂质;包含
～ compounds 包含化合物
～ of air 气泡〈塑料制品缺陷〉
～ polymerization 包接聚合(作用)
incoherent 不相干的
～ scattering 不相干散射
incoherentness 不相干性;不粘结性,无内聚性
incohesive 无黏聚力的
incombustibility 不燃性
incombustible 不燃的
～ adhesive 不燃性黏合剂
～ fabric 不燃性织物
～ mixture 不燃混合物
～ substance 不燃物
incoming 进入,引入,输入
～ sheet 进入片料〈造粒〉
～ web 进入片片;进入卷材
incompatibility 不相容性
incompatible 不相容的,不相混的
incomplete 不完全的
～ adhesion 欠黏
～ combustion 未完全燃烧
～ fusion 未完全熔焊;未完全塑化
～ mixing 拌和不均
～ reaction 不完全反应
incompressibility 非压缩性,不可压缩性
incompressible 不可压缩的
～ fluid 非压缩性流体,不可压缩流体
～ material 非压缩材料,不可压缩材料
incorporate 掺合,掺入;混合
incorrect 错误的,不正确的
incorrodible 不腐蚀的,耐腐蚀的
incorrosive 不腐蚀的
increase 增加,增长
～ in thickness 厚度增加
～ in volume 体积增加
～ in weight 重量增加
increasing of the surface 表面增加

increment 增加,增长;增加的量
incubation test 固化度试验〈热固性塑料〉
indene 茚
- ~ polymer 茚聚合物
- ~ resin 茚树脂

indentation 压痕,压陷〈硬度试验〉;凹陷,凹穴
- ~ depth 压痕深度
- ~ hardness 压痕硬度
- ~ hardness index 压痕硬度指数
- ~ index 压痕指数
- ~ load 压痕负荷
- ~ load deflection [ILD] 压痕负荷值;压痕负荷挠度
- ~ load deformation 压痕负荷变形
- ~ residual gage load [IRGL - value of foam] 压入残余度负荷
- ~ resistance 耐压陷性
- ~ test 压痕试验

indenting ball 压痕球〈塑料硬度试验〉
independent 独立的,单独的
- ~ adjustment 单独调整
- ~ control 独立控制,独立操纵
- ~ drive 独立传动,单独传动

index 指数;指标;索引,目录
- ~ chuck 分度卡盘
- ~ disc 指示盘,刻度盘
- ~ error 分度误差,指示误差
- ~ finger 指针
- ~ hand 指针
- ~ mark 刻度,分度
- ~ number 指数
- ~ of correlation 相关指数,相关指标
- ~ of creep 蠕变指数
- ~ of crystallinity 结晶度指数
- ~ of inertia 惯性指数
- ~ of notch sensitivity [notch factor] 切口脆性指数
- ~ of quality 质量指标
- ~ of refraction [refractive index] 折光指数;折射指数
- ~ of speciality 特性指数
- ~ of stability 稳定率
- ~ plate 分度盘
- ~ time 转位时间
- ~ wheel 刻度盘
- ~ yarn 标志纱线

indexer 指数测定仪;分度器
indexing 指数,分度;转位
- ~ disc 分度盘
- ~ mo(u)ld 旋转式模具,转位模具

indication 指示,显示;指标;标记
indicative value 近似值
indicator 指示器,显示器;指示剂
- ~ pressure 指示压力
- ~ range 指示剂变色范围

indicia 标记
indifferent equilibrium 惰性平衡
indirect 间接的,非直接的
- ~ analysis 间接分析
- ~ contact 间接接触
- ~ cure 间接硫化
- ~ drive 间接驱动
- ~ extrusion 间接挤塑
- ~ hardening 间接硬化
- ~ heat exchange 间接热交换
- ~ - heated mould 间接加热模具〈通过加热板由外部加热模具〉
- ~ heating 间接加热
- ~ injection 非直接注塑
- ~ sizing 间接上浆
- ~ take-up 间接卷取
- ~ welding 单面点焊

Indium [In] 铟
individual 个别的,单独的;特殊的
- ~ adjustment 单独调整
- ~ cast 单独铸塑,分开铸塑
- ~ construction 单件结构
- ~ drive 独立传动,单独传动
- ~ injection system 独自注塑系统
- ~ fibre 单根纤维
- ~ production 单件生产
- ~ structural unit (单独)结构单元
- ~ winding 单绕〈缠绕增强材料〉

indoor 室内,户内
- ~ exposure 室内暴露〈试样〉
- ~ paint 室内用涂料

~ test 室内试验,实验室试验
induced 诱导的;感应的
~ colo(u)r 诱导色,感应色
~ crystallization 诱导结晶
~ decomposition 诱导分解
~ nuclei 诱导晶核
~ reaction 诱导反应,感生反应
inducing colo(u)r 诱导色,感应色
induction 诱导;感应
~ effect 诱导效应
~ field 感应场
~ forces 感应力,德拜力
~ hardening 高频硬化,感应淬火
~ heated 感应加热的
~ – heated extruder 感应加热挤塑机
~ – heated extrusion machine 感应加热挤塑机
~ heating 感应加热
~ period 诱导期,感应期
~ switch 感应开关
~ time 诱导期
~ velocity 诱导速率
~ welding 感应焊接
inductive 诱导的;感应的
~ action 诱导效应
~ effect 诱导效应
inductivity 诱导率
indurated 硬化的
~ fiber 硬纤维
~ wood 浸胶压缩木材;高温高压改性木材
industrial 工业的,工业用的,工业生产的
~ chemist 工业化学家
~ chemistry 工业化学
~ fabric 工业用织物
~ furnace 工业炉
~ instrumentation [process instrumentation] 工业设备
~ laminate 工业用层合制品;工业层压板
~ moulding 工业用模塑件
~ packaging 工业包装
~ plastic 工业用塑料,工程塑料,商品塑料
~ pollution 工业污染
~ process 工业生产过程
~ product 工业制品
~ production 工业生产
~ resin 工业用树脂
~ robot 工业自动机
~ scale 工业规模,大规模
~ sewage 工业污水
~ sheet 工业用板材
~ standards 工业标准
~ television 工业电视
~ textiles 工业用纺织品
~ waste 工业废料
~ waste water 工业废水
~ water 工业用水
industry 工业
~ standard 工业标准
~ standard specifications [ISS] 工业标准规范
~ – wide 工业中广泛应用的
inelastic 无弹力的,无弹性的
inert 惰性的
~ additive 惰性添加剂
~ adhesive 惰性黏合剂
~ black 惰性炭黑
~ constituents 惰性组分
~ filler 惰性填料
~ gas 惰性气体;保护气
~ gas – shielded welding 惰性气体保护焊接
~ ingredient 惰性组分
~ material 惰性物料
~ pigment 惰性颜料
~ plasticizer 惰性增塑剂
~ powder 惰性粉料
~ resin 惰性树脂
~ substance 惰性物质
~ to deformation 耐变形性
inertance 惰性
inertia 惯性,惰性
~ coefficient 惯性系数
~ effect 惯性效应
~ force 惯性力

~ pressure 惯性压力
~ resistance 惯性阻力
~ torque 惯性扭矩
~ turbulence 惯性湍流
inertness 惰性;惯性
~ to acids 耐酸
inferior limit 下限
infestation 蔓延〈霉菌〉
infinite 无限的,无级的
~ adjustability 无级可调节性
~ dilution 无限稀释
~ life 无限寿命
~ variable 无级变速的
~ variable speed mechanism 无级变速装置
~ variety 无级变速
infinitely variable gear 无级变速齿轮
infinitive variable gear box 无级变速齿轮箱
inflame 燃烧
inflammability 易燃性,可燃性
~ limit 着火极限
~ point 燃点,发火点;着火温度
~ test 易燃性试验
inflammable 易燃的
inflammableness 易燃性
inflammation 着火,燃烧
~ temperature 着火温度
inflatable (可)充气的,(可)吹胀的,(可)膨胀的
~ bag 充气袋
~ bag method 充气袋法
- **bag technique** 充气袋法
~ cushion 充气衬垫〈包装〉
~ double - walled tent 充气双层壁帐篷
~ flexible bag mo(u)lding 充气(软)袋成型法
~ skin 充气外表层
~ structure 充气式结构〈塑料〉
inflate 充气,吹胀,膨胀
inflated 充气了的,膨胀了的
~ article 充气制品
~ cushion roof 充气缓冲顶

~ double - walled structures 充气双层壁结构
~ film 吹塑薄膜
~ hose 吹塑软管
~ tent 充气帐篷
~ tube structure 充气软管结构
inflating 充气;膨胀
~ agent 发泡剂;发气剂
inflation 充气,吹胀;膨胀
~ blow process 吹胀法
~ die 吹塑薄膜模头,吹塑薄膜机头
~ film process 吹塑薄膜法,管膜法
~ mandrel 吹胀杆;进气芯棒
~ pressure 充气压力
inflection 弯曲
inflexion 弯曲
inflow 流入
influent 流体;液体
information 信息,情报,资料
~ processing 信息处理,情报处理
~ retrieval [IR] 信号[情报,资料]检索
infrared 红外(线)的
~ absorption 红外线吸收
~ absorption spectrascopy 红外吸收光谱(法)
~ absorption spectrum 红外吸收光谱
~ analysis 红外线分析
~ attenuated total reflectance spectroscopy 红外衰减全反射光谱学
~ detection 红外检测
~ die glazing 口模红外上光〈塑料半成品〉
~ dryer 红外线干燥器
~ drying 红外线烘燥
~ heater 红外线加热器
~ heating 红外线加热
~ lamp 红外线灯
~ microspectrometer 红外线显微分光仪
~ moisture check 红外检测温度
~ oven 红外线烘箱;红外炉
~ polymerization index 红外聚合指数
~ preheater 红外预热器

~ preheating 红外线预热
~ radiant heating 红外(辐射)加热〈美〉
~ radiation 红外辐射
~ radiator 红外辐射器
~ range 红外线范围
~ ray 红外线
~ reflector lamp 红外反射灯
~ reflector radiator 红外反射辐射器
~ spectrogram 红外光谱图
~ spectrometry 红外光谱法
~ spectrophotometer 红外分光光度计
~ spectrophotometry 红外分光光度测定法
~ spectroscopic study 红外光谱研究
~ spectroscopy 红外光谱学
~ spectrum 红外光谱
~ thermometer 红外线温度计
~ welding[IR-welding] 红外线焊接
infrasonics 亚音频
infusibility 不熔性,难熔性
infusible 不熔的,难熔的
infusion 浸渍;浸渍剂;注入〈合成树脂〉
infusorial earth 硅藻土
ingrain 原纤染色
~ colo(u)rs 显色染料
~ dyes 显色染料
ingredient 成分;组分;配料
inherent 原有的,固有的
~ characteristic 固有特性
~ colour 固有颜色,本身颜色
~ vice 固有缺陷
~ viscosity 特性黏度,固有黏度,比浓对数黏度
inhibiter 抑制剂;阻聚剂;缓蚀剂
inhibiting 抑制
~ degradation 抑制降解
inhibition 抑制
~ constant 阻聚常数
~ effect 抑制效应
~ of oxidation 氧化的抑制
~ phenomena 阻抑现象
inhibitive substance 阻聚剂;抑制剂;防老剂

inhibitor 抑制剂;阻聚剂
~ dye 抑制剂染料
inhibitory coating 保护层,防护层
inhomogeneity 不均匀性;多相性
inimical impurity 有害杂质
initial 最初的,初始的,起始的
~ colo(u)r 初色,最初颜色
~ condition 原始条件,起始条件
~ cost 原价;原始成本;基建投资
~ creep 初级蠕变
~ crusher 初级压碎机
~ data 原始数据
~ elasticity 初弹性
~ friability 初始脆性
~ heating 预热
~ heating time 起始加热时间,初始升温时间
~ modulus 起始模量,初始模量
~ pressure 初压;表面压力
~ product 原料;最初产品,起始产品
~ reliability 初始可靠性〈产品〉
~ set 初形变,初变定〈反应性树脂〉;初反应
~ state 起始状态
~ strain 初应变
~ strength 起始强度
~ stress 起始应力,初应力
~ stress in stress relaxation 应力松弛的初始应力
~ - tangent modulus 起始正切模量,原切模量
~ tear 初撕裂
~ temperature 初始温度
~ tension 初张力
~ value 初值,始值
~ viscosity 起始黏度
initiate 引发;起始
initiating 引发
~ agent 引发剂
initiation 引发
~ point 引发点
~ reaction 引发反应〈聚氨酯泡沫塑料〉
initiatomer 单体引发剂

initiator 引发剂
　~ of polymerization　聚合引发剂
inject 注射;注塑
　~ into the air　自由注射;无模注塑
　~ time　注塑时间
injectable phenolic resin 注塑酚醛树脂
injected 注射的;注塑的
　~ area　注塑面积
　~ parison　注塑型坯〈吹塑〉
injection 注射,注塑
　~ ability　注塑能力,注塑性能
　~ and vacuum method　真空注塑(成型)法
　~ and vacuum mo(u)lding　真空注塑成型(法)
　~ area　注塑面积
　~ blow mould　注吹模
　~ blow mo(u)lding　注射吹塑(成型),坯吹塑(成型)
　~ blow mo(u)lding machine　注坯吹塑成型机
　~ - blow - stretch process [IBS - process]　注拉吹成型法
　~ boost time　注塑增压时间〈注塑〉
　~ capacity　注塑能力,注塑量
　~ carriage transverse adjustment　注塑部滑座横向调节
　~ carriage vertical adjustment　注塑部滑座纵向调节
　~ composition　注塑混合料
　~ compound　注塑料
　~ compression moulding　注射压缩成型
　~ control system　注射控制系统
　~ core　注塑型芯
　~ cycle　注塑周期;注塑循环
　~ cylinder　注塑料筒〈注塑机〉
　~ cylinder carriage　注塑料筒架
　~ efficiency　注塑效率
　~ end　注射端
　~ error　注射误差
　~ - extrusion blow mo(u)lding technique　螺杆式注坯吹塑(成型)法;注挤吹成型法

　~ extrusion machine　螺杆式注塑机
　~ extrusion mo(u)lding　螺杆式注塑成型
　~ flow number　注塑流动值
　~ force　注射力
　~ forward stroke　注射杆行程
　~ gun　注射枪
　~ heater　注塑加热器
　~ hole　注塑孔
　~ hydraulic controls　注塑液压控制
　~ machine　注塑机
　~ machine control　注塑机控制
　~ mandrel　注塑模芯,注塑型芯
　~ mixing　注射混合(法)〈熔体中注入交联剂〉
　~ mixing foam moulding　注射混合发泡成型
　~ mo(u)ld　注射模具,注射成型模具
　~ moulded　注塑(成型)的
　~ mo(u)lded adhesive composite　注塑成型粘接复合材料
　~ mo(u)lded article [part, piece]　注塑成型制件,注塑成型制品
　~ mo(u)lded foam　注塑泡沫塑料
　~ moulded heel　注塑(成型)跟〈鞋〉
　~ mo(u)lded parison　注塑型坯
　~ moulded part　注塑(成型)制件
　~ moulded part with sprue　有流道的注塑(成型)制件
　~ mo(u)lded plastics　注塑(成型)塑料
　~ mo(u)lded precision article　注塑精密件
　~ moulded sole　注塑(成型)底〈鞋〉
　~ mo(u)lded thread　注塑螺纹
　~ moulder　注塑(成型)机;注塑工
　~ mo(u)lding　注塑(成型);注塑制品
　~ mo(u)lding compound　注塑料
　~ mo(u)lding conditions　注塑(成型)条件
　~ mo(u)lding cycle　注塑周期
　~ mo(u)lding defect　注塑缺陷
　~ mo(u)lding gun　注塑喷枪
　~ mo(u)lding machine　注塑(成型)机

inlet

- ~ moulding machine with screw preplastication 螺杆预塑化式注塑(成型)机
- ~ mo(u)lding material 注塑料
- ~ mo(u)lding nozzle 注塑料嘴,注射喷嘴
- ~ moulding of large area 大面积注塑制品
- ~ mo(u)lding parameters 注塑参数
- ~ mo(u)lding plant 注塑装置
- ~ mo(u)lding pressure 注塑压力
- ~ mo(u)lding process 注塑成型法
- ~ – mo(u)lding technique 注塑技术
- ~ mo(u)lding technology 注塑工艺
- ~ mo(u)lding with screw plasticizing 螺杆(塑化)式注塑
- ~ nozzle 注塑料嘴
- ~ of sole 鞋底注塑
- ~ of sole direct to the upper 鞋底至鞋帮直接注塑
- ~ of thermoplasts [TP injection] 热塑性塑料注塑
- ~ order 注塑顺序
- ~ orifice 注塑孔
- ~ phase 注射相;注射阶段
- ~ piston 注射柱塞;注料杆;注塑杆
- ~ plunger 注射柱塞;注料杆,注塑杆(注塑机)
- ~ port 注射口
- ~ press 注射压机
- ~ pressure 注射压力,注塑压力
- ~ pressure control 注塑压力控制
- ~ pressure remote-control valve 注塑压力遥控阀
- ~ principle 注塑原理
- ~ process 注塑法
- ~ process with air pressure 气压注塑法
- ~ process with vacuum 真空注塑法
- ~ property 注塑性能
- ~ ram [injection piston or plunger] 注塑活塞,注料杆〈注塑机〉
- ~ ram pressure 注塑活塞压力
- ~ rate 注射速率
- ~ return stroke 注射杆回程
- ~ screw 注塑螺杆
- ~ sequence 注塑程序
- ~ shot 注料量
- ~ side 注射侧
- ~ sleeve 注射杆套
- ~ speed 注射速率[度],注塑速率[度]
- ~ speed control 注射速率[度]控制
- ~ stamping 注塑压花(法)
- ~ stretch blow moulding machine 注拉吹成型机
- ~ stroke 注料杆行程
- ~ technique 注塑技术
- ~ temperature 注塑温度
- ~ time 注塑时间
- ~ timer 注塑计时器
- ~ transfer blow moulding 注塑传递吹塑成型
- ~ transfer moulding 注塑传递成型
- ~ unit 注射装置
- ~ velocity 注塑速度
- ~ welding [unitized moulding] 注塑焊接

injector 注射器;注料器;喷射器
- ~ nozzle [jet mixer] 注料器嘴

injurious 有害的
- ~ ingredient 有害成分

ink 油墨
- ~ adhesion 印墨黏着力
- ~ duct 通墨道〈印刷〉
- ~ embossing 油墨印花
- ~ pan 油墨槽
- ~ transfer roller 传墨辊

inker roll bearing housing 墨辊轴承箱〈印刷〉

inking roller 墨印辊筒

inlay 嵌入;镶嵌件
- ~ printing 镶印

inlet 入口;浇口
- ~ diffusor 入口喷雾器
- ~ fitting [wall fitting] 入口配件
- ~ nozzle 入口喷嘴
- ~ pressure correction 入口压力校准〈注料嘴〉

inner

~ screen 进口滤网
~ temperature 进口温度
~ tube 进口管
~ velocity 进入速度
~ zone 进口区;供料口

inner 内部(的),里面(的)
~ bearing race 内轴承圈
~ diameter 内径[I.D.]
~ face 内表面
~ fender 内挡泥板
~ gear 内齿轮
~ liner 内衬
~ screw 内螺纹,阴螺纹
~ sizing unit 内定径设备〈内径〉
~ slide 内滑板〈模具〉
~ stress 内应力
~ structure 内部结构
~ viscosity 结构黏度
~ wall 内壁
~ width [I.W.] 空隙,内隙

innocuous 无害的
innocuousness 无害性
innoxious 无害的
innoxiousness 无害性
inoculating crystal 晶种
inodo(u)rous 无气味的,无臭的
inorganic 无机的
~ adhesive 无机黏合剂
~ chemistry 无机化学
~ fiber 无机纤维
~ filler 无机填料
~ ion exchange membrane 无机离子交换膜
~ ion exchanger 无机离子交换剂
~ pigments 无机颜料
~ plastic 无机塑料
~ polymer 无机聚合物
~ resin 无机树脂

input 输入;进料
~ material 进料
~ shaft 输入轴
~ speed 进料速度

insect 昆虫
~ - resistant treatment 防虫蛀处理

〈塑料,美〉

insecticidal 杀虫的
~ plasticizer 杀虫增塑剂

inseparable joints 不可分的连接
insert 嵌件;嵌入;插入;插入件
~ adapter 嵌入接头
~ block system 嵌块系统
~ - carring pin 嵌件支承销
~ holding pin 嵌件固定销
~ hole 嵌件孔
~ in bottom 底部(金属)嵌件〈模具〉
~ in top 顶部(金属)嵌件〈模具〉
~ mo(u)lding 嵌件模塑;深拉膜的热压花
~ of metal 金属嵌件
~ pin [carrier pin] 嵌件销〈模塑品嵌件〉
~ socket 承插接头〈塑料管线〉;嵌件插座
~ to support tube 夹持坯管嵌件〈吹塑〉

inserted 嵌入的
~ mandrel 嵌入芯模
~ mandrel technique 嵌入芯模技术〈吹塑〉

inserter 插件
inserting 嵌入,插入
insertion 嵌入,插入
~ piece 插入件,垫片

inset 嵌件
inside 内部(的),里面(的)
~ clearance 内间隙,内余隙
~ corner weld 内角焊缝
~ diameter 内径[I.D.]
~ diameter calibration 内径标定〈管材〉
~ dimension, [I.D.] 内尺寸
~ lead ga(u)ge 内螺纹导程
~ space 内空间
~ surface 内表面
~ taper 内圆锥体
~ thread 内螺纹
~ width 净宽

insol. [insoluble] 不溶的

insole, in-sole 内鞋底
insoluble 不溶(解)的
 ~ colorant 不溶着色剂
 ~ matter 不溶物
 ~ Particles 不溶颗粒
 ~ resin 不溶树脂
inspect 检查;检验
inspecting 检查;检验
 ~ standard 检验标准
inspection 检查;检验
 ~ bench 检验工作台
 ~ certificate 检验证书,检验合格证
 ~ department 检验部门
 ~ ga(u)ge 检验规
 ~ glass 观察镜,观察孔
 ~ hole 检查孔,观察孔
 ~ jig 检验夹具
 ~ schedule 检查程序
 ~ sheet 检验单
 ~ specification 验收技术条件
 ~ technique 检验技术
 ~ test 检查性试验
 ~ window 观察镜,窥镜
instability 不稳定性
 ~ constant 不稳定常数
installation 安装;装置,设备
 ~ cost 设备投资,设备费
 ~ diagram 安装图
 ~ procedure 安装工序
instant 瞬时,立即
 ~ adhesive 瞬时黏合剂
 ~ set polymer[ISP] 瞬时固化聚合物〈聚氨酯泡沫塑料〉
instantaneous 瞬时的,瞬间的
 ~ adhesive 瞬间黏合剂
 ~ bending strength 瞬时弯曲强度
 ~ combustion 瞬时燃烧
 ~ compliance 瞬时柔量
 ~ condition 瞬时条件
 ~ deflection 瞬时挠曲
 ~ deformation 瞬间形变
 ~ elastic difformation 瞬间弹性形变
 ~ elasticity 瞬时弹性
 ~ extension 瞬时延伸

 ~ load 瞬时负荷
 ~ modulus 瞬时模量
 ~ nucleation 瞬时成核
 ~ number average degree of polymerization 瞬时数均聚合度
 ~ output 瞬时产量
 ~ power 瞬时功率
 ~ rate of creep 瞬时蠕变速率
 ~ recovery 瞬时复原
 ~ recovery in creep 蠕变瞬间回复
 ~ set 瞬时变定
 ~ strain in creep 蠕变瞬时应变
 ~ strength 瞬时强度
 ~ stress 瞬时应力
 ~ stress-strain curve 瞬时应力-应变曲线
 ~ surface tension 瞬时表面张力
 ~ tangent modulus 瞬时正切模量
 ~ value 瞬时值
 ~ velocity 瞬时速度
instantaneously elastic spring 瞬时弹性弹簧
Instron 英斯特朗
 ~ capillary rheometer 英斯特朗毛细管流变仪
 ~ tensile tester 英斯特朗张力试验仪
 ~ type testing machine 英斯特朗试验机
 ~ type universal tester 英斯特朗型万能(材料)试验机
instruction 指导,规程;说明书,须知,指南
 ~ book 说明书
 ~ for use 使用指南
 ~ sheet 说明书,样本
instrument 仪器,仪表;工具;手段,方法
 ~ board[panel] 仪表板
 ~ error 仪器误差
 ~ of ratification 批准书
 ~ panel bracket 仪表板支架
 ~ with locking device 具有锁紧装置的调节器
instrumental match prediction 仪器配色测示法

instrumentation 检测仪
　~ engineer 检测工程师
insulant 绝缘材料
insulate 绝缘;保温,隔热;隔音
insulated 绝缘的;保温的,隔热的;隔音的
　~ feed bush mo(u)ld 流道绝热衬套模具〈注塑模具〉
　~ materials 绝缘材料;保温材料
　~ -runner(-type) mo(u)ld 保温流道型模具
　~ runner moulding 保温流道模塑
　~ wire 绝缘导线
insulating 绝缘的;保温的,绝热的
　~ ability 绝缘能力;保温性能
　~ board 绝缘板
　~ bush 绝缘(衬)套,绝缘套管
　~ charateristics 绝缘特性
　~ coat 绝缘涂层
　~ coating 绝缘涂料
　~ compound 绝缘物,绝缘料
　~ effect 绝缘作用,绝热效应
　~ efficiency 保温效率
　~ foam 保温泡沫塑料
　~ gasket 绝缘衬垫
　~ grid 绝缘板
　~ interlayer 绝缘夹层
　~ jacket 绝热夹套
　~ lacquer 绝缘漆
　~ layer 绝缘层
　~ lining 绝缘衬里;保温衬里
　~ material 绝缘材料;保温材料
　~ medium 绝缘介质
　~ paper 绝缘纸
　~ power 绝缘本领
　~ primer 绝缘底漆
　~ refractory 绝热耐火材料
　~ resin 绝缘树脂
　~ runner 保温流道〈热流道模具〉
　~ sheath 绝缘护套
　~ sheet 绝缘片材;绝缘板材
　~ support 绝缘支座
　~ strip 绝缘带
　~ tape 绝缘带
　~ tube 绝缘(软)管,绝缘(套)管
　~ varnish 绝缘清漆
　~ washer 绝缘垫圈
insulation 绝缘;保温,绝热;隔音
　~ board 绝缘板
　~ engineering 绝缘工程;绝缘技术
　~ material 绝缘材料;保温材料
　~ resistance 绝缘电阻
　~ resistance between plugs 插头之间绝缘电阻
　~ resistivity 绝缘电阻率
　~ sheath 绝缘护套(电缆)
　~ tape 绝缘带
　~ test 绝缘试验
　~ tubing 绝缘(软)管,绝缘(套)管
　~ washer 绝缘垫圈
insulator 绝缘材料,绝缘体
intaglio 凹(纹)雕(刻);凹版印刷
　~ copper plate printing 铜凹版印刷
　~ printing 凹版印刷
　~ printing cylinder 凹版印刷辊筒
intake 吸入;进气;入口;进口
　~ charge 进气,充气
　~ pressure 进气压力
　~ screen 进料滤网
　~ temperature 进口温度
　~ zone 进料区
integral 整体(的),组成的;积分(的)
　~ colo(u)r 整体颜色
　~ distribution curve 积分分布曲线
　~ extrusion 整体挤出
　~ foam 整体泡沫塑料
　~ foam interior 整体泡沫塑料内部〈多孔塑料结构〉
　~ foam structure 整体泡沫塑料结构
　~ hinge 连体铰链
　~ joint 整体连接
　~ mandrel 整体芯型,整体芯棒
　~ mo(u)ld 组合模具
　~ plasticization 全增塑
　~ skin 硬化表面层,结皮〈整体泡沫塑料〉
　~ skin foam [structural foam] 结皮泡沫塑料

~ skin foam moulding 结皮泡沫塑料成型
~ – skin mo（u）lding polyurethane foams 结皮模塑聚氨酯泡沫塑料
~ skin rigid foam 结皮硬质泡沫塑料
~ skin rigid urethane foam 结皮硬质聚氨酯泡沫塑料

integrate 完整(的),集成(的)
integrated 完整的,集成的
~ circuit 集成电路〈电〉
~ valve 控制阀
integrating circuit 积分电路〈电〉
intensified pressure resin injection 增压树脂注塑法
intensifier 增强器;增压器
intensity 强度;强烈
~ of a chemical reaction 化学反应强烈
~ of noise 噪声强烈
~ of radiation 辐射强度
~ of stimulus 刺激强烈
~ of X – ray diffraction X – 射线衍射强度

intensive 强烈的;彻底的,充分的
~ drying 充分干燥
~ mixer 强力混炼机;强力混合器
~ mixing 强力混合
~ setting 彻底定型〈薄膜拉伸时〉
~ sigma type mixer Σ型强力混炼机
~ test condition 强力试验条件
intensively cooled injection mould 充分冷却的注塑模具
interact 相互作用
interacting mechanism 相互作用机理
interaction 相互作用
~ coefficient 相互作用系数
interbedded 夹层的
interblend 相互掺混
interchain 链间的
interchange 互换
interchangeability 互换性
interchangeable [removable] 可互换的
~ can [change – can] 可互换槽
~ cavity mould 可换模腔模具
~ parts 可互换零件

Interchemical rotational viscometer 英特凯米卡尔旋转黏度计
intercoat adhesion 涂层间的黏合〈底层与面层间〉
intercommunicating 互通的〈泡孔〉
intercondensation polymer 共缩聚聚合物
interconnecting 相互连接
~ cells 互通泡孔
~ fibre sheet 交联纤维片层
~ foam 泡孔互连泡沫塑料,联孔泡沫塑料
intercrystalline 晶间的
~ fracture 晶间破裂
~ rupture 晶粒间断裂
interdependent 互相依赖的
interelectrode distance [gap] 电极间距〈高频焊接〉
interface 焊缝接触面〈焊接连接时〉,界面
~ layer 界面层,中间层
~ liquid/liquid 液 – 液界面
~ liquid/solid 液 – 固界面
~ structure 界面结构
~ temperature 界面温度
~ transfer coefficient 界面传递系数
interfacial 界面的
~ aging 界面老化
~ angle 界面角〈固 – 液之间〉
~ area 界面(面)积
~ bonding 界面粘接〈胶接〉
~ contact 界面接触
~ corrosion 界面腐蚀
~ energy 界面能
~ failure 界面破坏
~ film 界面薄膜
~ flow instability 面间流动稳定性,两相流动稳定性〈共挤塑〉
~ forces 界面力〈胶接〉
~ graft polymerization 界面接枝聚合
~ layer 界面层
~ polycondensation 界面缩聚(作用)
~ polycondensation polymerisation 界面缩聚(作用)

interference

~ polymerisation [polymerization] 界面聚合(作用)
~ pressure 界面压力
~ reflection[reflexion] 界面反射
~ region 界面(区域)
~ shear strength 界面剪切强度〈纤维,塑料〉
~ shear stress 界面剪切应力〈纤维,塑料〉
~ sliding friction 界面间滑动摩擦〈纤维,塑料〉
~ stress 界面应力
~ surface energy 界面能
~ tensimeter 界面张力计
~ tension [surface tension] 界面张力
~ traction 界面拉伸力[牵引力]〈纤维,塑料〉
~ unbonding 界面不黏合
~ viscosity 界面黏度
~ work 界面功

interference 干扰;干涉
~ colo(u)r 干涉色
~ fit 压紧配合〈如:管接〉
~ fringe 干涉条纹
~ microscope 干涉显微镜
~ spectrum 干涉光谱
~ suppressor 干扰消除器

interfibre bonding force 纤维间黏合力
intergranular 颗粒状;颗粒间的
interim 临时的,暂时的
interior 内部(的),里面(的)
~ coating of paint 内用涂漆
~ container 内包装物
~ decoration 内装饰
~ diameter 内直径
~ jaw 模内夹具
~ heating 内部加热
~ packing 内包装
~ paint 室内涂料
~ partitioning 内隔板
~ pressure 内压力
~ thickness 中部厚度
~ -type plywood 室内用胶合板

interlaced 交织的,交叉的,交错的;交联的;夹层的
~ polyethylene 交联聚乙烯
interlacing 交织;交错:交联
~ agent [crosslinking agent] 交联剂
interlaminar 层间的
~ bonding 层间黏合〈层合〉
~ bonding strength 层间黏合强度
~ peeling 层间剥离
~ shear 层间剪切
~ shear strength 层间剪切强度
~ strength 层间(剥离)强度〈层合材料〉
~ stresses 层间应力
~ tensile strength 层间拉伸强度
interlayer 中间层,夹层;层间的〈层压〉
~ adhesion 层间黏合〈层压材料〉
~ paper 中间层纸
~ temperature 层间温度
interleaf 中间层;夹层
interleaving paper [film] 衬垫纸
interlining 间夹层,防渗层
interlock 联锁(安全装置),安全开关
~ system 联锁系统
interlocker 联锁装置
interlocking 联锁(装置)
~ device 联锁装置
~ relay 联锁继电器
~ signals 联锁信号
intermediate 中间体,中间物,半成品;中间的
~ colo(u)r 中间色
~ compound 中间化合物,中间体
~ density polyethylene 中密度聚乙烯
~ die plate 中间模板
~ draft 中区牵伸
~ drive 中间传动
~ layer 中间层
~ material 中间产品,中间产物
~ moiecular weight [IMW] 中等相对分子质量
~ preparation 中间品制备
~ pressure 中等压力
~ product 中间产品,中间产物,半成品

~ temperature glue 中温固化胶
~ temperature setting 中温固化
~ temperature setting adhesive 中温固化黏合剂〈31~99℃固化的黏合剂〉
intermeshing 啮合的
~ co-rotating twin-screws 同向旋转啮合式双螺杆〈挤塑机〉
~ counter-rotating twin-screws 反向旋转啮合式双螺杆〈挤塑机〉
~ feed screw 啮合式供料螺杆
~ fingers 啮合式指形桨叶〈混合器,搅拌器〉
~ flight 啮合式螺纹
~ paddles mixer 啮合桨叶式混合器
~ screws 啮合螺杆
~ screw devolatilizer 啮合螺杆挥发器
~ threads 啮合式螺纹
~ twin screw extruder 啮合双螺杆挤塑机
intermiscibility 相(互)混性
intermittence 间歇性
intermittent 间歇的;间断的
~ bead 间断焊珠
~ drive 间歇式传动装置
~ duty 间歇工作状态
~ extrusion blowing 间断挤吹
~ feed 间歇供给
~ fluid bed dryer 间歇流动床干燥器
~ gas injections 间歇注气
~ handling 间歇操作
~ head 间歇(混合与注塑)机头〈泡沫塑料〉
~ heat 间断加热
~ operation 间歇操作
~ output 间歇产量
~ process 间歇操作法
~ seam welding 间断线焊
~ stress relaxation 间歇应力松弛
~ test 间歇性试验
~ weld 间断焊接缝
~ welding 间断焊接
intermix 混合,混炼;搅拌
intermixer 密闭式混炼机,密炼机
intermixture 混合;混合料

intermolecular 分子间(的)
~ condensation 分子间缩合(作用)
~ cross-linking 分子间交联
~ force 分子间力
~ -intramolecular polymerization 分子间-分子内聚合
~ structure 分子间结构
~ transfer 分子间转移
internal 内(部)的
~ air-pressure 内气压
~ and external surface 内外表面
~ antistat 内用抗静电剂
~ antistatic agent 内用抗静电剂
~ caliper gauge 内(径)卡规
~ condensation 内缩合(作用)
~ cooling 内冷却
~ diameter 内径
~ damping 内阻尼
~ die pressure 口模内压力
~ dimensions 内部尺寸
~ flame retardant 内(用)阻燃剂
~ fracture 内部破碎
~ friction 内摩擦
~ gauge 内径规
~ gears 内齿轮咬合
~ grinding machine 内研磨机
~ haze 内混浊,内光雾;内光雾度
~ heating 内加热
~ hexagon 内六角
~ impregnation 内浸渍
~ lubricant 内润滑剂
~ mixer 密闭式混炼机,密炼机
~ mixer with floating weight 浮动栓密炼机
~ mixing 密炼
~ mo(u)ld lubricant 内脱模剂
~ mo(u)ld release agent 内脱模剂
~ plasticization 内增塑(作用)
~ plasticized polymer 内增塑聚合物
~ plasticizer 内增塑剂
~ plug 内栓塞
~ polymerization 内聚合
~ pressure 内压力
~ pressure system 内压系统

- ~ reflectance 内反射
- ~ rotation 内旋转
- ~ screw thread 内螺纹
- ~ sizing 内径,定径;内施胶
- ~ stabilization 内稳定
- ~ stabilizer 内稳定剂
- ~ standard 内标(准)
- ~ standard sample 内标试样
- ~ strain 内应变
- ~ stress 内应力
- ~ stress relaxation 内应力松弛
- ~ structure 内部结构
- ~ surface 内表面
- ~ surface cooling [ISC process] 内表面冷却法
- ~ thread [female thread] 阴螺纹,内螺纹〈美〉
- ~ viscosity 内黏性;内黏度
- ~ waviness 内波纹〈透明塑料缺陷〉
- ~ welding flash [upset] 内焊脊〈管〉

internally heated runner 内加热流道
international 国际的
- ~ rubber hardness degree [IRHD] 国际橡胶硬度(等级)
- ~ standard 国际标准
- ~ unit 国标单位

International Standardiza–tion Organization [ISO] 国际标准化组织
International Standards Organization [ISO] 国际标准协会
International Union of Pure and Applied Chemistry (IUPAC) 国际纯化学和应用化学联合会
interparticle 粒间的
interpenetrating 互相贯穿的
- ~ elastomeric network [IEN] 互穿弹性体网络
- ~ homopolymer network [IHN] 互穿均聚物网络
- ~ network 互穿网络
- ~ polymer network [IPN] 互穿聚合物网络
- ~ thickening process [ITP] 互穿增稠方法

- ~ thickness control 互穿增稠控制

interplanar distance 晶面间距
interplay 相互作用
- ~ adhesion 中间层黏合,层间黏合

interpolymer 共聚物〈美〉
interpolymerization 共聚作用
interpretation (试验结果)数据分析
interrupted 断续的;间歇的
- ~ injection 间歇注塑
- ~ oscillation 间歇振荡
- ~ polar wind [dual pular wind] 间断极向卷绕
- ~ polymerization 间断聚合(作用)
- ~ thread 间断螺纹

interstice 间隙,空隙;裂缝
interstitial polymerization 填隙聚合,间充聚合〈一种制取聚合物共混物的化学方法〉
interstructure 组分,分配;隔板;挡板
intertwining 缠绕〈塑料分子〉
interval 间隔,间距;间隙;间歇
- ~ injection moulding 间隔注塑
- ~ mo(u)lding 间隔成型〈不同色料的注塑成型〉

intimate mixing 均匀混合
intra 〈词头〉内
- ~ –chain trans double bond (分子)链内反式双键
- ~ –intermolecular polymerization 分子内–分子间聚合

intracavity 腔内
intrachain 链内的
intractability 难处理,难加工
intractable 难处理的,难加工的
intrafibre 纤维内的
intragranular 颗粒内的
intramolecular 分子内的
- ~ condensation 分子内缩合(作用)
- ~ crosslinking 分子内交联
- ~ cyclization reaction 分子内成环反应
- ~ force 分子内力
- ~ reaction 分子内反应
- ~ transfer 分子内转移

intraresin reaction 树脂内反应

intricate 复杂的
- ~ casting 复杂铸塑件
- ~ mo(u)lded part 复杂(形状)模塑件
- ~ piece 复杂构件
- ~ plastic article 复杂(形状)塑料件

intrinsic 内在的;固有的;特性的
- ~ birefringence 特性双折射
- ~ breakdown 特性击穿
- ~ breaking energy 特性断裂能
- ~ constant 固有常数
- ~ contaminant 内在杂质
- ~ dynamic viscosity 特性动态黏度
- ~ heating 内热,内部加热
- ~ pressure 内压
- ~ properties 固有性质,本征特性
- ~ rigidity 固有刚性
- ~ strength 内在强度
- ~ viscosity 特性黏度
- ~ viscosity number 特性黏数
- ~ viscosity of polymer 聚合物特性黏度

introfaction 加速浸泡作用,促浸作用
introfier 加速浸渍剂
intrusion 螺杆挤料注塑
- ~ mo(u)lding 螺杆挤料注塑成型

intumescence 熔胀,膨胀
intumescent 膨胀的;熔胀的;发泡的
- ~ coating 熔胀性涂层,发泡性防火涂料
- ~ flam retardent [IFR] 发泡阻燃剂
- ~ paint 发泡性防火涂料

inventory 货单,清单;存料量〈注塑,挤塑料筒或机头内〉
inverse 相反,反向,逆向
- ~ cooling-rate curve 逆冷却速率曲线
- ~ heating-rate curve 逆加热速率曲线〈热分析〉
- ~ lamination 反向层合法〈层合泡沫塑料制品〉
- ~ proportion 反比例
- ~ ratio 反比例
- ~ stress 逆应力

inversed 反向的,逆向的

- ~ knife coater 反刀涂布机
- ~ L calender 倒 L 形压延机

inversion 颠倒,倒置,反向;转化
- ~ point 转化点
- ~ temperature 转化温度

invert 颠倒,反向;转化
inverted 颠倒的,反向的;转化的
- ~ blade coater 反刀涂布机
- ~ blister 凹痕〈制品缺陷〉
- ~ knife 逆刀,反刀
- ~ knife coater 反刀涂布机〈美〉
- ~ knife coating 反刀涂布
- ~ L calender 倒 L 形压延机
- ~ L-type [F-Type] 倒 L 形
- ~ mo(u)ld 反模,倒模
- ~ mo(u)ld moulding 倒模模塑,反模成型
- ~ plasticity 反塑性
- ~ pseudo plastic 反假塑性
- ~ ram press 反柱塞压机

investigation (材料)试验研究
investigative test 研究性试验,探索性试验

inviscid 非黏(性)的
- ~ fluid 非黏流体

iodine 碘
- ~ value [iodine number] 碘价;碘值

ion 离子
- ~ -beam etching 离子束蚀刻
- ~ bombardment 离子轰击〈使塑料非极性粘着面极化〉
- ~ chromatography 离子色谱法
- ~ exchange 离子交换(作用)
- ~ exchange capacity 离子交换容量
- ~ exchange chromatography 离子交换色谱法
- ~ exchange column 离子交换柱
- ~ exchange fiber 离子交换纤维
- ~ exchange filter 离子交换过滤器
- ~ exchange hollow yarn 离子交换中空丝
- ~ exchange materials 离子交换材料
- ~ exchange membrane 离子交换膜
- ~ exchange membrane diffusion dialysis

离子交换膜扩散渗析
~ exchange membrane electrodialysis　离子交换膜电渗析
~ exchange membrane separation technology　离子交换膜分离技术
~ exchange mixed bed　离子交换混合床
~ exchange resin　离子交换树脂
~ exchange separation　离子交换分离
~ exchange softener　离子交换软化剂
~ exchanger　离子交换剂
~ etching　离子蚀刻
~ microprobe mass spectrometer　离子微探针质谱计
~ plating　离子镀
~ - probe　离子探针
ionex catalyst　离子交换树脂催化剂
ionic　离子的
~ bond　离子键
~ copolymerization　离子型共聚
~ grafting　离子接枝
~ initiator　离子引发剂
~ polymer　离子聚合物
~ polymerization　离子型聚合(作用)
~ rubber　离子(化)橡胶
~ strength　离子强度
~ urethanes　离子型聚氨酯
ionitriding　离子氮化;电离渗氮
ionization　离子作用
~ foaming　离子辐射发泡
~ polymerization　离子型聚合(作用)
ionizing radiation　离子辐射
ionogenic polymer　离子(化)聚合物
ionomer　离子键聚合物,离子交联聚合物
~ resin　离子键树脂
IPN [interpenetrating polymer network]　互穿聚合物网络
IPP [isotactic polypropylene]　全同立构聚丙烯,等规聚丙烯
IR [infrared]　红外(线)的
Ir [iridium]　铱
IR [isoprene rubber]　异戊二烯橡胶
IR heater [infrared heater]　红外(线)加热器
IR - welding [infrared welding]　红外线焊接
iridescent　彩虹,闪光〈织物〉
~ fabric　闪光织物
~ film　彩虹薄膜
iridium [Ir]　铱
iron [Fe]　铁
~ disk mill　铸铁盘式磨粉机
~ loss　铁损
~ separator　磁选机
IRPI [infra - red polymerization index]　红外聚合指数
irradiate　照射,辐射
irradiated　照射的,辐射的
~ curing　辐射固化
~ foam　辐照泡沫塑料
~ plastic　辐射(交联)塑料
~ polyethylene　辐射(交联)聚乙烯
~ thermoplastic　辐射(交联)热塑性塑料
irradiation　照射,辐射
~ cross - linking　辐射交联
~ effect　辐射效应
~ grafting　辐射接枝
~ resistance　耐辐射性
~ time　辐照时间
Irrathene　〈商〉伊拉森〈辐射交联聚乙烯〉
irrecoverable　不可恢复的
~ compliance　不可恢复柔量
~ creep　不可恢复蠕变
irregular　不规则的
~ block　非规整嵌段,无规嵌段(聚合物)
~ checking　不规则细裂
~ curve　不规则曲线
~ folding　不规则折叠
~ network　不规则网络
~ pattern type cracking　不规则形开裂
~ polymer　非规整聚合物,无规聚合物
~ porosity　不规则孔率
~ width　门幅不齐

irregularity 不规则性;不均度,不匀率;凹凸不平
- ~ control 不匀率控制
- ~ index 不匀率指数,不规则性指数
- ~ test 不匀率试验

irreversible 不可逆(转)的
- ~ gel 不可逆凝胶

irrigation 灌溉,灌注

irritant 刺激(性)的;刺激剂〈生物〉
- ~ action 刺激作用〈生物〉

irt 防虫蛀处理〈美〉

isano oil 衣散油〈阻燃剂〉

ISC process [internal surface cooling] 内表面冷却法

ISO [International Standardization Organization, International Standards Organization] 国际标准化组织,国际标准化协会

ISO recommendation 国际标准化协会推荐标准

ISO standard 国际标准化协会标准

iso 〈词头〉异;等,同
- ~ - trans - tactic 反式全同立构

isoamyl 异戊基
- ~ acetate 乙(醋)酸异戊酯〈溶剂〉
- ~ butyrate 丁酸异戊酯〈增塑剂〉

isobaric 等压的;同量异序的
- ~ mass - change determination 等压质量变化测定
- ~ weight - change determination 等压重量变化测定〈热分析〉

isobutene [isobutylene] 异丁烯

isobutyl 异丁基
- ~ acetate 乙酸[醋酸]异丁酯〈溶剂〉
- ~ alcohol 异丁醇
- ~ isobutyrate 异丁酸异丁酯
- ~ vinyl ether [IVE] 异丁基乙烯基醚

isobutylene 异丁烯
- ~ - isoprene copolymer 异丁烯 - 异戊二烯共聚物,异丁橡胶
- ~ - isoprene rubber [IIR] 异丁烯 - 异戊二烯橡胶,异丁橡胶

isochromate 等色线

isochromatic 等色的

isochronous 等时的
- ~ stress - strain curve [diagram] 等时应力 - 应变曲线(图)
- ~ tensile creep modulus 等时拉伸蠕变模量〈塑料〉

isocyanate 异氰酸盐[酯]
- ~ adhesive 异氰酸酯(类)黏合剂
- ~ compound 异氰酸酯(类)化合物
- ~ content 异氰酸基含量
- ~ equivalent 异氰酸基当量
- ~ foams 异氰酸酯(类)泡沫塑料
- ~ generator 异氰酸酯(类)发生物
- ~ index 异氰酸基指数
- ~ plastic 异氰酸酯(类)塑料
- ~ polymer 异氰酸酯(类)聚合物
- ~ prepolymer 异氰酸酯(类)预聚体
- ~ raw material 异氰酸酯(类)原料〈制聚氨酯用〉
- ~ resin 异氰酸酯(类)树脂

isocyanates 异氰酸酯

isocyanurate 异氰脲酯
- ~ foam 异氰脲酯(类)泡沫塑料
- ~ plastic 异氰脲酯(类)塑料

isodecyl diphenyl phosphate 磷酸异癸二苯酯(增塑剂)

Isoderm process 艾索德姆发泡法

isoelectric 等电(位)的
- ~ heating 电热
- ~ point 等电点

isolate 隔离;离析

isolated 隔离的,绝缘的
- ~ cell foam 隔离泡孔泡沫塑料
- ~ double bond 孤立双键
- ~ feed bush mo(u)ld 绝热流道模具

isomer (同分)异构体

cis - ~ 顺式异构体

trans - ~ 反式异构体

isomers 异构体

isomeric (同分)异构的
- ~ polymer (同分)异构聚合物

isomerism 同分异构(现象)

isomerization 异构化
- ~ polymerization 异构化聚合(作用)
- ~ process 异构化过程

isomorph

~ reaction 异构化反应

isomorph 类质同象体,同晶型体

iso－octyl 异辛基

~ －decyl adipate[ODA] 己二酸异辛(基)癸(基)酯〈聚氯乙烯增塑剂〉

~ －decyl phthalate[ODP] 邻苯二甲酸异辛(基)癸(基)酯〈聚氯乙烯增塑剂〉

~ epoxystearate 环氧硬脂酸异辛酯〈增塑剂、稳定剂〉

~ ester of tall oil 妥尔油(酸)异辛酯〈增塑剂〉

~ isodecyl adipate 己二酸异辛·异癸酯〈增塑剂〉

~ isodecyl phthalate 邻苯二甲酸异辛·异癸酯

~ palmitate 棕榈酸异辛酯,软脂酸异辛酯〈增塑剂〉

isoosmotic solution 等渗压溶液

isophorone 异佛尔酮

~ diamine 异佛尔酮二胺〈固化剂〉

isophthalic acid 间苯二甲酸

isoprene 异戊二烯

~ rubber 异戊二烯橡胶

p－isopropoxy diphenylamine 对异丙氧基二苯胺〈抗氧剂〉

isopropyl 异丙基

~ acetate 乙酸异丙酯〈溶剂、脱水剂〉

~ 4－aminobenzenesulfonyldi(dodecyl－benzenesulfonyl) titanate 4－氨基苯磺酰基双(十二烷基苯磺酰基)钛酸异丙酯〈偶联剂〉

~ benzol 异丙(基)苯

~ myristate 肉豆蔻酸异丙酯〈增塑剂〉

~ myristate－palmitate 肉豆蔻酸棕榈酸异丙酯〈增塑剂〉

~ palmitate 棕榈酸异丙酯〈增塑剂、软润剂〉

~ percarbonate 过碳酸异丙酯

N－~ －N′－phenyl－p－phenylenediamine N－异丙基—N′－苯基对苯二胺,抗氧剂4010NA

~ tridodecyl benzenesulfonyl titanate 三(十二烷基苯磺酰基)钛酸异丙酯〈偶联剂〉

~ tri(dioctylphosphato) titanate 三(二辛基磷酰氧基)钛酸异丙酯〈偶联剂〉

~ tri(dioctyl pyrophosphato) titanate 三(二辛基焦磷酰氧基)钛酸异丙酯〈偶联剂〉

~ triisostearoyl titanate 三异硬脂酰基钛酸异丙酯〈偶联剂〉

~ trioleoyl titanate 三油酰基钛酸异丙酯〈偶联剂〉

~ xanthogen disulfide 二硫化二异丙基黄原酸酯〈促进剂〉

isopropylidene 异亚丙基

4,4′－~ bis(2,6－dibromophenol) 4,4′－异丙基叉双(2,6－二溴苯酚)〈阻燃剂〉

4,4′－~ bis(2,6－dichlorophenol) 4,4′－异丙基叉双(2,6－二氯苯酚)〈阻燃剂〉

2,2′－[~ bis(2,6－dichloro－P－phe－noxyl)] polypropyleneoxidediol 2,2′[异丙基叉双(2,6－二氯对酚氧基)]聚环氧丙烷二醇〈阻燃剂〉

4,4′－~ bisphenol 4,4′－异丙又双酚,双酚A〈抗氧剂〉

p,p′－~ bisphenol disalicylate 4,4′－异丙又双酚双水杨酸酯〈紫外线吸收剂〉

2－isopropylimidazole 2－异丙基咪唑〈固化剂〉

isorefractive mixture 等折射率混合物,等折光指数混合物

isostatic 等压的

~ compaction 等压压实

~ moulding 等压成型〈氟塑料〉

~ pressing 等压压制

isotactic 全同立构(的);等规立构(的)

~ block 等规嵌段,全同立构嵌段

~ chain 等规立构链

~ configuration 等规立构构型

~ index 等规指数,全同立构指数

~ moulding 等压成型(在加工聚四氟乙烯等材料时)

- ~ polymer 全同立构聚合物，等规聚合物
- ~ polyolefin 全同立构聚烯烃，等规聚烯烃
- ~ polypropylene [IPP] 全同立构聚丙烯，等规聚丙烯
- ~ polypropylene fiber 等规聚丙烯纤维
- ~ polystyrene 等规聚苯乙烯
- ~ pressing 全同加压，各向均压压塑
- ~ sequence 全同立构序列

isotacticity 等规立构性，等规度，全同立构规整度
- ~ index 等规指数

isotherm 等温线
- ~ creep curve 等温蠕变曲线

isothermal 等温的
- ~ change 等温变化
- ~ compression 等温压缩
- ~ crystallization 等温结晶
- ~ efficiency 等温效率
- ~ expansion 等温膨胀
- ~ extrudex 等温挤塑机
- ~ extrusion 等温挤塑
- ~ heating 等温加热
- ~ mass - change determination 等温质量变化测定
- ~ parameter 等温参数
- ~ pressure ratio 等温压力比
- ~ shear rate 等温剪切速率
- ~ thermogravimetric analysis 等温热失重(量)分析
- ~ weight - change determination 等温重量变化测定〈热分析〉

isothermic 等温(的)
- ~ dyeing 等温染色
- ~ flow 等温流动

isothymol 异百里酚
isotope 同位素
isotropic 各向同性的
- ~ carbon fibre 各向同性碳纤维
- ~ fiber 各向同性纤维
- ~ laminate 各向同性层压制品
- ~ material 各向同性材料
- ~ plate 各向同性板
- ~ pressure 各向同性压力
- ~ stress [hydrostatic stress] 各向同性应力

isotropy 各向同性(现象)
isoviscous 等黏的
- ~ state 等黏态
- ~ temperature 等黏温度

ISP [instant set polymer] 快速固化聚合物
ISRUF [integral skin rigid urethane foam] 结皮硬质聚氨酯泡沫塑料
ISS [industry standard specifications] 工业标准规范
issue 流出〈液体〉
ISTM [International Sociсty for Testing Materials] 国际测试材料协会
itaconic acid 衣康酸〈助剂〉
ITP [interpenetrating thickening process] 互穿增稠方法
I - type calender I 型压延机
IU [International Unit] 国际单位
IUPAC [International Union of Pure and Applied Chemistry] 国际纯化学和应用化学联合会
IVE [vinyl isobutyl ether] 异丁基乙烯醚
I. W. [inner width] 内宽
Izod 艾佐德
- ~ impact machine 艾佐德冲击机，悬臂梁式冲击机
- ~ impact strength 艾佐德冲击强度，悬臂梁式冲击强度
- ~ impact strength, notched 艾佐德缺口冲击强度
- ~ impact strength, unnotched 艾佐德无缺口冲击强度
- ~ impact test 艾佐德冲击试验，悬臂梁式冲击试验〈测冲击弯曲强度〉
- ~ impact tester 艾佐德冲击试验仪，悬臂梁式冲击试验仪
- ~ method 艾佐德方法
- ~ test 艾佐德试验，悬臂梁式试验
- ~ value 艾佐德值，悬臂梁式值

J

J. [journal] 杂志
jack 起重器;传动装置;插座
　~ arrangement　无规则排列〈玻璃纤维〉
　~ shaft　中间轴,副轴
jacket 护套,套管;定型套;夹套
　~ cooling　夹套冷却
　~ temperature　夹套温度
jacketed (带有)夹套的
　~ mixer　夹套式混合器
　~ reactor　夹套反应器
　~ rotary shelf dryer　夹套回转式搁板干燥机
　~ shelf　夹套搁板
　~ shelf dryer　夹套搁板式干燥机
　~ trough　夹套槽
Jamming [blocking] 阻塞
　~ against　挤压;反压
　~ in　固定楔住
　~ into　塞进,压入,夹紧,嵌入
Japan 日本;日本漆
　~ colour　日本色
Japanese Industrial Standart [JIS] 日本工业标准
jar 罐,缸;瓶,容器
jaw 颚板〈模具夹紧装置〉;模唇〈挤板模头〉;夹头〈材料试验机〉
　~ actuation　(模具)滑扑操纵机构
　~ breaker [jaw crusher]　颚式破碎机
　~ chuck　夹头盘,卡盘〈机〉
　~ clamp　夹紧钳
　~ crusher　颚式破碎机
　~ injection moulding tool　颚式注塑模〈下模可移动〉
　~ plate　颚板
　~ tool　颚式模具〈下模可移动〉
jectruder 注(塑)挤(塑)塑料成型
jellification 胶凝(作用);凝胶化(作用)
jelly 胶冻;冻胶

　~ mould　胶模
　~ strength　胶冻强度
jet 注嘴〈热固性塑料柱塞注塑机〉;喷射;射流;喷嘴;喷丝头
　~ agitator　喷射搅拌器
　~ blender　喷射混合器〈液体〉
　~ cell [air–diffusion chamber]　喷射室〈纺织〉
　~ condenser　喷射式冷凝器
　~ cooling　喷液冷却
　~ cutter　喷射切割机〈塑料〉
　~ cutting　喷射切割
　~ – cutting system　喷射切割装置
　~ dryer　气流干燥器
　~ filling　喷射充填〈注塑模〉
　~ forming　喷射成型
　~ hole　喷孔
　~ lip　喷嘴唇
　~ micronizer　喷射粉碎机
　~ mill　射流磨,喷射研磨机
　~ mixer　喷射混合器
　~ mixing　喷射混合
　~ mo(u)ld　喷射模〈热固性塑料〉
　~ mo(u)lding　喷射模塑〈热固性塑料〉
　~ mo(u)lding nozzle　喷射模型嘴〈热固性塑料〉
　~ orifice　喷孔,喷嘴
　~ out　喷射出
　~ pulverizer [nozzle pulverizer]　喷射粉碎机
　~ ring　喷射环(喷射研磨机)
　~ spinning [flash spinning type of melt spinning]　喷射流纺丝
　~ stretch　喷(丝)头拉伸
　~ stretch ratio　喷(丝)头拉伸比
　~ velocity　喷射速度
jetting 漩纹;喷射
jig [fixture] 夹具,钻模;样板,模板

~ electrode 焊条夹具;异型焊极
~ saw 摆动锯
~ welder 衬胎焊接机
~ welding 衬胎焊接;异型胎模焊
JIS [Japanese Industrial Standart] 日本工业标准
job 工作,作业;工件,零件;成品;工地,现场;职务
~ engineer 项目工程师
~ location 施工现场
~ mix 现场搅拌
~ operation 加工方法
~ production 单件生产
~ programme 工作程序,施工程序
~ rates 生产定额
~ schedule 工作进度(表)
~ sequence 加工序列
~ shop 加工车间
~ -site repair 就地修理
~ specification 施工规范,工作说明
jockey pulley 张力辊,导轮
jockey roller 导辊;张力辊
joggle 摇摆;木準接,啮合;偏斜
~ lap joint 压肩接头,啮合榫搭接
join 联接,接合
joined part 接合件,连接件
joining 连接,接合
~ by welding 连接焊
~ force 连接力;焊接力
~ glue 粘接胶
~ of plastics 塑料连接
~ pressure [welding pressure] 焊缝压力
~ process 连接方法
~ zone 连接区
joint 接头;连接;接合;黏合;接缝
~ aging time 黏固时间(黏合的后固化时间)
~ area 粘接面,黏合面
~ conditioning time 黏固时间(黏合的后固化时间)
~ coupling 万向接头
~ efficiency 粘接效率;焊接效率
~ face 接合面

~ factor 粘接系数
~ filler casting material 接缝浇注料,填缝料
~ filling 填缝
~ form 连接型式,连接形状
~ glue 接合胶,粘接胶
~ hinge 连接铰链
~ line 焊接缝
~ of metals 金属连接
~ packing [flange packing] 连接填衬料
~ part 连接件,连接部分
~ pin 连接销,铰链销
~ pipe 连接管
~ seam 接缝;接头
~ shape 连接型式;连接形状
~ sleeve 连接套
~ strength 粘接强度
~ surface(area) 粘接表面(积)
~ washer 密封垫
~ weld 连接焊,焊缝
~ welding 连接焊
jointing 接合,连接;接缝,填缝;填缝料
~ compound 填缝料;灌缝料
~ force 连接力
~ piece 连接件〈管〉
~ pressure 连接压力;焊接压力
~ sleeve 连接套(筒)
jointless 无接缝的,无接头的
~ fleximer flooring 无接缝柔软铺地材料
~ flooring 无接缝铺地材料
journal 轴颈;期刊
~ bearing 轴颈轴承,径向轴承
~ bearing sleeve 轴颈轴承套
~ box 轴颈箱
~ width 轴颈宽
JPC [Japan Patent Center] 日本专利中心
judgement 判断力,鉴定,评价
jumbo 巨大的
~ mixer 巨型混合器
Jumex extruder 捷米克斯挤塑机,四螺杆挤塑机

jumper wire　跨接线
junction　接合,连接;接头,接点〈邻近分子链〉
　~ box　接线盒
　~ piece　连接件
　~ point　接点;交联点
　~ welding　连接焊接
　~ zone　连接区段
juncture　接合点,连接点
just　正确的,确切的,合理的
　~ size　正确尺寸
jute　黄麻

K

K[kelvin]　开尔文〈绝对温度单位〉
K[potassium]　钾
k-factor　增殖系数
K-resin[styrene-butandien copolymer]　K-树脂,苯乙烯-丁二烯共聚物
K-value　K值〈按菲肯切尔(H. Fikentscher)公式计算的值,与聚合物分子大小及聚合度有关〉
k-value　K值;黏度值;增殖系数〈热导系数〉
kadin　高岭土
Kady mill　卡代磨
Kanavec　卡纳维克
　~ method　卡纳维克(塑性试验)法〈测定热固性塑料成型料流动性〉
　~ plastometer　卡纳维克塑性仪
kaolin[kaoline]　高岭土,陶瓷黏土
kaolinite　高岭石
kapillary[capillary]　毛细管
Kapton　〈商〉卡普顿〈聚酰亚胺薄膜〉
Karl Fisher reagent　卡尔·费歇试剂
katastatic stress　弛存应力
katharometer[catharometer]　导热析气计,热导计
Kcal kilocalorie　千卡,大卡〈热量非SI单位〉
Kelvin　开尔文〈人名〉;绝对温度单位〈符号K〉
　~ body　开尔文(固)体,黏弹(固)体
　~ degree[K]　开尔文温度

　~ scale　开尔文温标,绝对温标
　~ temperature[K]　开尔文温度,开氏温度,绝对温度
　~-Voigt body　开尔文-福格特体
　~-Voigt model　开尔文-福格特模型〈形变机械性能〉
kauri　栲树脂〈天然树脂〉
Kautex process　考代克斯法〈升台吹塑〉
keep　保持,维持
　~ dry　保持干燥〈包装〉
　~-fresh-container　保鲜容器〈包装〉
Kek mill　开克磨〈单转轮磨〉
keratin　角蛋白
　~ fiber　角蛋白纤维
kerosine　煤油
ksternich test　耐蚀试验
Kestner evaporator　凯斯特纳蒸发器,薄膜蒸发器
ketene　乙烯酮;烯酮〈类名〉
ketimine　酮亚胺〈固化剂〉
2-keto-1,7,7-trimethylnorcamphane　2-酮基-1,7,7-三甲基降莰烷〈增塑剂〉
ketone　酮〈类名〉;(某)甲酮
　~ resin　酮类树脂
kettle　锅;釜
Kevlar(aramid)fibre　〈商〉凯芙拉纤维〈聚对苯二甲酰对苯二胺纤维〉
key　楔;键;钥匙;主要的,关键词

~ band 特征谱带
~ board 键盘
~ colo(u)r 基本色
~ dimension 主要尺寸,关键尺寸
~ factor 关键因素
~ lock type system 闸板式锁模系统
~ panel 键盘
~ point 要点,关键
~ way 键槽,榫槽,销子槽
~ work 关键字
~ wrench 套筒板手
keyseat 键槽
kg [kilogram] 千克,公斤
kgf [kilogram force] 千克力,公斤力
kHz [kilohertz] 千赫(兹)
locker 引发剂,活化剂,促进剂
kieselguhr 硅藻土
kiln 窑;烘房;干燥装置
kilogram[kg] 千克,公斤
kilometer[km] 千米,公里
kiloton[kt] 千吨
land 种类,品种;品质,性质,特性
~ of sprue 浇口类型
~ of stresses 应力种类
kindling 着火,燃烧
~ point 着火点,燃点
~ temperature 着火温度
kinematic 运动(学)的
~ coefficient of viscosity 运动黏度系数
~ viscosity 运动黏度,比密黏度
~ viscosity coefficient 运动黏度系数
kinetic 运动的,动力(学)的
~ cnergy 动能
~ friction 动摩擦
~ friction coefficient 动摩擦系数
~ of polycondensation 缩聚反应动力学
~ of polymerization 聚合动力学
~ pressure 动压
~ theory of elasticity 弹性分子运动论
~ viscosity 动力学黏度,比密黏度

kirksite 铝锌合金〈吹塑模具用〉
kiss 轻触,接触
~ applicator 辊舐涂布机
~ coater 辊舐涂布机,挂胶机
~ coating 辊舐涂布,挂胶
~ embossing 低压轧花
~ roll 舐辊
~ roll coater 辊舐式涂布机
~ roll coating 辊舐涂布〈美〉
~ roller 舐涂辊
kit 小桶;箱,用具箱;成套工具
~ body [car body] 汽车车身
knead 捏合,捏和
kneadable material 可捏合材料;塑性材料
kneader 捏合机
~ blade 捏合叶片
~ mixer 捏合混合机
~ with masticator blades 带有割碎桨叶的捏合机
kneading 捏合,捏和
~ arm 捏合桨叶
~ blade 捏合桨叶
~ block 捏合段
~ disk 捏合盘〈盘式塑性仪〉
~ machine 捏合机
~ mill 捏和机
~ set 捏和机;捏合段
~ shaft 捏和轴
~ stock 捏合料
~ tool 捏和工具
~ zone 捏和段
knee 肘管,弯管;弯曲点〈曲线〉
~ bend 弯头,弯管接头
~ of curve 曲线的转折处
~ of S-N-curve 疲劳曲线的转折处
~ pipe 弯管,肘管,弯头
knife 刀;切刀;刮刀,刮板
~ carrier 刀架
~ carrying shaft 刀辊,刀轴
~ coat 刮刀涂布

knit

- ~ coater 刮刀涂布机〈美〉
- ~ coating 刮刀涂布
- ~ edge 刀刃
- ~ edge die 带有切刀的模头〈挤塑机〉
- ~ frame 刀框
- ~ line 刀纹,刀痕,刀路,刨纹〈涂布缺陷〉
- ~ mill 切碎机
- ~ – on – blanket coating 垫带式刮刀涂布
- ~ – on – roll coating [knife – over – roll coating] 辊式刮刀涂布
- ~ – over – roll coater 辊式刮刀涂布机
- ~ – over – roll coating 辊式刮刀涂布
- ~ roller 刀辊
- ~ rotor 圆盘式剪刀片〈造粒〉
- ~ spreader 刮刀〈涂布〉
- ~ switch 闸刀开关
- ~ wheel chips 刨花,木刨花;碎木片;碎料

knit 编织;汇合〈流料〉;接合
- ~ article 针织品
- ~ fabric construction 针织物结构
- ~ fabric shrinkage gauge 针织物收缩试验仪
- ~ line 接缝线,汇流纹;熔合纹;刀路
- ~ sack 编织袋

knitted 针织的
- ~ carpet 针织地毯
- ~ fabric 针织物,针织坯布

knitter 编织机,针织机;针织工

knitting 针织;针织品
- ~ foamed leather 针织泡沫革
- ~ length 编织长度
- ~ stiffness 编织密度
- ~ tightness 编织密度
- ~ width 编织宽度

knob 节;按钮,旋钮

knock 敲打,碰撞;顶销,脱模

knockout 顶出,脱模,推顶器
- ~ bar 顶出联动棒[杆]
- ~ connecting bar 脱模连接杆
- ~ die 顶出式冲模,冲切模
- ~ frame 推顶框架,脱模拉杆架,脱模板支架
- ~ hole 脱模销孔
- ~ machine 脱模机
- ~ pad 脱模垫
- ~ pin 顶出杆,脱模销
- ~ pin mark 脱模销痕
- ~ pin plate 脱模销托板
- ~ pin travel 脱模销行程
- ~ plate 脱模板,推顶板
- ~ plate mo(u)ld 有顶出板的模具
- ~ press 脱模力压机

Knoop 克诺普〈旧译努普〉
- ~ hardness 克诺普硬度
- ~ hardness number [KHN] 克诺普硬度值
- ~ microhardness test 克诺普显微硬度试验
- ~ scale 克诺普（硬度）标度

knot （绳）结
- ~ strength 结节强度,打结强度
- ~ strength ratio 结节强度比
- ~ tenacity 结节强度,打结强度

knotted net 打结网

know – how （技术）决窍,专门技术〈非专利或未申请专利的〉

knuckle thread [round thread] 圆螺纹

knurl 滚花
- ~ nut 滚花螺母

knurled – head screw 滚花头螺钉

knurling wheel 滚花轮

Kohinoor test 柯希诺尔试验〈划痕硬度〉

Ko – kneader Ko型捏合机,蜗杆捏合机〈带有动蜗杆和固定捏和齿的捏和机〉

koka flow tester 高化型流动试验仪

Kolloplex mill 柯洛普勒克斯磨

kopal 化石树脂

KO pin [knock – out pin] 顶出杆

Kovar alloys 柯发合金〈铁、镍、钴合金，用于高真空工程〉
Kr [Krypton] 氪
Krafft point 克拉夫特点，克拉夫特温度
kraft paper 牛皮纸
Kress expander 克莱斯扩幅机
Kritchever process 克利切凡尔法〈气体火焰处理聚乙烯表面便于印刷〉
krypton [Kr] 氪
kV [kilovolt] 千伏(特)
KV/cm [kilovolts centimetre] 千伏/厘米〈介电强度〉
kW [kilowatt] 千瓦，配
kW/h. [kilowatt - hour(s)] 千瓦/小时

L

l [levorotatory] 左旋
l;L [liter] 升
L - type L型
　~ calender L型压延机
　~ injection moulding machine L型注塑成型机
lab. [laboratoy] 实验室，试验室
lab - examined 试验研究的
label 标签;贴商标;示踪
　~ adhesive 标签黏合剂
　~ gummer 标签涂胶机
　~ panel 贴商标凹面〈包装〉
labeling equipment 贴标签装置
labeller 贴标签机
laboratory 实验室，试验室
　~ appliance 实验室仪器
　~ ball cock 实验室用球阀
　~ extruder 试验用挤塑机
　~ mill [cutting mill] 实验室用磨
　~ report 实验报告
　~ scale 试验室规模
　~ [lab]size 试验室规模
　~ test 实验室试验
labour capacity 劳动生产率
lac 紫胶,虫胶,虫漆;高强力细绳
　~ resin 虫胶树脂
　~ smearing machine 浸渍机
lace 带子, 花边;编带;饰带
lack 不足, 缺乏
　~ of adhesion 欠黏
　~ of fill - out 未填满〈增强塑料〉
　~ of fusion 欠焊〈焊接缺陷〉
lacquer (大)漆;挥发性漆;硝基(纤维素)漆,喷漆,涂漆
　~ base 底漆
　~ coating 漆涂(层)
　~ dope 硝棉清漆
　~ enamel 挥发性磁漆;硝基磁漆
　~ sealer 漆封剂
lactams 内酰胺
lactace 乳酸;乳酸盐[酯]
lactic acid 乳酸
lactide 交酯;丙交酯
lactim 内酰亚胺
lactone 内酯
lacto nitrile 乳腈;2 - 羟基丙腈;丙醇腈
ladder 梯型
　~ chain 梯型键
　~ fibre 梯型纤维
　~ polymer 梯型聚合物
lag 迟延;滞后;加保温套
　~ - up mo(u)lding 手糊成型
lagged 外罩的〈机器〉;护热的
lagging material 保温材料
lake 花斑〈表面〉;色淀〈遮盖性颜料〉,沉淀色料;缩坑〈制品缺陷〉;湖
　~ pigment 色淀颜料
lamella 薄片;薄板;薄层;片晶
lamellar 片状的,层状的
　~ buff 片状抛光轮

lamina

- ~ crystal　片晶,片状晶体
- ~ roll　层合辊,叠层辊
- ~ structures　片晶结构
- ~ thickness　片晶厚度

lamina [**laminae**]　薄层,薄片,薄板

laminal filler　片状填料

laminar　片状的,层状的
- ~ barrier container　层状阻隔型容器
- ~ bond　片层粘接
- ~ composite　片层复合材料
- ~ film　层状薄膜
- ~ flow　层流
- ~ flow adaptor　层流式模头接套
- ~ flow coextrusion　层流式共挤塑
- ~ mixing　层流混合
- ~ solid lubricant　层状固体润滑剂

laminate　层压,层合,叠层;层状的;层压塑料,层压制品;层压板
- ~ base　层合基材
- ~ composite　层合复合材料
- ~ layer　层压层
- ~ mo(u)lding　层压模塑,层压模塑法;层压制品
- ~ ply　层压层
- ~ substrate　层合基材

laminated　层压的,层合的
- ~ angle section　层合角型材
- ~ bearing　叠层(滑动)轴承
- ~ board　层压板,复合板
- ~ channel profile　层合槽型材
- ~ channel section　层合槽型材;U形层合型材
- ~ cloth　层合布板(用树脂浸渍织物制备的)
- ~ composite　层压(复合)材料
- ~ compressed wood　层压木材
- ~ fabric　叠层织物,层合布板
- ~ fabric bearing　层压布板制的轴承
- ~ fabric gear　层压布板制的齿轮
- ~ film　层合薄膜;复合薄膜
- ~ flooring　层合地板
- ~ foil　层合薄片,复合膜
- ~ gear　层压材料制的齿轮
- ~ mat　层合毡垫
- ~ material　层压材料,层合材料
- ~ mo(u)lded rood　层压模制棒材〈热固性塑料〉
- ~ mo(u)lded section　层合型材
- ~ mo(u)lded tube　层压模制管材〈热固性塑料〉
- ~ mo(u)lding　层压成型;层压模制品,层压材料
- ~ mo(u)lding die　层压模塑模具
- ~ mo(u)lding tube　层压模塑管
- ~ panel　层合条板
- ~ paper　层压纸板,层合纸板,复合纸板,硬纸板
- ~ phenolics　层合酚醛塑料
- ~ piece　层合制品
- ~ plastic　层合塑料
- ~ plastic material　层合塑料材料
- ~ plastics panel　(按尺寸切好的)层合塑料板
- ~ plate　层合平板〈层合结构的平板〉
- ~ plate theory　层合板理论
- ~ preform　层合型坯
- ~ product　层合模制品
- ~ profile　层合型材
- ~ rod　层合棒材
- ~ rolled tube　辊压层压管材,卷层管材〈热固性塑料〉
- ~ safety glass　层合安全玻璃,多层安全玻璃
- ~ section　层合型材
- ~ sheet　层压片材〈热固性塑料〉
- ~ shell　层合壳〈增强塑料〉
- ~ skin　层合表皮
- ~ structure　层合结构
- ~ synthetic resin bonded sheet　合成树脂黏合层压板
- ~ timber　层压木材,层压板
- ~ T-section　层合T形材

- ~ tube 层压管材
- ~ veneer 层压木材,层压胶合板
- ~ wood 层合木板,层压板
- ~ wood for exterior use 室外用层合木板

laminates 层压制品
- ~ for printed circuits 印刷电路用的层压制品

laminating 层压,层合;复合(层)
- ~ adhesive 层合黏合剂
- ~ agent 层合(粘结)剂
- ~ diagram 层压图
- ~ embossing 层合压花法
- ~ embossor 层合压花机〈层合与压花同时进行〉
- ~ film 层压薄膜,复合薄膜
- ~ material 层压材料
- ~ nip 层合间隙
- ~ press 层压机
- ~ resin 层合用树脂
- ~ rolls [squeezing tolls] 层合辊
- ~ sheet 层压片材
- ~ strength 层压强度
- ~ technique 层合技术
- ~ technology 层合技术;层合工艺
- ~ temperature 层合温度
- ~ wax 层合用蜡
- ~ web 层合基料

lamination 薄片,薄层;层;层合,层压,复合;层压结构
- ~ coating 贴合涂布,挤出贴合〈挤出物挤到基材上〉

laminator 层合机,复合机
- ~ roll 层合机辊

lamp 灯
- ~ basing cement 灯口胶(泥);灯泡封帽胶(泥)
- ~ black 灯黑〈一种炭黑〉;灯烟色
- ~ capping cement 灯泡封帽胶(泥)
- ~ shade 灯罩
- ~ socket 灯插座

lanara 酪素纤维

Lancester 兰塞斯特
- ~ muller type mixer 兰塞斯特研磨机型混合机,逆流圆盘混合机
- ~ plow type mixer 兰塞斯特桨叶式混合机,逆流桨式圆盘式混合机

land 合模面〈压塑模具和注塑模具〉;口模成型面;成型段〈挤塑机模头的〉;溢料面〈压塑模〉;齿顶端面〈螺杆〉
- ~ area 合模面面积;溢料面积;螺顶面积;成型面
- ~ face 合模面
- ~ length 成型段长度
- ~ mark 模面痕
- ~ pressure 旁压;两侧压力〈挤塑机螺杆〉
- ~ ratio 成型段比
- ~ surface 成型段表面
- ~ width 螺纹顶宽〈螺杆〉;合模面宽;溢料面宽〈压塑模〉;截坯宽度〈吹塑模〉

landed 突缘的
- ~ force 突缘模塞;有肩阳模〈压塑模〉
- ~ mo(u)ld 半溢料式模具;带台肩式模具
- ~ plunger 突缘模塞;有肩模塞
- ~ plunger mo(u)ld 半防溢式模具;带台肩模塞式模具
- ~ positive mo(u)ld 半防溢式模具;有肩全压式模具
- ~ scarf joint 钝头斜连接

lanthanium [La] 镧

lap 搭头,搭接
- ~ fillet weld 搭接角焊
- ~ joint 搭接;搭接接头;叠接;搭接缝
- ~ -joint strength 搭接强度,搭接头强度
- ~ shear joint 搭接剪切连接
- ~ shear strength 搭接剪切强度〈粘接〉

~ shear tensile 搭接剪切拉伸
~ – solvent sealing 搭接溶剂封合,化学搭接焊〈用溶剂或溶剂混合物〉
~ tab gate 搭接柄型浇口
~ time 搭接(起始)时间〈黏合剂〉;精研时间〈机械〉
~ weld 搭接焊接
~ – welded 搭接焊接
~ welding 搭接焊接
~ winding 重叠缠绕,叠绕
~ winding machine 重叠缠绕机
~ wrapped 重叠缠绕的
lapping 搭接,重叠;研磨,抛光
~ abrasive 研磨剂
~ time 搭接(起始)时间〈黏合剂〉;精研时间〈机械〉
larch wood 落叶松木材
large (巨)大的,大规模的
~ – area mo(u)lding 大型模塑制品,大尺寸模塑制品,大面积模塑制品
~ bore hose 大口径(软)管
~ diameter pipe 大口径管
~ output 大量生产
~ outside diameter pipe 大外径管
~ part 大规格模制品
~ particles 粗微粒
~ pitch 大螺距〈螺杆〉
~ scale 大规模的,工业规模的
~ scale experiment 大规模试验
~ scale field test 大规模野外试验
~ scale production 大量生产
~ scale test 大规模试验
~ – slot nozzle 宽缝式嘴
~ span 大跨度〈建筑学〉
Larson – Miller method 拉森-米勒法〈用高温,短时间蠕变断裂半理论法推断室温,长时间蠕变断裂状况〉
laser 激光;激光器
~ beam diffraction 激光束衍射
~ beam welding 激光束焊接
~ cutter 激光切割机〈切断塑料〉

~ cutting 激光切割
~ cutting unit 激光切割装置
~ fibre 激光纤维
~ flash photolysis 激光瞬间分解〈测定塑料结构〉
~ optoacoustic spectrum 激光光声光谱
~ radiation 激光辐射
~ Raman spectroscopy 激光拉曼光谱
~ sorter 激光探伤仪,激光缺陷检验仪
~ welder 激光焊接机
~ welding 激光焊接
~ welding equipment [system, unit] 激光焊接装置
last 最后(的);持续;耐久,楦〈鞋〉
~ runnings 后馏分〈蒸馏〉
lastic 弹性材料
lasting quality 耐久性
latch 固定销,插销,锁模钩;链环
~ plate 活模芯托板;嵌件固定板
latent 潜伏的
~ catalyst 潜伏催化剂
~ curing agent 潜伏性固化剂〈与树脂混合后,室温下有较长贮存期,一经加热就能使树脂固化的物质〉
~ curing resin 潜伏性固化树脂
~ heat 潜热
~ heat of solidification 固化潜热
~ solvent 潜溶剂;惰性溶剂
~ strain 潜应变
~ stress 潜应力
~ torque 潜在转矩
lateral 横向(的),侧向(的),侧面(的)
~ bending 横向弯曲
~ blow 侧吹
~ blowing 侧吹〈吹塑〉
~ buckling 横向翘曲
~ chain 侧链
~ contraction 横向收缩
~ deflection 横向挠曲
~ deformation 横向形变,侧向形变

~ face 侧面
~ flexibility 横向挠性
~ flexure 侧向挠曲
~ force 横向力
~ gating 侧向浇口〈注塑〉
~ injection 侧向注塑
~ movement 侧向运动
~ plan 侧视图
~ section 横断面
~ spread 侧面扩展,宽展
~ view 侧视图
~ waste 侧面废料〈泡沫塑料〉
laterally injected part 侧向注塑件
latex 乳胶;胶乳
~ adhesive 胶乳黏合剂
~ - albumen adhesive 胶乳-蛋白黏合剂
~ backed carpet 胶乳涂底地毯
~ bonded non-woven 胶乳黏合非纺织物
~ - casein adhesive 胶乳-酪蛋白黏合剂
~ film 胶乳膜
~ foam 泡沫胶乳
~ imterpenetrating polymer network [LIPN] 胶乳互穿聚合物网络
lath 板条
lathe 车床,旋床
~ type winding machine [traversing carriage winding machine] 车床型缠绕机
latices 胶乳类〈复〉
lattice 点阵;晶格
~ array [configuration] 晶格排列,晶格规整[构型]
~ constant 点阵常数;晶格常数
~ defect 点阵缺陷;晶格缺陷
~ fabric 格状织物
~ parameter 点阵参数
~ pattern 格子花纹
~ spacing 点阵间距;晶格间距
~ strain 晶格应变

~ structure 点阵结构;晶格结构
latticed plastics 格状塑料
Laue 劳厄
~ method 劳厄法
~ pattem 劳厄图〈单晶 X 射线衍射图像〉
launder 洗涤;洗涤槽;流水槽
laundry wheel 洗衣滚筒
3 - lauramidopropyl trimethyl ammonium methyl - sulfate 甲基硫酸(3-月桂酰胺丙基)三甲基铵〈抗静电剂 LS〉
laurate 月桂酸;月桂酸盐[酯]
lauroyl peroxide 过氧化月桂酰,过氧化十二烷酰〈交联剂,引发剂〉
lauryl stearyl thiodipropionate 硫代二丙酸月桂(基)十八(烷基)酯〈抗氧剂〉
law 定律
~ of radiation 辐射定律
~ of refraction 折射定律
Lawaczeck's viscometer 拉瓦萨克黏度计
lay 绕数;绕向(玻璃纤维);搁置;铺设;压平;层(次)
~ - decorating 敷饰
~ - down paper 覆盖层纸
~ - flat 平折;压平
~ - flat adhesive 平折黏合剂
~ - flat device 平折装置,压平装置〈管膜〉
~ - flat film 平折(吹塑)薄膜
~ - flat (tabular) film 平折(吹塑)薄膜
~ - flat tube 平折(吹塑)膜管
~ - flat tube die 平折膜管模头
~ - flat tubing 平折(吹塑)膜管
~ - flat tubing extrusion 平折膜管挤塑法
~ - flat width 折径,平折宽度〈吹塑薄膜〉
~ - out 布置;铺设;安装;涂敷
~ - up 叠铺;叠合坯料〈用于增强塑

layer

料〉;结构〈层压〉
- ~ – up mo(u)lding 叠铺成型〈增强塑料〉
- ~ up deformation 铺敷变形

layer 层(次);焊层;片坯料〈层压塑料〉;铺设
- ~ composite structure 层状复合结构
- ~ growth 层次增长
- ~ lattice lubricant 层状固体润滑剂
- ~ lay – up design 铺层设计〈增强塑料〉
- ~ mo(u)lding 多模层压塑,多层模塑
- ~ of adhesive in a joint 粘接胶合层
- ~ – on – layer foaming 层–层发泡
- ~ reinforcement 层状增强;层状增强材料
- ~ structure 层状结构
- ~ thickness 层状厚度

layering 层次;铺设

laying equipment 压平收卷装置〈吹塑管膜〉

layout table 成型台〈高频焊接〉;装配台

lb. [pound(s)] 磅〈0.453 公斤〉

lb. per sq. inch [psi – pounds pet square inch] 磅/英寸²

I. c. [loc. cit.] 在上述引文中

LC – machine [linear – contact high – frequency sealing machine] LC 机,线性接触高频焊接机

LCST [lower critical solution temperature] 较低临界溶解温度

LD [lethal dose] 致死剂量

LD50 [median lethal dose] 半致死剂量,致死中量〈50% 个体致死的剂量〉

L/D ratio [length/diameter ratio] 长径比〈螺杆〉

LDPE [low density polyethylene] 低密度聚乙烯

leach 沥滤;浸滤,浸提,浸沥

leaching 沥浸
- ~ tank 沥浸槽〈硬化纸板〉

lead 螺距;导程,导线;铅〈Pb〉
- ~ angle 螺纹导角
- ~ bell wire 电铃线
- ~ carbonate 碳酸铅
- ~ chlorophthalosilicate 氯代酞硅酸铅〈热稳定剂〉
- ~ chlorosilicate complex 氯硅酸铅络合物〈热稳定剂〉
- ~ chromate 铬酸铅
- ~ chrome pigment 铅铬颜料
- ~ cyanamide 氰氨化铅
- ~ 2 – ethyl hexoate 2 – 乙基己酸铅〈热稳定剂〉
- ~ error 螺距误差
- ~ monoxide 一氧化铅〈硫化活性剂和防焦剂〉
- ~ mould 铅模
- ~ of screw 螺旋导程
- ~ of screw flight 螺纹间距,螺距;螺杆螺纹导程
- ~ roll 导辊
- ~ salicylate 水杨酸铅〈光稳定剂,螯合剂〉
- ~ seal 铅封
- ~ silicate 硅酸铅〈热稳定剂〉
- ~ screw 导(螺)杆,推动螺杆
- ~ stabilizers 铅稳定剂
- ~ stearate 硬脂酸铅〈热稳定剂,润滑剂〉
- ~ sulfate 硫酸铅
- ~ time 设计拟定和实施时间
- ~ titanate 钛酸铅

leader 引导;导杆[轨,管]
- ~ pin 导柱,合模销;定位销
- ~ pin bushing 合模销套;定位销套
- ~ strip 导向板,导向器

leading 引导的,导向的
- ~ dimensions 轮廓尺寸,基本尺寸
- ~ edge 前缘;前沿
- ~ edge of flight 螺纹前缘

leaf for hot stamping 彩箔烫印

leafing 飘浮;浮层
leaflet 叶片;说明书
leak 泄漏
~ - blocking agent 堵漏剂
~ flow 漏流
~ free 不漏的;密封的
~ - proof 防漏的;密封的
proof closure 密封,防漏〈阀〉
- proof sealing 密封,防漏〈阀〉
proof under pressure 耐压的;紧密的
proof under vacuum 真空密封的
leakage 泄漏;漏失量
~ current 漏电流
~ factor 损失因数,损失系数
~ field 漏泄场;杂散场
~ flow 漏(泄)流〈挤塑〉
~ flow constant 漏流常数〈挤塑机〉
~ in the intensity of electric field 电场强度泄漏〈高频焊接时〉
~ resistance 漏泄电阻
~ test 泄漏试验;紧密性试验
leakless 不漏的
leather 皮革
~ adhesive 皮革黏合剂
~ bonding adhesive 皮革黏合剂
~ cement 皮革粘接剂
~ cloth 人造革;漆布
~ colo(u)r 皮革色
~ dressing 皮革整修
~ dye 皮革染料
~ finishing enamel 皮革涂饰色料
~ for bag making 制袋用革
~ for harnesses 马具用革
~ for saddle making 制马鞍用革
~ for upholstery 室内装饰用革
~ glue 皮革胶黏剂
~ paper 鞣革纸〈由合成树脂涂布或浸渍高强度纸制得〉
~ permeability value 皮革透湿值
~ substitute 人造革,假革
leatherette 人造革

leatherlike 皮革类的,类皮革的
~ material 皮革类材料
~ plastics sheeting 类皮革塑料薄片
leatheroid 纸皮,薄钢纸;人造革
left 左方的;向左
~ - handed machine 左转机器
~ - hand thread[LH] 左旋螺纹
leg gate 爪浇口
legend 图例;图表符号
lemon yellows 柠檬黄〈颜料〉
length 长度
~/diameter ratio [L/D ratio] 长径比
~ dimension 长度尺寸
~ distribution （纤维)长度分布
~ distribution graph （纤维)长度分布图
~ of extruder die 挤出机模头长度
~ of flat section 平坦部分长度〈试样〉
~ of heating - cooling channel 热 - 冷槽长度,恒温槽长度
~ of helical screw channel 螺旋槽长度
~ of life 使用寿命
~ of life test 使用寿命试验
~ of screw 螺杆长度
~ of tear 撕裂长度
~ of the lap joint 搭接长度
~ of travel 行程
~ of weld 焊缝长度
~ overall 总长(度),全长(度)
~ per chain link 每链环长度
~ - to - diameter ratio[L/D ratio] 长径比〈螺杆长度与直径比〉
lengthwise 纵向
~ cutting 纵向切割
~ direction 纵向
~ fold 纵向折叠
~ streaking 纵向条痕〈疵点〉
~ stripe pattern 纵条花纹
~ tear 纵向撕裂
lengthy test 长期试验,连续试验
Leno weave 列那型织物〈玻璃纤维织

物〉
lens 透镜
~ focus 透镜焦距〈激光焊接〉
Lenzing cutting bar 莱恩捷刀杆
Lessing ring 莱辛环〈蒸馏〉
let-go 脱胶;缺胶〈层合〉
let-off 基材输送装置;放卷,退卷
~ reel 退卷盘[筒,辊]
~ roll 放卷辊
~ unit 退卷装置
let-up 起层;成层
lethal dose[LD] 致死剂量
letterpress 活版印刷
~ printing 平板印刷
leuco vat dyes 还原染料隐色体
level 水平(面);等级,范围,程度;均匀的
~ controller 液面调节器
~ cut carpet 平绒地毯
~ indicator 液面指示器
~ line 水平线
~ of follow-up pressure 保压,保压力;续压压力〈注塑〉
~ of stress 应力分布
~ of stretch 拉伸程度
~ shade 均匀色泽
~ wind 均缠绕
~ winding [90 percent winding] 均缠绕
~ winding device 均缠装置
leveller 整平器,调平器
levelling 平整;均涂;匀饰性,流平性
~ agent 匀涂剂〈涂料〉;匀染剂〈染料〉
~ colo(u)r 匀染染料
~ dryer 平幅干燥器
~ machine 平整机
~ material 平整材料
~ property 匀涂性;平整性
~ roll 匀涂辊,整平辊
~ screw 校平螺钉
lever 杠杆;手把

levo [levorotatory]左旋;左旋的
Lexan 〈商〉莱克桑〈聚碳酸酯〉
LH [left-hand thread] 左旋螺纹
Li [lithium] 锂
lid 盖〈包装〉
life 寿命,使用期限,贮存期;周期产额〈模塑〉
~ belt 安全带,保险带
~ durability 使用寿命;耐用期
~ duration 使用寿命,耐用期
~ length 耐久性,使用期
~ period 寿命,存在期限
~ simulation test 使用寿命模拟试验
~ span 使用寿命,有效期限
~ test 使用寿命试验,耐久性试验
~ time 使用寿命,使用期限
lifetime 寿命,生存期;使用寿命:活性期
lift 冲压量,模压量,周期产额〈在模塑中的〉;提升
~ per hour 每小时(模塑)量
~ -type angle check valve 提升型角止回阀
~ -type check valve 提升型止回阀
lifting 提升(的);翘离〈层压制品〉;脱落〈涂膜〉
~ device 提升装置
~ gear 升降齿轮
~ piler 提升堆垛机
light 光,光线;光泽;轻的;浅色的;轻质的
~ absorbing 吸收光线的
~ absorption 光吸收
~ admitting walls 透光墙
~ ageing 光(致)老化
~ ageing test 光老化试验
~ ashes pl 烟(道)灰,飞扬灰尘
~ beam 光束
~ beam pyrolysis 光束致热解
~ -body 低黏度的;低稠(度)的〈涂料〉
~ coating 薄涂层

~ colo(u)r 浅色
~ coloured 浅色的,淡色的
~ degradable 光可降解的
~ degradation resistance 耐光降解性
~ - density construction 轻质结构〈美〉
~ - density material 轻质材料,轻质结构材料
~ - density structure 轻质结构件
~ exposure 曝光;照射
~ fabric 轻薄织物
~ fading test 光褪色试验
~ fast 耐晒的
~ fastness 耐光牢度〈颜色的〉耐晒性;耐光性
~ fastness rating 耐光牢度等级
~ fastness stadard 耐光牢度标准
~ filler 轻质填料
~ fittings 照明装备
~ focusing plastic fibre 光聚焦塑料纤维
~ fugitive 不耐光的
~ ga(u)ge sheet 薄(规格)板
~ - heat ageing 光热老化
~ industry 轻工业
~ - initiated polymerization 光引发聚合
~ intensity 光度,发光强度,光强
~ permanency 耐晒性;耐光性
~ permeability 透光性
~ polymerization 光聚合(作用)
~ proof 耐光的
~ radiator 光辐射器(用光辐射焊接)
~ resistance 耐光性,耐光度
~ resistant 耐光的
~ - resisting 耐光的
~ running 轻负荷运行
~ scattering 光散射
~ scattering coefficient 光散射系数
~ scattering photometer 光散射光度计
~ section 轻型型材

~ sensible 易感光的;光敏性的
~ sensitive surface 感光面
~ sheet 薄板材
~ sized 小尺寸的
~ stability 光稳定性,耐光性
~ stabilizer 光稳定剂,耐光剂
~ stabilizing agent 光稳定剂
~ stable 耐光照的,不感光的
~ stable unsaturated polyester resin 光稳定不饱和聚酯树脂
~ tight 不透光的
~ transmission [luminous transmittance] 光透射
~ transmittance 透光率
~ transmitting fibres 光导纤维
~ transmitting walls 透光墙
~ velocity 光速
~ wave 光波
~ weight 轻质的
~ weight building material 轻质建筑材料
~ - weight casting resih 轻质铸塑树脂
~ - weight concrete 轻质混凝土
~ - weight concrete panel 轻质混凝土板
~ weight construction 轻质结构
~ weight fabric 轻质织物
~ - weight foam concrete 轻质泡沫混凝土
~ - weight foam plaster 轻质泡沫石膏
~ weight metal 轻质金属
~ weight section 轻型型材
lightening 增白;照亮;减缓,减轻
lighter 更轻的;较浅色的
~ colour 浅色
~ shade 浅色调〈模塑件或涂层〉
lightfast 耐光的,不褪色的
lightly plasticized 稍微增塑的
lightmeter 光度计
lightness 光亮度

lightproof 不透光的,遮光的
lignin 木质素
- ~ plastic 木质素塑料
- ~ polyether 木质素聚醚
- ~ resin 木质素树脂

lignite wax 地蜡〈润滑〉
ligroin 里格罗英,石油英〈沸点约120~240℃的粗汽油馏分〉
lilac 淡雪青,丁香紫〈淡紫色〉
LIM [liquid impingement mo(u)lding] 液体浸渍模塑
LIM [liquid injection mo(u)lding] 液态料注塑
lime 石灰,氧化钙
limit 极限,界限;限度,限制;范围
- ~ breaking velocity 极限断裂速度
- ~ design 极限设计
- ~ load 极限载荷
- ~ moment 极限转矩
- ~ of accuracy 精度极限
- ~ of adhesion 附着力极限
- ~ of capacity 功率极限;功率范围
- ~ of compression 压缩限度
- ~ of deformation 形变极限
- ~ of detection 检测极限
- ~ of detection of impurities 杂质检测极限
- ~ of elasticity 弹性极限
- ~ of fatigue 疲劳极限
- ~ of flammability 可燃性极限〈气体〉
- ~ of inaccuracy 误差限度
- ~ of inflammability 可燃性极限
- ~ of micro-cracking 微龟裂极限,微裂纹极限
- ~ of mo(u)ld locking force 极限锁模力
- ~ of proportionality 比例极限
- ~ of stability 稳定性极限
- ~ of tolerance 容许极限
- ~ of wear 磨损极限
- ~ point 极限点

- ~ size 极限尺寸
- ~ speed 极限速率[度]
- ~ strength 极限强度
- ~ stress 极限应力
- ~ (stop) switch 限位(止动)开关,终端(止动)开关
- ~ value 极限值

limitation 极限,限度;局限性,缺陷[点]
limited 有限的,限定的
- ~ apparent flow 有限表观流动
- ~ deformation 有限形变
- ~ plastic flow 有限塑性流动
- ~ production 限量生产
- ~ swelling 有限溶胀

limiting 极限的;限制的,限定的,界限的
- ~ adhesive value 极限黏合值
- ~ compliance 极限柔量
- ~ concentration 极限浓度
- ~ condition 极限状态
- ~ curve 极限曲线
- ~ density 极限密度
- ~ diffusion coefficient 极限扩散系数
- ~ dimensions 限制尺寸
- ~ error 极限误差
- ~ factor 限制因素
- ~ fatigue stress 极限疲劳应力
- ~ friction 极限摩擦
- ~ intensity 极限强度
- ~ line 界线
- ~ load 极限负重
- ~ modulus 极限模量
- ~ oxygen index [LOI] 极限氧指数
- ~ point 极限点,限制点
- ~ pressure 极限压力
- ~ shearing stress 极限剪切应力
- ~ state 极限状态
- ~ stress 极限应力
- ~ surface 界面
- ~ temperature 极限温度
- ~ value 极限值
- ~ value of compressive strength 压缩强

度极限
- ~ viscosity 特性黏度
- ~ viscosity number 特性黏数,极限黏数
- ~ yield value 极限屈服值

line 衬(里);线(路),管路;生产线
- ~ belt 传送带
- ~ blender 管道掺混[混合,混配]器
- ~ capacity 流水线(生产)能力,生产线能力
- ~ fluid [transported fluid] 输送液体
- ~ linking element 传送带连接件
- ~ of centers 中心线
- ~ of deflection 挠曲线
- ~ of production 生产线,生产流程
- ~ of welding 熔接纹
- ~ pipe 输送管,总管
- ~ pressure 管路压力
- ~ production 流水线生产,流水作业
- ~ rate 生产线速率[度]
- ~ size 管道尺寸
- ~ speed 线速率[度]
- ~ transformer 线路变压器
- ~ -type horn [sonotrode] 线型超声波焊接工具〈用于细焊缝〉
- ~ welding 线焊接

linear (直)线型的
- ~ absorption coefficient 线吸收系数

C₇~C₉~ adipate 己二酸C₇~C₉直链烃基酯〈增塑剂〉
- ~ amorphous polymer 线型无定形聚合物
- ~ attenuation coefficient [linear extinction coefficient] 线性衰减系数〈光学〉
- ~ burning rate 直线燃烧速率
- ~ chain 直链,线型链
- ~ chain polymer 线型聚合物
- ~ compound 直链化合物
- ~ condensation 线型缩合
- ~ condensation polymerization 线型缩聚(作用)
- ~ -contact high-frequency sealing machine [LC-machine] 线性接触高频焊接机,LC(焊)机
- ~ copolymer 线型共聚物
- ~ deformation 线性形变
- ~ density 线密度
- ~ dilatometer 线(膨)胀测试仪
- ~ dilatometry 线(膨)胀测量法
- ~ dimensions 线性尺寸
- ~ expansion 线膨胀
- ~ expansion coefficient 线(膨)胀系数
- ~ expansibility 线膨胀性
- ~ expansivity 线膨胀性
- ~ extension 线性延长
- ~ extinction coefficient [linear attenuation coefficient] 线性衰减系数〈光学〉
- ~ extrusion speed 挤塑线速度
- ~ flow 直线流动,层流
- ~ gradient 线性梯度
- ~ high polymer 线型高聚物
- ~ hydrocarbon polymer 线型烃类聚合物
- ~ linkage 线型链〈苯核〉
- ~ low density polyethylene [LLDPE] 线型低密度聚乙烯
- ~ macromolecule 线型大分子,直链大分子
- ~ molecular chain 线型分子链
- ~ molecule 线型分子,直链分子
- ~ oligomer 线型低聚物
- ~ orientation 线型取向
- ~ phenomena 线型现象
- ~ polyester plasticizer 线型聚酯增塑剂
- ~ polyethylene 线型聚乙烯
- ~ polymer 线型聚合物
- ~ rate of flow 流动线速度
- ~ reactant 线型反应物
- ~ relationship 线性关系

linen

- ~ ring - chain polymer 线型环链聚合物
- ~ shrinkage 线型收缩率
- ~ speed 线性速率[度]
- ~ stretching 线性伸长
- ~ structure 线型结构
- ~ superpolymer 线型超高相对分子质量聚合物
- ~ superposition 线型叠加
- ~ surface velocity 表面线速度
- ~ traverse per rpm 每分钟转数的线型导丝动程〈长丝缠绕〉
- ~ unsaturated polyester 线型不饱和聚酯
- ~ viscoelasticity 线型黏弹性
- ~ winding 线性缠绕
- ~ zigzag - shaped polymer 线型锯齿形聚合物

linen 亚麻布;亚麻线
- ~ chips pl 亚麻碎片〈填料〉

liner 衬里,内衬;衬垫,衬套〈管或容器〉
- ~ bag 衬袋〈包装〉
- ~ keep plate 衬里夹板
- ~ of cylinder 料筒套筒
- ~ plate 衬板
- ~ tube 衬管

linerboard 天花板纸板,顶棚纸板

lining 衬里;衬砌;涂敷
- ~ board 衬板
- ~ compound 衬垫料
- ~ fabric 衬里织物
- ~ grade 衬里质量
- ~ material 衬里材料
- ~ of a pipe 管衬里
- ~ plate 衬板,饰面板

link 连接;连肘;链环,链节;键;键合
- ~ drive 连杆传动
- ~ lever 连杆
- ~ line 连络线
- ~ mechanism 连环机构,联杆机构
- ~ pin 连杆销,铰链销
- ~ system 连肘装置

linkage 连接;拉杆
- ~ assembly （模具）拉杆总成
- ~ system 联动装置

linked 连接(的)

linking 连接

linoleate 亚油酸;亚油酸盐[酯]

linoleic acid 亚油酸

linolenic acid 亚麻酸

linoleum 亚麻油,毡,油地毡,漆布
- ~ finish 漆布漆

linseed oil 亚麻籽油

linters 棉短绒,棉子绒〈填料〉

lip 模唇〈挤塑机嘴唇〉;边,缘
- ~ formation 溢边
- ~ heater 模唇加热器
- ~ mask 唇型罩
- ~ sealing 唇部密封

LIPN [latex interpenetrating polymer network] 胶乳互穿聚合物网络

liquefied 液化的
- ~ gas propellant 液化气燃料
- ~ natural gas [LNG] 液化天然气

liquefy 液化

liquefying point 液化点

liquid 液体;液态的
- ~ adhesive 液态黏合剂
- ~ bath orientation 液浴内取向
- ~ chromatogram 液相色谱(图)
- ~ chromatography 液相色谱(法)
- ~ chromatography analysis 液相色谱分析
- ~ colo(u)r 液态色料
- ~ container 液体容器
- ~ cooling 液体冷却;水冷
- ~ crystal 液晶
- ~ crystalline plastics 液晶塑料
- ~ dip coating 液态浸涂
- ~ epoxy injection mo(u)lding 液体环氧注塑
- ~ fed extruder 液体喂料挤塑机;熔料

挤塑机
~ friction 液体摩擦
~ honing 液体珩磨;液体搪磨〈金属表面加工方法〉
impingement mo(u)lding [LIM] 液体浸渍模塑
~ injection mo(u)lding [LIM] 液态料注塑
~ jet （混合）液体喷嘴
~ level 液面
~ membrane separation 液膜分离
~ organic barium – cadmium complex 液体有机钡镉络合物〈热稳定剂〉
~ nitrogen 液(态)氮
~ organic barium – cadmium – zinc complex 液体有机钡镉锌络合物〈热稳定剂〉
~ organic barium – zinc complex 液体有机钡锌络合物〈热稳定剂〉
~ organic calcium – zinc complex 液体有机钙锌络合物〈热稳定剂〉
~ paraffin 液体石蜡,流动石蜡,白〈润滑剂〉
~ phase 液相
~ polybutadine 液体聚丁二烯
~ polymerization 液相聚合(法)
~ prepolymer 液态预聚物
~ reaction moulding [LRM] 液体反应模塑
~ rigidity 液体刚性
~ resin 液态树脂
~ resin – injection processing 液态树脂注塑法
~ resistance 液体阻力
~ seal 液封
~ state 液态
~ thermostant 液体恒温器
~ tight 不透液体的
~ trap 液体捕集器
~ viscosimeter 液体黏度计
~ volume viscosity 液体体积黏度

~ waste 废液
~ waste disposal 废液处理
liquified gas propellant 液化气燃料
liquor 液体,(水)溶液
lit [little] 小的
liter. [literature] 文献
litharge 铅黄,黄丹,一氧化铅,密陀僧
lithium 锂
~ stearate 硬脂酸锂〈热稳定剂〉
Lithopone 〈商〉锌钡白,立德粉
litmus indicator 石蕊指示剂,石蕊试剂
litre [liter; l, L] 升(= dm^3)
little bubbles 小气泡〈在模塑件中〉
live 活的,有效的,新鲜的
~ catelyst 新鲜催化剂
~ centre 活性中心
~ gas 新鲜气体
~ lever 活动杠杆
~ steam 直接蒸汽,新鲜蒸汽〈用于高压釜的加压和加热〉
living 活的
~ polymer [telechelic polymers] 活性聚合物
~ polymerization 活性聚合(作用)
lixiviation 浸滤,浸析,浸取,浸出
LLDPE [linear low density polyethylene] 线型低密度聚乙烯
Lloyd's rules of shipping 劳埃德船级协会规则〈有关于增强塑料鱼船的结构〉
LMPE [low molecular weight polyethylene] 低相对分子质量聚乙烯
LNG [liquetied natural gas] 液化天然气
load 装模料,装料,加料〈压模〉;加入量;载荷;负载
~ at failure 断裂载荷,破裂载荷,破坏负荷
~ bearing 承载
~ – bearing capacity 承载[载重,负荷]能力
~ – bearing partition 承载隔板

- ~ bearing strength 承载强度,负载[载重]能力;载荷量
- ~ capacity 载荷量;装料[加料]量
- ~ cell 负荷[载荷]传感器
- ~ cell type transducer 负荷传感器型换能器
- ~ characteristic 负载特性
- ~ circuit 负载电路;加热电路〈塑料加工机〉
- ~ concentration 负荷集中
- ~ condition 载荷状态
- ~ curve 载荷曲线
- ~ cycles 负荷循环;负荷变换
- ~ – deflection curve 负荷变形曲线,负荷挠曲曲线
- ~ deflexion 载荷挠度
- ~ – deformation curve 负荷变形曲线
- ~ frame 试验机架
- ~ in bulk 散装(荷载)
- ~ indentation 压痕,受压凹痕
- ~ lever 载荷杆〈测试〉
- ~ limit 载荷限度
- ~ rate 荷载率
- ~ strain diagram 负荷应变曲线图
- ~ test 负荷试验
- ~ transducer conditioner 负荷传感调节器
- ~ unit 承载装置〈试验机架〉
- ~ verification 负荷校准

loadability 承载[载重,负荷]能力
loaded 负载的;树脂浸渍的
- ~ glass cloth[fabric] 树脂浸渍过的玻璃布
- ~ glass fibre mat 树脂浸渍过的玻璃纤维毡
- ~ roving 树脂浸渍过的粗纱
- ~ state 受载状态
- ~ stock 填料

loading 装模料,装料,加料;加入量;载荷,负载
- ~ agent 填充剂
- ~ area 加料面积
- ~ board 加料盘;加料板;承载板
- ~ capacity 装料量
- ~ cavity 装料腔,装料模槽
- ~ chamber 装料室,装料腔
- ~ chamber retainer plate 料腔托板,承载板,装载板
- ~ density 填料密度
- ~ fixture 装料工具
- ~ hopper 装料斗
- ~ in bulk 散装,堆装
- ~ level 负载程度
- ~ material 填料
- ~ pigment 颜料填充剂
- ~ plate 承载板,装载板,料腔(托)板
- ~ range 负荷范围
- ~ rate 装料速率[度]
- ~ shoe 装料底板;装料槽〈压模〉
- ~ shoe mould 加料底板式模具
- ~ space 装料腔,装料槽
- ~ test 负荷试验
- ~ time 装料时间;荷载时间
- ~ tray 加料盘
- ~ weight 装料量
- ~ well 装料槽,装料腔,装料室

loak foam 硬质聚氨酯泡沫塑料
lobster 弓形,弯曲
- ~ back 弯管,曲折管;管道补偿器,弓形弯管〈美〉
- ~ bend 弯管,曲折管;管道补偿器,弓形弯管

local 局部的
- ~ coefficient 局部系数
- ~ contraction 局部收缩
- ~ control 局部控制
- ~ defect 局部性疵点
- ~ deformation 局部形变
- ~ diffusion method 局部扩散法〈测定塑料结构〉
- ~ effect 局部效应
- ~ heater 局部加热器

~ heating 局部加热
~ mixture 局部混合
~ overcooling 局部过冷
~ overheating 局部过热
~ pressure 局部压力
~ rate of deformation 局部形变速率
~ relaxation mode 局部松弛模式
~ stiffening 局部加固
~ strain 局部应变
~ stress 局部应力
~ stretching 局部拉伸
localized 局部的
　~ bending 局部弯曲〈塑料形变件〉
　~ compression 局部压缩
　~ high temperature 局部高温
locate 设置,安排;固定;定位;探测,判明
locating 定位
　~ device 定位装置,安装夹具
　~ hole 定位孔
　~ notch 定位槽口
　~ pin 定位销〈模塑品嵌件〉
　~ plate 定位垫片
　~ ring 定位圈
　~ screw 定位螺钉
location of gate 浇口定位;浇口位置
loc. cit [l. c.] 在上述引文中
lock 锁,锁住,锁紧;闭锁
　~ bolt 锁紧螺栓,防松螺栓
　~ gate 闸门;装料闸门〈注塑机〉
　~ nut 锁紧螺母,防松螺母
　~ pin 锁销
　~ plate 锁紧垫片
　~ ring 锁环,密封圈
　~ screw 锁紧螺钉
　~ sleeve 锁紧套
　~ type clamping system 闭锁式锁模系统
　~ washer 锁紧垫圈,防松垫圈,保险垫圈
locking 闭锁;锁紧装置;锁模〈注塑模具〉
~ block 锁紧块〈注塑模具〉
~ bolt 锁紧螺栓
~ device 锁模装置
~ face 锁楔面
~ force 锁模力〈注塑机〉,合模力
~ hardware 锁件件,锁紧装置
~ key 锁模销
~ load 锁模力,锁模载荷〈注塑机〉
~ mechanism 锁模机构
~ pin 锁紧销
~ piston 锁模活塞
~ pressure 锁模压力,合模压力,闭模压力〈注塑机〉
~ ring 锁模圈
~ screw 锁紧螺钉
~ system 锁模系统
~ unit 锁模装置
~ valve 闭锁阀
~ washer 防松垫圈
logarithm 对数
logarithmic 对数的
　~ damping decrement 对数衰减,阻尼因子
　~ decrement 对数衰减率
　~ function 对数函数
　~ viscosity number 对数黏数,比浓对数黏数
LOI [limiting oxygen index] 极限氧指数
London 伦敦〈英国首都〉
　~ dispersion force 伦敦力,色散力
　~ smoke 暗灰色
long 长期(的),长久(的)长(的)
　~ barrel shell 长机筒壳
　~ -barrelled extruder 长料筒[机筒]挤塑机〈用于涂布〉
　~ chain 长链
　~ chain branch 长支链
　~ chain molecule 长链分子
　~ chain plastic 长分子链塑料
　~ chain polymer 长链聚合物

~ chain polymeric material 长分子链聚合材料
~ distance 长距离的
~ draft 大牵伸
~ duration cyclic test （长期）耐久性循环试验
~ duration test 耐久性试验
~ fatigue life 耐疲劳性,长期疲劳寿命〈美〉
~ feed 纵向供给
~ glass fibre mat 长玻璃纤维毡(片),长玻璃纤维垫子
~ lasting 耐久的
~ life test 耐久性试验
~ lived 经久耐用的
~ oil alkyd(resin) 长油度醇酸树脂〈含油60%以上〉
~ -oil varnish 长油清漆
~ period 长周期
~ pitch [large pitch] 长螺距〈螺杆〉
~ production run 大量生产,大批生产
~ range 远程的,广大范围的
~ range elasticity 大幅度弹性,广范围弹性;高弹性
~ range order 远程有序
~ run 长期运转
~ run test 长期(运转)试验
~ sequence copolymer 长序列共聚物
~ service stress 持续应力;疲劳应力
~ shaft weave 斜纹编织〈玻璃纤维织物〉
~ staple 长绒〈棉〉
~ stroke press 长行程压机
~ -stroke punch 长行程模冲
~ -term 长期的
~ -term behaviour 长期性能,长期行为
~ -term change 长期变化
~ -term creep 长期蠕变
~ -term development 远景发展,远景开发

~ -term durability 耐久性,耐用性
~ -term performance 长期性能,持久性能
~ -term properties 持久性能
~ term resistance 长期耐久性〈照射〉
~ -term run test 长期(运转)试验
~ -term stability 长期稳定性
~ -term storage 长期储存
~ -term strength 长期强度,持久强度
~ -term stressing 长期应力;长期受力
~ -term tensile stress 长期拉伸应力,持久拉伸应力
~ -term test 长期试验,持久试验
~ -term thermal stability 长期热稳定性
~ -term usage 长期使用
~ -term use 长期使用
~ -term welding factor 焊缝相对耐久性
~ test 长期试验
~ -time 长期的,持久的
~ -time burning test 长期燃烧试验
~ -time creep behaviour 长期蠕变行为〈塑料材料〉
~ -time deformation behaviour 长期形变行为
~ -time service behaviour 长期使用性
~ -time strength 持久强度
~ -time test[LTT] 长期试验
~ ton 长吨〈即1016.05kg〉
~ travel 长行程
~ trip 长行程
~ tube vertical evaporator 长管立式蒸发器
~ wearing 耐磨的
longevity 寿命,耐久性
~ test 耐久性试验
longitudinal 纵向的

~ component 纵向部分〈长丝缠绕〉
~ crack 纵向裂缝
~ dilatation 纵向膨胀
~ direction 纵向
~ dispersion 纵向渗透
~ displacement 纵向位移
~ elasticity 纵向弹性
~ expansion 纵向膨胀
~ extension 纵向延伸
~ fiber 纵向纤维
~ filaments 纵向长丝
~ force 纵向力
~ fracture 纵向断裂
~ groove [axial groove] 纵向槽纹
~ knife 纵切刀
~ laminating 纵向层压
~ load 纵向负载
~ mixing 纵向混合
~ modulus of elasticity 纵向弹性模量〈增强塑料〉
~ orientation 纵向取向
~ refractive index 纵向折射率
~ seam 纵向焊缝
~ section 纵断面,纵截面
~ shear strength 纵向剪切强度
~ shearing machine 纵向剪切机
~ shrinkage 纵向收缩(率)
~ slide 纵向滑板〈模具〉
~ stability 纵向稳定性
~ stiffener 纵向增强劲;纵向型材
~ strain 纵向应变
~ strength 纵向强度
~ stress 纵向应力
~ – transverse Poisson's ratio 纵横泊松比
~ – transverse shear modulus 纵横剪切模量〈增强塑料〉
~ – transverse shear modulus of elasticity 纵横剪切弹性模量
~ – transverse shear strength 纵横剪切强度〈增强塑料〉
~ travel 纵向位移
~ view 纵视图
~ wave 纵向波〈超声波焊接〉
~ wave modulus 纵波模量
~ weld seam 纵向焊缝
~ wind 纵向卷绕
~ winding 纵向缠绕〈增强塑料〉
~ wrapping 纵向卷绕
loom 主电缆〈电〉;织布机;圆织机〈纱〉
loop 环,圈
  conveyor 环形输送带
~ dryer 垂悬环形干燥器
~ expansion pipe 补偿器,膨胀管圈
~ forming 环成形
~ – pile carpet 毛圈式地毯
~ strength [loop tenacity] 弯曲强度;钩接强度〈纱〉
~ stress 圆周应力
~ tenacity 互扣强度,钩接强度
~ tensile test 钩接拉伸试验〈纱〉
~ test 折圈试验;弯环试验
~ – type clamp 环式夹板
~ – type dryer 垂悬环形干燥器
looped strand 加捻粗纱〈玻璃纤维织物〉
looping 成环,成圈;环绕
loose 松散的,疏松的;活动的;散装的
~ backing flange 松套法兰
~ black 粉末炭黑
~ block mould 活块模具
~ colo(u)r 浮色
~ cores 活(模)芯
~ density 散装密度
~ detail mould 活动件模具
~ fill type insulant 松散绝热材料
~ filler 疏松填料
~ fit 松配合
~ fitting knockout pin 松配脱模销
~ flange 活套法兰
~ full – disk buff 活全圆盘抛光轮
~ goods 散装货物

~ lining 用夹板固定的衬里
~ machine 撕松机〈用于预混纤维状模塑料的撕松〉
~ material 疏松材料
~ measure 松散体积
~ mo(u)ld 移动式模具,活动模具,可卸式模具
~ mo(u)ld wedge 松配模楔
~ packed 散装的
~ plate transfer mould 活板式压铸模
~ pressure block 活动压力板
~ punch 活络阳模
~ roll [dancer roll] 活动辊
~ roller 活动辊
~ stock 散装货
~ structure 松散结构
~ winding 松卷,松绕
loss 损耗,损失
~ angle 损耗角
~ by radiation 辐射损耗
~ coefficient 损失系数
~ compliance 损耗柔量
~ due to friction 摩擦损失
~ elasticity 弹性损失
~ energy (介电)损耗能
~ factor 损耗因素〈测定塑料动态弹性模量的指标〉
~ in efficiency 效率损失
~ in strength 强度损失
~ in weight 失重,重量损失
~ in weight on drying 干燥失重
~ index 损耗指数
~ maxima 损耗峰
~ modulus 损耗模量
~ of adhesion 失黏,脱黏(表面)
~ of adjustment 失调
~ of control 失控
~ of efficiency 效率损失
~ of life 寿命缩短,使用期缩短
~ of plasticizers 增塑剂损耗
~ of power 功率损失
~ of pressure 压力下降
~ of weight 失重,重量损失
~ on heating 加热损失;干燥损耗
~ on ignition 灼热损耗
~ peak 损耗峰
~ tangent 损耗正切,损耗角正切(值),损耗因素
lost 损失的,磨损的
~ head 切头
~ material 磨损的材料
~ sprue 损耗注道
~ wax process 失蜡(精密)铸造法;熔模铸造法
lot :批,堆
~ number 批数;批号
~ production 大批生产,批量生产
~ sample 批样
~ size 批量
louvre [louver] 百叶窗
~ dryer 百叶窗式干燥器
low 低的
~ alkali borosilicate glass 低碱硼硅(酸盐)玻璃
~ -angle neutron scattering 小角中子散射〈研究塑料结构〉
~ angle scattering 小角散射
~ angle wind 小角卷绕
~ bake coating 低温烘烤涂层
~ baking 低温烘烧
~ base extruder 低座挤塑机
~ boiler 低沸点物质
~ cost 低价,廉价
~ cost extender 廉价增量[增容,增充]剂
~ cost plasticizer 廉价增塑剂
~ -cycle fatigue 低频疲劳断裂
~ density 低密度
~ density foam 低密度泡沫塑料
~ density filler 低密度填料
~ density polyethylene [LDPE] 低密度聚乙烯

~ density polyethylene foam 低密度聚乙烯泡沫塑料
~ – distortion article 细小变形制品；收缩小的制品
~ draft 低倍牵伸
~ efficiency 低效率
~ elasticity 低弹性
~ – energy electron diffraction (LEED) 低能电子衍射
~ flow 低流动性〈塑料〉
~ foamer 低发泡剂
~ foaming 低发泡的
~ foaming PVC profiles 低发泡聚氯乙烯型材
~ fogging plasticizer 低雾(性)增塑剂
~ frequency 低频(率)的
~ frequency heating 低频加热
~ grade 低级的,低档的
~ level 低水平的,低标准的
~ limit 下限,最小限度
~ limit of size 下限尺寸
~ limit of tolerance 下限公差
~ loss 低(介电)损耗
~ loss polymer 低介电损耗聚合物
~ melting 低熔的
~ modulus elastic fibre 低模量弹性纤维
~ pitch [short pitch] 短螺距〈螺杆〉
~ plasticity 低可塑性
~ pressure moulding 低压成型
~ point 低点
~ polymer 低聚合物
~ polymerized precondensate 低聚合度预缩合物
~ pressure 低压(力)
~ pressure injection moulding 低压注塑(法)
~ – pressure laminate 低压层合
~ pressure mould closing 低压闭模〈吹塑〉
~ pressure moulding 低压模塑(法)

~ profile 低糙度〈不需要进行抛光成平整〉
~ profile resin [l. p. resin] 低收缩性树脂
~ ratio expanded foam 低度发泡塑料
~ ratio expansion 低度发泡
~ ratio expansion mo(u)lding 低发泡成型
~ ratio gear box 低比速齿轮箱
~ rubber compound 低橡胶混合物
~ shear strength 低剪切强度
~ shear viscosity 低剪切黏度
~ sheen agent 消光剂
~ shrink additive 低收缩性添加剂
~ shrink unsaturated polyester resin 低收缩不饱和聚酯树脂
~ shrinkage fibre 低收缩纤维
~ shrinkage film – tape 低收缩性膜带
~ shrinkage resin 低收缩性树脂
~ – slip 低爽滑性
~ smoke 低烟雾
~ speed adjustment 低速调整
~ speed stability 低速稳定性
~ strength 低强度
~ temperature brittleness 低温脆性〈板材〉
~ temperature flexibility 低温柔曲性
~ – tempering mo(u)ld [tool] 低温调温模具〈注塑〉
~ temperature physis 低温物理学
~ Vacuum [moderate vacuum] 低真空
~ viscosity 低黏性,低黏度
~ visibility 低能见度
~ voltage 低电压
~ – voltage heating 低电压加热
~ molecular 低分子的
~ molecular polyamide 低分子聚酰胺(固化剂)
~ molecular polymer 低分子聚合物,低聚物
~ molecular weight 低相对分子质量

~ molecular weight compound 低相对分子质量化合物
~ molecular weight polyethylene[LMPE] 低相对分子质量聚乙烯〈润滑剂,脱模剂〉
~ molecular weight polymer 低相对分子质量聚合物
~ molecular weight polypropylene 低相对分子质量聚丙烯〈润滑剂〉
~ molecular weight polystyrene 低相对分子质量聚苯乙烯〈增塑剂〉
~ pressure 低压
~ pressure casting 低压铸塑
~ pressure die casting 低压模铸
~ pressure discharge 低压辉光放电〈为非极性热塑性塑料极化〉
~ pressure foaming machine 低压发泡机
~ pressure injection mo(u)lding 低压注塑(成型)
~ pressure laminate 低压层合制品
~ pressure laminating 低压层合
~ pressure laminating resin 低压层合树脂
~ pressure mo(u)ld 低压模具
~ pressure mo(u)lding 低压模塑
~ pressure mo(u)lding resin 低压成型用树脂
~ pressure plastic 低压塑料
~ pressure polyethylene[LPPE] 低压聚乙烯
~ pressure polyethylene fibre 低压聚乙烯纤维
~ pressure polymerization 低压聚合
~ pressure press 低压压机
~ pressure resin 低压树脂
~ pressure sensing mo(u)ld protection device 低压敏感模具保护装置
~ pressure stage 低压段
~ temperature 低温
~ temperature bend test 低温弯曲试验
~ temperature brittle point 低温脆折点
~ temperature brittleness 低温脆性
~ temperature brittleness test 低温脆性试验
~ temperature cure 低温硬化
~ temperature curing 低温焙烘[硬化,固化]
~ temperature embrittlement 低温脆性
~ temperature extensibility 低温伸长率
~ temperature flexibility 低温柔曲性
~ temperature fluidity 低温流动性
~ temperature impact 低温冲击
~ temperature lacquer 低温漆
~ temperature mixing[LTM] 低温混合
~ temperature polymer 低温聚合物
~ temperature polymerization 低温聚合(作用)
~ temperature properties 低温性能
~ temperature reaction product of diphenylamine and acetone 二苯胺和丙酮的低温反应产物〈抗氧剂〉
~ temperature reaction product of phenyl-β-naphthyl amine and acetone 苯基β萘胺和丙酮的低温反应产物〈抗氧剂〉
~ temperature relaxation process 低温松弛过程
~ temperature resistance 耐寒性,耐低温性
~ temperature rubber 低温橡胶
~ temperature setting 低温固[硬]化
~ temperature solution polycondensation 低温溶液缩聚(作用)
~ temperature stiffening 低温固[硬]化
~ temperature test 耐寒试验,低温试验
~ temperature treatment 低温处理
~ temperature varnish 低温清漆
~ temperature welding 低温焊接

lower 下面的;低的

~ bulks 低膨松体
~ critical solution temperature 下限临界溶解温度
  die 下压模
~ cross head 下横梁
~ knock-out pin 下脱模销,下顶出杆
~ limit 下限
~ lip 下唇〈平片状模头〉
~ part of a mould 下模塞〈压塑〉
~ plastic limit 塑性下限
~ platen 下压板,下模板,下压台〈压机〉
~ ram 下活塞
~ ram transfer moulding 下活塞压铸
~ traverse 下横梁
~ yield value 低屈服值
~ yoke 下横梁
LPPE [low pressure polyethylene] 低压聚乙烯
LRM [liquid reaction mo(u)lding] 液体反应模塑
Ltd. [limited] 有限公司;有限的
LTM[low temperature mixing] 低温混合
LTF[long time test] 长期试验
lubricant 润滑剂;润滑的
~ bloom 润滑剂冒霜;油迹
~ exudation 润滑剂渗出
~ plasticizer 润滑性增塑剂
~ spray 润滑剂喷射
lubricated thermoplastic 自润滑热塑性塑料
lubricating 润滑的
~ ability 润滑能力
~ agent 润滑剂;脱模剂
~ effect 润滑作用,润滑效应
~ efficiency 润滑效率
~ grease 润滑酯
~ oil 润滑油
~ power 润滑能力,润滑效率
~ property 润滑性能

~ syringe 润滑剂注射器
lubrication 润滑(作用);涂油
~ failure 润滑失效
~ grove 润滑油槽
lubricator 润滑器,润滑装置;润滑剂
lug 把手;凸耳
luggage 行李,皮箱,皮包
~ carrier 行李箱,行李架
~ net 行李网
lumen 流明〈光通量单位;发光度单位〉;腔
luminance 亮度,发光度
luminescence 发光;冷光
luminescent 发光的,荧光的
~ dyes 荧光染料
~ effect 发光效应
~ material 发光材料
~ paint 发光涂料
~ pigments 发光颜料
~ plastics 发光塑料,荧光塑料
luminosity 亮度,发光度
luminous 发光的
~ apparent reflectance 光定向反射
~ body 发光体
~ colour 发光颜料
~ density 光密度
~ directional reflectance 光定向反射
~ efficacy 发光功率
~ efficiency 发光效率
~ exitance 光束发散度
~ fluorescent paint 荧光涂料
~ flux 光通量
~ fractional reflectance 光反射率
~ intensity 发光强度
~ paint 发光涂料
~ pigment 发光颜料
~ plastics 发光塑料
~ reflectance 光反射能力
~ reflection factor 光反射率
~ sensitivity 感光灵敏度
~ transmittance [light transmittance]

透光率,光透射率
lump 料团,块团〈不规则形状的块团〉
lustering 上光
  ~ agent 上光剂
lustre [luster] 光泽
  ~ buffing （高）光泽抛光
  ~ coating 上光涂料
  ~ finish 上光整理
  ~ lacking uniformity 光泽不匀
  ~ plate 上光板
lustreless 无光泽(的)
lustrous 光泽的;闪光的
  ~ filament 有光长丝
  ~ finish 有光整理,光泽整理
  ~ rayon 有光人造丝
  ~ yam 有光丝

lute 封泥
Luwa evaporator 卢瓦蒸发器
lux [lx] 勒(克司)〈照度单位〉
lye [*liquor*] 碱液
lyophile 亲液物;亲液胶体
lyophilic 亲液性;亲液的
  ~ polymer 亲液聚合物
  ~ sol 亲液溶胶
lyophobic 疏液的;憎液的
lyotropic liquid crystal polymer 亲液晶聚合物
lyre – type Ω 形
  ~ expansion Ω 形膨胀补偿器〈塑料管〉
  ~ expansion piece Ω 形膨胀补偿件〈管〉

M

M [mega] 兆,百万;大
m – [meta] 间(位);偏〈词头〉
m [meter] 米,公尺
M [mol] 摩〈摩尔,mole 符号〉
M – glass M – 玻璃
MA [maleic anhydridel] 马来酸酐,顺丁烯二酸酐,顺酐〈俗称〉〈固化剂〉
MABS [methyl methacrylate – acrylonitrile – butadiene – styrene copolymer] 甲基丙烯酸甲酯 – 丙烯腈 – 丁二烯 – 苯乙烯共聚物
MAC [maximum allowable concentration] 最大容许浓度
macaroni fiber 中空纤维
macerate 碎布;碎屑塑料;浸渍
  ~ mo(u)lding 碎屑压塑成型
  ~ mo(u)lding compound 碎屑模塑料
macerated 碎屑的;浸渍的
  ~ fabrics 布碎片,布屑〈填料〉
  ~ fabric mo(u)lding 碎布模制品
  ~ moulding compound 碎屑模塑料

  ~ plastic 碎屑[浸渍]塑料
maceration 浸渍(作用)
machinability 切削性;机械加工性
machinable 可机械加工的
machine 机器(的),机械(的);机械加工
  ~ attendance 机械保养
  ~ base 机器底座
  ~ capacity 机械能力
  ~ components 机器零件
  ~ construction 机器结构
  ~ cycle 机器工作周期
  ~ cycle monitoring 机械周期控制〈注塑机〉
  ~ dimensions 机器尺寸
  ~ direction 加工方向;轴向
  ~ efficiency 机器效率
  ~ element 机械元件
  ~ erection 机器安装
  ~ error 机器误差
  ~ for taking – off the burr 修边机

- ~ foundation 机座
- ~ frame 机架
- ~ glaze cylinder 机械上光滚筒
- ~ glazed finish 机械抛光
- ~ - glazed paper [machine finished paper] 机制上光纸
- ~ guards Pl 机械防护装置
- ~ hours 机器运转时间
- ~ industry 机器制造工业
- ~ knitting 机器编织
- ~ maintenance 机器保养
- ~ member 机械元件,机件
- ~ mo(u)lding 机械模塑,机械成型
- ~ mounting 机器安装
- ~ net 机制网
- ~ operation 机器操作
- ~ part 机器零件,机械部件,机件
- ~ performance 机械(运转)状态
- ~ property 机械加工性;切削性
- ~ shot capacity 机器注塑量
- ~ spares 机器备件
- ~ temperature profile 成型机温度分布
- ~ time 机器运转时间
- ~ tool 机械工具
- ~ unit 机组
- ~ upkeep 机器保护,机器维修
- ~ welding 机械化焊接
- ~ work 机械加工

machined 机械加工的
- ~ laminated tube 机制层合管材
- ~ part 机制件
- ~ rod 机制棒材
- ~ surface 加工面

machining 机械加工
- ~ allowance 机(械)加工余量
- ~ dimension 机械加工尺寸
- ~ of plastics 塑料的机械加工
- ~ marks 机械加工痕迹
- ~ precision (机械)加工精度
- ~ property 机械加工性能
- ~ quality 可机械加工性
- ~ time 机械加工时间

machineability 切削性;机械加工性
machineable 可机械加工的

macro 宏观的
- ~ - asperity 宏观粗糙度
- ~ - Brownian motion 宏观布朗运动
- ~ - crack 宏观裂缝
- ~ - creep 宏观蠕变
- ~ - dispersion 粗粒分散体
- ~ - flow 宏观流动

macrochain 大分子链
macroconfiguration 宏观结构
macrocycle 大环
macrocyclic 大环状的
- ~ compound 大环状化合物
- ~ polyetherpolyamide 大环状聚醚聚酰胺

macrography 宏观检查
macroheterogeneity 宏观不均匀性
macroion 大分子离子,高分子离子
macromechanics 宏观力学
macromer 大分子单体,高分子单体
macromolecular 大分子的
- ~ alignment 大分子排列
- ~ catalyst 大分子催化剂
- ~ chain 大分子链
- ~ compound 大分子化合物
- ~ dispersion 大分子分散体;大分子分散
- ~ material 高分子材料
- ~ resin network 大分子树脂网络

macromolecule 大分子,高分子
macromonomer 大分子单体,高分子单体
macronet ion exchange resin 大网状离子交换树脂
macroporous 大孔的
- ~ ion exchange resin 大孔离子交换树脂
- ~ polymer 大孔聚合物
- ~ structure 宏观多孔结构
- ~ resin 大孔树脂

macroradical 大分子自由基
macroreticular 大网状的
- ~ ion exchange resin 大网状离子交换树脂
- ~ resins 大网状树脂

~ resinous adsorbent 大网状树脂吸附剂
macrorheology 宏观流变学
macrosample 常量试样
macroscopic 宏观的
  ~ constant 宏观常数
  ~ deformation 宏观形变
  ~ elongation 宏观伸长
  ~ examination 宏观检验
  ~ irregularity 外观缺陷
macrostructure 宏观结构
magnesia 氧化镁,苦土
magnesium[Mg] 镁
  ~ carbonate 碳酸镁
  ~ hydroxide 氢氧化镁〈阻燃剂〉
  ~ oxide 氧化镁,苦土〈硫化剂〉
  ~ silicofluoride 氟硅酸镁
  ~ stearate 硬脂酸镁〈热稳定剂〉
  ~ sulfate 硫酸镁〈分散剂〉
  ~ sulfate heptahydrate 硫酸镁七水合物,泻利盐〈阻燃剂〉
magnet 磁铁;磁体
magnetic 磁(性)的
  ~ air-cushion vehicle 磁气垫运载工具
  ~ cushion 磁(缓冲)垫
  ~ declination 磁偏角
  ~ extracting device 磁选器,磁性分离器
  ~ extraction 磁力分离
  ~ field 磁场
  ~ flux 磁通量
  ~ grate [grate separator] 磁筛
  ~ grate separator 磁栅分离器〈吸除料斗中铁质〉
  ~ lens 磁透镜
  ~ nuclear resonance 磁核共振
  ~ nuclear resonance spectrum 磁核共振谱
  ~ permeability 磁导率
  ~ separation 磁力分离
  ~ separator 磁选器,磁力分离器
  ~ susceptibility 磁化率
  ~ tape 磁带

magneto-motive force [m.m.f.] 磁(动)势
magnetostrictive(sonic transducer) 磁致伸缩(声波)换能器
main 主要的,总的;干线;总管;电源
  ~ bearing 主轴承
  ~ body 主体,机身
  ~ chain [backbone chain] 主链〈聚合物〉
  ~ chain of polymer 聚合物主链
  ~ direction 主方向,总方向
  ~ drive 主传动
  ~ draft 主牵伸
  ~ driving shaft 主动轴
  ~ extruder 主挤塑机〈共挤塑〉
  ~ frame 主(机)架
  ~ frequency induction heating 工频感应加热
  ~ head 主横梁,顶梁
  ~ line 总管线〈管〉
  ~ pipe 主管,总管
  ~ plunger 主柱塞
  ~ product 主要产品
  ~ ram 主柱塞
  ~ reaction 主要反应
  ~ rule for construction 构型基本规则〈用于塑料计算结构〉
  ~ screw 主螺杆
  ~ shaft 主轴
  ~ stress 主应力
  ~ structure 主结构
  ~ switch 总开关
  ~ technical data 主要技术数据
  ~ test 主要检验
  ~ under pressure 受压总管
maintenance 保养,维护,维修
  ~ cost 维修费
  ~ efficiency 保养效率
  ~ free 不需维修(的),免保养(的)
  ~ -free operation 不需维修运行
  ~ manual 保养手册,维修手册
  ~ paint 维护涂料
  ~ work 维修工作,日常维修
major 较大的;主要的

~ adjustment 大调整
~ axis 主轴
~ diameter 大直径〈螺杆〉
~ industry 重点工业;大型工业
~ line 主线
~ overhaul 大修,总检修
~ parts 主要零件[部件]
~ repair 大修
~ service 大修
make – up 补偿;装配;制作;组成;编制
male 阳(性)的
~ die 阳模
~ drape forming 阳模包模成型
~ fitting 阳模配合
~ force 模塞,阳模
~ form [punch] 阳模
~ gauge 塞规
~ mo(u)ld 阳模,凸模;上半模
~ screw 阳螺纹,外螺纹
~ thread 阳螺纹,外螺纹
maleic 马来(酸的)
~ acid 马来酸,顺丁烯二酸
~ alkyd resin 马来酸醇酸树脂
~ anhydride [MA] 马来酸酐,顺丁烯二酸酐,顺酐〈俗称〉〉固化剂〉
~ anhydride addition compound modified unsaturated polyester resin 顺(丁烯二酸)酐加成物改性的不饱和聚酯树脂
~ resin 马来树脂;顺丁烯二酸树脂
malfunction 故障,不正常工作
~ alarm 故障警报
~ shutdown 故障停车
malleable 韧性的,可延展的;可锻的
~ iron 锻铁
malleability 延展的,可塑性
mallet 锤
~ handle die [dinking die] 冲孔模,冲切模,冲压模
~ perforator 锤式冲孔机
malonic acid 丙二酸
MAN [methacrylonitrile] 甲基丙烯腈
mandrel 芯模;芯型〈挤塑〉;芯棒〈吹型〉;管芯〈收卷〉

~ bend apparatus 芯材弯曲试验仪
~ bend flexibility 芯材弯曲柔顺性
~ bend test 芯材弯曲试验
~ carrier [mandrel support] 模芯支承架
~ diameter 模芯直径
~ drive shaft 管芯主动轴
~ forming [rolling] 芯型卷绕成型
~ heater 模芯加热器
~ holder 芯型支座,芯座
~ position 芯模位置
~ sizing 芯型定型
~ support [mandrel carrier] 模芯支承架
~ with cutting teeth 带切割钢齿的粗辊〈用于光泽塑料层消光〉
manganese acetate 乙[醋]酸锰〈催化剂〉
manganic oxide 三氧化二锰(填料)
man – hour 工时
manifold 歧管;复式接头;料道;多样的,各种各样的
~ chamber 料道室
~ clamp 歧管夹
~ die 歧管式模头;缝型模头
~ nozzle 歧管式注嘴
~ plate 歧管板;分配板
~ pressure 歧管压力
~ retainer 歧管固定架;分配器固定架
~ sheet die 歧管式料道挤板模头
Manila resin 马尼拉树脂
manipulated variable 操纵量,控制变量
manipulation 操纵,操作,控制
man – made 人造的
~ carrier material 合成载体材料〈用于涂布〉
~ fibre [artificial fibre] 人造纤维〈化学纤维的一类〉
man – power 人力
man – rated 适于人用的
manual 手动的,手工的
~ adjustment 手调
~ control 手动控制,手控,手动操纵
~ electric arc welding 手工电弧焊

manually actuated

~ finishing 手工修饰[整理,清理]
~ labor 手工劳动
~ loading 手工装料
~ manipulation 手工控制,手动操纵
~ mould [hand mould] 手动模具
~ operation 手工操作
~ regulation 手动调整
~ setting 手动调整,手调;手工装配
~ starter switch 手控起动开关
~ switch 人工开关,手控开关
~ welding 手工焊

manually actuated 手动操纵的〈阀〉
manuf. manufacture 制造;制造厂;制品
manufactory 制造厂
manufacture 制造;制品;制造厂
manufactured 制造的

~ adhesive unit 黏合剂制造装置
~ article 制品
~ product 制成品

manufacturer 制造者,制造厂
manufacturing 制造(的),生产(的)

~ cost 制造成本,生产成本
~ engineer 制造工程师
~ engineering 制造技术
~ errors 制造误差
~ facilities 生产设备
~ industry 制造工业
~ methods 生产方法
~ -oriented 从事生产的,与生产有关的
~ plant 制造厂
~ process 加工方法,加工过程
~ status 生产现状
~ technique 生产技术
~ tolerance 制造公差,制造裕度

mar 损坏;划痕,擦伤

~ proof 耐划痕的〈美〉
~ resistance 耐划痕性;耐擦伤性;划痕硬度
~ resistant 耐划痕的〈美〉

marble 大理石
Marco process 马可法;真空注塑成型法
marginal melt 边缘熔体,模具壁处的熔体

marine 船舶

~ finish coating 船舶修饰涂层
~ paint 船舶用漆
~ primer 船舶底漆
~ varnish 船舶清漆

mark 标志,标记的;痕迹

~ number 标号
~ off 划线;划痕

marked capacity 额定生产率,额定容量
marking 标志,标记;痕迹

~ pin 标志销〈为制品上带标志〉
~ plug 标志塞〈美〉

Marquardt index 马夸尔特指数〈表示酚醛树脂硬化程度的一种指标〉
Marrick blow moulding 马利克吹塑
Marrick process 马利克法〈英国马利克公司开发冷型坯吹塑法〉
marrying 配对
Martens 马丁

~ heat distortion temperature test 马丁热变形温度试验
~ heat resistance 马丁耐热性
~ scratch handness 马丁刻痕硬度
~ temperature 马丁耐热温度
~ test 马丁耐热试验

maser 微波量子放大器
mask 面罩,防护罩;遮蔽,掩蔽
masking 遮蔽;遮蔽物

~ agent 遮蔽剂
~ compound 遮蔽化合物
~ paint 遮蔽涂漆
~ paper 掩蔽纸
~ pressure sensitive adhesive tape 遮盖压敏胶带〈局部喷涂时〉
~ tape 遮蔽带

Mason horn 梅森焊接头,超声波焊接头
mass 物质;质量;块状物;大量,大批

~ absorption coefficient 质量吸收系数
~ burning rate 质量燃烧速率
~ colo(u)r 主色;浓色;覆盖色;本体颜色,体色
~ data 大量数据
~ density 物质密度,密度
~ detection 质量检测

~ detector 质量检测器
~ distribution function 质量分布函数
~ dyeing [spin dyeing] 纺前染色,本体染色
~ effect 质量效应
~ force 惯性力
~ law 质量定律
~ per unit area 每单位面积质量
~ percent 质量百分数
~ polymer 本体聚合物
~ polymerization [bulk polymerization] 本体聚合(作用),整体聚合(作用),块状聚合(作用)
~ polyvinyl chloride [MPVC] 本体法聚氯乙烯
~ produce 大量生产
~ produced plastic 大量生产的塑料;通用塑料,日常用塑料,大宗塑料
~ production 大量生产
~ range 质量范围
~ ratio 质量比,相对质量
~ resistivity 比电阻,质量电阻率
~ spectra 质谱
~ spectrograph 质谱仪
~ spectrometer 质谱仪
~ spectrometric thermal analysis [MTA,MSTA] 质谱热分析
~ spectrometry 质谱分析(法)
~ spectrometry and differential thermal analysis [MDTA] 质谱分析和差热分析
~ tone 主色;浓色
~ transfer 质量传递
Massey coater 马斯依涂布机〈涂布辊前带有计量和平整辊的辊式涂布机〉
master 主人;技师;主要的,基本的;熟练的
~ batch 母料;母体混合物,预混料;母炼胶〈橡胶〉
~ batch colouring 母料着色
~ batching 加母料;配母料
~ blank 标准坯件;模压坯件〈由软金属制〉
~ board 主控制(仪表)板

~ control 主控制,中央控制
~ control panel 主控制板〈塑料加工机〉
~ curve 总曲线;叠合曲线
~ drum 提花滚筒
~ forming process 原型成型法
~ model 主模型,原型;标准模型
~ mo(u)ld 母模,标准模
~ pack 大包装
~ pattern 原始图案;原模型
~ plan 总体规划,总布置图
~ powder 分散性粉末〈颜料〉
~ roll 导向辊,传动辊
~ sample 标准样品;标准试样
~ switch 主控开关,总开关
~ valve 总阀,主阀
~ viscometer 主黏度计;标准黏度计
~ workpiece 仿形样板,靠模样板
mastic 腻子,厚浆涂料〈建筑用〉;密封膏(剂)
~ gum 乳香
masticate 素炼;捏合
mastication 素炼(作用);捏合(作用)
masticator 素炼机;捏合机
~ blade 捏合机桨叶
mat 毡,毡料;无光泽的;褪光
~ binder 毡用黏合剂
~ clear varnish 半透明无光清漆
~ impregnation 毡料浸渍
~ in bonding 加毡黏合〈热固性增强塑料〉
~ mo(u)lding 毡料层压成型
~ surface 无光泽表面
match 匹配,配对
~ board 模板,(假)型板
~ die 成对模
~ -die mo(u)lding 对模成型
~ mark 配合符号
~ mo(u)ld thermoforming 对模[合模]热成型
~ plate 模板,分型板
~ up 配得上
matched 匹配的,配对的
~ bonding tool 靠模固定工具

matching

~ data　匹配数据
~ die　两半模,对模
~ die forming　对模成型
~ – die moulding　对模成型
~ (metal) die　成对(金属)模,对模
~ (metal) mo(u)ld　成对(金属)模,对模
~ metal mo(u)lding　对模成型
~ mould　对模
~ mo(u)ld forming　对模成型
~ – mo(u)ld thermoforming　对模热成型,合模热成型
~ pairs　配色染料
~ pressing dies　对压模
~ shaping dies　对成型模
~ tool　对模
~ tool forming　对模成型

matching　匹配,配对;拼色,配色

~ control　匹配控制,自动选配装置
~ mark　配对标记,装配标记
~ parts　配件
~ requirements　装配条件,配合要求
~ surface　配合面

material　材料,物料,物质;重要的

~ ablating effect　材料烧蚀效应〈电腐蚀加工,火花电蚀法〉
~ accumulation　材料积存;材料集中
~ balance　物料平衡
~ behavior　材料特性
~ being ground　(磨)碎料
~ certificate　材料(检验)合格证
~ characteristic values　材料特性值
~ damage　材料损坏〈型料成型时〉
~ damping　材料减振
~ difference　实质性的差别,本质的不同
~ distribution flaps　原料分配板
~ expenditure　材料耗损
~ flow　物料流动
~ handling　材料处理;材料加工〈美〉;物料装卸
~ list　材料单
~ mark　材料记号,材料代号
~ of construction　结构材料
~ order　材料单
~ particle　物质粒子
~ performance　材料性能
~ recycling　废料再生,材料回收再利用
~ standard　实物标准
~ substitution　材料代用品
~ supply　材料供应
~ testing　材料试验
~ testing equipment　材料试验装置
~ testing machine　材料试验机
~ to be ground　可磨碎料;待磨料
~ transfer　物质传递
~ usage　材料用量
~ well　料腔

mating　配套,配合

~ member　配合件
~ parts　配合件
~ surface　吻合面;配合面;合模面

matrix　母体,基料,母料;模型,阴模;原色,本色

~ fibre　基质型纤维,复合纤维
~ plate　网格板
~ polymerization　母体聚合
~ resin　母体树脂,基础树脂,基料树脂

matt　无光的,消光的,平光的,无(光)泽的

~ finishing　消光整理
~ paint　平光漆,无光漆
~ spot　无光斑,暗斑
~ surface　消光面,无泽面
~ yarn　无光纱

matte　无光的,消光的,无(光)泽的

~ coating　消光涂料;无光涂层
~ finish　消光
~ finish both sides　两侧面消光
~ finishing　消光加工;涂无光漆

matter　物质,物料,材料
matting　去光泽,消光
mattle(s)　斑点;混色斑纹
maturing　固化,熟化,陈化

~ temperature　熟化温度;固化温度;陈化温度

~ time 熟化时间,陈化时间
max. [maximum] 最大
maximal 最大的;最高的;最多的
　~ colo(u)r 最全色
　~ condition 最高条件
　~ draw ratio 最大拉伸比
　~ pressing force 最大压力
maximum 最大(的);最高(的);最多(的)
　~ allowable concentration [MAC(value)]最大允许浓度〈有害气体、蒸气或粉尘〉
　~ allowable operating temperature 最大允许操作温度
　~ allowable pressure 最大容许压力
　~ allowable pressure drop 最大容许压力降
　~ allowable working pressure 最高允许工作压力
　~ clearance 最大间隙
　~ clearance with mould open 最大开模度
　~ daylight opening 压板最大间距〈压机〉,最大模板间距
　~ dimensions 最大尺寸
　~ draft 最大拉伸
　~ elongation 最大伸长(率)
　~ error 最大误差
　~ ga(u)ge 最大规格
　~ height 最大高度
　~ injection volume 最大注塑体积
　~ limit of size 最大极限尺寸
　~ mo(u)ld opening 模具最大开距
　~ mo(u)ld size 模具最大尺寸
　~ load 最大载荷
　~ output 最高产率
　~ permissible concentration 最大允许浓度
　~ permissible service temperature 允许最高使用温度
　~ potential strength 最大潜在强度
　~ pressure 最大压力
　~ ratio 最大比例
　~ size 最大尺寸
　~ speed 最高速率,最大速度
　~ stress 最大应力
　~ surface stress in bend 最大弯曲表面应力
　~ temperature 最高温度
　~ torque 最大扭矩
　~ torque temperature 最大扭矩温度
　~ valence 最大值,极大值
　~ viscosity temperature 最高黏度温度
　~ weight of injection 最大注塑量
　~ weight per cycle (每次)最大注塑量
　~ width 最大宽度
Maxwell 麦克斯韦〈人名〉;麦克斯韦;Mx,〈磁通非SI单位. = 10 - 8Wb(韦伯)〉
　~ body 麦克斯韦黏弹体
　~ model 麦克斯韦模型〈形变力学性能〉
　~ equation of flow 麦克斯韦流动方程
Mayer bar 迈耶刮刀,螺旋刮刀
　~ coater 迈耶刮刀涂布机,辊式螺旋刮刀涂布机
MBI[mercaptobenzimidazole] 巯基苯并咪唑
MBS [methyl methacrylate - butadiene - styrene copolymer] 甲基丙烯酸甲酯 - 丁二烯 - 苯乙烯共聚物〈抗冲改性剂〉
MBT[mercaptobenzothiazole] 巯基苯并噻唑〈促进剂〉
MC[methyl cellulose] 甲基纤维素〈成膜剂,增稠剂,乳化剂,稳定剂,分散剂〉
mcf, m. c. f. [million cubic feet] 兆立方呎,百万立方呎
MDI[diphenyl methane diisocyanate] 二苯基甲烷二异氰酸酯
MDPE[medium density polyethylene] 中密度聚乙烯
MDS [morpholine disulphide] 二硫化吗啉
mean 平均(的)
　~ absolute deviation 平均绝对偏差
　~ absolute error 平均绝对误差
　~ annual 年平均

- ~ coefficient 平均系数
- ~ consumption 平均消耗(量)
- ~ degree of polymerization 平均聚合度
- ~ deviation 平均(偏)差
- ~ diameter 平均直径
- ~ difference 平均差
- ~ effective pressure 平均有效压力
- ~ effective value 平均有效值
- ~ effective viscosity 平均有效黏度
- ~ error 平均误差,标准误差
- ~ fatigue stress 平均疲劳应力
- ~ fibre length 纤维平均长度
- ~ length 平均长度
- ~ life 平均寿命
- ~ line 平均线,中线
- ~ load 平均负载
- ~ molecular weight 平均相对分子质量
- ~ molecule 平均分子
- ~ normal strain 平均法向应变
- ~ normal stress 平均法向应力
- ~ parameter 平均参数
- ~ piston speed 活塞平均速率[度]
- ~ pore size 平均孔径
- ~ pressure 平均压力
- ~ rate of stressing 平均应力速率
- ~ relative deviation 平均相对偏差
- ~ shearing stress 平均剪切应力
- ~ size 平均尺寸,平均大小
- ~ spherical (luminous) intensity 平均球面(发光)强度
- ~ strain 平均应变
- ~ strength 平均强度
- ~ stress 平均应力
- ~ temperature difference 平均温差
- ~ tension 平均张力
- ~ time between failures 平均故障间隔时间
- ~ value 平均值

means 手段,方法;装置,设备
- ~ of production 生产手段

meandering 曲折(的),弯曲(的)
- ~ dual-belt conveying unit 回曲形双带输送装置
- ~ dual-belt conveyor 回曲形双带输送机
- ~ flow 曲流

measling 斑点
measurability 可测性
measure 量度,测量;尺寸,大小;手段,措施
- ~ expansion 体积膨胀
- ~ of capacity 容量
- ~ of precision 精确度

measurement 测量,测定
- ~ accuracy 测量精度
- ~ of nuclear induction 核磁共振测定
- ~ transducer 测量传感器

measuring 测量
- ~ apparatus 测量仪器
- ~ bridge 测量电桥
- ~ cup 量杯
- ~ cylinder 量筒
- ~ error 测量误差
- ~ ga(u)ge 测厚计
- ~ glass (玻璃)量筒,量杯
- ~ load 测定载荷
- ~ means 测量方法
- ~ pick-up 遥测发送器
- ~ range 测量范围
- ~ tape 卷尺
- ~ technique 测量技术
- ~ tool 测量工具

mech [mechanical] 机械的;力学的
mechanical 机械的;力学的
- ~ accident 机械事故
- ~ adhesion 机械黏合
- ~ agitation 机械搅拌
- ~ agitator 机械搅拌器
- ~ anisotropy 力学各向异性
- ~ appliance 机械设备
- ~ automation 机械自动化
- ~ belt 传动带
- ~ blend 机械混合
- ~ blowing 机械发泡
- ~ breakdown 机械性损坏
- ~ buffing 机械抛光
- ~ characteristics 机械性能
- ~ clearance 机械间隙

~ constant 机械常数
~ control 机械控制,机械操纵
~ crimp 机械卷曲
~ cutting 机械切削
~ damage 机械损伤
~ decomposition 机械分解
~ deformation 机械变形
~ disintegration 机械粉碎
~ draft 机械牵伸
~ drawing〈for glassfibre〉 机械拉伸成型〈玻璃纤维〉
~ drawing process 机械拉伸法
~ drive 机械传动
~ efficiency 机械效率
~ ejection 机械脱模
~ embossing 机械轧花
~ energy 机械能
~ engineering 机械设计;机械工程
~ fabric 工业用织物
~ failure 机械故障,力学破坏
~ features 机械特性,力学特性
~ fibrillation 机械原纤化
~ finishing 机械整理
~ foam 机械发泡
~ foamed plastics 机械发泡塑料
~ frothing 机械起沫,机械发泡
~ handling 机械操作
~ hardening 机械硬化
~ industy 机械工业
~ interlocking 机械联锁
~ loss 力学损耗
~ loss factor 力学损耗因子
~ mixture 机械混合物
~ model 机械模型,力学模型
~ plastication 机械塑化
~ power (机械)驱动力,传动力
~ press 机械压机
~ properties 机械性能;力学性能
~ pulp 机械木纸浆
~ relaxation 力学松弛
~ sampling 机械取样
~ scanning 机动扫描
~ separation 机械分离
~ slicer 机械切片机;刨床

~ spectrum 力学谱
~ stirrer 机械搅拌器
~ stirring 机械搅拌
~ strength 机械强度
~ stress 机械应力
~ tensiometer 机械张力测定仪
~ test 仪器检验机械(性能)试验,力学试验
~ testing(of materials) (材料)机械试验;(材料)力学试验
~ treatment 机械加工
~ welding 机械焊接
~ working properties 机械加工性能
mechanically 机械地
~ blocked adhesive 物理法稳定型黏合剂
~ foamed 机械发泡的
~ foamed plastic 机械发泡塑料
mechanism 历程;机理;机构
~ of combustion 燃烧机理
~ of deformation 形变机理
~ of degradation 降解机理
~ of fracture 断裂机理,破坏机理
~ of polymerization 聚合历程,聚合机理
~ of reaction 反应历程,反应机理
mechanochemical 力化学的
~ crosslinkage 力化学交联
~ decomposition 力化学分解
~ degradation 力化学降解
~ grafting 力化学接枝
~ polycondensation 力化学缩聚
~ reaction 力化学反应
median 中等的;中间的;中值,中点,中线
~ fatigue life 中间疲劳寿命
~ fatigue strength 中间疲劳强度
~ line 中线
~ lethal dose[LD_{50}] 半致死剂量,致死中量〈50%个体致死的剂量〉
~ size 中等尺寸,中等大小
mediator 中间体;介质,媒介(物)
medical 医学的,医疗的
~ high polymer 医用高聚物

medium

~ polymer 医用聚合物
medium 介质,介体;媒介(物);手段,方法;中间的,平均的
~ accelerator 中等活性促进剂
~ alkali glass fiber 中碱性玻璃纤维
~ boiler 中级沸腾器〈沸点 90~115℃〉
~ counts 中支数〈纤维〉
~ crushing 中等破碎
~ density polyethylene[MDPE] 中密度聚乙烯
~ grain 中等颗粒,中等粒度
~ grained 中等颗粒的,中等粒度的
~ grainular 中等颗粒的
~ grinding 中等粉碎
~ hard,MH 中等硬度
~ (hard) board 中等硬度纤维板;半硬质纤维板
~ length fibre 中长纤维
~ - oil alkyd resin 中油度醇酸树脂
~ pressure 中等压力
~ pressure polyethylene 中压聚乙烯
~ section 中等型材
~ scale 中等规模
~ scale integration 中规模集成电路
~ shade 中等色调
~ size 中等尺寸,中等大小
~ soft,MS 半柔软
~ staple fibre 中长纤维
~ strength 中等强度
~ tenacity 中等强度
~ - thickness sheet(ing) 中等厚度片材
~ tone 中间色调,半色调
mega,[M] 兆,百万
MEK[methyl ethyl ketone] 甲乙酮
~ peroxide [methyl ethyl ketone peroxide] 过氧化甲乙酮
MEKP[methyl ethyl ketone peroxide] 过氧化甲乙酮
melamine 三聚氰胺,蜜胺
~ adhesive 三聚氰胺黏合剂
~ - alkyd resin 三聚氰胺-醇酸树脂
~ - alkyd resin coating 三聚氰胺-醇酸树脂涂料
~ benzoguanamine resin 三聚氰胺苯并哌胺树脂
~ decorative board 三聚氰胺装饰板
~ decorative laminate 三聚氰胺装饰层压板
~ - formaldehyde condensation resin 三聚氰胺-甲醛缩合树脂
~ formaldehyde glass fiber compound 三聚氰胺甲醛玻璃纤维模塑料
~ formaldehyde mo(u)lding compound 三聚氰胺甲醛模塑料
~ - formaldehyde plastics 三聚氰胺甲醛塑料
~ formaldehyde precondensates 三聚氰胺甲醛预缩合物
~ formaldehyde resin[MF] 密胺-醛树脂;三聚氰胺-甲醛树脂
~ formaldehyde resin adhesive 三聚氰胺-甲醛树脂黏合剂
~ glue 三聚氰胺胶(黏合剂)
~ mo(u)lding compound[material] 三聚氰胺模塑料
~ panel 三聚氰胺塑料板,密胺塑料板〈用作车身内覆盖材料〉
~ phenol formaldehyde resin[MPF] 三聚氰胺-酚醛树脂
~ phenolic resin 三聚氰胺酚醛树脂
~ plastic 三聚氰胺塑料,密胺塑料
~ resin 三聚氰胺树脂,密胺树脂
~ - resin adhesive[glue] 三聚氰胺树脂黏合剂[胶黏剂]
~ resin coating 三聚氰胺树脂涂料
~ - resin glue 三聚氰胺树脂胶(黏剂)〈美〉
~ resin mo(u)lding compound 三聚氰胺树脂模塑料
~ resin varnish 三聚氰胺树脂清漆
~ surfaced decorative laminate 三聚氰胺表面装饰板
~ - urea resin 三聚氰胺脲醛树脂
melamincplast 三聚氰胺塑料,蜜胺塑料
melanocratic 暗色的
melt 熔体,熔(融)料;熔融,熔化,塑化

- ~ accumulator 熔料储存器
- ~ additive brightener 熔体增白剂
- ~ adhesives 热熔型黏合剂
- ~ back flow 熔体回流
- ~ blend 熔体混合[掺混,混配]物
- ~ canal 熔体迪道
- ~ capacity 塑化能力,塑化量
- ~ characteristic 熔体特性
- ~ coater 熔融涂布机
- ~ coating 熔融涂布,热喷涂
- ~ coloration technique 熔体着色技术
- ~ crystallization 熔融结晶
- ~ elasticity 熔体弹性
- ~ embossing process 熔膜轧纹成形法
- ~ extractor 熔体分离器
- ~ extractor screw 熔体分离型螺杆
- ~ extruder 熔体(加料)挤入[塑]机
- ~ extrusion 熔体(加料)挤出[塑]
- ~ extrusion coating 熔体挤出涂布
- ~ feed extruder 熔体加料挤塑机
- ~ fed extruder 熔体加料挤塑机
- ~ film coating 熔膜涂布
- ~ filter 熔体过滤器
- ~ filtration 熔体过滤
- ~ flow behaviour 熔体流动性
- ~ flow index[MFI] 熔体流动指数,熔融指数
- ~ flow indexer 熔体流动指数测定仪
- ~ flow instability 熔体流动不稳定性
- ~ flow rate[MFR] 熔体流动速率
- ~ fluxed compound 熔体塑化料
- ~ fracture 熔体破裂〈挤塑物表面粗糙,出现"鲨鱼皮"、"螺旋丝、甚至破裂成碎块的现象〉
- ~ fracture stress 熔体破裂应力
- ~ grid 熔(炉)栅
- ~ grid type spinning machine 熔栅式纺丝机
- ~ head 熔体喷丝头〈纺丝〉
- ~ hopper 熔体料斗
- ~ index[MI] 熔体(流动)指数
- ~ index automaton 熔体(流动)指数自动测定仪
- ~ index plastometer 熔体(流动)指数塑性测定仪
- ~ indexer 熔体(流动)指数测定仪
- ~ inhomogeneity 熔体不均匀性
- ~ injection 熔体注塑(法)
- ~ instability 熔体不稳定性
- ~ line 熔合线,熔接线
- ~ metering pump 熔体计量泵
- ~ mix 熔体混合;熔体混合物
- ~ mixed pellet 熔体塑化粒料
- ~ pad 熔体余量(注塑时剩余量)
- ~ pigmentation 熔体着色法
- ~ polycondensation 熔体缩聚(作用)
- ~ polymerization 熔体聚合(作用)
- ~ pool 熔体池
- ~ pressure 熔体压力
- ~ pressure measurement transducer 熔体压力测定传感器〈在塑料加工机内〉
- ~ processing 熔体加工
- ~ residence time 熔体逗留时间
- ~ resistance 耐熔性
- ~ rheometer 熔体流变仪
- ~ roll 溶化辊,熔体辊〈在熔体辊压涂布机内〉
- ~ roll coater 熔体辊压涂布机
- ~ roll coating 熔体辊压涂布
- ~ runner 熔体流道
- ~ screw extruder 熔体螺杆挤塑机
- ~ section 熔融段,塑化段
- ~ slippage 熔体滑动
- ~ spinnable polymer 可熔纺聚合物
- ~ spinning 熔体纺丝,熔融纺丝〈合成纤维〉
- ~ spinning by extrusion method 熔体挤压纺丝法
- ~ spinning pump 熔融纺丝泵
- ~ spun〈filament〉 熔纺,熔融纺丝〈长丝〉
- ~ spun fiber 熔纺纤维
- ~ stage 熔融阶段;黏流态
- ~ store 熔料储存器
- ~ strength 熔体强度
- ~ strength at break 熔体断裂强度
- ~ swell 熔体挤出膨胀

meltability

~ temperature 熔融温度;熔体温度;塑化温度
~ thermocouple 熔体热电偶
~ transition 熔融转化
~ uniformity 熔体均匀性
~ viscometer 熔体黏度计
~ viscosity 熔体黏度
~ viscosity at extruder die 挤塑机模头处的熔体黏度
~ welding 熔融焊接
~ wheel 熔融轮〈熔融胶黏涂布机中〉
~ zone 熔融段,塑化段〈螺杆〉
meltability 可熔性
meltable 可熔化的
~ polyimide 可熔性聚酰亚胺
meltableness 可熔性
melten polymer 熔融聚合物
melting 熔融,熔化
~ behaviour 熔融行为,熔融性状
~ capacity [plasticizing capacity] 塑化能力,塑化量
~ centrifuge 熔体离心器
~ compound 熔融混合料
~ dilation 熔化膨胀
~ efficiency 熔化效率
~ flow 熔体流动
~ furnace 熔化炉〈纺丝〉
~ heat 熔化热,熔融热
~ pan 熔锅
~ plate extruder 熔化盘式挤塑机〈用于泡沫塑料挤出〉
~ point [mp.] 熔点
~ point apparatus 熔点测定器
~ point test 熔点试验
~ process 熔融过程
~ range 熔融范围,熔程
~ rate 熔化过程;塑料速率
~ salt 熔盐〈用于塑料加工时的温度测定〉
~ section 熔融段;塑化段
~ swamp 熔融槽
~ temperature 熔融温度;塑化温度
~ viscosity 熔体黏度
~ zone 熔融段,塑化段

member 构件,部件
membrane 膜,隔膜
~ equilibrium 膜(渗)平衡
~ permeability 膜的渗透性,膜的透气性
~ pressure 膜压
~ roof 薄膜棚顶
~ separation process 薄膜分离法
~ waterproofing 防水膜,防水层
memory 记忆;存储
~ capacity 存储量
~ effect 记忆效应;存储效应
mending 修理,修补
~ pressure-sensitive adhesive tape 压敏型修补胶带
menthane 蓋烷
~ diamine 蓋烷二胺〈固化剂〉
p-~ hydroperoxide 氢过氧化对蓋烷基〈交联剂,引发剂〉
mer 链节;基体
mercaptan 硫醇
mercapto 〈词头〉巯基
mercaptobenzimidazole [MBI] 巯基苯并咪唑〈抗氧剂 MB〉
mercaptobenzothiazole [MBT] 巯基苯并噻唑〈促进剂,防霉剂〉
2-mercaptoimidazoline 2-巯基咪唑啉〈促进剂〉
2-mercaptomethyl benzimidazole 2-巯基甲基苯并咪唑〈抗氧剂〉
γ-**mercaptopropyl trimethoxy silane** γ-巯基丙基三甲氧基(甲)硅烷〈偶联剂〉
mercerizer 丝光机〈纺织〉
merchantable life 商品使用期限
mercuric chloride 升汞,氯化汞
~ catalyst 氯化汞催化剂
mercury, Hg 汞,水银
~ pressure 水银柱,汞柱
Merlon 〈商〉麦尔龙〈美国 Mobay 化学公司产聚碳酸酯〉
Merrill-Brookfield visccmeter 梅里尔-布鲁克菲尔德黏度计
mesh 筛目,筛孔;网目,网眼

~ analysis 筛析
~ fabric 网眼织物
~ filter 筛网过滤器
~ number 筛网号
~ screening 网筛;过筛
~ size 筛目大小;粒度
mesocolloid 介胶体
meta 间位〈略作 m-〉;介;偏
~ -compound 间位化合物
~ cresol 间甲酚
~ -derivative 间位衍生物
metal 金属
~ adhesive 金属黏合剂
~ adhesive based on phosphate 磷酸盐型金属黏合剂
~ adhesive based on silicate 硅酸盐型金属黏合剂
~ adhesive joint 金属粘接
~ bonding 金属粘接,金属黏合
~ bonding joint 金属粘接
~ chelating dyes 金属螯合染料
~ clad laminate 金属覆盖层压材料
~ coating 金属涂层
~ complex dyes 金属络合染料
~ -containing adhesive 金属填充黏合剂
~ -containing plastic 金属填充塑料
~ cutting 金属切削
~ detector 金属探测器
~ fiber 金属纤维
~ -filled epoxy 金属填充环氧树脂
~ foil 金属箔
~ insert 金属嵌件
~ ion deactivating agent 金属离子钝化剂
~ -plated plastic 镀金属的塑料
~ plating 镀金属
~ -polymer-composite sheeting 金属-聚合物复合(材料)片材
~ powder 金属粉末
~ reinforcement 金属增强(材料)
~ roller coating 金属辊涂布
~ soap 金属皂
~ spray 金属喷镀

~ -sprayed injection mo(u)lding tool 喷镀金属的注塑模具
~ spraying 金属喷镀
~ stearate 硬脂酸金属盐〈稳定剂〉
~ strip 金属条
~ -to-metal adhesive 金属与金属黏合剂
~ -to-metal bond 金属与金属粘接
~ -to-metal bonding 金属与金属粘接[黏合]
~ -to-metal glue 金属与金属胶黏剂
metalation 金属化作用
metalescent coating 金属闪光粉涂层
metalizable 可敷金属的
metallic 金属的
~ coating 金属涂敷;金属涂料
~ colo(u)r 金属色
~ conducting polymer 金属性导体聚合物
~ embedding 金属件嵌铸〈模塑件中〉
~ fiber 金属纤维
~ finish 金属涂层
~ flake pigments 金属箔颜料
~ foreign body 金属杂质[夹杂物]
~ mo(u)ld 金属模具
~ paint 金属涂料
~ pigments 金属颜料
~ potting 金属件嵌铸〈美〉
~ powder 金属粉末〈填充剂〉
~ soap 金属皂〈稳定剂〉
~ stabilizer 金属盐类稳定剂
metallised plastic 镀金属塑料
metallization (表面)金属化,敷金属,喷镀金属,上金〈塑料〉
metallized 金属的;镀金属的
~ dyes 金属络合染料
~ plastic 镀金属的塑料,敷金属,上金塑料
metallizing (表面)金属化,敷金属,喷镀金属,上金〈塑料〉
~ coating 敷金膜层
~ of plastics 塑料(表面)金属化
metallo-organic 有机金属的

metalloscope

~ pigment [semi-mineral pigment] 有机金属颜料
metalloscope 金相显微镜
metamer 位变异构体
metameric colours 位变异构色
metamerism 条件配色〈颜色〉
metaphenylenediamine 间苯二胺〈环氧树脂固化剂〉
metastable 准稳的,亚稳的
metathesis 复分解(作用);置换(作用);易位(作用)
meter [metre, m] 米,公尺;计量;表〈测量仪〉
~ bar 计量棒
~ board 仪表板
~ case 仪表外壳
~ dial 仪表刻度盘
~ recorder 计量记录仪
~ screw 公制螺纹
metering 计量;测量;测定
~ ball valve 计量球阀
~ conveyor 计量传送带
~ cylinder 计量筒
~ device 计量装置
~ nipple 测量螺纹接管,定量轴套
~ point 计量点
~ pump 计量泵
~ rod 计量杆,计量棒
~ roll 计量辊
~ screw 计量螺杆
~ section 计量段
~ section of screw 螺杆的计量段
~ strength 计量强度〈螺杆内〉
~ tank 计量罐
~ zone 计量段〈螺杆〉
methacrylate 甲基丙烯酸酯
~ ester 甲基丙烯酸酯
~ plastic 甲基丙烯酸酯塑料
~ resin 甲基丙烯酸酯树脂
~ sirup 甲基丙烯酸酯浆料
methacrylic 甲基丙烯酸的
~ acid 甲基丙烯酸
~ ester 甲基丙烯酸酯
~ methyl ester [MMA, methyl methacrylate] 甲基丙烯酸甲酯
~ polymer 甲基丙烯酸聚合物
methacrylic resin 甲基丙烯酸树脂
methacrylonitrile [MAN] 甲基丙烯腈
N-(**methacryloyloxy**-2-hydroxypropyl)-N'-phenyl-p-phenylene diamine N-(甲基丙烯酰氧基代-2-羟基丙基)-N'-苯基对苯二胺〈防老剂〉
methacryloxymethyl triethoxysilane 甲基丙烯酰氧甲基三乙氧基(甲)硅烷
methanal [formaldehyde] 甲醛
methane 甲烷;沼气
methanol [methyl alcohol] 甲醇
~ extract 甲醇萃取物
~ soluble matter 甲醇可溶物质
method 方法,手段
~ of cylindrical film roll 薄膜卷筒法
~ of measurement 测量(方)法
~ of operation 操作法
~ of porosity analysis [determination] 多孔性分析测定法
~ of X-ray powder diffraction X射线粉末衍射法
methoxyethyl 甲氧乙基
~ acetoxystearate 乙酸基硬脂酸甲氧乙酯
~ acetyl ricinoleate 乙酰蓖麻醇酸甲氧乙酯〈增塑剂〉
~ oleate 油酸甲氧乙酯〈增塑剂〉
~ ricinoleate 蓖麻醇酸甲氧乙酯〈增塑剂〉
~ stearate 硬脂酸甲氧乙酯〈增塑剂〉
methyl 甲基
~ abietate 松香酸甲酯〈增塑剂、软化剂、增黏剂〉
~ acetate 乙酸甲酯,醋酸甲酯
~ acetyl ricinoleate [MAR] 乙酰蓖麻醇酸甲酯〈增塑剂〉
β-~ **acrolein** β-甲基丙烯醛
~ acrylate 丙烯酸甲酯
~ acrylate-ethyl methacrylate copolymer 丙烯酸甲酯-甲基丙烯酸乙酯共聚物

~ acrylate polymer 丙烯酸甲酯聚合物
α- ~ acrylic acid[MAA] α-甲基丙烯酸
~ alcohol[methanol] 甲醇
~ butyrate 丁酸甲酯
~ cellulose 甲基纤维素
~ α-cyanoacrylate α-氰基丙烯酸甲酯
2-(1- ~ cyclohexyl)-4,6-dimethyl phenol 2-(1-甲基环己基)-4,6-二甲基苯酚〈抗氧剂〉
~ dichlorosilane 甲基二氯(甲)硅烷
2- ~ -4,6-dinonyl phenol 2-甲基-4,6-二壬基苯酚〈抗氧剂〉
~ endomethylene tetrahydrophthalic anhydride 甲基桥亚甲基四氢邻苯二甲酸酐〈固化剂〉
~ ester of acrylic acid [methacrylate, methacrylic ester] 甲基丙烯酸酯
~ ester of tall oil 妥尔油酸甲酯,松浆油酸甲酯
~ ethyl ketone[MEK] 甲基乙基(甲)酮,甲乙酮,2-丁酮〈溶剂〉
~ ethyl ketone peroxide [MEKP] 甲(基)乙(基)酮过氧化物〈交联剂〉
~ hydrogen polysiloxane fluid 甲基含氢硅油〈脱膜剂、防黏剂、防锈剂〉
~ hydroxystearate 羟基硬脂酸甲酯〈润滑剂、脱模剂〉
~ -n-hexyl ketone 甲基正己基甲酮,2-辛酮〈溶剂〉
2- ~ -imidazole 2-甲基咪唑〈固化剂〉
~ -isoamyl ketone 甲基异戊基甲酮
~ -isobutyl ketone peroxide 甲基异丁基酮过氧化物〈交联剂〉
~ lactate 乳酸甲酯
~ methacrylate[MMA] 甲基丙烯酸甲酯
~ methacrylate-acrylonitrile-butadiene-styrene copolymer[MABS] 甲基丙烯酸甲酯-丙烯腈-丁二烯-苯乙烯共聚物
~ methacrylate-butadiene-styrene copolymer[MBS] 甲基丙烯酸甲酯-丁二烯-苯乙烯共聚物,MBS共聚物〈抗冲改性剂〉
~ methacrylate-butylacrylate copolymer 甲基丙烯酸甲酯-丙烯酸丁酯共聚物
~ oleate 油酸甲酯〈增塑剂〉
~ palmitate 棕榈酸甲酯,软脂酸甲酯〈增塑剂〉
~ pentachloro-stearate[MPCS] 五氯硬脂酸甲酯〈增塑剂〉
~ pentene polymer[TPX] 甲基戊烯聚合物
~ pentene resin 甲基戊烯树脂
N-4- ~ -2-pentyl-N'-phenyl-p-pheny-lene diamine N-4-甲基-2-戊基-N'-苯基对苯二胺〈抗氧剂〉
~ -phenyl silicone oil 甲基苯基硅油
~ phthalyl ethyl glycolate[MPEG] 甲基邻苯二甲酰基乙醇酸乙酯〈增塑剂〉
~ -polysiloxane 甲基聚硅氧烷
N- ~ -2-pyrrolidone N-甲基-2-吡咯烷酮〈溶剂〉
~ ricinoleate 蓖麻醇酸甲酯〈增塑剂〉
~ silicone oil 甲基硅油〈消泡剂、脱模剂、润滑剂〉
α- ~ -styrene-methyl methacrlate copolymer α-甲基苯乙烯-甲基丙烯酸甲酯共聚物
3-(or-4) ~ -1,2,3,6-tetrahydrophthalic anhydride 3-(或4)-甲基-1,2,3,6-四氢邻苯二甲酸酐〈固化剂〉
~ triethoxysilane 甲基三乙氧基(甲)硅烷
~ vinyl ether 甲基乙烯基醚
methylcellulose[MC] 甲基纤维素〈成膜剂,增稠剂,乳化剂,稳定剂,分散剂〉
methylene 亚甲基,〈旧称〉甲撑
2,2'- ~ bis(6-tert-butyl-4-cresol) 2,2'-亚甲双(6-叔丁基-4-甲酚)〈抗氧剂〉

methylol

4,4'- ~ bis(2,6-di-tert-butyl phenol) 4,4'-亚甲基双(2,6-二叔丁基苯酚)〈抗氧剂〉
2,2'- ~ bis(4-ethyl-6-tert-butyl phenol) 2,2'-亚甲基双(4-乙基-6-叔丁基苯酚)〈抗氧剂〉
2,2'- ~ bis(6-α-methylbenzyl-p-cresol) 2,2'-亚甲基双(6-α-甲基苄基对甲酚)〈抗氧剂〉
2,2'- ~ bis(4-methyl-6-tert-butylphenol) 2,2'-亚甲基双(4-甲基-6-叔丁基苯酚)〈抗氧剂2246〉
2,2'- ~ bis(4-methyl-6-cyclohexylphenol) 2,2'-亚甲基双(4-甲基-6-环己基苯酚)〈抗氧剂〉
2,2'- ~ bis[4-methyl-6-(α-methyl-cyclohexyl)phenol] 2,2'-亚甲基双[4-甲基-6-(α-甲基环己基)苯酚]〈抗氧剂〉
~ bis(stear-behenamine) 亚甲基双(硬脂酰-山嵛酸酰胺)〈润滑剂〉
~ bis(stearamide) 亚甲基双硬脂酰胺〈润滑剂〉
~ di-p-phenylene diisocyanate[MDI] 亚甲基双(对苯基二异氰酸酯)
~ chloride 二氯甲烷〈溶剂〉
4,4'- ~ -dianiline 4,4'-亚甲基二苯胺〈固化剂〉
~ -succinic acid 2-亚甲基丁二酸〈助剂〉

methylol 羟甲基
~ amide 羟甲基酰胺〈润滑剂〉
~ dyes 羟甲基型染料
~ group 羟甲基
~ melamine 羟甲基三聚氰胺,羟甲基密胺
~ urea 脲基甲醇,羟甲基脲

metre[meter] 米,公尺;计量;表〈测量仪〉

metric 米制的,公制的,公尺的;测量的
~ conversion 公制换算
~ measure 公制计量
~ size 公制公寸
~ thread 公制螺纹

~ ton 公吨,千公斤
~ units 公制单位

MF [melamine-formaldehyde resin] 三聚氰胺-甲醛树脂,密胺-甲醛树脂
MFE [mineral filler high electric] 改善电性能的无机填料,高电性能的无机填料
mfg. [manufacturing] 制造的
MFG [mineral filler best general] 改善一般性能的无机填料
MFH [mineral filler best heat resistance] 改善耐热性的无机填料
MFI [melt flow index] 熔体流动指数
MFM [mineral filler best moisture resistance] 改善耐湿性的无机填料
MFR [melt flow rate] 熔体流动速率
Mg [magnesium] 镁
mg [milligramme] 毫克
MH [medium hard] 中等硬(度)的
Ml [melt index] 熔体指数
mica 云母〈填料〉
~ -filled 云母填充的
~ flake 云母(鳞)片〈填料〉
~ flour 云母粉
~ plate 云母板,云母片
~ powder 云母粉〈填料剂〉
~ sheet 云母片〈片材〉
~ specks 云母斑点
~ talc 云母滑石

Micanite 云母层合材料,云母板
Micatherm 云母绝缘材料
micellar 胶束的;微胞的
~ catalysis 胶束催化
~ dispersion 胶束分散体
~ emulsion 胶束乳液
~ net structure 胶束网状结构,微胞网状结构
~ orientation 胶束定向
~ phase 胶束相
~ solubility 胶束溶解度
~ structure 胶束结构
~ surface 胶束表面
~ theory 胶束理论

micelle 胶束;微胞

~ formation 胶束形成
~ model 胶束模型
micro 微(的),显微(的)
~ – analysis 微量分析
~ and semimicro viscometer 微量和半微量黏度计
~ – balance 微量天平
~ – Brownian motion 微布朗运动
~ – capsule 微胶囊〈含有胶黏剂〉
~ – constituent 微量成分
~ – control 精密调节器
~ – crack 微裂
~ – crystal 微晶
~ – dispersed filler 微粒分散填充剂
~ – gel 微粒凝胶
~ glass beads (微)玻璃珠
~ – hardness tester 微型硬度计
~ – indentation hardness testing 微量压痕硬度试验
~ – indentation test 微型硬度计
~ – injection 微型零件注塑
~ injection moulding 微型零件注塑
~ – jet 微喷气刷
~ – jet roll coater 微喷气刷辊涂布机
~ – jet roll coating 微喷气刷辊涂布
~ – mechanics 微观力学
~ – mesh 微型小网眼
~ – miniaturization 微型化
~ – organisms resistance 耐微生物性〈塑料材料〉
~ – packing 微型填料
~ – phase structure 微观相结构
~ – porous rubber 微孔泡沫胶
~ probe analyzer 微探针分析器
~ – rheology 微观流变学
~ – sample 微量取样
~ – size particles 超细粒子
~ – stretching 微拉伸
~ – suspension polyvinyl chloride [MS – PVC] 微悬浮法聚氯乙烯
~ – test 微量试验;显微检验
~ – welding 微型焊接
microadding 微量填加
microadjustment 微量调整

microautoradiography 微射线自动照相术
microballoon 微球
microbe 微生物;细菌
microbial 微生物的;细菌的
~ resistance 抗微生物性
microbiological 微生物的
~ degradation 微生物降解
~ resistance 耐[抗]微生物性
microbulking 微膨化
microcapillarity 微毛细管作用
microcapsulary 微胶囊的
microcapsule 微胶囊
~ adhesive 微胶囊型黏合剂
~ dyes 微胶囊染料
microcellular 微孔的
~ foam 微孔泡沫塑料
~ rubber 微孔橡胶
~ structure 微孔结构,微泡沫结构
microchromatography 微量色谱法
microcolorimeter 微量比色计
microcomponent 微量组分
microcomputer 微型计算机
microconfiguration 微观结构
microcrack 显微裂纹,微裂纹
microcreep 微观蠕变
microcrystal 微晶
microcrystalline 微晶的
~ cellulose 微晶纤维素
~ structure 微晶结构
~ wax 微晶石蜡〈润滑剂〉
microcrystallite 微晶
microdensitometry 微密度测定(法)
microdetection 微量测定
microdetermination 微量测定
microdiffusion 微量扩散
microdyn tester 微型强力试验仪
microelements 微型元素;微型元件
microencapsulated 微型囊封的
microencapsulation 微型囊封,微型胶囊包封
microequivalent 微(克)当量
microexamination 显微检验,微观研究
microfeeder 微量加料器

microfibre 微纤维
microfibrillar structure 微纤结构
microfiltration membranes 微滤膜, 精密滤膜
microfissure 显微裂纹, 微裂纹
microflaw 显微裂纹, 发裂纹
microfoam 微孔
~ dyeing 微孔染色
~ rubber 微孔橡胶
microfocus X-ray tube 细聚焦 X-射线管
microfusion 微量熔化
microgap 微间隙
microgel 微粒凝胶
microgel particle 凝胶微粒
microgram [μg] 微克 $\langle 10^{-9}g \rangle$
microgranular 微晶粒状的
microhardness 微(型)硬度
~ ball indentor 球压痕微型硬度计
~ test (微)型硬度试验
~ tester 微型硬度测试仪
microheater 微型加热器
microheterogeneity 微观(结构)不均匀性, 微观(结构)多相性
microheterogeneous polyblend 微观非均相共混物
microhole 微孔, 细孔
microhomogeneity 微观(结构)均匀性
microlite 微晶
microliter [μL, μl] 微升 $\langle 10^{-9}L \rangle$
microlitic structure 微晶结构
microlustre method 微光泽测定法
micromatic setting 微调装置
micromechanical analysis 微观力学分析
micromeritics *pl* 粉粒学; 粉流学
micromesh 微孔筛
micrometer 测微计
~ thickness gauge 厚度测定器, 测厚计
micromolecule 微分子
micromorphology 微观形态学
micron mill 微粉磨机
micronized aluminum 微粒铝
micronizer [microniser] (喷射)研磨机, 微磨机, 微粒化机
micronotch 微切口; 微缺口
microorganism 微生物
microparticle support 微粒状载体
microphotograph 显微照相; 显微照片
microphotometer 测微光度计
microplastometer 微量塑性计
micropore 微孔
~ structure 微孔结构
microporosity 微孔性
microporous 微孔性的
~ materials on the base of polyurethane 微孔聚氨酯材料
~ plastic 微孔塑料, 细泡沫塑料
~ plastic film 微孔塑料薄膜
~ plastic sheet 微孔塑料板材
~ polymer 微孔聚合物
~ rubber 微孔橡胶
micropowder 超细粉
micropressure 微压
micropulverizer 微粉磨机
micropulverizing 精微粉碎
microrheology 微观流变学
microscope 显微镜
~ camera 显微照相机
~ slide 载物玻璃片, 滑动片(显微镜)
~ testing 显微镜检验
microsection (显微)磨片
microspheres [microballoons] 微球〈填料〉
microspheric 微球状的
microstrain 微应变
microstress 微应力
microstructure 微观结构
~ study 微观结构研究
microswelling 微膨胀
microtacticity 微观规整性[度]
microthin section 超薄切片
microtome 超微切片机
~ cut 微型切片
~ knife 细微切片刀
microviscosimeter 微型黏度计
microviscosity 微黏性
microwave 微波

~ curing 微波固化
~ discharge 微波放电
~ drying 微波烘燥
~ dyeing 微波染色
~ heater 微波加热器
~ heating 微波加热
~ heating cabinet 微波加热箱
~ inspection 微波检查
~ method 微波方法
~ radiation 微波辐射
~ vulcanization 微波硫化
middle 中间(的)
~ break 中断
~ coating 中间涂层
~ distillate 中间馏分
~ fraction 中间馏分
~ shade 中间色
midfibre 中长纤维
midget 微型(的),极小型的
~ motor 微型电动机
~ mo(u)lder 微型注塑机
migrate 迁移,移动;渗移〈增塑剂〉
migrating 迁移,移动
~ dye 泳移性染料
~ plasticizer 渗移性增塑剂
migration 迁移,移动;渗移
~ fastness 色移牢度
~ of colorant 着色剂渗移
~ of organic pigments in plastics 塑料中有机颜料渗移
~ of plasticizers 增塑剂渗移
~ stain 渗移斑
migratory plasticizer 渗移性增塑剂
Mikromat 球研磨机
mill 密耳〈0.001吋,0.254毫米〉
mil. [million] 兆,百万
mild 温和的,轻微的
~ cracking 轻度破裂
~ steel 软钢,低碳钢
mildew 霉
~ proofing 防霉处理
~ resistance 防霉变性
~ resistant 防霉的
~ retarding agent 抑霉剂

mileage 涂料涂敷量
milkiness 乳状,乳白色〈涂层〉
mill 研磨机,磨;机(器)
~ base 研磨基料
~ board 封面纸板,书面纸板
~ grain 磨粒
~ grinder 研磨机
~ line 研磨机组;辊压纹
~ mixing 辊炼
~ roll opening 辊隙
~ shrinkage 辊轧收缩〈塑料半成品加工后〉
~ surface 磨碎面
milled 磨碎了的;滚花的
~ fibres 磨碎纤维,研磨纤维〈玻璃纤维〉
~ glass fibre 磨碎的玻璃纤维,短玻璃纤维
~ head screw 滚花头螺钉
milling 研磨,磨碎;辊炼;铣削
~ ball 球磨机的球
~ cutter 切口铣刀,槽铣刀〈用于塑料加工〉
~ machine 研磨机;铣床
~ section 研磨室
~ time 研磨时间;滚压时间;辊炼时间
Mills–Pirelli 米尔斯·皮雷利
~ needle blowing 米尔斯·皮雷利气针吹塑
~ process 米尔斯·皮雷利法,空心针吹塑成型法
Millstruder 连续混合挤塑机
min. [minimum] 最小;最小的
min [minute] 分钟
mineral 矿物(的);无机的
~ additive 无机[矿物]添加剂
~ binder bond 无机粘结剂
~ black 石墨;矿黑
~ carbon 石墨
~ colour 矿物颜料,无机颜料
~ compound 无机化合物
~ dyeing 矿物染料染色
~ dyes 矿物染料

- ~ fibre 矿物纤维,无机纤维
- ~ fibre filler 矿物纤维填料
- ~ filled 矿物填充的
- ~ filler 矿物质填料,无机填料
- ~ filler best general［MFG］ 改善一般性能的无机填料
- ~ filler best heat resistance［MFH］ 改善耐热性的无机填料
- ~ filler best moisture resistance 改善耐湿性的无机填料
- ~ filler high electric［MFE］ 改善电性能的无机填料,高电性能的无机填料
- ~ loading 无机物填充量
- ~ matter 矿物质
- ~ paint 无机涂料
- ~ pigment 无机颜料
- ~ - reinforced thermoplastic［MRTP］ 矿物质增强热塑性塑料,无机物增强热塑性塑料
- ~ reinforcement 矿物质增强材料,无机物增强材料
- ~ resin 矿物树脂
- ~ spirit 石油溶剂油,200号溶剂油,〈俗称〉松香水,白节油〈漆用汽油〉

miniature 微型:小型的
- ~ car 小型汽车,微型汽车
- ~ limit switch 微型限位开关〈注塑机〉
- ~ socket 微型套筒

miniaturization 小型化,微型化
minifoam 微泡沫体
minimal 最小的,最低的,极小的
- ~ condition 最低条件
- ~ down time 最短停产时间
- ~ path 最短程
- ~ value 最小值,极小值

minimum 最小的,最少的,最低的
- ~ bend radius 最小弯曲半径
- ~ blister - free cure time 最短无泡固化时间
- ~ diameter 最小直径
- ~ diameter of thread 螺纹最小直径
- ~ film - forming temperature 最低成膜温度
- ~ function 最低函数
- ~ ignition time 最短点燃时间
- ~ problem 极小问题
- ~ size 最小尺寸
- ~ relaxation time 极小松弛时间
- ~ speed 最低速率［度］
- ~ standard 最低标准
- ~ stress 最小应力
- ~ temperature 最低温度
- ~ value 最小值,最低值
- ~ weather 最劣天气

minnow net machine 编网机
minor 较小的;较少的;较短的
- ~ adjustment 小调整
- ~ component 微量组分
- ~ cycle 短周期,小循环
- ~ defect 不重要缺陷,小毛病
- ~ detail 细节
- ~ diameter （螺纹）内径
- ~ diameter of thread 螺纹内径
- ~ repair 小修

minus 负;减的
- ~ effect 不良效果,副作用
- ~ material 次品

minute 分(钟);微小的;精密的
- ~ adjustment 精密调节,精调
- ~ bubbles 小气泡
- ~ crack 细裂缝
- ~ quantity 极少量,微量

mipor 微孔的
- ~ plastic 微孔塑料
- ~ rubber 微孔橡胶

mirror 镜,反射镜
- ~ finish 镜面光泽度
- ~ - image 镜象
- ~ polishing 镜面研磨;镜面抛光

mirrorstone 白云母;云母
misadjustment 错调,误调,失调
miscibility 溶混性;渗混性;可混性
miscible 可(溶)混的
mist 雾
- ~ cooling 喷雾冷却
- ~ separator 湿气分离器
- ~ spinning 喷雾纺丝

- ~ spray 喷雾
- **mitre** [miter] 斜接
- ~ joint 斜接头,斜接合
- ~ fillet weld 斜角焊接〈焊接〉
- **mix** [mixture] 混合;混合物;混合料
- ~ design 配料设计
- **mixed** 混合的
- ~ adhesive 混合黏合剂
- ~ adipates of $C_6 \sim C_{10}$ normal alcohols 己二酸 $C_6 \sim C_{10}$ 正构醇混合酯〈增塑剂〉
- ~ cell structure 混合泡孔结构
- ~ chromatogram 混合色谱(图)
- ~ coefficient 混合系数
- ~ col(o)ur effect 混色效应
- ~ colo(u)r method 混色法
- ~ complex 混合络合物
- ~ constant 混合常数
- ~ crystal 混合晶
- ~ dyes 混合染料
- ~ fabric 混纺织物
- ~ feeding 混合供料
- ~ fibre 混合纤维
- ~ flow [complex flow] 混合流动
- ~ flow impeller 混流式混合器
- ~ glass fibre cloth 混合玻璃纤维(滤)布
- ~ glue 预混胶,混合型黏合剂,混合胶
- ~ grain size 混合粒度
- ~ -in-place 就地混合
- ~ injection moulding 混合注塑
- ~ melting point 混合熔点
- ~ packing 混合填充(物)
- ~ phase 混合相
- ~ phthalates of $C_7 \sim C_9$ alcohols 邻苯二甲酸 $C_7 \sim C_9$ 醇混合酯〈增塑剂〉
- ~ phthalates of $C_9 \sim C_{11}$ alcohols 邻苯二酸 $C_9 \sim C_{11}$ 醇混合酯,911 酯〈增塑剂〉
- ~ phthalates of $C_6 \sim C_{10}$ normal alcohols 邻苯二甲酸 $C_6 \sim C_{10}$ 正构醇混合酯,610 酯〈增塑剂〉
- ~ phthalates of $C_8 \sim C_{10}$ normal alcohols 邻苯二甲酸 $C_8 \sim C_{10}$ 正构醇混合酯,810 酯〈增塑剂〉
- ~ polyester 混合聚酯
- ~ polymer 混合聚合物
- ~ polymerization 混合聚合(作用)
- ~ reflection 混合反射〈光学〉
- ~ sample 混合试样
- ~ sampling 混合取样
- ~ spinning 混纺
- ~ tensor 混合张量
- ~ transmission 混合透射〈光学〉
- ~ type 混合型
- ~ yarn 混纺纱;混色纱
- **mixer** 混合机
- ~ barrel 混合筒,搅拌筒
- ~ -dispenser 混合配料器
- ~ extruder 挤出混料机
- **mixing** 混合,掺混
- ~ agitator 混合搅拌器
- ~ at site 就地混合
- ~ beater 混合打浆机
- ~ chamber 混合室,搅拌室
- ~ colo(u)r 混合色,拼色
- ~ cone 混合圆锥体;混料喷嘴
- ~ container 混合容器
- ~ device 混合设备
- ~ drum 混合鼓
- ~ efficiency 混合效率,混合程度〈挤塑物〉
- ~ element 混合元件,混合部件〈螺杆〉
- ~ extruder 混炼挤塑机
- ~ feed 混合供给
- ~ formula 混合配方
- ~ gun 混合喷枪
- ~ head 混合头,混料喷头,混炼头
- ~ head screw 混合头螺杆
- ~ hollander 混合打浆机
- ~ machine 混合机
- ~ length 混合程
- ~ mill 辊式混合器,辊炼机
- ~ nozzle 混料喷嘴
- ~ order 混合加料顺序

~ paddle 混合浆叶
~ process 混合过程；混合方法
~ proportion 混合比，配合比
~ ratio 混合比(率)，配合比(率)，配料比
~ rod 混合棒；混料棒
~ roll 混合辊；混炼辊
~ rotor 密炼转子
~ screw 混合螺杆，混料螺杆，螺旋混合器
~ section 混合段
~ section and smear head 混合段和塑化头〈挤塑机〉
~ thoroughly 彻底混合；拌透
~ time 混合时间；混料时间；混炼时间
~ torpedo 混料鱼雷头，鱼雷型混炼头
~ tube 混合管
~ vessel 混合容器
~ zone 混合段
mixplaster 混合塑化机〈混合－熔融－反应器装置〉
mixtuder 混炼挤塑机
mixture 混合；混合物，混合料
~ composition 混合物成分
~ control 混合控制，混合比调节
~ density 混合物密度
~ dyes 混合染料
~ fabric 混纺织物
~ heat 混合热
~ of bis(hexamethylene) triamine and homologues 双六亚甲基三胺及其同系物的混合物〈固化剂〉
~ optimum 最佳混合物
~ ratio 混合比(率)，配料比
~ rule 混合定则
~ section 混合段
~ temperature 混合物温度
mL [milliliter] 毫升
mm [millimeter] 毫米
MM [molecular mass] 分子量
MMA [methyl methacrylate] 甲基丙烯酸甲酯〈有机玻璃单体〉
MMD [molecular mass distribution] 分子量分布
m. m. f. [magneto-motive force] 磁(动)势
Mn [manganese] 锰
Mn [molecular weight number average] 数均分子量
Mn [number-average molecular weight] 数(量平)均分子量
Mo [molybdenum] 钼
mobile 汽车；易动的，可移动的
~ bottom plate 可动底板〈模具〉
~ phase 流动相
~ shutter 闸板〈关闭装置〉
~ welding unit 移动式焊接装置
mocha 深咖啡，深褐色
mock-up 模型；模板
modality 模态
mode 方式，样式，形状；模式，模样
~ of operation 工作方式，操作方法
~ pattern 模型图案
~ shades 流行色泽，时髦色泽
model 模型(的)
~ assumption 模型假设
~ experiment 模型试验
~ fiber 模型纤维
~ law 模型定律
~ machine 样机
~ number 型号
~ set (成套)模型
~ test 模型试验
~ theory 模型理论〈注塑〉
modelling 模型化；模型形成〈注塑过程〉
moderate 中等的
~ cracking 中等破裂
~ foaming 中泡的
~ pressure polyethylene 中压聚乙烯
~ speed 中等速率[度]
~ strength 中等强度
~ vacuum [low vacuum] 低度真空
moderator 调节剂
modification 改性，变性；改良，改进
~ agent 改性剂
modified 改性的，变性的；改良的，改进的

~ acrylics 改性丙烯酸系塑料
~ alked resin 改性醇酸树脂
~ branched polyethlene 变性支链聚乙烯
~ cellulose 改性纤维素
~ cellulose fibre 改性纤维素纤维
~ continuous filament 变形长丝
~ cross-section fibre 异形截面纤维
~ fibre 变性纤维
~ hexamethylene diamine 改性的六亚甲基二胺〈固化剂〉
~ insulated-runner-type mould 改型保温流道式模具
~ intermittent sealing 改良间断热合
~ ion exchange membrane 改性离子交换膜
~ linear polyethylene 变性直链聚乙烯
~ machine 改型机〈注塑机〉
~ natural polymer 改性天然聚合物
~ Ostwald viscomeder 改良型奥氏黏度计
~ plastic 改性塑料
~ plastisol 改性增塑糊,改性增塑溶胶
~ polyphenylene oxide [MPPO] 改性聚苯醚
~ rayon 变性人造丝
~ resin 改性树脂
~ rubber 改性橡胶
~ staple fibre 变性短纤维
~ wood 改性木材
modifier 改性剂;调节剂
modul 系数;模数,模量
modular 模(数)的,系数的,比率的;由标准件组合的
~ chilling system 组合式冷却系统
~ concept 组合概念,组合原则
~ connector 组合式接插件
~ construction [principle, system] 组合结构 [原理,系统],单元结构
~ design 组合式设计
~ mo(u)ld 组合模具
~ principle 模数原理
~ screw 组合式螺杆

module 模件;组件;模量,模数,系数
~ of compliance 柔性模量
~ of elasticity 弹性模数
~ of resilience 回弹模数
~ of rigidity 刚性模数
~ units 组装单元
modulus 模量,模数;模件,组件
~ at 300% 300%定伸模量
~ at 500% 500%定伸模量
~ in compression 压缩模量
~ in flexure 挠曲模量
~ in shear 剪切模量
~ in torsion 扭转弹性模量
~ increase 模量增加
~ of cross elasticity 横向弹性模量
~ of direct elasticity 纵向弹性模量
~ of elasticity 弹性模量
~ of elasticity in flexure 弯曲弹性模量
~ of elasticity in shear 剪切性模量
~ of elasticity in tension 拉伸弹性模量
~ of flexural rigidity 弯曲刚性模量
~ of longitudinal elasticity 纵向弹性模量〈垂直应力与垂直变形关系的模量〉
~ of resilience 回弹模量
~ of rigidity 刚性模量
~ of rupture 断裂模量
~ of section 截面模量,断面系数
~ of shear 剪切模量
~ of shearing elasticity 剪切弹性模量
~ of stretch 拉伸模量
~ of torsion 扭转(弹性)模量
~ of transverse elasticity 横向弹性模量
~ of volume elactivity 体积弹性模量
~ of volume expansion 体积膨胀模量
~ -temperature curve 模量-温度曲线
~ tester 模量测定仪
Mohs' 莫斯(的),莫氏(的)
~ hardness 莫斯硬度,莫氏硬度
~ hardness number 莫斯硬度值
moire 石纹(的),波纹(的)
~ effect 波纹效应
~ fringes 条纹;波动光栅

moist
- ~ method 波纹法〈波带变形的测定〉
- ~ patter 波纹图像

moist 潮湿的,湿润的
- ~ atmosphere 潮湿空气
- ~ material 潮湿材料,湿料
- ~ product (潮)湿产品
- ~ room conditions 湿室环境

moisten 湿润,湿湿;变湿
moistener 湿润剂;湿润器
moisture 湿气,水分,(潮)湿度
- ~ absorption 吸湿;吸湿性
- ~ absorption resistance 防吸湿性
- ~ absorption test 吸湿试验
- ~ apparatus 测湿计,湿度计
- ~ barrier 防湿层
- ~ barrier property 防湿性,隔湿性
- ~ blow 湿气泡,潮湿生成的小水泡〈塑料制品或涂层〉
- ~ content 含湿量
- ~ content control 湿度控制
- ~ - curing urethane prepolymer 湿气固化聚氨酯预聚体
- ~ curve 湿度(变化)曲线
- ~ equilibrium 湿平衡
- ~ factor 湿系数
- ~ - free weight 脱湿重,干重
- ~ gradient 湿气梯度
- ~ meter 湿度计
- ~ percentage 含水率
- ~ permeability 透湿性
- ~ pickup 吸湿(率)
- ~ - proof 防湿的
- ~ proof cellophane 防潮玻璃纸〈涂有硝酸纤维素或偏氯乙烯共聚物等,用于食品及其他包装材料〉
- ~ regain 吸湿,吸湿性;回潮率
- ~ resistance 防湿性
- ~ resistant 防湿的
- ~ tester 测湿器
- ~ - tight 防湿的,不透水的
- ~ vapor transmission [MVT] 透湿性,透湿率

mol 摩〈摩尔 mole 的符号〉
mol. [molecular] 分子的
mol. [molecule] 分子
mol. wt [molecular weight] 分子量
molal 分子的:摩尔的.〈旧称〉克分子的;质量摩尔浓度的
- ~ concentration 质量摩尔浓度,mol/kg〈旧称重量克分子浓度〉
- ~ depression constant 摩尔冰点下降常数
- ~ elevation constant 摩尔沸点升高常数
- ~ solution 摩尔溶液
- ~ volume 摩尔体积
- ~ weight 摩尔质量

molality 重量摩尔浓度;重模
molar 摩尔的〈mole 的形容词〉;克分子的〈旧定义〉;摩尔浓度〈按体积计〉的;用于研磨的,能磨碎的
- ~ concentration 体积摩尔浓度〈旧称体积克分子浓度,现中英文均已废除,而代之以物质的量浓度 c〉
- ~ concentration 摩尔浓度
- ~ heat 摩尔热
- ~ mass 摩尔质量
- ~ mass average 平均摩尔质量
- ~ ratio 摩尔比
- ~ solution 摩尔溶液
- ~ volume 摩尔体积
- ~ weight 摩尔量;克分子量〈旧称〉

molarity 摩尔浓度;容模
mold [mo(u)ld] 模具,塑模,模型;霉,霉菌〈美〉
mole [mol] 摩尔,摩〈物质的量的 SI 单位,SI 七个基本单位之一。它是一系统的物质的量,该系统中包含的基本单元数与 0.012kg 碳-12 的原子数目相等〉;〈旧定义为〉克分子,克分子量

molecular 分子的
- ~ adhesion 分子(间)附着
- ~ arrangement 分子排列
- ~ asymmetry 分子不对称(性)
- ~ beam 分子束
- ~ bond 分子键
- ~ calculation 分子量测定,分子量计算

- ~ chain　分子链
- ~ chain axis　分子链轴,分子链轴线
- ~ chain length　分子链长
- ~ chain unit　分子链结构单元
- ~ cohesion　分子内聚(作用)
- ~ compounds pl　分子化合物
- ~ configuration　分子构型
- ~ conjugation　分子共轭
- ~ design　分子设计
- ~ disentanglement　分子(链)解开,分子(链)解缠结
- ~ entanglement　分子(链)缠绕
- ~ force　分子力
- ~ formula　分子式
- ~ heat　分子热
- ~ length　分子(链)长度
- ~ link　分子键
- ~ mass　分子量
- ~ mass distribution [MMD]　分子质量分布
- ~ mobility　分子流动性
- ~ motion　分子运动
- ~ network　分子网络
- ~ number　分子序(数)
- ~ orbital　分子轨道
- ~ orientation　分子取向
- ~ orientation of fibre　纤维的分子取向
- ~ packing density　分子堆积密度
- ~ polarization　分子极化(作用)
- ~ rearrangement　分子重排
- ~ sieves　分子筛⟨催化剂⟩
- ~ specific heat　分子比热
- ~ spectrum　分子光谱
- ~ still　分子蒸发器
- ~ structure　分子结构
- ~ symmetry　分子对称(性)
- ~ volume　分子体积
- ~ weight [MW]　分子量
- ~ weight average　平均分子量
- ~ weight dependent　分子量依赖性
- ~ - weight distribution [MWD]　分子量分布
- ~ weight distribution curve　分子量分布曲线
- ~ weight distribution measurement　分子量分布测定
- ~ weight fractionation　分子量分级[分离]
- ~ weight modifier　分子量调节剂
- ~ weight-viscosity constant　分子量-黏度常数

molecularly compatible　分子相容的
molecule　分子
mollification　软化(作用)
mollifier　软化剂;软化器
molten　熔化的,熔融的
- ~ bead　熔焊条;挤塑焊条
- ~ bead sealing　熔条封焊
- ~ bead welding　熔条焊接
- ~ bubble　熔融膜泡⟨吹塑薄膜⟩
- ~ extrudate　熔态挤塑物
- ~ film　熔融薄膜
- ~ film coating　熔膜涂布
- ~ glass　熔融玻璃
- ~ layer　熔层
- ~ mass　熔体,熔融物质
- ~ state　熔融状态

molybdate　钼酸盐⟨或酯⟩
molybdenum [Mo]　钼
- ~ disulfide　二硫化钼⟨润滑剂,填料⟩
- ~ sulfide　硫化钼

molybdic　钼
- ~ oxide　三氧化钼⟨阻燃剂⟩
- ~ sulfide　三硫化二钼

moment　瞬间,片刻;(力,弯,挠)矩
- ~ of deflection　弯曲矩
- ~ of flexure　挠曲矩
- ~ of force　力矩
- ~ of inertia　惯性矩
- ~ of rotation　转(动)矩

momentary state　暂时状态
momentum　动量
- ~ of gyration　旋转动量

monel metal　蒙乃尔合金⟨铜镍合金⟩
mono⟨字头⟩　单,一
- ~ - axial stretching　单轴拉伸
- ~ - crystalline　单晶的
- ~ - hole flow test　单孔流动试验

monoaxially

~ – molecular layer 单分子层
~ – rotation system powder moulding 单轴旋转粉末成型
monoaxially 单轴地
~ drawn film 单轴拉伸薄膜
~ drawn film strip 单轴拉伸膜条
~ stretched film tape 单轴拉伸膜带
monoazo 单偶氮
~ dyes 单偶氮染料
~ pigment 单偶氮颜料
monobasic 一(碱)价的;一元的;一代的
monoblock 整体
~ cast 整体铸塑
~ casting 整体铸塑件
monobutyl itaconate 衣康酸单丁酯〈增塑剂〉
monochromatic[monochroic] 单色的
~ analysis 单色分析
~ analyzer 单色分析器
~ colo(u)r 单色的
~ filter 单色滤光片,单色滤色片
~ harmony 单色协调
~ light 单色光
monochromator[monochrometer] 单色器
monoclinic 单斜晶的
~ form 单斜晶形
~ macrolattice 单斜大晶格〈部分结晶热塑性塑料〉
~ prisms 单斜棱晶
~ system 单斜晶系
monocolo(u)r 单色
monocomponent 单组分
~ fibre 单组分纤维
monocrystal 单晶
~ whisker 单晶须晶
monodisperse 单分散(性)
~ polymer 单分散聚合物
~ system 单分散系
monodispersity 单分散法
monifil[monofilament] 单丝,单纤丝
monofilament[monofil] 单丝,单纤丝
~ extrusion 单丝挤出
~ netting 单丝(渔)网
monofilm 单分子膜

monofluid spray gun 单喷射流枪
monolayer 单(分子)层
monolithic 整体的;单块(的),单片(的)
monomer 单体
~ cast nylon 单体浇铸尼龙,铸型尼龙
~ casting 单体浇铸塑
~ – plasticizer 单体型增塑剂
~ – polymer 单体聚合物
~ – polymer casting 单体–聚合物浇铸
~ – polymer particle 单体–聚合物粒子
~ ratio 单体配比
~ reactivity 单体(聚合)活性
~ reactivity ratio 单体竞聚率
~ relative reactivity 单体相对活性
~ sequence distribution 单体序列分布
monomeric 单体的
~ cement 单体粘接剂
~ cement bonding 单体粘接
~ substance 单体物质
~ unit 链节;单体单元;结构单元〈高聚物〉
monomethyl itaconate 衣康酸单甲酯〈增塑剂〉
monomolecular 单分子的
~ film 单分子膜
~ layer 单分子层
~ membrane 单分子膜
~ reaction 单分子反应
monotone 单色的
monovalent 一价的;单价的
monoxide 一氧化物
montanic ester wax 褐煤酯蜡,蒙旦蜡〈润滑剂〉
Mooney 门尼
~ index 门尼指数
~ plasticity 门尼塑性
~ plastometer 门尼塑度计
~ shearing disk plastometer 门尼剪切圆盘式塑度计
~ shearing disk viscometer 门尼剪切圆盘式黏度计
~ value 门尼值〈黏度〉

mould

~ viscosimeter 门尼黏度计
~ viscosity 门尼粘度
mop 布轮
mordant 蚀刻剂,浸蚀剂;媒染剂;媒染的
　　~ colour 媒介染料
　　~ dye 媒介染料
　　~ dyeing 媒染染色
morpholine disulphide［MDS］二硫化吗啉
morpholinyl 吗啉基
2－(4－~－dithio) benzothiazole　2－(4－吗啡啉基二硫代)苯并噻唑〈促进剂 MDB〉
2－(4－~ mercapto) benzothiazole　2－(4－吗啡啉基硫代)苯并噻唑〈促进剂 NOBS〉
morphological structure 形态结构
moth－proofing 防蛀;防蛀处理
mother 母;母的;本国的
　　~ liquor 母液
　　~－of－pearl effect 珍珠母效应;珠光效应
motion 运动;移动
motionless 不动的;固定的,静止的
　　~ mixer 静态混合器
　　~ mixing unit 静态混合装置
motive fluid 活动流体
motor 电动机
　　~ operator 电动机控制器
　　~ ratings 电动机功率
　　~ truck 载重汽车,运货汽车
mottle 斑点;混色斑纹,色斑
mottling 斑点,麻点
mould［mold〈美〉］ 模具;模(型);模塑;压模;霉菌〈英〉
　　~ accuracy 模具准确度
　　~ adjustment 模具调整
　　~ alignment 模具对准;模具定心
　　~ amortization 模具折旧
　　~ area 模具面积
　　~ attack 模具腐蚀
　　~ base 模型底座
　　~ block 模块

~ breating 模具放气
~ bumping 模具放气
~ carrier block 载模座
~ carrier unit 载模装置,模固定部件
~ cavity 模腔,型腔;模槽;阴模
~ cavity capacity 模腔容量;模槽容量
~ cavity depth 模腔深度;模槽深度
~ cavity pressure 模腔压力;模槽压力
~ chamber 模腔
~ changing time 模具更换时间
~ channel 模内沟道
~ charge 装模;装模料
~ chilling 模具冷却
~ clamp(ing) 合模
~ clamping capacity 合模力,锁模力
~ clamping compound 合模料
~ clamping device 合模装置,锁模装置
~ clamping force 合模力,锁模力
~ clamping frame 合模框架
~ clamping mechanism 合模机构,锁模机构
~ clamping plate 合模板
~ clamping pressure 合模压力,锁模压力
~ clamping stroke 合模行程,锁模行程
~ clamping unit 合模装置,锁模装置;夹持装置〈注塑机〉
~ cleaning 清模
~ cleaning gun 清模枪
~ close 闭模
~ closed time 闭模时间,合模时间
~ closing 闭模,合模
~ closing force 闭模力,合模力,锁模力;模具夹持力
~ closing ram 闭模活塞,合模活塞
~ closing speed control 闭模速度控制
~ closing station 闭模装置
~ closing time 闭模时间,合模时间
~ closing unit 闭模装置,合模装置,锁模装置〈注塑机〉
~ closure 闭模,合模
~ component 模具部件

mould

- ~ compound 模塑料
- ~ conditioning 模具调温
- ~ conditioning oven 模具调温箱
- ~ construction 模具结构
- ~ cooling 模具冷却
- ~ core 模芯
- ~ cover 模具盖〈发泡成型〉
- ~ cycle 模塑周期,成型周期
- ~ design 模具设计
- ~ designer 模具设计人员
- ~ die 模具
- ~ dimension 模具尺寸
- ~ discharging agent 脱模剂
- ~ efficiency 模塑效率
- ~ engraving machine 模具刻花纹机
- ~ extension 模具延长
- ~ extractor 脱模器
- ~ feeding 充模
- ~ feeding time 模具填充时间,充模时间
- ~ fill volume 模具装料容积,充模容积
- ~ filled short 充模缺料,不完全充模〈注塑〉
- ~ filling 充模;充模过程
- ~ filling cycle 充模周期〈注塑时〉
- ~ filling efficiency 充膜效率
- ~ -filling process 充模过程
- ~ -fill(ing) time 模具填充(料)时间,充模时间
- ~ fixing platen 模具固定板;
- ~ flow properties 模内流动性
- ~ for bottom ram press 下压式压机用模具
- ~ for up-stroke press 上压式压机用模具
- ~ force 模塑力〈注塑〉;阳模
- ~ forcer 阳模
- ~ form 模腔;型腔
- ~ frame 模架
- ~ gating 模浇口
- ~ gating nozzle 模具注嘴
- ~ heater-cooler 模具控温器,模具温度自动调节器
- ~ height 模具高度,装模空挡
- ~ height adjustment screw 模具高度调节螺丝
- ~ -holder 持模器,模具架〈用于聚氨酯泡沫塑料反应注塑〉
- ~ impression 模腔,型腔;阴模
- ~ indexing speed 模具轮回速度,模具转位速度
- ~ indexing table 模具轮回台,模具转盘
- ~ inhibitor 抑霉剂
- ~ insert 模腔嵌件,型腔嵌件
- ~ insert for marking 有标志的模腔嵌件
- ~ installation height 模具安装高度
- ~ jaw 模腔夹持器
- ~ life 模具寿命,模型寿命
- ~ load 装模料
- ~ loading 装模(料)
- ~ loading temperature 装模料温度
- ~ locking 锁模,合模
- ~ locking force 锁模力,合模力
- ~ locking unit 锁模装置〈注塑机〉
- ~ lubricant 润模剂,脱模剂
- ~ lubricant sprayer 润模剂喷雾器,脱模剂喷雾器
- ~ maker 模具制造者
- ~ making 模具制造
- ~ mark 模具痕;合模线迹;模具标志
- ~ mat 模压毡料
- ~ materials 模具材料
- ~ meeting surface 合模面
- ~ misalignment 模具未对准
- ~ mounting crane 模具安装用吊车
- ~ mounting platen 载模板,装模板
- ~ open time 开模时间
- ~ opening 模具开距,模板间距〈模具〉
- ~ opening force 开模力
- ~ opening stroke 开模行程〈注塑机〉
- ~ part tolerance 成型件公差
- ~ parting agent 脱模剂
- ~ parting face 合模面;分型面
- ~ parting force 开模力

~ parting line 合模线;拼缝线;分型线
~ parts 模具部件
~ pin 模具用销
~ plate 模板
~ platen measurements 模板尺寸
~ plating 模具电镀
~ plunger 阳模,凸模
~ polishing 模具抛光
~ preparation station 模具准备工作
~ pressure 模塑压力,成型压力
~ pressure sensor 模塑压力传感器
~ protection device 模具保护装置
~ register 模具对准
~ release 脱模,脱模剂
~ - release agent 脱模剂,润模剂
~ release equipment 脱模装置〈注塑机〉
~ release medium 脱模剂
~ residence time 模具滞留时间
~ resistance 防霉菌性
~ retaining flange 模具固定用法兰,定模法兰
~ - rotating injection mo(u)lding 模具旋转注塑
~ seam 合模脊缝;模缝
~ shrinkage 模塑收缩率,成型收缩,离模后收缩
~ shrinkage allowance 模塑收缩留量
~ size 模具尺寸
~ slide 模具滑板
~ station 载模装置
~ steel 模具用钢材
~ sticking 粘模
~ stripping 脱模
~ support 模具(支)架
~ supporting plate 模具支承板
~ surface area 模具表面积
~ temperature 模具温度
~ temperature controller 模温控制器
~ time 模塑时间
~ train 模具组
~ transfer type blow moulding machine 模具移动式吹塑机
~ turnover rate 模具周转速率

~ unloading 卸模,脱模
~ wash 模制件上黏附的脱模剂
~ wax 脱模蜡
~ wedge 模楔
~ wiper 脱模器
~ with conical splits 具有脱模锥度的模具〈便于模塑脱模〉
~ with injection ram and mo(u)ld movement along different axis 带有注塑柱塞并沿不同轴向动作的模具
~ with injection ram and mo(u)ld movement along the same axis 带有注塑柱塞并沿相同轴向动作的模具
mo(u)ldability 模塑性,成型性
mo(u)ldable 可模塑的
~ prepreg 可模塑预浸料
mo(u)lded 模塑的,模制的
~ article 模塑件,成型件
~ article cooling 模塑件冷却
~ brake lining 模塑制动垫
~ casting 模铸
~ fibre board 模压纤维板
~ foam block 模塑泡沫块
~ goods 模制品
~ hose 模塑软管
~ impregnated wood 模制浸渍木材
~ - in - foil decoration 模内敷箔装饰
~ in place 现场模塑
~ - in stress 模塑内应力〈模塑件内的残余应力〉
~ insulating material 绝缘模塑料
~ laminate 模塑层压制品,层合成型制品
~ laminated plastic tube 模塑层压塑料管,层合成型塑料管
~ laminated rod 模塑层压棒
~ laminated section 层合成型型材
~ laminated tube 层合成型管材
~ material 模塑料,成型料
~ part 模塑件,成型件
~ part from chips 木屑成型件〈由浸渍过树脂的木屑制的模塑件〉
~ piece 模塑件
~ plastics 模塑用塑料

mo(u)lder

~ plastic pirn 模塑的塑料纤管[筒管]
~ plywood 模塑胶合板
~ resin 模塑用树脂
~ rod 模塑棒材,成型棒材
~ screw 模塑螺丝,模制螺丝
~ specimen 模塑试样
~ structural foam 模塑结构泡沫塑料
~ thermoplastic article 热塑性塑料模塑件
~ thermosetting article 热固性塑料模塑件
~ thread 模塑螺纹,模制螺纹
~ tube 模塑软管
mo(u)lder 成型机;塑料加工者
mo(u)lding 模塑,成型;模制品,模塑件〈产品〉
~ abrasion 成型磨耗,模具磨耗〈使用时表面〉
~ article 模塑件,成型件
~ blank 模塑坯料,模塑半成品
~ board 模塑板,成型板
~ box 模塑箱,成型箱
~ breakthrough 模制品破裂
~ capacity 模塑量
~ chamber 模腔
~ charge 填模量
~ composite 模塑料,模塑混合料
~ composition 模塑料,模塑混合料
~ compound 模塑料,成型料
~ conditions 模塑条件;成型条件;模塑参数
~ cycle 模塑循环,模塑周期,成型周期
~ cycle time 模塑周期时间;成型周期时间
~ defect[fault] 模塑缺陷;成型缺陷〈模制品〉
~ efficiency 模塑能力,模塑效率
~ extractor 脱模器
~ gloss 塑件光泽度
~ in the first heat 模型首次加热
~ index 模塑指数
~ lubricant 润模剂,脱模剂

~ machine 模塑机,成型机
~ material 模塑料,成型料
~ mixture 模塑混合料
~ nest material 模腔材料
~ operation 模塑操作,成型操作
~ part 模塑件,成型件
~ plant 模塑装置,成型装置;模塑车间,成型车间
~ plate 装模板;装模压板
~ platen 装模压板
~ polytetrafluoroethylene 聚四氟乙烯模塑
~ powder 模塑粉,成型粉
~ powder density 模塑粉密度
~ press 模压机,压塑压机,成型压机
~ pressure 模塑压力,成型压力
~ process 模塑过程,成型过程;模塑方法,成型方法
~ product 模制品
~ resin 模塑用树脂,成型用树脂
~ shrinkage 模塑收缩(率),成型收缩(率)
~ skin 模塑表面层
~ strain 模塑应变
~ stripping 脱模
~ technique 模塑技术,成型技术
~ temperature 模塑温度,成型温度
~ time 模塑时间,成型时间
~ tolerance 模塑公差,成型公差
~ tool 模具,塑料加工成型模具
~ variables 模塑参数
~ viscosity 模塑黏度,成型黏度
~ wax 滑模蜡
~ with rotation[MWR] 旋转模塑
~ without sprue residues 无流道残料模塑件
~ zone 模塑段,成型段
mouldproof 防霉的
mount 架,座;安装,装配
mounting 安装;装配
~ clip 装配夹
~ diagram 装配图
~ plate 装模板
mouth 模口;孔;嘴

~ - piece 口模
movable 可移动的,移动式的
~ bottom plate 动模托板
~ crane 桥式吊车,移动式起重机
~ extruder 移动式挤塑机
~ part of the mould 模的可动部分
~ platen 动模板,移动板;动压板
~ platen side 动模板侧;动压板侧
~ retainer plate 可动托模板
~ side 动模侧
movement 活动,移动
moving 活动(的)
~ base mould 可移动芯棒的吹塑模具
~ bed 活动台架;移动床
~ bed process 移动床法
~ bed reactor 移动床反应器
~ belt production 传送带生产;流水线生产
~ blow mandrel 活动吹塑(气)芯
~ boundary 移动界面
~ bridge crane 移动桥式吊车
~ die platen 动模板
~ head 动模板〈注塑机〉;动夹头〈试验机〉
~ insert 活动嵌件
~ jaw 活动夹具
~ mo(u)ld 动模
~ mo(u)ld blow moulding 动模式吹塑成型
~ parts 活动部件
~ plate 动压板
~ pressure roll 动压辊〈焊接机〉
~ pressure roller 动压辊
~ side 动模侧
~ stop 移动挡板
~ table 动压台
~ veneer 薄层板〈制备层合板用〉
m. p. [melting point] 熔点
MPCS [methyl pentachlorostearate] 五氯硬脂酸甲酯〈增塑剂〉
MPEG [methyl phthalyl - ethyl - glycolate] 邻苯二甲酰基乙基乙醇酸甲酯〈增塑剂〉
MPVC [mass polyvinyl chloride] 本体法聚氯乙烯
MRTP [mineral - reinforced thermoplastic] 矿物质增强热塑性塑料,无机物增强热塑性塑料
MS [medium soft] 半柔软
MSPVC [microsuspension polyvinyl chloride] 微悬浮法聚氯乙烯
mucilage 黏液,黏胶;胶乳;胶乳黏合剂
mud 泥浆
~ baffle 挡泥板
~ - cracking 网状微细裂纹,龟裂
~ shield 挡泥板
muff joint 套接,套筒接头
muffle furnace 马弗炉,隔焰炉
mulch 覆盖(物)
~ film 农膜〈包括地膜、棚膜、篷膜等〉
muller 研磨机
~ mixer 研磨混合器,逆流转盘式混合器
multi 〈字头〉多
~ - axial drawing 多轴向拉伸
~ - axial stretching 多轴向拉伸
~ - bank press 多层压机
~ - cavity injection mould 多模腔注塑模具
~ - cavity mo(u)ld [multi - impression mould] 多模腔模具
~ - cellular structure 多孔结构
~ - chain polymer 多链聚合物
~ - chain condensation polymer 多链缩聚物,星形缩聚物
~ - channel die 多槽缝型模头
~ - channel sheet die 多槽缝片材机头
~ - circuit 多线路的
~ - circuit pattern 多线路缠绕
~ - circuit winding 多线路缠绕
~ - daylight press 多开距压机,多层压机
~ - directional stress 多方向应力
~ - element model 多元模型
~ - fibre 复型纤维
~ - flighted screw 多(头)螺纹螺杆

〈美〉
~ - fold tool 多型腔模具
~ - gate 多浇口
~ - gate mo(u)ld 多浇口模具
~ - gate system 多浇口系统
~ - gated 复式浇口(的)
~ - gating 多浇口
~ head blow mo(u)lding machine 多模头式吹塑机
~ - head plant 多头注塑装置〈用于聚氨酯泡沫塑料的反应注塑〉
~ - hole die plate 多孔口模板
~ - hole jet 多孔喷丝头
~ - impression injection mould 多模腔注塑模具
~ - impression mo(u)ld 多模腔模具,群腔模
~ - knife coating 多刮刀涂布
~ manifold die 多料道模头〈共挤塑〉
~ - nozzle 多头-注口嘴
~ - pack package 组合式包装
~ - platen press 多层板压机〈美〉
~ - ply 多层压机
~ - ply plywood 多层胶合板
~ - ply yarn 多股纱;多绞线
~ - section melt screw extruder 多段熔融螺杆挤塑机
~ - shell curing 分层固化〈增强塑料〉
~ - slot die [multi-channel die] 多缝型模头〈共挤塑〉
~ - stacked mo(u)ld 多层模具
~ - station blow moulding machine 多工位吹塑机
~ - station injection mo(u)lding machine 多工位注塑机
~ - station mo(u)lding machine 多工位模塑机
~ - station welding machine 多工位焊机
~ - tier injection mo(u)ld 多层注塑模具
~ - tier mould 多层(注)模具
~ - tier tool 多层注塑模具

~ - way package 多种方式包装
~ - zone screw 多段螺杆
multiblock copolymer 高嵌段共聚物,多嵌段共聚物
multibranched polymer 多支链聚合物
multicolo(u)r 多色的
~ cloth 多色织物
~ effect 多色效应
~ extruder 多色挤塑机
~ injection mo(u)ld 多色注塑模具
~ injection mo(u)lding 多色注塑(成型)
~ injection mo(u)lding machine 多色注塑机
~ print 多色印刷
~ profile extrusion 多色型材挤塑
multicolo(u)red 多色的
~ mo(u)lding 多色注塑件;多色成型件
~ plastic component 多色塑料件
multi-component 多组分的
~ catalyst 多组分催化剂
~ fibre 多组分(复合)纤维
~ film fabre 多组分薄膜纤维
~ injection 多组分注injection
~ mixing machine 多组分混合机
~ mixture 多组分混合料
~ polymer fibre 多组分共聚物纤维
~ polymerization 多元聚合
~ reactive liquids 多组分反应液
~ spraying 多组分喷涂
multichromatic spectrophotometry 多色分光光度计
multicylinder dryer 多筒干燥器
multifeeder machine 多喂料机
multifilament 复丝
~ roving 复丝粗纱
~ yarn 复丝
multifoam 多元泡沫塑料
multilaminate 多层层合制品
multilayer 多层
~ blow mo(u)lding 多层吹塑
~ cation exchange membrane 多层阳离子交换膜

multiple

- ~ compound 多层复合物
- ~ copolymer 多层共聚物
- ~ extrusion 多层(共)挤塑
- ~ fabric 多层织物
- ~ film 多层复合薄膜
- ~ film extrusion 多层薄膜(共)挤塑
- ~ sheet 多层片材
- ~ tubular film 多层复合管状薄膜
- ~ welding 多层焊接

multilevel 多水平面的,多层的
- ~ dryer 多层干燥机
- ~ injection mould 多层注塑模具
- ~ mould 多层模具

multiorifice die 多孔模头

multipass 多途径的,多(行)程的
- ~ dip coater 多道蘸涂布
- ~ dryer 多程干燥器;片材干燥器
- ~ heat exchanger 多通道热交换器,多程换热器

multiphase 多相的
- ~ plastics 多相塑料
- ~ structure 多相结构
- ~ system 多相系统

multiplaten press 多层板压机〈美〉

multiple 多(层)的;复合的
- ~ bead 多层焊道
- ~ beam agitator 多桨叶搅拌器
- ~ beam paddle mixer 多桨叶混合器
- ~ bond 重键
- ~ bonding 多重结合;重键
- ~ -cage mill 多笼式磨
- ~ cavity die 多模腔式模头
- ~ cavity mould 多模腔模具
- ~ -circuit 多线路的
- ~ circuit helical winding 多线路螺旋形卷绕
- ~ -circuit pattern 多路花纹
- ~ -circuit polar winding 交叉极向卷绕
- ~ -circuit winding 多路卷绕
- ~ -component injection moulding 多组分注塑(成型)
- ~ crossing 复式交叉
- ~ cutting tool 组合切削工具,组合刀具
- ~ deck press 多层压机
- ~ die 多位模
- ~ die tubing machine 多模(头)挤管机
- ~ dip coater 多道蘸涂机
- ~ disk clutch 多盘式离合器
- ~ dispersion 多分散性
- ~ draft 多倍牵伸
- ~ expansion 多级膨胀
- ~ fabric 多层织物
- ~ feed 多浇口进料
- ~ fibre material 多种纤维材料,复合纤维材料
- ~ flighted screw 多(头)螺纹螺杆
- ~ gating system 多点浇口系统
- ~ granulation 多次造粒,反复造粒
- ~ layer adhesive 复合层黏合剂
- ~ -lip die 多模唇模具
- ~ manifold die 多料道模头
- ~ metering 多位计量,多位配料
- ~ mo(u)ld 多腔模具
- ~ -opening platen press 多层板压机〈美〉
- ~ paddle mixer 多桨叶混合器
- ~ part adhesive 多组分黏合剂
- ~ pass 多程,多路
- ~ platen press 多层板压机
- ~ ply yarn 多股线
- ~ -point contact 多点接触
- ~ polar wind 交叉极向卷绕
- ~ processing 再次加工,多次加工〈热塑性塑料回收加工〉
- ~ purpose machine 多功能机,通用机,万能机
- ~ roll crusher 多辊破碎机
- ~ sampling 多次取样
- ~ -screw extruder 多螺杆挤塑机
- ~ shot moulding 多色注塑(成型)
- ~ -spindle 复式锭子
- ~ spinneret 组合喷丝板
- ~ stage press 多级压机
- ~ strand yarn 多股纱;合股纱
- ~ stretch 多次拉伸

multipoint

~ tab gate　复式柄型浇口
~ thread　多(头)螺纹
~ thread screw　多(头)螺纹螺杆
~ – unit press　多层压机
~ wound glass filament yarn　多股玻璃纤维长丝纱
~ wound glass staple fibre yarn　多股玻璃短纤维纱
~ wound yarn　多股络纱
multipoint　多点
~ injection　多点注塑
~ metering　多点计量
multipolymer　多元聚合物,共聚物
~ sheet　多元聚合物片材
multiport valve　多路阀
multi – purpose　通用[万能]的,多功能[多性能,多方面,多用途]的
~ adhesive　通用黏合剂;万能黏合剂
~ extruder　通用挤塑机;万能挤塑机
~ screw　通用螺杆〈挤塑机〉
~ tester　万能试验机,万能测试仪
multirun welding　多层焊接
multiscrew　多螺杆
~ extruder　多螺杆挤塑机
~ pipe extruder　多螺杆管材挤塑机,多螺杆挤管机
multispoked piece　多脚支架圈
multistage　多级的,多段的
~ compounding　多级混料
~ core[multi – stage cylindrical core]　多段(圆柱形)模芯
~ core screw　多级空心螺杆
~ cylindrical core　多段圆柱形模芯
~ drawing　多级拉伸
~ extrusion die　多级挤塑模头

~ extrusion screw with degassing　多级排气式挤塑螺杆
~ single – screw extrusion　多级单螺杆挤塑
~ stretching　多级拉伸
multistrand　多股
~ chain　多股链
~ roving　多股粗纱;多股纱条
multitrum weld　多层焊缝
multivalent　多价的
muntin　窗格条
muriatic acid　盐酸
murrey　桑果紫红〈深暗紫红色〉
muscovite　白云母,钾云母〈优质云母,填料〉
mushroom mixer　蘑菇状混合器,倾斜转筒式混合器
mutual solubility　互溶性
mV[millivolt(s)]　毫伏(特)
Mv[viscosity – average molecular weight]　黏(度平)均分子量
MVT[moisture vapor transmission]　透湿性,透湿率
Mw[molecular weight]　分子量
Mw[weight – average molecular weight]　重(量平)均分子量
MWD[molecular weight distribution]　分子量分布
MWR[mo(u)lding with rotation]　旋转模塑
Mylar film[polyethylene glycol terephthalate film]　米拉薄膜〈聚对苯二甲酸乙二(醇)酯薄膜,聚酯薄膜,公司〉
DuPont

N

N[Newton]　牛[顿]
N[nitrogen]　氮
N; n[normal]　正常的;标准的;规定的

n.[note]　摘录;注解
Na[sodium]　钠
nacarat　鲜艳橘红色

nacreous 有珍珠的
- ~ effect 珍珠母效应
- ~ pigments 有珠光的颜料

nailable 可打钉的
- ~ plastic 可钉塑料

naked 裸(体)的;无遮蔽的
- ~ eye 肉眼
- ~ flame 无遮蔽的火焰

name 名称
- ~ of part 品名
- ~ plate 铭牌

nap 绒毛;起绒

naphtha 石脑油,粗汽油,汽油馏分,溶剂(汽)油

naphthalene 萘

naphthenate 环烷酸盐

naphthol 萘酚;〈大写〉色酚
- ~ dye 萘酚类染料
- ~ yellow 萘酚黄

naphthyl group 萘基

napkin ring test 黏合薄壁圆筒的扭曲试验,黏合薄钢筒的剪切试验〈为测定耐剪切强度〉

narrow 窄的
- ~ fabric 带;窄幅织物
- ~ fabric with selvage 有织边的窄幅织物,有边织带
- ~ fabric without selvage 无织边的窄幅织物,毛边织带
- ~ restriction 注嘴;喷嘴
- ~ stitch buff [close stitch buff] 密针抛光轮
- ~ tube 小直径管

nascent 初生的,新生的

national 国家的
- ~ standard 国家标准

National Bureau of Standards[NBS] 国家标准局(美国)

Natta,Giulio G.纳塔
- ~ catalyst 纳塔催化剂

nattier blue 淡蓝色

natural 天然的,自然的;固有的;正常的
- ~ adhesive 天然黏合剂
- ~ aging 自然老化
- ~ cellulose 天然纤维素
- ~ circulation 自然循环;自然通风
- ~ circulation evaporator 自然循环蒸发器
- ~ colour 天然色;本色
- ~ condition test 天然条件试验
- ~ convection 自然对流
- ~ cooling 自然冷却
- ~ crack 自然裂纹,天然裂纹
- ~ draw ratio 自然拉伸比
- ~ durability test 天然耐久性试验
- ~ dye 天然染料
- ~ fiber 天然纤维
- ~ frequency 固有频率,自然频率
- ~ gas 天然气
- ~ glue 天然胶黏剂
- ~ gum 天然(树)胶
- ~ gutta percha 天然杜仲胶
- ~ high polymer 天然高聚物
- ~ parameter 特性参数
- ~ plastic 天然塑料;本色塑料
- ~ polymer 天然聚合物
- ~ resin 天然树脂
- ~ rubber[NR] 天然橡胶
- ~ rubber adhesive 天然橡胶黏合剂
- ~ rubber latex adhesive 天然胶乳黏合剂
- ~ size 原尺寸
- ~ time 松弛时间
- ~ wear 自然磨损

NB[nominal bore] 公称内径〈管材〉

N. B.[nota bene]〈拉〉 注意

NBR[acrylonitrile butadiene rubber] 丙烯腈-丁二烯橡胶,丁腈橡胶

NBR[nitrile butadiene rubber] 丙烯腈-丁二烯橡胶,丁腈橡胶

NBR – modified phenolic mo(u)lding powder 丁腈橡胶改性酚醛模塑粉

NC[numerically controlled] 数字控制的

NCR[nitrile chloroprene rubber] 腈基氯丁橡胶

NDOP[n–decyl, n–octyl phthalate] 邻苯二甲酸正癸·正辛酯

neat 纯的,净的

neatness 光洁性,光洁度;干净性
necessary condition 必要条件
neck（-in） 颈缩;缩幅〈吹塑薄膜〉;垂缩〈吹塑型坯〉;瓶颈〈瓶〉;辊颈〈压延机〉;细颈
- ~ bead 环形颈缘
- ~ -down 颈缩
- ~ finish 颈部(螺纹等)〈瓶〉
- ~ flash 颈部溢料〈瓶〉
- ~ -in 缩幅;缩颈
- ~ insert 颈部嵌件
- ~ ring 颈环,带环形槽的环
- ~ ring calibration 颈口定型
- ~ ring process 颈环法〈为使中空容器开口部分颈的内面和外侧精度高的一种成型方法〉
- ~ -shaping tool 颈部成型模
- ~ -thread finishing 颈部螺纹定型
- ~ tool 颈部(成型)模

necked-in container 缩颈容器
necking 颈状收缩,颈缩
- ~ behaviour 颈缩性
- ~ down 颈缩
- ~ zone 颈缩区

needle 针,针状物
- ~ bar[pin rod] 针床;针杆〈原纤化〉
- ~ blow 针吹法
- ~ blow moulding 针吹塑
- ~ blowing 插针吹塑,气针吹塑
- ~[needled] cylinder 针滚筒,针辊
- ~ crystal 针状结晶
- ~ felt 针刺毡
- ~ flame test 针状焰色试验
- ~ gate 针型浇口
- ~ -loomed felt 针刺毡
- ~ mat 绗缝毡(垫),针刺毡片
- ~ nozzle 针状关闭喷嘴
- ~ plate 针板
- ~ -punched batt 针刺毛毡
- ~ punched carpet 针刺地毯
- ~ -punched felt 针刺毡
- ~ punching 针刺;针刺法
- ~ punching machine 针刺机
- ~ roll 针辊
- ~ roller 针辊
- ~ row 针刺行列
- ~ seal nozzle 针阀关闭喷嘴
- ~ shut-off 针阀关闭
- ~ shut-off nozzle 针阀关闭喷嘴
- ~ splitting 针状编织
- ~ stich tear strength 针刺撕裂强度
- ~ tear resistance 耐针刺撕裂性,针刺撕裂强度,耐穿孔性
- ~ tear resistance test 耐针刺撕裂试验,针刺撕裂强度试验
- ~ tear strength 针刺撕裂强度
- ~ type(shut-off)nozzle 针阀关闭喷嘴
- ~ valve 针阀

needled 针状的,针刺的
- ~ buff 针刺抛光轮
- ~ mat 针刺毡
- ~ non-woven fabric 针刺无纺织物

negative 阴的,负的
- ~ catalyst 负催化剂
- ~ die 阴模,凹模,雌模
- ~ direction 逆向,反方向
- ~ forming 负压成型
- ~ gage pressure 负表压
- ~ mould 阴模
- ~ pressure 负压;真空
- ~ stain 负染色法,负反衬法
- ~ stretch 负拉伸
- ~ structural viscosity 负结构黏度
- ~ temperature 负温度,零下温度
- ~ thermoforming 负压热成型;压模热成型
- ~ vacuum forming 真空负压成型

nematic liquid crystal 向列型液晶
neon indicator 氖指示灯
neopentyl glycol 新戊二醇
- ~ dibenzoate 新戊二醇二苯甲酸酯〈增塑剂〉

Neoprene 〈商〉氯丁橡胶,氯丁二烯橡胶
- ~ adhesive 氯丁橡胶黏合剂
- ~ cement 氯丁胶浆

~ faced laminate 氯丁胶面层合板
~ latex 氯丁橡胶浆
~ rubber 氯丁橡胶
~ rubber – phenolic resin adhesive 氯丁橡胶,酚醛树脂黏合剂
~ synthetic rubber 氯丁合成橡胶
nerve 复原性;回缩性
Nesslers 内斯勒
~ colo(u)r comparsion 内斯勒比色法
~ reagent 内斯勒试剂
~ tube 内斯勒比色管
nest 模腔;模槽;阴模
~ mould 多巢模
~ plate 模穴套板,圈板
nested cloth 经纬织物
nesting 嵌置
~ cloth 经纬织物
~ fixture anvil 自动固定架
net 净的,纯的;网状的
~ area efficiency 净面效率〈有孔带状板,由幅宽减去孔径后的拉伸强度与板材拉伸强度之比〉
~ density 净密度
~ effect 综合效果
~ efficiency 净效率
~ extrusion 软网挤塑
~ film 网格薄膜
~ flow 净流
~ high – polymer 网状高聚物
~ knotting machine 结网机
~ – like structure 网状结构
~ machine 编网机
~ polymer 网状聚合物
~ sack 网眼袋
~ section 有效断面,净断面
~ structure 网状结构
~ tolerance 净公差
~ weight 净重
netting 网(状)
~ analysis 网络[格]分析〈增强塑料〉
~ twine 结网线,渔网丝
network 网络,体形结构
~ chain 网链

~ density 网络密度
~ fibrillated yarn 网状纤化纱
~ fibrillation 网状纤化(作用)
~ model 网状模型
~ point 交联点
~ polymer [crosslinked polymer] 网状聚合物,体形聚合物
~ structure 网状结构
~ yarn 网状纱〈玻璃纤维〉
neutral 中性的,中和的,中间的
~ colo(u)r 中和色,灰色,不鲜明的颜色
~ conductor 中(性)线,零线〈电〉
~ dyes 中性染料
~ equilibrium 随遇平衡,中性平衡
~ fiber 中性纤维
~ screw 中空螺杆
~ step wedge 中性阶梯楔〈光学〉
~ tone 中和色调〈灰色度〉
~ wedge 中和楔〈光学〉
neutralize 中和,平稳;抵消
neutrography [neutron radiography] 中子射线照相法
neutron 中子
~ beam 中子(射)束
~ diffraction 中子衍射
~ radiography 中子射线照相法
neutronography 中子照相法
new 新的
~ mode 新样式
~ model 新式样
Newton[N] 牛顿[力的单位]
~ liquid 牛顿液体,牛顿流体
Newton's law 牛顿定律
Newtonian 牛顿的
~ behaviour[character] 牛顿行为[特性]
~ flow 牛顿流动
~ fluid 牛顿流体
~ liminting viscosity 牛顿极限黏度
~ shear viscosity 牛顿剪切黏度
~ viscosity 牛顿黏度
NHDP [n – hexyl, n – decyl phthalate]

邻苯二甲酸正己·正癸酯
Ni [nickel] 镍
nib 疙瘩〈涂料〉
nibs *pl* 粒料〈挤塑〉;毛料〈纱疵〉
nick 刻痕,缺口;裂纹,裂缝
nickel[Ni] 镍
~ N,N-di-n-butyl dithiocarbamate N,N′-二正丁基二硫代氨基甲酸镍〈光稳定剂〉
~ 3,5-di-*tert*-butyl-4-hydroxybenzylphosphonate monoethylate 双(3,5-二叔丁基-4-羟基苄基膦酸单乙酯)镍〈光稳定剂2002〉
~ dimethyl dithio carbamate 二甲基二硫代氨基甲酸镍〈光稳定剂〉
~ dioctyl dithio carbamate 二辛基二硫代氨基甲酸镍〈光稳定剂〉
~ naphthenate 环烷酸镍〈催化剂〉
~ sesquioxide 三氧化二镍〈填料〉
night strip 导向带,磁带导带
nine-ply structure 九层结构〈胶合板〉
niobium [Nb;columbium] 铌
nip 辊隙〈两辊间隙〉,辊间装料隙
~ angle 辊隙角;咬入角;夹角;夹料辊
~ feed 辊隙进料
~ pressure 辊隙压力,辊压紧力
~ roll 夹辊,牵引辊;层压辊
~ roller 支承辊,受力辊
nipping of unit [squeeze rollers] 挤压装置,夹紧装置〈薄膜吹塑〉
nipple 螺纹接口〈管件〉
~ joint 螺纹接管;管接头接合
NIR[acrylonitrile isoprene rubber] 丙烯腈-异戊二烯橡胶
niton,Nt 氡
nitrate 硝酸盐;硝酸酯;硝化
nitrating plant 硝化装置
nitration 硝化(作用)
nitric acid 硝酸
nitriding 渗氮的,氮化的〈钢〉
nitrile 腈
~ chloroprene rubber[NCR] 腈基氯丁橡胶
~ -butadiene rubber [NBR] 丁腈橡胶
~ group 腈基,氰基
~ silicone rubber 腈硅橡胶
~ rubber 丁腈橡胶
~ rubber adhesive 丁腈橡胶黏合剂
~ rubber-phenolic resin adhesive 丁腈橡胶-酚醛树脂黏合剂
~ rubber PVC blend 丁腈橡胶-聚氯乙烯共混物
nitro 〈词头〉硝基
~ -colo(u)r 硝基色料
~ -group 硝基
~ urea 硝基脲
nitrobenzene 硝基苯
nitrocellulose [cellulose nitrate] 硝酸纤维素,硝化纤维素
~ coating 硝酸纤维素涂料
~ lacquer 硝酸纤维素漆
~ plastics 硝酸纤维素塑料
nitrogen N 氮
~ -pressurized rheometer 充氮流变仪;氮压流变仪
nitroglycerine 硝化甘油,甘油三硝酸酯
nitromethane 硝基甲烷
nitropropane 硝基丙烷
nitroso rubber 亚硝基橡胶
N-nitrosodiphenylamine N-亚硝基二苯胺〈高效阻聚剂N-NO,防焦剂〉
NMR[nuclear magnetic resonance] 核磁共振
~ measurement 核磁共振测定
~ spectrograph 核磁共振摄谱仪
~ spectrometer 核磁共振波谱仪
~ spectroscopy 核磁共振波谱法
~ study 核磁共振研究
no 无,不,非
~ -account 没价值的
~ -clamp bonding process 非夹紧性胶接法;快速粘接法
~ -effect level 无效剂量〈毒性〉
~ effect level for test animals 对(谷

椎)动物的容许剂量
- ~ - effect value　无效剂量值〈毒性〉
- ~ - load　无荷载,空载
- ~ - load running　空载运转
- ~ - load test　空载试验
- ~ - load work　空载运转
- ~ - matrix resin　非基料树脂
- ~ - mo(u)ld mo(u)lding　无模成型
- ~ - pressure laminate　常压层合(材料),常压层合制品,接触层合(材料)
- ~ - pressure resin　常压树脂;无压固化树脂
- ~ - rub wax　防滑蜡〈用于地板〉
- ~ - strength temperature　失强温度〈用于测定热变形性〉
- ~ - strength temperature tester　零强温度试验机〈简称 NST 试验机〉
- ~ - twist roving　无捻粗纱

No.; no. [number]　数;值;数字;号码

NODA [n-octyl n-decyl adipate]　己二酸正辛・正癸酯

NODP [n-octyl n-decyl phthalate]　邻苯二甲酸正辛・正癸酯

NODTM [tri(n-octyl, n-decyl) trimellitate]　偏苯三酸正辛・正癸酯

nodular　结节(状)的;粒状的

noise　噪声
- ~ abatement　消声,减声
- ~ absorbing　吸声,吸音
- ~ barrier　吸声板
- ~ barrier panel　隔声板,隔音板
- ~ damping　噪声衰减(的),隔音(的),消音(的)
- ~ deadener　隔声[音]材料
- ~ deadening foam　隔声[音]泡沫材料
- ~ level　噪声级
- ~ meter　噪声测试器
- ~ pollution　噪声污染
- ~ survey　噪声测定

NOL ring　诺尔环〈用浸过树脂的连续纤维环向缠绕而制成规定尺寸的环形试样〉

Nom. [nominal]　标称的;公称的

Nomex　〈商〉诺曼克斯〈杜邦公司产芳香族聚酰胺〉

nominal　公称的,标称的,标定的,铭牌的
- ~ bore [NB; DN]　公称内径
- ~ capacity　额定功率;额定生产率
- ~ density　公称密度〈泡沫塑料〉
- ~ diameter　公称直径
- ~ diameter of filaments or staple fibres　长丝或短纤维的公称直径
- ~ dimension　公称尺寸
- ~ error　公称误差
- ~ horsepower　额定马力,标定功率
- ~ inside diameter　公称内径
- ~ load　额定负荷
- ~ output　额定输出,标称输出
- ~ pipe size　公称管径
- ~ pressure [NP;PN]　额定压力,标称压力
- ~ rating　额定值
- ~ size　公称尺寸,公称规格
- ~ size of pipes　管材的公称尺寸
- ~ stress　公称应力
- ~ value　额定值,标称值
- ~ voltag　额定电压
- ~ weight　标称重量
- ~ width　标称宽度

non　非,不,无,未
- ~ - abrasive quality　非耐磨性
- ~ - abrasiveness [non-abrasive quality]　不磨损性
- ~ - absorbency　不吸水性
- ~ - aerated　无孔,未加发泡剂的〈泡沫塑料〉
- ~ aging　未老化的
- ~ alkali glass fibre　无碱玻璃纤维
- ~ - aqueous adhesive　非水溶液型黏合剂
- ~ - aqueous polymerization　非水溶液聚合,非水相聚合
- ~ - assembly adhesive　非装配用黏合剂
- ~ - bibulous paper　不吸水纸

~ - bleeding 不洇色(的),不渗色(的)
~ - blocking film 非粘连性薄膜
~ - blooming 不起霜
~ - blooming gradient 非渗出性组分
~ - breakable 不易破碎的,不破裂的,不易断裂的
~ - brittle 非脆(性)的
~ - buckling 不皱缩的
~ - burning type foamed plastics 耐[不]燃性泡沫塑料
~ - burning type plastics 耐[不]燃性塑料
~ - catalytic polymerization 非催化聚合
~ - cellular mo(u)lded parl 无孔模塑件
~ - coherent 无黏聚力的,松散的,无黏性的
~ - combustible 不易燃的
~ - combustible fabric 不燃性织物
~ - combustible material 不燃性材料
~ - compatible plasticizer 不相容增塑剂
~ - conductive 不传导的,绝缘的
~ - conductive fibre 绝缘纤维
~ - constant lead screw 不等距螺杆
~ - continuous 不连续的,间断的
~ - convertible coating 不可转化涂层
~ - copolymerizable monomeric additive 非共聚用单体添加剂
~ - corrodible 耐腐的,非腐蚀性的
~ - corrosive 无腐蚀(性)的;不锈的
~ - crystalline 非晶性的
~ - crystalline chain orientation 非晶性链段取向
~ - crystalline region 非晶区
~ - crystallizable region 非晶区,无定形区
~ - cutting shaping 非切削成型法
~ - defective 合格品,无缺陷物品
~ - degradable waste 不可降解废料
~ - destructive 非破坏性

~ - destructive evaluation 非破坏性评价
~ - destructive examination 非破坏性测验
~ - destructive inspection 非破坏性检验,无损探伤,无损检验
~ - destructive materials testing 非破坏性材料试验
~ - destructive method 非破坏性测试法
~ - destructive test 非破坏性试验
~ - diffusing plate 非散射板
~ - discolouring 不褪色的;不变色的
~ - drying oil 非干性油
~ - elastic 无弹性的,非弹性的
~ - elasticity 无弹性,非弹性
~ - equilibrium polycondensation 非平衡缩聚
~ - expansion 不膨胀
~ - extractable plasticizer 耐萃取的增塑剂
~ - fading 不褪色
~ - felting 不毡合性;不毡合的
~ - ferrous metal 非铁金属;有色金属
~ - flam 不燃的
~ - flame properties 耐燃性
~ - flammability 耐燃性
~ - flammable 不易燃的,耐燃的
~ - flammable celluloid 耐燃性赛璐珞
~ - flammable foam 耐燃性泡沫塑料
~ - flocculating resin 非絮凝树脂
~ - fraying 非磨损的;不散边的〈纺织〉
~ - friable 不易破碎的
~ - glare light 不眩光的
~ - homogeneous 不均匀的
~ - homogeneous deformation 非均匀形变
~ - homogeneous ion exchange membrane 非均相离子交换膜
~ - hygroscopic 不吸潮的
~ - ignitability 耐燃性,不可点燃性

~ – inflammability 不可燃性
~ – inflammable 不易燃的
~ – intercommunicating 不互通的〈泡沫塑料〉
~ – intercommunicating cell 闭孔不互相连通的泡孔〈泡沫塑料〉
~ – interconnecting cell 不互连泡孔,闭孔
~ – intermeshing screws 非啮合式螺杆
~ – intermeshing twinscrew extruder 非啮合式双螺杆挤塑机
~ – ionic 非离子的
~ – ionic dyes 非离子型染料
ionic emulsifier 非离子型乳化剂
~ – ionic surfactant 非离子型表面活性剂
~ – ionic surfaceactive agent 非离子型表面活性剂
~ – isothermal condition 非等温条件
~ – isothermal process 非等温过程
~ – joint glue 非连接性胶黏剂〈美〉
~ – linear 非线性的
~ – linear elastic relation 非线型弹性关系
~ – linear phenomena 非线性现象
~ – linear polymer 非线型聚合物
~ – linear viscoelasticity 非线性黏弹性
~ – linear winding 非线性缠绕〈增强塑料〉
~ – loadbearing 不能承载(的),不载重(的)
~ – magnetic coating 非磁性涂层
~ – magnetic material 非磁性材料
~ – marring 耐划痕的;非刮伤的〈美〉
~ – metallic 非金属的
~ – metallic insert 非金属嵌件
~ – metallic material 非金属材料
~ – migratory plasticizer 非迁移性增塑剂
~ – odo(u)rous 无气味的,无臭的

~ – oriented film 非定向薄膜,非拉伸薄膜
~ – polar 非极性的
~ – polar bond 非极性键
~ – polar linkage 非极性键
~ – polar plastic 非极性塑料
~ – polar polymer 非极性聚合物
~ – polar resin 非极性树脂
~ – pressure main 无压总管
~ – pressure tapping 无压管开孔
~ – pressurized main 无压总管
~ – reactive diluent 非反应性稀释剂,非活性稀释剂
~ – reactivity 无反应性
~ – reciprocating extruder screw 非往复式挤塑机螺杆
~ – recoverable creep 不可恢复蠕变
~ – recoverable deformation 不可恢复形变
~ – register printing 不对齐印刷
~ – regulatory products 不正规产品,不标准产品
~ – reinforced 未增强的
~ – reinforcing filler 非增强填料
~ – resonant forced vibration technique 无共振强制的振动技术
~ – return flow valve 止回流阀
~ – return valve [check valve] 止回阀,单向阀〈注塑机〉
~ – return valve stop 止回阀
~ – returnable 不可回收的
~ – returnable bottle 不可回收瓶
~ – rewound pirn 非复绕纬纱管〈织造〉
~ – rigid 非刚性的,软的
~ – rigid film 软质薄膜
~ – rigid plastic 非刚性塑料,软质塑料
~ – rigid polyvinyl chloride 非刚性聚氯乙烯,软质聚氯乙烯
~ – rigid sheeting 非刚性片材,软质片材
~ – rotting 防腐的

~ - scaling 不起鳞的
~ - selective filter 非选择性滤光器〈光学〉
~ - shattering 不碎的;防震的
~ - shrinkable 防缩的〈纺织品〉
~ - skid 防滑的
~ - skid coating 防滑涂层;防滑涂料
~ - skid deck coating 防滑舱面涂层
~ - skid flooring 防滑地板(材料),防滑铺地材料
~ - slip 防滑的
~ - slip finish 防滑整理
~ - slip surface 不滑爽表面;防滑面;摩擦面〈塑料铺面〉
~ - shrink 防缩
~ - shrink treatment 防缩处理
~ - soluble 不溶解的
~ - solvent 非溶剂
~ - solvent lamination 无溶剂层合
~ - staining 不污染的
~ - staining antioxidant 非污染性抗氧剂;无着色的抗氧剂;
~ - staining property 防污染性;不着色性
~ - stationary flow 非稳定流动
~ - steady flow 非稳定流动
~ - steady state 非稳定态
~ - stereospecific polymer 无规立构聚合物
~ - sticking characteristics 无黏性
~ - stop 不停的,不间断的
~ - stop operation 连续操作
~ - stop roll transfer 不停机换卷辊
~ - stop run 连续运转
~ - stop screen pack changer 不停车换滤网器;自动换滤网器〈挤塑机〉
~ - stressed 无应力的
~ - stretch 无伸缩
~ - structural adhesive 非结构型黏合剂
~ - structural covering 非结构式覆盖层
~ - swollen 不膨胀的

~ - symmetric 不对称的
~ - toxic 无毒(性)的
~ - toxic plasticizers 无毒增塑剂
~ - toxicity 无毒性
~ - tracking 无径迹
~ - transparency 不透明性
~ - transparent 不透明的
~ - twist roving 无捻粗纱
~ - uniform polymer 非均匀聚合物
~ - uniformity 不均匀性
~ - viscous 非黏性的,非黏滞的
~ - volatile matter 不挥发性物质
~ - volatile vehicle 不挥发性载色剂
~ - warp adhesive 不卷缩黏合剂,无皱纹黏合剂
~ - woven 无纺的,非织造的
~ - woven carpet 非织造地毯
~ - woven fabric 无纺布,无纺织物,非织布,非织造织物
~ - woven mat 无纺毡
~ - woven scrim 无纺平纹织物,非织造稀松布
~ - woven surface mat 无纺面层毡
~ - woven unidirectional [fibreglass mat (NUF mat)] 无纺单向(玻璃纤维)毡
~ - yellowing 不泛黄的

nonamer 九聚物
nonanoic acid 壬酸
nonplastic material 非塑性材料
nonplasticized - polyvinyl chloride 非增塑聚氯乙烯,硬聚氯乙烯
norbornene 降冰烯片
norm 定额,限额;标准,规格
normal 正常的,法向的,标准(的)
~ chain 正构链,直链
~ combustion 完全燃烧
~ condition 正常条件,标准状态
~ cooling 正常冷却
~ deformation 法向形变
~ density 正常密度
~ distribution 正态分布,高斯分布,常态分布

- ~ elastic modulus 正弹性模量
- ~ extension 法向拉伸
- ~ flight land width 法线螺棱宽度
- ~ horsepower 额定功率,标称功率
- ~ inspection 常规检查
- ~ load 常规载荷,额定载荷
- ~ mixture 常规混合
- ~ nylon 普通耐纶〈耐纶 66,尼龙 66〉
- ~ operating condition 正常操作条件
- ~ pipe thread,[NPT] 标准管螺纹
- ~ plywood 普通胶合板
- ~ pressure 常压,正常压力,标准压力
- ~ production programme 正常生产程序
- ~ rated power 额定功率
- ~ running 正常运转,正常操作
- ~ running temperature 正常运转温度,正常操作温度
- ~ screw channel width 法线螺槽宽度
- ~ size 标准尺寸,正常尺寸
- ~ speed 正常速率[度]
- ~ state 标准(状)态;正常状态
- ~ strain 法向应变
- ~ stress 法向应力
- ~ stress difference 法向应力差
- ~ stress effect 法向应力效应
- ~ viscosity 正常黏度,标准黏度
- ~ weld 正常焊接,标准焊接
- ~ width 标准宽度

normalization 标准化,规格化
normalize 正常,正规;解除内应力〈金属〉
Noryl resin 〈商〉诺里尔,〈美国通用电气公司产改性聚苯醚树脂〉
NR[natural rubber] 天然橡胶
nose key 钩头楔;鼻形楔
not 不,未,非,无
- ~ auto-crosslinking 非自交联的〈聚合物〉
- ~ self-crosslinking 非自交联的〈聚合物〉
- ~ -standardized moulding compound 非标准模塑料

notch 切口,刻槽,缺口
- ~ bar 缺口试棒
- ~ bar bending test 缺口试棒弯曲试验
- ~ bar impact test 缺口试棒冲击试验
- ~ bar pull test 缺口试棒拉伸试验
- ~ bar test 缺口试棒试验
- ~ brittleness 切口脆性,缺口脆性
- ~ effect 切口效应,缺口效应
- ~ factor 切口因素
- ~ impact resistance[strength] 耐切口冲击性[强度]
- ~ impact test 缺口冲击试验
- ~ sensitivity 切口敏感性
- ~ sensitivity test 缺口脆性试验
- ~ shape 切口形状
- ~ toughness 切口韧性

notched 切口的,缺口的
- ~ bar impact test 缺口试棒冲击试验
- ~ impact bar 切口冲击棒
- ~ impact resistance 缺口耐冲击性
- ~ impact strength 缺口冲击强度
- ~ Izod impact strength 缺口艾佐德冲击强度,缺口悬臂梁式冲击强度
- ~ knife dicer 齿刀式切粒机
- ~ specimen 切口试样
- ~ test bar 缺口试棒

notching press 缺口冲压机
novel 新型的,新颖的,新奇的
novelty yarn 花色纱,装饰纱
novolak[novolac] 线型酚醛树脂,可溶可熔酚醛树脂
- ~ epoxy 酚醛环氧树脂

nozzle 注料嘴,注嘴;喷嘴;喷丝板
- ~ adapter 注嘴接头
- ~ adaptor 注嘴接头〈注塑〉
- ~ aperture 喷丝板孔
- ~ approaching speed 注嘴靠模速率[度]
- ~ atomization machine 喷嘴喷雾机〈聚氨酯泡沫〉
- ~ attachment 喷嘴夹持器〈热气焊接仪〉
- ~ band heater 注嘴加热圈

~ block 注嘴体〈注塑机〉;纺丝头组件
~ casing 喷嘴罩
~ chamber 注嘴料腔〈注塑〉
~ clamp 喷嘴夹
~ clamping speed 喷嘴料供给速度
~ clamping stroke 喷嘴料供给行程
~ extension 注嘴接头
~ filler 喷丝头滤器
~ head 注嘴头;纺丝头喷嘴
~ heater 注嘴电热器
~ hole 纺丝孔
~ insert 注嘴嵌件
~ jet 注嘴喷射
~ manifold 分支注嘴;多头注嘴
~ mixer 喷嘴混合器
~ opening 纺丝孔
~ orifice 注嘴孔
~ outlet 注嘴出口
~ plate 注嘴固定板〈注塑机〉;纺丝板
~ position 注嘴位置
~ pressing force 注嘴压紧力
~ processing 喷雾法〈制备聚氨酯泡沫塑料〉
~ pull－back 注嘴退回
~ pulverizer 喷射粉碎机
~ －register 喷嘴分流器
~ return speed 注嘴返回速度
~ shoe 焊嘴套〈快速焊接嘴〉
~ shutt－off 注嘴止流;注嘴关闭
~ shutt－off valve 注嘴止流阀
~ side 注嘴侧〈注塑〉
~ －side mould part 注嘴侧模具部件〈注塑〉
~ stroke 注嘴冲程
~ thread 焊嘴螺纹
~ tip 注嘴尖
~ top 注嘴顶路
~ touch 注嘴与注嘴衬套紧密接触
~ touching 喷嘴接触
~ valve 注嘴阀
~ with flat contact 平头注嘴〈热流道模〉

NP［name plate］ 名牌
n. p. ;NP［normal pressure］ 正常压力,常压
NPL［Non－Patent Literature］ 非专利文献
NR［natural rubber］ 天然橡胶
NSA［National Security Agency］ 国家安全机构
NSA［National Standards Association］ 国家标准协会
NSR spectroscopy［nuclear spin resonance spectroscopy］ 核旋共振光谱学
Nt［niton］ 氡
nuclear 核的
~ energy 核能
~ fission 核裂变;核聚变
~ grade ion exchange resin 核子级离子交换树脂
~ magnetic resonance［NMR］ 核磁共振
~ magnetic resonance absorption 核磁共振吸收
~ magnetic resonance measurement［NMR measurement］ 核磁共振测定
~ magnetic resonance spectrograph［NMR spectrograph］ 核磁共振摄谱仪［光谱仪］
~ magnetic resonance spectrometer［NMR spectrometer］ 核磁共振波谱仪
~ magnetic resonance spectroscopy［NMR spectroscopy］ 核磁共振波谱法
~ magnetic resonance spectrum 核磁共振谱
~ magnetic resonance study［NMR study］ 核磁共振研究
~ physics［nucleonics］ 核子物理学
~ X－rays 核X射线
nucleated 核的
~ polyamide 有核聚酰胺
~ polypropylene 有核聚丙烯
~ resin 有核树脂
nucleating agent 成核剂

nucleation 成核作用;晶核形成;起泡（作用）;产生泡沫
~ agent 成核剂
~ of bubbles 气泡成核(作用)〈泡沫塑料〉
~ process 成核过程
~ rate 成核速率[度]
nucleator 成核剂
nucleonics [nuclear physics] 核子学
nucleus 核;环;晶核
~ formation 晶核形成
~ of crystallization 结晶核
NUF mat [non-woven unidirectional(fibreglass) mat] 无纺单向(玻璃纤维)毡
number 数目,数字
~ - average degree of polymerization 数均聚合度
~ - average kinetic chain length 数均动力学链长
~ - average molecular weight [M_n] 数均分子量
~ distribution function 分子数分布函数
~ of cycles of load stressing 应力负荷循环数
~ of daylight 压板档数(层压机)
~ of parallel screw flights 平行的螺纹数
~ of shots 注射次数
~ of teeth 齿数
~ of threads 螺纹数;纱线数
~ of turns of flight 螺纹总圈数〈螺杆〉
~ range 数值范围
numerical data 数据〈数字〉
nun union 锁紧螺母,外套螺母
nursing Ware 护理物品
nut 螺母
~ blocking adhesive 锁紧螺母黏合剂;螺母防松黏合剂
~ crest 内螺纹顶
~ root 内螺纹根
~ union 螺帽,连接螺套

NW [net weight] 净重
nylon 尼龙,耐纶,锦纶〈聚酰胺〉
~ basc laminate 尼龙布基层合塑料
~ bearing 尼龙轴承
~ canes 尼龙制饮用麦杆吸管
~ coated fabric 耐纶涂覆织物
~ coating 耐纶涂层
~ fabric 尼龙织物
~ fishing net 耐纶渔网,尼龙渔网
~ hose 耐纶袜,尼龙袜
~ modified phenolic mo(u)lding 尼龙改性酚醛模塑制品
~ monofilament 尼龙单丝
~ mo(u)lding powder 尼龙模塑粉
~ mousse 耐纶弹力丝〈低旦〉
~ plastics 聚酰胺塑料
~ 4 [polybutyrolactam, PA-4] 聚丁内酰胺,尼龙4
~ 6 [polycaprolactam, PA-6] 聚已内酰胺,尼龙6,锦纶
~ 8 [polyoctanoyllactam, PA-8] 聚辛内酰胺,尼龙8
~ 9 [polynonanoylamide, PA-9] 聚壬酰胺,尼龙9
~ 10 [polydecalactam, PA-10] 聚癸内酰胺,尼龙10
~ 11 [polyundecanoylamide, PA-11] 聚十一酰胺,尼龙11
~ 12 [polylaurylactam, PA-12] 聚十二内酰胺,尼龙12
~ 13 [polytridecanoylactam, PA-13] 聚十三酰胺,尼龙13
~ 46 [polyhexamethylene butydiamine, PA-46] 聚已二酰丁二胺,尼龙46
~ 66 [polyhexamethylene adipamide, PA-66] 聚已二酰己二胺,尼龙66
~ 610 [polyhexamethylene sebacamide, PA-610] 聚癸二酰已二胺,尼龙610
~ 612 [polyhexamethylene dodecanamide, PA-612] 聚十二烷二酰已二胺,尼龙612
~ 1010 [polydecamethylene sebacamide,

PA - 1010] 聚癸二酰癸二胺,尼龙1010
~ reinforced 聚酰胺增强(塑料)
~ resin 聚酰胺树脂
~ tube 尼龙管
~ tyre 尼龙帘线轮胎
~ yarn 尼龙线

O

O 正;原;邻(位)
~ - cresol [*ortho cresol*] 邻甲酚
O [oxygen] 氧;氧气
O ring O形环,O圈,密封圈
Ob. [minimum light transmission of plastics] 塑料最小透光率
object colour 物体色
oblate 扁圆的,扁球状的
~ ellipsoid head contour 扁椭球封头曲面
~ spheroid 扁球
oblique 倾斜(的)
~ ball bearing 斜滚珠轴承
~ crossing 斜交叉
~ crystal 单斜晶
~ design 斜纹图案,斜纹花型
~ drawing 斜视图
~ extruder head 斜向挤塑机机头
~ head 斜(向)机头〈挤塑机〉
~ joint 斜搭接头
~ plane 斜面
~ rail 斜轨〈长丝缠绕〉
~ section 斜断面
~ stress 斜应力
~ to grain 斜纹
~ - to - grain wood bonding 木纹斜黏合
obliterating power 遮盖力
oblong 长方形的
~ hole 槽;缝;窄口
OBP [octyl benzyl phthalate] 邻苯二甲酸辛·苄酯〈增塑剂〉
observation check 外部检查
obt. [obtain] 获得,得到

Oc [maximum haze of plastics] 塑料最大混浊度
occupational 职业的
~ disease 职业病
~ safety 生产安全;安全生产
OCF [Owens - Corning Fiberglas] 欧文斯 - 科宁玻璃纤维
~ test 欧文斯 - 科宁玻璃纤维试验
ochre [ocher] 赭石〈着色剂〉;赭色
ochreous 赭色的
octabromodiphenyl oxide 八溴二苯醚〈阻燃剂〉
n - octadecyl β - (4 - hydoxy - 3,5 - di - tert - butylphenyl) propionate β - (4 - 羟基 - 3,5 - 二叔丁基苯基)丙酸正十八碳醇酯〈即抗氧剂1076〉
octahedrite 八面石
octakis - (2 - hydroxypropyl) sucrose 八(2 - 羟丙基)蔗糖〈交联剂、增塑剂〉
octamer 八聚物
octanoic acid 辛酸〈增塑剂、杀虫剂、杀菌剂〉
octyl 辛基
~ benzyl phthalate [OBP] 邻苯二甲酸辛·苄酯〈增塑剂〉
~ decyl adipate 己二酸辛·癸酯〈增塑剂〉
~ decyl phthalate [ODP] 邻苯二甲酸辛·癸酯〈增塑剂〉
~ diphenyl phosphite 亚磷酸二苯·辛酯〈抗氧剂〉
~ diphenyl phthalate [ODPP] 邻苯二甲酸辛基二苯酯

~ epoxy stearate 环氧硬脂酸辛酯〈增塑剂、稳定剂〉
~ ether 二辛醚,正辛醚〈防水剂,塑料润滑剂,抗静电剂〉
n - ~ methacrylate[OMA] 甲基丙烯酸正辛酯
p - tert - ~ phenol - formaldehyde resin 对叔辛基苯酚甲醛树脂〈硫化剂〉
P - ~ phenyl salicylate 水杨酸对辛基苯酯〈紫外线吸收剂〉
~ tridecyl phthalate 邻苯二甲酸辛基·十三烷基酯〈增塑剂〉
octylated phenol 辛基化苯酚〈抗氧剂〉
octyltin stabilizer 辛基锡类稳定剂
OD[outside diameter; outside dimensions] 外径;外部尺寸
ODA[isooctyldecyl adipate] 己二酸异辛基·癸基酯〈PVC增塑剂〉
ODD[isooctyldecyl phthalate] 邻苯二甲酸异辛基·癸基酯〈PVC增塑剂〉
odorimetry 气味测定法
odo(u)r 气味,香气,臭味
~ - free 无气味的,无臭的
~ threshold value 气味临界值〈毒性〉
odo(u)rless 无气味的,无臭的
ODP[octyl decyl phthalate] 邻苯二甲酸辛癸酯〈增塑剂〉
ODPP[octyl diphenyl phthalate] 邻苯二甲酸辛基二苯酯
off 关,闭;偏离
~ center 偏心
~ center sprue 偏心注口
~ colo(u)r 变色,不正常颜色,色差
~ colo(u)r product 变色产品
~ cut 边角料
~ - grade 等外品
~ - print 特别印刷
~ shade 色差
~ - size 不合尺寸,尺寸不合格
~ sorts 等外品
~ standard 等外级
~ - the - shelf 现用的;现存的
~ - white 近于白色(的),灰白色(的)

offset 偏移;补偿;倾斜;胶版
~ cavity 偏置模槽
~ circuits 偏移缠绕
~ - coupling 偏管接头
~ die 侧向模头,偏心机头〈挤塑机〉
~ die - head 偏模头〈挤塑〉
~ gravure coater 补偿槽辊涂布机
~ gravure print coating 胶凹版印刷涂布
~ - gravure roll coater 补偿槽辊涂布机
~ grooves 偏槽沟纹
~ guide pin 侧向导销,侧向定位销
~ heart - shaped curves 偏心形弯曲
~ injection mo(u)lding 偏置注塑成型
~ lap joint 压肩接头,啮合榫搭接
~ modulus 补偿模量
~ moulding 侧位模塑,偏position成型
~ - mounted 偏位装配的
~ needles on a roller 辊上偏位针
~ pipe die 弯头挤管模
~ polar winding 偏心极性缠绕〈缠绕容器底〉
~ printing 胶(版)印(刷)
~ roll 侧辊
~ rotary tumbler 偏心转动混合器
~ spiders 偏位支架
~ stress 补偿应力
~ transfer mo(u)ld 侧位压铸模
~ yield point 偏置屈服点
~ yield strength 偏置屈服强度,补偿屈服强度,残余变形屈服强度
~ yield stress 偏置屈服应力
Oh; oh[ohm] 欧姆[Ω]
ogee S形曲线
OI[oxygen - index] 氧指数
OIDP[octyl - isodecyl phthalate] 邻苯二甲酸辛异癸酯
oil 油
~ absorption 吸油量;吸油性
~ base paint 油涂料
~ capacity 油含量〈液压系统〉
~ circulation lubrication 油循环润滑
~ cloth 油布;漆布

oiliness

- ~ color 油溶性染料
- ~ extended rubber 充油橡胶
- ~ gauge 油位表,油量计;油规;油比重计
- ~ - immersed vane pump 油浸式叶轮泵〈塑料加工机〉
- ~ - in - water emulsion, OW 油水乳化液;油水乳化剂
- ~ length 含油量,油含量
- ~ - modified alkyd resin 油改性醇酸树脂
- ~ - modified polyurethane 油改性聚氨酯
- ~ - modified resin 油改性树脂
- ~ paint 油性漆,调和漆,油性色漆,油涂料
- ~ permeabilily 渗油性
- ~ pigment - type paint 颜料性油涂料,油色料
- ~ plasticized 油增塑的
- ~ port 油室
- ~ - reactive resin 油改性树脂
- ~ - reactive phenolic resin 油改性酚醛树脂
- ~ resistance 耐油性
- ~ seal 油封
- ~ soluble phenolic resin 油溶性酚醛树脂
- ~ soluble resin 油溶性树脂
- ~ surfacer 油性二道漆,油性中层漆
- ~ varnish 油性清漆,油基清漆

oiliness 含油性;润滑性

oleamide 油酰胺,油酸酰胺〈脱模剂,润滑剂,爽滑剂,防粘剂,抗粘连剂〉

olefin 烯烃
- ~ compound 烯烃化合物
- ~ copolymer 烯烃共聚物
- ~ fibre 烯烃类纤维
- ~ hydrocarbon 烯烃
- ~ oligomerization 烯烃[低聚]
- ~ plastic 烯烃类塑料
- ~ polymer 烯烃类聚合物
- ~ resin 烯烃类树脂

olefinic 烯(属)的
- ~ bond 烯键
- ~ constituents 烯组分
- ~ link 烯键
- ~ polymerization 烯烃聚合
- ~ thermoplastic elastomers [OTE] 烯烃热塑性弹性体

oleoresin 含油树脂
oleoresinous 油性树脂
- ~ materials 油性树脂材料
- ~ varnish 油性树脂清漆

oleyl nitrile 油酰基腈〈增塑剂、软化剂〉
oligomer 低聚物,齐聚物
- ~ molecule 低聚物分子

oligomerization 低聚(合);齐聚
Olsen 奥尔森
- ~ flow tester 奥尔森流动试验机
- ~ type hardness tester 奥尔森式硬度试验机

omnidirectional pressure 各向(均匀)压力
on - line 在线,在生产线上;联机
- ~ data 联机数据
- ~ inspection 生产线上检查
- ~ maintenance 不停产检修
- ~ operation 联机操作
- ~ viscosimetry 生产线中黏度测定(法)

on - off lever 开关操纵杆
on shade 近似颜色
on - roll coater [on roll knife coater] 辊刮式涂布机
on - site 现场的,就地的
- ~ inspection 现场检查
- ~ test 现场试验
- ~ winding 现场缠绕成型〈特大型增强塑料制品〉

on - strud joint 轴颈和槽上下接缝的粘接
ondulation 波纹,起皱〈经纬纱〉
one 一(个)的;单(独)的
- and - one - half - screw machine [single - screw extruder with double - screw feed] 双螺杆进料单螺杆挤塑机
- ~ - circuit pattern 单环路缠绕〈长丝〉

~ coat paint 单层涂漆,底面合一漆,一道成活漆
~ - component adhesive 单组分黏合剂
~ - floor - line machine 单(楼)层生产线机器
~ helical circuit 单螺旋线绕程〈长丝缠绕〉
~ minute strain 一分钟应变〈蠕变试验〉
~ - par adhesive 单组分黏合剂
~ pass bonding 单程黏合
~ - piece 整体的
~ - piece casting 整体铸塑件
~ - piece construction 整体式结构
~ - piece screen 整体式滤网
~ - shot 一次使用的,一次完成的
~ - shot - adhesive 一次性用黏合剂
~ shot container 一次性容器(包装)
~ - shot method 一步法
~ - shot mo(u)lding process 一次成型法
~ - shot operation 一步操作
~ - shot process[technique] 一步(发泡)法〈聚氨酯泡沫塑料〉
~ - shot urethane foam 一步法(成型)聚氨酯泡沫塑料
~ - shot usage 一次性使用
~ - side 平面(的)
~ - side coating [single side coating] 单面涂布
~ - side double coating 单面双涂布
~ - side welding 单面焊
~ - stage 一步(的),单级
~ - stage moulding 一步成型
~ - stage phenol resin 一步法酚醛树脂
~ - stage process 一步法
~ - stage resin 一步法酚醛树脂;甲阶酚醛树脂,可溶可熔酚醛树脂〈美〉
~ - step 一步(的)
~ - step mixing 一次混料法〈各组分一次加入并混合的方法〉
~ - step operation 单步操作

~ - step resin 一步法树脂;甲阶酚醛树脂,可溶可熔酚醛树脂
~ - step synthesis 一步法合成
~ - way 单向的,单程的
~ - way - adhesive 一次性用黏合剂,反应性黏合剂
~ - way bottle 一次性用瓶
~ - way - cement 一次性用黏合剂〈美〉
~ - way circulation 单向循环
~ - way drawing 单轴拉伸
~ - way packing 一次性用包装
~ - way rivet 单向铆钉
~ - way stretch 单向拉伸
~ - way valve 单向阀,止回阀
opacifying 不透明的
~ agent 不透明剂,遮光剂
~ pigment 不透明颜料
opacisation 消光,闷光(处理);磨砂;无表面光泽;不透明的
opacity 不透明性;不透明度;浑浊度
opal [translucent coloured] 乳白的
opalescence 乳白光,乳白色
opaque 不透明的;无光泽的
~ body 不透明体
~ coating 不透明涂层[涂料],遮光涂料
~ colo(u)r 不透明色料,遮光色料
~ coloured 不透明色的
~ pigment 不透明颜料,遮光颜料
opaqueness 不透明性;不透明度
open 开(的),断开(的),敞开的
~ - air atmosphere 室外气候
~ - air exposure 室外曝晒
~ - air roofing 露天屋顶(材料)
~ - air weathering 室外耐气候老化〈测试〉
~ - air weathering test 室外耐气候牢度试验
~ area 筛孔面积〈筛〉
~ assembly 晾胶〈黏合〉
~ assembly time 晾胶时间〈黏合〉
~ bubble 敞开式气泡〈模压件缺陷〉
~ casting [casting into open moulds]

敞口铸塑
~ cell 开孔
~ cell cellular material 开孔泡沫材料
~ cell cellular material with closed skin 表层密闭式的开孔泡沫材料
~ cell cellular plastic 开孔微孔塑料
~ cell foam 开孔泡沫塑料
~ cell foamed plastics 开孔泡沫塑料
~ cell plastics 开孔泡沫塑料
~ cell product 开孔泡沫制品
~ cell structure 开孔结构
~ cure 无模固化
~ die 敞口模头
~ - end preform 开口型坯〈吹塑〉
~ - ended sack 开口袋〈包装〉
~ fabric 组织稀松织物
~ face mo(u)ld 敞口模具
~ - faced die 敞口冲坯模,手动冲模
~ gate 宽浇口
~ - grained 粗粒的,大粒的
~ - hole 空心的
~ - hole insert 空心嵌件
~ joint[weld] 开缝连接,开缝焊接
~ - mesh bag 网眼袋
~ mill 开放式辊炼机,开放式塑炼机
~ mop [loose full - disk buff] 敞开抛光轮
~ mo(u)ld 开放式模具
~ - mo(u)ld process 敞模成型法
~ mo(u)lding 敞模成型
~ - mouth sack 开口袋〈包装〉
~ nozzle 开口型注料嘴
~ porosity 表面多孔性
~ preform 敞模预成型
~ preform process 敞模预成型法〈增强塑料〉
~ preforming 敞模预成型法
~ roll mill 开炼机,开放式塑炼机
~ side machine 单柱式机;侧开式机
~ side press 敞边式压机;侧开档压机
~ structure 松散结构
~ - weave material 稀松组织
~ whole disc mop [loose full - disk buff] 敞开抛光轮

openability tester 开口性测试仪〈测定平折管膜粘连性〉
opening 孔,缝;开孔;开启;开放
~ die 可拆模
~ force 开模力
~ period 开模周期
~ pressure 开启压力
~ ram 开启活塞
~ speed 开模速率[度]
~ stroke 开启行程〈注塑模具〉
~ surface 开模面
~ traverse 开启行程〈塑料压机〉
operate 操作,操纵;开动,启动;运转;运算,计算
operating 操作的,运转的
~ arm 操纵杆
~ board 操纵盘
~ button 操作按钮
~ characteristic 运转特性
~ conditions 操作条件;工作条件
~ control 操作控制
~ cost 操作费用,运转费用
~ cycle 操作周期,运转周期
~ data 操作数据,运转数据
~ efficiency 操作效率
~ end 操作端,操作侧
~ engineer 工厂工程师
~ forces 有效力,作用力
~ handle 操纵手柄
~ instruction 操作说明
~ instruction manual 操作说明书
~ life 工作寿命
~ line 操作线,作业线
~ maintenance 运行维护,小修
~ manual 操作手册,操作规程
~ mechanism 操纵机构
~ nip 运转辊隙
~ panel 操纵台,操纵盘
~ parameter 操纵参数
~ pressure 操作压力
~ quietness 运转平静
~ scrap 生产废料,工序残料
~ sequence 操作程序
~ side 操作端,操作侧

~ speed 运转速率[度]
~ stress 工作应力
~ stroke 操作行程
~ supplies 动力燃料
~ temperature 操作温度
~ time 运转时间
~ valve 操纵阀
~ variables 操作变量
~ voltage 工作电压
operation 操作,运转
~ by hand 手工操作
~ frequency 运转频率
~ instruction 使用说明书,操作说明书
~ parameter 操纵参数
~ rate 运转率,开工率
~ sequence control 操作程序控制
~ specifications 操作规范,操作规程
~ test 操作试验
operational 操作的,运转的
~ data 运转数据
~ development 产品改进性研制
~ factor 运行参数
~ life 运转寿命
~ order 操作次序
~ performance 操作性能
~ tension 工作电压
operator 操作者
OPP [oriented polypropylene (film)] 定向聚丙烯(薄膜)
opposition 对抗;反对
OPS [oriented polystyrene (film)] 定向聚苯乙烯(薄膜)
optical 光(学)的;光导的;旋光的
~ anisotropy 光学各向异性
~ axis 光轴
~ brightener 荧光增白剂
~ character 光学特性
~ densitometer 光密度计
~ density 光密度
~ density of smoke 烟的光密度,光学烟密度
~ dispersion 光色散
~ distortion 光学畸变

~ fiber 光学纤维,光导纤维
~ frame 光学仪构架
~ isomer 旋光异构体
~ isomerism 光学异构现象
~ microscope 光学显微镜
~ path difference 光程差
~ plastics 光学塑料
~ properties 光学性质(材料)
~ rotatory dispersion 旋光色散
~ spectrum 光谱
~ scanner 光扫描仪
~ strain 光学应变
~ testing 光学性能试验〈材料〉
~ whitening agent 荧光增白剂
optimization 最优化,最佳化
optimum 最佳的
~ channel depth 最佳螺槽深度
~ condition 最佳条件,最佳状态
~ control 最佳控制
~ cure 最适硫化;最佳固化
~ drying 最佳干燥
~ efficiency 最佳效率
~ feed location 最佳进料位置
~ fibre orientation 纤维最佳取向
~ performance 最佳性能
~ performance parameters 最佳性能参数
~ pitch 最佳螺距
~ programming 最佳程序设计
~ seeking method 优选法
~ speed 最佳速率[度]
~ temperature 最佳温度
~ value 最佳值
optional 任意的,任选的
~ equipment 附加设备,备用设备
OPVC [oriented polyvinyl chloride] 定向聚氯乙烯
orange 橙(子);橙色(的);橙色颜料
~ dye 橙色染料
~ peel 桔皮纹,橙皮〈成型件面〉
~ pigment 橙色颜料
order 程序,有序;订购,定(货)单
~ - disorder distribution 有序-无序分布(状态)

ordered

~ - disorder transformatson 有序-无序转变
~ - disorder transition 有序-无序转变
~ distribution 序态分布
~ of assembly 装配顺序

ordered 有序的
~ aromatic copolyamide 有序共聚芳酰胺
~ arrangement 有序排列
~ copolymerization 有序共聚(作用)
~ fold surface 规则折叠面
~ linear copolymer 有规线型共聚物
~ region 有序区
~ sequence 有序序列
~ state 有序状态
~ structure 有序结构

ordnance 军用品
~ paint 军用色漆

org. [organic] 有机的

organic 有机的
~ acid 有机酸
~ adhesive 有机胶黏剂,有机黏合剂
~ base 有机碱
~ bentonite 有机(改性)膨润土;有机(改性)皂土
~ binder bond 有机粘结剂
~ chemist 有机化学家
~ chemistry 有机化学
~ coating 有机涂层
~ compound 有机化合物
~ diluent 有机稀释剂
~ dye 有机染料
~ fiber 有机纤维
~ fiber reinforced plastics 有机纤维增强塑料
~ filler 有机填料
~ glass 有机玻璃
~ ion exchanger 有机离子交换剂
~ isocyanate 有机异氰酸酯
~ laboratory 有机化学实验室
~ material 有机材料
~ peroxide 有机过氧化物
~ pigment 有机颜料
~ promoter 有机助催化剂,有机助聚剂,有机促进剂
~ protective coating 有机防护性涂层
~ reinforcing agent 有机增强剂
~ stabilizer 有机稳定剂
~ synthesis 有机合成

organo 〈词头〉有机
~ additive 有机添加剂
~ - functional group 有机官能基团
~ - mat 有机玻璃毡〈由玻璃短纤维与黄麻制备〉

organobentonite 有机(改性)膨润土,有机(改性)皂土

organochlorosilane 有机氯(甲)硅烷

organogel 有机凝胶

organoleptic 器官感觉的
~ evaluation 感官检验
~ property 感官性能
~ test 感官试验

organometallic 金属有机的,有机金属的
~ catalyst 金属有机[有机金属]催化剂
~ compounds 金属有机[有机金属]化合物
~ polymer 金属有机[有机金属]聚合物
~ stabilizer 金属有机[有机金属]稳定剂

organophilic 亲有机物的
~ bentonite 亲有机物质的(改性)膨润土

organophosphate 有机磷酸酯

organophosphite 有机亚磷酸酯

organosilicon 有机硅
~ compound 有机硅化合物

organosilicone 有机硅氧烷
~ polymer 有机硅聚合物
~ rubber 有机硅橡胶

organosiloxane 有机硅氧烷
~ polymer 有机硅氧烷聚合物

organosol 有机溶胶;稀释增塑糊
~ resin 稀释增塑糊树脂

organotin 有机锡
~ compound 有机锡化合物

~ stabilizers 有机锡稳定剂
orientate 取向;定向;拉伸
orientated short – fibre reinforcement 取向短纤维增强材料
orientation 取向;定向
　~ birefringence 取向双折射
　~ blow moulding [stretch – blow moulding] 取向吹塑
　~ distribution function 取向分布函数〈分子结构〉
　~ effect 取向效应
　~ factor 取向因子
　~ force 取向力
　~ function 取向函数
　~ index 取向指数
　~ of filaments 纤维拉伸
　~ of films 薄膜拉伸
　~ of molecules 分子取向
　~ polarization 取向极化(作用)
　~ release stress 取向松弛应力,解取向应力
　~ stage 取向阶段〈拉吹〉
　~ uniformity 取向均一性
oriented 取向的,定向的
　~ adsorption 取向吸收
　~ crystalization 取向结晶
　~ film 定向薄膜;拉伸薄膜
　~ polymer 定向聚合物
　~ polymerization 定向聚合,有规立构聚合
　~ polypropylene(film)[OPP] 定向聚丙烯(薄膜)
　~ polystyrene(film)[OPS] 定向聚苯乙烯(薄膜)
　~ polyvinyl chloride[OPVC] 定向聚氯乙烯
　~ region 取向区
　~ structure 有规立构结构
　~ webs 取向纤维网,定向纤维网
orienting 取向,定向
　~ effect 定向效应
　~ feeder 定向加料器
　~ roll 导(向)辊
orifice 孔;模孔,模口;喷嘴;喷丝孔

~ bushing die bushing 模头外套〈挤塑〉
~ flowmeter 孔板流量计
~ land [die land] 口模成型面
~ meter 孔板流量计
~ of spinneret 喷丝孔
~ plate 孔板
~ plate flowmeter 孔板流量计
~ relief [die relief] 模缝;口模缘〈美〉
~ viscometers 锐孔黏度计
original 最初的,原来的
~ cross – section 原始(横)断面
~ design 原设计,原结构
~ detail 原有的零件
~ drawing 原图
~ inspection 初步检查
~ mo(u)ld 原型;样模
~ record 原始记录
~ sample 原(试)样
~ scheme 原始方案
~ shape 原样
~ size 原始尺寸
~ specification 原始技术条件
~ state 初始状态
~ stock 原始混合物
~ treatment 初次处理
Ornamin process 奥纳敏成型法(在三聚氰胺或脲醛树脂压塑的半固化阶段打开模具,放上浸渍的印花纸而压塑为一体的过程)
ortho 〈词头〉正;原;邻位〈略作 o – 〉
~ cresol 邻甲酚
orthogonal 正交的,垂直的
~ body 正交体
~ cone 正交锥面
~ involution 正交对合
~ ply laminate 正交层合板,正交铺敷层压板
~ surface 正交曲面
~ viscoelastic body 正交黏弹体
orthographic 直线的,直角的;正视的;正交的
~ view 正视图

orthopaedic 矫形术
orthophthalate plasticizers 邻苯二甲酸酯类增塑剂〈包括 DOP,DIDP,DTDP〉
orthotropic 正交(各向)异性(的)
~ anisotropy 正交各向异性
~ elastic material 正交异性弹性材料
~ material 正交异性材料
~ plate 正交异性板
orthotropy 正交各向异性;直交
Os [osmium] 锇
oscillate 摆动;振动
oscillating 摆动的;振动的
~ agitator 摆动搅拌器
~ circuit 振荡电路
~ curve 振动曲线
~ cylinder viscometer 振动圆柱黏度计
~ detector 振荡检测器
~ die 振荡式口模
~ dipole 振动偶极
~ disk rheometer 振荡盘式流变仪
~ disk viscometer 振荡盘式黏度计
~ feeder 摆动进料器
~ friction welding 振动摩擦焊接
~ function 摆动函数;振动函数
~ hot-wire slicer 振动热丝切片机
~ impeller mixer 振动桨式混合器
~ load 波动载荷
~ mill 振动磨;振动粉碎机
~ roller 振动辊筒
~ screen 振动筛
~ screw motion 螺杆振动运动
~ stress 振荡应力
~ traverse motion 往复摆动运动
~ twisting-machine 振动加捻机
oscillation 摆动,振动,振荡,波动
~ constant 振荡带数
~ counter 振动计数器
~ damping 摆动阻尼
~ energy 振荡能
~ frequency 振荡频率
~ period 摆动周期,振荡周期
~ pickup 振荡传感器
oscillator 振动器,振荡器
oscilloscope 振荡器;示波器

osmium [Os] 锇
osmometer 渗透压力计
osmometry 渗透压测定法
osmosis 渗透(性,作用)
osmosize 渗透
osmotic 渗透的
~ coefficient 渗透系数
~ pressure 渗透压
Ostwald viscometer 奥氏黏度计,奥斯特瓦尔德黏度计
Ostwald's colourmeter 奥氏比色计,奥斯特瓦尔德比色计
OTE [olefinic thermoplastic elastomers] 烯烃热塑性弹性体
out 外部的,外表的;断开,脱离;输出
~ -draw 外拉伸
~ -draw textured yam 外拉伸变形丝
~ -feed rolls 溢流辊
out of 失去
~ -action 不能工作的,不能运转的
~ -adjustment 失调的
~ -alignment 未对准的
~ center 偏心
~ -date 过时的,陈旧的,落后的
~ control 失去控制
~ doors 户外,室外,露天
~ -fashion 不合时尚的,不流行的
~ -flat 不平直的
~ level 倾斜
~ -line 偏移的
~ -miter [out-of-register] 不对齐〈模具〉
~ -operation 不工作的;断开的
~ -order 故障,不正常
-position 偏移的
~ -reach 达不到
~ register 不对齐
~ repair 失修
~ -round 不圆
~ -run 不运转的
~ shape 走样,失去正常形状
~ size 不正常大小,失去正确尺寸的
~ time 不按时的
~ use 无用,报废

~ – service 失效的,不工作的
~ – service inspection 停运转检查
~ – service record 故障记录
~ – service time 停工作时间
outage 预留容量〈包装〉
outboard 外部的,外面的
~ bearing 外装轴承
outbreak 破裂;断裂
outcome 结果;产量,输出(量)
outcoming signal 输出信号
outdoor 户外的,室外的,露天的
~ durability 耐室外用,耐户外用,室外耐久性,户外寿命,耐气候,耐老化
~ exposure 户外曝晒,露天曝露
~ exposure test 室外曝晒试验
~ life 室外曝露寿命
~ paint 户外用漆
~ resistance 耐室外用,耐户外用,室外耐久性,耐室外性
~ – seacoast weathering 海滨室外大气老化
~ test 室外试验
~ – tropical weathering 热带室外大气老化
~ use 室外(使)用
~ weathering 室外大气老化
~ weathering test 室外大气老化试验
outer 外部的,外面的
~ barrier layer 外表层,外部阻隔层〈整体泡沫塑料〉
~ casing 外壳,外罩
~ coating 外部敷层
~ cover 外罩,外套
~ curvature 外拱形〈成型〉
~ defect 外观疵点
~ diameter 外径
~ die 口模套
~ die ring 外模圈
~ extremities 外形尺寸极限
~ face 外端面
~ inspection 外观检查
~ layer 外层
~ ring [die bushing] 外模套〈挤塑〉
~ ring die 外模套模头

~ shaft 外轴
~ sizing unit 定外径装置
~ skin 外表层,外皮
~ support 外支架
~ tube 外管
outfit (成套)设备,装置,装备
outflow 流出
outgassing 渗气
outlay 经费,费用,支付
outlet 出口;排气口;出口管;排泄阀
~ adapter 出口(异径)接头
~ area size 出口(横)截面
~ branch 排出支管
~ connection 出口接管
~ connector 出口连接器
~ diffuser 出口扩散器
~ elbow 排出管弯管
~ nozzle 出料嘴
~ pipe 出口管,排出管
~ roller 输出辊
~ valve 排放阀,排泄阀
outline 外形,轮廓
~ blanking 外形模压(件),外形冲压(件)
~ blanking die 外形模压模,外形冲压模
~ cutting [outline blanking] 外形切割
~ die 冲切模
~ drawing 外形图,轮廓图
~ plan 初步计划,提纲
outerpack [overpack] 外包装
outproduce 超量,超额生产
output 出口;产量;排出量;输出(量);功率;工作能力
~ coefficient 利用系数
~ growth 产量增加
~ instruction 输出指令
~ per day 每日产量
~ per shift 每班产量
~ power 输出功率
~ quota 产量定额
~ rate 产量,生产率
~ writer (文件)输出改写程序
outrigger bearing 外架轴承

outsert 金属嵌件
- ~ moulding 金属嵌件注塑
- ~ technology 金属嵌件注塑法

outshot 等外品(原料)

outside 外部,外表,外观
- ~ circle 外圆周,齿顶圆
- ~ coating 外表涂层
- ~ corner weld 外角焊
- ~ diameter 外径
- ~ diameter of a screw 螺杆外径
- ~ diameter of thread 螺纹外径
- ~ dimension 外径尺寸
- ~ drawing 外视图,外形图
- ~ face 外面
- ~ shell 外壳
- ~ surface 外表面
- ~ thread 外螺纹
- ~ view 外观图,外形图

outsize 特大(的)

oval 椭圆(形的)
- ~ belt conveyor 卧式输送带
- ~ filler rod 椭圆焊条

oven 烘箱
- ~ ageing 烘箱老化;热致老化
- ~ ageing test 烘箱老化试验
- ~ baked enamel 烘烤漆层
- ~ drying 烘干
- ~ dry weight 烘干重量,干重
- ~ test 耐热试验

over 超,过度
- ~ charge 超载
- ~ damping 过度阻尼
- ~ -dimensioned 超尺寸的
- ~ drive 超速传动
- ~ elongation 过度伸长
- ~ -end winding 轴向卷绕
- ~ -end withdrawal 轴向退绕
- ~ etching 过度腐蚀
- ~ exposure 过度曝晒
- ~ -length fibre 超长纤维
- ~ -molding [double-shot moulding] 双注塑法
- ~ -temperature protector [switch] 超温保护器

overage 过度老化
overageing 过度老化
overall 总的,全部的
- ~ accuracy 总准确度
- ~ coefficient 总系数
- ~ composition 总成分
- ~ density 总密度,粗密度
- ~ design 总体设计
- ~ diameter 最大直径,外形直径
- ~ dimension 总尺寸,外形尺寸,轮廓尺寸
- ~ efficiency 总效率,综合效率
- ~ heat transfer coefficient 热通导率,热通过系数
- ~ height 总高,外廓高度
- ~ length 总长度,外形长度
- ~ mo(u)ld height 模具最大开距
- ~ performance 综合性能
- ~ pressure ratio 总压比
- ~ project 综合方案,全面计划
- ~ properties 综合性能
- ~ quantity 总量
- ~ reliability 综合可靠性
- ~ size 总尺寸,外廓大小
- ~ test 总试验,综合试验
- ~ view 全貌图
- ~ width 总宽(度)
- ~ wrapper 全(标签)包装
- ~ yield 总产量

overblow 过度发泡
overcapacity 超负荷
overcoat process 多层涂刷法
overcoating 罩涂,外敷层,保护涂层;超幅宽贴合〈挤塑贴合〉
overcompression 过度压缩
overcure 过度固化,过度熟化
- ~ resistance 耐过度固化性

overcuring 过度固化
overcut fibre 超长纤维
overfeed 超喂,过量加料
- ~ roller 超喂辊
- ~ tenter 超喂拉幅机

overfeeding 加量加料,过量进料
overfilled 过量填装料,装料过满

overflow 溢流,溢料;焊渣〈高频焊接〉
- ~ capacity 溢流量
- ~ groove 溢流槽,溢料槽
- ~ mould 溢流式模具
- ~ outlet 满溢出口
- ~ pipe 溢流管;排水管
- ~ space 溢料间隙
- ~ trap 溢流阀
- ~ trimming machine 修边机
- ~ tube 溢流管
- ~ valve 溢流阀
- ~ viscometer 溢流黏度计

overfulfil the quota 超过定额
overgassing 过多放气
overgrind 过度粉碎
overhang roll 悬辊
overhaul 大修;彻底检查(帐目等);彻底改革,全面修订
- ~ life 大修周期
- ~ period 大修周期

overhauling 检修〈设备〉
overhead 过顶的,高架的;管理费
- ~ cost 管理费
- ~ tank 压力槽
- ~ view 俯视图
- ~ welding 仰焊

overheating 加热过度,过热
overlacquer 罩光喷漆
overladen 过载的,过负荷的
overlap 搭接;重叠;焊瘤〈焊接〉
- ~ area 搭接面,重叠面积
- ~ coefficient 重叠系数
- ~ fillet weld 搭接角焊
- ~ gate 搭接浇口
- ~ joint 搭接(接头),叠接
- ~ length 搭接长度
- ~ ratio 搭接程度;搭接比(例)
- ~ seam 搭接缝
- ~ welding 搭接焊

overlapped joint weld 搭接焊缝
overlapping 搭接;重叠
- ~ area 重叠面积
- ~ blades 啮合式桨叶;叠接桨叶〈捏和机〉
- ~ flap 塔接活盖〈包装〉
- ~ louvres pl 塔接(通风)天窗

overlay 敷面(层),贴面(层);重叠
- ~ mat 表面毡,敷面(玻璃)毡
- ~ paper 贴面纸〈浸渍」热固性树脂预聚物的于无色透明纸〉
- ~ sheet 贴面片材
- ~ up 铺叠(增强塑料)
- ~ welding 堆焊

overload 过负荷,超载
overmixing 过度混合
overpack [outerpackck] 外包装,重复包装
overpackaging 太昂贵包装
overpacking 过量充模;过量灌料
overressure [excess pressure] 过压;超压
overprint (ing) 套色印刷
overproduce 生产过剩;超额生产
overruns 溢流
overs 筛余物〈过筛后的剩余物〉
oversize 过大,超过尺寸
- ~ material 过大物料;筛上物料

overpeed 超速;高速运转
overspray 过喷
overstrain 过度应变,残余应变
overstress 过度应力
overstretch 过度拉伸〈薄膜,带材,纤维〉
overtemperature 超温,过热温度
- ~ protector 超温保护器
- ~ sensor 超温感传器
- ~ switch 超温保护开关

overtime work 超时工作,加班工作
overtone 涂盖色;保护色;浓色;无光泽色
overturnable hopper 可翻料斗
overvoltage 超额电压
overweight 超重
overwrap 外包装
- ~ film 外包装薄膜

overwrapping machine 外包装机
OW [oil-in-water emulsion] 油水乳化液;油水乳剂
oxalic ester 草酸乙二酸酯

oxamide 乙二酰二胺,草酰胺〈稳定剂〉
oxazine dyes 噁嗪型染料
oxichlorination 氧氯化
oxidable 可氧化的
oxidant 氧化剂
oxidation [oxidization] 氧化(作用)
- ~ behaviour 氧化行为
- ~ colourising 氧化显色
- ~ colo(u)rs 氧化染料
- ~ coupling 氧化偶联
- ~ degradation 氧化降解〈塑料材料〉
- ~ dyes 氧化染料
- ~ inhibitor 氧化抑制剂,抗氧剂
- ~ polymerization 氧化聚合
- ~ resistance 耐氧化性,抗氧化能力
- ~ susceptibility 氧化敏感性
- ~ reduction 氧化还原
- ~ reduction polymer 氧化还原聚合物
- ~ reduction polymerization 氧化还原聚合(作用)
- ~ reduction resin 氧化还原树脂
- ~ reduction system 氧化还原体系
- ~ reduction type ion exchange resin 氧化还原型离子交换树脂

oxidative 氧化的
- ~ ageing 氧化老化
- ~ cationic polymerization 阳离子型氧化催化聚合
- ~ coupling 氧化交联,氧化偶联
- ~ degradation 氧化降解〈作用〉
- ~ dehydrogenation 氧化脱氢
- ~ polycondensation 氧化缩聚,脱氢缩聚
- ~ polymerization 氧化聚合(作用),脱氢聚合
- ~ stability 氧化稳定性〈热〉

oxidatively degradable plastic 氧化降解性塑料
oxide 氧化物
- ~ pigment 氧化物颜料

oxidizable 可氧化的
oxidize 氧化
oxidized 氧化(了)的
- ~ cellulose 氧化纤维素
- ~ colour 氧化(显色的)染料

oxidizer 氧化剂
oxidizing 氧化的
- ~ action 氧化作用
- ~ agent 氧化剂
- ~ dyes 氧化染料

oxirane 环氧;环氧乙烷
- ~ group 环氧基
- ~ oxygen content 环氧氧(含)量

oxy 〈词头〉氧化;羟基
1-~-3-methyl-4-isopropylbenzene 1-羟基-3-甲基4-异丙基苯〈抗氧剂〉

oxybis 氧代双
4,4′-~(benzene sulfonyl hydrazide)[OBSH] 4,4′-氧代双(苯磺酰肼)〈发泡剂〉
4,4′-~(benzene sulfonyl semicarbazide) 4,4′-氧代双(苯磺酰氢基脲)〈发泡剂〉

oxycellulose 氧化纤维素
- ~ ester 氧化纤维素酯
N-oxydiethylene-2-benzothiazole sulphenamide N-氧双-1,2-亚乙基-2-苯并噻唑次磺酰胺〈促进剂〉

oxygen [O] 氧
- ~ ageing 氧(化)老化
- ~ containing 含氧(气)的
- ~ convertible resin 氧化可逆性树脂
- ~ index 氧指数〈塑料烧燃性〉
- ~ index flammabity test 氧指数可燃性试验
- ~ index test 氧指数试验〈测定塑料材料可燃性〉
- ~ tubing 氧气用软管

oxygenant 氧化剂
oxygenolysis 氧化分解(作用)
oxypolymerization 氧化聚合,脱氢聚合
oz. [ounce(s)] 英两,唡〈已废〉,盎斯
ozone [O₃] 臭氧
- ~ ageing 臭氧老化
- ~ box 臭氧老化试验箱
- ~ crack 臭氧致龟裂
- ~ cracking resistance 耐臭氧龟裂性

~ degradation 臭氧(致)降解(作用)
~ fading 臭氧(致)褪色
~ resistance 耐臭氧性
~ resistance test 耐臭氧试验
~ treatment 臭氧处理
ozonized air 臭氧化空气〈非极性热塑性塑料电极化时〉

P

p;pp [page] 页数
p [para] 对位;仲;副
p. a. [per annual] 每年;每年的
PA [phthalic anhydride] 邻苯二甲酸酐,苯酐〈俗称〉
PA [polyamide] 聚酰胺
PA-4 [polybutyrolactam] 聚丁内酰胺,尼龙4
PA-6 [polycaprolactam, nylon 6] 聚己内酰胺,尼龙6
PA-8 [polyoctanoyllactam, nylon 8] 聚辛内酰胺,尼龙8
PA-9 [polynonanoylamide nylon 9] 聚壬酰胺,尼龙9
PA-10 [polydecalactam, nylon 10] 聚癸内酰胺,尼龙10
PA-11 [polyundecanoylamide, nylon 11] 聚十一酰胺,尼龙11
PA-12 [polylauryllactam, nylon 12] 聚十二内酰胺,尼龙12
PA-13 [polytridecanoyllactam, nylon 13] 聚十三内酰胺,尼龙13
PA-46 [polyhexamethylene butydiamine, nylon 46] 聚己二酰丁二胺,尼龙46
PA-66 [polyhexamethylene adipamide, nylon 66] 聚己二酰己二胺,尼龙66
PA-610 [polyhexamethylene sebacamide, nylon 610] 聚癸二酰己二胺,尼龙610
PA-612 [polyhexamethylene dodecanamide, nylon 612] 聚十二烷二酰己二胺,尼龙612
PA-1010 [polydecamethylene sebacamide, nylon 1010] 聚癸二酰己二胺,尼龙1010

PAA [polyacrylate] 聚丙烯酸酯
PAA [polyacrylic acid] 聚丙烯酸
PABM [polyaminobismaleimide] 聚氨基双马来酰亚胺
pacin factor 基本因素,基本条件
pack 包装,包捆;组合(件);装填,填充;保压
~ hardening 表面硬化处理〈渗碳,渗氮等〉
pack. [packing] 填料
package 包装,包捆;组件;外壳;卷装;纱筒
~ build 卷装成形
~ density 卷装密度;包装密度
~ drawing 组装草图
~ engineer 包装技术员
~ engineering 包装技术
~ shelf 小件行李架
~ shell 组装外壳
~ size 卷装大小
packaging 包装
~ adhesive 包装(用)黏合剂
~ bag 包装袋
~ density 封装密度
~ equipment 包装设备
~ film 包装薄膜
~ insert 包装填充物
~ line 包装生产线
~ machine 包装机
~ material 包装材料
~ table 包装带,打包带
~ unit 包装单元
packed column 填充粒;填充塔;填料塔
picker 包装机,打包机
packing 包装,打包;装填;衬垫;保压;填

pad

　　充物;填料
~ box　装料箱,货箱;填料箱
~ density　堆积密度;填充密度〈填料〉;存储密度
~ diamension　包装尺寸
~ factor　填充因子
~ film　包装薄膜
~ gland　填料函,填料(压)盖
~ inspection　包装检验
~ line　包装生产线
~ list　装箱单
~ machine　包装机
~ material　包装材料
~ of products　产品包装
~ paper　包装纸
~ press　包装压机
~ pressure　保压压力
~ ring　密封环;垫圈
~ room　包装室,包装间
~ scale　包装秤
~ sheet　垫片
~ specification　包装规格
~ support　填充物支架
~ tape　包装带;打包带
~ test　镶嵌试验〈牙〉
~ twine　打包绳

pad　衬垫,底座;垫片,垫料
padding　填料;衬垫
paddle　桨叶;搅棒
~ agitators　桨式搅拌器
~ agitator with blade　桨式搅拌器
~ dryer　搅拌式干燥器
~ dyeing machine　桨叶式染色机
~ mixer　桨式混合器
~ mixer with crossed beams　十字形(桨式)搅拌器
~ mixer with flat blade　扁平桨式搅拌器
~ mixer with multiple beams　多桨式混合器
~ mixer with single beam　单桨式混合器
~ stirrer　桨式搅拌机
PAEK [polyaryl ether ketone]　聚芳醚酮

PAI [polyamide – imide]　聚酰胺–酰亚胺
PAI [polyaminotriazole]　聚氨基三唑
paint　漆;涂料;喷涂
~ and varnish industry　油漆工业
~ binder　涂料黏合剂
~ coat　涂层
~ coating　涂漆
~ compound　漆混合料
~ film　漆膜
~ for cocooning　蚕丝漆〈可剥性保护漆〉
~ for cocoonization　蚕丝漆
~ grinder mill　涂料研磨机
~ mixer　涂料混合器
~ pouring　浇注着色
~ remover　脱漆剂
~ resin　漆用树脂
~ roller mill　辊式研漆机
~ spray booth　喷漆室
~ spraying plant　喷漆器
~ surface　涂漆面
~ vehicle　漆料
painter　油漆工;涂敷机,喷漆机
painting　涂漆,涂装;着色
~ defect　涂漆层缺陷
~ gun　喷(涂)枪
~ of plastics　塑料(用)涂漆
~ surface　涂漆面
pair of rollers　成对滚筒;成对辊
palatable　可口的
pale　浅色的,苍白的
~ colour　浅色
~ shade　淡色调〈模塑件或涂层〉
palladium [Pd]　钯
pallet　转盘;衬垫〈机器〉;托板,托盘〈运输〉
palletized unit load　码垛设备装载量
palletizer　码垛机
palmitate　棕榈酸酯
PAMS [polyα – methylstyrene]　聚α–甲基乙烯〈增塑剂〉
PAN [polyacrylonitrile]　聚丙烯腈
pan　平盘;平[板]面;集料器;槽

~ grinder 盘式研磨机
~ mill [mixer] 碾盘式碾磨机[混合器]
panchromatic 全色的
~ film 全色胶片
pane 玻璃板〈窗〉；嵌板〈门〉
panel 板材；板条，仪表板，控制板，面板
~ board 仪表板，操纵板
~ core density 板芯密度〈发泡〉
~ holder 板条夹
~ planing machine 刨板机
~ size 板材尺寸
paneling 瘪陷；镶板
pantachromatic 多色的
PAPA [polyazelaic polyanhydride] 聚壬二酸酐
paper 纸
~ base laminate 纸基层合制品
~ board 纸板
~ carrier 纸载体〈黏合膜〉
~ chromatography 纸色谱（法）
~ cloth 纸布
~ composite 纸复合材料
~ converting 纸整理加工
~ electrophoresis 纸电泳法
~ fibre 造纸纤维
~ filler 纸质填料
~ film 纸形薄膜
~ finishing 纸精饰
~ honeycomb core 浸渍树脂的蜂窝状纸芯
~ imitated film 仿真纸膜
~ interlayer 纸夹层，中间层纸
~ label 纸标签
~ -like (plastic) film 仿真纸（塑料）膜
~ machine 造纸厂
~ mill 造纸厂
~ mill beater 纸张打浆机
~ number 纸号〈平方米重〉
~ roll 纸辊，纸卷
~ shredding machine 碎纸机
~ sizing 纸上胶
~ weight 纸重〈纸号〉

papercore plywood 纸心胶合板
PAPI [polymethylene polyphenylene isocyanate] 聚亚甲基聚苯基异氰酸酯〈参见 PMPPI〉
papyrolen 纸形聚乙烯薄膜
para 对位〈略作 p-〉；仲；副
~ chloro-meta-xylenol 对氯间二甲苯酚〈防霉剂〉
~ -compound 对位化合物
~ -cresol 对甲酚
~ dyes 偶合染料
~ -formaldehyde 仲甲醛；多聚甲醛
~ -isomer 对位异构体
~ -linkage 对位键
~ -octylphenyl salicylate 水杨酸对辛基苯酯
~ -orientation 对位取向
~ -oriented ring 对位取向环
~ -xylene 对二甲苯
parabanic acid 仲班酸；乙二酰脲
paracrystal 次晶
paracrystalline 次晶的
~ lattice 次晶晶格
~ region 次晶晶区
~ structure 次晶结构
paracrystallinity 次晶性
~ model 次晶模型
paraffin [paraffine] 石蜡；链烷烃
~ oil 石蜡油
~ wax 固体石蜡，天然石蜡〈润滑剂，脱模剂〉
paraformaldehye 仲甲醛；多聚甲醛
parallel 平行的，并联的
~ arrangement 平行排列
~ build 平行成形，平行卷绕
~ build package 平行成形卷装
~ connection 并联
~ dielectric heating 平行介电加热
~ folding 平行折叠
~ laminate 顺纹层压板
~ laminated 同向层压的，平行层压的；顺纹层压的
~ -lamination 平行层压，顺纹层压
~ mat 顺纹（玻璃）毡

parallelising

- ~ – plate plastometer 平行板塑性计
- ~ plate viscometer [viscosimeter] 平行板黏度计
- ~ rod 条状试棒
- ~ slide gate valve 平滑闸门
- ~ spinning 平行纺丝
- ~ system 并联系统
- ~ test 平行试验
- ~ to grain 顺纹〈胶接点〉
- ~ – to – grain gluing 顺纹粘接
- ~ wound bobbin 平行卷绕筒子〈纺丝〉

parallelising 平行度〈纺丝〉

paramagnetic 顺磁(性)的

- ~ nickel chelate 顺磁性镍螯合物〈光稳定剂〉
- ~ rasonance 顺磁共振
- ~ resonance absorption 顺磁共振吸收

paramagnetism 顺磁性

parameter 参数,系数

- ~ of adjustment 调整参数〈塑料加工机〉
- ~ value 参数值

parametric optimization 最佳参数选择

parchment paper 仿羊皮纸,硫酸纸

parchmyn 仿羊皮纸

parent 母体的

- ~ material 原材料,母料
- ~ molecule 母体分子
- ~ strock 母体原料

parison 型坯

- ~ accumulator 型坯蓄料器
- ~ control 型坯控制
- ~ curl 型坯蜷缩〈挤压吹塑〉
- ~ cutter 型坯切切;型坯切断器
- ~ cutting knife 型坯切刀
- ~ die swell 型坯离模膨胀
- ~ drop time 型坯下垂时间
- ~ gripping device 型坯夹具〈挤吹〉
- ~ head 型坯〈挤出〉机头〈吹塑机〉
- ~ holder 型坯夹持器
- ~ mould 型坯模
- ~ programmer 型坯程序控制器
- ~ programming 型坯程序控制
- ~ – shift type blow moulding machine 型坯移动式吹塑机
- ~ swell 型坯膨胀
- ~ tube 型坯管;管坯

parquet – type polymer 镶嵌型聚合物

parquetry 嵌木地板

- ~ finishing 木地板涂漆装饰〈用合成树脂漆〉
- ~ sealing 木地板密封涂敷〈用合成树脂漆〉

part 部分;部件,制件,制品

- ~ cure 部分硫化
- ~ extractor 模塑件脱模装置〈注塑机〉
- ~ inspection 模塑件检验
- ~ number 零件号码

part. [partial]部分的,局部的

part. [particle]粒子;微粒

partial 部分的,局部的;条件的

- ~ assembly drawing 零件装配图
- ~ automatic 半自动的
- ~ automatization 部分自动化
- ~ cracking 局部龟裂
- ~ foaming restraint 局部发泡防溢器
- ~ hydrolysis 部分水解
- ~ melting 部分熔融
- ~ overlap seal end 部分搭接封端
- ~ oxidation 部分氧化
- ~ pressure 分压(力),部分压力
- ~ quantity 分量
- ~ re – closing 部分再闭模
- ~ vacuum 部分真空

partially 部分地,局部地

- ~ crosslinked thermoplastic 部分交联热塑性塑料
- ~ crystalline 部分结晶
- ~ crystalline plastic 部分结晶塑料
- ~ expanded 部分发泡的

particle 粒子,颗粒

- ~ board 刨花板;木屑板
- ~ board binder 木屑板粘结剂
- ~ composite structure 颗粒复合结构
- ~ diameter 粒径
- ~ distribution 粒子分布
- ~ filter 微粒过滤器

- ~ fineness 颗粒细度
- ~ flow process 颗粒流程
- ~ orientation 粒子取向
- ~ shape 颗粒形状
- ~ size 粒度,粒径,粒子大小
- ~ size analysis 粒度分析,粒径分析
- ~ size distribution 粒度分布,粒径分布,粒子大小分
- ~ size range 粒度范围,粒径范围
- ~ structure 粒状结构

particulars pl 详细资料,详细数据,摘要

particulate 粒状的
- ~ composite 粒状复合材料
- ~ solid 固体粒子

parting 分离(的),离别(的)
- ~ agent 脱模剂
- ~ compound 脱模剂
- ~ line 合模线;模缝痕;拼缝线
- ~ line mark 合模线痕〈模塑件上〉
- ~ plane 分模面,界面
- ~ surface 分模面

partition 分配,分隔
- ~ board 隔板
- ~ chromatography 分配色谱法
- ~ coefficient 分配系数
- ~ gas chromatography 分配气相色谱法
- ~ insulator 绝缘导管
- ~ line 合模线;模缝痕;拼缝线
- ~ panel 隔板
- ~ panel lining 隔板衬里
- ~ wall 隔壁

partitioned mo(u)ld cooling 塑模间隔冷却

partitioning insert 间隔填入物〈包装〉

partly 部分地,局部地
- ~ crystalline plastic 部分结晶塑料
- ~ risen foam 部分发泡
- ~ wall 界墙,共同墙

parylene 聚对亚苯基二甲基

PAS[polyaryl sulfone] 聚芳砜

pass 通过;经过;合格
- ~ the test 通过试验
- ~ through rolls 滚筒行程,滚筒通过

passage 通过;通道

passive 钝(态)的;被动的
- ~ anvil 钝化砧〈超声波焊〉
- ~ hardness 钝态硬度,耐磨硬度,耐磨性

paste 糊(料);膏
- ~ adhesive 糊状黏合剂
- ~ board 胶合纸板
- ~ coo(u)r 浆色;色浆
- ~ dyes 浆状染料,染料糊
- ~ extrusion 糊料挤出〈聚四氟乙烯〉
- ~ filer 膏状填料
- ~ like 糊状的
- ~ – making resin 制糊树脂
- ~ mill[mixer] 磨浆机,调糊机
- ~ mo(u)lding 糊料成型
- ~ paint 厚漆
- ~ processing 糊料加工
- ~ resin 糊状树脂
- ~ roller mill 磨浆机
- ~ technology 糊料加工

pastel 轻淡的;菘蓝染料
- ~ colo(u)r 淡色;粉彩色
- ~ shade 浅色,淡色

pat.[patent] 专利

PAT[polyaminotriazole] 聚氨基三唑

parent 母体的
- ~ material 原材料,母料
- ~ molecule 母体分子
- ~ strock 母体原料

parison 型坯
- ~ accumulator 型坯蓄料器
- ~ control 型坯控制
- ~ curl 型坯蜷缩〈挤坯吹塑〉
- ~ cutter 型坯切刀;型坯切断器
- ~ cutting knife 型坯切刀
- ~ die swell 型坯离模膨胀
- ~ drop time 型坯下垂时间
- ~ gripping device 型坯夹具〈挤吹〉
- ~ head 型坯(挤出)机头〈吹塑机〉
- ~ holder 型坯夹持器
- ~ mould 型坯模
- ~ programmer 型坯程序控制器
- ~ programming 型坯程序控制

~ – shift type blow moulding machine 型坯移动式吹塑机
~ swell 型坯膨胀
~ tube 型坯管;管坯
parquet – type polymer 镶嵌型聚合物
parquetry 嵌木地板
~ finishing 木地板涂漆装饰〈用合成树脂漆〉
~ sealing 木地板密封涂敷〈用合成树脂漆〉
part 部分;部件;制件,制品
~ cure 部分硫化
~ extractor 模塑件脱模装置〈注塑机〉
~ inspection 模塑件检验
~ number 零件号码
part. [partial] 部分的,局部的
part. [particle] 粒子;微粒
partial 部分的,局部的;条件的
~ assembly drawing 零件装配图
~ automatic 半自动的
~ automatization 部分自动化
~ cracking 局部龟裂
~ foaming restraint 局部发泡防溢器
~ hydrolysis 部分水解
~ melting 部分熔融
~ overlap seal end 部分搭接封端
~ oxidation 部分氧化
~ pressure 分压(力),部分压力
~ quantity 分量
~ re – closing 部分再闭模
– vacuum 部分真空
partially 部分地,局部地
~ crosslinked thermoplastic 部分交联热塑性塑料
~ crystalline 部分结晶
~ crstalline plastic 部分结晶塑料
~ expanded 部分发泡的
patch 补片;斑点
patching 修补
~ compound 修补材料
patent 专利(的);专利权
~ appeal 专利要求
~ application 专利申报,专利申请
~ claim 专利要求

fee 专利费用
infringement 专利侵犯
~ leather 漆皮
~ specification 专利说明
patentability 可专利(性)
patentable 可取得专利的
path 路径,途径
patrix 母料;阳模;上模
pattern 花样;图案;模型
~ area 花纹面积,花型范围
~ barrel 提花滚筒,纹板滚筒
~ coater (熔胶)图形涂布机
~ design 花纹设计
~ die 型模
~ drum 提花滚筒,纹板滚筒
~ paper 花纹板;木纹纸
~ plate 提花板;模板
~ sheet 花纹贴面层,塑料贴面板
~ varnish 美术漆
patterned surface 花纹表面
patterning mechanism 提花机构〈纺织〉
pay – off 松卷装置,退卷〈塑料薄膜〉;基材输送装置
paying load 有效负载
Pb [lead] 铅
PB [polybutene,polybutylene] 聚丁烯
PBAN [poly(butadiene – acrylonitrile)] 聚丁二烯 – 丙烯腈,丁腈橡胶
PBD [polybutadiene] 聚丁二烯
PBI [polybenzimidazole] 聚苯并咪唑
PBMA [poly(n – butyl methacrylate)] 聚甲基丙烯酸正丁酯
PBO [polybenzoxazole] 聚苯并噁唑
PBS [poly(butadiene – styrene)] 聚丁二烯 – 苯乙烯,丁苯橡胶
PBT [polybenzothiazolel 聚苯并噻唑
PBTP [polybutylene terephtalate] 聚对苯二甲酸丁二(醇)酯
p.c. [per cent] 百分之几
pc. [piece] 件;片;块
PC [polycarbonate] 聚碳酸酯
PCB [polychlorinated bisphenyl] 多氯代联苯
PCD [polycarbodiimide] 聚碳二酰亚胺

PCDT [poly(1,4 - cyclohexylene dimethylene terephthalate)] 聚(1,4 - 亚环己基二亚甲基对苯二甲酸酯)
PCL [polycaprolactone] 聚己酸内酯
PCTFE [polychlorotrifluoroethylene] 聚三氟氯乙烯〈增塑剂〉
Pd [palladium] 钯
PDAIP [poly diallyl isophthalate] 聚间苯二甲酸二烯丙酯
PDAP [polydiallyl phthalate] 聚邻苯二甲酸二烯丙酯
PDESF [polydiethylsiloxane fluid] 聚二乙基硅氧烷液体,二乙基硅油〈清泡剂,防水剂,抛光剂,润滑剂,脱模剂〉
PDMS [polydimethylsiloxane] 聚二甲基硅氧烷〈脱模剂〉
PDMSF [polydimethylsiloxane fluid] 聚二甲基硅氧烷液体,二甲基硅油〈脱模剂,润滑剂,添加剂,清泡剂〉
p. e. [par example] 举例,例如
PE [polyethylene] 聚乙烯
PEA [polyethyl acrylate] 聚丙烯酸乙酯
peak 峰;峰值;最大值
~ distortion 峰畸衰
~ efficiency 最高效率
~ effcientcy of unit 设备的最大生产率
~ exothermic temperature 最高的放热温度〈热固性树脂〉
~ load 峰值复载,高峰负载,最大荷载
~ load operation 最高负荷操作
~ pressure 最高压力
~ temperature 峰值温度
~ - to - valley height 峰谷间高度;最大不平度〈连接件表面粗糙不平〉
~ - to - valley ratio 峰谷比,峰值 - 谷值比
peanut hull flour 花生壳粉〈填料〉
pear - shaped channel 梨形料道
pearl 珍珠;珠灰色;蓝灰色
~ essence 珠光粉
~ filler 碳酸钙填料
~ mill 珍珠球磨机
~ polymerization 珠状聚合

pearlescent 珠光般的
~ conting 珠光涂层;珠光涂料
~ pigment 珠光颜料
pearlized polymethyl methacrylate 珠光有机玻璃,珠光聚甲基丙烯酸甲酯
pebble 卵石
~ mill 卵石球磨机
PEC [chlorinated polyethylene] 氯化聚乙烯 [CPE.]
Peco [Thennox] torpedo 皮克型鱼雷头;内加热的鱼雷头
pedestal 架,底座〈机器〉
~ - type impeller mixer 架式高速混合器
PEEK [polyether ether ketone] 聚醚醚酮
peel 皮;剥皮;剥离
~ adhesion (皮层)剥离黏着力
~ adhesion test 黏合剥离试验
~ back 铸模剥落
~ bond strength 剥离黏合强度
~ - off 剥离
~ - off plastics coating 可剥离塑料涂层
~ - off velocity 剥离速度
~ paint [print] 可剥离涂层;可剥离性涂料〈用于临时性防腐〉
~ ply 剥离层
~ resistance 耐剥离性
~ strength 剥离强度
~ test 剥离试验
peelable 可剥(性)的
~ bond 可剥性胶合
~ coating 可剥性(保护)涂层
~ protective coating 可剥性保护涂层
~ seal 可剥封口
peeler 剥皮机,剥片机;切片机
peeling 剥离;剥皮;脱落,剥落〈膜、片、涂层〉
~ flexure test 挠曲剥离试验
~ from the block 剥离成膜
~ process 剥离方法
~ strength 剥离强度
~ stress 剥离应力
~ technique 剥离技术

~ test 剥离试验
~ tester 剥离试验仪
PEG [polyethylene glycol] 聚乙二醇
peg 栓钉
 ~ stirrer 钩状搅拌器
 ~ -type baffle 钩状(搅拌)桨叶
PEK [polyether ketone] 聚醚酮
pellet 粒料;切粒
 ~ head 造粒机头
 ~ mill 线粒料辊炼机
 ~ size 颗粒尺寸,颗粒大小
pelleted adhesives 粒料黏合剂;片状黏合剂
pelleter 切粒机,造粒机
pelleting 切粒,线料造粒,造粒
 ~ machine 切粒机,造粒机
 ~ press 切粒机
 ~ property 切粒能力
pelletization 切粒,线料造粒
pelletize 切粒
pelletized 粒状的
 ~ compound 粒状料
 ~ dyes 粒状染料
pelletzer 切粒机,造粒机
pelletizing 切粒,造粒
 ~ die 造粒模具
 ~ machine 切粒机,造粒机
 ~ process 切粒过程,造粒过程
 ~ rod 切粒棒条
 ~ system 切粒系统
pencil hardness 铅笔(刮痕)硬度
pendulum 摆锤
 ~ damping test 摆锤阻尼试验
 ~ hardness 摆测硬度
 ~ hardness test 摆测硬度试验
 ~ hardness tester 摆测硬度(试验)仪
 ~ hardness testing machine 摆测硬度仪
 ~ impact strength 摆锤冲击强度
 ~ impact test 摆锤(式)冲击试验
 ~ plastometer 摆锤(式)塑性计
 ~ strength tester 摆锤(式)(冲击)强度(试验)仪
 ~ test 摆锤(式)(冲击)试验

~ tester 摆锤(式)试验仪
~ type 摆锤型
penetrability 穿透性;渗透性
penetrable 可渗透的,可穿透的
penetrant 渗透剂
penetrating 穿透的
 ~ power 穿透力,渗透性
 ~ property 渗透性
penetration 渗透;渗透性〈黏合剂渗透粘接件的表面毛细管〉;针入度;熔焊
 ~ depth 渗透深度〈物料对塑料的渗透程度〉;熔焊深度
 ~ number 针入值
 ~ of adhesive 黏合剂渗透
 ~ test 渗透性试验;针入度试验
 ~ tester 渗透测定仪;针入度测定仪
penetrometer 渗透测定仪;针入度测定仪
pentabromochlorocyclohexane 五溴氯环己烷〈阻燃剂〉
pentabromodiphenyl oxide 五溴二苯醚〈阻燃剂〉
pentabromoethylbenzene 五溴乙苯〈阻燃剂〉
pentabromophenol 五溴苯酚〈阻燃剂〉
pentachlorophenol [PCP] 五氯苯酚〈防霉剂〉
pentachlorophenyl laurate 五氯苯(酚)月桂酸酯〈防霉剂〉
pentachlorothiophenol 五氯硫酚〈塑解剂、再生活化剂〉
pentacites 季戊四醇树脂
pentaerythritol [pentaerythrite] 季戊四醇〈热稳定剂〉
 ~ fatty acid ester 季戊四醇脂肪酸酯〈增塑剂〉
 ~ tetrabenzoate 季戊四醇四苯甲酸酯〈增塑剂〉
 ~ tetracaprylate caprate 季戊四醇四辛酸癸酸酯〈增塑剂〉
2,4-pentanedione 2,4-戊二酮〈螯合剂、溶剂、金属络合剂〉
Penton〈商〉片通,〈Hercules Powder 公司产氯化聚醚〉

PEO [polyethylene oxide] 聚环氧乙烷,聚氧化乙烯〈俗称〉
peony 牡丹红〈暗红色〉
PEOX [polyethylene oxide] 聚环氧乙烷,聚氧化乙烯〈俗称〉
peptide bond 肽键
peptize 胶溶;塑解
peptizer 胶溶剂;塑解剂
peracetic acid 过乙酸
percent 百分率,百分数
~ by volume 体积百分数
~ by weight 重量百分数
~ conversion 转化率
~ error 百分误差,误差百分率
~ test 挑选试验
~ transmittance 透光百分率
~ swell 膨胀百分率
percentage 百分率,百分数
~ concentration 百分浓度,浓度百分率
~ content 含量百分率
~ elongation 伸长率,延伸率
~ elongation at break 断裂伸长率
~ elongation at yield 屈服伸长率
~ moisture content 含水率
~ of deviation 偏差率
~ of error 误差率,误差百分率
~ of reject 废品率,废品百分率
~ relative humidity 百分相对湿度;相对湿度百分数
~ transmission 透射百分率
~ variation in rate 速率变化〈挤塑机〉
perchloroethane 全氯乙烷,六氯乙烷〈俗称〉
perchlorocyclopentadecane 全氯环十五烷,三十氯(代)环十五烷〈阻燃剂〉
percolate 渗滤;渗滤液
percolation 渗透,渗滤
percolator 渗滤器
perfect 完全的,理想的
~ combustion 完全燃烧
~ elastic material 理想弹性材料
~ elasticity 理想弹性
~ mixing 完全混和

~ parison 理想型坯,理想管坯〈吹模〉
~ plasticity 理想塑性
perfectly 完全地
~ elastic 理想弹性的;完全弹性的
~ reflecting diffuser 全反射漫射体
~ transmitting diffuser 全发射漫射体
perfluorinated 全氟化的
~ ethylene propylene copolymer 全氟乙-丙共聚物
~ ion exchange membrane 全氟离子交换膜
~ sulfonic acid resin 全氟磺酸树脂
perfluoroalkoxy resins [PFA] 全氟烷氧基树脂
perfluorocaprylic acid 全氟辛酸〈分散剂〉
perfluoroethylene 全氟乙烯,四氯乙烯〈俗称〉
~ propylene 全氟乙丙共聚物
~ propylene copolymer 全氟乙丙共聚物
~ propylene polymer [FEP] 全氟乙丙聚合物
~ propylene plastic [FEP] 全氟乙丙塑料
perfluorooctanoic acid 全氟辛酸〈分散剂〉
perfluoropropylene 全氟丙烯,六氟丙烯〈俗称〉
perfluorotrimethylene diisocyanate 全氟三亚甲基二异氰酸酯
perfluosulfonic acid type cation exchange membrane 全氟磺酸型阳离子交换膜
perforate 穿孔,冲孔
perforated 穿孔的;多孔的
~ bottom plate 多孔底板
~ cylinder 多孔滚筒
~ drum dryer 多孔转鼓烘干[干燥]机
~ plate 多孔板
~ plywood 穿孔胶合板
perforating 打孔,穿孔
~ die 冲孔模

~ machine 打孔机,穿孔机
perforation 打孔,穿孔
performance 操作;运行;性能,特性
　~ analysis 性能分析
　~ characteristics 使用性能,应用性能,工作特性
　~ curve 特性曲线
　~ data 使用数据,性能数据
　~ density 表观密度
　~ figure 性能数字,质量指标
　~ index 性能指数,性能指标
　~ parameter 性能参数
　~ range 性能范围
　~ standards 技术性能标准
　~ test 运转试验,性能试验,应用试验
period 周期;时期
　~ of ageing 老化期〈塑料材料〉
　~ of dwelling 停压时间;保压时间;放气时间
periodic 周期性的,定期的
　~ change 周期变化
　~ copolymer 嵌段共聚物,周期(性)共聚物
　~ copolymerization 嵌段共聚,周期(性)共聚
　~ error 周期性误差
　~ inspection 定期检修,定期检查
　~ overhaul 定期大(检)修
　~ sample 周期性取样
　~ test 定期试验,周期性试验
peripheral 周围的,外面的
　~ equipment 外围设备,外部设备
　~ fibre 周缘纤维
　~ layer 外层;边缘层
　~ speed 圆周速率[度]
　~ winder 圆周缠绕机
peripherally drilled roll 内有加热槽的辊;周期钻孔辊
perishable food 易腐食品
permanence 耐久性;永久性;稳定性;安定性
　~ condition 不变条件
permanent 永久的,持久的
　~ antistatic fibre 持久性防静电纤维
　~ compression set 永久压缩变形
　~ curl 永久卷曲〈长丝〉
　~ deformation 永久变形,残余变形
　~ distortion 永久变形
　~ dyes 永固染料,持久染料,不变色染料
　~ elongation 永久伸长,固定伸长
　~ exparsion 持久膨胀,残余膨胀
　~ extension 永久伸长,残余伸长
　~ folding test 持久弯折试验
　~ hardness 永久硬度
　~ joint 永久性连接〈如焊接,粘接〉
　~ label 持久标签〈包装〉
　~ magnet 永久磁铁
　~ output 长期生产率
　~ plasticizer 永久性增塑剂
　~ repair 小修,日常修理
　~ resistance to swelling 耐持久性溶胀
　~ set 永久变形;固定伸长
　~ stability 耐久性
　~ storage 长期储存
　~ strain 永久应变
　~ stress 永久应力
　~ tackiness 持久黏性〈黏合〉
　~ use 长期应用
permanently tacky adhesive 永久性黏合剂
permeability 渗透性;穿透性;磁导率
　~ apparatus 透气率测定仪
　~ coefficient 渗透系数
　~ test 渗透性试验
　~ to air 透气性
　~ to gas 透气性
　~ to light 透光性
　~ to liquids 透液性
　~ to oil 透油性
　~ to water 透水性
　~ to water vapor 透水气性
permeable 可渗透的
　~ plastics 可透塑料
　~ to vapour 可透蒸气的
permeameter 渗透计
permeation 渗透(作用)
　~ barrier 阻渗层

~ coefficient 渗透系数
permissible 容许的
~ circumferential stress 容许圆周应力
~ deformation 容许变形
~ deviation 容许偏差
~ dose 容许剂量
~ error 容许误差
~ explosives 安全炸药
~ load 容许负荷
~ stress 容许应力
~ tolerance 容许公差
~ variation 容许偏差
permittivity 介电常数,介电系数;电容率
peroxide 过氧化物
~ catalyst 过氧化物催化剂;过氧化物固化促进剂
~ catalyzer 过氧化物催化剂
~ crosslinking 过氧化物交联
~ initiation 过氧化物引发
~ initiator 过氧化物引发剂
~ paste 过氧化物糊〈不饱和聚酯树脂固化剂〉
~ vulcanization 过氧化物硫化
peroxyketal 过缩酮〈交联剂〉
perpendicular 垂直(的);垂直线
persimmon 柿棕〈棕色〉
persistence 持久(性)
persistent 持久的,不变的
~ characteristic 持久特性
~ strength 长期性强度
persisting elongation 永久伸长
persorption 吸混作用;渗入吸附(气孔)
perspective 透视
~ drawing 透视图
~ view 透视图
PES(U) [polyether sulfone] 聚醚砜
PET(P) [polyethylene terephthalate] 聚对苯二甲酸乙二(醇)酯
petrochemical 石油化工产品;石油化学的
~ industry 石油化学工业
~ material 石油化工原料
~ plant 石油化工厂

petrochemistry 石油化学
petrol 汽油
petroleum 石油
~ chemical plant 石油化工厂
~ chemicals 石油化工产品
~ cracking 石油裂解
~ cracking process 石油裂化过程
~ crude 原油
~ ether 石油醚
~ hydrocarbons *pl* 石油烃
~ hydrocarbon resin 石油烃(基)树脂
~ naphtha 石脑油,粗汽油,汽油馏分,溶剂(汽)油
~ processing 石油加工
~ product 石油产品
~ resin 石油树脂〈软经剂,增黏剂〉
petticoat 球形绝缘器
PF [phenol formaldehyde resin] 酚醛树脂
PFA [perfluoroalkoxy resins] 全氟烷氧基树脂
PGC [pyrolytic gas chromatography] 热解气相色谱法
PGE [phenyl glycidyl ether] 苯基缩水甘油醚
pH [concentration of hydrogenions] 氢离子浓度
pH value pH 值
pharmaceutical polymer 药用聚合物
phase 相(位)
~ angle 相角
~ boundary 相界
~ boundary charge 界面电荷
~ boundary crosslinking 相界交联
~ boundary layer 相界层
~ change 相变
~ contrast microscope 相衬显微镜
~ diagram 相图
~ difference 相位差
~ distortion 相位畸变
~ interface 相界面
~ inversion 相逆转〈聚合过程中的〉
~ inversion temperature 相转变温度
~ microscope 相衬显微镜

- ~ plate 相位板
- ~ separation 相分离, 析相作用
- ~ surface 相表面
- ~ transition 相转变

PHEMA [poly(2 - hydroxyethyl meth - acrlate)] 聚(甲基丙烯酸 - 2 - 羟乙酯)

phengite 多硅白云母

phenol 酚〈类名〉;苯酚
- ~ aldehyde resin 酚醛树脂
- ~ - base adhesive 酚醛黏合剂
- ~ fibre 酚醛纤维
- ~ formaldehyde (苯)酚(甲)醛(缩合物, 树脂)
- ~ formaldehyde condensation resin 酚醛缩合树脂
- ~ formaldehyde oligomer 酚醛低聚物
- ~ formaldehyde resin [PF] 酚醛树脂
- ~ - furfural resin 苯酚糠醛树脂
- ~ furfural moulding powder 苯酚糠醛模塑粉
- ~ glue 酚醛胶(黏剂)
- ~ modified melamine moulding powde 苯酚改性三聚氰胺模塑粉
- ~ resin 酚醛树脂
- ~ resin glue 酚醛树脂胶(黏剂)〈美〉

phenolic (苯)酚的; 酚类的
- ~ adhesive 酚醛黏合剂
- ~ aniline resin 苯酚苯胺树脂
- ~ antioxidant 酚类抗氧剂
- ~ cement 酚醛胶泥粘接剂
- ~ epoxy resin 酚醛环氧树脂
- ~ fibre 酚醛纤维
- ~ foam 酚醛泡沫塑料
- ~ foams plastics 酚醛泡沫塑料
- ~ formaldehyde resin 酚醛树脂
- ~ glue 酚醛胶(黏剂)〈美〉
- ~ glue film 酚醛胶黏膜
- ~ laminate 酚醛塑料层合板
- ~ laminated roll 酚醛树脂层卷棒条
- ~ laminated sheet 酚醛树脂层合片材
- ~ modified alkyd resin 酚醛改性醇酸树脂
- ~ moulding composition 酚醛模塑料〈美〉
- ~ moulding compound [material] 酚醛模塑料
- ~ mo(u)lding powder 酚醛模塑粉
- ~ novolac 线型酚醛树脂, 酚醛清漆
- ~ plastic 酚醛塑料
- ~ - polyvinyl acetal adhesive 酚醛 - 聚乙烯醇缩醛黏合剂
- ~ resin 酚醛树脂
- ~ resin adhesive 酚醛树脂黏合剂
- ~ resin and plastics 酚醛树脂及塑料
- ~ resin coating 酚醛树脂涂料
- ~ resin for shell moulding 用于壳模成型的酚醛树脂
- ~ resin paint 酚醛树脂漆
- ~ resin varnish 酚醛树脂清漆
- ~ - rubber adhesive 酚醛 - 橡胶黏合剂
- ~ silicone adhesive 酚醛树脂 - 有机硅黏合剂
- ~ system weakly acidic cation exchange resin 酚醛系弱酸性阳离子交换树脂
- ~ varnish 酚醛树脂清漆

phenolics 酚醛树脂; 酚醛塑料

phenolplast 酚醛塑料

pbenomenological 现象的, 唯象的
- ~ effect 唯象效应〈塑料形变的力学性能〉
- ~ rheology 唯象流变学

phenomenon 现象

phenoplast 酚醛塑料

phenoplastic 酚醛塑料
- ~ mo(u)lding material 酚醛模塑料

phenoxy resin 苯氧树脂

phenyl 苯基
- ~ acrlate 丙烯酸苯酯
- ~ alkylsulfonate 烷基磺酸苯酯〈增塑剂〉
- ~ diisodecyl phosphate 亚磷酸苯二异癸酯〈抗氧剂〉
- ~ glycidyl ether [PGE] 苯基缩水甘油醚〈反应性稀释剂〉
- ~ group 苯基

2 ~ indole   2-苯基吲哚〈热稳定剂〉
2 ~ -4-methylimidazole   2-苯基-4-甲基咪唑〈固化剂〉
N-~-α-naphthylanune   N-苯基-α-苯胺,防老剂甲,尼奥宗A〈抗氧剂〉
N-~-β-naphthylamine   N-苯基-β-萘胺,防老剂丁,尼宗奥D〈抗氧剂〉
N-~-N'-octyl-p-phenylene diam-ine   N-苯基-N'-辛基对苯二胺,防老剂688〈抗氧剂〉
N-~-N'-sec-octyl-p-phenylene-dia-mine   N-苯基-N'-仲辛基-对苯二胺〈抗氧剂〉
~ salicylate   水杨酸苯酯〈紫外线吸收剂〉
~ silane resins   苯基硅烷树脂
phenylated 苯基的
  ~ polypyromellitimidine   均苯型聚酰亚胺;苯基化梁均苯四甲酰亚胺
phenylcellulose 苯基纤维素
phenylene 亚苯基
  m-~ diamine [MPD]   间苯二胺〈固化剂〉
phenylethylene 苯乙烯
2-**phenylimidazole**   2-苯基咪唑〈固化促进剂〉
5-**phenyltetrazole**   5-苯基四唑〈发泡剂〉
phenyltrichlorosilane   苯基三氯(甲)硅烷
Phillips injection moulding   菲利浦法注塑;菲利浦法熔融发泡
phlogopite 金云母
phonometer 声强计
phoryl resins 磷酰酚树脂
phosgene 光气;碳酰氯
phosphate 磷酸酯〈或盐〉
  ~ plasticizer   磷酸酯类增塑剂
phosphating   磷酸酯化;磷酸盐化
phospbation   磷酸酯化;磷酸盐化
phosphorescence   磷光(现象)
phosphorescent   磷光(性)的

~ pigment   磷光颜料
~ paint   磷光涂料
phosphoric 磷的
  ~ acid   磷酸
  ~ ester   膦酸酯
phosphorus, P   磷
photoacoustic spectroscopy   光声谱(学)
photoactivate   光敏化
photoactivated oxidation   光致氧化
photoaddition   光(致)加成
photoageing   光(致)老化
photochemical   光化学的
  ~ decomposition   光化分解
  ~ degradation   光化降解
  ~ destruction   光化破坏
  ~ oxidant   光化氧化剂
  ~ polymerization   光化聚合
  ~ sensitivity   光化敏性
  ~ stability   光化稳定性
photochemistry   光化学
photochromic   光致变色的
  ~ dyes   光致变色染料
  ~ fibre   光致变色纤维
photochromism   光致变色
photocolorimeter   光比色计
photocolorimetry   光比色法
photoconduction   光电导
photoconductive   光电导的
  ~ film   光敏薄膜
  ~ polymer   光电导聚合物
photoconductivity   光电导性
photocopolymerization   光致共聚(作用)
photocrosslinking   光致交联
photodecomposition   光分解
photodegradable   可光降解的
  ~ plastic   可光降解的塑料,光降解性塑料
  ~ polymer   可光降解的聚合物
photodegradation   光降解
  ~ of fibre   纤维的光降解
  ~ of plastic   塑料的光降解
photodepolymerization   光解聚(作用)
photodeterioration   光致劣化,光致变质,光(致)老化

photoelastic 光弹性的
- ~ analysis 光弹性分析
- ~ effect 光弹性效应
- ~ material 光弹性材料
- ~ measurement 光弹性测定
- ~ plastic 光弹性塑料
- ~ test 光弹性试验
- ~ varnish 光弹性清漆

photoelasticity 光弹性
photoelectric 光电的
- ~ cell 光电池
- ~ colorimeter 光电比色计
- ~ effect 光电效应

photografting method 光接枝法〈改良聚合物表面,如:可印性〉
photographic reference standard 感光参考标准
photogravure printing machine 照相凹版印刷机
photoinitiated polymerization 光引发聚合
photoinitiation 光(化)引发(作用)
photoluminescence 光致发光,荧光
photolysis 光解(作用)
photomicrograph 显微照相
photomultiplier 光电倍增管
photooxidant stabilizer 防光老化剂
photooxidation 光氧化(作用)
photooxidative degradation 光氧化降解
photoplastic 光敏塑料
photopolymer 感光聚合物
photopolymerisation [photopolymerization] 光致聚合(作用)
photorearrangement 光致重排
photoresist 光敏抗蚀剂
photoresistance 光敏电阻
photoresponsive polymer 光敏聚合物
photosensibilized reaction 光敏化反应
photosensitised polymerirazion 光敏聚合
photosensitive 光敏的
- ~ adhesive 光敏胶黏剂
- ~ control film 光敏控制薄膜
- ~ initiator 光敏引发剂
- ~ layer 光敏层,感光层
- ~ polymer 光敏聚合物
- ~ resin 光敏树脂
- ~ tape 光敏磁带

photosensitiveness 光敏性
photosensitivity 光敏性,感光灵敏度
photosensitization 光敏作用
photosensitized degradation 光敏降解
photosensitizer 光敏剂
photostability 耐光性,光稳定性
photostabilizer 耐光剂,光稳定剂
photostabilization 耐光性
phototransformation of polymers 聚合物的光转变
phototropic dyes 光变性染料
phthalate 邻苯二甲酸酯[盐]
- ~ ester plasticizer 邻苯二甲酸酯(类)增塑剂
- ~ plasticizer 邻苯二甲酸酯(类)增塑剂

phtalic 邻苯二甲(酸)的
- ~ acid 邻苯二甲酸
- ~ alkyd resin 邻苯二甲醇酸树脂
- ~ anhydride [PA] 邻苯二甲酸酐,〈俗称〉苯酐〈固化剂〉
- ~ ester 邻苯二甲酸酯
- ~ resin 邻苯二甲酸树脂

phthalo cyanine 酞菁〈染料〉
- ~ blue 酞菁蓝
- ~ green 酞菁绿

phys. [physical] 物理的
physical 物理的
- ~ adsorption 物理吸附
- ~ antioxidant 物理性防老化剂
- ~ blowing agent 物理发泡剂
- ~ catalyst 物理催化剂
- ~ change 物理变化
- ~ chemical propaties 理化性能,(物)理化(学)性质
- ~ condition 物理状态
- ~ cross-linking 物理交联
- ~ dimension 外形尺寸,实际尺寸
- ~ indes 物性指标
- ~ isomerism 物理异构
- ~ modification 物理改性

~ property 物理性能
~ relaxation 物理松弛
~ test 物理性试验
physicist 物理学家
physicochemical surface properties 物理化学的表面活性〈粘接面经表面预处理〉
physico – mechanical testing 物理机械试验
physics laboratory 物理实验室
PI〔polyimide〕聚酰亚胺
PI〔polyisoprene〕聚异戊二烯
P. I. A.〔Plastics Institute Australia〕澳大利亚塑料协会
PIA〔Plastics Institute of America〕美国塑料协会
pi bond π键
PIB〔polyisobutene〕聚异丁烯
PIC – foam〔polyisocyanurate foam〕聚异氰脲酸酯泡沫塑料
PIC – rigid foam〔polyisocyanurate rigid foam〕聚异氰脲酸酯硬质泡沫塑料
pick 纬纱〈织物〉
pick – off 引离;剥离
~ roll 引离辊
pick – up 模塑件夹具
~ groove 挂模槽;侧位料槽
~ roll 胶辊,蘸料辊
picking strength 黏板力
picking – up 咬模;啃模〈模具〉
pickle time 酸洗时间,酸蚀时间,酸浸时间
pickling 酰洗;浸渍
~ process 酸浸(法);浸渍(法)〈金属件粘接前〉
piece 块,件;零件;毛坯
~ cutter 冲切机
~ production 单件生产
~ work 计件工作
pierce 穿孔,冲孔
~ and blank tool 冲切膜
~ and die 冲孔模
piercer 冲头
piercing 冲孔

~ die 冲孔模
~ mandrel 穿孔芯棒
~ press 冲孔压机
piezo – crystal 压晶〈超声波焊接机〉
piezoelectric 压电的
~ crystal 压电晶体
~ resin 压电树脂〈具有压电特性的树脂〉
piezoelectrical pressure transducer 压电压力传感器〈为掌握塑料加工机中的物料压力〉
pigeon 鸽灰〈紫光灰色〉
~ blood 鸽血红〈暗红色〉
pigment 颜料;色料
~ cake 颜料块,颜料滤饼
~ carrier 颜料载体
~ dispersant 颜料分散剂
~ flushing 颜料泛色;颜料洗出
~ paste 颜料糊;漆浆
~ vehicle 颜料载色剂
~ volume 颜料体积
~ volume concentration〔PVC〕颜料体积浓度
pigmentation 着色
pigmented 着色的
~ layer 着色层
~ resin 着色树脂
~ surface 着色面
pigmenting 着色的
~ capacity 着色力
~ property 着色性能
pile 绒头,绒毛
~ carpet 绒头地毯
~ fabric 绒头织物,起绒织物,割绒织物
pill 料丸,球料
pilling (织物表面)起球〈疵点〉
pilot 引导的;辅助的,检查的;实验性的,中间规模的
a ~ run 小批试产
~ batch 试验批量
~ flame 点火火焰〈燃烧性试验〉
~ house 操纵室
~ lamp 信号灯,指示灯

pimelic acid

- ~ model　试选样品
- ~ pack　试生产包装
- ~ package　试生产包装
- ~ plant　中间试验厂,中试装置
- ~ plant scale　中间规模厂
- ~ process　试验性生产过程
- ~ production　试产,试制
- ~ roller　导辊
- ~ scale　中间试验规模,中试规模
- ~ scheme [project]　小规模试验计划
- ~ studies　探索性研究
- ~ test　小规模试验,典型试验
- ~ valve　导向阀

pimelic acid　庚二酸
pimple　疙瘩〈模压缺陷〉
pin　销,杆,针
- ~ gate [gating]　针形浇口
- ~ plate　脱模销板
- ~ -point gating [pin gate]　针点浇口
- ~ -point gating with conical sprue bushing　带锥形浇道套的针点浇口
- ~ rod　针杆
- ~ roll　针辊
- ~ test　扩口试验〈塑料管〉

pinch　夹紧,收缩
- ~ effect　收缩效应
- ~ off　切口
- ~ off blades　切口片
- ~ off clearance　切坯口间距
- ~ off dam　切坯闸
- ~ off edge　切坯刀口
- ~ off land　切坯刀口厚度
- ~ off tail　切坯尾料
- ~ pass mill　平整机
- ~ roll　夹辊;牵引辊
- ~ tube blow moulding　切管吹塑

pine　松(树)
- ~ needle　松针绿〈暗绿色〉
- ~ trees　松树状花纹〈压延薄膜缺陷〉

pineapple　菠萝的
- ~ cone　菠萝筒子,双锥形筒子〈纺织〉

torpedo [torpedo spreader]　鱼雷形分流酸
- ~ -type mixing nozzle　菠萝型混合喷嘴

pinhole　针眼,针孔〈模压缺陷〉
- ~ gate　针孔型浇口
- ~ plotter　打孔器
- ~ test　针孔试验
- ~ tester　针孔试验机

pinholing　针眼,针孔〈涂层〉
pinion　齿轮
pink discolouraiton　发红
pinned　装有销钉群的
- ~ disc [dowelled disc]　钢销圆盘
- ~ disc mill　棒磨机
- ~ rotor　装有销钉群的转筒
- ~ stator　装有销钉群的定子

pinpoint　针尖;针孔形
- ~ closing nozzle　针孔形止料注嘴
- ~ gate [gating]　针孔浇口,点浇口
- ~ gate mo(u)ld　点浇口模具
- ~ gating with conical sprue bushing　有圆锥形主流道衬套的针孔浇口

Piotrowski's process　彼特洛夫斯基法〈法塑吹塑成型方法的一种〉

pipe　管;管材;装管子
- ~ adaptor　管接件
- ~ bell [belled pipe end]　管套接口
- ~ bend　弯管;管肘

bend of　90°管弯头
- ~ bending　弯管;管弯曲
- ~ bracket　管托架
- ~ branch　支管;管接头
- ~ bundle　管束
- ~ cap　管用密封盖
- ~ clamp [clip]　管夹
- ~ coil　蛇管,旋管
- ~ condenser　管式冷凝器
- ~ connection　管连接;管接头
- ~ coupling　管连接;管接头
- ~ cramp　管吊钩;固定钩
- ~ cross　十字管,四通管
- ~ cutter　切管机
- ~ diameter　管径
- ~ die　管模
- ~ drain　排水管
- ~ elbow　肘管;弯管

~ end 管端
~ exhaust 排气管
~ expander 扩管器
~ extruder 管材挤塑机,挤管机
~ extrusion 管材挤塑
~ extrusion die 管材挤塑模头
~ extrusion (die) head 管材挤塑机头
~ extrusion line 挤管生产线,挤管装置
~ fitter 管安装工
~ fitting 管配件,管件
~ flange 管子法兰
~ flexibility 管子韧性
~ former 管材定径器;管材定型模
~ forming 管材成型
~ fracture 管的损坏
~ frame 管架
~ hanger 吊管钩
~ hanger buffer 管夹垫
~ haul-off machine 管材牵引机
~ holder 管支架
~ hook 吊管钩,固定钩〈塑料管线〉
~ -in-pipe system 管套管系统〈金属管内衬塑料管〉
~ installation 管线安装,管线铺设
~ joint 管连接;管接头
~ jointing 管连接,管焊接
~ line 管道,管线
~ line stopper 管道塞
~ line work 管道安装
~ lining 管道衬里
~ reducer 渐缩管
~ section 管段,管节(一节管子)
~ spigot 管接头,套管
~ strand 管束
~ support 管架
~ tapping 管攻丝
~ Tee T形管节
~ thread 管子螺纹
~ -to-flange joint 管与法兰连接〈塑料管线〉
~ -to-pipe joint 管与管连接〈塑料管线〉
~ train 管材挤塑机

~ valve 管阀
~ wall 管壁
~ water cooled former 管材水冷定型器
~ with helical reinforcing corrugations 螺旋增强波纹管
~ with helical reinforcing web 织物增强螺旋形管
pipeline 管线;管道
~ mixer 管道式混合器
piperazine 哌嗪
pipette 吸移管,移液管
piping 管道,管道系统
~ head 挤管机头
lay-out 管道布置
PIR [polyisocyanurate] 聚异氰脲酸酯
pirn 纬纱管;纤丝
piston 阳模,模塞;活塞
~ accumulator 活塞式储料缸
~ barrel 活塞筒体
~ pressure 活塞压力
~ pump 活塞泵
~ ring 活塞环,活塞圈
~ rod 活塞杆
~ spinning pump 活塞纺丝泵
~ stroke 活塞冲程
~ travel 活塞行程,活塞冲程
~ type 活塞式
~ type injection moulding machine 活塞式注塑机
pit 麻点,痘痕〈模压件缺陷〉
pitch 螺距,节距
~ circle (齿轮)节圆
~ diameter (螺纹)中径,节径
~ line 节线,齿距线
~ of screw 螺距
~ of screw thread 螺距
~ of spindle 螺杆螺距
pitched blade 弯浆,弯搅拌浆
pitting 点蚀;凹痕
- corrosion 点状腐蚀,穴状腐蚀
- corrosion test 点蚀试验
PIV-gear [positive infinitely variable gear] 无级变速齿轮,PIV型齿轮

pivalolactone [PVL] 新戊内酯
pivot 轴颈,枢轴
pivotable injection unit 可旋转的注塑装置
pkg. [package] 包装
pl. [plural] 复数(的)
plain 简单的;平的
 ~ bearing 普通轴承,滑动轴承
 ~ bending 平面弯曲
 ~ butt joint [weld] 对接焊(缝),I形焊缝
 ~ cloth 平布,平纹织物〈增强材料〉
 ~ core pin 平面成孔销〈模具〉
 ~ die 简单模
 ~ dyed 单色的,素色的
 ~ face 光面
 ~ flange 平面法兰
 ~ flash line 平合模线
 ~ pin ejector 平面脱模销
 ~ pipe 直管
 ~ roller 光面(印花)辊筒
 ~ scarf joint 平(面)斜接
 ~ seam 平缝
 ~ shade 单色,素色
 ~ weave 平纹织物〈增强材料〉
 ~ woven fabric 平纹织物,平纹布
planar 平(面)的
 ~ helix winding 平面螺旋线缠绕法
 ~ molecule 平面(排列)分子
 ~ orientation 平面取向,沿面取向
 ~ winding 平面缠绕
 ~ wrap 平缠绕
plane 平面
 ~ deformation 平面形变
 ~ diagram 平面图
 ~ end bottom 平底〈反应釜〉
 ~ fracture 平面破裂
 ~ load-bearing structure 平面支承结构
 ~ of a laminate 层压面
 ~ of bending 弯曲平面
 ~ of flexure 挠曲平面
 ~ of rupture 裂断面
 ~ -parallel 平行平面的

 ~ plate 平板
 ~ section 剖面
 ~ shear strength 平面剪切强度
 ~ strain 平面应变,平面变形状态
 ~ stress 平面应力;双轴应力
 ~ view 平面图;俯视图
 ~ wave 平面波
planetary 行星式的
 ~ change-can mixer 行星式可替换筒的混合器
 ~ dissolve 行星式溶解器
 ~ extruder 行星式挤塑机
 ~ gear 行星齿轮
 ~ motion 行星运动
 ~ (paddle) mixer 行星(桨叶)混合器
 ~ roller 行星滚柱
 ~ roller extruder 行星滚柱式挤塑机,四螺杆行星式挤塑机
 ~ screws 行星螺杆
 ~ screw extruder 行星螺杆式挤塑机
 ~ stirrer 行星式搅拌器
 ~ transmission 行星变速器
 ~ winding machine 行星式缠绕机
planing 刨削
 ~ fixture [machine] 刨削固定装置,刨床
 ~ method 刨削法
 ~ powder 刨削粉末
planishing 压光
planning 计划,规划;设计
 ~ engineer 规划工程
 ~ stage 规划阶段
plant 工厂,车间;装置,设备;植物
 ~ capacity 设备能力,工厂生产能力
 ~ cellulose 植物纤维素
 ~ conditions 生产条件
 ~ engineer 工厂工程师
 ~ equipment 固定设备
 ~ factor 设备利用率
 ~ fiber 植物纤维
 ~ for plastics machines 塑料机械厂
 ~ manager 企业领导人
 ~ mucilage 植物黏质物;植物胶浆
 ~ operation 工业生产

~ parameter 装置参数
~ scale 装置规模,工厂规模
~ scale equipment 工厂规模设备
~ scale operation 大规模生产,工业生产
~ set-up 装置安装
plaque 成型板〈用于塑料测试〉
~ mo(u)ld 成型板模具
plasma 等离子体
~ jet coating 等离子喷涂
~ polymerization 等离子聚合
~ spray coating 等离子喷涂
~ spray process 等离子喷涂法
plastainer 塑料容器
plaster 熟石膏
~ board 石膏板
~ cast 石膏模型;石膏铸件
~ mo(u)ld 石膏模
plastelast 塑弹性物;弹性塑料
plastic 塑料;塑性的
~ additive (agent)塑料添加剂
~ adherend 塑料粘接件
~ adhesive 塑料黏合剂
~ after effect 塑性后效
~ alloy 塑料"合金"
~ anisotropy 塑性各向异性
~ ashing test 塑料灰化试验
~ -backed magnetic tape 塑料磁带
~ barn cloche 塑料(膜)棚
~ base protective layer 塑料保护层
~ bearing 塑料轴承
~ beverage crate 塑料饮料瓶周转箱
~ binder 塑料粘接剂
~ -bodied automobile 塑料车身汽车
~ bonding 塑料粘接,塑料黏合
~ bonding adhesive 塑料粘接剂
~ building material 塑料建筑材料
~ cake 塑料圆形料坯〈用于唱片〉
~ can 塑料盒〈包装〉
~ capsule 塑料封壳
~ cement 塑料粘接剂,塑胶
~ clay 易塑黏土;陶土
~ -coated fiber 塑料涂层纤维,涂塑料纤维

~ -coated paper 塑料涂层纸,涂塑料纸
~ -coated textiles 塑料涂层织物,涂塑料织物
~ coating 塑料涂层,塑料涂膜
~ coefficient 塑性系数
~ composite film 塑料复合薄膜
~ composition 塑料混合料
~ compounding 塑料混料
~ concrete 塑料混凝土
~ consistency 塑性稠度
~ containers 塑料容器
~ cover 塑料罩
~ covering 塑料覆盖
~ crate 塑料周转箱
~ crystal 塑性晶体
~ cutting 塑料切削
~ deformation 塑性形变,永久形变,不可逆形变
~ dispersion 塑性分散系
~ elongation 塑性伸长,永久伸长
~ engineering 塑料工程(学)
~ film 塑料薄膜
~ film condenser 塑料薄膜电容器
~ fitting 塑料配件
~ flow 塑性流动
~ flow curve 塑性流动曲线
~ flow model 塑性流动模型
~ fluid 塑性流体
~ fluidity 塑性流动性
~ foam 泡沫塑料
~ foil 塑料箔,塑料彩箔
~ forming 塑性加工;塑料成型
~ friction 塑性摩擦
~ fuel tank 塑料燃料油箱
~ gasket 塑料衬垫
~ gear 塑料齿轮
~ granulation 塑料造粒
~ -impregnated paper core 浸渍塑料的纸芯
~ incinerator 塑料焚烧炉
~ -insulated cable 塑料绝缘电缆
~ laminate 层压塑料;塑料层压板
~ leather cloth 塑料人造革

plastic

- ~ lens 塑料透镜
- ~ light filter 塑料滤光器
- ~ limit 塑性极限
- ~ – lined 衬塑料的
- ~ – lined paper 衬有塑料的纸,塑料衬纸
- ~ lining 塑料衬里
- ~ material 可塑物质,塑料
- ~ matrix 塑料基料,塑料母料
- ~ melt 塑料熔体
- ~ – metal adhesive joint 塑料–金属粘接
- ~ memory 塑性记忆
- ~ microballs 塑料粉球
- ~ mill [rolling mill] 塑料研磨机
- ~ mixer and kneader 塑料用混合器和捏和机
- ~ mobility 塑性浓度;塑性流动性
- ~ modified concrete 塑料改性混凝土
- ~ modified mortar 塑料改性砂浆
- ~ monofilament 塑料单丝
- ~ mortar 塑料泥浆
- ~ mo(u)lding press 塑料成型(压力)机
- ~ net 塑料网
- ~ package 塑料包装
- ~ packaging 塑料包装
- ~ packing 塑料填密,塑料垫圈
- ~ parts 塑料零件
- ~ paste 塑料糊
- ~ pattern 塑料模型
- ~ pinched – off 塑料截坯口〈挤吹成型〉
- ~ pipe 塑料管
- ~ plating 塑料金属喷镀
- ~ precision article 塑料精密制品
- ~ press 塑料成型压机
- ~ processing 塑料(成型)加工
- ~ processing machine 塑料加工(成型)机
- ~ processing tool 塑料加工(成型)模具
- ~ processor 塑料加工厂;塑料加工者
- ~ product 塑料制品,塑料产品
- ~ property 塑性
- ~ range 塑性范围
- ~ raw material 塑料原料
- ~ recovery 塑料回收
- ~ refractory 塑性耐火材料
- ~ resilience 塑性回弹
- ~ resin tank 塑料树脂储罐〈涂布机〉
- ~ resistance [internal friction] 塑性阻力
- ~ ribbed pipe 塑料波纹管,塑料肋管
- ~ sandal 塑料凉鞋
- ~ scrap 塑料废料
- ~ semiproduct 塑料半成品
- ~ set – up temperature 塑料固化温度
- ~ setting 塑性定形
- ~ shaping technique 塑料成型工艺
- ~ – sheathed cable 塑料护套电缆
- ~ sheeting 塑料片材
- ~ size 塑料浸润剂
- ~ slide bearing 塑料滑动轴承
- ~ slideway 塑料滑道
- ~ slip 塑料滑移
- ~ slipper 塑料拖鞋
- ~ solid 塑性固体
- ~ sponge 塑料海绵
- ~ stage 塑性阶段,可塑阶段
- ~ state 塑性状态
- ~ stopper 塑料塞
- ~ strain 塑性应变
- ~ strength 塑料强度
- ~ structural component 塑料结构元件
- ~ substances 可塑物质
- ~ tailored for the job 特殊用途的塑料
- ~ technology 塑料工艺(学)
- ~ – to – metal bond 塑料与金属粘接
- ~ tool 塑料工具;塑料加工模具
- ~ tooling 塑料工具;塑料模
- ~ – to – plastic adhesive [glue] 塑料与塑料黏合剂
- ~ tube 塑料(软)管
- ~ tubing 塑料(软)管
- ~ valves 塑料阀
- ~ viscosity 塑性黏度
- ~ waste 废塑料,塑料废物

- web 塑料带;塑料网
- weld 塑料焊接
- welder 塑料焊接机;塑料焊工
- welding 塑料焊接
- welding engineering 塑料焊接工程
- welding equipment 塑料焊接设备
- welding technique 塑料焊接技术
- window frame 塑料窗框
- yield 塑性屈服
- yield point 塑料流点;塑性屈服点
- yield temperature 塑料屈服温度;塑性屈服温度
- yield test 塑流试验
- yield value 塑流值,塑性屈服值
- yield with temperature 塑料屈服温度,热稳定性

plasticate 塑炼,塑化
plasticating 塑炼,塑化
- ~ capacity 塑化能力,塑炼能力〈挤塑机的〉
- ~ chamber 塑炼室
- ~ cylinder 塑炼料筒
- ~ efficiency 塑炼,效率〈挤塑机〉
- ~ extruder 挤出塑炼机
- ~ extrusion 挤出塑炼
- ~ pressure 塑化压力,塑炼压力

plastication 塑炼,塑化
plasticator 塑炼机,塑化器
plasticimeter 塑性计
plasticised fabric 塑性化织物
plasticity 塑性,可塑性,塑性力学
- ~ agent 增塑剂
- ~ coefficient 塑性系数
- ~ index 可塑性指数
- ~ number 可塑度;可塑值
- ~ range 塑性范围
- ~ recovery number 塑性恢复值
- ~ retention 塑性保持率
- ~ retention index 塑性保持指数
- ~ stability 塑性稳定性
- ~ test 可塑性试验
- ~ theory 塑性理论
- ~ value 塑性值

plasticization 增塑作用

plasticize 增塑,塑化
plasticized 增塑的
- ~ compound 增塑混合料
- ~ material 增塑(塑)料
- ~ melt-spinning 塑化溶体纺丝
- ~ plastic 增塑塑料
- ~ polyvinyl chloride 增塑聚氯乙烯,软质聚氯乙烯

plasticizer 增塑剂,增韧剂
- ~ -adhesives 增塑-黏合剂
- ~ blend 增塑共混料,混配增塑剂
- ~ content 增塑剂含量
- ~ efficiency 增塑剂效率
- ~ -extender 增塑稀释剂,充填增塑剂〈低质量增塑剂〉
- ~ extraction 增塑剂萃取
- ~ limit 增塑剂限量,增塑剂极限
- ~ migration 增塑剂渗移
- ~ mixer 增塑剂混合器
- ~ resistance 耐增塑剂性
- ~ unit 塑化装置〈注塑机〉
- ~ volatility 增塑剂挥发度

plasticizing 增塑;塑化
- ~ agent 增塑剂
- ~ bath 塑化浴
- ~ capacity 塑化能力
- ~ cylinder 塑化料筒
- ~ device 塑化装置
- ~ efficiency 塑化效率
- ~ injection cylinder 注塑的塑化料筒
- ~ oil 增塑油
- ~ rate 塑化程度;塑化定额
- ~ section 塑化段
- ~ system 塑化系统
- ~ unit 塑化装置〈注塑机〉
- ~ zone 塑化区,塑化段〈挤塑机〉

plastification 增塑(作用),塑化(作用),塑炼
- ~ bath 塑化浴
- ~ temperature 塑化温度
- ~ zone 塑化区,塑化段

plastificator 塑化器,塑炼机〈螺杆〉;增塑剂
plastifier 增塑剂

plastify 增塑；塑化
plastifying 增塑的，塑化的
  ~ capacity 塑化能力
  ~ extruder 塑化挤塑机
  ~ extrusion 塑化挤塑
plastigel 增塑凝胶
  ~ paste 增塑凝胶糊
plastimeter 塑性仪，塑度计
plastimetry 塑性测定法
plastisol 塑料溶液，增塑溶胶，增塑糊
  ~ fusion 增塑糊塑化，增塑溶胶塑化
  ~ mo(u)lding 增塑糊模塑（件）
  ~ paste 增塑溶胶糊
plastoelastic 塑弹性的
  ~ body 塑弹性物体
  ~ deformation 塑弹性形变
plastoelasticity 塑弹性
plastograph 塑度仪；塑性计
plastography 塑度（形变）分析法
plastomer 塑性体，塑料
plastometer 塑性计；塑度仪
plastometry 塑性测定法
plastosoluble dye 塑料可溶性染料
plastpeel 塑料袋〈美〉
platability 可镀性
platable 可电镀
  ~ plastics 可电镀塑料
  ~ resin 可电镀树脂
plate 板材，平板；极板；电镀
  ~ and frame filter press 板框式压滤机
  ~ bender 弯板机
  ~ calender 板材压延机
  ~ crystal 片晶
  ~ dispersion plug 色料分散板
  ~ ejection 板式脱模
  ~ ejector 脱模板
  ~ electrode 板极
  ~ hear exchanger 板式热交换器
  ~ -like pigment 板状颜料
  ~ mark 压板印，压板痕，模板痕
  ~ mill 轧板机
  ~ out 沉析贴层，结垢〈塑料加工装置上所形成塑料加工助剂沉积物〉
  ~ roller 滚板机
  ~ shearing machine 剪板机
  ~ thickness 板厚度
platen 压板，模板；压力台；电镀的
  ~ area 压板面积
  ~ hanger and guide 板悬吊和导轨装置（多层平板压机）
  ~ plastics 电镀塑料
  ~ press 平板压机
  ~ size 压板尺寸
  ~ speed 压板闭合速度
platform blowing 托坯吹塑
plating （电）镀；金属镀层
  ~ bath 电镀浴，电镀槽
  ~ liquid 电镀液〈高真空蒸发用预涂料〉
  ~ machine 折叠机，卷板机，压光机
  ~ on plastics 塑料电镀
  ~ solution 电镀液
  ~ temperature 电镀温度
  ~ time 电镀时间
platinum [Pt] 铂，白金
  ~ bushing 铂金轴套；铂金喷丝嘴〈美〉
pleating 褶皱
plenum 送气通风；充实
  ~ chamber 喷混室；送气通风室
  ~ chamber method [process] 喷混室法
  ~ chamber preforming 喷混室式预成型
pleochroic 多色的
pleochroism 多色性，多色（现象）
pleochromatic 多色的
pleochromatism 多向色性，多色（现象）
plexiglass 有机玻璃，聚甲基丙烯酸甲酯
  ~ safety screen 有机玻璃安全护板
pliability 柔韧性，揉曲性，可挠弯性
  ~ test 弯曲试验；韧性试验
pliable 易挠弯的，柔韧的
  ~ sheeting 易弯板材
plied 合股的
  ~ (filament) yarn 合股线
Pliers pl 钳子〈焊接〉
plotting 测绘，标绘；制图

~ paper 方格(绘图)纸,制图纸
ploughing 沟槽纹〈摩擦磨损〉
plow blade blender [mixer] 犁片式混合器,对流盘式混合器〈美〉
plug 柱塞,模塞,阳模
~ – and – ring forming 模塞 – 夹圈(热)成型
~ – and socket connection 插承连接〈塑料管〉
~ – assist 柱塞助压,助压模塞
~ – assist air – press forming 柱塞助压压气成型
~ – assist air slip vacuum forming 柱模塞助压气滑真空成型
~ – assist blowing 柱模塞助压压气成型
~ – assist forming 柱塞辅助成型
~ – assist pressure forming 柱模塞助压压力成型
~ – assist reverse draw forming 柱塞辅助回吸成型
~ – assist reverse draw vacuum forming 柱塞辅助回吸真空成型
~ – assist vacuum forming 柱塞辅助真空成型
~ – assist vacuum thermoforming 柱塞助压真空热成型
~ box 插座,插接线盒〈电〉
~ continuous flow mixer 柱塞连续流动混合器,间歇流动混合器
~ depth of action 柱模塞压入深度
~ ejector 脱模栓
~ extruder 推杆式挤出机
~ flow 活塞式流动
– forming 模头成型,柱模塞助压成型
– gate 塞型浇口
~ ga(u)ge 塞规
~ – in unit 插件
~ shape 柱模塞外形
~ speed 柱模塞行程速率[度]
~ surface 柱模塞表面
~ temperature 柱模塞温度
~ – to – mo(u)ld clearance 柱模塞间隙

~ – up 冷料栓〈柱塞注塑时形成〉
~ valve 旋塞阀
plunger 柱塞,模塞;阳模;射料杆
~ bushing 柱塞套
~ cylinder 射料杆驱动油缸
~ die 上模,阳模,冲头
~ dwell 栓塞停留时间
~ extruder 柱塞挤塑机
~ extrusion type rheometer 柱塞挤出型流变仪
~ face 射料杆面积
~ injection 柱塞式注塑
~ injection mo(u)lding 柱塞式注塑成型
~ injection mo(u)lding machine 柱塞式注塑(成型)机
~ injection press 柱塞式注塑机
~ mo(u)ld 活塞式(压铸)模具
~ mo(u)lding 活塞式铸压,柱塞成型辅助活塞式压铸
~ mo(u)lding press 活塞式铸压机,柱塞模塑压机
~ plat 模塞托板
~ preplasticating injection mo(u)lding machine 柱塞预塑化式注塑(成型)机
~ preplasticiser 柱塞式预塑化器
~ preplasticising 柱塞式预塑化
~ press 柱塞式压机
~ pressure 射料杆压力
~ retainer 阳模底板,模塞底板〈模具〉
~ retainer plate 柱塞底板
~ rod 柱塞杆
~ speed 射料杆速率[度]
~ transfer moulding 柱塞式压铸
~ – type 柱塞式
~ – type injection 柱塞式注塑成型
~ – type injection machine 柱塞式注塑(成型)机
~ – type injection moulding machine 柱塞式注塑成型机
~ – type plasticizing equipment [unit] 柱塞式塑化设备〈注塑机〉

~ - type transfer mo(u)ld 活塞式铸压模

~ - type transfer mo(u)lding 活塞式铸压

~ - type viscometer 柱塞式黏度计

ply 板片,层片;折叠;线纱股数〈纱〉;合股

~ elevator 层状提升机

~ structure 层片结构

~ yarn 合股线

plyglass 合成纤维各向异性材料

plywood 胶合板

~ adhesive 胶合板用黏合剂

~ glue 胶合板用胶

p. m. per mimute 每分钟

FMA [polymethyl acrylate] 聚丙烯酸甲酯

PMAC [polymethoxy acetal] 聚甲氧基缩醛

PMAN [polymethacrylonitrile] 聚甲基丙烯腈

PMCA [polymethylα - chloroacrylate] 聚α - 氯代丙烯酸甲酯

PMI [polymathacryl imide] 聚甲基丙烯酰亚胺

PMMA [polymethyl methacrylate] 聚甲基丙烯酸甲酯,有机玻璃

PMP [poly - 4 - methyl - 1 - pentene] 聚4 - 甲基 - 1 - 戊烯

PMPPI [polymethylene polyphenylene isocyanate] 聚亚甲基聚苯基异氰酸酯〈参见 PAPI〉

PMPSF [polymethylphenyl siloxane fluid] 聚甲基苯基硅氧烷液体〈润滑剂,脱模剂,浸渍剂,处理剂〉

PN [nominal pressure] 额定压力,标称压力

pneumatic 气体的;空气的;气动的,气压的

~ atomizer 气动喷雾器

~ bumper 充气保险杠,气保险杠

~ circuit 气动回路,压缩空气循环系统

~ control 气动控制,气动调节

~ conveyance 气动输送〈塑料物料〉

~ conveyer 气动输送机

~ conveying 气动输送

~ conveying dryer 气流干燥器〈粒料〉

~ conveyor 气动输送机

~ cushion film 气垫膜

~ ejection 气动脱模

~ hammer 气锤

~ hopper loader 气动斗式加料器〈粒料〉

~ press 气动压机

~ prestretch 气动预拉伸

~ test 气压试验

pneumatically driven press 气动压机

pneumatize 气动化

pneumohydraulic 气动液压的

pneupack 缓冲包装

Po [polonium] 钋

PO [polyolefin] 聚烯烃

pock 麻点,痘痕

~ mark 痘斑〈中空制品表面的浅痕〉

~ markings 麻点

pocket 口袋;穴,坑;小型的,袖珍的

~ air 夹气

~ chill roll 小型激冷辊,有网纹面的冷却辊

~ package 袋式包装,小型包装

pocking 痘痕

point 点;尖端;针尖

~ brilliance 光点亮度

~ by point method 逐点测定法

~ by point test 逐步试验

~ gate 针尖型浇口

~ gate mo(u)ld 点浇口模具

~ imperfection 疵点

~ of chain rupture 链破裂点

~ of contact 接触点

~ of fracture 断裂处

~ of fusion 熔点

~ of ignition 着火点

~ of impact 冲击点〈摆锤撞击试验〉

~ of inflection 转折点

~ of inflexion 弯曲点

~ of intersection 交叉点,交会点〈织物

编织〉
- ~ of inversion 转化点
- ~ of support 支点
- ~ welding 点焊

poise 泊〈动力黏度非 SI 单位,等于 10^{-1} Pa·s〉

poisonous substance 有毒物质
poisonousness 毒,毒性
Poisson's 泊松
- ~ distribution 泊松分布
- ~ index 泊松指数,泊松常数
- ~ law 泊松定律
- ~ number 泊松数
- ~ ratio 泊松比

polar 极性的;极化的
- ~ absorption 极性吸收
- ~ activation 极性活化
- ~ additive 极性添加剂
- ~ adhesive 极性黏合剂
- ~ bond 极性键
- ~ chain molecule 极链分子
- ~ colo(u)rs 极性染料
- ~ crystal 极性晶体
- ~ dome 缠卷筒的凸度〈容器底〉
- ~ dyes 极性染料
- ~ group 极性基团
- ~ hole 极孔;端拱开孔〈增强塑料〉
- ~ link 极性键
- ~ molecule 极性分子
- ~ monomer 极性单体
- ~ opening 极性孔
- ~ pigment 极性颜料
- ~ plastic 极性塑料
- ~ polymer 极性聚合物
- ~ port 端拱顶开口
- ~ reaction 极性反应
- ~ rubber 极性橡胶
- ~ solvent 极性溶剂
- ~ wind 极性缠绕
- ~ winding 极性缠绕(法)
- ~ winding machine 极性缠绕机
- ~ wrapping 极性卷绕

polarimeter 偏振计,旋光仪
polariscope 偏光仪
polarity 极性
- ~ effect 极化效应
- ~ sign 极性符号

polarizability 极化性,极化度
polarization 极化(作用),偏振现象
- ~ colorimeter 偏振比色计
- ~ in electrodialysis 电渗析过程极化
- ~ imcroscope 偏光显微镜
- ~ photometer 偏光光度计
- ~ process 极化(作用)〈非极性表面〉;偏振现象

polarized 偏振的,极化的
- ~ fluorescence 偏振荧光
- ~ infrared 偏振红外线
- ~ light 偏振光;极化光
- ~ light method 偏振光法
- ~ light microscope 偏光显微镜
- ~ radiation 极化辐射

polarizer 偏振器,起偏振器
polarizing 偏振,极化
- ~ angle 偏振角
- ~ filter 偏振滤光片,偏振片
- ~ fluorescence 偏振荧光
- ~ microscope 偏光显微镜
- ~ plate 偏振片

polarogram 极谱图
polarograph 极谱仪
polarographic analysis 极谱分析(法)
polarography 极谱法
polaroid 人造偏振片,偏振片;〈大写〉宝丽来一次成像(照相机,照片)
pole 极,极点,磁极
- ~ piece (磁)极片

polish 抛光,磨光
- ~ up 修饰;改善

polished 抛光的,磨光的
- ~ cylinder roll 抛光滚筒,抛光辊,磨光辊
- ~ face 抛光面,磨光面
- ~ finish 抛光处理
- ~ plate 抛光板
- ~ specimen 抛光试样
- ~ surface 抛光面,精加工面

polisher 磨光器,抛光器

polishing 抛光
~ band 抛光带
~ barrel [polishing drum] 抛光滚筒
~ belt 抛光带
~ between plates 压板抛光,压光板
~ buff 抛光轮
~ by tumbling [barrel polishing, drum polishing] 滚筒抛光
~ composition 抛光剂
~ compound 抛光剂
~ disc 抛光盘
~ drum 抛光滚筒
~ lathe 抛光机
~ machine 抛[磨]光机
~ mop 抛光轮
~ oil 抛光油
~ paper 研磨纸;抛光纸
~ paste 抛光膏
~ plate 抛光板
~ powder 抛光粉
~ ring [ring buff, polishing mop] 抛光球
~ roll 抛光辊
~ roller 抛[磨]光辊
~ rolls 轧光辊;抛光辊;轧板机
~ roll stack 轧光机
~ stack 轧光辊;抛光辊
~ tool 抛光工具,抛光仪
~ varnish 抛光漆[膏]
~ wheel 抛[擦]光轮
~ with polishing rolls 用抛光辊抛光
~ with polishing wheels 用抛光轮抛光
politure 抛光,光泽
pollution 污染,沾染;浑浊
~ control 污染[沾染,浑浊]控制
~ - free 无污染的,无公害的
polonium [Po] 钋
poly 〈字头〉多;聚
~ (n - butyl methaciylate) [PBMA] 聚甲基丙烯酸正丁酯
~ (1,4 - cyclohexylene dimethylene suberamide) 聚(1,4 - 亚环己基二亚甲基辛酰胺)
~ (1,4 - cyclohexylene dimethylene terephthalate) [PCDT] 聚(1,4 - 亚环己基二亚甲基对苯二甲酸酯)
~ (ethylene - propylene) terpolymer [EPDM] 乙烯 - 丙烯三元聚合物
~ (2 - hydroxyethyl methacrylate) [PHEMA] 聚(甲基丙烯酸2 - 羟乙酯)
~ (4 - methyl - 1 - pentene) [PMP] 聚(4 - 甲基 - 1 - 戊烯)
~ a - methyl styrene [PAMS] 聚α - 甲基苯乙烯〈增塑剂〉
~ - Paper 涂塑料纸,塑料涂层纸
~ - para - xylene 聚对亚苯基二甲基
~ (m - phenylene isophthalamide) 聚间苯二甲酰间苯二胺
~ (p - phenylene terephthalamide) 聚对苯二甲酰对苯二胺
~ (1,2_propylene glycol sebacate) 聚癸二酸1,2 - 丙二醇酯〈增塑剂〉
~ - p - xylene 聚对二甲苯
pdyacetal 聚甲醛;聚缩醛
polyacetylene 多炔
polyacryl 聚丙烯(酰基)
~ ether 聚丙烯醚
~ sulfone 聚丙烯砜
polyacrylamide 聚丙烯酰胺
polyacrylate [PAA] 聚丙烯酸酯
~ adhesive 聚丙烯酸酯黏合剂
~ dispersion 聚丙烯酸酯分散体
~ plastics 聚丙烯酸酯塑料
~ resin 聚丙烯酸酯树脂
~ rubber 聚丙烯酸酯橡胶
polyacrylic 聚丙烯(酸酯)的
~ acid [PAA] 聚丙烯酸〈织物处理剂〉
~ acid ester 聚丙烯酸酯
~ fibre 聚丙烯酸酯纤维
~ plastic 聚丙烯酸酯塑料
~ polymer 聚丙烯酸酯聚合物
~ resins 聚丙烯酸酯树脂
~ rubber 聚丙烯酸酯橡胶
polyacrylonitrile [PAN] 聚丙烯腈
~ fibre 聚丙烯腈纤维,腈纶〈俗称〉
polyaddition 加聚,加成聚

~ product 加聚产物
~ reaction 加聚反应
polyalcohol 多元醇
polyalkyl 聚烷基
 ~ acrylate 聚丙烯酸烷基酯
 ~ vinyl ether 聚烷基乙烯基醚
polyalkylene 聚亚烷基
 ~ amides 聚亚烷基酰胺
 ~ oxide 聚亚烷基氧化物;聚环氧烷
 ~ oxide rubber 聚亚烷基环氧橡胶
 ~ terephthalate 聚对苯二甲酸亚烷基酯
polyallomer 异质同晶聚合物
 ~ copolymer 异质同晶共聚物
 ~ resin 聚异质同晶树脂
poly(alpha-methylstyrene),[PAMS] 聚α-甲基苯乙烯〈增塑剂〉
polyamide [PA] 聚酰胺〈固化剂〉
 ~ curing agent 聚酰胺固化剂〈环氧树脂〉
 ~ fibre 聚酰胺纤维
 ~ -imide [PAI] 聚酰胺-酰亚胺
 ~ -imide resins 聚酰胺-酰亚胺树脂
 ~ multipolymer 多元共聚尼龙,聚酰胺多元共聚物
 ~ plastics 聚酰胺塑料
 ~ resin 聚酰胺树脂
 ~ sulfonamide 聚酰胺磺酰胺
polyamine 多胺;聚胺
 ~ -methylene resin 聚胺甲基树脂
polyaminobismaleimide [PABM] 聚氨基双马来酰亚胺
 ~ resin 聚氨基双马来酰亚胺树脂
polyaminoresin 聚胺树脂
polyaminotriazole [PAT] 聚氨基三唑
 ~ fibre 聚氨基三唑纤维
polyampholyte 聚两性电解质
polyamphoteric electrolyte 聚两性电解质
polyanhydride 聚酐
polyaryl 聚芳基
 ~ acetylene 聚芳基乙炔
 ~ ester 聚芳酯

~ ether ketone 聚芳醚酮
~ sulfone [PAS] 聚芳砜
polyaryiamide [PARA] 聚芳酰胺
polyarylate 聚芳酯;多芳基化合物
polyarylic ester 聚芳酯
polyaryloxysilane 聚芳氧基(甲)硅烷
polyazelaic polyanhydride [PAPA] 聚壬二酸酐
polyazo dye 多偶氮染料
polybasic 多元的,多碱(价)的,多盐基的〈俗〉;多代的
 ~ acid 多价酸;多元酸
polybenzimidazole [PBI] 聚苯并咪唑
 ~ fibre 聚苯并咪唑纤维
 ~ resin 聚苯并咪唑树脂
polybenzothiazole [PBT] 聚苯并噻唑
polybenzoxazinedione 聚苯并噁嗪二酮
polybenzoxazinone 聚苯并噁嗪酮
polybenzoxazole [PBO] 聚苯并噁唑〈胶黏剂〉
polyblend 聚合物共混料,聚合物混配料;掺混料
 ~ fibre 聚合物混纺纤维
polybrominated biphenyl [PBB] 多溴代联苯
polybutadiene [PBD] 聚丁二烯
 ~ acrylic acid copolymer 丁二烯-丙烯酸共聚物
 ~ -acrylonitrile [PBAN] 聚丁二烯-丙烯腈
 ~ rubber 聚丁二烯橡胶
 ~ styrene [PBS] 聚丁二烯-苯乙烯
 ~ -type resin 聚丁二烯型树脂
poly-1-butene 聚-1-丁烯
polybutene [PB] 聚丁烯
polybutyl methacrylate,[FBMA] 聚甲基丙烯酸正丁酯
polybutylene [polybutene] 聚丁烯
 ~ plastic 聚丁烯塑料
 ~ resins pl 聚丁烯树脂
 ~ terephthalate [PBT、PBTP] 聚对苯二甲酸丁二(醇)酯
polybutyrolactam [PA-4] 聚丁内酰胺,尼龙4

polycaproamide [PA-6] 聚己内酰胺，尼龙6
polycaprolactam [PA-6] 聚己内酰胺，尼龙6
polycaprolactone, [PCL] 聚己酸内酯〈可作为生物降解添加剂〉
polycarbodiimide [PCD] 聚碳酰二亚胺
polycarbonate [PC] 聚碳酸酯
  ~ fibre 聚碳酸酯纤维
  ~ of hologenated bisphenol A type 双酚A型卤代聚碳酸酯
  ~ plastics 聚碳酸酯塑料
polycarborane siloxane 聚碳硼烷硅氧烷
polycarbosilane 聚碳硅烷〈金属有机聚合物〉
polycarboxylic acid 多羧酸
polychlorinated bipheny [PCB] 多氯代联苯
polychloroether [chlorinated polyether] 聚氯醚
polychlorofluorocarbon 聚氟氯碳
polychlorofluorohydrocarbon 聚氟氯烃
polychloroprene 聚氯丁二烯，氯丁橡胶
  ~ adhesive 聚氯丁二烯黏合剂
  ~ rubber [CR] 聚氯丁二烯橡胶，氯丁橡胶
polychlorotrifluoroethylene [PCTFE] 聚三氟氯乙烯〈增塑剂〉
polychroism 多色(现象)
polychromatic 多色的
  ~ fibre 热敏变色纤维
  ~ finish 多色修饰；有金属光泽的涂层
polychromatophile 亲不同色的
polychrome 多色的
polycondensate 缩聚物
polycondensation 缩聚(作用)
  ~ product 缩聚物
polycyclopentadiene 聚环戊二烯
  ~ resin 聚环戊二烯树脂〈增稠剂、分散剂、保养剂、浸渍剂〉
polycrystal 多晶体
polycrystalline 多晶的
  ~ polymer 多晶型聚合物
  ~ structure 多晶结构
polycrystallinity 多晶性
polydecalactam [PA-10] 聚癸内酰胺，尼龙10
polydecamethylene sebacamide [PA-1010] 聚癸二酰癸二胺，尼龙1010
polydiallyl 聚二烯丙基
  ~ isophthalate [PDAIP] 聚间苯二甲酸二烯丙酯
  ~ phthalate [PDAP] 聚邻苯二甲酸二烯丙酯
polydiene 聚二烯
  ~ rubber 聚二烯橡胶
polydiethyl siloxane 聚二乙基硅氧烷，二乙基硅油〈脱模剂〉
  ~ fluid [PDESF] 聚二乙基硅氧烷液体，二乙基硅油〈润滑剂,脱模剂,消泡剂,防水剂,抛光剂〉
polydimethyl siloxane [PDMS] 聚二甲基硅氧烷，二甲基硅油，硅油〈脱模剂〉
  ~ fluid [PDMSF] 聚二甲基硅氧烷液体，二甲基硅油〈润滑剂,脱模剂,添加剂,消泡剂〉
polydiolefin 聚二烯烃
polydisperse 多分散
  ~ polymer 多分散(性)聚合物
polydispersibility 多分散性
polydispersity 多分散性
  ~ index 多分散指数
polyelectrolyte 高分子电解质，聚合物电解质
polyene 多烯
polyenic 多烯的
polyester 聚酯
  ~ capacitor 聚酯电容器
  ~ ether 聚醚酯
  ~ fibre 聚酯纤维，涤纶〈俗称〉
  ~ - fibre reinforced plastic 聚酯纤维增强塑料
  ~ film 聚酯薄膜；聚酯胶卷，聚酯软片
  ~ foam 聚酯类泡沫塑料
  ~ imide 聚酯酰亚胺
  ~ mo(u)lding compound 聚酯模塑料

~ photosensitive film 聚酯光敏胶片[胶卷],聚酯感光胶片[胶卷]
~ plastic 聚酯塑料
~ plasticizers 聚酯增塑剂
~ polyols 聚酯多元醇
~ powder coating 聚酯粉末涂料
~ - prepreg 聚酯预浸料(坯)
~ resin 聚酯树脂
~ rubber 聚酯型橡胶
~ terephthalate 聚对苯二甲酸酯
~ urethane 聚酯型聚氨酯
polyesteramide 聚酯酰胺
~ fibre 聚酯酰胺纤维
polyesterification 聚酯化(作用)
polyethene 聚乙烯
polyether 聚醚
~ - based polyurethane foams 聚醚类聚氨酯泡沫塑料
~ based polyurethane rubber 聚醚型聚氨酯橡胶
~ ester fibre 聚醚酯纤维
~ ether ketone [PEEK] 聚醚醚酮
~ - fluorinated polyurethane prepolymer 聚醚型含氟聚氨酯预聚物
~ foam 聚醚类泡沫塑料
~ hexols 聚醚六元醇
~ imide 聚醚酰亚胺
~ ketone [PEK] 聚醚酮
~ octols 聚醚八元醇
~ pentols 聚醚五元醇
~ - polyester elastomer 聚醚-聚酯弹性体
~ polyurethane foams 聚醚类聚氨酯泡沫塑料
~ rubber 聚醚型橡胶
~ sulphon [PES] 聚醚砜
~ tetrols 聚醚四元醇
~ triols 聚醚三元醇
~ urethane elastomer 聚醚型聚氨酯弹性体
polyethyl acrylate [PEA] 聚丙烯酸乙酯
polyethylene [PE] [polyethene] 聚乙烯
~ bag 聚乙烯袋

~ by irradiation 辐射聚乙烯
~ - chlorotrifluoroethylene [E/CTFE] 聚乙烯-三氟氯乙烯
~ component 聚乙烯部件〈用于焊接或粘接〉
~ fibre 聚乙烯纤维,乙纶〈俗称〉
~ film 聚乙烯薄膜
~ foam 聚乙烯泡沫塑料
~ glycol [PEG] 聚乙二醇
~ glycol esters 聚乙二醇酯类〈抗静电剂,增塑剂,润滑剂〉
~ glycol raw material 聚乙二醇原料
~ glycol terephthalate 聚对苯二甲酸乙二(醇)酯
~ imine 聚乙烯亚胺
~ insulated cable 聚乙烯绝缘电缆
~ monofilament 聚乙烯单丝
~ of high density 高密度聚乙烯
~ of low density 低密度聚乙烯
~ oxide [PEO] 聚环氧乙烷,聚氧化乙烯〈俗称〉
~ plastics 聚乙烯塑料
~ polyamine 多亚乙基多胺,〈俗称〉多乙撑多胺〈固化剂〉
~ powder 粉末聚乙烯〈改性剂、黏合剂、添加剂〉
~ - propylene plastics 乙丙塑料
~ terephthalate [PET、PETP] 聚对苯二甲酸乙二(醇)酯
~ terephthalate fibre 聚苯二甲酸乙二(醇)酯纤维,聚酯纤维,涤纶〈俗称〉
~ wax 聚乙烯蜡〈润滑剂〉
γ - (**polyethyleneamino**) **propyltrimethoxy silane** γ-(多乙撑氨基)丙基三甲氧基(甲)硅烷〈偶联剂〉
polyfluorocarbon 聚氟碳
polyfluorohydrocarbon 聚氟烃
polyformaldehyde 聚甲醛
~ plastic 聚甲醛塑料
~ resin 聚甲醛树脂
polyfunctional 多官能团的;多功能的
~ compound 多官能化合物
~ group 多官能团
~ isocyanate 多官能异氰酸酯

~ monomer 多官能单体
~ structure 多官能结构
~ structure unit 多官能结构单元
polyfunctionality 多官能性;多官能度
polygenetic 多色的
~ dye 多色染料
polyglycol 聚乙二醇
~ distearate 聚乙二醇二硬脂酸酯
polyglycolylurea 聚乙内酰脲,聚脲基乙酸内酰胺,聚海因
polyhexafluoropropylene 聚六氟丙烯
polyhexamethylene adipamide [PA-66] 聚己二酰己二胺,尼龙66
polyhexamethyleae butyldiamine [PA-46] 聚己二酰丁二胺,尼龙46
polyhexamethylene dodecanamide [PA-612] 聚十二烷二酰己二胺,尼龙612
polyhexamethylene sebacamide [PA-610] 聚癸二酰己二胺,尼龙610
polyhydantoin 聚乙内酰脲,聚海因〈俗称〉
polyhydrazide 聚酰肼
polyhydric alcohol 多元醇
polyhydrocarbon 聚烃类
polyhydroxy alcohol 多羟基醇,多元醇
polyhydroxy ether 聚羟基醚
~ resin 聚羟基醚树脂
polyimidazopyrrazolone 聚咪唑吡咯啉酮
polyimide [PI] 聚酰亚胺
~ -ester 聚酰亚胺酯
~ fibre 聚酰亚胺纤维
~ film 聚酰亚胺薄膜
~ foam 聚酰亚胺泡沫塑料
~ plastic 聚酰亚胺塑料
polyion 聚离子
polyisobutene [PIB] 聚异丁烯
~ rubber 聚异丁烯橡胶
polyisobutylene [PIB] 聚异丁烯
polyisocyanate 聚异氰酸酯
polyisocyanurate [PIR] 聚异氰脲酸酯
~ -based plastic 聚异氰脲酸酯塑料
~ foam [PIC-foam] 聚异氰脲酸酯泡沫塑料
~ rigid foam [PIC-rigid foam] 聚异氰脲酸酯硬质泡沫塑料
polyisoprene [PI] 聚异戊二烯
~ deutero 氘化聚异戊二烯
polylaminated structure 多层结构〈增强塑料〉
polylaurylatctam [PA-12] 聚十二内酰胺,尼龙12
polyliner 多孔分流梭
~ torpedo 多孔分流梭鱼雷头
ω-polymer "卷心菜式"聚合物,ω-聚合物
polymer 聚合物
~ -additive mixture 聚合物-添加剂混合物
~ absorbent 聚合物吸附剂
~ alloy 聚合物"合金"
~ binder 聚合物黏合剂
~ blend 共混聚合物,聚合物共混物,混配聚合物
~ blend fibre 聚合物混纺纤维
~ blending 聚合物共混,聚合物混配
~ chain 聚合物链
~ chips 聚合物切片
~ compound 聚合物混配料,聚合物混合料
~ dispersion 聚合物分散体
~ emulsion 聚合物乳液
~ film 聚合物膜
~ flow constant 聚合物流动常数
~ gel 聚合物凝胶
~ homologue 同系聚合物
~ melt 聚合物熔体
~ melt temperature 聚合物熔体温度
~ modification 聚合物改性
~ modified 聚合物改性的
~ modifier 聚合物改性剂
~ morphology 聚合物形态学
~ network 聚合物网络
~ -poor phase 贫聚合物相
~ processing 聚合物加工
~ radical 聚合物自由基
~ reaction 聚合物反应

~ rheology 聚合物流变学
~ -rich phase 富聚合物相
~ science 聚合物科学
~ semiconductor 聚合物半导体
~ series 聚合物系列
~ solution 聚合物溶液
~ stabilizer 聚合物稳定剂
~ structure 聚合物结构
~ transition 聚合物(相)转变
~ of 2,2,4-trimethyl-1,2-dihydroquinoline 2,2,4-二甲基-1,2-二氢化喹啉聚合物〈抗氧剂〉
~ waste 聚合物废料

polymeric 聚合(物)的
~ acceptor 聚合物接受体
~ additive 聚合(物)添加剂
~ adhesive 聚合物型黏合剂
~ alloy 聚合物"合金"
~ catalyst 聚合物催化剂
~ chelate 聚合螯合物
~ coating 聚合物涂层
~ compound 聚合物
~ dyes 聚合染料
~ material 聚合(物)材料
~ melt 聚合物熔体
~ modification 聚合物改性
~ modifier 聚合物改性剂
~ organotin mercaptide 聚合型有机锡硫醇化合物〈热稳定剂〉
~ plasticizer 聚合型增塑剂〈增塑剂〉
~ polyisocyanate [PPI] 聚异氰酸酯
~ resin 聚合物树脂
~ segment 聚合物链段

polymericular weight 聚合物分子量
polymerizable 可聚合的
~ plasticizer [reactive plasticizer]
polyxmerism 聚合(现象)
polymerizate 聚合物
polymerization 聚合(作用)
~ accelerator 聚合促进剂
~ activator 聚合活化剂
~ activity 聚合活性
~ agent 聚合剂
~ catalyst 聚合催化剂
~ coupling reactant 聚合偶联剂
~ degree 聚合度
~ in homogeneous phase 均相聚合
~ -in-situ 反应注塑造;现场聚合
~ inhibitor 阻聚剂
~ initiation reaction 聚合引发反应
~ initiator 聚合引发剂
~ kinetics 聚合动力学
~ mechanism 聚合机理
~ plant 聚合装置
~ process 聚合过程
~ product 聚合产品
~ promotor 聚合促进剂
~ rate 聚合速率
~ regulator 聚合调节剂
~ resins 聚合型树脂
~ retainer 聚合抑制剂,阻聚剂
~ stabilizer 聚合稳定剂

polymerize 聚合
polymerized 聚合(了)的
~ oil 聚合油
~ substance 聚合物
~ 2,2,4-trimethyl-1,2-dihydroquinoline powder 2,2,4-三甲基-1,2-二氢化喹啉聚合体〈高分子量粉末状〉,防老剂 124
~ 2,2,4-trimethyl-1,2-dihydroquinoline resin 2,2,4-三甲基-1,2-二氢化喹啉聚合体〈低聚合度树脂状〉,防老剂 RD

polymerizer 聚合剂;聚合器
polymerizing 聚合
~ power 聚合能力
polymethacrylate 聚甲基丙烯酸酯
polymethacrylimide [PMI] 聚甲基丙烯酰亚胺
polymethacrylonitrile [PMAN] 聚甲基丙烯腈
polymethoxy acetal [PMAC] 聚甲氧基缩醛
polymethyl 聚甲基
~ acrylate [PMA] 聚丙烯酸甲酯
~ α-chloroacrylate [PMCA] 聚 α-氯代丙烯酸甲酯

~ α-chloromethacrylate [PMCA] 聚α-氯甲基丙烯酸甲酯
~ methacrylate [PMMA] 聚甲基丙烯酸甲酯,有机玻璃
~ methacrylate molding powder 聚甲基丙烯酸甲酯模塑粉
~ methacrylate optical fiber 聚甲基丙烯酸甲酯光学纤维〈光导纤维〉
~ methacrylate plastics 聚甲基丙烯酸甲酯塑料
~ siloxane 聚甲基硅氧烷
polymethyiene 聚亚甲基
~ polyphenylene isocyanate [PAPI] 聚亚甲基聚苯基异氰酸酯
polymethylenic 聚亚甲基的
polymethylphenyl 聚甲基苯基
~ siloxane [PSI] 聚甲基苯基硅氧烷
~ siloxane fluid [PMPSF] 聚甲基苯基硅氧烷液体,苯甲基硅油〈润滑剂、脱模剂、浸渍剂、处理剂〉
polymethylstyrene 聚甲基苯乙烯
Polymin [polyethylenimine] 聚乙烯亚胺
polymolecular 多分子的
~ layer 多分子层
polymolecularity 多分子性
polymorph 多晶型物
polymorphism (同质)多晶(现象)
polynonanoylamide [PA-9] 聚壬酰胺,尼龙9
polynuclear 多核的;多环的
polyoctanoyllactam [PA-8] 聚辛内酰胺,尼龙8
polyol 多元醇,多羟基醇
~ raw material 多元醇原料〈制造聚氨酯的组分〉
polyolefin [PO] 聚烯烃
~ fibre 聚烯烃纤维
~ foam concentrate 泡沫聚烯烃母料
~ -modified high-impact polystyrene 聚烯烃改性的耐高冲击聚苯乙烯
~ -plastic 聚烯烃塑料
~ resin 聚烯烃树脂
polyorganosiloxane 聚有机硅氧烷
polyorganostannosiloxane 聚有机锡硅氧烷
polyorganotitanosiloxane 聚有机钛硅氧烷
polyoxadiazole 聚噁二唑
~ fibre 聚噁二唑纤维
polyoxamide 聚乙二酰胺,聚草酰胺
~ fibre 聚草酰胺纤维
polyoxazole 聚噁唑
polyoxyethylene 聚氧化乙烯,聚环氧乙烷
~ alkyl ether 聚氧乙烯烷基醚〈乳化剂、匀染剂〉
~ (5) glycerin monooleate 聚氧乙烯(5)甘油单油酸酯〈防雾剂、抗静电剂〉
~ (20) glycerin monostearate 聚环氧乙烷(20)甘油单硬脂酸酯〈防雾剂〉
~ (9) monooleate 聚氧乙烯(9)单油酸酯〈防雾剂〉
~ nonyl phenol ether 聚氧乙烯壬基酚醚〈乳化剂〉
~ octyl phenol ether 聚氧乙烯辛烷基酚醚〈乳化剂 OP〉
~ (20) sorbitan monooleate 聚环氧乙烷(20)山梨糖醇酐单油酸酯〈防雾剂〉
~ stearate 聚氧乙烯硬脂酸酯〈柔软剂 SG〉
~ xylitol stearate 聚氧乙烯硬脂酸木糖醇酯〈油剂〉
polyoxymethylene [POM] 聚甲醛,聚氧化甲烯
~ plastic 聚甲醛塑料,聚氧化甲烯塑料
polyoxypropylene glycol 聚氧化丙二醇
polyparabanic acid [PPA] 聚乙二酰脲
polyparamethyl styrene 聚对甲基苯乙烯
polypeptides 多肽
polyperfluorotriazine 聚全氟三嗪
polyphase 多相(的)
~ current 多相电流
~ emulsion 多相乳状液
polyphenol-aldehyde fibre 聚酚醛纤维

polyphenyl sulfide [PPS] 聚苯硫醚
polyphenylacetylene 聚苯乙炔
polyphenylene 聚亚苯基
~ ether [PPE] 聚苯醚
~ oxide [PPO] 聚苯醚,聚苯撑氧〈俗称〉
~ sulfone 聚苯砜
~ sulphide [PPS] 聚苯硫醚
polyphenylquinoxaline 聚苯基喹喔啉
polyphenylsulfone [PPSU] 聚苯基砜
polyphosphmiitrilic chloride 聚氯化膦腈
polyphosphoric acid 多(缩)磷酸
polyphthalate carbonate [PPC] 聚邻苯二甲酸-碳酸酯
polypivatolactone, [PPVL] 聚新戊内酯
polypropene [polypropylene] 聚丙烯
~ adipate 聚己二酸丙二醇酯〈增塑剂〉
~ condenser 聚丙烯电容器
~ /ethylene - propylene rubber alloys 聚丙烯-乙丙橡胶合金
~ fibre 聚丙烯纤维,丙纶〈俗称〉
~ film 聚丙烯薄膜
~ film capacitor 聚丙烯薄膜电容器
~ foams 聚丙烯泡沫塑料
~ glycol 聚丙二醇
~ oxide [PP OX] 聚丙醚;聚环氧丙烷
~ plastic 聚丙烯塑料
~ sulphide 聚硫化丙烯;聚丙(撑)硫醚
~ terephthalate [PPT] 聚对苯二酸丙二醇酯
polypropylene [PP, polypropene] 聚丙烯
polypyromellitimide 聚均苯四(甲)酰亚胺
polypyrrolidone 聚吡咯烷酮
polyquinazolinedione 聚喹唑啉二酮
Polyquinazolone 聚喹唑啉酮
Polyquinoxaline 聚喹喔啉〈胶黏剂、涂料〉
polysaccharide 多糖
polysebacic polyanhydride 聚癸二酸多酐〈固化剂〉
polysilicone rubber 聚硅橡胶
polysiloxane 聚硅氧烷
~ emulsion 有机硅乳剂〈消泡剂、脱模剂、织物整理剂〉
~ rubber 聚硅氧烷橡胶;硅橡胶
polysorbate 聚山梨酯
polystamianediol 聚锡二醇
~ dilaurate 聚锡二醇二月桂酸酯〈热稳定剂〉
~ ether ester 聚锡二醇醚酯〈热稳定剂〉
~ lauro - maleate ester 聚锡二醇月桂酸-马来酸酯〈热稳定剂〉
polystyrene [PS] 聚苯乙烯
~ beads 聚苯乙烯珠粒料
~ compound 聚苯乙烯混合料
~ fibre 聚苯乙烯纤维
~ film 聚苯乙烯薄膜
~ film capacitor 聚苯乙烯薄膜电容器
~ foam 聚苯乙烯泡沫塑料
~ homopolymer 聚苯乙烯均聚物
~ maleic anhydride resin 聚苯乙烯-顺(丁烯二酸)酐树脂〈固化剂〉
~ modified by rubber 橡胶改性聚苯乙烯
~ moulding material 聚苯乙烯模塑料;聚苯乙烯成型料
~ paper 聚苯乙烯(泡沫)纸
~ plastic 聚苯乙烯塑料
~ resin 聚苯乙烯树脂
polystyrol 聚苯乙烯
polysulfonate copolymer 聚磺酸酯共聚物
polysulphide [polysulfide] 聚硫化物;多硫化物
~ polymer 多硫聚合物
polysulphone [polysulfone] 聚砜
polytainer 聚乙烯容器
polyterephthalate [PIP] 聚对苯二甲酸酯
polyteipene 聚萜(烯);多萜(烯)
~ resin 聚萜树脂;多萜树脂〈增黏剂,胶黏剂〉
polytetrafluoroethylene [PTFE] 聚四氟

乙烯
- ~ aqueous dispersion 聚四氟乙烯水分散液
- ~ capacitor 聚四氟乙烯电容器
- ~ fibre 聚四氟乙烯纤维
- ~ plastics 聚四氟乙烯塑料
- ~ wax 聚四氟乙烯蜡〈润滑剂〉

polytetramethylene terephthalate [PTMT] 聚对苯二甲酸丁二醇酯

polythene [polyethylene] 聚乙烯

polythiazole 聚噻唑

polytridecanoylactam [PA-13] 聚十三酰胺,尼龙13

polytrifluorochloroethylene [polychlorotrifluoroethylene] 聚三氟氯乙烯
- ~ fibre 聚三氟氯乙烯纤维

polytrifluorostyrene 聚三氟苯乙烯

polytrimethylene terephthalate 聚对苯二甲酸丙二(醇)酯

polytrimethylhexamethylene terephthalamide 聚对苯二甲酰三甲基己二胺

polytrioxane 聚三噁烷

polytropic extruder 多热源挤塑机

polyundecanoylamide [PA-11] 聚十一酰胺,尼龙11

polyurea [PU] 聚脲
- ~ fiber 聚脲纤维

polyurethane [PU, PUR] 聚氨酯,聚氨基甲酸酯;聚氨基甲酸乙酯
- ~ adhesive 聚氨酯黏合剂
- ~ backing 聚氨酯衬(里料)
- ~ block foam 聚氨酯块状泡沫塑料
- ~ block production 聚氨酯块状泡沫塑料生产
- ~ - cast elastomer [PU - cast elastomer] 聚氨酯铸塑弹性体
- ~ coating 聚氨酯涂料
- ~ elastomer [UE] 聚氨酯弹性体
- ~ fibre 聚氨酯纤维
- ~ foam 聚氨酯泡沫塑料
- ~ lacquer 聚氨酯漆
- ~ leather 聚氨酯合成革
- ~ lining 聚氨酯衬(里)
- ~ plastic 聚氨酯塑料
- ~ - polymethyl methacrylate copolymer 聚氨酯-聚甲基丙烯酸甲酯共聚物
- ~ prepolymer 聚氨酯预聚物
- ~ reaction injection moulding technique [polyurethane RIM technique] 聚氨酯反应注塑(成型)法
- ~ reaction resin (forming) material 聚氨酯反应性树脂(模塑)料
- ~ resin 聚氨酯树脂
- ~ resin adhesive 聚氨酯树脂黏合剂
- ~ RIM - process [RIM - technique, RIM - technology] 聚氨酯反应注塑(成型)法
- ~ rubber 聚氨酯橡胶

polyvalent 多价的

polyvinyl 聚乙烯基
- ~ acetal 聚乙烯醇缩(乙)醛
- ~ acetal/phenolic resin adhesive 聚乙烯醇缩醛/酚醛树脂黏合剂
- ~ acetate [PVAC] 聚乙酸乙烯(酯),聚醋酸乙烯(酯)
- ~ acetate plastics 聚乙酸乙烯塑料
- ~ acetate emulsions 聚乙酸乙烯(酯)乳液
- ~ acetate latex 聚乙酸乙烯(酯)胶乳
- ~ acetate resin adhesive 聚乙酸乙烯(酯)树脂黏合剂
- ~ alcohol [PVA, PVAL] 聚乙烯醇〈分散剂〉
- ~ alcohol fiber 聚乙烯醇纤维〈填料〉
- ~ alkyl ether 聚乙烯基烷基醚
- ~ amine 聚乙烯胺
- ~ butyral [PVB] 聚乙烯醇缩丁醛
- ~ butyral emulsion 聚乙烯醇缩丁醛乳液〈织物处理剂、胶黏剂、纸张处理剂〉
- ~ butyrate [PVBu] 聚乙烯丁酸酯
- ~ carbazole [PVK] 聚乙烯咔唑
- ~ carbazole resin 聚乙烯咔唑树脂
- ~ chloride [PVC] 聚氯乙烯
- ~ chloride - acetate [PVCA] 聚氯乙烯-乙酸乙烯酯

~ chloride acetate copolymer 聚氯乙烯乙酸乙烯酯共聚物
~ chloride acetate fibre 聚氯乙烯-乙酸乙烯酯(共聚)纤维
~ chloride/acrylate rubber blend [PVC/ACR] 聚氯乙烯与丙烯酸酯类橡胶的共混物
~ chloride/acrylonitrile - butadinestyrene terpolymer blends [PVC/ABS] 聚氯乙烯与丙烯腈-丁二烯-苯乙烯三元共聚物的共混物,聚氯乙烯-ABS树脂共混物
~ chloride acrylonitrile fibre 聚氯乙烯-丙烯腈(共聚)纤维
~ chloride/chlorinated polyethylene blends [PVC/CPE] 聚氯乙烯与氯化聚乙烯的共混物
~ chloride coated fabric 聚氯乙烯涂覆织物
~ chloride/ethylene - vinyl acetate copolymer blends [PVC/EVA] 聚氯乙烯与乙烯-乙酸乙烯共聚物的共混物
~ chloride fibre 聚氯乙烯纤维,氯纶〈俗称〉
~ chloride film 聚氯乙烯薄膜
~ chloride foam 聚氯乙烯泡沫塑料
~ chloride latex 聚氯乙烯胶乳
~ chloride leather 聚氯乙烯人造革
~ chloride/methyl methacrylate - butadine - styrene terpolymer blends [PVC/MBS] 聚氯乙烯与甲基丙烯酸甲酯-丁二烯-苯乙烯三元共聚物的共混物
~ chloride mo(u)lding compound 聚氯乙烯模塑料,聚氯乙烯成型料
~ chloride mo(u)lding material 聚氯乙烯模塑料,聚氯乙烯成型料
~ chloride/NBR blends 聚氯乙烯与丁腈橡胶的共混物
~ chloride paint 聚氯乙烯涂料
~ chloride paste [PVC past] 聚氯乙烯糊
~ chloride plastic 聚氯乙烯塑料
~ chloride polymerized in emulsion [E-PVC] 乳液(聚合的)聚氯乙烯
~ chloride polymerized in supension [S-PVC] 悬浮(聚合的)聚氯乙烯
~ chloride - vinyl methyl ether [PVM] 聚氯乙烯-甲基乙烯基醚
~ dichloride 聚二氯乙烯
~ ester 聚乙烯酯
~ ether 聚乙烯醚
~ fluoride [PVF] 聚氟乙烯
~ fluoride sheet 聚氟乙烯板材
~ formal [PVFM] 聚乙烯醇缩甲醛
~ formal acetal 聚乙烯醇缩甲乙醛〈黏合剂〉
~ imidazole 聚乙烯基咪唑
~ isobutyl ether [PVI, PVIE] 聚乙烯异丁醚〈黏合剂、增塑剂、添加剂〉
~ methyl ether [PVME] 聚乙烯甲醚
methyl ether/maleic anhydride, [PVM-MA, PVM/MA] 聚乙烯甲醚-马来酐
~ methyl ketone [PVMK] 聚乙烯基甲基酮
~ nicotinate 聚烟酸乙烯酯
~ oleyl ether 聚乙烯基油醇基醚
~ propionate [PVPr] 聚乙烯丙酸酯
~ pyrrolidone [PVP] 聚乙烯吡咯烷酮
~ resin 聚乙烯(基)类树脂
~ stearate 聚硬脂酸乙烯
~ synthetic fibre 聚乙烯(基)类合成纤维
polyvinylidene chloride [PVDC] 聚偏(二)氯乙烯
polyvinylidene chloride fibre 聚偏(二)氯乙烯纤维
polyvinylidene chloride [PVDC] **plastic** 聚偏(二)氯乙烯塑料
polyvinylidene cyanide 聚偏(二)氰乙烯
polyvinylidene cyanide fibre 聚偏(二)氰乙烯纤维
polyvinylidene fluoride [PVDF] 聚偏(二)氟乙烯
polyxylenol 聚二甲酚

POM [polyoxymethylene] 聚甲醛;聚氧化甲烯
pomegranate 石榴红〈暗红色〉
pommel 推料杆,柱塞〈柱塞式挤塑机〉
pontianac 玷玳型树脂〈天然树脂〉
pony mixer 行星式搅拌机,换罐式混合器
Pony type change – can mixer 换罐式混合器
poor 不良的
~ conductor 不良导体
~ feeling 手感不良
~ flow quality 流动性不良
~ penetration 渗透不良
~ weld 不良焊缝
~ wet 浸渍不好,润滑不好
POP [printing – Out – Paper] 印相纸〈一种感光纸〉
popcorn （爆）玉米花
~ plastic 米花状塑料
~ polymer 米花状聚合物
~ polymerization 米花状聚合,膨体聚合
pore （微）孔,细孔
~ diameter 孔径
~ distribution 孔径分布
~ filler 填料,填隙料
~ foaming agent 微孔发泡剂
~ size 孔径;孔径大小
~ size distribution 孔径分布
porolated film 微孔薄膜〈透气性〉
porolating machine 微孔成型机〈透气性薄膜〉
poromaster process 波洛玛斯特法〈采用放电法使塑料薄膜具有透气性、透湿性微孔的方法〉
poromer 透气性合成率
poromeric 透湿不透水性;透湿不透水的
~ material 合成多孔材料,透气性（人造革）材料
poromerics 微孔材料,透气性人造革
porosimeter 孔率计,孔度计

porosimetry 孔隙率测定
porosint 多孔材料
porosity 多孔性;孔隙率,多孔度
~ of cellular plastics 微孔塑料孔隙率
~ test 透气性试验
~ testing machine 透气性试验机
porous 多孔的,疏松的
~ earner 多孔载体
~ cells 开孔微泡〈泡沫塑料〉
~ cloth 多孔织物
~ glue line 多孔胶膜层
~ membrane 多孔膜
~ mo(u)ld 多孔模具
~ plastics 多孔性塑料
~ plate 多孔板
~ polymer 多孔聚合物
~ polymeric material 透气性聚合材料
~ structure 多孔结构
~ surface 多孔表面
~ water – proofing 透气性防水（整理）
porousness 孔隙率,孔隙度,多孔性
port 孔,口
portable 手提的;轻便的
~ bag 手提包
~ grinder 手摇钻
~ machine 移动式机器
~ mixer 手动混合器
~ mo(u)ld 手提式模具,移动式模具
~ press 手动压机
~ tester 便携式试验仪
portion 部分;份;分配
~ pack 份包装
~ package 份包装
pos. position 位置,地位
pos. positive 正的;阳的
position 位置;地位,状态;安置;定位
~ indicator 位置指示器
positioner 定位器;转动换位器
positive 正的,阳性的
~ catalyst 正催化剂
~ cooling 人工冷却,补助冷却
~ crystal 正晶体

~ die 阳模；上半模
~ ideal adjustable gear 无级变速齿轮
~ mixer 强制混合器
~ mo(u)ld 全压式模具，不溢式模具
~ pressure 正压
~ thermoforming 阳模热成型
positron annihilation 正电子湮没
post 〈词头〉后，续
~ - bonding 后黏合（工艺）
~ - cementing 后粘接，后胶接〈由于添加剂渗移使固化胶层而进行粘接〉
~ - chlorination 后氯化（作用）
~ - condenser 后冷凝器；后凝缩器
~ consumer recycling 消费后回收
~ cure 后固化
~ curing 后固化
~ decoration 后装饰
~ - demold shrinkage 脱模后收缩
~ - drawing 后拉伸
~ - elastic behavior 弹性后效
~ embossing 后压花
~ expansion 后发泡
~ - formable laminate 可后成型的层合板
~ forming 后成型
~ - heating 后加热；热后处理〈产品〉
~ injection pressure 注塑持续压力，保压〈注塑〉
~ - mo(u)ld shrinkage 出模收缩
~ - mo(u)lding assembly 模压后组装
~ - mo(u)lding crystallization 模压后结晶
~ - mo(u)lding insert technique 成型后嵌件置入法
~ - mo(u)lding treatment 成型后处理〈精加工，修整〉
~ processing 后加工
~ - shrinkage 后收缩；后收缩率〈塑料成型件〉
~ softening 后软化
~ stove 后烘烤

~ - treatment 后处理
~ - type change - can mixer 立式换筒式混合器
~ - type mixer 立式搅拌机
postbaking 后烘烤
postchlorinated 氯化过的
~ polyethylene 氯化聚乙烯
~ polyvinyl chloride 氯化聚氯乙烯
postcompression stage 续压期〈注塑时〉
postcrystallization 后结晶
postextrusion 挤出后
~ filament swelling 挤出丝膨胀
~ swelling 挤后膨胀
postformed 后成型的
~ die 后成型模
~ laminate 后成型层压料〈板材〉，后成型层压料板
~ laminated section 后成型层压型材
~ mo(u)lding 后成型制品
postforming 后成型
~ sheet 后成型片材
postreticulated polymer 网化后聚合物；交联后聚合物
postwelding treatment 焊接后处理
pot 料槽；罐；釜
~ crusher 罐式压碎机
~ heater 热压锅，罐式加热器
~ life 适用期
~ mill 球磨机，瓷罐球磨机
~ mo(u)lding 料槽式传递成型；料槽式压铸
~ plunger 加料室柱塞，压料柱塞
~ retainer 料槽托模板〈美〉
~ slug 料槽，余料
~ spianing 离心纺纱；罐式纺纱
~ - type collar 碗形垫圈
~ - type transfer mo(u)lding 罐式传递模塑
potash mica 钾云母，白云母〈优质云母，填料〉
potassium [K] 钾
~ bromide 溴化钾〈催化剂〉

potential

~ chloride 氯化钾
~ ω-hydroperfluoroheptylate ω-含氢全氟庚酸钾〈分散剂〉
~ hydroxide 氢氧化钾；苛性钾
~ persulfate 过硫酸钾〈引发剂〉

potential 势；位；潜(在)的
~ energy 势能
~ shrinkage 潜在收缩
~ temperature 潜在温度

potentiality 可能性；潜力
potentially frothing 潜在发泡
potentiometric titration 电位滴定法
potted article 封装件
potting 铸封，灌封，嵌铸
~ compound 灌封浆料，铸封料
~ material 铸封料
~ syrups 灌封浆料，铸封料

pouch 盒，(小)袋
~ packaging of liquids 袋式液体包装

pound 磅；英磅
~ per square inch [psi] 磅每平方英寸
~ per square inch absolute 绝对磅每平方英寸
~ per square inch gauge [psig] 磅每平方英寸表压

pour 倾注
~ -in-place 现场浇灌〈发现〉
~ -in-place foam 现场浇灌发泡
~ -off 蒸馏出来；倾出，倒出，流出，溢出
~ point 倾倒点；浇灌点；流动点

pourability 可灌注性
pourable 可浇注的，可灌入的
~ plastics 液态塑料
~ sealing compound 可浇注封口料，灌封料

poured in-place 现场浇灌〈发泡〉
pouring 灌注；铸封
~ compound 铸封料
~ defect 浇铸缺陷
~ mo(u)ld 浇注模

~ spout 浇注
~ temperature 浇铸温度
~ time 浇注时间〈发泡〉

Poval 聚乙烯醇〈日本对聚乙烯醇的通称〉
pow. powder 粉末
powd. [powdered] (研成)粉末(状)的
powder 粉料，粉末，粉状
~ adhesive 粉状黏合剂
~ bath 粉浴〈流化〉
~ binder 粉状粘结剂
~ blend 共混粉料，混配粉料
~ blender 粉料混合器，粉料混配器
~ bonding 粉末黏合
~ coating 粉料涂布；粉末涂装
~ compact 压塑粉，压型粉，粉粒料；粉末坯块
~ density 松密度，粉密度
~ dispenser 粉末撒布器
~ extrusion 粉料挤塑
~ filler 粉状填料
~ grain density 粉粒密度
~ grain size 粉粒大小
~ graphite 粉状石墨
~ lacquer 粉末(喷)涂料
~ layer 粉末层
~ load 粉料加料
~ mixer 粉料混合机
~ mixing 粉料混合
~ moulding 粉料模塑
~ shaker 粉料振动机
~ sintering 粉末烧结
~ sintering moulding 粉末烧结成型
~ sintering process 粉末烧结法
~ sprayer 喷粉器
~ spraying 粉料喷涂
~ technique 粉末技术
~ varnish 粉末状磁漆

Powdered 粉(末)状的
~ catalyst 粉(末)状催化剂
~ glue 粉(末)状胶黏剂
~ high-temperature adhesive 粉(末)

状耐高温黏合剂
~ pigment   粉(末)状颜料
~ plastic   粉(末)状塑料
~ resin   粉(末)状树脂
~ rubber   粉(末)状橡胶
~ scrap   粉(末)状废料,粉屑
power   本领;力;功率;动力;电源
~ board   配电板
~ cable   电力电缆
~ capability   功率
~ constant   功率常数
~ consumption   功率消耗;动力消耗
~ control   功率调整
~ current   动力电流
~ cut-off   关车;结束工作
~ fabric   弹性织物
~ factor   功率因数
~ feed   自动进料,机械进料
~ loss   功率损耗
~ output   动力输出量,电能效率
~ pack   驱动装置;液压装置〈注塑机〉
~ panel   配电盘
~ piping   高压管道
~ range   功率范围
~ rating   额定功率
~ saw   电锯
~ station   动力厂,发电站
~ supply   供电电源
~ water   水力;高压水
PP [polypropylene]   聚丙烯
PPA [polyparabanic acid]   聚乙二酰脲
ppb. [parts per billion]   十亿分之(几)
PPC [chlorinated polypropylene]   氯化聚丙烯
PPC [polyphthalate carbonate]   聚(邻)苯二甲酸-碳酸酯
PPE [polyphenylene ether]   聚苯醚
PPI [Plastics Pipe Institute]   塑料管材协会
PPI [Polymeric Polyisocyanate]   聚异氰酸酯

ppm [parts per million] 10^{-6},百万分之(几)
PPO [polyphenylene oxide]   聚苯醚,聚苯撑氧〈俗称〉
PPOX [polypropylene oxide]   聚丙醚;聚环氧丙烷
PPS [polyphenyl sulfide]   聚苯硫醚
PPSU [polyphenylene sulfone]   聚苯基砜
PPT [polypropylene terephthalate]   聚对苯二甲酸丙二(醇)酯
pptd. [precipitated]   沉淀的
pptn. [precipitation]   沉淀(作用)
PPVL [polypivalolactone]   聚新戊内酯
pr [pair]   偶,对
prac. [practically]   实际上
practical   实用的,实际的
~ chemist   有实际经验的化学家
~ chemistry   实用化学
~ experience   应用经验,实用经验
~ formulation   实用配方
~ life   实际使用寿命
~ viscosity measurement   实际黏度测定
~ unit   实用单位
practice   实践,实用,实验
~ injection mo(u)lding of plastics   实用塑料注射成型
pre 〈词头〉预,先,前
~ -accelerated resin   预加速反应树脂
~ -aging   预老化,人工老化
~ -assemble   预装配
~ -blowing   预吹胀法,预吹塑
~ -blowing technique   预吹胀法,预吹塑法
~ -dewatering chute   预脱水槽
~ -evaporated   预蒸发的
~ -expanded   预发泡的
~ -expanded bead   预发泡珠粒料
~ -expanded bead steam mo(u)lding   (因蒸汽)预发泡珠粒料蒸汽成型〈聚苯乙烯泡沫塑料〉
~ -expanded particles   预发泡颗粒

~ - expander 预发泡机
~ - expander autoclave 预发泡热压釜
~ - expansion 预发泡
~ - expansion method 预发泡法
~ - pinch 预夹坯〈吹塑〉
~ - programmed 预先程序设计的
preageing period 人工老化期
preassemble 预装配
prebend 预弯曲
preblend 预混配,预混合
prebody 预浓缩〈漆〉
prebond treatment 预粘结处理
precalculate 预先计算
precast 预浇铸
precipitability 临界沉淀点,沉淀度
precipitant 沉淀剂
precipitate 沉淀
precipitated 沉淀(了)的
~ and coldground powder 沉淀和低温研磨(塑料)粉料〈用于金属涂布〉
~ barium sulphate 沉淀硫酸钡
~ calcium carbonate 沉淀碳酸钙〈填料〉
~ filament [coagulated filament] 凝结长丝
~ pigment 沉淀颜料
~ silica 沉淀二氧化硅
precipitating 沉淀的
~ agent [precipitant] 沉淀剂
~ tank 沉淀槽
precipitation 沉淀;析出
~ bath [coagulation bath] 沉淀浴
~ chromatography 沉淀色谱法
~ - exchange resin 沉淀交换树脂
~ polymerization 沉淀聚合(作用)
~ temperature 沉淀温度
~ precipitator 沉淀器
precise 精密的,精确的
~ adjustment 精密调整
~ form 精密成型
~ instrument 精密仪器
~ measurements 精确的尺寸

precision 精密;精确度
~ accuracy 精度
~ article by injection moulding 精密注塑(成型)制件
~ casting 精密铸塑件
~ component 精密零件,精密构件
~ cut flock 定长切断纤维
~ injection machine 精密注塑机
~ injection mo(u)ld 精密注塑模(具)
~ injection mo(u)lding 精密注塑(成型)
~ instrument 精密仪器
~ mo(u)lding 精密模塑
~ plastic article 精密塑料制件
~ prescribed 要求精度
~ regulator 精密调节器
~ setting 精密调节,微调
~ standard 精密标准
~ winding 精细缠绕(法)
~ work 精密加工
precoat 预涂(层)
precoating 预涂,预涂层
precompounded 预混的
precompressed 预压缩的,预压塑的
~ fluoroplastic 预压(缩)氟塑料〈用于烧结〉
~ mixture 预压混合
~ moulding 预压塑
precornpression 预压缩;预压塑
preconditioning 预处理,预调整
precooling 预冷却
precrack 划痕;裂缝
precracking test 破裂试验
precreping 预压花
~ calender 预压花机
precrushing 预粉碎
precure 预固化〈缺陷〉
precuring 预固化
precut blank 冲切坯料,预切坯料
precutter 预切断机
predecomposition 预分解
predecoration 预装饰

predetermined 预定的
- ~ characteristics 预定性能
- ~ pressure 预定压力
- ~ size 预定尺寸,给定尺寸

prediction of service life 预测使用寿命〈涂层〉

predispersing 预分散

predistributor roll 布料辊,辊式布料器〈涂布机〉

predraft 预拉伸,预牵伸

predrawing 预拉伸
- ~ process 预拉伸过程

predry 预干燥

predryer 预烘机

predrying 预干燥〈模塑料〉
- ~ treatment 预干燥处理

prefabricate 预制品,预制,装配

prefabricated parts 预制构件

prefabrication 预制〈毛坯〉

preference 选择,优先
- ~ test 选择试验,优选试验

preferential value 优先值,最佳值

preferred 优先的
- ~ orientation 择优取向,最佳取向

prefilling press 预压实机〈粉状压实〉

prefoam 预发泡

prefoamed 预发泡的
- ~ beads 预发泡珠粒料

prefoamer 预发泡机

prefoaming 预发泡
- ~ container 预发泡容器
- ~ time 预发泡时间

preform 预压(料)锭;预制品,料坯,型坯;预成型
- ~ binder 料坯粘结剂
- ~ blow mo(u)lding 预制型坯吹塑
- ~ density 型坯密度
- ~ die 型坯模
- ~ extruder 型坯挤塑机
- ~ gripper 型坯夹持器
- ~ insert 预发泡;预制品;料板;预成型嵌件
- ~ machine 预成型机
- ~ mandrel 型坯芯棒
- ~ matched (metal) die method 预成型(金属)对模法
- ~ mo(u)lding 预成型模塑
- ~ process 预成型法〈热固性塑料层合制品〉
- ~ screen 预成型网

preformed part 预成型件

preformer 预成型机

preforming 预成型
- ~ nozzle 预成型嘴
- ~ piston 预成型活塞
- ~ press 预成型压机〈用于粉状或粒状热固性模塑料〉
- ~ process 预成型模塑
- ~ tool 预成型压模〈热固性塑料〉

pregassed 含发泡剂
- ~ beads 含发泡剂的颗粒
- ~ granules 含发泡剂的颗粒

pregassing method 预加发泡剂法

pregel 树脂淤积层;预凝胶

pregwood 浸胶木材

preheat 预热
- ~ cycle 预热周期
- ~ roll 预热辊

preheated granules 预热颗粒料

preheater 预热器
- ~ by dielectric 高频预热器

preheating 预热
- ~ cabinet 预热柜
- ~ cylinder 预热料筒
- ~ hopper 预热料斗
- ~ mill 预热研磨机
- ~ oven 预热烘箱
- ~ roll 预热辊
- ~ rolling mill 预热辊式研磨机
- ~ section 预热段
- ~ treatment 预热处理
- ~ tube 预热管〈快速焊接焊嘴〉
- ~ zone 预热区

preimpregnated 预浸渍的

preimpregnation
~ mat 预浸渍毡
~ of fabrics 织物预浸渍
~ of mats 毡片预浸渍
~ of rovings 粗纱预浸渍
preimpregnation 预浸渍
preirradiation polymerization 预辐射聚合
preliminary 初步的,预备的
~ ageing 预先老化
~ breaker 预先破碎机
~ breaking 预先破碎
~ compression 预压缩
~ crushing 预先压碎
~ dimensions 预定尺寸
~ dip 预浸渍
~ draft 预牵伸,预拉伸
~ drying 预干燥
~ filter 初步过滤
~ heating 预热
~ investigation 初步调研
~ measures 初步措施
~ sketch 草图
~ test 初步试验
~ test data 初步试验数据
~ treatment 预处理
~ treatment of the base 粘接件预处理
~ winding 预备卷绕
preload 初载荷,预载荷,初始负荷
~ equipment 预负荷装置
preloaded 预加料的,预负荷的
~ roll 预负荷辊
preloading of rolls 辊预负荷
premature 过早的
~ cure 过早固化
~ failure 过早损坏
~ oxidation 过早氧化
~ sintering 早期烧结
premelt 预熔化
premium 优质,高级
~ -grade 优质的,高级的
~ grades 优等品
~ properties 优良性能

premix 预混合,预混料〈树脂与其他组分〉
~ melt 预混料熔体
~ moulding 预混料成型
~ moulding compound 预混模塑料
~ ratio 预混合比
premixed application 预混法〈黏合剂〉
premixer 预混机
premixing 预混
~ method [process] 预混法
premould combination 预成型前配料
premo(u)lding 预成型
pre-operational test 交付使用前的试验
preoriented yarn 预取向丝
prep. prepare 制备
preparation 制备
~ and compounding 制备和混合
~ of laminate 层合板制备
~ of sample 样品的制备
~ of semi-finished products 半成品制备
~ of test specimen 试样制备
~ of the parts to be welded 焊接件制备
~ of waste paper 废纸处理
~ test 调制试验,预备试验
prepare 制备
prepared for use, p.f.u 准备好的
prepinch 预夹坯
preplastication 预塑化
preplasticator 预塑化器
preplasticization 预塑化
preplasticizer 预塑化器;预增塑剂
preplasticizing 预塑化
~ injection mo(u)lding machine 预塑化注塑成型机
~ method 预塑化法
~ mo(u)lding compound 预塑化模塑料,预塑化成型料
~ screw 预塑化螺杆
~ unit 预塑化装置
prcplnstification [preplastication] 预

塑化
prepolycondensate 预缩聚物
prepolymer 预聚物
- ~ foaming 预聚物发泡
- ~ gel 预聚物凝胶
- ~ mo(u)lding 预聚物模塑,预聚物(发泡)成型
- ~ process 预聚合(泡沫)法;预聚物发泡成型;二步聚合法〈制造聚氨酯泡沫塑料〉
- ~ technique 预聚物发泡成型

prepolymerization 预聚合
prepreg 预浸料〈浸了树脂的玻璃纤维材料〉
- ~ laminating 用预浸料层合
- ~ mat 预浸料的毡片
- ~ method 预浸料法
- ~ mo(u)lding 预浸料模塑,预浸料成型
- ~ process 预浸料坯法
- ~ winding 预浸料坯缠绕,干法缠绕

prepregnated mat 预浸料的毡片,预浸玻璃纤维毡
prepress lamination 预压层合
preprinting 预印刷
preproduction 试制,试生产
- ~ adhesive test 黏合剂试生产试验
- ~ heating 预热
- ~ model 试制模型
- ~ package 按照合同包装;试制包装
- ~ test 试制性试验

prescribed 规定的
- ~ limit 规定极限
- ~ properties *pl* 规定性能

preselectable steps 可预选的步骤
preselecting programme 预选程序控制
preselector 预选器〈塑料加工机〉
presentation 装潢〈包装〉
preserve 保护;防护;罐头食物
preserved plywood 防腐胶合板
preserving package 防腐包装
preset (预)调整,给定程序的

- ~ adjustment 预调整
- ~ control 程序控制

preshrinkage 预收缩
press 压机,压力机;模压;冲压;打包机;印刷机
- ~ bag method 压袋(成型)法
- ~ bonding 压力粘接,压黏
- ~ button 按钮
- ~ button control 按钮控制,按钮操纵
- ~ capacity 压机能力
- ~ control 按钮控制器
- ~ cure 加压硫化,加压固化
- ~ -curing time 加压固化时间〈美〉
- ~ die 冲头,冲模;阳模
- ~ ejector rod 压机脱模栓
- ~ ejector rod crosshead 压机脱模栓横梁
- ~ ejector system 压机脱模系统
- ~ factor 压制因素
- ~ finishing 压光,滚光
- ~ fit 压配合
- ~ -fit insert 压配嵌件
- ~ -fitting tip nozzle 压紧喷嘴〈热流道模〉
- ~ forming 模压成型
- ~ head 压机顶板
- ~ jig 模压工具[装置]
- ~ maker 模压工人
- ~ mo(u)ld 模塑模具,压模
- ~ -moulded 压塑
- ~ -on pressure 压紧力〈喷嘴〉
- ~ overhang 压机外伸
- ~ platen 压板,模板
- ~ polishing 压光
- ~ ram 冲头,冲模;阳模
- ~ roll 压辊
- ~ table 压台〈压机〉
- ~ through pack 深拉包装
- ~ -to-flow 有弹性垫层的压紧装置
- ~ tool 压塑模具
- ~ welding 加压焊接

press[pressure] 压力

pressed 压制的,压缩的
- ~ billet 预压料锭
- ~ -fibre board 纤维板
- ~ film 压膜
- ~ plastics sheet 压制塑料板材
- ~ sheet 压制板材

pressing 加压;压制;冲压
- ~ die 冲压模
- ~ plate 压制板,压塑板
- ~ power 加压力;压制力
- ~ roll 压辊
- ~ stage 加压阶段
- ~ time 压制时间

pressure 压力
- ~ accumulator 储压器
- ~ air 压缩空气
- ~ at expulsion 出口压力
- ~ back-flow 压力逆流
- ~ bag 加压充气袋
- ~ bag mo(u)lding 软袋施压成型,加压袋压法成型
- ~ belt 承压皮带
- ~ block 受压垫块
- ~ -blocking 受压黏连
- ~ blowing in free space 无模吹塑成型,自由吹塑成型
- ~ bottle 耐压瓶
- ~ break 压裂
- ~ bubble-plug assist vacuum forming 气胀模塞助压真空成型
- ~ bubble vacuum snap-back forming 气胀阳模真空回吸成型
- ~ build-up 压力构成;压力增长
- ~ butt weld 对接压焊(缝)
- ~ casting 加压力铸塑
- ~ coating 加压涂布
- ~ cone 锥形压力头
- ~ control 压力调节,加压控制
- ~ control lever 加压控制杆
- ~ course 加压过程
- ~ curing 加压固化
- ~ dependence 压力关系
- ~ diecasting 压铸
- ~ difference 压力差
- ~ differential meter 差压式流量计
- ~ digester 加压蒸煮器
- ~ distribution 压力分布
- ~ distribution port 压力分布孔
- ~ drop 压(力)降
- ~ during heating time 加热期间的压力
- ~ dwell 保压,续压
- ~ embossing 压花
- ~ equalization 压力均衡,均压
- ~ -equalizing plate 均压板
- ~ expansion 加压发泡
- ~ face 压力面
- ~ feed 加压供料
- ~ flow 压流;背流,回流,逆流
- ~ flow constant 回流常数〈挤塑机〉
- ~ forming 冲压(成型),模冲;模压
- ~ gauge [gage] 压力计
- ~ gelation process 加压凝胶化法
- ~ gelling 加压凝胶化
- ~ gradient 压力梯度
- ~ holding phase 压力保持相位;压力保持周期
- ~ inside the mo(u)ld 模内压力
- ~ intensifying cylinder system 增压式油缸合模装置
- ~ level 压力级
- ~ line 耐压管线
- ~ loss 压力损失
- ~ lubrication 加压润滑
- ~ manometer 压力计
- ~ marking 压痕
- ~ measuring cell 测压仪
- ~ meter 压力计
- ~ modulus 压缩模量
- ~ mo(u)lding 加压成型
- ~ nozzle 压力喷嘴
- ~ of the rolls 辊压力
- ~ opportunity 加压时机
- ~ outer sizing unit 加压外定径装置〈管材〉

~ pack [aerosol package] 加压包装,密封包装
~ package 加压包装,密封包装
~ pad 传压垫,加压垫,承压垫
~ pattern 加压(分布)曲线
~ pipe 耐压管
~ piston 推料活塞;加压活塞
~ profile 加压(分布)曲线
~ ram 推料活塞;加压活塞
~ ratio 压力比
~ recorder 压力记录器
~ reducer 减压器
~ regulating valve 压力调节阀
~ regulator 调压器
~ relief volve 卸压阀;安全阀
~ resistance 耐压性
~ roll 压辊;压力辊,压紧辊
~ roller 加压辊,压紧辊
~ sensing device 压力传感仪
~ -sensitive 压敏的
~ -sensitive adhesion 压敏黏合
~ -sensitive adhesive 压敏黏合剂
~ -sensitive adhesive tape 压敏胶黏带
~ -sensitive paper 压敏纸
~ -sensitive tack 压敏黏着性
~ -sensitive tape 压敏胶黏带
~ -sensitive tape test 压敏胶黏带试验
~ sensitives 压敏材料
~ sensor 压力传感器
~ sequence 加压顺序
~ -set ink 吸附性印刷油墨
~ shear strength 加[受]压剪切强度〈粘接〉
~ shear test 加[受]压剪切试验
~ sintering 加压烧结〈聚四氟乙烯〉
~ -specific volume-temperature diagram 压力-比容-温度图
~ -speed governed extruder 压力和转速调节的挤塑机
~ spray nozzle 压力喷涂嘴

~ surface 受压面
~ switch 压力开关
~ tap 测压接嘴,测压孔
~ test(ing) 压力试验,试压
~ thermoforming 压力热成型,气压热成型
~ -tight 不泄压的,耐压密封的
~ -tight seal 耐压密封
~ time 受压时间
~ transducer 压力传感器
~ transmission 传压,压力传递
~ tubing 耐压(软)管
~ type transfer moulding 加压型传递成型
~ unit 压力单位
~ -vacuum forming 加压真空成型
~ vessel 高压容器;压力容器
~ viscosimeter 压力黏度计
~ welding 压焊
pressureless 无压(的)
~ rotational casting 无压旋转铸塑
pressurization 增压;气密,(高压)密封
pressurize 增压
pressurized 增压的,加压的;耐压的
~ main 受压总管
~ packing 加压包装;气溶胶包装
prestrain 预应变
prestress 预应力
prestretch 预拉伸〈拉伸成型〉
~ zone 预拉伸区
prestretching 预拉伸
~ process 预拉伸过程
~ roll 预拉伸辊
pretension 预应力,预张力,预拉伸
~ roll 预伸张辊
pretreating 预处理
~ finish 预先整理
~ roll 预处理辊;预浸渍辊
pretreatment 预处理
~ time 预处理时间
preventive 预防的
~ against corrosion 防腐蚀剂

~ maintenance 预防性保养,预防性维护
pre wetting roll 预润湿辊
pre wiring 预配线,预接线
P. R. I. [Plastics and Rubber Institute] 塑料和橡胶协会〈美国〉
prim. [primary] 初级的,原始的
primary 最初的,初级的;基本的,主要的
~ accelerator 主促进剂
~ backing 基底(布)
~ carpet backing 地毯基底(布)
~ coat 底涂层
~ colours 原色,基色〈红、黄、蓝三色〉
~ creep 初级蠕变
~ crusher 粗碎机
~ crystallization 主结晶
~ degradation 初级降解
~ dispersion 主分散
~ extinction 初级消光
~ gluing 一次胶合
~ high polymer 初级高聚物,初始聚合物
~ material 原料
~ mole 基本分子
~ package 初级包装
~ plastics 原塑料
~ plasticizer 主增塑剂
~ processing 初级加工
~ product 初级产品
~ qualitites 基本特性;基本性质
~ runner 主流道
~ standard 基本标准
~ structure 原结构,一级结构
~ treatment 一次处理,初级处理
prime 最初的,原始的
~ coat 底漆层;底胶
~ cost 生产成本,原价
~ grade material [virgin material] 纯净物料,原料
~ grade polymer 原始聚合物
primer 底漆;底胶;底层涂料;腻子
~ coat 底涂层
~ coating 底涂层;底余料;底层涂布
~ layer 底层〈共挤塑〉
priming 打底
~ coat 底涂层
primrose 樱草;淡黄色的
prin [principal] 主要的,基本的
prin [principle] 原理;原则
principal 主要的,基本的
~ axis 主轴
~ axis of strain 应变主轴
~ axis of stress 应力主轴
~ bond 主键
~ chain 主链
~ characteristic data 主要特性数据
~ direction 主方向
~ length 主体长度
~ link 主键
~ normal stress 法向主应力
~ stress 主应力
~ use 主要用途
~ valency 主(要化合)价
principle 原理;规律
~ layout 总布置图
~ of action 作用原理
~ of operation 工作原理,操作原理
~ of superposition 叠加原理
~ of winding 缠绕规律〈增强塑料〉
print 印刷,印制;印花
~ coat 印花涂层
~ embossing 印花
~ free 指检干硬化
~ roll 印辊
~ -roll kiss coating 印辊吻涂
~ roller 印染辊,印花辊
~ stain 印花沾污
printability 印刷性,可印性〈塑料〉
~ treatment 可印性处理
printed 印刷的;印花的
~ board 印刷线路板
~ circuit 印刷线路〈导电涂料,导电环氧树脂〉

~ circuit board 印刷线路板
~ foil 印花贴面纸,彩印箔
~ plywood 印刷胶合板,彩印胶合板
~ sheeting 印花薄膜
~ wiring board 印刷线路板
printer 印刷机,打印机;印刷工
~ terminal 打印机终端
printing 印刷;印花;印染
~ blotches 印花不匀〈疵点〉
~ foil 印刷彩印箔
~ ink 印刷墨
~ on plastics 塑料印刷
~ - out - paper [POP] 印相纸〈一种感光纸〉
~ paste 印染浆
~ plate 印刷板
~ roll 印花辊筒
~ unit 彩印厂
probability 概率;几率
~ density 概率密度
~ distribution 几率分布
~ of error 误差概率
~ of failure 故障概率
~ of fracture 断裂概率
~ of success 成功概率
~ sampling 概率取样法
probe 试探;探测;试样
~ inspection 取样检验
~ trace 探针轨迹
procedure 程度;手续;过程;方法
proceeding 程序,方法
~ of melting 溶化方法
proceedings (会议)记录,会报
process 程序,方法;操作,作业;加工,处理;工艺;生产
~ automation (生产)过程自动化
~ capacity 加工能力
~ chart 工艺流程图,制造程序图
~ condition 加工条件,操作条件
~ control 程序控制;加工控制
~ control setting 程序控制调整
~ control system 程序控制系统

~ - controlled injection mo(u)lding 程序控制注塑(成型)
~ - controlled injection mo(u)lding machine 程序控制注塑(成型)机
~ controlled system 程序控制系统
~ cost 加工费;操作费用
~ design 工艺设计
~ drift 给定加工参数的偏差
~ engineer 工艺工程师
~ flow 生产流程
~ flow diagram 工艺流程图
~ instrumentation 生产工艺控制仪
~ line 生产过程流水线
~ of feeding the tool 充模过程
~ of manufacture 制造过程
~ optimization 加工最佳化,最佳加工选择
~ printing 彩色套印
~ redundancy 加工余量
~ variables 加工变量
~ variation 加工偏差
processability 可加工性;加工性能
processing 加工,处理
~ aid 加工助剂
~ characteristics 加工特性,加工性能
~ condition 加工条件
~ diagram 加工(顺序)图;工艺流程图
~ ease 操作容易;便于加工
~ error 加工误差
~ loss 加工损耗
~ machine 加工机械
~ method 加工方法
~ monitoring 加工监控
~ of reinforced unsaturated polyesters 增强不饱和聚酯的成型加工
~ of data 加工数据处理
~ operation 加工操作
~ parameter 加工参数
~ plant 加工装置,加工车间
~ plastics 塑料加工
~ process 加工方法,加工过程

~ properties 加工性能,加工特性
~ technology 生产工艺;加工技术
~ temperature 成型加工温度
processor 加工者;加工程序;信息处理机
prod [produce] 生产,制造
prods [products] 产物,产品
produce 生产,制造;产品,产物
producer 生产厂,制造厂;生产者
product 产品,产物
~ design 产品设计
~ finishing 产品修整;产品精加工
~ of quality 优质产品
~ performance 产品特性
~ properties 产品性能
~ stock 产品储存
~ styling 产品外形设计
~ testing 产品试验
~ uniformity 产品均匀性
~ - use 制品用途
~ verification test 产品检验试验
~ yield 产品收率
production 生产,制造;产品;产量
~ capacity 生产能力
~ - caused dimensional inaccuracies 加工过程中产生的尺寸偏差
~ constant 生产常数
~ control 生产控制
~ cost 生产费用,生产成本
~ efficiency 生产效率
~ engineering 生产工程,生产技术
~ facility 生产装置,生产设备
~ flow 生产流程
~ formula 生产配方
~ line (流水)生产线
~ line for film strips 膜带生产线
~ line work (流水)生产线
~ model 生产模型
~ per hour 每小时生产量
~ piece 产品样品
~ process 生产过程
~ quality control 产品质量控制

~ rate 生产率
~ run 生产过程;成批生产
~ - scale operation 生产规模操作
~ sample 产品样品
~ scheme 生产流程图
~ specifications 生产技术条件
~ street (流水)生产线
~ targets 生产指标
~ test 生产性试验
~ typical test 产品定型试验
~ unit 生产装置
~ yield 生产率
productive 生产的
~ capacity 生产能力
~ force 生产力
~ output 生产量
~ rate 生产率
productivity 生产率,生产能力
profile 型材;外形;截面(图),断面(图);分布图
~ board 侧板,模板
~ dies 型材挤塑口模
~ extrusion 型材挤塑
~ extrusion head 型材挤塑机头
~ extrusion line 型材挤塑生产线
~ fibre 异型纤维
~ haul - off(take - off) machine 型材牵引机〈挤塑制品〉
~ hole 异型孔〈纺丝〉
~ section 异型件
~ spinneret 异型喷丝板
profiled 异型的
~ casting 异型铸件
~ fibre cross - section 异型纤维横截面
~ filament 异型长丝
~ flash line 异型合模线
- film extrusion 异型薄膜挤塑
~ heated tool 异型电加热器
~ joint 异型密封,异型连接
~ spinning 异型纺丝
profiling 压型;异型材挤塑

~ roll 压型;压花辊
program [programme] 程序
  ~ control 程序控制
  ~ controller 程序控制器
  ~ cycle controller 周期程序控制器
  ~ parameter 程序参数
programmed (编了)程序的
  ~ heating 程序控制加热,按程序加热〈塑料加工机〉
  ~ injection control 程序控制注塑
  ~ materials testing 按程序材料试验
  ~ parison 程序控制型坯
programmer 程序设计器
programming 程序设计(的)程序编制(的)程序控制(的)
progressive 进步的,逐步的,逐渐的
  ~ core (螺杆)渐变芯
  ~ cutting 逐步切割
  ~ drafts 递增牵伸
  ~ drier 逐步干燥器
  ~ failure 逐步损坏
  ~ gluing 逐步胶合
  ~ pitch 渐变螺距
  ~ plasticization 逐步塑化
  ~ screw 渐变型螺杆
  ~ thread 渐变螺纹
project 计划,设计;工程;投射;投影;突出
  ~ engineer 主管工程师,项目工程师
  ~ planning 工程规划
  ~ programming 工程程序设计
projected 投影的;凸出的;计划的
  ~ area 投影面积,投影成型面积
  ~ mo(u)lding area 投影成型面积
projection 投射(物);投影(物);凸出(物);计划
  ~ of compounds 混合物体积增加
  ~ welding 凸焊
prolonged action of heat 长期加热效应
promotion 促进,助长
promotional package 促销包装
promoter [promoter] 促进剂,活化剂,助催化剂,助聚剂
proof 耐,防,抗;检验,试验;不透……的
  ~ fabric 防水织物
  ~ load 保险负荷
  ~ pressure 保证压力
  ~ pressure test 耐压试验
  ~ resilience 保证弹力,弹性极限应力
  ~ sample 试样,样品
  ~ strength 保证强度,弹性极限强度
  ~ stress 保证应力,试验应力,弹性极限应力
  ~ test 验收试验,安全试验
proofing 防护;涂胶
propagating chain end 使链末端增长
propagation 传播;增长
  ~ crosslink (链)增长交联
  ~ of heat 热传播,热传导,传热
  ~ velocity 传播速度
propellant 发泡剂;膨胀剂
  ~ -filled melt 含发泡剂熔体
  ~ -type injection moulding 含发泡剂注塑法
propeller 螺旋桨,推进器
  ~ agitator 螺旋桨式搅拌器
  ~ mixer 螺旋桨式混合器
  ~ mixing element 螺旋桨式混合(器的)元件
  ~ stirrer 螺旋桨搅拌机
propene [propylene] 丙烯
  ~ plastics 丙烯类塑料
propenoic acid 丙烯酸
property 性质,性能;特性
  ~ change 性质的变化
  ~ test 性能试验
propionate 丙酸酯
propionic acid 丙酸
proportion 比例;配合
proportional 比例的
  ~ action controller [proportional position action controller] 比例(位置)作用控制器
  ~ counter 正比计数器

proportionating

- ~ error 相对误差
- ~ limit 比例极限
- ~ position action controller 比例位置作用控制器
- ~ sampling 比例取样
- ~ valve 比例阀,定量阀

proportionating 比例;配料
- ~ control 比例控制
- ~ pump 配料泵,比例泵

proportioning 配合〈按比例定量〉,配料;比例
- ~ and handling device 配料和输送装置
- ~ piston 配料活塞
- ~ pump 配料泵,比例泵
- ~ rotary piston device 旋转活塞配料装置

proprietary 专利的;所有的
- ~ articles 专利品
- ~ formula 专用配方,有专利权的配方
- ~ material 专用材料
- ~ name 商品名(称),专利商标名

propyl 丙基
- ~ acetate 乙酸(正)丙酯,醋酸(正)丙酯
- ~ trimethoxysilan,[PTMO] 丙基三甲氧基硅烷

propylene [propene] 丙烯
- ~ aldehyde 丙醛
- ~ -ethylene block copolymer 丙烯-乙烯嵌段共聚物
- ~ -ethylene random copolymer 丙烯-乙烯无规共聚物
- ~ glycol 丙二醇
- ~ glycol dibenzoate 二苯甲酸丙二醇酯〈增塑剂〉
- 1,2-~ glycol monolaurate 1,2-丙二醇单月桂酸酯〈增塑剂〉
- ~ glycol monoricinoleate 丙二醇单蓖麻醇酸酯〈增塑剂、湿润剂〉
- 1,2-~ glycol monostearate 1,2-丙二醇单硬脂酸酯〈增塑剂〉

- ~ oxide rubber 环氧丙烷橡胶
- ~ plastics 丙烯类塑料
- ~ product 丙烯类产品
- ~ resin 丙烯类树脂

protect 保护,防护
protected enclosure 防护罩壳,安全机壳
protecting 保护(的),防护(的)
- ~ band 保护带,安全带
- ~ cap 防护罩
- ~ casing 保护罩外壳
- ~ cover 防护盖
- ~ device 保护装置
- ~ film 防护膜
- ~ jacket 护套
- ~ piece 护板
- ~ sleeve 保护套
- ~ tube 防护管

protection 保护,防护
- ~ face 防护面,覆盖面
- ~ from corrosion 防腐蚀
- ~ test 防护试验

protective 保护的,防护的
- ~ agent 防护剂
- ~ boots 保护罩
- ~ cap 护帽,安全帽
- ~ case casing 保护罩,防护罩
- ~ clothing 防护衣
- ~ coat 护层
- ~ coating 保护涂料;保护涂层
- ~ colloid 保护胶体
- ~ colo(u)r 保护色
- ~ cover 安全罩,防护外壳;保护层
- ~ film 保护膜
- ~ glove 防护手套
- ~ grate [grid] 护栅,安全围栏〈注塑机〉
- ~ helmet 防护帽,安全帽
- ~ lacquer 防护漆
- ~ layer 防护层
- ~ measure 防护措施
- ~ package 防护性包装
- ~ painting 防护涂料

~ sheet 防护片材
~ surface coating 防护表面涂层;防腐蚀涂层
~ varnish 防护清漆
protector 保护器
protein 蛋白质
~ fibre 蛋白质纤维
~ plastic 蛋白(质)塑料
~ resin 蛋白(质)树脂
proteitiaceous 蛋白酶
prototype 模型;原模;样模;母模
~ mo(u)ld 试验模具;原型模具
~ protracted heating 长期加热
protrude 突出,凸起
protruding – type insert 突露型嵌件
prove 证明;检验
provisional 假定的,暂时的,临时的
~ patent specification 暂时专利说明书
~ specification 暂时说明书〈专利〉
PS [polystyrene] 聚苯乙烯
pseudo 〈词头〉假,伪;准
~ – affine deformation 准仿射形变
~ – compound 假化合物
~ – Newtonian flow 假牛顿流动
~ – program 伪程序,模程序
~ – reinforcement 假增强
~ – steady plastic flow 假稳定塑性流
~ – thermostatic 伪恒温的
~ – viscosity 假黏度
~ – viscosity behaviour 假黏度性能
pseudocomponent 假组分
pseudocrystal 假晶
psenaofinite strain 假有限应变
pseudomer 假异构件
pseudoplastic 假塑性(的)
~ flow 假塑性流动
~ fluid 假塑性流体
~ material 假塑性材料
pseudoplasticity 假塑性
~ index 假塑性指数
PSI [polymethylphenyl siloxane] 聚甲基苯基硅氧烷

psi [pounds per square inch] 磅每平方英寸
psig [negative gage pressure] 负表压
psig [pounds per square inch gauge] 磅每平方英寸表压
psycho rhcology 心理流变学
pt. [point] 点
PTDQ [polymerized trimethyl dihydroxyquinoline] 三甲基二氢化喹啉聚合体
PTFE [polytetrafluoroethylene] 聚四氟乙烯
PTMO [propyltrimetholysilan] 丙基三甲氧基硅烷
PTMT [polytetramethylene terephthalate] 聚对苯二甲酸丁二醇酯
PTP [polyteraphthalate] 聚对苯二甲酸酯
pts [parts] 份
PU [polyurea] 聚脲
PU [polyurethane] 聚氨酯,聚氨基甲酸酯;聚氨基甲酸乙酯
PU – cast elastoner [PU – cast elastomer] 聚氨酯铸塑弹性体
pub. [public] 公共的
public 公共的,公有的
~ cleansing 公共清洁
~ hazard 公害
~ nuisance 公害
puce 暗红色(的)
puckering 皱纹,皱褶
puff – up 膨胀
puffing agent 增稠剂,浓缩剂
putty 阵阵地喷(漆)
pug mill 捏和碾磨机
pull 拉,牵,拔
~ back 拉回
~ back bar 拉杆
~ back cylinder 回程液压缸,后拉压缸,后拉油缸
~ back cylinder type double – action press 后拉油缸式双动压机
~ back pin 后拉杆;注口冷料固定销

pulled surface

~ back ram　回程(液压)活塞
~ back system　后拉装置
~ back traverse　回程,返程
~ cord bag　用加捻绳捆扎的袋
~ direction　牵引方向
~ lever　拉杆
~ lid　拉拔盖〈包装〉
~ - off　扯开;拧开;撬开;断开
~ - off yarn over - end　轴向退绕
~ - off yarn tangentially　切向退绕
~ pin　脱模销
~ rod　拉杆
~ roll　牵引辊
~ roll stand　牵引辊架;卷绕装置〈挤塑物〉
~ strength　黏附强度,剥离强度〈黏合〉
~ strip　拉伸窄条,拉伸带
~ tape　拉伸带
~ test　拉伸试验
~ - up　脱层

pulled surface　皱裂面;起皱表面;粗糙表面;表面酥脆
puller　牵引机
pulley　滑轮;皮带轮;滚筒
~ guide　导向辊〈涂布机〉
pulling　拉,牵引
~ device　牵引装置
~ lever　拉杆
~ rate　牵引速率
~ speed　牵引率［速度］
~ test　拉伸试验
~ velocity　牵引速度
pulp　浆料,纸浆
~ board　纸浆板
~ colour　水色浆
~ engine　打浆机;(纸浆)捣碎机〈美〉
~ mould　浸胶纸浆成型模
~ moulding　浸胶纸浆模塑(成型)
~ slurry　液体浆
~ stock　浆液料
~ strainer　纸浆筛滤器

pulper　碎浆机
pulping process　打浆法
pulsatile pressure　压力波动
pulsating　脉动的,波动的
~ behaviour　脉动性能,膨胀性能
~ effect　脉动效应
~ fatigue strength　脉动疲劳强度〈屈挠疲劳试验〉
~ pressure　脉冲压力
pulsation　脉动,脉冲,波动
~ - diffusion method　脉动扩散法〈分子量测定〉
~ dryer　脉冲干燥器
~ of follow - up pressure　保压波动〈注塑〉
~ tension strength　脉冲拉伸强度
~ tension test　脉冲拉伸试验
~ test　脉冲试验;恒定周期性疲劳试验
pulsator　振动筛;脉动器
pulse　脉冲
~ column　脉冲柱
~ controlled　脉冲控制的
~ counter　脉冲计数器
~ frequency　脉冲频率
~ generator　脉冲发生器
~ injection　脉动注射
~ NMR method　脉冲核磁共振法
~ phase conveying　脉冲固相输送
pulsed　脉冲的
~ NMR - gradient method　脉冲核磁共振梯度法
~ voltage　脉冲电压
pulsing timer　脉冲计时器
pultrusion　挤拉成型,拉出成型
~ machine　挤拉成型机
pulverize　磨碎,磨(成)粉,粉碎;喷(成)雾〈将液体〉
pulverizer　粉磨机
~ mill　粉磨机
pulverizing　研磨,磨(成)粉;喷雾
~ equipment　粉碎设备

~ mill 粉磨机
~ mixer 粉碎拌搅机
pulverulent 粉状的
~ plastic 粉状塑料，粉状模塑料
pumice 浮石
pumicing 浮石抛光
pump 泵
~ line 泵吸管线
~ packing 泵填料
~ pressure [piston pressure] 泵压力，活塞压力
~ shifting 泵调速
~ speed 泵速，泵功率
pumping 泵；泵送，泵唧
~ capacity 挤塑量
~ efficency 挤塑效率
~ section 泵吸管断面
~ zone 挤塑段
punch 冲头；冲模，冲切，阳模；冲孔器
~ and die 有上下模的片材冲模
~ and die with guide pillars 装有导柱的冲模
~ pin 冲头
~ press 冲压机；冲床
– ratio 冲压比
~ retainer plate 阳模托板
punched 冲孔的，冲切的
~ card 冲孔卡（片）
~ felt [needle felt] 针刺毡
~ hole 冲孔
~ – out 冲切
~ piece 冲压件
~ plastic tape 穿孔塑料带
~ roll 纹板滚筒
punching 冲孔；冲制，冲切，冲模
~ die 冲压模，冲切模；冲孔模
~ hole diameter 冲孔直径
~ machine 冲压机，冲切机；冲孔机
~ pad 冲压垫
~ pelleter 冲压片机，冲压粒机
~ pelleting press 冲压成片机
~ press 冲压机；冲床

~ pressure 冲孔压力
~ quality 冲切性（能）
~ stock 冲切料
~ test 冲切试验，穿孔试验
~ tool 冲切工具，冲压模
puncture 穿孔，刺孔，扎孔；穿穿，刺透
~ penetration rate 穿透度
~ resistance 耐穿刺性
~ test 穿刺试验
puppet 坯条；坯料，滚压坯料
PUR [polyurethane] 聚氨酯，聚氨基甲酸酯；聚氨基甲酸乙酯
pur. purified 精制的
pure 纯的
~ colo(u)r 纯色
~ deformation 纯形变
~ research 纯研究
~ viscosity 纯黏度
purely elastic deformation 纯弹性形变
purge gas 吹扫气体〈激光切割塑料时〉
purging 清洗，吹洗；清机
~ compound 清机料
~ the cylinder 清洗料筒〈注塑〉；清洗机筒〈挤塑〉
purification 提纯，纯化；净化，精制
~ method 提纯方法，净化方法
purified gas 净化气
purifying tank 净化槽
puring vat 纯化槽
purity 纯度，纯净
~ degree 纯度〈化学药品〉；清洁度〈连接件表面〉
~ grade 纯度（等）级
purple 红紫色
purpose 目的；用途
purree 印度黄
push 推，搡
~ – back 复位；回程
~ – back cylinder type double – action press 回程油缸式双动压机
~ – back pins 开模杆，顶回杆，复位销

~ - back ram 回程活塞〈压机〉
~ button 按钮
~ button control 按钮控制,按钮操纵
~ button operation 按钮操作
~ button plate 按钮板
~ button switch 按钮开关
~ - out ram 推料柱塞
~ - out ram accumulator 推料柱塞式储料器
~ - but ram unit 推断柱塞机构
~ - pull oscillator 推挽式振荡器
~ - up 向上推;容器底达稳定〈瓶底内凹〉
pushing face of flight 螺槽推进面
putrefaction resistance 耐腐烂性
put up 卷装;包装〈成品〉
putty 油灰,腻子
~ chaser 探腻子碾
PVA [polyvinyl acetate] 聚醋酸乙烯酯,聚乙酸乙烯酯
PVA [polyvinyl alcohol] 聚乙烯醇
PVAC [polyvinyl acetate] 聚醋酸乙烯酯,聚乙酸乙烯酯
PVAL [polyvinyl alcohol] 聚乙烯醇〈分散剂〉
PVB [polyvinyl butyral] 聚乙烯醇缩丁醛
PVC [pigment volume concentration] 颜料体积浓度〈漆〉
PVC [polyvinyl chloride] 聚氯乙烯
~ integral door 聚氯乙烯组装门;聚氯乙烯全塑门
~ modified phenolic moulding powder 聚氯乙烯改性酚醛模塑粉
~ system macroreticular weakly basic anion exchange resin 大网状弱碱性聚氯乙烯型阴离子交换树脂
PVC/ABS [polyvinyl chloride/acrylonitrile - butadiene - styrene teipolymer ble - nds] 聚氯乙烯与丙烯腈 - 丁二烯 - 苯乙烯三元共聚物的共混物,聚氯乙烯 - ABS 共混物
PVC/ACR [polyvinyl chloride/acrylate rubber blends] 聚氯乙烯与丙烯酸酯类橡胶的共混物
PVC/CPE [polyvinyl chloride/chlorinated polyethylene blends] 聚氯乙烯 - 氯化聚乙烯的共混物
PVC/EVA [polyvinyl chloride/ethylene - vinyl acetate copolymer blends] 聚氯乙烯与乙烯 - 乙酸乙烯酯共聚物的共混物
PVC/MBS [polyvinyl chloride/methyl methacrylate - butadiene - styrene terpolymer blends] 聚氯乙烯与甲基丙烯酸甲酯 - 丁二烯 - 苯乙烯三元共聚物的共混物
PVC/NBR [polyvinyl chloride/NBR blends] 聚氯乙烯 - 丁腈橡胶共混物〈即丁腈橡胶改性聚氯乙烯〉
PVCA [polyvinyl chloride - acetate] 氯乙烯 - 乙酸乙烯酯共聚物
PVCC [chlorinated polyvinyl chloride] 氯化聚氯乙烯
PVD [polyvinyl dichloride] 聚二氯乙烯
PVDC [polyvinylidene chloride] 聚偏(二)氯乙烯
PVDF [polyvinylidene fluoride] 聚偏(二)氟乙烯
PVF [polyvinyl fluoride] 聚氟乙烯〈涂层,脱模剂〉
PVFM [polyvinyl formal] 聚乙烯醇缩甲醛〈现用〉
PVFO [polyvinyl formal] 聚乙烯醇缩甲醛〈过去用〉
PVI [polyvinyl isobutyl ether] 聚乙烯(基)异丁(基)醚〈黏合剂,增塑剂,添加剂〉
PVIE [polyvinyl isobutyl ether] 聚乙烯异丁醚
PVK [polyvinyl carbazole] 聚乙烯咔唑
PVL [pivalolaone] 新戊内酯
PVM [polyvinyl chloride - vinyl methyl ether] 聚氯乙烯 - 甲基乙烯基醚

PVME [polyvinyl methyl ether] 聚乙烯甲醚

PVMK [polyvinyl methyl ketone] 聚乙烯基甲基酮

PVMMA, PVM/MA [polyvinyl methyl ether/maleic anhydride] 聚乙烯甲醚-马来酐

PVOH [polyvinyl alcohol] 聚乙烯醇〈现用 PVAL〉

PVP [polyvinyl pyrrolidone] 聚乙烯吡咯烷酮

PVPr [polyvinylpropionate] 聚乙烯丙酸酯

p-v-t diagram [pressure-specific volume-temperature diagram] 压力-比容-温度图〈测定热塑性塑料收缩率〉

pyenometer [pyknometer] 比重瓶,比重计

pyramidal 棱锥形的;角锥形的
~ clement 角锥形元件

pyranyl 吡喃基
~ resin 吡喃基树脂

pyrazolone 吡唑啉酮

pyrex glass 派热其斯耐热玻璃

pyridine 吡啶
~ dyes 吡啶染料,氮苯染料
~ -zinc chloride complex 吡啶-氯化锌络合物〈活性剂 711〉

pyrogen 热原质;〈大写〉焦精〈硫化染料商品名〉

pyrogenic 热解的
~ decomposition 高温分解
~ silica 热解法氧化硅〈增稠剂〉

pyrogram 热解色谱图

pyrolysis 热解(作用),高温分解〈塑料受热而降解〉
~ capillary gas chromatography 热解毛细管气体色谱法
~ -molecular-weight-chromatography 热解分子量色谱法

pyrolytic 热解的
~ degradation 热降解
~ gas chromatography [PGC] 热解气相色谱法
~ polymer 热聚物

pyromellitic 苯四甲酸
~ dianhydride [PMA] 均苯四甲酸二酐〈固化剂〉

pyrometer 高温计

pyronine dyes 焦宁染料

〈含 =R〉$\begin{matrix}O\\C\end{matrix}$ R=基团的染料〉

Q

Q [quantity] 量,数量

QC [quality control] 质量控制,质量管理,质量检查

quadripolymer 四元聚合物

qual. [qualitative] 定性的;质量的

qualification test 质量鉴定试验,合格试验;粘接性质量(检测)试验
~ specification 质量鉴定试验规范

qualitative 定性的;质量的
~ analysis 定性分析
~ assessment 质量评定
~ data 质量数据
~ test 定性试验

quality 质量,品质
~ assurance 质量保证
~ classification 品质分级
~ coefficient 质量特性,质量系数
~ control [QC] 质量控制,质量管理,质量检查
~ control data 质量控制数据

quant.
- ~ control standards 质量检验标准,质量管理标准
- ~ control testing 质量控制试验
- ~ factor 品质因数〈焊接或粘接〉
- ~ grade 质量等级
- ~ inspection 质量检验
- ~ level 质量水平,质量等级
- ~ of aspect 外观质量
- ~ of mixing 混合的质量〈物料〉
- ~ of weld 焊接质量
- ~ requirements 质量要求,质量规格
- ~ specification 品质规格,质量标准,技术规范,质量说明书
- ~ standard 质量标准
- ~ test 质量试验,质量检验

quant. [quantitative] 定量的,数量的

quantitative 定量的,数量的
- ~ analysis 定量分析
- ~ data 数量数据
- ~ differential thermal analysis 定量差热分析
- ~ transformation 量变

quantity 数量;参数
- ~ applied 应用量
- ~ of heat 热量
- ~ of light 光量
- ~ of state 状态参数
- ~ production 大量生产,成批生产

quartered wood 四开木材

quartering 割料;四分取样法

quartz 石英,二氧化硅
- ~ crystal pressure recorder 石英晶体压力记录器
- ~ fibre 石英纤维
- ~ fibre-reinforced phenolics 石英纤维增强酚醛树脂
- ~ sand 石英砂

quasi 〈词头〉准;似;半
- ~ -adiabatic extrusion 准绝热挤塑
- ~ -brittle fracture 准脆性断裂
- ~ -crosslink 准交联
- ~ -crystalline lattice 似晶格
- ~ -isotropic fiber reinforced plastics 准各向同性纤维增强塑料
- ~ -prepolymer process 半预聚物法
- ~ -single-strand chain 准单股链
- ~ -static loading 准静态负荷
- ~ -viscous creep 似黏性蠕变

quater polymer 四元聚合物

quaternary 四元的
- ~ ammonium ion exchange resin 季铵型离子交换树脂
- ~ exchange resin 四组分离子交换树脂

quaterpolymer 四元聚合物

quench 骤冷,淬火
- ~ bath 骤冷浴,淬火浴
- ~ tank 骤冷槽
- ~ tank extrusion 槽式骤冷挤塑

quencher 骤冷器,淬火器;猝灭剂

quenching 骤冷,淬火
- ~ bath 骤冷槽,骤冷浴
- ~ effect 骤冷效应
- ~ medium 骤冷剂
- ~ stress 骤冷应力,淬火应力
- ~ tank 骤冷槽〈挤塑〉
- ~ temperature 骤冷温度,淬火温度
- ~ water 骤冷水

quick 快速的
- ~ -acting 快速作用(的),速动的
- ~ ageing 快速老化〈塑料材料〉
- ~ ash 烟灰,飞扬的灰尘
- ~ -change pick-off gear 快速变换传动齿轮
- ~ check 快速试验,快速检验
- ~ -cleaning extruder 可快速清理的挤塑机〈螺杆和料筒〉
- ~ coke 焦炭,焦煤
- ~ -curing 快速固化的
- ~ -curing moulding compound 快速固化模塑料;快速固化成型料
- ~ deep freezing 快速深度冷冻
- ~ -drying coating 快干涂料
- ~ -drying primer 快干底漆
- ~ motion test 加速试验〈材料〉
- ~ setting adhesive 快速固化黏合剂
- ~ speed 快速,高速
- ~ -stick 快黏

~ – stick test 快黏试验
~ test 快速试验
quicklime 生石灰
quiescent fluidized bed 平稳流化床
quieter material 消声材料,消音材料
quilted 绗缝的
~ mat 绗缝毡〈美〉
quilting 绗缝
~ machine 绗缝机
quinacridone red 喹吖啶酮红,酞菁红〈着色剂〉
quinaldine red 喹哪啶红
quinoline 喹啉

~ dyes 喹啉染料
quinone (苯)醌
p – ~ **dioxime** 对(苯)醌二肟〈硫化剂〉
~ pigments 醌颜料
quinonimine dyes 醌亚胺染料
quinoxaline 喹噁啉,对二氮萘
~ colo(u)ring matters 喹噁啉染料
~ – phenylquinoxaline copolymer 喹噁啉–苯基喹噁啉共聚物
~ plastic 喹噁啉塑料
quoin 楔子,模楔
quota 份额,定额
q. v. [quod vide〈拉〉] 参照,参考

R

R;r [radius] 半径
R [right] 右;正确
rabbling 搅拌,耙动
~ mechanism 搅拌机构;耙式结构〈多层干燥器〉
race(rotation) 空转
rack 齿条;框架;导轨
~ and pinion 齿条齿轮传动
~ drive 齿条传动
~ packaging 机架包装
radial 径向的;辐射的;半径的;光线的
~ clearance (螺杆)径向间隙
~ cracking 径向裂纹
~ crown 径向中高度
~ deflecting louvres 径向导向隔板
~ direction 径向
~ distortion 径向畸变
~ distribution 径向分布
~ fill – out 径向充模
~ flow impeller 径向流动叶轮
~ flow mixer 径向流动混合器
~ impeller 径向叶轮
~ installation 径向安装
~ piston [plunger] pump 径向活塞泵
~ removal 径向移动

~ screw clearance 径向螺杆间隙
~ shrinkage 径向收缩
~ stiffness 径向劲度
~ strain 径向应变
~ stress 径向应力
~ thrust 径向推力
~ wear 径向磨损
radiant 辐射的,放射的
~ curing process [RCP] 辐射固化法
~ diyer 辐射干燥器
~ energy 辐射能
~ heat 辐射热
~ heat seal 辐射热合
~ heat welding 辐射热焊接
~ heater 辐射加热器
~ intensity 辐射强度
radiation 辐射;放射
~ boiler 辐射锅炉
~ – chemical determination of short – chain branching 辐射化学短链支化测定
~ – chemically chlorinated polyvinyl – chloride 辐射化学氯化聚氯乙烯
~ – cross – linking 辐射交联〈塑料〉
~ curing, [RC] 辐射固化

~ damage 辐射损伤〈塑料材料〉
~ degradation 辐射降解
~ dosimetry 辐射剂量测定法
~ dryer 辐射干燥器
~ - grafted 辐射接枝的
~ grafting 辐射接枝〈热塑性塑料表面〉
~ heating 辐射加热
~ induced 辐射诱导的
~ - induced cross - linking 辖射诱导交联
~ - induced grafting 辐射诱导接枝
~ - induced polymerization 辐射诱导聚合
~ - initiated 辐射引发的
~ - initiated polymerization 辐射引发聚合
~ - initiated polymerization of ethylene 辐射引发乙烯聚合
~ intensity 辐射强度
~ ion polymerization 辐射离子聚合(作用)
~ meter 辐射测量仪
~ polymerization 辐射聚合
~ preserved food 辐射防腐食品
~ processing 放射加工,辐射加工
~ resistance 耐辐射性
~ - shielding acrylate sheet 防射线有机玻璃
~ temperature 辐射温度〈红外焊接时〉
~ treatment 辐射处理
radiator 辐射器,辐射体;散热器,暖气片;冷却器,水箱〈汽车〉
radical 基,根;自由基,游离基;原子团
~ absorber 自由基吸收剂〈反应性树脂〉
~ chain reaction 自由基链反应
~ copolymerization 自由基共聚合,游离基共聚合
~ polymerization 自由基(引发)聚合,游离基(引发)聚合
~ polymerization initiator 自由基聚合引发剂

~ transfer 自由基转移
radioactive 放射性的
~ decay 放射性衰变
~ detector 放射性检测器
~ disintegration 放射性脱变
~ dust 放射性微尘
~ static eliminator 放射性静电消除器
~ - tracer - fibre 放射性示踪纤维
radio engineering 元线电工程
radiofrequency 射频,高频
~ curing 射频固化
~ drying 射频烘干[干燥],高频烘干[干燥]
~ gluing 射频胶黏
~ heating [RF - heating] 射频加热,高频加热
~ mo(u)lding [RF - mo(u)lding] 射频(电感加热)成型
~ preheating [RF - preheating] 射频预热,高频预热
~ sealer 射频热合机
~ - tarpaulin welder [RF - tarpaulin welder] 高频帆布焊接机
~ welding [RF - welding] 射频焊接,高频焊接
radiographic inspection 射线照相检查
radiography 射线照相法〈用于塑料结构的研究〉
radiolucency 射线可透过,X射线阻碍
radiometer 放射科,辐射计
radiopacity 辐射不透明度
radiopaque 不透射线的
radium emanation [RaEm] 镭射气,氡
radius 半径
~ of bend 弯曲半径
~ of fillet 倒圆半径
~ of gyration 旋转半径
radome 雷达天线罩,天线屏蔽器
RaEm [radium emanation] 镭射气,氡
rail 导轨铁路
~ shipment 铁路装运
railroad shipment 铁路装运
rain 雨(水)
~ gutter 雨水口,排水沟

~ gutter pipe 雨水管,雨水落下管
~ wear 雨衣
rainbow dyeing 虹彩染色
rainwater head 雨水排水口
raising 厚度增高〈涂层〉
rake 倾斜
raking mechanism 耙式结构〈多层干燥器〉
RAM [restricted area mo(u)lding] 窄面模塑
ram 活塞,柱塞;模塞,阳模;射料杆
~ accumulator 柱塞式储料缸
~ area 活塞面积
~ extruder 柱塞挤塑(出)机,液压挤塑(出)机
~ extrusion 柱塞挤塑(出);挤塑烧结〈氟塑料〉
~ flow 挂塞流〈通过柱塞压力造成的二维流动〉
~ force 活塞推力
~ injection mo(u)lding 柱塞式注塑成型
~ injection mo(u)lding technique 柱塞式注塑成型法
~ motion time 柱塞行程时间
~ preplasticator 柱塞式预塑化器
~ press 柱塞式压机;冲压机
~ pressure 柱塞压力,射料杆压力
~ pump 柱塞泵
~ retraction time 活塞返回时间〈注塑〉
~ acrew 柱塞螺杆,往复螺杆〈吹塑〉
~ screw accumulator 柱塞式螺杆储料器,往复式螺杆储料器〈吹塑机〉
~ screw accumulator preplasticator 柱塞螺杆储料式预塑化机
~ screw accumulator transfer plasticizing 柱塞螺杆储料式挤压塑化
~ [piston] travel 柱塞行程,射料杆行程
~ travel time 柱塞行程时间
~ type continuous extruder 柱塞式连续挤塑机
~ -type preplasticizing aggregate 柱塞式预塑化机组
~ weight 柱塞杆重量
Raman spectrum 喇曼光谱
ramie 苎麻,青麻
rammer 冲头;冲压;推料器
rand 卡圈,边,缘;沿条;贴边料
randall-fenton joint 二次胶接件〈榫槽连接〉
random 无规的;任意的
~ arrangment 无规排列
~ chain 无规链
~ chain scission 链无规断裂
~ chemo-mechanical fibrillation 无规化学机械原纤维化
~ colo(u)r 任意色
~ copolymer 无序共聚物,无规共聚物
~ copolymerization 无规共聚(作用)
~ crack 不规则裂缝
~ crosslinking 无规交联
~ degradation 无规降解
~ distribution 无规分布
~ drafts 无规牵伸,任意牵伸
~ lamination 无规层合;无序层合
~ linear copolymer 无规线型共聚物
~ mechanical fibrillation 无规机械原纤维化
~ noise program 无规噪声程序
~ orientation 无规取向
~ packing 无规填充
~ pattern 无规晶格;无规排布;无规图案
~ polymer 无规聚合物
~ program 无规则程序
~ sample 抽样,任意取样
~ structure 无规结构〈聚合物〉
randomly coiled chain 无规旋卷链
Raney nickel 雷尼镍〈催化剂〉
range 范围
~ inspection 抽查
~ of adjustment 调节范围
~ of application 应用范围
~ of products 产品种类
~ of shades 色泽分布范围
~ of sizes 尺寸范围

rapid

~ of strain 应变范围
~ of stress 应力范围
~ of temperatures 温度范围
ofuse 应用范围,使用范围
~ of work 加工范围

rapid 快速的
~ ageing 快速老化
~ cycle mo(u)lding [RCM] 快速(周期)模塑
~ - mounting 快速装配
~ polymerization 快速聚合
~ travel 快移
~ (power)traverse 快移,高速运动

rapidogcn dyes pl 快速着色剂

rare 稀有的,稀少的
~ earth naphthenate 环烷酸稀土(金属)〈催化剂〉
~ mixture 稀有混合物

Raschig ring 拉西环〈蒸馏〉

raspberry 木莓色(暗色),紫绛色

ratchet 棘轮;棘齿
~ tooth dicer 棘齿刀式切粒机

rate (速)率,速度
~ of absorption 吸收率
~ - of - blow controller 吹胀比控制装置
~ of burning 燃烧速率
~ of burning test 燃烧速率试验
~ of consumption 消耗率
~ of cooling 冷却速率
~ of creep 蠕变速率
~ of crystallization 结晶速率
~ of cure 固化速率
~ of decay 衰变速率,衰减速率
~ of deflection 挠曲速率
~ of deformation 形变速率
~ of discharge 出料速度
- **of dyeing** 上色率
~ of elongation 伸长速率
~ of extension 拉伸速率;伸长率;延伸率
~ of extrusion 挤塑速率
~ of feed 供料速率
~ of finished products 成品率
~ of flow 流量;流率
~ of hardening 硬化率
~ of heat transfer 传热率;传热系数
~ of impulse 脉冲速率
~ of injection 注射速率
~ of loading 负荷速率[度]
~ of oxidation 氧化速率
~ of polymerization 聚合速率[度]
~ of production 生产率
~ of reaction 反应速率,反应速度
~ of scintillation 闪烁速率
~ of shear 剪切速率
~ of slip 滑动速率[度]
~ of stirring 搅拌速率
~ of strain 应变速率
~ of stressing 应力变化速率
~ of temperature 温度上升速度
~ of vapour deposition 蒸镀速率[度]
~ of vulcanization 硫化速率
~ of wear 磨耗率

rated 额定的,标称的
~ capacity 额定能力
~ consumption 额定消耗量
~ life 额定寿命
~ load 额定载荷
~ operating pressure 额定工作压力
~ output 额定产量;额定输出;额定功率
~ power 额定功率,设计功率
~ size 额定尺寸
~ value 额定值

rating 额定(值)
~ plate 额定铭牌,定额牌,标牌

ratings pl 发动机功率

ratio 比(率)
~ bar 计量杆
~ of components 成分比
~ of compression 压缩比
~ of drawing 牵伸比;拉伸比
~ of elongation 伸长比
~ of flow path and thickness 流程与厚度比
~ of flow path to wall thickness 流程与

reactivation

　壁厚之比〈成型制件〉
~ of mixture　混合比
~ of rigidity　刚性化
~ of the fibre length to the diameter　纤维长径比
~ of viscosity　黏度比
raw　生的;原始的;粗的
~ casting　粗铸塑件
~ cotton　原棉
~ data　原始数据,原始资料
~ fabric　本色织物
~ feed　新进料,原料
~ fibres　纤维原料
~ gas　粗气
~ material　原料,原材料
~ oil　原料油
~ polymer　原始聚合物
~ rubber　生橡胶
~ seam　粗缝
~ stock　原料
ray　射线;放射,辐射
　α-~ fluorescent spetroscopy　α-射线荧光光谱
　γ-~ irradiation polymerization　γ-射线照射聚合(作用)
~ of light　光线
~ seal　辐射热合
~ sealing　辐射热合
Rayleigh scattering　雷利散射
Raymond bowl mill　雷蒙特球磨球
rayon　人造纤维,人造丝
~ cut staple　人造短纤维
RC［radiation curing］　辐射固化
RCM［rapid cycle mo(u)lding］　快速(周期)模塑
RCP［radiant curing process］　辐射固化法〈聚合物〉
RD［diffuse reflectance］　漫反射
rd.［round］　圆周;圆的
rc［referring to］　涉及(到),关系到,归功于
Re［Reynold's number］　雷诺数
Re［rhenium］　铼
Re number［Reynold's number］　雷诺数

react　反应;(反)作用
reactant　反应物;反应体
~ resin　活性树脂
reacting　反应的
~ capacity　反应能力
~ substance　反应物
~ weight　反应量
reaction　反应;(反)作用
~ capacity　反应能力
~ casting　反应铸塑
~ cement　活性粘接剂
~ control　反应控制
~ foam casting　反应发泡铸塑
~ foam rao(u)lding　反应发泡成型法
~ foaming　反应发泡铸塑法
~ heat　反应热〈反应性树脂〉
~ injection mo(u)lding［RIM］　反应(发泡)注塑
~ injetion mo(u)lding technique［RIM technique］　反应(发泡)注塑(成型)法
~ intermedicate　反应中间体
~ mechanism　反应机理
~ mixture　反应混合物
~ inoulded piece　反应模塑件
~ mo(u)lding　反应模塑,反应成型
~ mo(u)lding technique　反应成型法
~ product　反应产物
~ product of diphenylamine and diisobutylene　二苯胺和二异丁烯的反应产物〈抗氧剂〉
~ promoter　反应促进剂
~ rate　反应速度
~ resin　反应性树脂
~ sensitive adhesive　反应性黏合剂
~ time　反应时间
~ to fire　燃烧反应
~ -type phenolic foam　反应型酚醛泡沫塑料
~ velocity　反应速度
~ vessel　反应(容)器
~ zone　反应区
reactionless　无反应的,惰性的
reactivation　再活化,再生

reactive

~ by induction heating 感应加热再活化
~ by infrared heating 红外加热再活化
~ of catalyst 催化剂再生
~ of the film of adhesive 胶黏膜的再活化

reactive 反应的;活性的
~ adhesive 反应性黏合剂
~ curing agent 反应性固化剂
~ diluent 反应性稀释剂
~ dye 活性染料
~ disperse dyes 活性分散染料
~ fibre 反应性纤维
~ filler 活性填料
~ flame retardant 反应性阻燃剂
~ intermediate 活性中间体
~ monomer 反应性单体,活性单体
~ pigment 活性颜料
~ plasticizer 反应性增塑剂
~ polymer 反应性聚合物,活性聚合物
~ polystyrene 反应性聚苯乙烯
~ resin 活性树脂,反应性树脂
~ thinner 活性稀释剂;反应性稀释剂
~ time 反应时间〈聚氨酯泡沫塑料反应注塑时〉

reactivity 反应性,活(化)性
~ of monomer 单体反应性
~ ratio 竞聚率;反应率,活性率

reactor 反应器;反应剂
~ bottom 反应器底

reading 读数
~ microscope 读数显微镜

readout 读出;数字显示装置

ready 有准备的
~ for operation 准备运转
~ for service 操作准备
~ for use 备用的
~ – mixed paint 调和漆

reagent 试剂;试液
~ resistance 耐试剂性,耐化学药品性

real 真实的;有效的
~ density 真实密度
~ power 有效功率

~ surface 作用面〈粘接件〉
realignment （分子）重排列,重组合
ream 雾状条纹〈透明塑料〉;令〈纸张计量单位,=500张纸〉;胀口〈管〉
rear 后部;后面(的)
~ apron 后挡板
~ barrel temperature 机筒后部温度
~ bottom radius 后螺底半径
~ bumper 后保险杠〈汽车〉
~ cavity·plate 后模槽板
~ clamping plate 后锁模板
~ end plate 后端盖板
~ face of flight 后螺面
~ heat zone 后加热段
~ panel 后挡板
~ shoe 后模板〈注塑模具〉
~ valving 后部排气
~ view 后视图

re-arrangement 重排列〈分子〉
re–assembly 重新装配
re–boiler 重沸器,再煮器

rebound 回弹
~ characteristic 回弹特性
~ degree 回弹率
~ elasticity 回弹性
~ hardness 回弹硬度
~ performance 回弹性能
~ resilience 回弹性
~ resiliometer 回弹性测定计
~ strain 回弹应变
~ stroke 回弹行程
~ test 回弹试验
~ travel 回弹行程

receiver 接收器;容器
receiving 接收,接受
~ bin 接收器;接料仓
~ inspection 验收检验
~ tray 承接盘
~ vessel 接料器,收集盘

reception 接收
~ test 验收试验

receptive 易接收的
receptivity 接收性能;吸收性
receptor 接收器

recess 切口;凹口,凹穴;嵌入
recessed 凹穴的
- ~ mould 凹框模
- ~ panel 贴标签凹面板
- ~ sleeve 凹槽套筒

recipe 配方
- ~ of mix 混料配方

reciprocal 相互的,可逆的,倒数
reciprocating 往复的
- ~ action 往复动作
- ~ brush 往得式电刷,往复式涂刷
- ~ feeder 往复式加料器
- ~ hot wire 往复式热丝
- ~ hot wire sealing 往复式热丝封合
- ~ hot wire welding 往复式热丝焊接
- ~ mixing head 摆动式混合头〈反应系统〉
- ~ motion 往复运动
- ~ pump 往复泵
- ~ screw 往复式螺杆
- ~ screw accumulator 往复式螺杆储料器
- ~ screw injection machine 往复式螺杆注塑机
- ~ screw injection mo(u)lding 往复式螺杆注塑成型
- ~ screw injection mo(u)lding machine 往复式螺杆注塑成型机
- ~ screw press 往复式螺杆注塑机
- ~ screw unit 往复式螺杆机组
- ~ single-screw machine 往复式单螺杆成型机
- ~ table 往复平移式工作台

reciprocction feeder 往复式加料器
recirculating (再)循环
- ~ air 循环空气
- ~ air dryer 循环空气干燥器
- ~ pump 循环泵

reclaim 回收;再生;再生料
- ~ mill 精研;精磨机

reclaimed 再生的
- ~ fibre 再生纤维
- ~ plastics 再生塑料
- ~ rubber 再生橡胶

reclaimer 回收设备
reclaiming 回收;再生
- ~ by plastification 再塑化
- ~ process 再生工艺
- ~ scrap 回收废料

reclosable package 可再密封包装
reposing (模具)再闭合
recoating 再涂
- ~ machine 罩漆机,再涂机

recoUer 卷取机;绕机;重卷机
recombination 复合,再化合
- ~ rate 复合率

recommendation 推荐,建议
- ~ standard 推荐标准

recompression mo(u)lding test 再压缩成型试验
recondition 再调节;修理;回收
record 记录,录音;唱片
- ~ compound 唱片料〈聚氯乙烯〉
- ~ press 唱片压制机

recorder 记录器;录音机
recording 记录,录音;唱片,录音磁带
- ~ disc 录音盘
- ~ film 录音胶片
- ~ flow controller 记录式流量调节器
- ~ hygrometer 记录式温度计
- ~ impact-torsion test 记录式冲击扭曲试验
- ~ microphotometer 记录式显微光度计
- ~ spectrophotometer 自动记录分光光度计
- ~ tape 录音带
- ~ thermometer 记录式温度计,温度记录仪

recover 恢复;再生,利用〈废料〉
recoverability 回复性能
recoverable 可恢复的,可回收的
- ~ extension 可复伸长,弹性伸长
- ~ strain 可恢复应变;可塑变形

recovered 回收的
- ~ fibre 回收纤维
- ~ plastic 回收塑料
- ~ solvent 回收溶剂

recovery 恢复,复原;回收,回复;再生;

recrystallization

利用〈废料〉
- ~ capacity 恢复量;回收能力
- ~ characteristic 回复特性
- ~ creep 回复蠕变〈塑料材料〉
- ~ curve 弹性回复曲线
- ~ index 回复指数
- ~ of colo(u)r 色泽复原
- ~ of elasticity 弹性复原
- ~ of plastic waste 塑料废料的回收
- ~ of shape 形状回复〈塑料滞后弹性形变〉
- ~ percent 回收(百分)率
- ~ processing 回收加工
- ~ ratio 回收率
- ~ temperature 回复温度
- ~ test 恢复(性能)试验
- ~ time 回复时间〈塑料材料的滞后-回弹变形〉

recrystallization 再结晶(作用)
recrystd. [recrystallized] 再结晶的
rectangular 长方形的,矩形的
- ~ bar [horn, sonotrode] 矩形电极;矩形发生器〈用于超声波焊接〉
- ~ distribution 均匀分布
- ~ runner 矩形流道
- ~ tab gating 矩形柄形浇口

rectifier 整流器
rectilinear polymer 直线型聚合物
recuperability 可回收性
recuperate 恢复;再生,回收
recuperation 恢复;再生,回收
- ~ tank 回收容器,收集槽

recurring 重复;再熟化
- ~ unit 重复单元,重复链节

recycle (再)循环;回收
- ~ gas 循环气体
- ~ monomer 循环单体
- ~ of air 循环空气
- ~ stock 再循环物料;回收料

recycled plastic 回收塑料
recycling (再)循环;回收,再利用〈生产过程中的废塑料等〉
red 红色(的);赤热的
- ~ brittleness 热脆性
- ~ iron oxide 氧化铁红,铁红,铁丹
- ~ lead 铅丹,红丹,四氧化三铅
- ~ mud filled polyvinyl chloride plas-tics 赤泥填充聚氯乙烯塑料
- ~ shortness 热脆性

reddish 微红的
red flg. [reducing flange] 异径法兰
redox [reduction-oxidation] 氧化还原(作用)
- ~ cawlyst 氧化还原催化剂
- ~ initiation 氧化还原引发(作用)
- ~ initiator 氧化还原引发剂
- ~ polymer 氧化还原聚合物
- ~ polymerization 氧化还原聚合
- ~ reaction 氧化还原反应
- ~ resin 氧化还原树脂

redress 矫正,调整;弹性复原
reduce 减少,降低;减径,缩径;还原,对比
reduced 还原的;减低的,降低的;缩减的;对比的
- ~ colo(u)r 还原色;还原染料
- ~ factor of stress concentration 降低应力集中的因数
- ~ gear [reduction geat] 减速齿轮
- ~ in scale 缩小比例
- ~ inherent viscosity 比浓对数黏度
- ~ osmotic pressure 比浓渗透压
- ~ pigment 还原颜料
- ~ product 还原产物
- ~ temperature 对比温度
- ~ transparency 对比透明度
- ~ viscosity 比浓黏度

reducer 还原剂;还原器;异径管;减速器;减压阀;退黏剂
- ~ assembly 异径管配件

reducing 还原的;减低的,降低的;缩减的,渐缩的;异径的
- ~ adaptor socket 异径接套
- ~ agent 还原剂
- ~ blish 异径接套
- ~ coupling 异径连接〈塑料管线〉
- ~ elbow 异径弯头
- ~ fitting 异径管件

~ flange 异径法兰
~ gears 减速齿轮;减速器
~ nipple 异径螺纹接套〈管〉
~ pipe 渐缩管
~ power 还原能力
~ power of a white pigment 白颜料掩盖力
~ socket 异径套节〈管〉
~ Tec 异径 T 形件
reductant 还原剂
reduction 还原;减低,降低;缩径;粉碎
~ accelerator 还原促进剂
~ activator 还原活化剂
~ gear(ing) 减速齿轮;减速器
~ of area 断面收缩率
~ of rigidity 刚性降低
~ - oxidation [redox] 还原氧化
~ piece 异径件
redective 减少的;还原的;还原剂
Redwood viscometer 雷德伍德黏度计
reefer car [refrigerator car] 冷藏车
reel 卷轴,卷筒;卷取
~ drum 卷筒
~ of yarn 纱线卷轴
~ off 退卷
~ - off machine 松卷机,退卷机
~ spinning 漏斗法纺丝
~ up 缠绕;卷取
reelability 可绕性
reeled film 卷绕的薄膜;膜卷
reeler 卷取机
reeling machine 卷取机;绕丝机
reentrant 凹入(的),凹腔(的)
~ base of a bottle 凹状瓶底
~ mo(u)ld 凹腔模具
re - entry heating 重返大气层加热
re - evaporating 再蒸发
re - evaporation 再蒸发
ref. [reference] 参考;参考文献
ref. temp. [reference temperature] 参考温度
reference 参考(文献);参照;基准,标准
~ atmosphere 基准环境
~ book 参考书,手册

~ condition 参考条件,基准条件
~ data 参考数据
~ marks 参考标记
~ material 参考材料
~ operator 定值器;调节值控制器
~ sample 参考试样
~ standard 参考标准
~ substance 参照物
~ temperature 基准温度
~ test block 标准试样
~ time 基准时间
refill 再装填
refine 精制;提纯
refined weld zone 焊缝过渡区〈塑料焊接〉
refinement 精制;提纯;改进
~ in technique 技术改进
refiner 精研机
refining 精制;提纯;改进
~ of plastics 塑料件修整
reflect 反射
reflectance 反射;反射度;反射系数
reflected light 反射光
reflecting 反射的
~ microscope 反光显微镜
~ power 反射能力
reflection [reflexion] 反射;反光
~ coefficient 反射系数
~ density [reflection (optical) density] 反射光密度
~ factor 反射系数
reflectivity 反射性;反射系数
reflector 反射器〈红外焊接〉
reflex reflection [retro - reflection] 折回,反射〈光学〉
refract 折射
refracted ray 折射线
refracting angle 折射角
refraction 折射(度)
~ constant 折射常数
~ index 折射率;折射指数
refractive 折射的
~ index 折射率;折射指数
~ index detector 折射指数检测器

refractory 耐火材料;耐火的;难熔的;耐熔的
- ~ fiber 耐火纤维
- ~ fibre reinforced plastic 耐火纤维增强塑料
- ~ lining 耐火衬里
- ~ material 耐火材料
- ~ oils pl 耐热油
- ~ plastics 耐火塑料
- ~ products 耐火产品

refrigerant 冷冻剂,致冷剂

refrigerating 制冷的,致冷的,冷冻的
- ~ exchanger 致冷交换器,冷却器
- ~ pipe 制冷管,冷却管

refrigerator 致冷器,(电)冰箱,冷柜,冷藏库,冷藏箱,
- ~ car 冷藏车
- ~ lining 冷柜衬里

refuse 废物,废料;下脚料;再熔化,再塑化
- ~ bag 废物袋,垃圾袋
- ~ plastics 废塑料
- ~ - polymer composite 废料-聚合物复合材料

regain 回收

regenerate 再生;再生料;回收

regenerated 再生的;回收的
- ~ bath 回收槽
- ~ cellulose 再生纤维素
- ~ energy 回收能(量);回弹,弹性复原〈泡沫塑料〉
- ~ fibre 再生纤维
- ~ plastic 再生塑料
- ~ polymer 再生聚合物
- ~ rubber 再生橡胶

regeneration 再生;回收
- ~ efficiency 再生效率
- ~ of ion exchange resin 离子交换树脂的再生

region 区域,范围

register 记录;定位,对准;注册
- ~ hole 定位孔
- ~ pin 定位销

registered 登记的;记录的;注册的
- ~ trade name 注册商品名称
- ~ trademark 注册商标

regranulate 再次造粒,二次造粒

regranulating 再次造粒〈塑料废料〉
- ~ line 再次造粒生产线

regrind 回用料,粉碎再生料;再次磨碎

regrinding 回用料;再次磨碎〈塑料废料〉;废料再生;再生料

reground material 回收料;再生料

regular 正规的,规整的,习惯的;照例的;
- ~ block 规整嵌段
- ~ checking 定期检查
- ~ convention formula 惯用配方
- ~ crosslinking 规整交联
- ~ folding 规整折叠
- ~ overhauling 定期(大)检修,定期大修
- ~ packing 规整填料
- ~ Polymer 规整聚合物,有规聚合物
- ~ production 正规生产
- ~ reflection [specular reflection] 镜面反射,正常反射〈光学〉
- ~ service conditions 正常使用条件
- ~ size 正规尺寸,标准尺寸
- ~ symbol 正规符号,正规记号,正规代号
- ~ transmission [direct transmission] 直接传送,正规传送〈光学〉

regularity 规整性

regulate 调整,调节;规整

regulated polymer 有规聚合物

regulating 调整的,调节的
- ~ unit 调整机构
- ~ valve 调节阀

regulation 调节,调整,控制;规整
- ~ of the tool temperature 模温调节
- ~ size 常规尺寸

regulator 调节剂;调节器

regulatory products [dangerous articles] 有危险的产品,受规章限制的产品

reinforce 增强,加强

reinforced 增强的,加强的
- ~ butt joint [weld] 加强的对接焊缝
- ~ composite 增强复合材料

~ conctete 增强混凝土
~ cord 增强帘子线
~ expanded plastic 增强的发泡塑料
~ filler 增强填料,补强填料
~ fillet weld 加强角焊缝
~ liquid reaction mo(u)lding [LRM-R] 增强的液体反应模塑
~ mat 增强玻璃纤维毡
~ material 增强材料
~ mo(u)lding compound 增强成型材料,增强模塑料
~ pipe [armed pipe] 增强管
~ plastics [RP] 增强塑料
~ plastics/composite [RP/C] 增强塑料/复合材料
~ plastics moulding 增强塑料模塑
~ plywood 增强胶合板
~ polystylene plastics [RFPS] 增强聚苯乙烯类塑料
~ reaction injection mo(u)lding [RRIM] 增强反应注塑成型,增强反应注塑模塑
~ reaction mo(u)lding 增强反应成型
~ rib 加强肋
~ rubber 增强橡胶
~ structural foam 增强结构泡沫材料
~ thermoplastics [RTP] 增强热塑性塑料

reinforcement 增强材料,补强材料
~ bandwidth 增强(材料)带宽
~ bumper 增强保险杆[杠]
~ cord 增强帘子线
~ fibres 增强纤维制品
~ mat 增强毡片
~ tape 增强带材
~ web 增强筋条

reinforcer 增强材料;增强剂
reinforcing 增强,加强
~ agent 增强剂
~ band 增强带
~ corrugntion 增强筋条,增强波纹〈管〉
~ effect 增强效应〈填料或增强材料〉
~ fiber 增强纤维

~ filler 增强填料
~ material 增强材料
~ pigments 增强颜料
~ rib 加强筋
~ strip 加强带
~ wob 增强筋条,增强波纹〈管〉

reject 废品;疵品,不合格产品;等外品;下脚料
~ bin 废料仓
~ chute 废料槽
~ loss 废品耗损
~ rate 废品率

rejected material 废料,废品
rejection 等外品;废品,废物
relation 关系
relationship 相互关系
relative 相对的,相关的
~ accuracy 相对准确度
~ deformation 相对形变
~ density 相对密度〈旧称比重〉
~ dielectric constant 相对介电常数
~ efficiency 相对效率
~ elongation 相对伸长
~ error 相对误差
~ flammability 相对可燃性
~ flow 相对流动
~ frequency 相对频率
~ humidity [RH] 相对湿度
~ impact strength 相对冲击强度
~ molecular mass 相对分子质量
~ molecular-mass average 相对平均分子质量
~ peel strength 相对剥离强度
~ permeability 相对透气性
~ permittivity 相对介电常数;〈电〉相对电容率
~ pressure 相对压力
~ rigidity 相对刚性
~ speed 相对速率[度]
~ tack 相对黏性
~ velocity 相对速度
~ viscosity 相对黏度
~ viscosity increment 相对黏度增量

relaxation 松弛(作用)

~ behaviour 松弛性能
~ mechanism 松弛机理
~ modulus 松弛模量
~ of stresses 应力松弛
~ phenomenon 松弛现象
~ rate 松弛速率
~ shrinkage 松弛回缩
~ spectrometiy 松弛光谱测定(法)〈测定塑料松弛转变〉
~ spectrum 松弛谱
~ temperature 松弛温度
~ test 松弛试验
~ time 松弛时间〈恒定形变时应力的调整时间〉
~ transition 松弛转变
~ velocity 松弛速度
release 释放,解除,放松
~ agent 脱模剂〈在模塑时〉
~ catch 掣子,卡子,爪
~ -coated paper 脱膜涂层纸
~ film 剥离膜
~ mark 剥离痕迹
~ of entrapped air 截留空气释放
~ the stress 消除应力
~ time 弛缓时间
releasing 释放,解除,放松
~ a vacuum 解除真空,停止真空
~ agent 脱模剂
~ tension 消除张力
reliability 可靠性〈塑料结构件〉
~ coefficient 可靠性系数
~ design 可靠性设计
~ index 可靠性指数
~ test 可靠性试验
reliable (in operation) （操作中)可靠的
relief 减轻;释放;凸纹;凸起的
~ angle 切口角
~ cylinder 凸纹印花辊筒
~ lines 切片刀痕
~ map 模型图
~ pattern 凸纹花样
~ pressure valve 减压阀,安全阀
~ printing 凸版印刷,凸纹印花
~ valve 安全阀,保险阀

relieve 退切〈在模具中〉;解除,放气
reloading 重复负荷;重新加载;换料;再装料
reluctivity 磁阻率
remainder 剩余物
remanent 残余的,剩余的
~ strain 残余应变
remelt 再熔
remoistenable adhesive 再湿性黏合剂
remoistening adhesive 再湿性黏合剂
remote control 遥控
removable 可移动的,可更换的
~ bin 活动式料仓
~ plate mould 可换模板式模具;活模板式模具
~ plunger mould 可换柱塞式模具,活动模塞式模具
~ suspension blade 活(悬)浮板
~ tray 可拆卸板,活板
~ tray mo(u)ld 活板模具
~ wedge mo(u)ld 活模楔模具
removal 移去;排除;脱离;脱模
~ from the mo(u)ld 脱模
~ of volntile constituents 脱挥发分
~ time 脱模时间
remove 移去;排除;脱离;脱模
remover 脱离器;脱模剂
removing from the mo(u)ld 脱模
re-operate 重新运转,重新操作
reorientation 再取向
rep. [repeat] 反复,重复
repack 重新包装,改装
repair 修理,检修
~ parts 修理部分;备用零件
~ piece 配件;备件
~ size 修理尺寸
~ welding 修补焊
repassivation behaviour 再钝化性能
repeat 重复
~ feed 重复供给
repeatability 重复性
repeated 重复的,屡次的
~ bend test 重复弯曲试验,弯曲疲劳试验

~ crystallization 重复结晶
~ deformation 重复变形
~ flexural fatigue 重复挠曲疲劳,挠曲疲劳寿命
~ flexural strength 重复挠曲强度,挠曲疲劳强度
~ flexural stress 重复挠曲应力,弯曲疲劳应力
~ flexure 重复挠曲
~ impact test 重复冲击试验,冲击疲劳试验
~ load 重复载荷,反复载荷
~ stress 重复应力,疲劳应力〈美〉
~ structural unit 重复结构单元〈聚合物〉
~ test 重复试验,疲劳试验
~ trials 重复试验,疲劳试验
~ toreion test 扭转疲劳试验
~ use 重复使用
repeater 放大器;增强器;增强剂
repeating unit 重复单元〈聚合物〉
tepellent 排斥的,防水的;防水剂
replace 替换
replaceable 可替换的
~ head 可换式机头
replacement 替换,取代,代替
~ design 等代设计〈增强塑料〉
~ part 备件
~ time 更换时间
replasticize 再增塑,再塑化
replenish (再)装满,(再)填满;补充,补足
replenishment 再装满;补给;充填
replica 复制品
~ of mould details 模型件复制品
~ polymerization 模板聚合,复型聚合
~ technique 复制技术
replicability 复制性
replicate sample 复制样品
report 报告
re-precipitation 再沉淀
representative (有)代表性的,典型的
~ formula 典型配方
~ sample 有代表性的样品

~ viscosity-temperature curve 代表性黏度-温度曲线
reprint 再版,翻印
reprocess 再加工,重新处理,后处理;再生,回收
reprocessed 再加工的,后处理的
~ material 再加工(材)料
~ plastic 再加工塑料
reprocessing 再加工;再生,回收
reproducibility 再现性
reproducible 可再现的;可再生产的
reproduction 再生产
rept [report] 报告,报告书
repulsion 推斥,排斥
~ energy 排斥能力
~ forces 排斥力
req. requisition 需要,必要条件
res. research 研究;调查
res. resolution 分解;溶解
resample 再取样
research 研究
~ and development [R and D] 研究和开发
~ and development test 研究和开发试验
~ chemist 研究化学家
~ scheme 研究方案
~ technique 研究方法,研究手段
~ test 研究性试验
reseda green 木犀草绿,灰光淡绿色
reserve 保留,储备,储存;备用的
~ capacity 储备能力
~ parts 备件
reservoir 储器器〈吹塑机上的料筒〉
~ capacity 储料器容量
reset 重新安置;重新调节
~ button 复位按钮
reshape 再成型
residence 滞留,停留
~ time 滞留时间,停留时间,持续时间〈在模具中〉
residual 剩余的,残余的
~ acid 残留酸
~ deformaiion 残余变形,永久变形

- ~ effect 残余效应
- ~ elasticity 残余弹性,剩余弹性
- ~ elongation 残余伸长,永久伸长
- ~ elongation at break 断裂残余伸长
- ~ error 残留误差
- ~ extension 残余伸长,剩余伸长
- ~ flash 残留飞边
- ~ hardness 残余硬度
- ~ heat 残余热
- ~ induction 剩余磁感应
- ~ moisture content 残留含水量
- ~ monomer 残留单体
- ~ oil 渣油;残油
- ~ phenomena 残留现象
- ~ pressure 剩余压力
- ~ products 残余产物;副产品;残油〈石油化工〉
- ~ quantity 残余量
- ~ shrinkage 剩余收缩
- ~ solvent 残余溶剂
- ~ state of stress 残余应力状态
- ~ strains 残余应变
- ~ stress 残余应力
- ~ stretch 剩余拉伸,剩余伸张
- ~ styrene 残余苯乙烯
- ~ vinyl chloride monomer [RVCM] 残余的氯乙烯单体
- ~ volatile-matter 残余挥发物

residue 残余物
- ~ on distillation 蒸馏残余物

resilience 回弹,回弹性
- ~ energy 回弹能量
- ~ factor 回弹系数
- ~ test 弹性试验
- ~ testing machine 弹性试验机

resiliency 回弹,回弹性

resilient 回弹的;弹性的
- ~ -elasticity recovery 弹性变形恢复
- ~ fabric 弹性织物
- ~ floor covering 弹性地板覆盖材料
- ~ floor tile adhesive 弹性地板砖用黏合剂
- ~ roll 弹性辊

resiliometer 弹性测定计

resin 树脂;成型料,模塑料;用树脂浸渍
- ~ absorption 吸树脂量,树脂吸收率
- ~ acid 树脂酸
- ~ adhesive 树脂型黏合剂
- ~ alloy 树脂"合金";聚合物"合金"
- ~ applicator 树脂涂布器
- ~ bearing (合成)树脂轴承
- ~ bearing galss fiber 浸过树脂的玻璃纤维轴承
- ~ binder 树脂粘结剂
- ~ blender 树脂掺混器,树脂混配器
- ~ board 树脂纸板〈用树脂浸渍过的纸板〉
- ~ bonded 树脂粘结的
- ~ -bonded board 树脂胶合板
- ~ -bonded grinding wheel 树脂黏合砂轮
- ~ -bonded laminate 树脂胶合层压板
- ~ -bonded plywood 树脂胶合板
- ~ -bonded wood fibre product 树脂黏合木纤维制品
- ~ -bonding 树脂黏合
- ~ casting body 绕铸体
- ~ -coated 树脂浸渍的,树脂涂布的
- ~ coated sand 浸渍[涂布]树脂砂
- ~ collector 树脂收集器
- ~ concrete 树脂混凝土
- ~ containing extractants 萃取树脂
- ~ content 树脂含量
- ~ cure 树脂固化
- ~ -curing agent mixture 树脂与固化剂混合物
- ~ demand 树脂用量
- ~ dispensing unit 树脂分配装置
- ~ duct 树脂导管通道
- ~ emulsion 树脂乳液
- ~ entrapping powder of an inorganicion-exchange 无机离子交换包埋树脂
- ~ ester 酯化树脂
- ~ exchanger 离子交换树脂
- ~ -filled rubber 树脂填充橡胶
- ~ filling 树脂填充
- ~ finishing 树脂涂饰,树脂整理
- ~ flakes 树脂薄片

- ~ flow 树脂流动
- ~ for abrasive disks 砂轮用树脂
- ~ for glue and adhesive 胶黏剂用树脂
- ~ for grinding wheels 砂轮用树脂
- ~ for technical purpose 工业用树脂,工程用树脂
- ~ for the adhesive 黏合剂用树脂
- ~ - free 不含树脂的
- ~ - glass fabric laminate 树脂和玻璃纤维织物层压材料
- ~ glue 树脂胶
- ~ - impregnated 树脂浸渍的
- ~ impregnated densified wood 树脂浸压层合木材
- ~ impregnated abric 树脂浸渍织物
- ~ impregnated paper 树脂浸渍纸
- ~ impregnated wood 树脂浸渍木材
- ~ impregnation 树脂浸渍
- ~ injection 树脂注塑〈真空注塑法〉
- ~ injection mo(u)lding 树脂注塑成型
- ~ injection point 树脂注塑点,树脂注塑孔
- ~ injection processing 树脂注塑法
- ~ injection system 树脂注塑系统
- ~ injection unit 树脂注塑装置
- ~ kettle 树脂反应釜
- ~ latex 树脂乳剂
- ~ line 树脂软管线
- ~ loaded glass cloth 树脂浸渍的玻璃布
- ~ matrix 树脂本体,树脂母体
- ~ mixture 树脂混合物
- ~ moat 树脂槽
- ~ modified 树脂改性的
- ~ overflow 树脂溢流
- ~ paper 树脂浸渍纸
- ~ paste 树脂糊
- ~ pick - up 树脂吸收〈浸渍布,浸渍纸〉
- ~ - pigment ratio 树脂颜料比
- ~ pocket 树脂淤积
- ~ - rich area 多树脂区,富树脂区
- ~ - rich layer 富树脂层
- ~ - rubber blend 树脂-橡胶共混料,树脂-橡胶混配料
- ~ slurry 树脂淤浆
- ~ solution 树脂胶液
- ~ - starved area 贫树脂区,树脂缺胶面
- ~ streak 树脂流纹,树脂条纹
- ~ supply tube 树脂供料软管
- ~ syrup 树脂胶液
- ~ tank 树脂储料槽;浸渍槽
- ~ - treated laminated compressed wood 树脂浸渍层合压缩木材
- ~ treatment 树脂处理
- ~ trough 树脂槽
- ~ varnish 树脂清漆
- ~ wear 树脂磨损

resinate 树脂酸酯;用树脂浸渍
resinification 树脂化(作用)
resinify 树脂化
resinography 显微树脂学〈用光学方法检验固化的合成树脂或粘接〉
resinoid 热固性树脂;热固性黏合剂;似树脂的,树脂状的
- ~ bond 树脂粘结剂
- ~ grinding wheel 树脂粘结砂轮
- ~ plastic 热固性塑料;树脂状的塑料

resinology 树脂学
resinous 树脂的
- ~ composition 树脂混合物
- ~ plasticizer 树脂型增塑剂
- ~ varnish 树脂系清漆

resintance 抵抗,耐…性;电阻;阻力
- ~ butt welding 电阻对接焊
- ~ coupling 电阻耦合
- ~ heating 电阻加热
- ~ of ageing 耐老化性,抗老化性
- ~ of plastics to colour 塑料耐色性
- ~ of plastics to fungi 塑料防霉性
- ~ seam welding 电阻缝接焊
- ~ spot welding 电阻点焊
- ~ thermometer 电阻温度计
- ~ to abrasion 耐磨损性
- ~ to acids 耐酸性
- ~ to ageing 耐老化性
- ~ to alkali 耐碱性

~ to alternating moist and dry conditions 耐干湿(交替)性
~ to artificial weathering 耐人工气候老化性,耐人工自然老化性
~ to bacteria attack 防菌浸蚀性
~ to bending 耐弯曲性
~ to boiling 耐蒸煮性
~ to brittleness 耐脆性
~ to chalking 耐白垩化性〈涂敷技术〉
~ to chemical attack 耐化学浸蚀性
~ to chemical substances 耐化学物质性
~ to chemicals 耐化学药品性
~ to cigarette burns 耐香烟烧伤性
~ to cold 耐寒性
~ to cold and tropic conditions 耐寒冷和耐热带气候性
~ to colo(u)r change 耐色牢度性
~ to compression 耐压缩性,耐压强性
~ to corrosion 耐腐蚀性
~ to cracking 耐开裂性
~ to creasing 防皱性
~ to crushing 防皱性
~ to cuts 耐切割性
~ to deformation 耐变形性
~ to fatigue 耐疲劳性
~ to fire 耐火性,耐燃性
~ to flow 流动阻力
~ fo fiiels 耐燃料油性
~ to fungi 防霉性
~ to glow heat 耐炽热性
~ to heat 耐热性
~ to heat and humidity 耐湿热性
~ to high-energy radiation 耐高能辐射性
~ to hot water 耐热水性
~ to impact 耐冲击性
~ to indentation 耐压痕性,耐压痕强度
~ to insects 防蛀性
~ to irradiation 耐辐射性
~ to light 耐光性
~ to liquids 耐液体性;耐溶液性
~ to microorganisms etc 防微生物性等

~ to organic liquids 耐有机溶液性
~ to pressure 耐压(力)性
~ to ripping 耐撕裂性
~ to rupture 耐破裂性
~ to salt solutions 耐盐溶液性
~ to salt spray 耐盐水喷雾性
~ to scratching 耐刮痕性
~ to shear 耐剪切性,耐剪强度
~ to shock 耐冲击性,耐震击性
~ to shrinkage 防缩性
~ to solvents 耐溶剂性
~ to sparking 击穿电阻
~ to staining 耐污染性
~ to surface leakage 耐表面漏泄性
~ to swelling 耐溶胀性
to tear propagation 耐撕裂蔓延性
~ to tearing 耐撕裂性
~ to torsion 耐扭曲性,耐扭强度
~ to tracking 耐电弧径迹性
~ to tropical conditions 耐热带气候条件性
~ to ultra violet discoloration 耐紫外线变色性
~ to washing 耐洗涤性
~ to water 耐水性
~ to wear 耐磨性
~ to weathering 耐气候老化性
~ welding 接触焊,电阻焊,电阻热焊
~ wire 电阻丝
resistant 抵抗的,耐久的;耐…的
~ to ageing 耐老化的
~ to compression set 耐压缩永久变形
~ to corrosion 耐腐蚀的
~ to oil 耐油的
~ to oxidation 耐氧化
resistivity 抵抗性;比电阻,电阻率,电阻系数
~ against fire 耐火性
~ against water 耐水性
resistor 电阻器
resit [resite] 丙阶酚醛树脂,不溶不熔酚醛树脂
resitol [resilite] 乙阶酚醛树脂,半熔酚醛树脂,不溶可熔酚醛树脂

resoften 再软化
resol [resole] 甲阶酚醛树脂,可溶酚醛树脂,可溶可熔酚醛树脂,线型酚醛树脂
resolution 分辨,鉴别;分解
- ~ capacity 分辨能力,鉴别能力
- ~ of tensor 张量分解
- ~ power 分辨本领
resolving 分辨,鉴别;分解
- ~ power 分辨率,分辨本领,分辨能力,鉴别力
resonance 共振
- ~ condition 共振状态
- ~ curve 共振曲线
- ~ dispersion 共振色散
- ~ effect 共振效应
- ~ energy 共振能
- ~ frequency 共振频率
- ~ line 共振线
- ~ test 共振试验
resonant 共振的
- ~ forced vibration technique 共振强制振动技术
- ~ frequency 共振频率
- ~ machine 振动试验机
resorcin [resoreinol] 间苯二酚
- ~ adhesive 间苯二酚黏合剂
- ~ -formaldehyde resin [RF] 间苯二酚甲醛树脂
- ~ glue 间苯二酚胶黏剂〈美〉
- ~ monobenzoate [RMB] 间苯二酚单苯甲酸酯〈光稳定剂〉
- ~ -resin adhesive [glue] 间苯二酚树脂黏合剂〈美〉
resource 资源;原料
resp. [respectively] 各自地,分别地
restrained 限制的,约束的
- ~ foaming 限制发泡
- ~ press platen 固定压板〈压机〉
restraint foaming 抑制发泡
restricted (受)限制的,有限的
- ~ area mo(u)lding [RAM] 窄面模塑
- ~ gate [gating] 窄浇口
- ~ runner 窄流道

restrictor 节流阀,节流圈
- ~ bar 节流栓
- ~ ring 节流环
resultant 反应产物;综合的;有效果的
- ~ draft 实际牵伸
- ~ error 综合误差
- ~ of reaction 反应产物
- ~ stress 合成应力
retail package 零售包装
retained on sieve 筛余渣
retainer 托模板;挡板;保持架,定位器
- ~ pin 固定销
- ~ plate 模板,托板,托模板〈模腔〉
- ~ ring 固定环
retaining 保持,固定
- ~ claw 固定爪〈顶出杆〉
- ~ clip 固定夹
- ~ pin 嵌件定位销
- ~ plate 挡板,制动板
- ~ ring 固定环〈挤塑机模〉
retard spectrum 推迟时间谱
retardance 抑制
- ~ elasticity 推迟弹性
retardant 阻滞剂,抑制剂
retardation 阻滞;光程差
- ~ period 延缓时间
- ~ spectrum 推迟谱
- ~ time 滞留时间,推迟时间
retarded 推迟的;减速的
- ~ action 推迟作用
- ~ closing 延迟闭合
- ~ elasticity 推迟弹性
- ~ oxidation 缓慢氧化
- ~ spontaneous recovery 延迟自然恢复
- ~ velocity 减速度
retarder 阻滞剂,抑制剂;缓聚剂,阻聚剂
retention 保留,滞留;记忆
- ~ bar 节流元件
- ~ flex 挠曲强度保持率
- ~ mixer 储存混合器
- ~ modulus 弹性模量保持率
- ~ tank 储存罐〈在混合器上〉
- ~ time 保留时间,保压时间

retentivity 保持性
retest 重复试验,再试验
reticle [reticule] 交叉线,十字线;网袋
- ~ plastic film 网纹塑料膜片
- ~ plastic [sheet] 网纹塑料片材

reticulate 网状的
- ~ structure 网状结构

reticulated 网状的
- ~ foams pl 网状泡沫塑料
- ~ structure 网状结构
- ~ urethane foam 网状聚氨酯泡沫塑料

retort 蒸馏器,蒸馏
- ~ pouch 蒸煮袋

retouching injection press 触摸式注压机
retract 收缩
retractable 可缩(性)的
- ~ blow pipe 回缩吹气管
- ~ fibre 收缩性纤维
- ~ film 回缩膜,收缩性膜

retracting process 收缩过程〈挤塑物〉
retraction 收缩,回缩;回复
- ~ modulus 回复模数
- ~ stress 收缩应力
- ~ velocity 回缩速度

retroactivity 复原能力,〈塑料卸载后形变〉恢复能力
retro-reflection 折回,反射
retrogradation 退减作用;变稠〈黏合剂储存时〉
return 恢复,返回
- ~ conveyor 循环运送带
- ~ current 返流
- ~ curve 回复曲线
- ~ filter 回流过滤器
- ~ flow 回流
- ~ flow blocking device 止回流装置
- ~ line 回流管线
- ~ pin 回程杆
- ~ product 返料;废料
- ~ pulley 变向滑轮
- ~ speed 回程速率[度],回复速率[度]
- ~ spring 回复弹簧
- ~ travel 回程

- ~ valve 回流阀

reuse 重复使用,再使用
- ~ package 重复使用包装

rev. [review] 评论
rev. [revolution] 旋转;转数
reverse 相反;反向;逆向
- ~ bend 反向弯曲,槽型弯曲
- ~ bend test 交变弯曲试验

bubble forming 反向气胀成型
- ~ colo(u)ring 倒色,互换色
- ~ curvature 负曲率
- ~ cylinder 倒转滚筒
- ~ draw forming 反向拉伸成型
- ~ drive 反向传动
- ~ -flighted screw 反螺纹螺杆
- ~ flow baffle 回流挡板
- ~ gear 反向齿轮;逆转装置
- ~ impact 反向冲击
- ~ impact test 反向冲击试验,反面冲击试验
- ~ motion 返回运动回流〈注塑机〉
- ~ mo(u)lding 倒模压成型〈有下置阳模的成型机〉
- ~ osmosis 反渗透
- ~ osmosis membranes 反渗透膜
- ~ -printed film 反面印刷薄膜
- ~ printing 反面印刷
- ~ polymerization 可逆聚合
- ~ reaction 可逆反应
- ~ roll 逆辊〈涂布〉
- ~ roll coater 反辊式涂布机,逆辊涂布机
- ~ -roll coating 反辊涂布,逆辊涂布
- ~ -rotation 反向旋转
- ~ shrinkage 回缩;缩
- ~ taper 倒锥度
- ~ -taper ream 倒锥孔
- ~ tapered nozzle 倒锥孔喷嘴
- ~ valve 回动阀

reversed 反向的,相反的
- ~ bending 反弯曲
- ~ flow 回流,逆流
- ~ pyramids 反棱锥体
- ~ stress 逆向应力

reversibility 可逆性
reversible 可逆的
~ change 可逆变化
~ deformation 可逆形变
~ elongation 可逆伸长
~ gel 可逆凝胶
~ polymerization 可逆聚合
~ reaction 可逆反应
~ rotation 可逆转动
reversing 反向;回动
~ bevel 换向伞形齿轮
~ drive 反向传动
~ gear 反向齿轮,回动装置
~ handle 换向手柄,反向柄
~ motion 反向运动,换向运动
~ roll 逆转辊
~ shaft 逆转轴
~ slide 换向滑板
~ spring 回动弹簧
~ valve 回流阀
reversion 反转;复原;返原〈硫化〉
revert 复原,倒转
revertible 可逆的,可返转的
revolution 旋转
~ perminute [rpm] 转数每分钟,每分钟转数
~ per second [rps] 转数每秒,每秒钟转数
~ per unit of time 单位时间转数
revolve 旋转,转动
revolver 旋转器;滚筒
revolving 旋转的
~ blade 旋转刮板
~ cone mixer 旋转锥形混合器
~ cylinder 回转滚筒
~ -disk feeder 转盘式供料器
~ door 转门
~ drier [dryer] 转筒干燥器
~ drum 转鼓,滚筒
~ feed table 旋转供料台
~ filter 转滤器
~ helical blade 旋转螺旋型叶片
~ knife 旋转切割刀
~ table 旋转式台,转台

rewind 复卷
~ tension 复卷张力
rewinder 复卷机,卷取机
rewinding 复卷
~ cutting machine 复卷切割机
~ machine 重绕机,复卷机
reworked 再加工的
~ material 再生料
~ plastic 再生塑料
reworking 再次加工;再生加工
Reynold's number [Re] 雷诺准数
RF [resorcinol-formaldehyde resin] 间苯二酚-甲醛树脂
rf;RF [radio frequency] 射频;高频
RF-heating [radiofrequency heating] 射频加热,高频加热
RF-moulding [radio-frequency mo(u)lding] 射频(电感加热)成型
RF-preheating [radiofrequency preheating] 射频预热
RF-sealing [radiofrequency sealing] 射频熔合
RF-tarpaulin welder [radiofrequency-tarpaulin welder] 射频帆布焊接机
RF-welding [radiofrequency welding] 射频焊接
RH;R.H;r.h. [relative humidity] 相对湿度
rhenium [RE] 铼
rheodynamic stability 流变动力学稳定性
rheological 流变(学的)
~ behaviour 流变特性
~ breakdown 流变破坏〈塑料熔体〉
~ equation of state 流变(态)方程(式)〈塑料熔体〉
~ law 流变学定律
~ model 流变学模型
~ parameter 流变参数
~ process 流变过程
~ property 流变性质
rheology 流变学
~ modifier 流变改进剂
~ model 流变学模型

rheometer 流变仪
rheometric test 流变性试验
rheopectic 震凝的;流凝的;耐流变的
rheopecticity ［rheopexy］ 震凝性;震凝度;流凝性;耐流变性
rheostat 变阻器
rheoviscometer 流变黏度计
rhodamines 若丹明〈有机染料〉
rhombic 正交(晶)的;斜方晶体的;菱形的
~ mica 金石母
~ system 斜方晶系
r. h. p ［rated horse power］ 额定马力
Rhus lacquer 漆树漆,天然漆,生漆
rib 肋条,加强肋〈塑料成型件〉
~ arch 肋拱
~ stiffener 加强肋
~ -type sheet 肋条形片材
ribbed (带)肋的,呈肋状的
~ construction 加强结构
~ panel 加肋板
~ plate 加肋板
~ sheet roller 型材轧辊,加肋辊
~ sleeve 加肋套筒
~ surface 有肋表面
~ tube 加肋管
ribbing 肋条,加强肋
ribbon 条,带
~ - blade agitator 螺带式搅拌器
~ blender 螺带式掺混机,螺带混配机
~ conveyor 带式运输机
~ die 挤带模头
~ extruder 带状物挤塑机
~ gate ［film gate］ 带状浇口,膜状浇口
~ - like fibre 带状纤维,扁纤维
~ - like filament 带形丝,扁丝
~ mixer 螺带式混合机
~ of steel 带钢
~ polymer 带状聚合物
~ screw 螺带
~ spinning 带状纺丝法
~ spinning head 带状纺丝头,带状喷丝头

rich 富有的;浓厚的
~ colour 富丽色泽;浓色
~ oil 富油,饱和吸收油
~ shade 浓色调,浓色相,强色
ricinol amide 蓖麻醇酸酰胺〈脱模剂〉
ricinoleate 蓖麻醇酸酯
ridnus oil 蓖麻油
ridge 脊(状);螺脊,螺纹
~ forming 脊架成型
~ mo(u)ld 脊架成型模具
rifled 来复线的
~ barrel 来复线机筒
~ liner 来复线衬套
right 正确(的);正常(的);右面(的);向右
~ angle 直角
~ elevation 右视图
~ - hand 右的,右手的
~ - hand thread 右旋螺纹
~ - hand ［rightward］welding 右向焊接,向后焊〈热气焊〉
~ - handed rotation 右旋
rigid 刚性的,坚硬的;固定的
~ block 刚性嵌段
~ - body displacement 刚体位移
~ cellular material ［plastic］ 硬(质)泡沫材料［塑料］,硬(质)多孔材料［塑料］
~ chain 刚性链
~ charging machine 固定装料机
~ contruction 刚性结构
~ control 严格控制
~ fit 紧固配合
~ female mo(u)ld 硬(质)阴模,坚阴模
~ fibre 刚性纤维
~ foam 硬(质)泡沫塑料
~ foamed plastics 硬(质)泡沫塑料
~ frame 刚性构架
~ joint 刚性连接
~ material 刚性材料
~ packing 刚性填料
~ plastic 刚性塑料,硬质塑料
~ plastic sheet 硬质塑料片材

~ platen 硬板
~ polyester polyurethane foams 硬质聚酯型聚氨酯泡沫塑料
~ polyethylene 硬质聚乙烯
~ polymer 刚性聚合物
~ polyurethane foam 硬质聚氨酯泡沫塑料
~ polyvinyl chloride 硬质聚氯乙烯
~ polyvinyl chlorid mo(u)lding compound 硬质聚氯乙烯模塑料
~ polyvinyl chloride structural foam 硬质聚氯乙烯结构泡沫塑料
~ PVC film [RPVC] 硬质聚氯乙烯薄膜
~ PVC structural foam 硬质聚氯乙烯结构泡沫塑料
~ - rod like polymer 刚棒形聚合物
~ section 硬质型材
~ shaft 刚性轴
~ sheet 硬质片材
~ structure 刚性结构
~ surface 硬(质)表面
~ urethane 硬质聚氨酯泡沫塑料
~ waveguide 刚性波导管

rigidity 刚性[刚度],硬度[性]
~ index 刚性指数
~ modulus 刚性模量

rigidsol 硬质增塑糊

RIM [reaction injection mo(u)lding] 反应发泡注塑

RIM – technique [reaction injection mo(u)lding technique] 反应(发泡)注塑(成型)法

rim 边(缘)
~ gear 齿环,齿轮缘,齿轮圈
~ pressure 边缘压力

ring 环(形的);圈
~ and ball softening point 环球法软化点
~ baffle 环形挡板
~ buff 抛[磨] 光环〈压制件去除倒刺〉
~ chain 环链〈聚合物〉
~ clamp 环夹,夹箍

~ crusher 环式压碎机
~ cylinder 环形料筒
~ die 环形口模,环形试片冲模
~ – feed multi – gating 环形供料多浇口
~ follower 模具(内螺纹)圈,螺纹环〈成型件上有外螺纹〉
~ follower mo(u)ld 螺纹环模具〈有内螺纹环的模具〉
~ for deep drawing 深度拉伸用环
~ forming addition polymerization 环形加成聚合(作用)
~ gate 环形浇口
~ groove 环槽
~ heater 环形加热器
~ molecule 环状分子
~ nozzle 环形口模;环形注嘴
~ nut 环形螺母
~ – opening 开环
~ – opening polyaddition reaction 开环加成聚合反应
~ opening polymerization 开环聚合(作用)
~ packing 环形密封件,密封环
~ piston 环形活塞
~ plate method 环板法〈测定热固性塑料成型塑料的流动性〉
~ roll mill 环形辊磨机
~ – shaped heater 环形电热器
~ specimen 环形试样
~ structure 环状结构
~ surface 环面
~ test piece 环形试片
~ type mo(u)ld 环状模型
~ viscometer 环式黏度计

rinse 漂洗;冲洗;涮;漱洗
~ tempering 调温冲洗

rinser 冲洗器

RIP [resin injection processing] 树脂注塑法

ripening 熟化,陈化
~ index 成熟指数
~ time 熟化时间,陈化时间

ripping 纵向槽纹〈塑料成型件〉;剥;

ripple

撕;折
~ strength 撕裂强度
~ strength tester 撕裂强度试验仪
ripple 波纹,皱纹,网纹〈挤塑物〉
~ factor 波纹系数
~ finish 波纹整理
~ frequency 波纹频率
~ lacquer 波纹漆,皱纹漆
~ marks 波痕,麻点
~ varnish 皱纹漆
ripples pl 制网纹,制成皱纹
rise 上升;增长
~ of foam 泡沫发泡〈制聚氨酯泡沫材料时〉
~ rate 起发速度,发泡速度〈聚氨酯泡沫塑料〉
~ time 起发时间;发泡时间〈聚氨酯泡沫塑料〉
riser 提升管〈塑料制〉,升降器
rising 上升的;增长的;发泡的
~ ability 发泡能力,膨胀能力〈聚氨酯泡沫塑料〉
~ - bubble viscometers 上升气泡黏度计
~ casting 膨胀铸塑
~ - film evaporator 升薄膜式蒸发器
~ floor 升降台
~ table 提升台
~ table blow mo(u)lding 提升台架式吹塑成型
~ time 发泡时间〈聚氨酯泡沫塑料〉
rivel varnish 条纹漆
rivet 铆钉;铆接
~ bar 铆钉杆
~ - bonding 铆钉连接,铆接
~ clipper 铆钉钳
~ head knockout pin 铆钉形脱模销
~ head 铆钉头
~ hole 铆(钉)孔
~ lap joint 铆钉搭接
~ rod 铆钉杆
~ shank 铆钉杆体
~ type insert 铆钉形嵌件
riveted joint 铆钉连接

riveting 铆接
~ die 铆钉模,铆接模
~ machine 铆接机
~ tool 铆接工具
r. m. reversed motion 逆动,反向作用
r/m revolutions per minute 转/分
r. m. running meter 延米,纵长米
RMB [resoreinol monobenzoate] 间苯二酚单苯甲酸酯〈光稳定剂〉
robot 自动机,机械手;机器人
~ extractor 自动脱模机[器]
rocker 摇摆器,振动器;平衡杆〈涂布机〉
~ arm 摇臂,摇杆
rocking 摇摆的,摇动的
~ movement [rocking motion] 往复运动,摆动
~ sieve 摇摆筛
Rockwell 洛克韦尔,洛氏
~ haidness 洛氏硬度
~ hardness tester 洛尔硬度计
~ indentation hardness 洛氏压痕硬度
~ superficial hardness equipment 落氏表面硬度计
rod 棒(材);棒条
~ coater 辊式涂布机
~ crack 纵裂纹
~ mill 棒磨机
~ stock 棒状材料
~ type insert 棒形嵌件
rodent resistance 耐啃咬性
roentgenography X 射线照相法
Rolinx process 洛林克斯法,注塑压花法
roll 辊(筒);卷(筒);滚动,滚压
~ adjusting gear 辊隙调节机构
~ adjustment 辊隙调节
~ arrangement 辊排列,滚筒配置
~ bearing 滚柱轴承〈压延机〉
~ bending 辊弯曲〈压延机辊〉
~ bowl 辊筒
~ bowl width 辊筒宽
~ calender 辊筒式压延机
~ camber 辊中高,辊外凸
~ centres pl 辊距〈辊中心间距〉

~ chamber 辊腔
~ changing rigs 换辊设备
~ clearance 辊隙(距)
~ coater 辊筒涂布机
~ coating 辊式涂布,辊涂
~ cover [roll coverage] 辊覆盖层
~ coverage [roll cover] 辊覆盖层
~ – covering 辊涂敷
~ conveyor 辊式输送机
~ crossing 辊交叉
~ crossing alignment 辊轴交叉调整
~ crown 辊凸面,辊中高,辊外凸
~ crusher 辊式压碎机
~ deflection 辊挠曲
~ diameter 辊直径
~ die extruder 辊模挤塑机
~ dimension 压延辊尺寸
~ embossing 纹辊
~ end 辊端
~ face 辊面
~ feed 料卷供料
~ feed theimofoimer 片卷供料式热成型机
~ flooring 卷材地板
~ former 辊压成型机
~ frame 辊压机架
~ gap 辊距,辊(间)隙
~ glazing 辊压抛光
~ glazing unit 辊压抛光机构
~ grinder 滚筒研磨机,轧辊磨床;卷带式抛光机
~ heating 加热辊
~ kiss coater 辊舐涂布机
~ kiss coating 辊舐涂布
~ laminating 辊压层合
~ leaf stamping 箔卷烫印
~ – leaf stamping press 箔卷烫压印机
~ – loading 辊负荷
~ mark 辊痕〈半成品上〉
~ mill 辊磨机;辊炼机
~ mouth 辊间装料隙
~ neck 辊轴颈
~ nip [roll mouth or roll throat] 辊隙;辊间隙,进料口

~ nip adjustment 辊距调节(装置)
~ nip pressure 辊间隙压力
~ of fabric 布卷
~ opening 辊隙;轧辊开度
~ out 胀管,扩口
~ plating 滚镀
~ preloading 辊预负荷
~ pressure 辊压力
~ pull back 辊式拉伸装置
~ release agent 防黏滚剂
~ reverse coater 逆辊涂布机
~ roofing 卷铺屋面材料
~ separating force 辊分离力
~ separation 辊分离,辊距
~ setting 辊隙调节
~ shell [roller shell] 轴套
~ skewing 辊交叉
~ slitting 滚刀切割
~ slitting and rewinding machine 滚刀切割和复卷机
~ slitting machine 滚刀切割机〈用于膜片〉
~ speed ratio 辊速比
~ spew 卷裹渗出
~ spew test 卷裹渗出试验
~ spot welding 滚动点焊接
~ stand 滚轧机座
~ stock 滚轧件〈美〉;卷材
~ supporting forces 滚辊支承力
~ surface temperature 辊面温度
~ temperature 辊温
~ throat 滚辊料垄,滚辊进料口
~ type 辊种类
~ up 卷起,卷绕
~ wave 滚波
~ welding 滚压焊
~ width [roll bowl width] 辊宽
~ with adjustnble curvature 可调曲率辊
~ with drilled channels 开沟槽的辊
~ with heating chamber 有加热内腔的辊
~ with inner partition plates 有内隔板的辊

rolled 滚压的
- ~ and moulded laminated section 滚压和模塑层合型材
- ~ down 冷轧
- ~ laminated tube 卷制层压管,滚轧层合管
- ~ material 压延材料
- ~ pipe 卷制管材
- ~ section 滚压型材
- ~ sheet 辊轧片材
- ~ strand 轧制线
- ~ tube 滚制管,轧制管

roller 辊,辊轴,滚筒;滚轧机
- ~ adjustment 辊隙调节
- ~ bearing 滚柱轴承
- ~ body 辊体
- ~ coat 辊式涂布
- ~ coating 辊式涂布
- ~ conveyor 辊式运输机
- ~ crusher 滚筒式破碎机
- ~ cutter for pipe 滑轮割管器
- ~ cutting and wind-up machine 滑轮切割和缠卷机,复卷切割机
- ~ die 辊式模头〈挤塑机〉
- ~ die extruder 辊模挤塑机
- ~ draft 滚筒牵伸
- ~ draw fame 滚筒拉伸机架
- ~ drawing mechanism 滚筒拉伸机构〈生产膜片或带〉
- ~ drier [diyer] 辊筒式干燥器
- ~ electrode 滚动电极〈塑料焊接〉
- ~ electrode type high frequency pre-heater 滚动电极式高频预热机
- ~ end 辊端
- ~ feed 滚筒供料
- ~ gear bed 辊道,辊式传送带;滚床
- ~ head 辊式模头
- ~ leveller 辊式矫直机
- ~ machine 辊压机
- ~ mill 辊磨机
- ~ press 辊压机
- ~ presser 压辊〈美〉
- ~ pressure 辊压力
- ~ printing 滚筒印花
- ~ sander 磨光辊,砂磨辊
- ~ spot welding 滚动电极点焊接
- ~ spreader 辊筒涂敷机
- ~ support 滚柱支架,滚筒架
- ~ table [train] 辊道,辊式传送带
- ~ thrust bearing 滚柱止推轴承

rolling 滚动,滚压;卷边
- ~ -ball viscometer 滚球式黏度计
- ~ bank 滚动料垄
- ~ bearing 滚动轴承
- ~ compression mill 滚压磨
- ~ direction 轴滚动方向
- ~ friction 滚动摩擦
- ~ hide 辊上薄膜皮
- ~ machine 卷管机〈由板卷成管〉
- ~ operation 滚动过程
- ~ pipe from sheel material 板材卷成管
- ~ press 滚压机
- ~ screw 滚动螺杆
- ~ sheet 辊轧片材
- ~ sphere viscometer 滚球式黏度计
- ~ stand 辊轧机座
- ~ stock 辊轧件

roof 屋顶
- ~ gully 屋顶(雨水)排水口
- ~ light sheet 屋顶轻型片材
- ~ membrane 屋顶填衬膜
- ~ shingle 屋顶板
- ~ surface 屋顶面
- ~ tile 屋顶瓦

roofing 盖屋顶材料;卷材;屋顶,屋面
- ~ board 屋顶板
- ~ card board 屋顶油纸板
- ~ felt 屋面油毡
- ~ membrance 屋顶填衬膜
- ~ tile 屋顶瓦

room 房间,室
- ~ conditions pl 室内环境
- ~ temperature 室温
- ~ temperature cure 室温固化
- ~ -temperature curing 室温固化
- ~ temperature gluing 室温粘接
- ~ temperature mo(u)lding 室温成型
- ~ temperature setting 室温固化

~ temperature setting adhesive 室温固化黏合剂
~ temperature stability 室温稳定性
~ temperature vulcanization [RTV] 室温硫[熟]化

root 根;螺纹根;齿根
~ circle 齿根圆
~ diameter 根直径〈螺杆〉
~ diameter plus thread depth 螺杆直径
~ distance 齿根距离
~ flank 齿腹,齿面〈节圆和齿根之间〉
~ gap 根部间隙〈焊缝〉
~ layer （焊缝）底层,根部焊道
~ - mean - square strain 均方根应变
~ - mean - square stress 均方根应力
~ of a tooth 齿根
~ opening 根部间隙
~ radius 齿根圆角半径
~ run 根部焊道,底层〈热气焊缝〉
~ toughness 根部韧性〈热气焊缝〉

rope 绳索
~ drive 绳索传动

roper 捻绳机

roping 条痕;纵向槽纹〈塑料成型件〉

rose wood 青龙木

rosin 松香;松脂
~ acid 松香酸〈乳化剂〉
~ modified alkyd resin 松香改性醇酸树脂
~ resin 松香树脂
~ unsaturated polyester resin 松香不饱和聚酯树脂

Rossi – Peakes tester 罗西•皮克斯试验仪

rosy 玫瑰色的;浅玫瑰色的;粉红色的

rot resistance 耐腐性

rotaiy 旋转的,转动的
~ blower 旋转式鼓风机
~ breaker 旋转压碎机
~ chopper 转刀切碎机
~ compression moulding press 旋转式压缩成型机
~ compressor 旋转式压缩机
~ crusher 旋转式破碎机
~ cut 旋转切割
~ - cut veneer 旋转切割薄木板
~ cutter 旋转式切断机,旋转切割机
~ disk 旋转盘
~ disk knife 旋转圆盘刀,转盘刀
~ drum 转鼓,滚筒
~ drum drier 转鼓式干燥器
~ drum mixer 转鼓式混合机
~ dryer 旋转式[转鼓式,转筒式]干燥器
~ feeder 旋转式加料器
~ flowmeter 旋转式流量计
~ injection machine 旋转式注塑机
~ joint 旋转连接,联轴
~ kiln （旋）转炉
~ kiln diyer 转炉式干燥器〈用于层合材料〉
~ knife 圆盘刀
~ mixer 转筒混合器,滚筒混合器
~ motion 转动
~ mo(u)lding 回转多模成型,回转多模模塑,转盘式模塑
~ pelleter 旋转压片机,旋转造粒机
~ pelleting machine 旋转压片机
~ preforming press 旋转式预压片机,旋转式预成型压机
~ press 旋转压机,滚压机
~ printing machine 滚筒印花机
~ pump 回转泵,旋转式泵
~ screen 旋转筛
~ screen dryer 圆网干燥机
~ shelf 旋转架,转盘
~ shelf dryer with circulating air 循环空气转盘式干燥器
~ sieve 旋转筛
~ slide valve 旋转式滑阀
~ spraying machine 旋转式喷涂机
~ spreader 旋转分流芯
~ switch 旋转开关
~ table 转台,转盘
~ thermoformer 旋转热成型机
~ type automatic compression mo(u)lding press 旋转式自动压缩成型机

rotate

　塑压
　~ type blow mo(u)lding　旋转式吹塑成型
　~ type injection mo(u)lding machine　旋转式注塑成型机
　~ type mo(u)lding machine　旋转式成型机
　~ - type structural foam injection mo(u)lding machine　旋转式结构泡沫塑料注塑成型机
　~ vane feeder　旋转叶轮式加料器
　~ veneer　镟制层板
　~ viscosimeter　旋转黏度计

rotate　旋转,转动,回转;循环;变换

rotating　旋转
　~ arm　旋转臂
　~ arm machine　转臂式卷绕机
　~ blade　转刀;旋转桨叶
　~ bowl　转筒
　~ cross cutter　旋转式横切刀
　~ cylinder　旋转圆筒
　~ die　旋转口模
　~ disk　转盘
　~ disk viscometer　旋转盘式黏度计
　~ drum　转鼓
　~ dryer　旋转干燥器
　~ expander　回转式展幅机
　~ feed　旋转供给
　~ friction element　旋转摩擦元件〈摩擦焊接机〉
　~ in opposite direction　反向旋转
　~ in the same sense　同方向旋转
　~ inwardly　向内旋转
　~ knife　转刀
　~ outwardly　向外旋转
　~ oven　转炉
　~ pan mixer　转盘混合器
　~ screen　旋转筛
　~ shutter - type separator　旋转闸板式分离器,旋转百叶窗式分选器
　~ sieve　回转筛
　~ speed　转速
　~ sphere viscosimeter　旋转滚珠黏度计
　~ spreader　旋转式分流梭

　~ table　转台
　~ tensioning device　转动式张力装置

rotation　旋转,转动
　~ axis　旋转轴
　~ mo(u)lding　旋转成型,滚塑,回转熔塑
　~ per minute　每分钟转(数)
　~ plastometer　旋转式塑性仪
　~ rate　旋转速率[度]
　~ speed　旋转速率[度],转速
　~ tensor　旋转张量
　~ through 180°　180°旋转
　~ viscometer　转动黏度计

rotational　旋转的,转动的
　~ cast foam　旋转铸塑泡沫塑料
　~ casting　旋转铸塑,滚铸
　~ efficiency　转动效率
　~ foaming　旋转发泡
　~ force　旋转力
　~ forming　旋转成型
　~ frequency　转动频率,转数
　~ friction welding　旋转摩擦煤接
　~ inertia　转动惯量
　~ isomer　旋转异构体
　~ motion　转动
　~ mo(u)kling　旋转成型,滚塑,回转熔塑
　~ powder mo(u)lding　旋转粉料成型
　~ - resin/fibre - mo(u)lding process　树脂/纤维旋转成型法;树脂/纤维旋转铸塑法
　~ resistance　旋转阻力
　~ shell　圆环形壳层
　~ spectrum　转动光谱
　~ speed of the roll　辊转速
　~ - symmetric cross - cutter　同步旋转横割机,轴对称横割机
　~ vacuum embossing　旋转式真空压花,真空离心压纹
　~ vacuum embossing machine　旋转式真空压花机
　~ velocity　转动速度
　~ viscometer　旋转黏度计,转动黏度计

~ viscosimeter 旋转黏度计,转动黏度计
rotationally mo(u)lded part 旋转成型件
rotatory 旋转的;旋光的
~ diffiision 转动扩散
~ dispersion 旋光色散
~ drier [dryer] 旋转干燥器
- instrument 旋光测定计
~ isomer 旋转异构体
rotoforming 旋转成型法
rotogravure 转轮凹版印刷〈塑料印刷〉
~ press 转轮凹版印刷机
rotomo(u)lding 旋转成型,滚塑,回转熔塑
~ casting 旋转铸塑成型件
rotor 转子〈混合机〉;转动件;滚筒,转子;捏和桨叶
~ assembly 转子装配
~ blade 转子叶片
~ - cage 转笼
~ - cage impeller 转旋笼式搅拌机
~ cage impeller mixer 滚筒混合器,转笼式混合器
~ knife 滚刀
~ speed 转子速率[度]
rouge 铁丹;胭脂色
rough 粗(糙)的
~ adjustment 粗调
~ approximation 粗略近似
~ calculation 粗略计算
~ casting 铸塑毛坯
~ draught 草图
~ finish 粗修(整),粗加工
~ grain 粗粒(料)
~ grinding 粗磨
~ machining 粗加工
~ material 原料
~ package drawing 总体布置示意图
~ plan 初步计划,设计草图
~ roll 糙粗轧辊
~ sheet on the rolls (粗)滚轧片材
~ - surface 粗糙面
~ test 经验试验

~ tuning 粗调
~ vacuum 低度真空
~ weather 恶劣气候
~ weight 粗略重量,毛重
~ working 粗加工
roughened 变粗糙的
roughness 糙度;粗度
~ coefficient 粗糙系数
round 圆(形)的
~ blank 圆形毛坯
~ block(for gear wheel manufacture) (齿轮)粗坯〈用于制造齿轮〉
~ botlom bag 圆形底袋
~ die 圆形模头
~ ejection pin 圆型脱模销
~ flask 圆底烧瓶
~ head rivet 圆头铆钉
~ - headed screw 圆头螺杆
~ hole 圆孔
~ hole sieve 圆孔蹄
~ mo(u)ld insert for marking 标志用圆形模具嵌件
~ piece 圆形工件
~ plug for marking 标志用圆形柱销〈美〉
~ rod 圆杆,圆棒
~ runner 圆形流道
~ stock 圆料
~ thread 圆形螺纹
rounded
~ corner 圆角
~ crest 圆弧顶〈螺纹〉
~ nose (螺杆)圆头
~ root 圆根〈螺纹〉
routine 程序,常规,例行
~ control 常规控制
~ inspection 常规[例行]检查
~ quality control 程序质量控制
~ test 例行试验,常规试验
routing 程序安排;挖刨;研磨
roving (无捻)纱线;(无捻)纱条〈玻纤〉
~ band winding 粗纱缠绕法
~ bobbin 粗纱管
~ chopper 粗纱切断机

~ cloth （无捻）粗纱布〈美〉
~ fabric （无捻）粗纱布,（无捻）粗纱织物〈玻纤〉
~ package 粗纱筒
~ winding 粗纱缠绕法
RP ［reinforced plastics］ 增强塑料
rpm, r. p. m. ［revolutions per minute］转/分
rps; r. p. s ［revolutions per second］转/秒
KRIM ［reinforced reaction injection mo(u)lding］ 增强反应注塑成型,增强反应注塑模型
RT ［room temperature］ 室温
RTP ［reinforced thermoplastics］ 增强热塑性塑料
RTV ［room-temperature vulcanizing］ 室温硫化
Ru ［ruthenium］ 钌
rub 摩擦；磨损；研磨
~ -off test 擦拭试验
rub-off 擦净；磨去
rubbed joint 磨平滑的粘接面的粘接
rubber 橡胶
~ adhesive 橡胶型黏合剂
~ back-up roll 胶面托辊
~ bag 橡胶袋
~ bag mo(u)lding 橡胶袋施压成型,胶囊施压成型
~ -based adhesive 橡胶型黏合剂
~ belt 胶带
~ blanket 橡胶垫板
~ blanket coater ［spreader］ 橡胶垫板涂布机
~ block 橡胶块
~ bonding 橡胶粘接
~ bowl 橡皮辊筒
~ cement 橡胶型胶黏剂,橡胶泥
~ coated fabric 涂（橡）胶织物
~ coating 涂（橡）胶
~ compound 橡胶混合料
~ cork 橡胶塞
~ covered pressure roll 胶面压实辊
~ die 橡胶印模

~ dispersion 橡胶分散体
~ -elastic network polymer 橡胶弹性网状聚合物
~ elasticity 橡胶弹性
~ foam 泡沫橡胶
~ gasket 橡胶密封垫
~ joint 橡胶密封
~ kneader 橡胶捏合机
~ latex 胶乳
~ -metal bond 橡胶与金属的粘接
~ modified plastics 橡胶改性塑料
~ -modified polymer 橡胶改性聚合物
~ pad 橡皮垫
~ plate printing 胶版印刷
~ plunger mo(u)lding 橡皮柱塞成型
~ -reinforced polypropylene 橡胶增强聚丙烯
~ -reinforced polystyrene 橡胶增强聚苯乙烯
~ -reinforcing resin 橡胶增强树脂
~ resin 橡胶树脂
~ resin alloy 橡胶树脂"合金"
~ -resin blend 橡胶树脂共混料
~ ring 橡胶垫圈
~ -ring forming 橡胶环成型法
~ roll 橡皮辊
~ sheet cutter 橡胶切片机
~ sole 橡胶底〈鞋〉
~ solvent 橡胶溶剂
~ spreader 橡胶刮涂,辊涂胶机
~ stud 橡皮闸瓦
~ -to-metal bonding 橡胶与金属粘接
~ transition 橡胶态转化,玻璃态转变
rubberlike 橡胶状的
~ behaviour 橡胶态性能
~ elastic deformation 橡胶状弹性形变
~ elasticity 橡胶状弹性
~ material 橡胶状材料
~ polymer 橡胶状聚合物
rubbery 橡胶状的
~ copolymer 橡胶状共聚物
~ network 橡胶网络

- ~ polymer 橡胶状聚合物
- ~ shear modulus 橡胶态剪切模量
- ~ state 橡胶状态
- ~ substance 橡胶性物质
- ~ tensile modulus 橡胶拉伸模量
- ~ transition 橡胶态转变

rubbing 摩擦;磨损
- ~ abrasion 磨损
- ~ fastness 耐摩擦牢度
- ~ finish 可打磨漆,磨光漆
- ~ leathers 搓条皮圈,搓条皮板
- ~ machine 摩擦机

rubine 红宝石色,玉红
ruby 红宝石色,玉红
ruffler 弯曲装置
rufous 赤褐色的
rugged 粗糙的,不平的
ruggedising 加固,增强
ruggedness 坚固性;耐久性;强度
rule 规则,规律;定律,章程;尺
- ~ of combination 组合规则
- ~ of mixture 混合定律,混合规则〈复合材料〉

run 运行;运转
- ~ -down equipment 磨损了的设备
- ~ -down pipe 溢流管
- ~ number 批号,批数
- ~ -out trumpet 喇叭形管端
- ~ time 运转时间
- ~ through 贯通
- ~ -up 起动,试车
- ~ -up time 起动时间

runner 流道;流道(冷)料
- ~ cloth 导布
- ~ diameter 流道直径
- ~ hook 流道料钩
- ~ length 流道长度
- ~ mo(u)ld 无分流道模具
- ~ profile 流道形状
- ~ removal 流道凝料切除
- ~ stripper plate 流道冷料顶出板

runnerless 无流道
- ~ injection mo(u)ld 无流道注塑模具
- ~ injection mo(u)lding 无流道注塑成型
- ~ mo(u)ld 无流道模具
- ~ mo(u)lding 无流道模塑

running 运行,运转;流动
- ~ a blank 空白试验
- ~ characteristic 运转特性
- ~ clearance 运转间隙
- ~ condition 运转状态
- ~ cycle 工作循环
- ~ efficiency 运转效率
- ~ experience 运转经验
- ~ free 空转
- ~ head 排出接管〈半圆形雨水排水沟〉
- ~ hours 运转时间
- ~ -in of mo(u)lds 模具试运转
- ~ -in test 空转试验
- ~ life 运转寿命
- ~ maintenance 运行保养
- ~ meter, r.m. 延来,纵长来
- ~ outlet 排出口〈雨水排水沟〉
- ~ points 蒸馏点
- ~ program 运转程序,操作程序
- ~ short of 不够用〈原料〉
- ~ surface 波状表面;(滑雪)滑动面
- ~ temperature 运转温度
- ~ test 运转试验
- ~ time 运转时间,操作时间

runnings pl 馏出物
runway 跑道〈塑料〉
rupture 破裂;破坏,断裂
- ~ disc 安全防爆膜
- ~ of the metal 金属断裂〈粘接时〉
- ~ point 断裂点
- ~ strength 断裂强度
- ~ stress 断裂应力
- ~ test 破坏试验,断裂试验

rust (铁)锈;铁锈色〈棕色〉
- ~ creep 腐蚀蠕变
- ~ -proof steel 防锈钢
- ~ -protective paint 防锈漆,防锈涂料
- ~ -resisting paint 防锈漆,防锈涂料

radienium, Ru 钌
rutile 金红石,二氧化钛

RVCM [residual vinyl chloride monomer] 残余的氯乙烯单体

Ryton [商] 雷通〈菲利浦公司制聚苯硫醚〉

S

s [second] 秒
S [soft] 柔软的
s. [soluhle] 可溶的
S [sulphur] 硫
s [symmetrical structure] 对称结构〈化学〉
S – glass S – 玻璃〈高强玻璃〉
S glass fibre S 玻璃纤维,高强玻璃纤维
S – N – curve [Wohler stress – cycle diagram] 应力 – 疲劳曲线,应力数曲线,韦勒疲劳曲线
S – PVC [suspension polyvinyl chloride] 悬浮法聚氯乙烯
S – twist S 捻,顺手捻〈纱〉
S – type calender S 型压延机
SA [styrene acrylonitrile copolymer] 苯乙烯 – 丙烯腈共聚物
SABRA [stirface activation beneath reaction adhesivis] 在反应性黏合剂下表面活化
sack 装,装袋
~ weigher 装袋秤量器
sad 深暗的
saddle 脊架
SAE [Society of Automotive Engineers] 汽车工程师协会
SAET [Transactions of the Society of Automotive Engineers] 汽车工程师协会会报
safe 安全的,可靠的
~ allowable stress 安全容许应力
~ coefficient 安全系数
~ in operation 安全操作
~ to operate 安全操作,安全运行
~ working pressure 安全工作压力
safety 安全(性),可靠性
~ belt 安全带,保险带
~ cutout 安全切断器
~ door 安全门〈注塑机〉
~ device 安全防护装置
~ factor 安全因素
~ flim 保护膜
~ gate 安全栅门
~ glass 安全玻璃
~ glass interlayer 安全玻璃中间层
~ glazing 装配安全玻璃
~ guaid 防护装置〈注塑机〉
~ helmet 安全帽
~ in food 食用安全性,食品级〈食品接触的塑料〉
~ in production 安全生产
~ index 安全指数
~ measure 安全措施
~ mechanism 安全装置
~ net 安全网〈高空作业〉
~ pin 安全销,保险销
~ stop 安全停止装置,保险停止器,行程限制器
~ strap 安全带
~ valve 安全阀
sag 流挂;下垂;熔塌
~ rate 下垂率,下垂速度
sagging 流挂;下垂;熔塌;帘流
salable 可出售的
~ product 合格品;正品
sale 卖,出售;销路,销量
~ – oriented 从事销售的
~ package 销售包装
~ promotion 推销
~ stimulating 促销
salicylanilide N – 水杨酰苯胺〈防霉剂〉
salicylic acid 水杨酸〈助发泡剂〉

N – salicylidene N – salicylhydrazide　N – 亚水杨基 – N′ – 水杨酰肼〈抗氧剂〉
sallow　灰黄色
salmon pink　鲑鱼粉红〈肉色〉
salpetre　硝(石);钾硝;硝酸钾
salt　盐
~ fog resistance　耐盐雾性
~ spray cabinet　盐水喷雾室
~ spray jet　喷盐水嘴
~ spray resistance　耐盐水喷雾性
~ spray test　盐水喷雾试验
salvage　废品
Sambay evaporator　沙姆拜蒸发器
Samesarion [electrostatic spraying]　静电喷涂
sample　样品;取样;试样;试料
~ card　样本
~ collection　试样收集
~ container　样品容器
~ degradation　试样降解
~ design　样品设计
~ piece　试样;样件
~ preparation　试样准备
~ pretreatment　试样预处理
~ punch　冲(切试)样
~ puncher　冲(切试)样器
~ recovery　试样回收
~ size　试样量
~ survey　样品鉴定
sampler　取样器;模型,样板
~ device　脱模装置
sampling　采样,取样
~ condition　取样条件
~ inspection　取样检查
~ mo(u)ld　试样压模
~ rate　取样率
~ ratio　取样比例
san. [sanitary]　卫生的
SAN [styrene acrylonitrile copolymer]　苯乙烯 – 丙烯腈共聚物
sand　砂(子);沙灰〈浅黄灰色〉
~ blast　喷砂
~ blast nozzle　喷砂嘴
~ blasted　喷砂的
~ blaster　喷砂机
~ catcher　捕砂器,砂捕集器〈喷砂装置〉
~ grain　砂粒
~ grinding stone　磨石;砂石
~ mill　混砂机
~ paper　砂纸,金刚砂纸
~ shell mo(u)lding　壳模成型法,克罗宁法
~ sifter　砂筛,滤砂器
sandblast　喷砂(法)
~ nozzle　喷砂嘴
sandblasted　喷砂的
~ finish　喷砂修整
sandblasting　喷砂
~ machine　喷砂机
sanding　砂磨;喷砂;砂纸打磨
~ finish　无光漆,磨光漆
~ scaler　掺砂涂料
sandwich　夹芯,夹层;复合板
~ construction　夹层结构,夹芯结构
~ copolymer　嵌段共聚物
– effect　夹层效应
~ foam process　夹芯发泡成型法
~ foaming　夹芯发泡成型
~ heating　双面加热,夹芯加热
~ injection mo(u)lding　夹层注塑成型
~ laminate　夹芯层压板
– layer　夹层;芯层
~ material　夹层材料;芯材
~ mo(u)ld　夹层材料压塑模具
~ mo(u)lding　夹层泡沫成型
~ moulding technique　夹层泡沫成型技术
~ panel　夹层板;复合材料板
~ plate　夹层板
~ printing　夹层印刷
~ roofing panel　屋面复合板
~ structure　夹层结构,复合结构
sanitary　卫生的
~ cell　卫生池
~ cubicle　卫生室
~ installation　卫生设备

sanitation

~ non-wovens 卫生无纺布
sanitation 卫生设备
sap (树)汁;液;浆
~ wood 液材
saponification 皂化
~ number 皂化值〈美〉
~ value 皂化值
sapomly 皂化
sapphire 蓝宝石(色的)
Sarati plastics 沙拉思塑料〈由偏二氯乙烯聚合或偏二氯乙烯与少量其他不饱和化合物共聚制成的〉
sat. [saturate] 饱和
satellite 卫星
~ extruder 卫星挤塑机,辅助挤塑机
~ gear 行星齿轮
satin 缎纹
~ crape finish 缎纹整理〈使之具有缎纹光泽〉
~ crape like finish 缎纹整理
~ film 半透明薄膜
~ weave 缎纹编织〈玻璃纤维织物〉
satinizing 缎纹整理;消光处理;粗糙化处理〈表面〉
satisfy 使饱和
satn. [saturation] 饱和
saturant 浸渍剂,饱和剂
saturate 使饱和
saturated 饱和的
~ colo(u)r 饱和色
~ compounds 饱和化合物
~ humidity 饱和湿度
~ humidity ratio 饱和绝对湿度
~ mixture 饱和混合物
~ polyester 饱和聚酯
~ steam 饱和蒸汽
~ vapour 饱和蒸气
saturation 饱和,饱和度
~ induction 〈磁性〉饱和感应
saturator 浸胶机;饱和器
saw 锯
~ cutting 锯切割
~ dust 锯末,锯屑
~ dust mo(u)lding 锯末成型件

~ teeth profile 锯齿状型材
sawn veneer 锯成的薄木板
sawing machine 锯机,动力锯
Saybolt 塞波尔特
~ chromometer 塞波尔特比色计
~ viscometer 塞波尔特黏度计
Sb [antimony] 锑
SB [styrene butadiene copolymer] 苯乙烯-丁二烯共聚物
SBR [styrene-butadiene rubber] 苯乙烯-丁二烯橡胶,丁苯橡胶
SBR-adhesive [styrene-butadienerubber-adhesive] 苯乙烯-丁二烯橡胶黏合剂,丁苯橡胶黏合剂
SBS [styrene butadine styrene block copolymer] 苯乙烯-丁二烯-苯乙烯嵌段共聚物
~ rubber [SBS] 橡胶
scale 标度;秤;规模;鳞片
~ deflection 标度偏转
~ drawing 比例图
~ flaking 鳞片剥落
~ hopper 称量加料斗
~ model 按比例缩小的模型
~ of hardness 硬度标度
~ of roughness 粗糙标度
~ operation (大)规模生产
~ -up 放大比例
~ -up model 放大模型
scaler (脉冲)计数器;定标器
scaling 计数;分层,起鳞,剥落,剥离;结垢;定标
~ machine 剥皮机
~ up 扩大〈由实验室到工业生产〉;按比例放大
scaly 鳞片状
scanning 扫描
~ electron dMraction 扫描电子衍射
~ electron microscope [SEM] 扫描电子显微镜
~ speed 扫描速率[度]
scarf 嵌接,镶接
~ connection 斜接连接
~ joint 斜接接头,嵌接

~ - ring torsion test 嵌接环形扭曲试验〈用6个依次连续对接的筒形扭曲试验测定粘接的强度〉
~ welding 斜面焊接,嵌接焊
scarlet 大红,猩红
scattered 散射的
~ light 散射光
~ light measurement 散射光测定
scattering 散射〈试验值〉
~ amplitude 散射振幅
~ power 散射能力
scavenged sprue 废注口料,注料口废料
Schautnex machine 沙曼克斯机〈直接排气法〉
schedule 目录,清单,一览表,图表;计划表;时间表
schematic 示意的,概略的,图解的
~ diagram 原理图;示意图;简图
scheme 草图,平面图;图表;方案,规划
Schopper
~ folding tester 邵珀尔折叠试验仪
~ tensile tester 邵珀尔拉伸试验机
science 科学
scientific 科学的
~ experment 科学实验;科学仪器设备
~ glassware 科学玻璃仪器
~ research 科学研究
scintillation 闪烁(现象)
~ counter 闪烁计数器
~ rate 闪烁率
scintillator 闪烁器
scintillometer 闪烁计
scission 裂开,切断;分离
~ of link 断键
~ of the polymer chain 聚合物链断裂
~ reaction 裂解反应
sclerometer 肖氏硬度计,回跳硬度计
scleroscope 肖氏硬度计,回跳硬度计
~ hardness 肖氏硬度,回跳硬度
~ hardness test 肖氏硬度试验
scorch 烧焦;早期硫化
~ safety 防焦性;焦烧安全性
scorching 焦烧;早期硫化;引发;激发〈交联,固化〉

score 擦伤,伤痕〈模面〉
scoring 划痕;擦伤;咬模,啃模
~ resistance 耐擦伤性
~ test 划痕试验
scotch 刻痕,擦伤;压碎
scourer 擦洗仪
scouring 擦洗
SCP [steam curing process] 蒸汽交联法〈聚合物〉
scram botton 紧急停车按钮
scrap 切屑,边角料,废料,废品,残渣;碎片
~ chopper 废料切碎机
~ cutter 边角料切碎机,修边刀
~ granulater 边角料造粒机
~ grinder 边角料粉碎机,废料粉碎机
~ grinding 边角料粉碎
~ plastic 废塑料;再生塑料
~ rate 废品率
~ recyling 边角料回收
~ reduction 边角料粉碎
~ regrinder 回收料破碎机
~ removal system 切除边角料设备
~ reprocessing 回收料再加工
~ returns 废品料;废料回收
~ roll 落丝筒
scraper 刮刀,刮板〈混合机〉
~ bar 刮板〈涂布机〉
~ bar clearance 刮板间距〈涂布机〉
~ bar treater 刮板式处理器
~ blade 刮料板
~ knife 刮刀
~ rod 刮棒,刮板
scrapless forming process, **SFP** 无屑成型法
scratch 刮痕
~ board 刮板
~ hardness 刮痕硬度,耐刮硬度
~ hardness test 刮痕硬度试验
~ hardness tester 刮痕硬度试验仪
~ proof 耐刮的,耐划痕的
~ resistance 抗刮性,耐刮痕性
~ resistant 耐刮痕的
~ - resistant film 耐刮伤薄膜

screen
- ~ test 刮痕试验
- ~ testing machine 刮痕试验机
- ~ width 刮痕宽度

screen 筛网,滤料网;屏蔽
- ~ agent 屏蔽剂,遮盖剂
- ~ analysis 筛析
- ~ aperture 筛孔
- ~ assembly 筛网组件
- ~ beater mill 带有筛的锤磨机
- ~ belt dryer 筛带式干燥机
- ~ cage 筛篮筐
- ~ casing 筛网外壳
- ~ changer 换网器〈挤塑机〉
- ~ changing 筛网更换
- ~ classifier 筛分器
- ~ cloth 筛滤布
- ~ filter 筛滤板
- ~ floating 筛余物
- ~ fractionation 筛分,筛析
- ~ head 滤网头〈挤塑机〉
- ~ element 筛网组件
- ~ mat 筛分材料毡
- ~ mesh 筛眼,筛孔
- ~ opening 筛孔
- ~ pack 过滤网,滤网叠〈挤塑机孔板〉
- ~ pack changer 换网器
- ~ pillar 筛网立柱
- ~ plate 筛板,过滤板
- ~ printing 网板印刷
- ~ printing ink 网板印刷油墨
- ~ residue 筛余物
- ~ roll 筛网辊
- ~ separation 筛分
- ~ size 筛号
- ~ surface 筛面
- ~ tailings 筛余物
- ~ unit 筛网机组
- ~ varnish （筛)网漆〈印刷〉

screened 遮蔽的;筛过的
screener 筛选机
screening 屏蔽〈高频焊接装置〉;筛分,筛选
- ~ area 筛网面积

- ~ capacity 筛分能力;遮蔽能力
- ~ effect 屏蔽效应
- ~ machine 筛分机
- ~ material 筛分材料;屏蔽材料

screw 螺杆;螺丝;螺钉;螺旋
- ~ advance speed 螺杆推进速度,螺杆进料速度〈注塑时〉
- ~ antechamber 螺杆前腔
- ~ antirotation lock 防螺杆反转装置
- ~ axis 螺杆轴
- ~ back pressure 螺杆背压
- ~ barrel 螺旋式滚筒
- ~ blending element 螺杆混料元件
- ~ bolt 螺杆,螺栓
- ~ bushing 螺杆套筒
- ~ cap 螺丝帽;螺旋盖〈包装〉
- ~ channel 螺槽
- ~ channel depth 螺槽深度
- ~ channel width 螺槽宽度
- ~ character 螺杆特性
- ~ characteristic curve 螺杆特性曲线
- ~ cleaning machine 螺杆清理机
- ~ clearance 螺杆工作间隙
- ~ configuration 螺杆构型
- ~ conveyor 螺旋输送机
- ~ cooling 螺杆冷却
- ~ core 螺杆芯孔
- ~ core pin 螺杆芯孔成型销
- ~ core plug 螺杆芯孔塞
- ~ core tube 螺杆芯孔管
- ~ cutting machine 攻螺纹机
- ~ cylinder 螺筒
- ~ design 螺杆的设计
- ~ diameter 螺杆直径
- ~ die 螺杆式模头,螺杆分配机头
- ~ displacement 螺杆位移
- ~ drive 螺杆传动
- ~ drive power 螺杆传动功率
- ~ drum 螺旋式转鼓
- ~ dryer 螺旋式干燥器
- ~ efficiency 螺杆效率
- ~ clearance 螺杆间隙
- ~ extruder 螺杆式挤塑机
- ~ extrusion 螺杆挤塑机

screw

- ~ – extrusion hot – melt adhesive applicator 螺杆式挤出热熔胶涂布机
- ~ extrusion press 螺杆挤压机
- ~ feed rate 螺杆进料速率[度]
- ~ feeder 螺杆加料器;计量螺杆
- ~ flight 螺纹,螺杆螺纹
- ~ flight depth 螺纹深度
- ~ geometry 螺杆几何(形状)
- ~ heating – cooling 螺杆温度调节
- ~ horsepower 螺杆(传动)功率
- ~ hub 螺杆尾突台
- ~ hub seal 螺杆尾突台密封
- ~ impeller 螺旋式搅拌叶轮
- ~ in – line type injection mo(u)lding machine 同轴螺杆式注塑成型机
- ~ – in tube 旋拧入管
- ~ injection mo(u)lding 螺杆式注塑(成型)
- ~ injection mo(u)lding machine 螺杆式注塑(成型)机
- ~ kneader 螺杆捏合机
- ~ length 螺杆长度
- ~ machine 制螺纹机,制螺钉机;螺杆挤塑机
- ~ mixer 螺杆式混合器
- ~ mixer with vertical screw 立式螺旋混合器
- ~ – out – type injection mo(u)ld 旋拧出式注塑模具〈用于制作螺纹件〉
- ~ pelletizer 螺杆造粒机
- ~ piston 螺杆式活塞
- ~ piston injection mo(u)lding 螺杆 – 活塞式注塑成型
- ~ – piston plasticizing 螺杆 – 活塞塑化
- ~ – piston shaft 螺杆 – 活塞体
- ~ – plasticating injection mo(u)lding 螺杆塑化式注塑成型
- ~ plasticizing 螺杆式塑化
- ~ plasticizing unit 螺杆式塑化装置〈注塑机〉
- ~ plastification 螺杆式塑化
- ~ plug 螺旋塞
- ~ plunger[screw piston] 螺杆式活塞
- ~ preplasticaling injection moulding 螺杆预塑化式注塑成型
- ~ preplasticating injection mo(u)lding machine 螺杆预塑化式注塑成型机
- ~ preplasticator 螺杆式预塑化器
- ~ preplasticisation 螺杆预塑化
- ~ preplasticiser 螺杆预塑化器
- ~ press 螺杆式挤压机
- ~ push – out device 螺杆顶出装置,螺杆脱模装置
- ~ ram 螺杆式柱塞
- ~ ram injection unit 螺杆柱塞式注塑装置
- ~ reactor 螺杆式反应器
- ~ reducing gear 螺杆减速齿轮〈挤塑机〉
- ~ restriction section 螺杆增压段
- ~ revolution 螺杆转数
- ~ ribbon 螺带
- ~ ribbon extruder 螺带式挤塑机
- ~ root 螺杆芯杆,螺纹齿根
- ~ rotating in opposite direction 反(方向)旋转螺杆
- ~ rotation in the same direction 同方向旋转螺杆
- ~ shaft 螺杆传动轴
- ~ shank 螺杆尾〈无螺纹部分〉
- ~ shank seal 螺杆尾密封
- ~ sleeve 螺纹套管
- ~ speed 螺杆速率[度],螺杆转速
- ~ speed indicator 螺杆转速指示器
- ~ speed range 螺杆转速范围〈挤塑机〉
- ~ spindle 螺纹心轴
- ~ stem 螺杆芯杆
- ~ stirrer 螺旋搅拌器
- ~ tachometer 螺杆转速表
- ~ thread 螺(杆螺)纹
- ~ thread nomenclature 螺纹名称
- ~ thread types 螺纹种类,螺纹类型
- ~ thrust 螺杆,螺杆推力
- ~ tip 螺杆梢
- ~ torque 螺杆扭矩
- ~ transfer mo(u)lding 螺杆式压铸

~ transfer plasticizing 螺杆挤压塑化
~ travel indicator 螺杆行程指示器
~ truncation 螺杆尖端钝化
~ turn 螺杆旋转
~ – type extruder 螺杆式挤塑机
~ – type extrusion machine 螺杆式挤塑机
~ – type injection mo(u)lding machine 螺杆式注塑成型机
~ – type mixer 螺杆式混合器
~ – type plasticizing 螺杆式塑化
~ – type plasticizing equipment [unit] 螺杆式塑化装置〈注塑机〉
~ – type plunger 螺杆式柱塞
~ vent 螺杆排气口
~ wear 螺杆磨损
~ with armoured flight lands 螺纹棱面铠装的螺杆
~ with constant (depth of) thread 等螺槽深度螺杆
~ with countersunk 埋头螺钉
~ with decreasing depth (of thread) 螺槽深度递减螺杆
~ with decreasing pitch 螺距递减螺杆
~ with decreasing thread 螺槽深度递减螺杆
~ with equal pitch 等螺距螺杆
~ working clearance 螺杆工作间隙
~ zone 螺杆段

screwed 螺纹的,螺旋的
~ bayonet joint 螺纹卡口式连接〈吹塑瓶时〉
~ pipe joint 螺纹管连接
~ sleeve 螺纹套管
~ tip 螺杆梢

screwing 螺纹的,螺旋的
~ machine 攻螺纹机
~ – out action 旋出,挤出

screwless 无螺杆的
~ extruder 无螺杆挤塑机
~ plasticator 无螺杆塑化机

scrim 稀松无纺(的)织物
scroll conveyor 螺旋输送机
scrubbing tower 洗涤塔

scuff 擦伤,划伤
~ mark 擦伤痕
~ resistance 耐擦伤性,耐磨性〈塑料地板〉
~ resistant 耐擦伤的,耐磨性的

scuffing 擦伤,划伤;磨损
~ test 擦伤试验

scum 泡沫
~ plastic 泡沫塑料
~ rubber 泡沫橡胶

scumbling [antiquing] 薄涂不透明色〈在压花塑料制品表面上产生双色效果〉

SDS [solids draining screw] 固体加工螺杆

Se [selenium] 硒
SE [sound emission] 声发射
SEA [sound emission analysis] 声发射分析

seal 封闭;密封;垫圈
~ area 密封面积
~ coat 封闭涂层
~ gum 密封胶
~ joint 密封接合
~ ring 密封环
~ up 密封
~ weld 密封焊缝

sealability 密封性;可焊接性
sealable flat rim 可焊接的平边〈待加工成型件〉
sealant 密封材料;密封胶;热焊辅料
sealed 密封的,封口的
~ end 封端
~ package 密封包装

sealer 热合机,熔封机;封焊机
sealing 热合,熔封,密封
~ adhesive 密封黏合剂,密封胶
~ area 热合面,熔封面;捍接面
~ bar 熔接棒
~ band 密封带
~ bead 封口焊道
~ bushing 密封衬套
~ coat(ing) 密封涂层
~ compound 密封料

~ cycle 热合周期
~ diameter 封闭直径
~ die 热合模
~ disk 密封垫圈
~ edge （模具）密封边缘
~ efficiency 密封效率
~ element 密封件,密封体
~ film 热合薄膜;密封膜;防护覆盖膜〈用于层压制造〉
~ flange 封闭法兰
~ frequency 热合频率
~ joint 热合接头;密封接头;密封面
~ ledge 密封边缘
~ material 密封材料
~ needle 封接注射针
~ of the gate 浇口密封
~ paint 封闭漆
~ paste 密封糊剂,密封（软）膏
~ press 热合压力机;封焊压机
~ pressure 热合压力;封焊压力
~ plug 密封塞
~ profile 密封型材
~ ring 密封环
~ run 封底焊道
~ solution 密封液〈用于多孔表面〉
~ strip 密封条
~ surface 密封面,热合面
~ tape 密封带;胶黏带
~ time 热合时间
~ washer 密封垫圈
~ web 密封带〈塑料制〉
~ weld 封焊

seam 焊;接缝;合模缝;焊缝
~ mark 缝痕;伤痕
~ sealant 灌缝料,密封料
~ sealing 缝焊
~ weld 滚焊（缝）,缝焊
~ welder 线缝焊机,焊缝焊机
~ welding 线焊,缝焊,滚焊
~ width 焊缝宽度

seaming 接缝,卷边接缝
~ die 合缝模

seamless 无缝的
~ bottle mo(u)lding technique 无缝瓶成型法
~ tube 无缝管

season colour 流行色,应时色
seat cover 座罩
sebacate 癸二酸盐[酯]
sebacic acid 癸二酸
~ ester plasticizer 癸二酸酯增塑剂
sec. [second] 秒
sec. [secondary] 仲
secant 正割,割线
~ curve 正割曲线
~ modulus 正割模量〈塑料材料〉
~ modulus of elasticity 弹性正割模量

second 秒;片刻;第二的;副的,辅助的
~ cure 再固化,再硬化
~ - generation synthetic fibre 第二代合成纤维
~ moment of inertia of area 断面二次转动惯量
~ operation 修整,补充加工,精加工
~ - order reaction 二级反应
~ - order splitting 二级分裂
~ - order theory 二次流动理论
~ order transition 二次转变,玻璃化转变
~ - order transition point〈temperature〉二次转变点〈指温度〉
~ order transition temperature 二次转变温度,玻璃化温度
~ reaction 补充反应,再反应
~ stage injection 两级注塑
~ surface decorating 背面装饰
~ - surface decoration 底面装璜法

secondary 第二的;副的,辅助的
~ accelerator 辅助促进剂
~ antioxidant 辅助抗氧剂
~ breaker 二次破碎机
~ carpet backing 地毯第二层底布
~ characteristics 次要性能
~ colo(u)r 次色,复色,混合色,调和色
~ creep 二级蠕变
~ cross - linked 后交联;二次交联
~ crystallization 再结晶,后期结晶,重

结晶
　~ decomposition　再度分解
　~ dispersion　次级分散
　~ gluing　二次黏合;装配黏合
　~ heating　后加热,再加热
　~ ion mass spectroscopy [SIMS]　二次离子质谱仪
　~ material　再生料
　~ member　副构件
　~ particle　次级粒子
　~ plasticizer　辅助增塑剂
　~ plastics　辅助塑料
　~ polymerization reaction　二次聚合反应;后期聚合反应
　~ processing　二次加工
　~ product　次要产品,副产品
　~ raw material　再生原料〈塑料〉
　~ reaction　副反应,次级反应〈反应性树脂〉
　~ runner　分流道
　~ shades　次色,混合色,调和色
　~ stress　二次应力,副应力
　~ structure　二级结构
　~ treatment　后处理
　~ valence bond　副价键
section　型材;截面,断面;(分)段
　~ bar　型材棒
　~ construction　组合式结构
　~ cutter　切片机
　~ cutting　切片,切割
　~ drawing　剖面图
　~ modulus　断面模量
　~ shape　型材
　~ strip　异型带条
sectional　部分的;分级的,分段的,组合的;截面的
　~ area　截面积
　~ drive　分段驱动
　~ mo(u)ld　镶合模具,组合模具
　~ pattern　组合式模型
　~ screw　镶段螺杆
　~ view　断面图,剖视图
　~ width　断面宽度
sediment　沉降物,沉积物

sedimentation　沉积,沉淀
　~ average molecular weight　沉降平均分子量
　~ of fillers　填料沉积
　~ rate　沉降速率[度]〈树脂配方中填料〉
see-through-clarity　透明度
Seebeck effect　塞贝克效应,热电效应
seediness　粒度;成粒性
seeding　加晶种;成团块〈涂料〉;成细粒〈在涂层中〉;接种
　~ agen[nucleating agent]　晶核剂
　~ polymerization　接种聚合
　~ technique　晶核结晶法
segment　(分割)部分;段;链段〈高聚物〉;弓形(体)
　~ bend　弓形弯曲;弯形管道,弓形弯管〈补偿器〉
　~ die　组合模,可拆卸模
　~ motion　链段运动
segmental　部分的,引形的
　~ arch welding machine　燥扇形状物的焊接机
　~ bearing　弓形轴承〈层压塑料轴承〉
　~ bonding　局部黏合法
　~ die　组合模
segmented　嵌[镶]　段的
　~ copolymer　嵌段共聚物
　~ extruder barrel　镶段式挤塑机筒
　~ polymer　(多)链段聚合物
segmer　链段
segregate　分离;条纹色调
segregation　色条纹〈成品表面缺陷〉;色调〈制品表面〉;分离;离析
seizing　咬模〈模具〉
　~ of mo(u)ld　咬模,卡模
seizure　咬模;卡模〈模具〉
selection grid　分选筛
selective　选择的
　~ assembly　选择装配,选配
　~ detection　选择性检测
　~ fit　选择配合
　~ heating　局部加热
　~ permeability of ion exchange membrane

离子交换膜选择透过性
~ polymer　选择聚合物
~ polymerization　选择聚合,立构有择聚合,不对称立体选择性聚合
~ sampling　选择性采样
~ solvent　选择性溶剂
~ test　选择性试验
~ treatment　选择处理
selectivity　选择性
selenium [SE]　硒
~ cell　硒电池
self　本身,自己
~ acting　自动的
~ acting control　自动控制
~ acting machine work　机器自动工作时间
~ - acting scale　自动秤
~ - actuated controller　自动控制器
~ - adherent　压敏的;自黏性的
~ - adherent film　自黏膜
~ - adherent tape　自黏胶带
~ - adhering　压敏的;自黏性的
~ - adhesion　自黏(性)
~ adhesive　自黏合剂
~ adhesive film　自黏膜
~ - adhesive tape　自黏胶带
~ - adjusting　自动调整,自动调节
~ - adjusting weigher　自动秤量器
~ - adjustment　自动调整
~ ageing　自然老化
~ - aligning ball bearing　自动调整滚珠轴承
~ - aligning double row ball bearing　自动调整双排滚珠轴承
~ - ligning grip　自调夹紧装置〈试验机〉
~ - carrying mo(u)ld lubricant　本身含有脱模剂
~ - cleaning [selt - wiping]　自动清理,自动清除
~ - cleaning effect　自动清理效应
~ - cleaning screw　自动清理螺杆,自净式螺杆〈挤塑机〉
~ - cleansing　自动清洗

~ - closing　自闭合
~ - closing nozzle　自闭注料嘴
~ - colour　原色,本色,天然色
~ coloured　原色的,天然色的
~ - combustion　自燃
~ - compensating　自动补偿的
~ - compensating dust seal　自动补偿式防尘圈
~ - condensation　自身缩合
~ - contained　自主的,独立的;整装的
~ - contained drive　单独传动
~ - contained mo(u)ld　整装模具
~ - contained press　单独传动的压机
~ - control　自动控制
~ - cooling　自然冷却
~ - crimping　自动卷曲
~ - crosslinking [auto = crosslinking] 自交联〈聚合物〉
~ - curing　自动固化的;自动硫化的
~ - curing adhesive　自固化黏合剂
~ - curing organic binder　自固化有机粘结剂
~ - decomposition　自分解
~ - destmet plastics　自毁性塑料
~ diffusion　自扩散
~ - edge　布边,织边〈纺织品〉
~ - etching primer　自蚀底漆
~ - exting [self - cxtinguishing]　自熄性
~ - extinguishing　自熄;自熄性;自熄性的
~ - extinguishing fibre　自熄性纤维
~ - extinguishing plastic　自熄性塑料
~ - extinguishing resin　自熄性树脂
~ - extinguishing time　自熄时间
~ - extinguishing unsaturated polyester resin　自熄性不饱和聚酯树脂
~ - feed　自动供料,自动加料
~ - feeder　自动加料器
~ - hardening　自硬的
~ - hardening organic binder　自固化有机粘结剂
~ - heal break　自愈合裂缝〈在介质

SEM

作用下〉
~ - heating 自热,自动加热
~ - heating temperature 自动加热温度
~ - ignition 自燃
~ - ignition point 自燃点
~ - ignition temperature [SIT] 自燃温度
~ - indicating weighing machine 自动指示计量机,自动显示磅秤
~ - inflammable 自燃的
~ - inflammable mixture 自燃混合物
~ - lubricating 自动润滑的
~ - lubricating bearing 自润滑轴承
~ - lubricating property 自润滑性能
~ - operated controller 自动控制器
~ - plasticizing action 自增塑(作用)
~ - polymerization 自动聚合(作用)
~ - purging effect 自净化效应
~ - purification 自净化
~ - quenching 自熄灭〈塑料燃烧性能〉
~ - regulation 自动调节
~ - sealing 自密封,自封合
~ - sealing profile 自封型材
~ - sealing tank 自封槽
~ - sealing tape [SST] 自封黏胶带
~ - shade 单色,本色
~ - shearing gate 自断式浇口
~ - skinned foam 自结皮泡沫塑料
~ - skinning 自结皮
~ - skinning rigid foam 自结皮硬质泡沫塑料〈具有较高密度表皮层的泡沫塑料〉
~ - stick 自黏附
~ - supporting 自承重的;自架的
~ - tapping screw 自攻丝螺纹
~ - tone 单色调
~ - vulcanizing 自动硫化的
~ - wiping 自动擦净,自动清除
SEM [scanning electron microscope] 扫描电子显微镜
semi 〈词头〉半,部分,不完全的
~ - arid 半干燥的

~ - automatic 半自动的
~ - automatic control 半自动控制
~ - automatic mould 半自动模具
~ - automatic mo(u)lding machine 半自动成型机
~ - automatic operation 半自动操作
~ - automatic press 半自动压机
~ - automatic welding 半自动焊接
~ - automation 半自动化
~ - batch reactor 半间歇反应器
circular baffle 半圆防护板
~ - circular gate 半圆浇口
~ - commerical production 半工业性生产
~ - commercial scale 半工业规模〈生产〉
~ - compreg 半浸胶压缩木材
~ - conducting polymer 半导体聚合物
~ - conduction property 半导体性
~ - conductive polymer 半导体聚合物
~ - conductivity 半导电性
~ - conductor 半导体
~ - continuous 半连续的
~ - crystal polymer 半结晶聚合物
~ - crystalline thermoplastic 半结晶热塑性塑料
~ - diaphanous 半透明的
~ - diffraction angle 半衍射角
~ - dry winding 半干法缠绕
~ - drying oil 半干性油
~ - durable adhesive 半耐久性黏合剂
~ - finish parts 半成品零件,半成品配件
~ - finished 半加工的,半成品
~ - finished goods 半成品
~ - finished material 半成品,半制品
~ - finished piece 半制品,半制件
~ - finished product 半成品
~ flash mo(u)ld 半溢料式模具
~ - gloss 半光泽的
~ - hard 半硬的
~ - hardboard 半硬质(纤维)板
~ - homogeneous 半均匀

~ – homogeneous ion exchange membrane 半均相离子交换膜
~ – industrial 半工业的
~ – interpenetrating 半互贯〈网络聚合物〉
~ – manufactures 半成品
~ – manufactured goods 半成品
~ – manufactured products 半成品
~ – mineral pigment [metallo organic pigment] 半无机颜料
~ – opaque 半透明的
~ – permeable 半渗透性的
~ – permeable membrane 半渗透膜
~ – plant 中间试验厂,中试装置
~ – plastic 半塑性的
~ – plastic state 半塑性状态
~ – polar 半极性的
~ – polar bond 半极性键
~ – polar link 半极性键
~ – positive mo(u)ld 半全压式模具,半溢料式模具
~ – prepolymer process [quasi – prepolymer process] 半预聚物法
~ – product 半成品
~ – production 半生产,间歇生产
~ – regenerated fibre 半再生纤维
~ – rigid 半硬的
~ – rigid cellular plastics 半硬质微孔塑料
~ – rigid foam 半硬质泡沫塑料
~ – rigid plastic 半刚性塑料,半硬质塑料,韧性塑料
~ – rigid polyurethane foams 半硬质聚氨酯泡沫塑料
~ – rigid sheet 半硬片材
~ – rigid urethane foam 半硬质聚氨酯泡沫塑料
~ – scale plant 半工业化装置
~ – scale production 半工业化生产
~ – sectional view 半断面图,半剖视图
~ – skilled worker 半熟练工人
~ – spherical 半球形的
~ – synthetic 半合成的
~ – synthetic fiber 半合成纤维
~ – synthetic polymer 半合成聚合物
~ – transparence 半透明性
~ – transparent 半透明的
~ – works 中间试验厂,中试装置
sense of rotation 旋转方向
sensitive 敏感的
~ film 感光胶片
~ to heat 热敏材料
sensitivity 敏感性,灵敏度;感光性
sensitize 敏感,敏化;感光
sensitizer 敏化剂;敏化器
sensitizing agent 敏化剂
sensor 传感器,敏感元件
sensory 敏感的,感觉的,灵敏的
~ adaptation 灵敏配合
~ evaluation 感觉评定
~ testing 敏感测试
sep. [separate] 分离;分开
separable 可分离的,可拆式的
~ joints 可拆式联轴节〈塑料〉
separate 分开,分离
~ application 分别涂布
~ application adhesive 分涂黏合剂
~ pot 成型料预塑化槽
~ pot mo(u)ld 预塑化槽压铸模,分离料巢式模具
~ streams 分料流〈共挤出〉
separating 分开的,分离的
~ calorimeter 分离量热器
~ edge 切削刃
~ efficiency 分离效率
~ electrode 切割电极〈高频焊接〉
~ force 分离力
separation 分开;分离;离析
~ hevel 分模线〈模具〉
~ of line dust 清除细尘
~ of hydrogen atoms 除氢原子
~ of sprues 分离浇口
separator 分离器;选分器;分流芯;型坯分离器;隔板
~ head 分配头;分流芯头
~ roll 分离辊
sepg. [separating] 分离的

sequence 顺序,程序
 ~ control 程序控制
 ~ controlled computer 程控计算机,程控电脑
 ~ switch 程序开关
sequential 顺序的,连贯的
 ~ arrangement 顺序排列〈在聚合物分子中〉
 ~ coating 顺序涂布
 ~ distribution 单体链节分布〈共聚物中〉
 ~ draw 顺序拉伸,外拉伸
 ~ impact moulding 顺序冲压成型
 ~ interpenetrating networks 顺序互穿网〈聚合物〉
  ~ interpenetrating polymer network [SIPN] 顺序互穿网状聚合物
 – mo(u)lding 连贯注塑成型
 ~ operation 时序操作
 ~ polar wind 顺序极性缠绕
 ~ sampling 顺序取样
 ~ wind 顺序缠绕〈容器底部〉
sequestering agent (多价)螯合剂;络合剂〈静电浸渍浴涂胶〉
serial 连续的,顺序的,串联的
 ~ production 成批生产,系列生产
series 连续,串联,系(列);级数
 ~ of materials 材料系
 ~ production 成批生产,批量生产
serrated 锯齿形的
 ~ blade 锯齿形叶片
 ~ blade impeller 锯齿形搅拌叶轮
 ~ heating element 锯齿形加热元件〈塑料焊接〉
 ~ scarf joint 锯齿形嵌接
serum 胶液〈树脂〉
service 服务;运行,使用;维护,保养
 ~ condition 使用条件
 ~ durability 耐用性;耐久性
 ~ life 运转寿命,使用寿命,使用期限
 ~ line 供给管线〈气,水〉
 ~ manual 维护手册,操作规程
 ~ parts 备用零件
 ~ pipe 供水管
 ~ pressure 供给压力〈管〉
 ~ property 使用性能
 ~ simulation test 运转模拟试验
 ~ speed 运行速率[度]
 ~ stress 使用应力
 ~ temperature 使用温度,操作温度
 ~ test 运转试验
 ~ time 使用期限
 ~ tubing 供水管
 ~ voltage 供给电压
 ~ wear 使用磨损
serviceability 适用性
serviceable life 适用期,使用期
services 能量
servo 伺服机构;伺服的
 ~ – acting 伺服作用的
 ~ – hydraulic testing machine 伺服液压试验机
 ~ mechanism 伺服机构
 ~ valve 伺服阀;电动液压阀
sesqui – threaded double – channel 一倍半螺纹双槽螺杆
set 套,组;固化,硬化,凝固;设定,给定,塑化;永久变形,残留变形;装置
 ~ bolt 固定螺栓
 ~ elongation 永久伸长
 ~ forming 固定成型
 ~ length 规定长度
 ~ of mo(u)ldings 成型件产量
 ~ of rolls 成套辊
 ~ of spare parts 成套备件
 ~ of weights 小砝码,成套砝码
 ~ out 开始,准备
 ~ point 固化点,凝固点;给定值
 ~ point control 设定点控制
 ~ point temperature 固化点温度
 ~ resistance 耐永变性
 ~ roller 定形辊
 ~ screw 固定[调节,定位,制动]螺钉
 ~ shaping 固定成型;冷却法热成型
 ~ speed 规定速率[度],额定速率[度]
 ~ stretching 拉伸定形

~ temperature 设定温度,给定温度
~ time 规定时间,调定时间;固化时间,硬化时间
~ - to - touch 快干使灰尘不黏附的〈清漆〉
set - up 固化,硬化;凝固;装置;装配;配套〈美〉
~ time 固化时间;安装时间
setting 安装,装配;调整,设定,给定;固化,硬化;凝结,凝实
~ accuracy 调整精度,定位精度
~ agent 硬化剂
~ dies 调整冲模
~ drawing 装配图
~ plan 装配平面图
~ point 调整点;凝结点,凝固点
~ pressure 设定压力,给定压力
~ range 调整范围
~ screw 调节螺钉,定位螺钉
~ spindle 调整心轴
~ temperature 固化温度,凝固温度;调节温度
~ time 调整时间;固化时间
~ to touch 触摸黏性〈黏合剂〉〈美〉
~ up 固定;硬化;凝结,凝固;装配〈美〉
~ value 设定值,给定值
settle 安排,整理;固定;沉降
~ rate 沉降速率［度］
sever 切断;分离
sewage 污水;下水道
~ pipe 下水管
~ piping 下水管道系统
~ sludge 污水污泥
sewed buff 针刺布轮
sewer 污水管;下水道
~ pipe 下水管,污水管
~ pipeline 下水管道线〈塑料制〉
sewing 缝纫;熔合〈塑料〉
~ thread 缝线
sewn mop 针刺布轮
SF ［safety factor］ 安全系数
SF ［structural foam］ 结构泡沫〈塑料〉
SFAM mat ［synthetic fibrous anisotropic material mat］ 各向异性合成纤维毡,SFAM 毡
SFP ［scrapless forming process］ 无边料成型法
SF/FP ［structural foam made of polypropylene］ 聚丙烯制结构泡沫塑料
SFRTP ［short glass fibre reinforced thermoplastic］ 短玻璃纤维增强热塑性塑料
shade 色调,色泽;遮光罩
~ card 配色样卡
~ deviation 色泽差异
~ pitching 打色样,拼色
shading 暗色;染［着,涂］色
shaft 中心轴,（传动）轴
~ coupling 联轴节,联轴器
~ drive 轴传动
~ of agitator 搅拌器轴
shake 摇动,振动
~ down 试运转;试验性的,临时的
~ - out ﹙用醚﹚提取;振摇出
shaker 振荡器
~ screen 摇动筛
shaking sieve 振荡筛
shallow pan 浅皿〈实验室〉
shank 柄;轴;镜身,镜筒;螺钉尾〈无螺纹部分〉
shape 模型;型材;形状;成型件
~ factor 形状系数
~ memory polymer 形状记忆聚合物
~ of test samples 试样形状
~ of the article 制品形状
~ of the weld 焊缝形状
~ retention 形状稳定性〈塑料成型件〉
shaped 成型的
~ by compression 压缩成型
~ casting 铸塑料
~ goods 定型产品;定型制件
~ sheet ［formed sheet］ 成型片材
shaping 成型,刨削〈加工〉
~ by gravity 重力成型
~ by rollers 滚压成型
~ cycle 成型周期成型模

~ jig 成型用具
~ mould cavity 成型模腔
~ mo(u)ld surface 成型模面〈模具内壁〉
~ operation 成型作业
~ plate 型板,样板
~ plug 成型模塞
~ pressure 成型压力
~ temperature 成型温度
shark skin 鲨皮斑〈塑料件疵点〉
sharp 锐利的,尖锐的;急速的,急转的;突然的
~ – corrugated roll 尖槽辊〈带V形槽的辊〉
~ – V thread 锐角螺纹
~ gate 急转式浇口;突变浇口
shatter crack 细裂纹,微裂纹
shatterproof 耐震的,防冲的,不碎的
shatterproofness 耐震性,防冲性,不碎性
shaving board 刨花(胶合)板
shavings 残屑,刨屑,刨花,薄片
Shaw 肖氏
~ hardness 肖氏硬度
~ process 肖氏法〈制高精度塑料模具方法〉
shear 切变,剪切
~ agate 切断浇口
~ action 剪切作用
~ breakage 剪切破裂
~ crack 剪切裂缝
~ creep recovery 剪切蠕变回复
~ deformation 剪切形变
~ degradation 切变降解
~ – dependence of viscosity 黏度的切变依赖性
~ edge 剪切边料［飞边］;切口〈模具〉
~ elasticity 剪切回弹,剪切弹性(模量)
~ energy 剪切能
~ flexure 剪切挠曲
~ force 剪切力
~ fracture 剪切破裂

~ gate 自断浇口
~ gradient 剪切梯度
~ initiation stress 剪切起始应力
~ marks 刮痕
~ modulus 剪切模量,切变模量
~ rate 剪切速率
~ rate thickening 切速稠化
~ resistance 耐剪切性
~ strain 剪切应变
strength (耐)剪切强度
~ stress 剪切应力
~ stress relaxation modulus 剪切应力松弛模量
~ stretch 切变拉伸
~ surface 剪切面
~ test 剪切试验
~ thickening 切变增稠
~ thinning 切变稀释,剪切稀化
~ torpedo 剪切鱼雷形分流梭
~ velocity 剪切速度［率］
~ viscosity 剪切黏度
~ viscosity coefficient 剪切黏性系数
~ zone 切变区域
shearing 剪切,切变
~ action 剪切动作
~ adhesion 耐剪黏着力
~ component 剪切分力
~ deformation 剪切形变
~ die 剪切模
~ disk viscometer 剪切圆盘式黏度计
~ failure 剪切破坏
~ fatigue tester 剪切疲劳试验仪
~ force 剪切力
~ fracture 剪切断裂
~ rate 剪切速率
~ rate range 剪切速率范围
~ resilience 剪切回弹
~ section 剪切断面
~ strength 剪切强度
~ stress 剪切应力
~ stress course 剪切应力过程;剪切应力分布
~ test 剪切试验
sheath 套,护套;外壳

sheathe 包复,覆盖;铠装
sheathed 带有护套的
sheathed heater 护套加热器
sheathing 套,护套;外壳,包皮;覆盖层
 ~ compound 护套料〈用于电缆〉
 ~ die 护套模头〈电缆〉
 ~ element 包覆元件
sheen (擦亮)光泽
sheet 片材,薄板,薄片,膜片
 ~ bar 薄板坯料,板料
 – base 压片机底座
 ~ blowing 膜片吹塑
 ~ blowing method 膜片吹塑法〈中空制品〉
 ~ calender 片材压延机
 ~ clamp 片材夹具
 ~ coating 片材涂布
 ~ cut 裁片
 ~ cutter 裁片机
 ~ die 片材口模
 ~ dimensions 片材尺寸〈片材:0.25mm及以上;薄膜:低于0.25mm〉
 ~ extruder 片材挤塑机
 ~ extrusion 片材挤塑
 ~ extrusion die 挤塑片材口模
 ~ extrusion equipment 片材挤塑设备
 ~ extrusion line 片材挤塑生产线
 ~ for deep drawing 深度拉伸用片材
 ~ for lining 衬里用片材
 ~ formation pass 成片辊隙
 ~ forming 片材成型
 ~ gasket 密封垫圈,密封垫片
 ~ g(u)ge 片材厚度,厚薄规,板规
 ~ gluer 片材胶合机
 ~ granulator 片材成粒机
 ~ haul-off machine 片材牵引机
 ~ high polymer 片状〈结构〉高聚物
 ~ metal 薄钢板
 ~ metal brake 薄钢板压弯成型机
 ~ mo(u)ld 片材成型模具
 ~ mo(u)lding 片材成型
 ~ mo(u)lding compound [SMC] 片状模塑料,片状成型料
 ~ on the rolls 辊压片材,压延片材

 ~ packing 片状密封件
 ~ parison 片状型坯〈吹塑〉
 ~ pasting machine 片材黏合机
 ~ plate 片材薄板
 ~ polymer 片状聚合物
 ~ reinforcement for laminates 层压板用片材增强料
 ~ resistance 薄膜电阻
 ~ rolls 薄片轧辊
 ~ stock roll 片材卷,成卷片材
 ~ structure 片状结构
 ~ -to-sheet bond 片材粘接,片材与片材粘接
 ~ train 片材生产线
sheeter 压片机
 ~ lines 刨痕,刨纹〈加工痕迹〉
 ~ mark 刨纹
sheeted coating compound 片状涂料
sheeting 片材,薄片,膜片;成片
 ~ calender 片材压延机
 ~ -die extruder 片状模头挤塑机
 ~ dryer 片材材料用干燥器,片材干燥器
 ~ line 片材生产线
 ~ stretcher 膜片拉伸机
shelf 搁置,储存;搁板
 ~ ageing 搁置[储存]老化
 ~ dryer 柜式干燥机
 ~ goods 货架商品
 ~ life 适用期,储存寿命,储存期
 ~ test 搁置试验
 ~ time 储存时间
shell 壳(体);浇铸模型;外套,轴套
 ~ -and-tube exchanger 管壳式换热器
 ~ casting 壳型铸塑
 ~ core 壳芯〈壳型法〉
 ~ coupling 套筒联轴节
 ~ cutter 果壳切碎机
 ~ fabric 面料
 ~ filler 壳粉填料
 ~ flour 果壳粉〈填料〉
 ~ flour filler 果壳粉填料
 ~ method 壳型制造方法

~ mo(u)ld 壳模;克罗宁法用模具
~ mo(u)ld over male mo(u)ld 阳模制壳模
~ mo(u)lding 壳体模塑,壳模铸塑
~ mo(u)lding resin 壳体模塑用树脂
shellac 紫胶,虫胶,虫漆
~ formaldehyde resins 虫胶甲醛树脂
~ plastics 虫胶塑料
~ varnish 紫胶清漆,虫胶清漆
shelve 置于架上,储存
shield 屏蔽;护罩;防护
shielded cable 护套电缆
shielding 防护(的),屏蔽(的)
~ case 保护罩
~ gas 保护气,惰性气体
shift 移位;变换;变速(器);方法;轮班
~ factor 移位因子
~ knob 开关按钮
shifter 分离器
shifting machine 筛分机
shingle of lamination 层合屋顶板
ship bottom paint 船底漆,船底涂料
shipping 海运,航运,装运;发货
~ container package 海运集装箱包装
~ package 装运包装
~ order 送货单,发货单
SHIPS [super-high impact polystyrene] 超高冲击聚苯乙烯
shish kebab 串型结晶结构〈由应力诱发结晶产生聚合物块状链结构〉
shock 震动;冲击;休克
~ absorbent 减震的,缓冲的
~ absorber 减震器,缓冲器
~ absorber bumper 防撞保险杠
~ absorbing insert 减震垫,弹性垫,弹性缓冲层垫
~ absorption 减震,缓冲
~ bending test 冲击弯曲试验
~ chill roll 骤冷辊
~ cooling 骤冷
~ elasticity 冲击弹性
~ factor 冲击系数
~ freezing process 骤冷法
~ heating 骤热
~ load 冲击负荷
~ mo(u)lding 冲压模塑
~ pendulum 冲击摆锤
~ proof 抗震的,防震的
~ proofing package 防震包装
~ proofness 耐冲击性,防震性〈塑料材料〉
~ resistance 耐冲击性,防震性
~ stress 冲击应力
~ strength 耐冲强度
~ test 冲击试验
~ wave 冲击波
shockless 无冲击的,无震动的
shoddy yarn 软再生毛纱
shoe 模套,模圈;导向板;鞋
~ hot melt adhesive 鞋用热熔胶
~ lining 鞋衬里
~ sole 鞋底
shooting 注射,注塑
~ cylinder 注塑料筒
~ piston 注塑料柱塞;注料杆
shop fascia 店面招牌
Shore 邵氏
~ durometer 邵氏(压痕)硬度计
~ durometer hardness 邵氏压痕硬度
~ hardness 邵氏硬度
~ hardness tester 邵氏硬度计
~ scleroscope 邵氏反弹式硬度计
~ scleroscope hardness 邵氏反弹式硬度
~ scleroscope hardness test 邵氏反弹式硬度试验
short 短的;脆的;缺料〈模塑件〉;废料
~ beam shear strength 短试片剪切强度
~ chain 短链
~ -chain branch 短支链
~ circuit 短路〈电〉
~ fibre 短纤维
~ -fibre reinforced thermoplastic 短纤维增强热塑性塑料
~ fining 欠料,给料不足;欠注〈注塑〉
~ glass fibre 短玻璃纤维
~ glass fibre-filled thermosetting ma-

terial 短玻璃纤维填充热固性材料
~ glass fibrereinforced thermoplastic [SFRTP] 短玻璃纤维增强热塑性塑料
~ grain 细粒
~ grained 细粒的
~ – life 短寿命的
~ mo(u)lding 注料不足成型件,欠注成型件
~ – neck bottle 短颈瓶
~ of raw material 原料不足
~ oil 短油度的;短油,聚合程度不大的油
~ – oil alkyd 短油度醇酸树脂,含油量低的醇酸树脂
~ – oil varnish 短油清漆
~ pitch 短螺距
~ run 小批量〈生产〉
~ run production 小批量生产
~ shot 缺料;欠注,注料不足〈模具〉
~ shot point 欠注点
~ slit drawing 微缝型拉伸〈薄膜〉
~ – spin [spinneret] extrusion 短纺挤塑
~ spin technology 短纺工艺
~ stopper 阻聚剂
~ stopping agent 速止剂;链终止剂〈高聚物〉
~ – stroke press 短冲程压机
~ – term 短期的
~ – term creep 短期蠕变
~ – term load 短期负荷
~ term resistance 短期隐定性〈辐射〉
~ – term test 短期试验,快速试验
~ test 快速试验
~ – time deformation behaviour 瞬间形变行为
~ time set 瞬间变形
~ time stability 瞬间稳定性
~ time tensile test 瞬间拉伸试验
~ time test [STT] 瞬间试验,快速试验
~ ton 短吨〈2000磅,合907.185kg〉
shortage 短缺,不足,缺乏

shortness 短;不足;脆性;难抽丝〈树脂〉
shot 注射,注塑;注塑量;注射量
~ bag mo(u)lding 沙袋成型,挤压成型
~ capacity 注塑能力;注塑量
~ cycle 注塑周期
~ – in mandrel [spigot] technique 嵌入芯棒技术〈吹塑〉
~ mo(u)lding 沙袋成型,挤压成型;注塑
~ rate 注塑速率;注塑周期
~ size 注塑量〈注塑机的〉
~ volume 注塑体积;注塑量
~ volume controller 注塑量调节器
~ weight 注塑量〈注塑机的〉
shotting cylinder 注塑料筒
shoulder 突缘;肩,台肩;钝边〈焊〉;瓶肩〈吹塑瓶〉
~ insert 凸缘嵌件
shouldered rod 肩状试棒.
show 显示,显露,展示
shredder 撕碎机;粉碎机
~ black 粉碎机叶轮
shredding 撕碎,粉碎
shrink 收缩
~ allowance 收缩余量
~ back 回缩,缩回,退缩
~ block 收缩胎模
~ coating 热收缩涂层
~ film 收缩薄膜
~ fit 冷缩配合
– **fixture** 防收缩模架,冷却模架
~ jig 防缩夹具,防缩模
~ mark 缩痕,缩纹
~ on 缩紧〈热状态〉
~ – on coating 热收缩涂层
~ packaging 收缩包装
~ – proof 不收缩的,防缩的
~ – proof finish 防缩整理
~ resistance 防缩性
~ resistant finish 防缩加工
~ ring 收缩环
~ stress 收缩应力
~ tension 收缩张力

shrinkable
~ tunnel 收缩烘道
~ wrapper 收缩包装；收缩包装机
~ wrapping 收缩包装；收缩包装材料
shrinkable 可收缩的
~ block jig 可收缩胎模
~ tube 可收缩管
shrinkage 收缩；收缩性，收缩量，收缩率
~ across flow 横向收缩，与流动方向垂直的收缩
~ allowance 收缩裕量
~ anisotropy 各向异性收缩
~ behaviour 收缩性能
~ block 收缩胎模，冷却胎模
~ cavity 收缩孔
~ control finish 防缩整理
~ crack 收缩开裂；收缩裂缝
~ crimping 收缩卷曲
~ curve 收缩曲线
~ factor 收缩系数
~ film [sheeting] 收缩薄膜[膜片]
~ fit 收缩配合
~ fixture 防缩模架，防缩檀
~ hole 缩孔
~ in length 经向收缩，长度收缩
~ in width 纬向收缩，宽度收缩
~ jig 防缩夹具，防缩模
~ level 收缩程度
~ limit 收缩极限
~ mark 缩痕，缩纹
~ of volume 体积收缩
~ pool 收缩坑〈成型件表面缺陷〉
~ potential 潜在缩率
~ ratio 收缩比，收缩系数
~ strain 收缩应变
~ stress 收缩应力
~ temperature 收缩温度
~ with flow 流向收缩
shrinking 收缩
~ effect 收缩效应
~ -on 缩紧〈热状态〉
~ treatment 收缩处理
shrivel varnish 皱纹漆
shrunk-on 收缩到……上，缩紧
~ coating 收缩涂层

~ flim 收缩薄膜
~ joint 收缩连接〈塑料管线〉
~ pipe joint 收缩管连接
~ sleeve 收缩磁筒
shuffling time 压合时间
shut-off 合模口，模缝脊；关闭〈管路〉；切断〈电源〉
~ cock 关闭旋塞，(切断)活栓〈美〉
~ needle 关闭式注针，关闭式注嘴
~ nozzle 关闭式注嘴
~ pressure 封闭压力
~ slide 关闭式闸板
~ valve 关闭阀，止流阀
shuttering 模板
shuttle carriage 导向滑板，穿梭式滑架〈缠绕法〉
~ riding on oblique rail 有倾斜滑轨的穿梭式滑架〈缠绕法〉
SI [silicone] 硅
SI [silicones in general] 有机硅通称
S. I. thread 单螺纹
SI unit [Standard International unit] 标准国际单位制单位，SI单位
Siamese blow 连体吹塑，夏姆吹塑
~ mo(u)lding 夏姆吹塑法，一模多穴吹塑法，连体吹塑法〈一个型坯同时吹塑多个中空容器〉
siccative 干燥剂；干燥的
side 旁边(的)，侧面(的)
~ bars 侧位活络销
~ bearing 轴向轴承，端轴承
~ blow 侧吹
~ blowing 侧吹
~ by side 并排地，并列地，相并
~ chain 侧链，支链
~ core 侧位模芯
~ cut 切边
~ cylinder type double-acting press 侧缸型双动压机〈合模力由主柱塞提供，开模力由侧位油缸提供〉
~ delivery head 斜向挤出机头〈注塑机〉
~ discharge chute 侧向卸料槽
~ edge gate 齐边浇口

~ effect 副作用
~ elevation 侧视图
~ entering 从侧面进入
~ - entering feed 侧向进料
~ entry 侧向进料
~ extrusion 侧向挤塑,角式挤塑,斜向挤塑
~ fed 侧向进料的
~ feed 侧向供料
~ feed [fed] head 侧向供料机头〈吹塑机〉
~ feed opening 侧向供料口
~ gate 侧向浇口
~ - gated injection mould 侧向浇口注塑模
~ group 侧基
~ hung door 双扇门,摺门
~ injection mo(u)lding 侧向注塑成型
~ pin 侧向销
~ plate 侧板
~ pressure 侧压
~ product 副产品
~ pull 侧抽芯〈模具〉
~ roll 侧辊〈倒 L 形四辊压延机〉
~ runner mixer 有侧料道的混合器〈穆勒混合器的一种〉
~ shears 切边机
~ strain 侧向应变
~ trim unit 两侧裁边装置
~ tube 支管,侧管
~ view 侧视图
~ weld sealing 纵向焊接;用电热丝边部热合〈制袋时〉

sieve 筛;筛分;滤网;过滤
~ analysis 筛分,筛析
~ aperture [sieve opening] 筛孔
~ drum drier [dryer] 筛网圆筒干燥机
~ grate 筛箅
~ mesh 筛眼,筛孔
~ number 筛号
~ opening 筛孔
~ plate 筛板
~ - plate column 筛板塔

~ residue 筛余物,筛渣
~ retention 筛分保留物
~ size 筛眼孔径
SIR [Swedisch Rubber Association] 瑞典橡胶协会
sift 筛分;过筛
sifter 筛,筛分机,筛选机
sifting 筛分,筛选;筛余物
~ machine 筛分机,筛选机
sight 视力;观察
~ control 目视检查
~ hole 观察孔,检视孔
~ glass 观察镜,视镜
sighting 着色,涂色,指色
sigma 〈希腊字母〉σ,∑,西格玛
~ bond σ 键
~ - shaped kneader mixer 西格玛型捏和混合机,Z 型捏和机
~ - type kneader [Z - shaped blade kneader] 西格玛型捏和机
signal 信号,指令
silane 硅烷
~ coupling agent 硅烷偶联剂〈偶联剂〉
~ finish 硅烷处理剂
~ treatment 硅烷处理
silanized 硅烷处理的
~ silica flour 硅烷处理的石英粉〈合成树脂专用填料〉
~ strand 硅烷处理的玻璃纤维束
Silica 二氧化硅〈填料〉
~ flour 石英粉
~ gel 硅胶〈催化剂〉
~ gel drier 硅胶干燥剂
~ soil 硅土
~ white 白炭黑;硅补强剂
silicate 硅酸盐 [酯]
siliceous earth 硅藻土
silicic acid 硅酸
~ gel 硅(酸)胶
silicon [Si] 硅
~ carbide 碳化硅
~ carbide fiber 碳化硅纤维
~ dioxide 二氧化硅

- ~ modified rubber 硅改性橡胶
- ~ mo(u)ld release agent 硅脱模剂
- ~ polymer 硅聚合物

silicone 聚硅氧烷;硅酮;硅树脂
- ~ adhesive 聚硅氧烷黏合剂,硅酮黏合剂,有机硅黏合剂
- ~ -bonded glass cloth 聚硅氧烷涂敷的玻璃布,硅酮涂敷的玻璃布
- ~ carbide [carborundum] 碳化硅
- ~ -ceramic mo(u)lding compound 硅酮-陶瓷模塑料,聚硅氧烷-陶瓷模塑料
- ~ diffusion pump fluid 有机硅扩散泵油
- ~ dioxide 二氧化硅
- ~ elastomer 硅酮弹性体,有机硅弹性体
- ~ fluid (有机)硅油
- ~ foam 有机硅泡沫塑料
- ~ glass resin 有机硅玻璃树脂
- ~ grease (有机)硅脂〈润脂剂〉
- ~ grease of high vacuum 高真空用(有机)硅脂
- ~ in general 有机桂通称
- ~ laminate 有机硅层压塑料
- ~ mo(u)lding compounds 有机硅模塑料
- ~ -nitrile rubber 硅腈橡胶
- ~ oil (有机)硅油
- ~ plastic 聚硅氧烷塑料,硅酮塑料,有机硅塑料
- ~ -polycarbonate block copolymer 有机硅-聚碳酸酯嵌段共聚物
- ~ polyether copolymer 硅酮聚醚共聚物,聚硅氧烷-聚醚共聚物
- ~ polymer 硅酮聚合物,有机硅聚合物
- ~ potting compound 硅酮浇铸料,聚硅氧烷浇铸料
- ~ release agent 聚硅氧烷脱模剂
- ~ release agent for paper 纸用有机硅脱模剂
- ~ resin 硅酮树脂,聚硅氧烷树脂,有机硅树脂
- ~ rubber 有机硅橡胶,硅(酮)橡胶,聚硅氧烷橡胶
- ~ surfactants 有机硅表面活性剂〈脱模剂、润滑剂、消泡剂、抗静电剂、添加剂〉
- ~ varnish 硅酮清漆,有机硅清漆,聚硅氧烷清漆

silicones 聚硅氧烷,(聚)硅酮

silk 丝;丝织品
- ~ screen printing 丝网印刷〈塑料表面〉
- ~ screen system 丝网印刷系统

sill profile 门槛型材
sillimanite 硅线石
siloxane (聚)硅氧烷
- ~ epoxide (聚)硅氧烷环氧物

silver [Ag] 银;镀银
- ~ lactate catalyst 乳酸银催化剂
- ~ spray process 喷银法
- ~ streak 银丝纹

simple 简单的,单纯的
- ~ beam impact machine 简支梁冲击(试验)机
- ~ beam impact test 简支梁冲击试验
- ~ bending 单纯弯曲
- ~ elongation 单向伸长
- ~ ester plasticizer 单酯类增塑剂
- ~ folding 单折叠
- ~ injection mould 单注塑模
- ~ -lap joint 单搭接〈粘接〉
- ~ -lapped adhesive joint 单搭接粘接
- ~ -overlapping adhesive joint 单搭接粘接
- ~ scarf joint 单斜接
- ~ spread 单面涂布
- ~ stress 单轴向应力

simplification 简化;合理化
simplified 简易的,单一的
- ~ mo(u)ld 简易模具
- ~ symbol 简化符号

Simpson 辛普森
- ~ mixer 辛普森混合机
- ~ mix-muller 辛普森混合研磨机

simulation 模拟,模仿

simultaneous 同时的
~ draw texturing machine 同时拉伸变形机
~ interpenetrating network [SIN] 同时互穿网状(聚合物)
~ thermogravimetry and differential thermal analysis 同时热失重分析和差热分析法
SIN [simultaneous interpenetrating network] 同时互穿网状(聚合物)
singe 烧焦
single 单(独)的
~ – action press 单动压机
~ – action ram 单动活塞
~ arm kneader 单桨捏和机
~ bar bumper 单杆保险杠
~ bead 单焊珠
~ – bead weld 单珠焊
~ beam paddle mixer 单(臂)桨叶式混合机
~ bevel groove 单斜坡口〈焊接〉
~ – bevel groove weld 单斜坡口焊接;半 V 形焊缝
~ – blade 单桨叶
~ blade kneader 单桨捏和机
~ blade kneading 单桨捏和
~ blade mixer 单桨搅拌机
~ – block shear test 单块试样剪切试验〈测定夹层结构层间粘接性〉
~ bond 单键
~ – bond polyheterocyclics 单键聚杂环;螺旋形聚合物
~ – cavity conical mo(u)ld 单模腔锥形模具〈测定材料热变形性能〉
~ – cavity cubical mo(u)ld 单模腔立体模具〈测定材料热变形性能〉
~ – cavity mo(u)ld 单模腔模具
~ cavity roll 空心滚筒
~ circuit circular hoop winding 单环路圆周缠绕
~ circuit helical winding 单环路螺旋缠绕
~ – circuit pattern 单环路缠绕
~ circuit polar winding 单环路极向缠绕
~ circuit winding 单环路缠绕
~ coating 单(层)涂层
~ column press 单柱压机
~ component epoxy adhesive 单组分环氧黏合剂
~ – core cable 单芯电缆,单线电缆
~ crystal 单晶
~ crystal fibre 单晶纤维,须状
~ curved 单弧形的〈壳〉
~ depth of thread 螺纹单齿深度
~ die tubing machine 单模头挤管机
~ drive 单独传动
~ drum drier [dryer] 单鼓干燥器
~ – faced 单面的
~ – fillet weld 单角焊
~ – flight [single flighted] 单螺纹的
~ – flighted screw 单(头)螺纹螺杆
~ gate mo(u)ld 单浇口模具
~ glass filament yarn 单玻璃纤维长丝纱
~ glass staple fibre yarn 单玻璃纤维短丝纱
~ groove 单面坡口
~ – head machine 单机头机〈聚氨酯泡沫塑料反应铸塑〉
~ helical structure 单螺旋形结构
~ – impression mo(u)ld 单型腔模具
~ lap joint 单搭接〈粘接〉
~ – lap weld 单搭接焊接
~ lapped adhesive joint 单搭接粘接
~ layer 单层的
~ – layer film 单层薄膜
~ – manifold die 单料道模头〈共挤塑〉
~ – material extrusion 单一原料挤塑
~ – overlapped adhesive joint 单搭接粘接
~ – part injection mould 单件注塑模具
~ – part production 单件生产
~ parting line mo(u)ld 单合模线模具
~ pass 单向;单程
~ pass bonding 单程黏合

V with root face

~ pass dip coater　单向蘸涂机
~ piece production　单件生产
~ piece work　单件加工
~ - point injection　单孔注塑
~ - property test method　单项性能试验方法
~ - property type test　单项性能试验
~ - roll breaker　单辊破碎机
~ - roll crusher　单辊压碎机
~ - roll extruder　单辊挤塑机
~ roll feeder　单辊供料器
~ - runner disc mill　单回转磨
~ - runner mill　单回转磨
~ - screw extruder　单螺杆挤塑机
~ - screw extruder with double - screw feed　双螺纹喂料的单螺杆挤塑机
~ - screw extruding machine　单螺杆挤塑机
~ - service package　单份包装
~ - shear lap joint　单剪搭接〈搭接试样作剪切强度试验〉
~ - shell　单壳(层)的
~ - sided coating　单面涂布
~ - slot die　单缝口〈共挤塑〉
~ - spindle　单轴的
~ - split mo(u)ld　简易对分式模具
~ spread　单面涂敷量；单面施胶，单面涂布
~ spread weight　单面涂布量
~ - stage　单级的
~ - stage compression　单级压缩
~ - stage plastication　一步塑化，直接塑化
~ - stage plunger injection　单级柱塞式注塑成型
~ - stage process　一步制备法
~ - stage resin [SS]　甲阶酚醛树脂；可溶性酚醛树脂
~ - stage screw　单级螺杆
~ - start screw [single - flight screw]　单螺纹螺杆
~ - start screw profile [single - flight screw profile]　单螺纹螺杆型材
~ - station injection mo(u)lding　单工位注塑成型
~ - station machine　单工位机
~ - station thermoformer　单工位热成型机
~ - step resin　甲阶酚醛树脂；可溶酚醛树脂
~ strand chain　单股链
~ - strand polymer　单链聚合物
~ stroke　单行程
~ - stroke machine　单冲程机
~ - stroke preforming press　单冲程预成型压机
~ tab gate　单柄型浇口
~ thread　单(头)螺纹
~ - thread screw　单(头)螺纹螺杆
~ - tier injection mould　单层多型腔注塑模具
~ toggle crusher　单极颚式破碎机
~ toggle locking system　单肘节销模系统
~ V [single vee]　单V形
~ V - butt joint　单V形对接
~ V - butt joint with sealing run　单V形封底对接焊缝
~ V - butt weld　单V形对接焊缝
~ V - butt weld with backing run　单V形封底对接焊缝〈美〉
~ V groove　单V形坡口；单V形槽
~ V weld　单V形焊缝
V with root face　单V形齿根面，Y形焊缝
~ yarn　单纱
sink　沉降，凹入
~ mark　凹痕，缩痕〈制品缺陷〉
~ of heat　受热器，吸热设备；保热，保温
sinter　烧结，熔结，粘结
~ coating　烧结涂布；熔丝涂层
~ forming　烧结成型
~ - fuse　烧结 - 熔化
~ - fuse coating　烧结涂布
~ mo(u)lding　烧结成型；熔结成型
sintered　烧结的
~ cake　烧结块

~ ceramics 烧结陶瓷
~ material 烧结材料
sintering 烧结,熔结
~ bath 烧结浴,液体化粉末浴
~ oven 烧结炉,熔结炉
~ point 烧结点,软化点
~ temperature 烧结温度
sinuous – type buff 波型抛光轮
siphon [syphon] 虹吸,虹吸管
~ tube 虹吸管
SIPN [sequential interpenetrating polymer network] 顺序互穿网状聚合物
SIR [styrene isopren elastomeric copolymer] 苯乙烯–异戊二烯共聚弹性体
sirup 胶液〈树脂〉
SIS [styrene – isoprene – styrene] 苯乙烯–异戊二烯–苯乙烯〈热塑性橡胶〉
SIS [Swedisch Standardization] 瑞典标准
sisal 剑麻
~ fiber 剑麻纤维
SIT [self – ignition temperature] 自燃温度
site 地点,场所,现场
~ assembly 现场装配
~ test 现场试验
six – impression injection mould 六型腔注塑模
size 尺寸,大小;值;容积;上浆,浆料,浸润剂,润滑剂;校准;规格;筛分;定径
~ distribution 粒径分布
~ finish 胶浆;铸型涂料
~ mixture 混合浆料
~ of grain 粒度,粒径,粒径度
~ of granulation 颗粒大小,颗粒度
~ of mesh 筛孔尺寸
~ of platen 压板面积;模具安装面积〈压机〉
~ press 上浆辊,双辊涂布机;黏合压机
~ – reducing machine 粉碎机
~ reduction 磨碎;粉碎;成粒,粒化
~ residue 浸润剂残留量
~ separation 粒析;颗料离析
~ spray [lubricant spray] 浸润剂喷洒

〈玻璃纤维〉
sized 上浆(过)的
~ fibre 上浆纤维
~ glass fiber 上浆玻璃纤维
sizing 上浆,填料,胶料;定径;定型;校准;表面修整[处理]
~ agent 浆料,上浆剂
~ bush （管材）定径套
~ control 尺寸控制
~ die 定型模,定径模
~ material 浆料,胶料
~ plate 定型板,定径板
~ press 上浆辊,双辊涂布机;黏合压机
~ roll 定型辊
~ roller 上浆辊
~ sleeve 定型套,定径套
~ system 定型系统
~ treatment 上装（处理）
~ tube 定型管,定径管
skein 绞纱
skeiner 缠绕器
skeletal matrix 骨架阳模
skeleton 骨架
~ construction 骨架结构
~ drawing 草图,轮廓图,示意图,结构图
~ electrode 骨架电极
~ forming 骨架成型,快速反吸真空成型
~ frame 骨架,框架
~ male die 框式阳模
~ mo(u)ld 骨架成型模具〈吹塑模〉
~ pattern 轮廓模型,构架模型
~ structure 骨架结构
~ tool forming 骨架模成型
skew 斜(的);弯曲的,挠曲的;不对称的
~ curve 挠曲线
~ distribution 不对称分布
~ symmetric 不对称的
skewed roll [crossed roll] 斜交辊
skewing 斜(交);弯曲
skid 滑动;过压溢料痕〈注塑疵点〉
~ drying 涂层表面干燥

~ resistance 防滑性
~ resistant 防滑的
billed worker 受过专门培训的工人
skim rubber 胶清橡胶
skin 皮层〈微孔塑料的〉;表层;皮肤
~ anti-toxic 皮肤消毒剂
~ breakage 表面破坏
~ coat 表层
~ covering 表层
~ density 皮层密度
~ -dry 表层干燥的
~ effect 表皮效应〈高频焊接时〉
~ friction force 表面摩擦力
~ formation 表皮形成
~ hardness 表面硬度
~ irritant 皮肤有刺激性的
~ layer 表层
~ package 表皮包装,贴体包装
~ reinforcement 表面增强材料
~ stress 表面应力
~ structure 平面承载结构
~ zone 表皮区
skinning 结皮;剥皮
~ machine 剥皮机〈薄片〉
skips 漏涂处〈黏合剂或漆涂敷时〉
skirt 边〈缘〉
skirting 踢脚板,护墙板
skive 磨光;削去,削去〈皮〉;切成薄片
sl. [slight] 轻微的;少量的〈化学浸蚀〉
slab (厚)片,板;板坯
~ of cement 混凝土板
~ of foam 泡沫塑料板
~ of slate 石板
~ production 板料生产,板块生产
~ rubber 橡胶块
~ stock (泡沫塑料)板块,板料
~ stock production 板块生产
slag wool 矿渣棉
slagging 料团堵塞〈料道〉
slat 板条
~ expander 板条式展幅机
slate 石板,板岩;石板色,蓝灰色
~ flour 板岩粉〈填料〉
sleeve 套筒

~ barrel 套筒
~ bearing 套筒轴承
~ collar 套筒〈美〉
~ ejection 套筒式脱模
~ ejector 套筒脱模销
~ joint 套接
~ knockout pin 套筒脱模销
~ nut 套筒螺母〈美〉
~ pin 套筒销
~ welding 套焊〈塑料管〉
sleeving 套管
slenderness ratio 细长比,长径比
slice 切片,切割〈块材〉;薄片
sliced 切成片的
~ film [seet] 切成薄片,刨成片材
~ veneer 切成薄木片,刨成片板〈胶合板用〉
slicer 刨片机;切片机
slicing 刨片,切片
~ knife 切片刀
~ machine (块料)刨片机
slick foil 润滑膜
slide 滑动;滑动阀;滑(动)片,承物玻璃片,载片〈显微镜〉
~ bar 滑杆
~ bearing 滑动轴承
~ block 滑块
~ box 滑动盒
~ carriage 滑座
~ control 滑动调节
~ face 滑动面
~ follower 滑动随动机构
~ gauge 滑尺,游标卡尺
~ glass 滑片,承物玻璃片,载片
~ guide 滑动导轨
~ lid 滑盖
~ lock lid 滑锁合盖
~ mould 滑动模具
~ nozzle 滑动喷嘴
~ plate 滑板
~ prevention 防止滑移,防滑
~ property 滑动性能
~ rail 滑轨
~ support 滑动架,滑动轨

~ surface （area）滑动面
~ valve 滑阀
slider 滑动器，滑板
slideway 滑道
sliding 滑动（的）
~ bar 滑杆
~ bearing 滑动轴承
~ bed 滑座
~ bed mo(u)ld 滑座式模具
~ balde 滑片
~ block 滑块
~ carriage 滑动架
~ carriage mo(u)ld 滑道式模具
~ clamp 滑动夹头
~ clutch 滑动离合器
~ contact 滑动接点〈电〉
~ core [moving insert] 滑动模芯,滑动型芯
~ core mo(u)ld 可滑动型芯的吹塑模
~ discharge door 滑动式卸料阀门
~ discharge equipment of banbury mixer 密炼机滑动式卸料装置
~ door 滑门
~ draft 滑溜牵伸
~ friction 滑动摩擦
~ guide 滑动导轨
~ joint 滑动接合
~ material 润滑材料
~ mould blow moulding 滑动型模具吹塑
~ mo(u)ld process 滑动型模具成型法〈挤吹〉
~ nozzle 滑动喷嘴
~ panel 滑动模槽板
~ plate 滑动模槽板
~ -plate metering device 滑板式计量装置
~ plate viscometer 滑板式黏度计
~ pressure bar 滑动压板
~ property 滑动性能,滑爽性
~ punch 滑动阳模,滑动冲头
~ resistance 防滑性
~ sleeve 滑动套筒
~ socket joint 滑动套筒接合

~ speed 滑动速率［度］
~ support 滑动支架
~ surface 滑动表面
~ switch 拨动开关,滑动开关
~ valve 滑阀
~ wedge 滑动模楔
~ weight balance 游码式天平
slight 轻微的,稍微的
~ attack 轻微化学浸蚀
sligthly 轻微地,稍微
~ incompatible 微不相容的,相容性不良的
~ soluble，[sl. s.] 微溶
slinger mo(u)lding 离心式成型（法）
slip 滑动,滑移；滑爽(性),滑润(性)
~ additive 滑爽添加剂,防粘连剂
~ agent 滑爽剂,防粘连剂
~ board 滑板
~ cap 滑动盖
~ coefficient 滑动系数
~ draft 滑溜牵伸
~ expansion joint 滑动膨胀接合
~ fittings 滑动配件
~ forming 滑片成型〈热成型〉
~ joint 滑动接头
~ -layer 滑层；减摩层〈自动形成〉
~ of cylinder 滚筒滑动
~ plane 滑动面,滑移面；断层；裂缝形成,开裂〈由于骤冷〉
~ proofing agent 防滑剂
~ resistance 防滑性
~ resistant 防滑的
~ ring forming 滑环式成型
~ sheet 防黏垫片
~ -stick 滑动黏附现象
~ -stick effect 滑动黏附效应
~ surface 滑动面
~ thermoforming （阳模）滑片热成型
slippage 滑动,滑移〈黏合剂固化时粘接件的相对移动〉
~ effect 滑动效应
~ factor 滑动系数
slipperiness 滑溜,滑动行为,减摩行为
slippery hand 手感光滑

slipping 滑动
　~ agent 滑爽剂
slit 隙缝,缝口〈挤塑模〉;撕裂
　~ die 缝口模头
　~ - die extrusion 缝口模挤塑
　~ die rheometer 缝口模流变仪
　~ fibre 裂膜纤维,扁丝,撕裂薄膜
　~ fibre yarn 裂膜纤维线
　~ film 撕裂(薄)膜
　~ film die 裂膜模头〈美〉
　~ film tape 裂膜带,裂膜扁丝
　~ - film tape technology 裂膜带工艺
　~ - film tape used for weaving 编织用裂膜带
　~ orifice 裂缝
　~ viscometer 缝隙式黏度计
　~ yam 撕裂纱,撕裂纤维,劈裂纤维
slitter 切条机,分切机
　~ - type cube cutter 分切型方粒切粒机
　~ type machine 切条型机
slitting 分切;纵向切条纵裂
　~ needle 成纤用针,纤化用针
　~ resistance 耐撕裂性
　~ saw 切割锯
　~ shear machine 剪切机
　~ strength 撕裂强度
sliver (长)条纱,薄片;碎料
　~ sheet 薄片
slope 倾斜;斜面;斜度
　~ angle 坡度角
　~ coefficient 斜率
　~ of curve 曲线斜度
　~ of thread 螺纹斜度
sloped feed throat 斜壁进料口
sloping roof 斜(屋)顶
slops tank 废馏分槽;蒸馏
slot 槽,缝;裂口,缝口
　~ cut die 切口模
　~ depth 槽深
　~ die 缝口模头
　~ die extruder 缝口模(头)挤塑机
　~ die extrusion 缝口模(头)挤塑
　~ - die film extrusion 缝口模薄膜挤塑
　~ extrusion 缝口挤塑
　~ extrusion flim 缝口挤塑薄膜
　~ gate 缝型浇口
　~ lining 槽衬里
　~ pitch 槽距
　~ spreading 缝口涂布
　~ type die 缝型模头
　~ weld 槽型焊缝,切口焊缝
　~ width 槽宽
slotted 裂缝的,开裂的,有槽的
　~ applicator 缝口模头涂布机
　~ plate 槽孔板
slow 缓慢的
　~ agitator 慢速搅拌器
　~ burning 缓慢燃烧
　~ burning fabric 耐烧织物
　~ burning test 缓慢燃烧试验
　~ - curing adhesive 缓慢固化黏合剂
　~ - curing resin 缓慢固化树脂
　~ - down 降速,减速〈模具开闭〉
　~ - elastic defoliation 缓弹性变形
　~ motion picture 快速拍摄影片
　~ oxidation 缓慢氧化
　~ running control 低速运转控制
　~ speed mixer 慢速混合器
sl. s. [slightly soluble] 微溶
slub 块,堆
sludge 污泥;残渣
　~ valve 污水阀
slug 冷料;流道残料
　~ well 冷料井
slugging fluidized bath [bed] 腾涌式流化床
slugwise 间歇
　~ continuous flow mixer 间歇流动混合器
　~ proportioning 间歇配料
sluice 闸门,水门
　~ valve 闸阀,滑板阀
slurry 淤浆
　~ polymerization process 淤浆聚合法
　~ preform process 料浆喷射预成型法
　~ preforming 料浆喷射预成型(法)

slush 糊料;胶泥
~ casting 搪铸;中空铸塑(件),糊料 慢塑
~ mo(u)ld 搪塑模具
~ mo(u)lding 搪塑
SM [styrene monomer] 苯乙烯单体
SMA [styrene maleicanhydride copolymer] 苯乙烯-马来酸酐共聚物,苯乙烯-顺酐共聚物
small 小型的,小规模的,细小的
~ - angle electron diffraction 电子小角衍射
~ angle scattering 小角散射
~ angle X - ray scattering 小角X-射线散射,X-射线小角散射
~ batch production 小批生产
~ batch production method 小批生产法
~ casting 小铸塑件
~ crack 细裂纹
~ foot 小成型件底座
~ - injection mo(u)lding 小型成型件注塑成型
~ - lot production 小规模生产
~ mo(u)lding 小型(塑料)成型件
~ part 小规格制品,小件模制品
~ pipe 小直径管
~ scale 小型的,小规模的
~ scale production 小规模生产
~ scale test 小规模试验,小型试验
~ serial production 小批生产
~ test 小型试验
SMC [sheet mo(u)lding compound] 片状模塑料,片状成型料
smear head 塑炼头
smearing 塑化
smell 臭味;气味;嗅觉
smelly 有臭味的
smoke 烟,烟尘;烟雾
~ detector 烟探测器;检烟器
~ emission 发烟
~ production 发烟
smooth 平滑的,光滑的,平稳的
~ - bored 平整钻孔的

~ combustion 稳定燃烧
~ contour 光滑外形
~ fibre 平滑纤维
~ finish 光洁加工
~ metering rod 光滑计量棒
~ mixing head 平滑混炼机头〈挤塑螺杆〉
~ operation 平稳运转
~ roll 压光辊
~ surface 光滑表面
~ - touch 手感光滑
smoothing 光滑的,平稳的
~ equipment 平整设备
~ press 压光压机
~ roll 压光辊
smoothness 光滑度;滑爽性
smouldering 闷烧,发烟燃烧
SMS [styreneα - methyl styrene copolymer] 苯乙烯-α-甲基苯乙烯共聚物
smudge 污迹,斑痕〈成型件缺陷〉
Sn [tin] 锡
snag 钩丝〈针织疵点〉
snap 迅速闭合;压钮,按扣
~ fitting 卡扣接头
~ package 按扣包装
snap - back 急速返回,迅速回复
~ coating 急速回缩涂层
~ fibre 弹性纤维,松紧纤维
~ forming 快速回吸成型
~ thermoforming 快速回吸热成型,阳模反吸热成型
snarl effect 卷缩效应
snubber rolls 牵伸辊,拉伸辊〈薄膜或丝拉伸〉
snug - fitting 紧密配合,紧贴配合
soak 浸湿,浸渍
~ into 浸渍,浸透
soaking 浸渍,浸透
~ machine 浸渍机
soap 肥皂
~ bubble leak detection 皂泡法检漏
~ content 皂含量;碱含量
~ fastness 耐皂牢度
socket 插管,套管;连接管;轴颈轴承

~ fitting 插管配件
~ fusion 插管熔焊
~ fusion fitting 插管熔焊配件
~ fusion joint 插管熔焊
~ fusion jointing 插管熔焊
~ joint 窝接
~ machine 套接机
~ tube 套管
soda 苏打;碱;碳酸钠
~ cellulose 碱纤维素
~ lime glass 钠钙玻璃〈增强材料〉
~ lye 碱液
~ paper 碱法纸粕
~ pulp 碱法浆粕
sodium [Na] 钠
~ acide 叠氮化钠
~ alginate 藻酸钠
~ alkyl sulfonate 烷基磺酸钠〈分散剂、乳化剂〉
~ bicarbonate 碳酸氢钠,酸式碳酸钠,重碳酸钠,重碱,小苏打,焙烧苏打〈发泡剂〉
~ borohydride 氢硼化钠
~ butyl oleate sulfate 油酸正丁酯硫酸钠盐〈油剂、润湿剂〉
~ carbonate 碳酸钠,纯碱
~ carboxymethyl cellulose 羧甲基纤维素钠盐〈泥浆处理剂,有机助洗剂,增黏剂,乳化剂,胶黏剂〉
~ carboxymethyl hydorxyethyl cellulose 羧甲基羟乙基纤维素钠盐〈糊剂〉
~ cellulose 纤维素钠盐
~ dibutyl dithiocarbamate 二丁基二代氨基甲酸钠 [促进剂 TP]
~ dibutyl naphthalene sulfonate 二丁基萘磺酸钠〈乳化剂〉
~ 2,2'-dihydroxy-4,4'-dimethoxy-5-sulfobenzophenone 2,2'-二羟基-4,4'-二甲氧基-5-磺基二苯甲酮的钠盐〈紫外线吸收剂〉
~ N-dimethyldithiocarbamate 二甲基二硫代氨基甲酸钠〈终止剂〉
~ hydroxide 氢氧化钠,苛性钠,烧碱
~ isopropyl xanthate 异丙基黄原酸钠,促进剂 SIP
~ lauryl sulfate 十二烷基硫酸钠〈乳化剂〉
~ metabisulfite 焦亚硫酸钠〈还原剂、试剂〉
~ paper 碱法纸粕
~ pentachlorophenate 五氯苯酚钠〈防霉剂〉
~ polyacrylate 聚丙烯酸钠
~ polybutadiene rubber 丁钠橡胶
~ polyoxyethylene alkyl ether sulphate 烷基聚氧乙烯醚硫酸酯钠盐〈表面活性剂〉
~ polysulfide 多硫化钠〈终止剂〉
~ pyrophosphate 焦磷酸钠〈分散剂、螯合剂、软水剂〉
~ rubber 丁钠橡胶
~ stearate 硬脂酸钠〈热稳定剂、润滑剂〉
~ sulfocyanide 硫氰化钠〈溶剂〉
~ sulfoxylate formaldehyde 甲醛合次硫酸氢钠,雕白块〈活化剂〉
~ tetraphenylborate 四苯硼酸钠〈催化剂〉
~ thiocyanate 硫氰酸钠〈溶剂〉
soft (柔)软的
~ agent 软化剂
~ board 软质纤维板
~ ductile material (软)韧性材料
~ -elastic foam 软质泡沫塑料
~ fiber 软纤维
~ fibre board 软质纤维板
~ film 软质薄膜
~ flow 低黏性,易流动性
~ foam 软质泡沫塑料
~ handle 柔软手感
~ metal bland 软金属坯料
~ packing 软填料
~ plastics 软塑料
~ polyvinyl chloride 软质聚氯乙烯
~ resin 软质树脂
~ rubber 软质橡胶
~ water 软水
~ wood 软木

~ X ray inspection 低能量X射线检查
soften 软化;增塑(处理)
softener 软化剂,增塑剂
softness 柔软(性,度)
　~ index [number, value] 软化指数
softening 软化;增塑
　~ agent 软化剂,柔软剂
　~ point 软化点
　~ range 软化范围
　~ region 软化区
　~ temperature 软化温度
software 软件
soil resistance 耐污性
soiling 污染,沾染,沾污
　~ behavior 易沾污性
sol 溶胶
　~ - gel transfermation 溶胶-凝胶变换
sol. [soluble] 可溶的
solar 太阳的
　~ collector 太阳能收集器
　~ - energy 太阳能
　~ plastics 太阳能装置用塑料
　~ power 太阳能
solder 钎焊,低温焊
　~ resist 阻焊剂
solderability 可焊接性
solderable 可焊的
　~ finish 可焊涂层
　~ plastic finish 可焊塑料(涂)层〈焊接时不必除掉塑料涂层〉
sole 底部,底基;鞋底
　~ attaching update 绱鞋〈把鞋底用注塑法固定在鞋帮上〉〈美〉
　~ bonding 鞋底黏合
solenoid 螺线管;感应线圈
　~ valve 电磁阀
solid 固体(的);突心(的);整体(的)
　~ bearing 整体轴承
　~ binder 固体粘结剂
　~ board 实心硬纸板
　~ body 固体
　~ casting 密实浇铸
　~ cement 固体粘接剂

~ colour 单色,素色
~ construction 整体结构
~ content 固体含量
~ conveyance 固体输送〈挤塑机螺杆进料段〉
~ dies 整体模
~ female mo(u)ld 坚硬阴模
~ fibreboard 整体纤维板
~ glass beads 实心玻璃珠
~ glass microsphere 实心玻璃微珠〈填料〉
~ heel 实心跟〈鞋〉
~ lubricant 固体润滑剂
~ matter 固体
~ mo(u)ld 坚硬模具
~ phase 固相
~ phase condensation 固相缩合
~ phase initiation 固相引发(作用)
~ phase polycondensation 固相缩聚(作用)
~ phase polymerization 固相聚合(作用)
~ - phase pressure forming, [SPPF] 固相压力成型法
~ phase reaction 固相反应
~ - phase wear 固相磨损
~ plate 固定板
~ plate mo(u)ld 实体板式模具
~ point 固化点;凝固点
~ polymer 固体聚合物
~ polymerization 固相聚合
~ residue 固体残渣
~ roller 实心皮辊
~ rubber 固体橡胶
~ section of the extruder (物料)在挤塑机区段中呈固态
~ shade 单色,素色
~ state 固态
~ - state polymerization 固相聚合
~ structure 实心结构
~ volume 实体积
~ waste 固体废物
~ wood 实心木材
solids draining screw, [SDS] 固体加工

螺杆
solidification 固化;凝固
- ~ heat 固化热
- ~ of a melt 熔体固化;熔体凝固
- ~ point 固化点;凝固点
- ~ process 固化过程〈熔体注塑时〉
- ~ rate 固化速率
- ~ temperature 固化温度

solidify 固化;凝固
solidifying agent 固化剂;凝固剂
solo fibre 单纯纤维,独种纤维
solubility 溶度;溶解性,可溶性
- ~ parameter 溶解度参数
- ~ parameter test 溶解度试验

solubilization 增溶;溶解作用
soluble 可溶的
- ~ cellulose 可溶性纤维素
- ~ colorant 可溶性着色剂
- ~ dispersion dye 可溶性分散染料
- ~ indigo 可溶性靛蓝
- ~ matter 可溶性物质
- ~ matter loss 可溶物损耗
- ~ resin 可溶性树脂
- ~ starch 可溶性淀粉
- ~ vat dyes 可溶性还原染料
- ~ vat ester dyes 可溶性酯化还原染料

solute 溶质,溶解物;溶解的黏合介质〈热塑性塑料〉
solution 溶液;溶解
- ~ adhesive 溶液黏合剂
- ~ block polymer 溶液嵌段聚合物
- ~ casting 溶液浇铸
- ~ coating 溶液涂布
- ~ complex 溶液络合物
- ~ polycondensation 溶液缩聚(作用)
- ~ polymerization 溶液聚合(作用)
- ~ /solvent viscosity ratio 溶液/溶剂黏度比,相对黏度
- ~ spinning 溶液纺丝
- ~ temperature 溶解温度
- ~ welding 溶液焊接

solvated resin 溶济化树脂
solvation 溶剂化(作用)
solvency 溶解本领

- ~ power 溶解力

solvent 溶剂;有溶解力的;溶剂的
- ~ action 溶解作用
- ~ - activated adhesive 溶剂活化黏合剂
- ~ adhesive 溶剂型黏合剂
- ~ - based adhesive 溶剂型黏合剂
- ~ - based coating 溶剂型涂料
- ~ bonding 溶剂粘接
- ~ - borne adhesive 溶剂型黏合剂
- ~ casting 溶剂铸塑,溶剂浇铸
- ~ cement 溶剂型胶黏剂
- ~ - cement spigot 溶黏套接(管)
- ~ - cemented fitting 溶黏配件
- ~ - cemented socket 溶黏套接(管)
- ~ cleaning 溶剂清洗
- ~ cracking 溶剂致龟裂
- ~ - degreased 溶剂脱脂的
- ~ drum dyeing machine 鼓筒式溶剂染色机
- ~ dyeing 溶剂染色
- ~ evaporation 溶剂蒸发
- ~ fluids 溶液
- ~ impregnated resin 溶剂浸渍树脂
- ~ laminating 溶剂层合,溶剂贴合
- ~ mo(u)lding 溶剂成型
- ~ phase 溶剂相
- ~ polishing 溶剂抛光
- ~ popping 溶剂气泡〈在固化涂层中〉
- ~ recovery 溶剂回收
- ~ release adhesive 溶剂挥发型黏合剂
- ~ resistance 耐溶剂性
- ~ sealing 溶剂溶接,溶剂封合
- ~ sensitive adhesive 溶剂敏感型黏合剂
- ~ socket 溶黏套接
- ~ soluble adhesive 溶剂型黏合剂
- ~ spinning 溶剂纺丝
- ~ - spray transfer printing 溶剂喷射转印
- ~ spun fibre 溶纺纤维
- ~ stress crazing 溶剂应力龟裂腐蚀
- ~ tolerance (溶剂)稀释容限
- ~ - type adhesive 溶剂型黏合剂

~ vapour 溶剂蒸气
~ welding 溶剂粘接,溶剂焊接
solventation 溶剂化作用
solvent-free 无溶剂的
~ adhesive 无溶剂黏合剂
~ impregnating lacquer 无溶剂浸渍漆
~ impregnating technique 无溶剂浸渍法
~ lamination 无溶剂层合
solventless 无溶剂的
~ coating 无溶剂涂布
~ bonding 无溶剂粘结
~ silicone mo(u)lding compounds 无溶剂有机硅模塑料
~ silicone resin 无溶剂有机桂树脂
sonic 声音的,声波的
~ converter 声波变换器〈超声波焊机〉
~ source 声波源,噪声源
sonotrode 超声波焊接工具;超声焊极
~ output face 超声波焊具工作面
soot 烟黑,烟炱,油烟〈炭黑的一种〉
~ carbon 烟黑型炭黑
~ chamber test 烟尘箱试验
sophisticated extruder 改良型挤塑机
sorbic acid 山梨酸
sorbitan 脱水山梨糖醇,山梨糖醇酐
~ monolaurate 山梨糖醇酐单月桂酸酯〈防雾剂〉
~ monooleate 山梨糖醇酐单油酸酯〈防雾剂〉
~ monopalmitate 山梨糖醇酐单棕榈酸酯〈防雾剂〉
~ monostearate 山梨糖醇酐单硬脂酸酯〈防雾剂〉
~ stearate 山梨糖醇酐硬脂酸酯,乳化剂 S-60
~ stearate polyoxyethylene ether 山梨糖醇酐硬脂酸聚氧乙烯醚,乳化剂 T-60
sorbitol 山梨糖醇〈热稳定剂〉
sorption 吸着(作用)
~ film 吸着膜
sort 种类,类别,分类

sound 声,声音;完善的
~ absorbend material 吸音材料
~ absorbing 吸音的
~ absorbing coating 吸音涂层
~ -absorbing foam 吸音泡沫塑料
~ -absorbing panel 吸音板
~ -absorbing plastic 吸音塑料
~ absorption 吸音
~ absorption coefficient 消音系数
~ absorption materials 吸音材料
~ attenuation 消音;吸音;弱音
~ baffle 隔音板,消声板
~ barrier 隔音板
~ cable 通信电缆
~ casting 优质铸塑件
~ damping 隔音的
~ -damping property 隔音性
~ deadening 消音,隔音
~ deadening material 消声材料隔音材料
~ emission,[SE] 声发射
~ emission analysis [SEA] 声发射分析〈塑料疵点检验〉
~ insulating material 隔音材料
~ insulating property 隔音性
~ insulation 隔音,消声
~ insulation board 隔音板,消声板
~ -proof 隔音的
~ proofing 隔音,消音
~ reveiver 声接受器〈透声检验塑料〉
~ track 声迹
~ transducer 声波变换器
~ transmission coefficient 声透射系数
~ transparency 透声性
~ velocity 音速
sounding 发响
source image 光源映象
soy(a) bean 大豆
space 空间;间距
~ cavity 空隙
~ -charge distribution 空间电荷分布
~ -dyeing 间隔染色,间段染色〈纱线〉
~ formula 立体式

~ grid　空间晶格;空间点阵;空间栅格
~ lattice　空间格子
~ length　间距
~ model　立体模型
~ network polymer　立体网状聚合物
~ polymer　立体形聚合物;立构聚合物
~ resolving power　空间分辨率
~ truss dome　空间桁架穹顶
spacer　隔板;垫板;调距模板
~ block　隔块,隔板
~ fork　垫模片
~ fork and spring box mo(u)ld　垫模片和弹簧箱式模具
~ plate　隔板,垫板,调距模板
~ ring　间隔圈,隔环;调距环
spacing　间隔,间距
~ clamp [clip]　间隔夹具
~ clip bracket　间隔固定夹;管夹
~ ring　间隔圈,隔环;调距环
~ sleeve　间隔套
~ washer　间隔垫圈
spaghetti　空心圆条;绝缘套管〈挤塑〉
spall　散裂,裂开
spalling　散裂,剥落
~ resistance　耐散裂强度,耐散裂性
~ test　散裂试验
span　跨距,间距
~ width　净宽
Spandex　斯潘德克斯纤维〈高弹纤维属名,一般指聚氨酯纤维〉
spangles　闪光颜料
spanishing　凹纹印刷
spar　晶石;桅杆
- varnish　桅杆清漆;船杆用漆
spare　备件;备用的
~ detail　备用零件;附件
~ parts　备件,备用零件
~ parts kit　备件箱,零件箱
~ piece　备件,配件
spark　火花;电火花
~ discharge　火花放电
~ discharge method　火花放电法
~ erosion　电火花腐蚀(法)〈模具加工〉
~ gap　电火花间隙
~ testing　电火花试验〈电缆外皮〉
~ tracking　电弧径迹
~ welding　电弧焊
sparking voltage　击穿电压
spatial　立体的;空间的
~ arrangement　立体排列,空间排列
~ configuration　立体构型
~ configuration of molecules　分子的空间结构
~ isomerism　立体异构
~ structure　立体结构
spatially cross-linked structure　立体交联结构
spatula　刮刀
speader　分流棱
spec. specification　规格;说明书
special　特别的,特殊的;专用的
~ dye　特种染料
~ edition　特刊,号外
~ fixture　专用夹具
~ -grade synthetic resin　专用合成树脂,特殊牌号合成树脂
~ purpose　专用,特殊用途
~ purpose extruder　专用挤塑机
~ purpose ingredient　专用配合剂
~ purpose plastic　专用塑料
~ purpose rubber　专用橡胶
~ purpose screw　专用螺杆〈挤塑机〉
speciality [specialty]　特种产品,特殊产品
~ elastomer　特种弹性体
~ products　特种产品,特殊产品
~ type　特种类型
specially-built　特殊构造〈机械〉
specific　特殊的;特定的;比(率)
~ adhesion　特性黏合;比黏附
~ area　比表面积
~ capacity　比容(量)
~ character　特性,特点
~ conductance　电导率,电导系数
~ density　比密度,相对密度
~ elongation　比延伸,延伸率

~ gravity 相对密度
~ gravity bottle 比重瓶
~ heat 比热容
~ heat capacity 比热容量
~ humidity 比湿度
~ induction capacity 介电常数;电容率
~ inductive capacity 介电常数;电容率
~ injection pressure 比注塑压力
~ insulation resistance 比绝缘电阻
~ mass 比密度,相对密度
~ materials 特殊材料
~ modulus 比模量
~ (mo(u)lding) pressure 比(成型)压
~ performance index 特性指数
~ power 比功率,功率系数
~ pressure [unit pressure of a press] 比压
~ property 特性
~ ratio 比率
~ reagent 特效试剂
~ refraction 折射系数,折射度
~ resistance 比电阻,电阻率
~ speed 比速
~ strength 比强度,强度系数
~ stress 比应力
~ surface 比表面,表面系数
~ surface area 比表面积
~ test 特殊试验
~ value 比值
~ viscosity (增)比黏度
~ volume 比容
~ weight 相对密度
specifics 详细说明书;细节
specification 规格,技术规范;说明书
~ chart 规范一览表
~ control 规范控制
~ of quality 质量规格
~ requirements 技术规格要求
~ sheets 说明书,样本
~ standards 规格标准
specified 规定的,指定的,额定的
~ horsepower 额定功率,铭牌功率

~ load 规定载荷
~ rate 额定量
~ speed 规定速率[度]
~ temperature 规定温度
~ time 规定时间,给定时间
~ value 给定值
~ weight 规定重量
specimen 试样,试片,样品
~ copy 样本
~ test 试样试验
~ with crack at edge 边缘破裂试样
speck 斑点〈缺陷〉
specking 花斑
spectral 光谱的
~ absorption 光谱吸收
~ analysis 光谱分析
~ band 光谱带
~ concentration 光谱密度
~ directional reflectance 光谱定向反射本领
~ distribution 光谱分布
~ luminous efficiency 光谱发光效率
~ range 光谱范围
~ reflectance 光谱反射本领
~ transmission 光谱渗透性
spectrochemistry 光谱化学
spectrogram 光谱图
spectrograph 摄谱仪,光谱仪
spectrophotometry 分光光度测定法
spectroscope 分光镜
spectroscopy 光谱学
spectrum 光谱
~ analyser 光谱分析仪
~ of relaxation time 松弛时间谱
~ of retardation time 推迟时间谱
specular 镜的,反射的
~ gloss 镜面光泽,反射光泽
~ reflection 镜面反射
~ reflection value 镜面反射值
speed 速率,速度
~ control 速率[度] 控制
~ gear 高速齿轮
~ of assembly 组装速率[度],装配速率[度]

~ of combustion 燃烧速率[度]
~ of cooling 冷却速率[度]
~ of impact 冲击速率[度]
~ of propagation 传播速率[度];增长速率[度]
~ of reaction 反应速率[度]
~ of revolution [torque] 旋转速率[度],扭矩速率[度]〈美〉
~ of testing 试验速率[度]
~ of torque 扭矩速度
~ of travel 前推进速度
~ of water absorption 吸水速率
~ range 速率[度] 范围
~ rate 速率
~ ratio 传动比,速比
reducer 减速器
 ~ - up roll 加速辊
 ~ variator 变速器
speedometer 速度计
speedwall joint 复式榫和槽粘接
spent 用完的,废的
 ~ material 废料
spew 溢料;毛刺,飞边
 ~ area 溢料面,溢料脊
 ~ groove 溢料槽,溢料缝
 ~ he 模脊缝,溢料线,合模线
 ~ relief 溢料缝隙
 ~ ridge 溢料脊
 ~ way 溢料缝
Spewing 渗出,流出,压出;溢料
spheric(al) 球形的
 ~ condenser 球形电容器
 ~ housing 球形外壳
 ~ joint 球形接合
 ~ lubrication head 球形润滑头
 ~ roller bearing 球形滚柱轴承
 ~ shell 球形壳
 ~ surface 球表面
 ~ valve 球阀
spheroid 球体,球状容器
spheroidal 球体的
 ~ particle 球形颗粒
spherulite 球晶,球粒
 ~ growth 球晶生长

~ structure 球粒结构
sp. gr. specific gravity 相对密度
SPI [Society of the Plastics - Industry] (New York) 塑料工业协会(纽约)
spider 模芯支架;星形板〈顶出器〉;旋转十字架
~ die 支架式口模
~ fin 星形模芯支架〈挤塑机〉
~ gate [spoke gate] 星形浇口
~ head 模芯支架式模头〈挤吹模具〉
~ leg 支架柱
~ - like spreader sprue 星形式分流主流道
~ - like sprue 星形式注流道
~ line 分流痕
spiderless head 无模芯支架模头
spigot 插口;尖形管端;套管
~ and socket joint 窝接,承插接合
~ discharge 卸料孔
~ joint 插管接头
spin 旋转;纺(丝)
~ - away centrifugal dryer 旋转式离心干燥器
~ - bonded fabric 纺黏型织物
~ - casting 旋转铸塑
~ - die manifold 纺丝模头箱
~ - draw ratio 纺丝拉伸比
~ - drawing 纺丝拉伸
~ dyeing 纺前染色,本体染色
~ extruder 纺丝挤出机
finish 纺丝整理
~ forming 旋转成型
~ off 有用副产品,衍生产品
~ roviing 纺粗纱
~ welding 旋转焊接,摩擦焊接〈美〉
~ welding disk 旋转[摩擦]焊接盘〈美〉
spindle 心轴;锭子〈纺织〉
~ cutter 卷筒纸芯切割器
~ mill 棒状研磨机
~ press 螺旋压力机
spinnability 可纺性
spinneret [spinnerette] 喷丝板,纺丝板
~ assembly 喷丝板组合件

~ draft 喷丝头拉伸
~ nozzle 喷丝板,纺丝板,喷丝嘴
~ orifice 喷丝(板细)孔
spinning 旋转;纺丝
~ bar 纺丝模头集料管
~ bath 纺丝浴
~ bath stretch 纺丝浴拉伸
~ bucket 纺丝罐
~ by extruder 挤塑纺丝
~ can 纺丝筒,纺纱筒
~ cell 纺丝仓,纺丝室
~ cep 纺丝管纱
~ die 喷丝模头,喷丝板
~ dope 纺丝浆液〈美〉
~ draft 纺丝头拉伸
~ extruder 纺丝用挤塑机
~ extrusion 纺丝挤塑
~ funnel 纺丝漏斗
~ head 喷丝头,纺丝头
~ hopper 纺丝料斗
~ machine 纺丝机
~ mandrel 旋转心轴〈离心铸塑〉
~ nozzle 喷丝板,纺丝板,纺丝头
~ pot 纺丝罐
~ process 纺丝过程
~ property 纺丝性能
~ solution 纺丝溶液
~ temperature 纺丝温度
spiral 螺旋(的),螺线(的),螺纹的
~ angle 螺旋角
~ agitator 螺旋搅拌器
~ barrier screw 螺旋屏障型螺杆
~ classifier 螺旋式分离机
~ coil 螺旋管
~ conveyor 螺旋输送机
~ current 旋流,涡流
~ die 螺旋式模头
~ dryer 螺旋式干燥器
~ flow 螺旋形流动
~ flow distributor 螺旋形流动分布器
~ flow test 螺旋形流动试验
~ gear 斜齿轮,螺旋形齿轮
~ groove 螺旋槽
~ guide 导向螺旋

~ knurled insert 螺纹滚花嵌件
~ mo(u)ld cooling 螺旋式模具冷却
~ pipe 蛇管
~ ribbon mixer 螺带式搅拌机
~ spindle 涡流器,旋流器;旋转体
~ stirrer 螺旋搅拌器
~ test 螺旋形流动试验〈测定塑料成型料流动性〉
~ test flow number 螺旋流动试验黏度指数
~ -type blown film die 螺旋型吹塑模头
~ winding 螺旋式卷绕
spirane polymer 螺旋环状聚合物
spirit 酒精,乙醇,醇
~ dye 醇溶染料
~ varnish 醇溶性清漆
Spirolite pipe 螺线缠绕管
spiropolymer 螺旋形聚合物
splash 飞溅,喷溅;斑点,污迹〈注塑件缺陷〉
~ apron 挡泥板
~ board 挡泥板〈汽车〉
~ shield 挡泥板
splashing 喷溅
~ plate 挡泥板
splay 斜面,斜度;倾斜(的);有倾斜度的
~ mark 喇叭形斑痕〈模腔在完全被填满之前注塑料过早冷却和固化而造成的注塑件缺陷〉
splice 拼接,绞接;加固〈纺织〉
splicing 拼接的
~ tape 黏胶带
~ ~ yarn 加固纱
spline (楔形)键槽,刻槽;楔形齿
splined feed zone 槽纹供料段
splint 开口销,开尾销;夹板;薄板
~ board 薄木板
splinter-proof glass 防碎玻璃,安全玻璃
split 破裂,撕裂,分裂;劈开,对开;裂缝;组合模〈模具〉
~ cavity 对开模槽

~ cavity block 对开阴模
~ cavity mo(u)ld 对开模槽式模具
~ chase mo(u)ld 对开模具
~ die 对开式模,组合式模
~ die plate 对开式模板
~ -feed hopper 分路供料漏斗
~ -feed technique 分开供给法,分部位供料法
~ fibre 裂膜纤维
~ fibre yam 裂膜(纤维)纱
~ film 裂膜
~ film fibre 裂膜纤维
~ -follower mo(u)ld 对开式模具
~ index 撕裂指数,纵裂指数
~ knitting 裂膜带编织(法)
~ knitting machine 裂膜带编织机
~ level 分层面〈注塑模〉
~ line 对开缝;分模线〈模具〉
~ mandrel 瓣合型芯,组合型芯
~ mo(u)ld 对开式模具,瓣合式模具,组合式模具
~ of mo(u)ld 模瓣
~ -off 分裂;裂开
~ pattern 组合模
~ pin 开口销
~ ring mo(u)ld 瓣合模圈式模具
~ -segment die 组合模
~ strength test 撕裂强度试验
~ taper mo(u)ld 瓣合锥度式模具
~ up into components 分解成各组分
~ weaving 编织用裂膜带;裂膜带织法
~ wedge mo(u)ld 瓣合楔式模具
~ yam 裂膜丝
splits 模瓣
splitting 分裂,劈裂
~ ability 撕裂性能
~ die 组合模
~ of chain 链断裂
~ of film 裂膜
~ -off 分裂,裂开
~ resistance 耐撕裂性
~ tendency 拼合倾向
splutter print 喷印

s.p.m [stroke per minute] 冲程/每分钟
spoilage (食品)腐败;损坏
spoiled 损坏的
~ casting 残铸塑件
~ product 废品
spoke 模芯支架;辐条,轮辐
~ gate 轮辐式浇口
sponge 海绵;泡沫材料
~ plastics 海绵塑料,泡沫塑料
~ product 海绵制品;海绵状材料制品
~ rubber 海绵橡胶
~ section 泡沫形成区
~ weave 海绵组织
spongelike structure 海绵状结构
ponging agent 发泡剂
spongy 海绵状的
~ material 海绵状材料
spontaneous 自发的,自然的
~ combustion 自燃
~ crimp 自发卷曲
~ curing 自然固化
~ elastic recovery 自发弹性恢复
~ ignition 自燃
~ ignition temperature 自燃温度
~ nucleation 自发成核
spool 卷(线)筒;线筒
spooler 卷取机,缠绕机;络筒机
spoon agitator 勺形搅拌器
sporadic nucleation 散现成核
spot 斑点,污点〈成型件或涂层上〉;点滴;局部;现象
~ analysis 点滴分析
~ bonding 局部粘接
~ check 抽样检查
~ cure 局部固化
~ gluing 局部胶接
~ heating 局部加热
~ press 快速加热压机,高压蒸汽加热压机
~ repair 现场修理
~ resistance 耐污染性
~ sample 局部试样;抽查
~ test 当场试验,当场测试
~ weld 点焊;点焊缝

~ welder 点焊机
~ welding 点焊(法);局部焊接
~ welding gun 点焊枪
~ welding machine 点焊机
spotting 斑点形成,斑痕
spout 槽道;喷嘴
SPPF - process [solid phase prensure forming] 固相压力成型法
SPR [Society for Physical Research](London) 物理研究协会〈伦敦〉
spray 喷射,喷涂;连枝毛坯〈在注塑中〉
~ - and - wipe painting 喷抹上漆;衍彩印刷〈彩色印刷〉
~ angle 喷射角
~ - applied wrapping 茧状包装(法)
~ bonding 喷涂黏合法
~ coating 喷涂;喷镀层
~ cooling 喷(水)零冷却
~ drawing 喷淋拉伸
~ drier [dryer] 喷雾干燥器
~ drum 内部雾化的转鼓
~ drying 喷雾干燥
~ foam 喷射法制(的)泡沫塑料
~ foaming 喷射发泡
~ foaming gun 泡沫喷枪〈制备聚氨酯泡沫塑料〉
~ granulator 喷雾造粒机
~ gun 喷枪
~ hole 喷孔
~ - in - place 就地喷涂,现场喷涂
~ in injection mo(u)lding 注塑毛坯连枝
~ metal coating 喷镀金属层
~ metal mo(u)ld 喷镀金属的模具
~ mo(u)lding 喷附成型,喷涂成型
~ nozzle 喷嘴,喷孔
~ packaging 茧状包装
~ sizing 喷雾上浆
~ spinning 喷纺成型(法)
~ test 喷淋法防水性试验〈织物〉
~ - up 喷射成型〈增强塑料〉;喷发〈发泡〉
~ - up method 喷射成型
~ - up moulding 喷涂成型法

~ - up technique 喷射成型法,喷涂成型法
~ - webbing 网状涂喷涂;茧状包装(法)〈美〉
sprayability 喷涂性
sprayable adhesive 可喷涂性黏合剂
sprayed 喷涂的
~ in - place 就地喷涂,现场喷涂
~ metal mould 喷涂的金属模具
sprayer 喷雾器
~ for mo(u)ld lubricant 喷润滑模剂用的喷雾器,脱模剂用喷雾器
spraying 喷雾;喷涂;喷射
~ jet 喷嘴
~ lacquer 喷漆
~ machine 喷雾机,喷涂机;发泡机
~ plant 喷射装置,喷雾装置,喷涂装置
~ process 喷雾法
spread 涂布;涂胶量;蔓延,扩散;流失〈着色〉
~ adhesive 涂布黏合剂,涂胶层
~ calender 涂胶压延机
~ coat [spread - caoting adhesive] 刮涂胶
~ coater 刮涂机
~ coating 刮刀涂布,刮涂
~ defects 散布性疵点
~ of fire 火焰蔓延
~ of flame 火焰蔓延〈塑料燃烧性〉
~ spontaneously 自发展涂
spreadability 覆盖性
spreadable life 可涂期,适用期〈黏合剂〉
spreader 分流梭〈注塑机〉;涂胶机,涂布机,刮涂机
~ calender 涂胶压延机
~ coating 涂胶机涂布
~ knife 刮涂刀
~ roll 涂胶辊,展幅辊
spreading 流散,扩展,溢流;刮涂
~ agent 铺展剂,涂铺剂,展涂剂
~ calender 涂胶压延机
~ device 涂胶装置〈中空吹塑模具〉
~ knife 涂布刮刀,刮涂刀

spring

~ machine 刮涂机,涂胶机,涂布机
~ mass 上涂料
~ of powder adhesive 粉状黏胶剂涂布
~ paste 涂布用糊剂
~ rate 涂布率
~ roller [expander roll] 刮涂辊;展幅辊
~ weight 涂布量

spring 弹簧;弹性
~ - absorber model 弹簧-减震器模型
~ action 弹簧作用
~ back 弹回,回弹,弹性后效〈塑料冷加工〉
~ balance pull 弹簧测力计
~ bearing 弹性轴承
~ box mo(u)ld 弹簧箱式模具
~ constant 弹性常数
~ ejector 弹簧顶出器
~ follower 弹簧垫圈
~ hammer 弹簧锤
~ linkage system 弹簧键系统
~ pad 弹簧垫
~ washer 弹簧垫圈

springiness 弹性

sprue 浇口,主流道,注料道;注口冷料
~ and runners 主流道和分流道〈注塑进料系统〉
~ break 注口冷料断脱
~ break controller 注口冷料断脱控制器
~ break limit switch 注口冷料断脱行程开关
~ bush 主流道衬套,注料口衬套〈注塑模〉
~ bushing 主流道衬套
~ channel 注料道
~ cone 锥形注道〈模具〉
~ ejector 注口冷料脱模销〈注塑模〉
~ ejector bar 注道顶出板
~ ejector pin 注口冷料脱模销
~ gate 直接浇口,浇道口
~ gating 直接浇口
~ granulator 注口冷料粉碎机

~ groove 注料沟,溢料槽,溢料腔
~ height 注料口深度
~ hook 注口冷料钩
~ knockout pin 注口冷料脱模销
~ location ring 注料口定位圈
~ lock 注口冷料钩,主流道冷料钩
~ lock pin 注口冷料固定销
~ mark 浇口痕
~ pin 注口冷料脱模销
~ plate 注道冷料顶板
~ puller 注口冷料脱模销,注道冷料脱模销,浇道拉出杆
~ runner 注道
~ slug 浇口;浇口冷料
~ sticking 注口冷料黏模
~ type 浇口类型
~ waste 浇铸耗损
~ well 注口冷料井

sprueless 无流道
~ injection 无流道注塑(成型)
~ injection moulding 无流道注塑成型
~ lateral direct injection 无流道侧面直接注塑
~ mo(u)ld 无注口料式模具,无流道模具
~ mo(u)lding 无流道(注塑)成型
~ nozzle 无流道喷嘴
~ - type runner system 无注口料式流道方式

spue 溢料
~ line 溢料线

spun 旋转;纺(丝)
~ - bonded 纺(丝)黏(合)的
~ - bonded fabric 纺黏型织物
~ - bonded fibre 纺黏纤维
~ fabrics 短纤维织物
~ filament 纺单丝
~ glass 玻璃纤维
~ orientation 纺丝取向
~ roving 定长毛纱;加捻粗纱
~ yarn 细纱;纺织纱

spur 直齿
~ gear 正齿轮
~ gear on parallel axes 平行轴上的正

齿轮,正齿轮传动
~ tooth(ing)　直齿咬合
sputtered chrome　喷镀铬〈真空蒸附〉
sq. square　平方
sq. m. square meter　平方米
square　（正）方形;平方;直角的
~ bend　直角弯头
~ bottom bag　方形底袋
~ box mo(u)ld　方箱式模具,正方箱式模具
~ butt(type of)joint　平头对接,工型对接〈塑料焊接〉
~ butt weld　平头对接焊(缝)
~ centimeter[cm²]　平方厘米
~ crossing　十字形交叉
~ edge　平面焊口
~ gate　正方断面浇口
~ meter[m²]　平方米
~ runner　正方(断面)流道
~ thread　方螺纹
~ toothed scarf joint　榫槽多次胶接件,齿形嵌接
~ -type expansion joint[piece]　矩形补偿节
squeegee　刮墨器;涂刷器,刮浆板
~ roll　挤压辊,橡皮辊子〈俗称〉
squeeze　挤压
~ bottle　挤瓶;喷洗瓶〈软塑料制〉
~ moulding　挤压模塑〈用于试验模具,模型模具由环氧树脂制造〉
~ nut　压制螺母,挤压螺母;螺纹插件
~ off　挤出,压出
~ out　溢胶;刮去〈剩余树脂〉;挤出
~ packing　鼓胀包装,拱形包装
~ ring　环形嵌槽
~ roll　挤压辊
~ roll coater　挤压辊式涂布机
~ rollers　夹辊〈制膜〉
~ stroke　挤压冲程
squeezing　挤压
~ dies　挤压模
squirrel cage　鼠笼
~ impeller　鼠笼式搅拌机
Sr.[strontium]　锶

SR[synthetic rubber]　合成橡胶
SRP[styrene/rubber plastics]　耐高冲击聚苯乙烯
s(-s-)　对称的(结构)〈聚合物〉
SS[single-stage resin]　甲阶酚醛树脂;可溶молярный酚醛树脂
SST[sell-sealing tape]　自封黏胶带
SST(test)[step-by-step test]　逐步试验
st. standard　标准(的)
stability　稳定性,稳定度;坚固性,耐久性
~ condition　稳定性条件
~ constant　稳定常数
~ factor　稳定系数
~ in forage　储存稳定性
~ in use　使用稳定性
~ test　稳定性试验
~ to light　光稳定性
~ to precessing　加工稳定性
~ to storage　储存稳定性
~ under load　荷载稳定性
stabilization　稳定(作用)
~ period　稳定期
stabilize　稳定;坚固
stabilized structure　稳定结构,坚固结构
stabilizer　稳定剂;稳定器
~ section　稳定型材,加筋型材
stabilizing　稳定
~ agent　稳定剂
~ effect　稳定效应
~ medium　稳定介质;加筋材料
~ structure　稳定结构,坚固结构
~ treatment　稳定性处理
~ zone　稳定区
stable　稳定的,坚固的
~ equilibrium　稳定平衡
~ state　稳定状态
stack　堆积,重叠
~ -a-box　堆叠箱
~ dryer　堆积式干燥机
~ layer　堆叠层
~ mo(u)ld　叠模压塑模具
~ mo(u)lding　叠模压塑
~ of lamellae　片晶叠层

~ type filter　叠片式过滤器
stacked　堆积的,重叠的
　　~ column packing　堆积填柱物
　　~ injection mo(u)lds　多层注塑模具
　　~ tower packing　堆积填塔料
stacker　堆垛机,码垛机
stacking　堆放,堆垛,堆积,堆叠
　　~ density　堆积密度,堆密度
　　~ pallet　堆叠托板
　　~ strength　堆叠强度
stafluid　静电流化床装置
stage　阶段,分段;时期
　　~ filter　分级过滤器
　　~ grafting　分段接枝
　　~ heating　分段加热
　　~ joint　梯形连接,阶梯连接
　　~ of experiments　实验阶段
　　~ of manufacture　制造阶段
　　~ of work　工作阶段,工作时期
staged reactor　多级反应器
daggered　交错(排列)的
stain　污斑;污染;着色(剂)
　　~ resistance　耐污染性
　　~ stained　染色的,涂漆的
　　~ glass　彩色玻璃
　　~ wood　着色木
staining　斑点;污染性;着色,染色
　　~ agent　染色剂;着色剂
　　~ power　染色本领,着色力
　　~ test　着色试验
stainless　不锈的
　　~ property　防锈性
　　~ steel　不锈钢
stair–step dicer　阶梯型切粒机
stajet　静电粉末喷涂装置
stalk　浇口,注口;注料口冷料
stalling point　凝固点;固化点
stamp crushing　摘碎
stamping　冲压,模冲;冲压件,模压件;烫印
　　~ calender　轧花压延机
　　~ die　冲压模;压花模
　　~ foil　压花箔,烫印箔
　　~ hammer　冲压锤

~ line　冲压生产线
~ machine　冲压机,压印机
~ mo(u)ld　冲压模
~ press　冲压机,压印机
~ tool　冲压工具
stand　站;台;台架
　　~ – by　备用(的)
　　~ – by plant　备用设备
　　~ for test tubes　试管台架
　　~ oil　熟油
　　~ pipe　竖管
　　~ test　台架试验
standard　标准
　　~ atmosphere　标准大气压,标准环境
　　~ bar　标准试样
　　~ colo(u)rs　标准色
　　~ condition　标准状态,标准条件
　　~ deviation　标准偏差
　　~ equipment　标准设备;标准附件
　　~ error　标准误差
　　~ film　标准薄膜
　　~ frame　(模具)标准框架
　　~ gate　标准浇口
　　~ grade　标准级
　　~ humidity　标准湿度
　　~ insert　标准嵌件
　　~ length　标准长度
　　~ line　连续生产线,标准化生产线
　　~ model　标准样品
　　~ mo(u)ld frame　标准模架
　　~ moulding compound　标准模塑料
　　~ nozzle　标准喷嘴
　　~ of comparison　对比标准样品
　　~ operation　标准操作
　　~ parts　标准零件
　　~ pitch　标准螺距
　　~ sample　标准试样
　　~ sieve　标准筛
　　~ size　标准尺寸
　　~ specification　标准规格
　　~ specimen　标准试样
　　~ state　标准状态
　　~ substance　标准物
　　~ symbol　标准符号

~ temperature   标准温度
~ test   标准试验
~ time   额定时间,标准时间
~ tolerance   标准公差
~ type   标准型,标准型式
~ unit   标准单位
~ volume   标准体积
~ weight   标准重量
standard. [standardization] 标准化
standardized 标准的
　~ mo(u)lding composition   标准成型料,标准模塑料
　~ moulding compound   标准成型料,标准模塑料
　~ products   标准化产品
standing 直立的;固定的;标准的
　~ storage   长期储藏
　~ time   停滞时间,停留时间
stannic (正)锡的,四价锡的
　~ oxide   氧化锡,氧化正锡
stannous 亚锡的
　~ 2 - ethylhexoate stearate   2 - 乙基己酸 - 硬脂酸亚锡〈热稳定剂、润滑剂〉
　~ octoate   辛酸亚锡〈聚氨酯泡沫塑料用催化剂〉
staple (定长)短纤维,切段纤维;大宗产品
　~ cutter   短纤维切断机
　~ cutting   切断成短纤维
　~ fibre   (定长)短纤维,切段纤维
　~ fibre cutting machine   短纤维切断机
　~ fibre glass yam   玻璃短纤维纱
　~ fibre woven fabric   短纤维织物
　~ fibre yam   定长短纤维纱
　~ glass fiber   玻璃短纤维
　~ goods   大宗商品
　~ - products   大宗商品
　~ spun yarn   短纤维纺纱
　~ yarn   短纤维(纺纱)纱
star 星
　~ chain   星型链
　~ connection   星型连接
　~ feeder   星型加料器
　~ gear   星型齿轮

~ pattern   星型网眼
~ polymer   星型聚合物
starch 淀粉,(淀粉)浆
~ adhesive   淀粉黏合剂,淀粉胶
~ glue   淀粉胶〈美〉
~ paste   浆糊
~ permanent   耐久架料
start 开始;开动,启动
~ button   启动按钮
~ - up   开动,启动
~ - up operation   开始运转
~ - up period   启动阶段
~ - up test   试运转
~ - up time   启动时间
starter strip 导向带
starting 开始;开动,启动
~ button   启动按钮
~ material   原材料
~ product   原料;起始产品
~ switch   启动开关
~ temperature   起始温度
~ time   启动时间
starvation 缺乏,不足;漏涂胶,欠胶,缺胶
starve 缺乏
starved 缺乏的;欠胶的
~ area   欠胶面
~ feeding   欠量加料,供料不足
~ joint   欠胶接头
starving 饿料,贫料
state 状态,情况,性能;国家,帮,州〈美〉
~ of cure   固化程度
~ of matter   物质状态
~ of orientation   取向状态
~ of oxidation   氧化状态
~ of plastic flow   塑性流动状态
~ of stress   应力状态
~ of surface   表面状态
~ - specified standards   国家(规定的)标准
~ verification   国家鉴定
static 静止的,固定的
~ adhesion   静态黏合
~ beaming test   静态弯曲试验

statics
- ~ behavior　静态
- ~ characteristics　静态特性
- ~ charge　静电荷
- ~ chaise accumulation　静电荷积累
- ~ charge dissipation　静电荷流散
- ~ charge gauge　静电荷测定仪
- ~ check　静电检验
- ~ content　静电量
- ~ deformation　静态形变
- ~ detector　静电检测器
- ~ dielectric constant　静电介电常数
- ~ discharge　静态放电
- ~ electricity　静电
- ~ electrification　静电
- ~ elimination　静电消除
- ~ eliminator　静电消除器;静电消除剂
- ~ endurance limit　静态持久极限
- ~ equilibrium　静态平衡
- ~ fatigue　静态疲劳
- ~ flock adhesive　静电植绒黏合剂
- ~ free　无静电
- ~ friction　静摩擦
- ~ friction coefficient　静摩擦系数
- ~ hold time　保持时间,保压时间
- ~ inhibitor　防静电剂
- ~ load(ing)　静负荷
- ~ long-term load　长期静负荷
- ~ mixer　静态混合器
- ~ modulus　静态模量
- ~ preload　静预负荷
- ~ pressure　静压
- ~ strain　静态应变
- ~ strength　静态强度
- ~ stress　静态应力
- ~ temperature　静态温度
- ~ test　静态试验
- ~ testing machine　静态试验机
- ~ torsion test　静态扭曲试验

statics　静电学;材料力学
station　站;位置;安置
stationary　静止的,固定的
- ~ bed　固定床
- ~ blade　固定叶片
- ~ cutter blade　固定切割刀片
- ~ disc　固定(圆)盘
- ~ dryer　静态干燥器
- ~ film　静态薄膜
- ~ knife　固定刀〈切粒机〉
- ~ lip　固定模唇
- ~ machine　固定机器
- ~ mandrel　固定模芯〈吹塑〉
- ~ membrane roofing　固定膜屋顶
- ~ part of the mould　固定半模〈注塑〉
- ~ phase　固定相
- ~ platen　定压板,定模板
- ~ platen side　定压板侧;定模板侧〈注塑机〉
- ~ screen　固定筛
- ~ state　固定状态
- ~ table　固定工作台
- ~ teeth　固定齿
- ~ temperature　恒定温度
- ~ type　固定式

statistical　统计的
- ~ assurance　统计可靠性
- ~ copolymer　统计(结构)共聚物
- ~ data　统计数据
- ~ distribution　统计分布

stator　定子;挡板
- ~ blade　静叶片,定片;固定刀〈切粒机〉
- ~ cage　固定罩
- ~ ring　固定环

Std.　[standard]　标准(的)
steady　稳定的,坚固的;平衡的
- ~ flow　稳定流动,稳流〈塑料熔体〉
- ~ flow process　稳流过程
- ~ flow viscosity　稳流黏度
- ~ load　稳定负荷,固定负荷
- ~ plastic flow　稳定塑性流动
- ~ pressure　稳定压力
- ~ shear melt viscosity　平衡剪切溶体黏度
- ~ state　稳态,恒定态
- ~ state characteristics　稳态特性
- ~ state compliance　稳态柔量
- ~ state condition　稳定状态;稳态条件
- ~ state creep　稳态蠕变

~ state current 稳态电流
~ state error 稳态误差
~ state flow 稳态流动,稳流
~ state model 稳态模型
~ state performance 稳态特性
~ state viscosity 稳态黏度
steam 蒸汽,水蒸气
~ ager 蒸汽硫化罐
~ bag 蒸汽袋
~ boiler 蒸汽锅炉
~ buffle 挡汽板
~ chamber 蒸汽室
~ channel 蒸汽通道,汽道〈模具〉
~ chest 蒸汽柜
~ chest mo(u)lding 蒸汽箱成型
~ coil 蒸汽蛇形管
~ conditioner 蒸汽调节器
~ core 蒸汽通道
~ -cored mo(u)ld 蒸汽加热式模具,蒸汽通道式模具
~ curing process,[SCP] 蒸汽交联法〈聚合物〉
~ drier 蒸汽干燥器
~ heat 蒸汽热
~ heat dryer 蒸汽加热干燥机
~ heated mo(u)ld 蒸汽加热式模具
~ heater 蒸汽加热器
~ heating 蒸汽加热(法)
~ injection orifice 喷汽孔
~ jacket 蒸汽套管,蒸汽夹层
~ jacketed 蒸汽夹套的
~ jacketed mo(u)ld 蒸汽夹套模
~ -jet 蒸汽喷射,蒸汽喷嘴
~ -jet compression 蒸汽喷射压塑法
~ mo(u)lding 蒸汽发泡成型
~ mo(u)lding of pre-expanded beads 预发泡珠粒料的蒸汽发泡成型〈聚苯乙烯发泡〉
~ -mo(u)lding process 蒸汽发泡成型法〈聚苯乙烯发泡〉
nozzle 蒸汽喷嘴
~ permeability 透气性
~ pipe 蒸汽管
~ plate 蒸汽加热板〈模具〉

~ platen 蒸汽加热板〈美〉
~ platen press 汽热板式压机
~ point 沸点
~ pressure 蒸汽压力
~ slot 蒸汽缝隙
~ tight 汽密(的)
~ treatment 蒸汽处理〈为可发性聚苯乙烯发泡〉
~ way 蒸汽通道
stearamide 硬脂酰胺〈润滑剂、脱模剂〉
stearamidopropyl dimethyl-β-hydroxy ethyl ammonium dihydrogen phosphate 硬脂酰胺丙基·二甲基·β-羟乙基铵二氢磷酸盐,抗静电剂SP
stearamidopropyl dimethyl-β-hydroxy ethyl ammonium nitrate 硬脂酰胺丙基·二甲基·β-羟乙基铵硝酸盐,抗静电剂SN
stearate 硬脂酸盐[酯]
~ of zinc 硬脂酸锌〈热稳定剂〉
stearic acid 硬脂酸〈润滑剂,硫化活性剂〉
stearoyl peroxide 过氧化硬脂酰〈交联剂,引发剂〉
stearyl 硬脂酰
~ alcohol 硬脂醇〈润滑剂〉
N-~-12-hydroxystearamide N-硬脂酰基-12-羟基硬脂酰胺〈润滑剂,脱模剂〉
steatite 块滑石
steel 钢;钢制的;刀口;冲切工具
~ ball 钢球〈测硬度〉
~ blade 钢叶片
~ dart 钢镖〈落球试验〉
~ disk 钢制圆盘
~ rule clicking die 刀口冲模
~ rule cutting 刀口模冲切
~ rule cutting die 刀口冲模
~ -rule die 刀口模;冲切模
~ structure 钢结构
steeping 浸渍
~ bath 浸渍浴
~ tank 浸渍槽
steering order 控制顺序,控制信号

stellite 钨铬钴合金
stem 杆,棒
stencil 型板〈印刷〉
~ process 型板印刷法;漏板印刷法
step 步骤;阶梯的,逐步的
~ addition polymer 逐步加成聚合物
~ -by-step test 逐步试验
~ gate 阶梯浇口
~ -growth polymerization 逐步增长聚合
~ joint 梯形连接,阶梯连接
~ ladder fibre 阶梯型纤维
~ ladder polymer 阶梯型聚合物
~ -line 阶段的,分段的;分段排列
~ -off construction 阶梯式结构
~ -over 跨越
~ pattern 阶梯型花纹
~ reaction polymerization 逐步聚合作用
~ sizing 连续筛选,连续筛分;筛分尺寸
~ -test procedure 逐步试验法
~ up cure 分段固化
stepped 分段的,阶梯的
~ cavity mo(u)ld 阶梯模腔式模具
~ cylindrical horn [sonotrode] 阶梯形圆柱焊头〈超声波焊接工具〉
~ drawing frame 分段拉伸机
~ ejector pin 阶梯形顶出销
~ flash line 阶梯形合模线
~ injection 分阶段注塑
~ joint 搭接(缝)
~ pin ejector 阶梯形脱模销
~ section 阶梯形截面
~ series multigating 阶梯式多浇口系列
~ -up production 加速生产
stepping test 逐步试验
stepwise 逐步的,分段的
~ continuous 按步连续
~ decomposition 逐步分解
~ polymerization 逐步聚合(作用)
~ synthesis 分步合成
stere [m^3] 立方米

stereo 立体的
~ -directed polymer 立体定向聚合物
~ -grinder machine 立体磨光机,立体研磨机
~ -homopolymer 立构均聚物
~ -rubber 有规立构橡胶
stereoblock 立构规整嵌段,定向嵌段
~ copolymer 立构规整嵌段共聚物
~ polymer 立构规整嵌段聚合物
stereograft polymer 立构接枝聚合物
stereoisomer 立体异构体
stereoisomeric 立体异构的
~ form 立体异构型
~ monomer 立体异构单体
~ unit 立体异构单元
stereoisomerism 立体异构现象
stereorandom copolymer 立构无规共聚物
stereoregular 有规立构的
~ crystalline polyvinyl chloride 有规立构的结晶性聚氯乙烯
~ polybutadiene 有规立构聚丁二烯橡胶
~ polymer 有规立构聚合物,定向聚合物
~ polymerization 有规立构聚合,定向聚合
~ rubber 有规立构橡胶
stereoregularity 立构规整性
stereorepeating unit 立体重复链节,立构重复单元
stereoselective 立体有择的
~ polymerization 立体有择聚合(作用)
stereospecific 立体有择的
~ catalyst 立体有择聚合催化剂
~ polymer 立体有择聚合物,有规立构聚合物
~ polymerization 定向聚合(作用),立体有择聚合(作用),有规立构聚合(作用)
stereospecificity 有规立构性,立体定向性
stereotactic 立体有规的

~ polymer 定向聚合物,立体有规聚合物
~ polymerization 定向聚合,立体有规聚合
steric 立体的,位的;空间的
~ configuration 空间构型
~ effect 位阻效应
~ factor 位阻因素
~ hindrance [inhibition] 位阻现象
~ isomer 立体异构体
~ isomerism 立体异构(现象)
~ order 位序
~ regularity 立构规整性
stick 棒,杆;黏附;黏模
~ force 黏附力
~ - in thermocouple 针形热电偶〈测塑料温度〉
~ line 黏附线;黏模线
~ mark 黏模痕
~ - slip effect 黏滑效应,滑动-黏附效应
sticker 黏着剂
stickiness 黏着性
sticking 黏模;粘接,粘结
~ agent 黏着剂
~ in the mo(u)ld 黏模
~ tape 黏胶带
~ temperature 黏着温度
~ to mo(u)ld 黏模
sticky 黏性的,胶黏的
~ point 黏点
stiff 刚性的,劲度的,坚硬的;困难的;浓稠的
~ fibre 硬纤维
~ flow 难流动(性)
~ paste 浓膏
~ reinforcement 加劲增强
stiffen 硬化加强,加劲,增强,补强
stiffened structure 加劲结构;加强结构
~ stiffener 硬化剂;增稠剂;套接管〈薄壁聚乙烯管〉
stiffening 硬化加强,加劲;补强,增强
~ band 加强带
~ member 加强件

~ plate 加筋板,加强板
~ rib 加强肋
stiffness 刚性,劲度,硬性
~ behaviour 刚性,劲度,硬性
~ in bend 弯曲刚度[劲度]
~ in flexure 挠曲刚度[劲度]
~ in torsion 扭曲劲度[刚度]
~ modulus 弯曲刚性模量
~ stest 刚度试验
stilbene azo dyes 二苯乙烯偶氮染料
still 静止的;蒸馏条〈蒸馏〉;(树脂)吹塑
stippling 形成斑点
stir 搅拌
~ - in resin 分散型乙烯基树脂,拌合型树脂
~ - in type paste 分散型乙烯基
~ up 搅拌
stirred 搅拌的
~ autoclave 带搅拌的高压釜
~ tank reactor 带搅拌的反应釜
~ tank reactor cascades 带搅拌的串联反应釜
stirrer 搅拌器
~ bar 搅(拌)棒
~ motor 搅拌器用电动机
stirring 搅拌
~ aim 搅拌器桨叶
~ rate 搅拌速率
~ rod 搅拌棒
~ screw conveyor 螺旋搅拌输送器
~ tub 搅拌桶
~ - type mixer 搅拌式混合器
stitch 穿刺,一针,缝;绑结;滚压
~ welding 针脚式焊接,定位焊
stitched mop 针刺抛光轮
stitching machine 缝合机
stock 原料;毛坯,坯料;备料;加工用料
~ allowance 坯料裕量
~ bin 料仓
~ blender 混器器,掺合器
~ distributor 物料分配器
~ dye 原染料,备用染料
~ guide 导料器;物料挡板

Stoddard solvent

- ~ house 料房,库房
- ~ pan 接料盘
- ~ production 大量生产,成批生产
- ~ roll 退卷辊
- ~ room 原料库;储藏室
- ~ sizes 标准尺寸;库存量〈半成品,成品的〉
- ~ temperature 物料温度,(塑料)熔体温度;加工温度
- ~ thermocouple 物料热电偶

Stoddard solvent 斯托达德溶剂
staff 材料,物料
stokes 泡[st]〈运动黏度非 SI 单位,等于 10^{-4} m/s〉
stone powder 岩石粉〈填料〉
stop 停止,制动;制动器;挡板
- ~ block 限位板,止动块
- ~ bolt 制动螺栓
- ~ button 制动按钮,停车按钮
- ~ end 密封罩,密封盖
- ~ nut 防松螺母
- ~ pin 制动销
- ~ plate 盲板;止动片
- ~ pod 限位垫板
- ~ valve 节流阀,断流阀

stopcock 活栓;活塞
stopper 制动器,挡板;阻聚剂
stopping 制动;填塞
- ~ device 制动装置
- ~ lac 填塞料,底层腻子
- ~ medium 填塞料,底层腻子

storability 耐储性,储存稳定性〈美〉
storable 可储存的,耐储藏的
storage 储藏,储存
- ~ bin 储料槽,料仓
- ~ box 储料箱
- ~ capacity 存储容量
- ~ chest 储料柜
- ~ compliance 储能柔量
- ~ container 储料容器
- ~ cylinder 储料筒〈吹塑机〉
- ~ head 储料机头〈吹塑机〉
- ~ hopper 储料斗
- ~ life 储存期

- ~ modulus 储能模量
- ~ polymer 耐储存聚合物
- ~ -proof 耐储的
- ~ property 耐储存性
- ~ stability 储存稳定性
- ~ systems 储料装置〈吹塑〉
- ~ tank 储罐,储槽
- ~ test 储存试验
- ~ vessel 储存容器

store 储存,储藏;储存器;商店
- ~ holder 储料器
- ~ room 储藏室
- ~ tank 储(存)槽

stored 储存的,储藏的
- ~ energy 储能,蓄能
- ~ heat 储热,蓄热
- ~ program 存储程序

storing property 耐储存性
stove 炉(子);烘干,烤干〈漆,涂料〉;硬化〈塑料表面〉;硬化炉〈树脂〉
stoving 烘干,烘烤;硬化〈塑料表层〉
- ~ enamel 烘漆,烤漆,烘干型磁漆
- ~ paint 烘漆,烤漆
- ~ schedules 烘烤程序
- ~ temperature 烘烤温度
- ~ varnish 清烘漆,烘干型清漆

straight (正)直的;纯的;直线的
- ~ -arm mixer 直臂混合器
- ~ -arm paddle mixer 直臂桨式混合器
- ~ -arm stirrer 直臂搅拌器
- ~ blade 直型桨叶
- ~ casting 简单铸塑
- ~ -chain 直链
- ~ -chain hydrocarbons 直链烃
- ~ -chain macromolecule 直链大分子
- ~ -chain molecule 直链分子
- ~ Couette flow 奎特型直线流
- ~ curved blade 直曲型桨叶
- ~ die 直口模,直模头
- ~ dipping process 直接浸渍法
- ~ extrusion head 直挤塑机头,轴向挤塑机头
- ~ fibre 直纤维,未卷曲纤维

~ fitting 直配接,同径配接〈塑料管〉
~ flow 直线流
　　flow annular piston accumulator 直流环形推料活塞式储料缸
~ forming 阴模成型,简易真空成型,直接(真空)成型
~ head 直机头〈挤塑机〉
~ hydraulic mo(u)ld clamping sys-tem 全液压合(锁)模装置,直压式合模装置
~ - in relationship 线性关系,直线关系
~ injection mo(u)lding 直接注塑成型〈热固性塑料〉
~ - line 直线
~ - line extrusion head 直挤塑机头,轴向挤塑机头
~ - line feed 直线供给
~ - line "herringbone" layout 直线鲱鱼骨形布置,直线人字形布置
~ linear polyethylene 直线型聚乙烯,直链聚乙烯
~ melamine 未改性的三聚氰胺树脂,未改性蜜胺
~ molecular orientation 线型分子取向
~ - notched pin 圆柱开槽销
~ overlapping gate 直接搭头浇口
~ paddle agitator 直桨叶搅拌器
~ paddle mixer 直桨叶混合器
~ pin 圆柱销
~ polymer 纯聚合物,均聚物
~ product 纯产品
~ reciprocating motion 直线往复运动
~ resin 纯树脂
~ tab gating 直片型浇口
~ thread 圆柱螺纹
~ - through 直通的
~ through die 直模头
~ through head 直机头
~ through manifold 直流道
~ vacuum forming 直接真空成型,简易真空成型
~ - way valve 直通阀
~ weld 直焊缝

straighten 弄平,矫直
straightening press 矫正压机
strain 应变,变形;伸长,张力;拉紧;弯曲;(过)滤
~ ageing 应变时效,弯曲老化〈热气焊接〉
~ amplitude 应变振幅,极限应变平均差
~ birefringence 应变双折射
~ crack 变形裂纹
~ crystallization 应变结晶
~ deviation 应变偏差
~ distribution 伸长分布
~ energy 应变能
~ - fatigue curve 应变-疲劳曲线
~ ga(u)ge 应变仪
~ ga(u)ge indicator 应变仪指示器
~ indicator 应变指示器
~ level 应变程度,应形程度
~ meter 应变仪;伸长计
~ of flexure 挠曲应变
~ path 变形区域
~ rate 应变速率
~ recovery 应变恢复
~ recovery curve 应变恢复曲线
~ relaxation 应变松弛
~ roll 张力辊;松紧辊
~ tensor 应变张量,伸长张量
~ tester 应变试验仪
~ velocity 应变速度
strained ring 张力环
strainer 滤料器〈挤塑机〉
~ head 滤料器头〈挤塑机〉
~ plate 滤料板〈挤塑机〉
straining of colo(u)r 滤色浆
strainless 未应变的
strains pl 剪切应力
strand 丝,丝束〈玻璃丝〉;股〈电缆,绳索〉;料〈挤塑〉
~ coating 线材涂布
~ cutter 线料切粒机
~ die 拉丝口模,线料口模
~ granulator 线料造粒机
~ integrity 丝束集结

stranded conductor

~ pelletizer 线料切粒机
stranded conductor 股合导线
strata pl 层
straticulate 薄层的,分层的
stratiffied 分层的,层压的
　~ mixture 分层混合料,层状混合物
　~ plastics 层压塑料
straw cellulose 草纤维素
stray 杂散;散射
　~ field 杂散场;漏磁场〈高频焊接〉
　~ voltage 杂散电压
　~ reflection 散射反射
strayfield dielectric heating 漏磁场高频加热
streak 条纹,条痕
　~ flaw 条状裂纹
　~ line 条纹线
　~ plate 条纹,条痕板
streakiness 条纹形成
streaking 条纹,条痕〈涂料〉;形成条痕
streaks 表面条纹〈缺陷〉
streaky surface 有条痕的表面
stream feeder 连续供料器
streaming birefringence 流动双折射;流线双折射
streamline 流线(型);流水线;连续的
　~ continuous flow mixer 流线型连续流动混合器
　~ flow 流线型流动
　~ flow pattern 流线型流动图型
　~ proportioning 连续配料,不间断配料
　~ proportioning of components 流线型的各组分配比
streamlined production 流水线生产,流线型生产
street 街道;异径
　~ elbow 长臂肘管,异径弯管
　~ fitting 异径管套筒
strength 强度;强力
　~ at rupture 断裂强度
　~ behaviour 强度性能
　~ criterion 强度准则
　~ of a dye 染料着色力

~ of welding seam 焊缝强度
~ parameter 强度参数
~ tester 强度试验仪
~ testing instrument 强度试验仪
~ - to - density ratio 强度密度比(率)
~ - to - weight ratio 比强度;强度重量比(率)
strengthening 加固,增强
~ agent 增强剂
~ mechanism 增强机理
stress 应力
~ amplitude 极限应力平均差
~ at break 断裂应力
~ build - up 应力构成,应力聚集
~ coat 应力涂料;脆性涂料
~ coat method 用脆性涂层做应力试验,用脆性涂层鉴别应力
~ component 应力分量;分应力
~ concentration 应力集中
~ concentration factor 应力集中因子[系数]
~ corrosion 应力腐蚀
~ crack 应力开裂;疲劳裂缝
~ crack corrosion 应力开裂腐蚀
~ crack formation 应力裂纹形成
~ crack ratio 应力开裂比例,疲劳裂缝质量因数
~ crack resistance 耐应力开裂性
~ cracking 应力开裂,应力龟裂
~ cracking behaviour 应力开裂行为
~ cycle 应力周期
~ decay 应力衰减
~ - deformation curve 应力变形曲线
~ deviation 应力(偏)差
~ direction 应力方向
~ disribution 应力分布
~ dissipation 应力散失
~ - elongation ratio 应力伸长比
~ - fatigue curve 应力-疲劳曲线
~ free 无应力的
~ - free mo(u)ldings 无内应力模塑制品
~ in bending 弯曲应力

~ in torsion 扭应力
~ - induced corrosion 应力诱导腐蚀
~ induced crystallization 应力诱导结晶(作用)
~ intensity 应力强度;裂纹强度
~ peak 最高应力〈应力集中〉
~ range 应力范围
~ rate 应力速率
~ ratio 应力比
~ recovery 应力回复
~ relaxation 应力松弛
~ relaxation test 应力松弛试验
~ relaxation tester 应力松弛测试仪
~ relaxometer 应力松弛仪
~ relief 应力消除
~ relief test 应力消除试验
~ relieving 应力消除
~ reversal 应力变换
~ rupture 应力破裂,应力断裂
~ rupture strength 应力断裂强度
~ sensitivity 应力敏感性
~ softening effect 应力软化效应
~ - strain curve 应力-应变曲线
~ - strain relations 应力-应变关系
~ strain test 应力-应变试验
~ surface 应力面
~ tensor 应力张量
~ to rupture 断裂应力
~ trajectories 应力轨迹
~ under compression 压缩应力
~ whitening 应力白化;折痕〈着色的硬塑料板返复折叠时产生的白色痕〉
~ wrinkle 应力皱纹
~ yield 屈服应力
stressed 负荷的,应力的
stressing under external pressure 外压应力
stressless 无应力的
stretch 拉伸;伸张
  ~ bar 展幅杆
  ~ bath 拉伸浴
  ~ blow mo(u)lding 拉吹成型
  ~ blow mo(u)lding machine 拉吹成型机

~ breaker 牵切机〈短纤维〉
~ die 拉伸(成型)模
~ elongation 拉伸弯形,伸长
~ fabric 弹力织物
~ forming 拉伸成型,包模成型
~ ratio 拉伸力比
~ roll 松紧辊;张力辊
~ - spinning 拉伸纺丝
stripper 脱模器;顶出器;模板;剥离器
~ bar 压缩膜模板
~ bolt 脱模顶杆
~ cutter roll mill 冲切辊式破碎机,切割辊式破碎机
~ ejection 脱模顶出
~ frame 脱模框
~ machine 脱模机,卸料机
~ mo(u)ld 脱模板式模具
~ pin 脱模销
~ plate 脱模板,开模板
~ plate ejection 膜模板式脱模,碰板脱模
~ plate mo(u)ld 脱模板式模具,碰板式模具
~ plate type injection mould 脱模板式注塑模
~ punch 脱模顶件
~ ring 脱模圈
~ roll 引离辊
~ roller 引离辊
stripping 脱模;剥离;脱除;汽提
~ agent 退模剂,剥模剂
~ coating 可剥(保护)涂层
~ compound 可剥防腐浸渍料
~ device 脱模装置
~ film 可剥膜
~ fixture 脱模工具
~ fork 脱模叉
~ frame 脱模框
~ from the mo(u)ld 脱模
~ of mo(u)ld 脱模
~ machine 脱模机,卸料机
~ strength 剥离强度
~ test 剥离试验
~ unit 脱模装置

stroke 行程,冲程,动程
- ~ cure 抚熟固体试验〈在钢板上测定热固性树脂固化速度〉
- ~ limitating 行程极限
- ~ of a piston 活塞冲程
- ~ of a screw 螺旋动程
- ~ per minute [s.p.m.] 冲程/每分钟〈活塞〉
- ~ retraction 倒转,逆行
- ~ volume 注料杆推料体积,冲程容积,工作容积

strong 强的,坚固的;浓(厚)的
- ~ acid 强酸
- ~ caustic liquor 强碱液
- ~ fibre 强力纤维
- ~ part 坚固部件
- ~ solvent 强溶剂

strongly 强烈地
- ~ acidic ion exchange resin 强酸性离子交换树脂
- ~ basic anion ion exchange fibre 强碱性阴离子交换纤维
- ~ basic ion exchange resin 强喊性离子交换树脂
- ~ basic vinyl pyridine anion exchange resin 强碱性乙烯吡啶阴离子交换树脂

strongmelt-process 辊式熔体涂布法〈熔融黏合剂涂布〉

strontium [Sr] 锶
- ~ ricinoleate 蓖麻醇酸锶〈稳定剂〉
- ~ stearate 硬脂酸锶〈热稳定剂〉

structural 结构的
- ~ adhesive 结构(型)黏合剂,结构胶
- ~ application 结构应用
- ~ board 结构板
- ~ bond 结构粘接件
- ~ breakdown 结构破裂
- ~ change 结构变化〈部分结晶的热塑性塑料〉
- ~ component 结构零件
- ~ defect 结构缺陷
- ~ development 结构改进
- ~ disorder 结构无序〈纤维增强塑料中未按最佳方向排列〉
- ~ element 结构元件;结构单元
- ~ fault 结构缺陷
- ~ feature 结构特征
- ~ foam 结构泡沫塑料
- ~ foam mo(u)lding 结构泡沫塑料成型;结构泡沫塑料成塑件〈夹芯注塑〉
- ~ foam mo(u)lding machine 结构泡沫塑料成型机〈注塑〉
- ~ foam mo(u)lding part 结构泡沫塑料成型件
- ~ tensor 拉伸张量
- ~ tester 拉伸试验仪
- ~ thermoforming 拉伸热成型
- ~ yam 弹力丝

stretchability (可)拉伸性,(可)伸长性,(可)延伸性

stretched 拉伸的
- ~ figments 拉伸长丝
- ~ nylon 弹力耐纶,弹力尼龙
- ~ orientation polymethyl methacrylate sheet 拉伸定向聚甲基丙烯酸甲酯板材
- ~ -out structure 引伸结构
- ~ tape 拉伸带

stretcher 拉伸器,延伸器
- ~ bar 拉幅辊,拉幅杆
- ~ forming 拉伸成型(法)
- ~ leveller 拉伸矫直机
- ~ -roller 拉伸辊〈薄膜〉
- ~ strain 拉伸变形

stretching 拉伸;延伸
- ~ device 拉伸装置
- ~ force 拉(伸)力,张力
- ~ frame 绷架,拉伸架
- ~ godet 拉伸导辊〈纺丝〉
- ~ machine 拉幅机
- ~ pedestal 拉伸架
- ~ strain 拉伸应变
- ~ stress 拉伸应力
- ~ unit 拉伸装置

stria 条纹〈在透明塑料中〉

striated structure 条纹结构

striation 条纹,条痕,波痕〈模芯支架脚

使熔料局部定向所引起〉
Strickland B winding 斯垂克兰德・贝缠绕成型法〈纵向和圆周方向配合缠绕制备玻璃纤维增强旋转壳体〉
striction 颈缩
strike – through 透底,击穿
striker 冲击摆锤
striking 冲击(的)
~ pendulum 冲击摆锤
~ pendulum apparatus 冲摆设备
string crystals 串晶,"羊肉串"结构
stringer 加劲条,增强带〈在薄板结构上〉
~ bead 焊珠,焊蚕;窄焊道〈焊条不横摆〉
stringiness 拉丝性;黏稠性〈黏合剂〉
strip 窄条,带,板条;剥离
~ – back peel test 剥离试验〈测定膜与基材粘接的剥离强度〉
~ breakdown 板坯
~ coating 带涂布;可剥涂层;板条镶面板涂装
~ – coating line 带涂布装置;板条镶面板涂装(生产)线
~ curtain 带条状窗帘〈PVC制〉
~ cutter 带料切割机;切带机
~ extrusion head 挤带机头
~ feed 带条状供料〈压延机〉可剥膜
~ fuse 片状保险丝,熔片
~ heater 带式加热器,电热圈
~ heating 电热丝加热
~ label 带条状标签〈包装〉
~ marker 路标〈交通〉
~ material 条状材料
~ of bias fabric 斜条纹织物
~ of carpeting 狭长地毯
~ packaging 可剥性包装
~ parison 带状型坯〈吹塑〉
~ shaped filler rod 带形焊条
~ specimen 带条形试样
~ tension 带条张力
~ thermocouple 条状热电偶
~ weld 带状焊;搭接焊缝
~ welding 带状焊;搭接焊

~ winder 缠带机,绕带机
~ winding 绕带
~ winding machine 绕带机,缠带机
strippable 可剥的
~ coating 可剥性涂层
~ lacquer 可剥涂层漆
~ protective coating 可剥保护涂层
~ foam part 结构泡沫塑料成型件
~ formation 结构形成
~ formula 结构式
~ material 结构材料;建筑材料
~ member 结构件
~ model 结构模型
~ order 结构有序〈纤维在增强塑料中按最佳方向排列〉
~ panel 结构板,支承板
~ part 结构件
~ plastics 结构塑料
~ purposes 结构目的
~ rearrangement 结构重排
~ regularity 结构有规性
~ relaxation 结构松弛
~ shape 结构形式
~ stability 结构稳定性
~ surface effect 结构表面效应
~ test 结构试验
~ unit 结构单元
~ viscosity 结构黏度,非牛顿黏度
~ viscosity index 结构黏度指数
structure 结构
~ design 结构设计
~ diagram 结构图
~ element 结构元件
~ element made from plastic 塑料结构(元)件
~ material 构造(用)材料;建筑(用)材料
~ model 结构模型
~ of polymers 聚合物结构
~ regularity 结构规整性
~ relaxation 结构松弛
~ test 结构试验
~ type 结构类型
~ with twist 加捻结构〈纱〉

strut 支柱;支撑
st. s [standard sample] 标准样品
STT [short time test] 短期试验,快速试验
stud 双头螺栓;销子
～ bar 剩余的材料,料头
～ bolt 双头螺栓
～ insert 螺柱嵌件
study 研究;分析
staffer 柱塞式挤塑机〈美〉
stuffing 填料,填塞
～ box 填料函,填料箱
～ cylinder 供料料筒
～ piston 供料活塞〈注塑机〉
～ plunger 供料柱塞
～ screw 供料螺杆
stylus 铁笔;描形针;触针;雕刻刀
～ printing 触针印刷
styrenated 苯乙烯化的
～ alkyd resin 苯乙烯改性醇酸树脂
～ phenols 苯乙烯化苯酚,防老剂 SP
styrene 苯乙烯
～ acrylonitrile copolymer [SAN] 苯乙烯－丙烯腈共聚物
～ acrylonitrile plastic 苯乙烯－丙烯腈塑料
～ acrylonitrile resin 苯乙烯－丙烯腈树脂
～ alloy 苯乙烯"合金"〈塑料〉
～ - alpha - methylstyrene [SMS] 苯乙烯-α-甲基苯乙烯共聚物
～ - alpha - methylstyrene plastics 苯乙烯-α-甲基苯乙烯塑料
～ butadien copolymer [SB] 苯乙烯－丁二烯共聚物
～ butadien elastomeric copolymer 苯乙烯－丁二烯共聚弹性体
～ butadien plastic 苯乙烯－丁二烯塑料
～ butadien styrene block copolymer [SBS] 苯乙烯－丁二烯－苯乙烯嵌段共聚物
～ butadien rubber [SBR] 苯乙烯－丁二烯橡胶,丁苯橡胶

～ butadien rubber - adhesive [SBR - ad - hesive] 苯乙烯－丁二烯橡胶黏合剂,丁苯橡胶黏合剂
～ butadien thermoplastics 苯乙烯－丁二烯热塑性塑料
～ chloroprene rubber 苯乙烯－氯丁二烯橡胶
～ copolymer 苯乙烯共聚物
～ die 聚苯乙烯板料挤塑口模
～ - divinylbenzene copolymer 苯乙烯－二乙烯苯共聚物
～ emulsion 苯乙烯类乳剂
～ insulation 苯乙烯类绝缘材料
～ isopren elastomeric copolymer 苯乙烯－异戊二烯共聚弹性体
～ - maleic anhydride, [SMA] 苯乙烯－马来酸酐共聚物
～ - maleimide copolymers 苯乙烯－马来酰亚胺共聚物
～ methylmethacrylate 苯乙烯－甲基丙烯酸甲酯
～ methylmethacrylate copolymers 苯乙烯－甲基丙烯酸甲酯共聚物
～ methylmethacrylate resins 苯乙烯－甲基丙烯酸甲酯树脂
～ α - methylstyrene copolyner [SMS] 苯乙烯-α-甲基苯乙烯共聚物
～ α - methylstyrene plastic 苯乙烯-α-甲基苯乙烯塑料
～ plastic 苯乙烯类塑料
～ polymers 苯乙烯类聚合物
～ resin 苯乙烯类树脂
～ - rubber plastic [SRP] 聚苯乙烯橡胶改性塑料,耐高冲击聚苯乙烯
～ - rubber plastic alloy 苯乙烯－橡胶塑料"合金"
～ system quaternary ammonium type Ⅰ strongly basic anion exchange resin 苯乙烯系强碱性季铵Ⅰ型阴离子交换树脂
～ system strogly acidic cation exchange resin 苯乙烯系强碱性阳离子交换树脂
～ system weakly bisic anion exchange

resin 苯乙烯系弱碱性阴离子交换树脂
styrofoam 苯乙烯泡沫塑料的俗称
styrol 苯乙烯
styryl disperse dye 苯乙烯基型分散染料
sub class 副族
subassembly 组合件,部件;组件装配
subcavity 多模腔
~ gang mo(u)ld 多模槽式模具
~ mo(u)ld 多模槽式模具,溢料式多槽模具
subcritical crack growth 亚临界的裂缝延伸
suberic acid 辛二酸
subfreezing temperature 冰点以下温度,零下温度
subject 题目,主题;科目;服务,以⋯为条件(to)
sublimate 升华,提纯,纯化,净化
submarine gate 潜伏式浇口,隧道型浇口
submerge 浸入;浸没
submersion 浸入;浸没
submicrofracture 亚微观断裂
submicroscopic damage 亚微观损伤
subnormal 正常以下的
subquality product 不合格产品,次级品
subsequent unit 后续装置
subsidiary 辅助的
~ experiment 辅助试验
substance 物质,实物;材料
substandard 非标准的
~ product 等外品
substantive 直接的
~ azo dyes 直接偶氮染料
~ dyes 直接染料
substituent 代用品,代用料;取代基〈聚合物〉
~ group 取代基团
substitute 代用品;替代的;取代
~ material 代用(材)料, substituted phenols 取代酚
substitution 代替,取代
~ method 取代法

~ product 取代产物
~ rate 取代程度
~ reaction 取代反应,置换反应
substrate 基material, 基料;底材,衬垫;底涂层;载体
~ web 基材
substratum 基材,底层
substructure 亚结构;基础结构
subsurface 表面下的;地下的
subterranean cable 地下电缆
subwater pipe line 水下管线
subzero 零下(的)
~ temperature 零下温度
~ test 零下(温度)试验
succeeding stretch 二次拉伸;继续拉伸
succinic acid 丁二酸;琥珀酸
~ peroxide 过氧化丁二酸〈交联剂,引发剂〉
sucker pin 液动或真空驱动的模具销
sucrose 蔗糖
~ acetate isobutyrate 乙酸异丁酸蔗糖酯〈增塑剂〉
~ benzoate 苯甲酸蔗糖酯〈增塑剂〉
~ octaacetate 八乙酸蔗糖酯,蔗糖八醋酸酯〈增塑剂〉
suction 吸取,空吸
~ bottle 吸滤瓶
~ box 吸收箱
~ chamber 吸引室
~ conveyor 真空输送机
~ couch roll (抽)吸辊
~ fan 排风扇,通风机;吸尘器
~ filter 吸滤器
~ hood 吸罩
~ inlet 吸入口
~ installation 通风系统;吸尘设备
~ line 吸入管线
~ machine 吸尘机
~ mo(u)ld 抽吸成型,真空成型
~ pipe 吸管
~ pipe socket 吸管插座
~ pressure 吸收压力
~ pump 真空泵,吸入泵,空吸泵
~ roll 吸辊

sulfamide

~ strength 吸收强度,耐吸收性
~ table 吸台〈涂布〉
~ valve 吸入阀
sulfamide 磺酰胺;硫酰胺
~ resins 磺酰胺树脂
~ - formaldehye resin 磺酰胺-甲醛树脂
sulfate [sulphate] 硫酸盐[酯]
~ pulp 硫酸盐纸浆
~ wood pulp paper 硫酸盐木浆纸
sulfide [sulphide] 硫化物
~ colo(u)rs 硫化染料
~ staining 硫化物污染
sulfite 亚硫酸盐[酯]
~ liquor 亚硫酸盐废液
~ pulp 亚硫酸盐纸浆
sulfo chloride 磺酰氯
sulfonamide dyes 磺酰胺型染料
sulfonate carboxylate copolymers 磺酸酯羧酸酯共聚物
sulfonated [sulphonated] 磺化的
~ coal cation exchanger 磺化煤阳离子交换剂
~ oil 磺化油
~ phenolic resin 磺化酚醛树脂
sulfonation 磺化(作用)
sulfone [sulphone] 砜
sulfonic acid 磺酸
sulfoxide [sulphoxide] 亚砜
sulfur [sulphur, S] 硫〈硫化剂〉
~ chloride vulcanized vegetable oils 氯化硫硫化的植物油〈软化剂〉
~ colo(u)rs 硫化染料
~ dye 硫化染料
~ nitride 硫-氮化合物
~ nitrogen polymers 硫-氮聚合物
sulfuric [sulphuric] 硫(的)
~ acid 硫酸
~ acid anodizing 硫酸阳极处理〈金属粘接件电化学表面处理〉
sulfurless cure 无硫硫化
sulphate cellulose [sulfate pulp] 硫酸盐纤维素
sulphite cellulose [sulfite pulp] 亚酸

盐纤维素
sulphur [S, sulfur] 硫
sun 太阳,日;日光,阳光
~ checking 晒裂
~ checking agent 耐日光致龟裂剂;光稳定剂
~ cracking 晒裂
~ crazing 日晒银纹裂;日晒细纹裂
~ discolo(u)ration 日晒变色
~ exposure test 日晒试验,曝晒试验
~ - hour 日照小时
~ - proof 耐晒的
~ - ray 太阳光线
~ test 日晒试验
sunfast 耐晒的
sunk 加深的,深入的,凹陷的
~ spot 缩孔;凹痕缩痕〈制品缺陷〉
sunken 沉下的,地下的;凹下的
~ joint 凹陷接缝〈不完善的粘接或焊接〉
~ pipe 地下管道
sunlight 日光
~ ageing 日光老化(作用)
~ degradation 日光降解
~ exposure 日光曝晒(试验)
~ resistance 耐光性,耐晒性
super 超,特级,优等
~ absorbent polymer 超吸附性聚合物
~ draft 超牵伸
~ drawing 超拉伸
~ high draft 超大牵伸
~ high impact polystyrene [SHIPS] 超高冲击聚苯乙烯
~ light foam 超轻质泡沫塑料
~ performance 优越性能
~ - pressed plywood 高压胶合板
~ sensitizing 超敏化作用〈塑料镀金属的前处理过程〉
~ soft flexble urethane foam 超柔软聚氨酯泡沫塑料
~ strength 超强度
supercalender 多辊压延机;超级[高度]压光机
supercompression 超压缩

supercontraction 过收缩
supercooling 过冷
supercritical 超临界的
superficial 表面的,外部的;面积的
　～ area　表面积
　～ coat　表层
　～ damage　表面损坏
　～ expansion　表面膨胀
　～ layer　表面层
　～ structure　表面结构
　～ unit　面积单位
superfluous 过剩的,多余的
　～ parameter　多余参数
superheat 过热
　～ temperature　过热温度
superheated 过热的
　～ bubble cap foaming　过热蒸汽泡罩发泡法
　～ steam　过热蒸汽
　～ steam autoclave foaming　过热蒸汽高压釜发泡法
　～ steam jet foaming　过热蒸汽喷射发泡成型法
　～ steam-prefoaming　过热蒸汽预发泡成型法
superhigh boiler 超高温沸腾器〈1350℃沸腾〉
superimposed 叠合的,叠加的
　～ initial modulus　叠加初始模量
　～ layer　叠加层
　～ seam　叠缝
　～ stresses　叠加应力
superior 高级的,优良的;在上的,较高的
　～ limit　上限
　～ rubber　超级橡胶
superlattice 超晶格
　～ structure　超晶格结构〈部分结晶热塑性塑料〉
　～ transformation　超晶格转变
supermicroscope 超级显微镜
supermolecular structure 超分子结构
supernatant clear liquid 沉清液;清液层
supernormal 超正常的

superplastic alloy 超级塑料"合金"
superplasticity 超塑性
superpolyamide 超高分子聚酰胺
superpolyester 超聚酯
superpolymer 超高聚物
superposable die 叠合模,复合模,组合模
superposed on 重叠,重合
supersede 取代,代替,更换
supersonic 超声(波)的
　～ flaw detector　超声探伤仪,超声探测裂缝仪
　～ horn　超声焊极
　～ wave　超声波
　～ welder　超声波焊接机
　～ welding machine　超声(波)焊(接)机
superstructure 超等结构;上层结构
supervision 监督;检测;管理
　～ of instrumentation　操作控制
superwood 高级木材,(树脂)浸渍木材
supplant 取代,代替,更换
supplemental 补充的,辅助的
supplementary 补充的,辅助的
　～ means　辅助手段
　～ parts　辅助零件
　～ test　补充试验
supplier 供应者,供给者
supply 供给;供料
　～ line　供给线
　～ mains　供给干线,供应总管
　～ required　需求供应
support 支承;支架;载体
　～ fins　支翼
　～ of catalyst　催化剂载体
　～ pillar　支模柱
　～ plate　托板,模底板,垫模板,底板
　～ post　支模柱
　～ rod　支承杆;载体棒
　～ roll　支承辊,托辊,垫辊〈压延机〉
　～ structure　承载结构
supported 支承的
　～ film　带垫衬薄膜
　～ film adhesive　带垫衬胶黏膜

supporter
~ flange joint 法兰夹紧连接
~ screwed joint 螺栓压紧连接
~ sheet 带垫衬片材
supporter 支架;载体
~ of combustion 燃烧载体
supporting 支承,承载
~ air 支撑空气,压缩空气〈挤塑软管膜吹塑成用〉
~ base 基材;底基;垫材;载体
~ bed 支承板,垫板,底板
~ core 支承模芯
~ layer 衬垫层
~ material 衬垫材料,基材
~ member 支承构件
~ plate 托板,垫模模,支承板
~ ring 支承圈,托圈;垫圈
~ roll 支承辊,承压辊,垫辊,托辊〈压延机〉
~ structure 支承结构
suppressant 抑制剂
suppressing agent 抑制剂
surface 面,表面
~ abrasion 表面磨耗
~ abrasion resistance 表面耐磨性
~ abrasion scratching 表面磨刮痕
~ abrasion test 表面磨蚀试验
~ absorber 表面吸收器
~ – active agent 表面活性剂
~ adhesion 表面黏附
~ adsorption 表面吸附
~ aftertreatment 表面后处理〈粘接面〉
~ ageing 表面老化
~ area 表面积;有效面积
~ area of catalyst 催化剂有效面积
~ asperity 表面粗糙度
~ blemish 表面缺陷
~ blowhole 表面气孔
~ blush 表面湿纹,表面湿浊;雾面
~ burn 表面燃烧
~ characteristic 表面特性
~ check 表面龟裂;表面检验
~ chelate compound 表面螯合物
~ coat 表面涂层,面漆
~ coating ion exchange membrane 表面涂层离子交换膜
~ coating powder 表面涂布粉
~ coating resin 表面涂饰树脂,面漆用树脂
~ coefficient 表面系数
~ colo(u)r 表面着色
~ combustion 表面燃烧
~ condenser 表面冷凝器
~ condition 表面状况;表面条件
~ conductance 表面电导
~ conductivity 表面导电性
~ contact 表面接触
~ cooling 表面冷却
~ crack 表面裂纹
~ damage 表面损伤
~ defect 表面缺陷
~ density 表面密度
~ diffusion 表面扩散
~ distortion 表面变形,表面扭曲
~ elasticity 表面弹性
~ element 表面元素
~ energy 表面能〈黏合面〉
~ film 表面膜
~ finish 表面涂饰剂;表面涂饰;表面抛光;表面光洁度
~ finishing 表面修饰,表面修整,表面加工
~ force 表面力
~ friction 表面摩擦
~ gloss 表面光泽
~ hardening 表面硬化
~ hardness 表面硬度
~ haze 表面上浑浊(度),表面光雾〈成型件上〉
~ heater 表面加热器
~ heterogeneity 表面不匀性
~ imperfection 表面缺陷
~ in contact 接触面
~ inspection 外部检查
~ irregularity 表面不平;表面粗糙;表面缺陷,表面凹陷
~ joint 表面连接,面接接头
~ layer 表面层〈粘接件〉
~ leakage 表面漏泄

~ level 水平面
~ maring 表面擦伤
~ mat 表面毡〈增强塑料〉
~ modification 表面改性
~ modified 表面改性的
~ moisture 表面水分
~ of contact 接触面
~ of fracture 断裂面
~ of revolution 回转曲面
~ of separation 界面
~ of the weld 焊接面
~ oxidation 表面氧化(作用)
~ peeling 表面剥离
~ phenomenon 表面现象
~ pins 平推顶杆,回程杆
~ planing machine 平面刨床〈塑料加工用〉
~ plate 平(面)板;平台
~ polishing 表面抛光
~ potential 表面(电)势,表面电位〈粘接件材料〉
~ preparation 表面预处理
~ preservation 表面防护〈粘接件表面〉
~ pressure 表面压力
~ pretreatment 表面预处理
~ printing [flat printing] 平版印刷
~ profile 表面粗糙断面
~ properties 表面性能
~ protection 表面防护(层)
~ quality 表面质量
~ reaction 表面反应
~ recombination 表面复合
~ residual stress 表面残余应力
~ resistance 表面电阻
~ resistivity 表面电阻率
~ rolling 表面滚化
~ roughness 表面粗糙度
~ shear viscosity 表面剪切黏性
~ sheet 面层片材,覆面片材〈装饰层压材料〉
~ sizing 表面孔封闭;上浆
~ skin 表(面皮)层〈泡沫塑料〉
~ smoothness 表面平滑度,表面整度
~ - specific adhesion 表面特性黏附
~ spread of flame 火焰表面蔓延
~ speed (辊)表面速度
~ stabilized 表面稳定的
~ stress 表面应力
~ structure 表面结构
~ tack 表面黏性
~ tack eliminator 防黏剂
~ tackiness 表面黏着性
~ temperature 表面温度
~ tension 表面张力
~ texture 表面网纹;表面纹理;表面结构
~ trace 表面痕迹
~ tracer 表面粗糙断面
~ traction 表面牵引力
~ - treated film 表面处理过的薄膜
~ treating 表面处理
~ treating agent 表面处理剂
~ treatment 表面处理
~ valley 表面凹陷
~ viscosimeter 表面黏度计
~ viscosity 表面黏度
~ waviness 表面纹面;表面波纹〈制品缺陷〉
~ wear 表面磨损
~ welding 表面焊接;熔焊,堆焊
~ width (辊)表面宽度
~ winder 表面卷取机
~ winding 表面卷取
surfacer 二道底漆,二道浆
surfacing 表面涂饰,敷面,面饰,表面平整;堆焊
  ~ mat 面层毡,表面毡片;敷面毡料〈增强塑料〉
  ~ material 铺面材料
  ~ veil 表面贴面毡
surfactant 表面活性剂
surgical use 外科应用
surging 波动,料涌〈挤塑〉
surplus 残料,余料;多余的,剩余的
  ~ heat 多余的热量,过剩的热量
  ~ material 溢料,余料,剩余材料

surveyor's report

~ pressure 剩余压力
~ stock 剩余原料
surveyor's report 鉴定证明书
susceptibility 敏感度[性];磁化率
~ to ……对……的敏感性
susceptible 敏感的
~ to corrosion 对腐蚀敏感,易受腐蚀
susceptiveness 敏感性
suspend 悬浮;吊挂
suspended 悬浮的;吊挂的
~ ceiling 吊顶
~ matter 悬浮物
~ particles 悬浮颗粒
~ particles dryer 悬浮粒干燥器
~ phase 悬浮相
suspending 悬浮
~ agent 悬浮剂
~ medium 悬浮介质
suspension 悬浮;悬浮体,悬浮液
~ bed 悬浮床
~ blade 悬挂叶片
~ bridge 悬桥,吊桥
~ cable 悬挂索缆
~ method 悬浮法
~ of fluidized powder 流化粉末悬浮体
~ point 悬挂点〈帐篷〉
~ polymer 悬浮聚合物
~ polymerizate 悬浮聚合物
~ polymerization 悬浮聚合(作用)
~ polytetrafluoroethylene 悬浮法聚四氟乙烯
~ polyvinyl chloride [S‑PVC] 悬浮法聚氯乙烯
suspensoid 悬胶体
sustain 承受;维持;遭受
s.w. [specific weight] 比重(旧称)相对密度
swabbing 擦涂,刷涂;刷色,刷浆
swaging 深拉,深冲;压花;卷边
~ roll 卷边辊
~ tool 卷边工具,折边模
swan neck 鹅颈管,弯管〈塑料落水管〉
~ press 曲柄立式压机
swarf 细屑

sweat out 盲汗,发出汗〈增塑剂渗出〉
sweating 发汗;渗出〈增塑剂渗出〉
sweep extractor 扫刮式脱模器
swell 膨胀;溶胀
~ factor 膨胀系数
~ ratio 膨胀[溶胀,泡胀] 比〈挤塑物〉
~ test 膨胀试验
~ up 膨胀,溶胀
swelling 膨胀,溶胀,泡胀
~ agent 膨胀剂,溶胀剂,泡胀剂
~ behaviour 膨胀[溶胀,泡胀] 性能〈挤塑物〉
~ determination 膨胀[溶胀,泡胀]度测定
~ heat 膨胀[溶胀] 热
– index 膨胀[溶胀、泡胀] 指数〈挤塑物〉
~ power 膨胀[溶胀、泡胀] 能力
~ property 膨胀性
~ rate 膨胀[溶胀、泡胀] 率
~ ratio 膨胀[溶胀、泡比
~ stage 膨胀阶段;膨胀相
~ tendency 膨胀[溶胀、泡胀]趋势
swept volume 推料体积,推进体积;冲程容积
swing 摇摆,摆动;旋转,转动
~ die 旋转式口模
~ jaw 活动颚板
~ mill [vibrating mill] 摆动磨,振动破碎机
~ mo(u)ld 转动式模具
~ sieve 摆动筛,振动筛
~ sledge 摆锤
~ sledge hammer crusher 摆锤破碎机
~ -sledge(hammer)mill 摆锤磨
~ table 转台
~ type check valve 回转式止逆阀
swinging 摆动;转动
~ ball bearing 摆动式滚珠轴承,自动调整滚珠轴承
~ disc type check valve 回转盘式止逆阀
~ gate 转动式联锁装置〈挤塑机模〉

~ scraper 旋转式刮刀
~ screen 摆动筛；振动筛
~ shovel 转动式铲刀
~ sieve 摆动筛；振动筛
swirl 旋涡，涡流；弯曲；粗糙表面，皱纹表面〈结构泡沫缺陷〉
~ mat 卷曲玻璃丝毡
switch 开关；转换
~ board 开关板，配电盘
~ board diagram 接线图，线路图，电路图
~ case 开关箱
~ off 关掉，断开
switchover point 转向点
swivel winder 转动缠绕机
swivelling head 可转位的注塑机头
swollen cellulose 溶胀纤维素
syenite 正长岩，硅酸钠 - 硅酸钾 - 硅酸铝复合物
syn. synthetic 合成的
synaeresis [syneresis] 脱水收缩〈胶体〉
synchro 同步(的)
~ draw 同步拉伸
synchronous operation 同步运行，同步操作
synclastic 同方向弯曲的〈壳〉
syndiotactic 间同(立构)的，间规的
~ 1,2 - polybutadiene 间规聚 1,2 - 丁二烯
~ polymer 间同立构聚合物，间规聚合物
syndiotacticity 间规度，间同立构规整度
syneresis 脱水收缩
synergism [synergy] 协同作用，协合作用，增效作用
synergist 协同剂，增效剂
synergistic 协同的，增效的
~ action 协同作用，协合作用，增效作用
~ additive 增效性添加剂
~ agent 增效剂
~ combination 增效性配合
~ effect 协同效应
~ system 增效性配方

synoptical table 一览表
syntactic 综合的；组合的，复合的
~ bubble 复合空泡
~ cellular plastic 组合微孔塑料
~ foam 复合泡沫塑料
~ foams plastics 复合泡沫塑料
synthesis 合成(法)
~ reactor 合成反应器
~ temperature 合成温度
synthesize 合成
synthetic 合成的，人造的；综合的
~ cement 合成粘接剂
~ dyes 合成染料
~ dyestuff 合成染料
~ elastomer 合成弹性体
~ fabrics 合成纤维织物
~ fatty acid 合成脂肪酸〈乳化剂〉
~ fibre 合成纤维
~ fibrous anisotropic material mat [SFAM mat] 各向异性合成纤维毡，SFAM 毡
~ flooring 合成地板〈塑料制〉
~ foam 合成泡沫塑料
~ leather 合成革，人造革
~ material 合成材料
~ method 合成方法
~ organic pigment 合成有机颜料
~ natural rubber 合成天然橡胶
~ paper 合成纸
~ plastic 合成塑料
~ polymer 合成聚合物
~ polymer dispersion 合成聚合物分散体
~ products 合成制品
~ resin 合成树脂
~ resin adhesive 合成树脂黏合剂
~ resin bonded - cloth sheet 合成树脂黏合布板
~ resin bonded - fabric sheet 合成树脂黏合布板
~ resin bonded laminate 合成树脂胶合层压材料
~ resin bonded - paper sheet 合成树脂黏合纸板
~ resin cement 合成树脂粘接剂

~ resin coating 合成树脂涂料
~ resin finish 合成树脂涂料
~ resin ion exchanger 合成树脂离子交换剂
~ resin mo(u)lding compound 塑料粉,合成树脂模塑料
~ resin varnish 合成树脂清漆
~ rubber 合成橡胶
~ rubber adhesive 合成橡胶黏合剂
~ segmented copolymer 合成嵌段共聚物
~ sponge material 合成海绵材料
~ study 综合研究
~ thermoplastics 合成热塑性塑料
~ utilization 综合利用
~ wool 合成羊毛,人造羊毛
syphon [siphon] 虹吸,虹吸管
syrups 糊浆;铸塑浆,浆料
syst. [system] 系统
system 系统,体系;装置,设备
~ pressure 操作压力
~ research 系统研究
SZ [size] 尺寸;大小

T

T,t [temperature] 温度
T [tentative standard] 暂行标准〈ASTM〉
t [thickness] 厚度
T [ton] 吨
T-cock 三路活塞,三路旋塞
T-die T形模头
~ film T形模头挤出的膜,平膜
T-joint T形接头;T形焊接
T-peel T形剥离
~ resistance T形耐剥离性
~ strength T形剥离强度
~ test T形剥离试验
T-piece T形接头,T形管件,三通管件,丁字形三通
T-shape manifold T形料道〈挤塑〉
T-slot groove T形槽,丁字形槽
T-type T形,丁字形
~ die T形模头
~ joint T形胶接接头
T-welding T形焊接,丁字形焊接
Ta [teantalum] 钽
tab 前室,预备室;薄片;标记,标签,标志牌
~ gate 柄形饶口,直角浇口
~ gating 柄形浇口,直角浇口
tab. table 表;表格;图表

TAB [Technical Abstracts Bulletin]
Taber 泰伯
~ abraser [machine] 泰伯磨试验机
~ abrasion index 泰伯磨耗指数
~ abrasion test method 泰伯磨损试验法
table 工作台,平板
~ area 工作台面积
~ feeder 平板供料器;台式供料器
~ press 台式压机
tablet 料板,料片;片剂
~ compressing press 压片机
~ density (压片料)料片密度
~ loader 装片器
~ press 压片机
tabletting 压片;切粒;造粒〈美〉
~ machine 造粒机;压片机
~ press 压片机
tabulation 制表(格)
TAC [triallyl cyanurate] 氰尿酸三烯丙酯
tack 黏性
~ coat 黏合层,粘接层
~ -free 无黏性;不粘手的,指触干燥的
~ -free state 消黏状态,表面无黏性

~ - free time 消黏时间〈聚氨酯泡沫塑料反应注塑时〉
~ instrument 黏性测试仪
~ life 黏性消失时间,黏性寿命
~ - maker 增黏剂〈弹性体黏合剂〉
~ nozzle 定位焊嘴
~ producer 增黏剂
~ - producing agent 增黏剂
~ range 黏性期〈压敏胶〉
~ retention charateristics 黏附性持续性
~ strength 黏附强度
~ temperature 发黏温度
~ welding 定位焊,点固焊,间断焊,平头焊接
tackifier 增黏剂
tackiness 黏性,黏合性,胶黏性
~ agent 黏合剂
~ tester 黏性试验机
tacking 定位焊
~ nozzle 定位焊嘴〈热气焊接仪〉
~ seam 定位焊缝〈焊接〉
tackmeter 黏性测试仪,黏性计
tacky 发黏的
~ dry 干(后)黏性
~ fibre 粘结纤维
~ range 黏性期
tackyness 黏附性
tactic 顺序的,规则的;有规(立构)的
~ block 有规(立构)嵌段
~ block polymer 有规(立构)嵌段聚合物
~ polymer 有规(立构)聚合物
~ polymerization 有规(立构)聚合
tacticity 立构规整度
tactile 触觉的
TAF foam mo(u)lding [Toshiba and Asahi foam mo(u)lding] TAF 泡沫塑料成型法,气体反压注塑法,东芝和旭公司泡沫塑料成型法
TAF process TAF 泡沫塑料成型法
tag 标签
~ cloth 标签布
Tag blask point test 泰格闪点试验

Taga process 田贺法,田贺成型法〈挤塑成型与注塑成型结合的一种成型方法〉
Tagliabue open and closed cup[TLC] 塔利亚贝氏开闭杯[闪点]试验
tail 尾(部),末端;飞边剩余料〈铸塑件〉
~ board 尾板〈卡车〉
~ flap 挡带〈汽车,自行车〉
~ lamp 尾灯,后灯
~ - to - tail bond 尾 - 尾连接
~ water 废水
tailings 粗粒填料;尾馏分〈蒸馏〉
tailor - made 定制的,特制的
~ adhesive 定制的黏合剂〈美〉
~ fibre 特制纤维
~ material 定制的材料
tailored polymer 特制聚合物
tails 熔合缺陷〈铸塑件〉
take - away 引出,牵引;牵引装置
~ roll 牵引辊
~ speed 牵引速率[度]
take - off 牵引装置
~ apron 牵引挡护板
~ conveyor 牵引装置
~ device 牵引装置;退绕装置
~ equipment 牵引装置
~ roll 牵引辊
~ speed 牵引速率[度]
~ tower 牵引塔架〈垂直吹薄装置〉
~ unit 牵引装置
take - out unit 脱模装置
take - up 牵引滚筒;卷取装置
~ bobbin 卷绕筒管
~ equipment 牵引装置;卷取装置
~ godet 卷丝导丝盘
~ machine 卷绕机
~ mechanism 牵引机构;卷取机构
~ package 卷取筒子
~ reel 卷取筒
~ roll 牵引辊;卷取辊
~ speed 牵引速率[度];卷取速率[度]
~ spool 牵引筒;卷取筒〈涂布带〉
~ winder 缠绕机

taking of samples 取样
tal total 总数;总计
talc [talk] 滑石
  ~ powder 滑石粉
talcum 滑石
  ~ powder 滑石粉
tall 高大的,巨大的
  ~ oil 妥尔油
  ~ oil varnish 妥尔清油漆
  ~ resin 妥尔油树脂,浮油树脂〈由木浆废液制得〉
tallate 树脂酸盐
tally 标签,名牌,标记牌
talus 废料
tamping 填塞;填塞料
tandem 串联(的),串列(的)
  ~ blow moulding machine 串联式吹塑机
  ~ dies 复式拉深模
  ~ connection 串联
  ~ drive 串联驱动
  ~ extrusion 串联式挤塑
  ~ extrusion coating line 串联式挤塑涂布生产线
  ~ line 串联装置
  ~ mixer 复式混合器
  ~ transfer mo(u)ld 积层式传递模具
tangent 正切
  ~ line 切线
  ~ modulus 正切模量
  ~ modulus of elastisity 弹性正切模量
  ~ of the dielectric loss angle 介电损耗角正切
  ~ of the loss angle 损耗角正切(值)
tangential 切向的,相切的
  ~ feed opening 切向供料口
  ~ force 切向力
  ~ mill 切相研磨机
  ~ screw 切向螺旋
  ~ strain 切向应变
  ~ stress 切向应力
tank 箱,桶,容器,储罐,槽罐;槽车
  ~ agitator 槽罐搅拌器
  ~ bottom 容器底
  ~ connector 槽罐连接器
  ~ lining 储罐衬里
  ~ mixer 罐式混合器
  ~ reactor 罐式反应器
  ~ wagon 油槽车
tantalum [Ta] 钽
tap 螺丝攻;旋塞,龙头,分布嘴;分流〈电〉;节流阀,调节阀
  ~ connector 排水栓旋接器
  ~ water 自来水
tape 条,带,带材;胶黏带,密封胶带;磁带,录音带
  ~ adhesive 胶黏带
  ~ cassette 磁带盒
  ~ fibre 原纤化膜带
  ~ line 带材生产线〈膜带生产〉
  ~ - producing plant 带材生产装置
  ~ slitting unit 切带装置〈薄膜生产〉
  ~ suction gun 带材吸入风枪〈薄膜生产装置〉
  ~ tension roller 皮圈张力辊
  ~ winder 缠带机〈薄膜带〉
  ~ winding 缠带
  ~ winding method 缠带法
  ~ with selvage 有织边带
  ~ without selvage 无织边带
  ~ yarn 扁丝
taper (圆)锥度;锥形的
  ~ pipe 异径管
  ~ pin 锥形销
  ~ reamer 锥孔铰刀
  ~ thread 圆锥形螺纹
  ~ winding 锥形卷绕
tapered 锥形的,斜
  ~ bobbin 锥形筒子
  ~ extruder 锥形挤塑机
  ~ flange 异径法兰
  ~ flange adaptor 异径法兰连接器
  ~ hole 锥(形)孔
  ~ insert 锥形嵌件
  ~ locating ring 锥形定位环
  ~ overlap 斜塔接
roller bearing 锥形滚柱轴承
  ~ secondary sprue 锥形分流道

~ single lap joint 单斜搭接
~ thread 锥形螺纹
tapering 锥形的
~ screw 挤压螺杆
~ spindle 锥形轴
tapping 攻螺丝;放出,泄出
~ attachment 攻丝夹具
~ clamp 卡箍〈塑料管〉
~ clip 卡箍
~ key 攻丝扳手
~ machine 攻丝机
~ pipe 泄水管
~ screw 自攻丝螺丝
~ tool 攻丝工具
tar 焦油
~ -base epoxide 焦油(基)环氧树脂
~ -base epoxide resin 焦油(基)环氧树脂
target 目标;指标
tarnish 失去光泽;使无光泽
~ film 失光泽膜
tasteless 无味的
taupe 灰褐色
tautness 拉紧
~ meter 伸长计
tawny 黄揭色
Tb [terbium] 铽
TBC [tetracarboxy butane] 四羧丁酯
TBP [tri-n-butyl phosphate] 磷酸三正丁酯〈增塑剂〉
TBPA [tetrabromophthalic anhydride] 四溴邻苯二甲酸酐,四溴苯酐〈阻燃剂〉
TBT [tetrabutyl titanate] 四丁基钛酸酯,钛酸四丁酯
TBTD [tetrabutyl thiuram disulfide] 二硫化四丁基秋兰姆〈促进剂〉
T.C. [take care] 小心;注意
TC [technical characteristic] 技术性能
TC [Technical Committee] 技术委员会
TC [total carbon] 总碳含量〈废水中〉
TCBO [trichlorobutylene oxide] 三氯环氧丁烷,三氯代丁撑氧
TCE [trichloroethylene] 三氯乙烯
TCEP [trichloroethyl phosphate] 磷酸三氯乙酯〈阻燃剂〉

TCP [tricresyl phosphate] 磷酸三甲苯酯〈增塑剂〉
TD [technical data] 技术资料;技术数据
TD [test data] 试验数据
TDI [toluene diisocyanate] 甲苯二异氰酸酯
TDR [technical data report] 技术数据报告
Te [tellurium] 碲
tear 撕裂,扯裂;裂缝,龟裂;磨损;泪
~ drops 泪珠,泪滴〈铸塑件缺陷〉
~ drop manifold 泪珠形分料道(挤塑)
~ off 撕裂,撕开;撕下
~ -off package 撕开包装
~ phenomena pl. 撕裂现象
~ propagation 撕裂延展
~ propagation force 撕裂增生力,撕裂延展力
~ propagation resistance 耐撕裂延展性
~ propagation strength 耐撕裂延展强度
~ propagation test 撕裂延展试验
~ resistance 耐撕裂性
~ resistance test 耐撕裂试验
~ resistance tester 耐撕裂试验仪
~ speed 撕裂速度
~ strength 撕裂强度
~ tape 撕裂带
~ test 撕裂试验
~ tester 撕裂试验仪
tearing 撕裂的
~ force 撕力
~ strain 扯裂应变
~ strength 撕裂强度
~ test 撕裂试验
TEC [triethyl citrate] 柠檬酸三乙酯〈增塑剂〉
tech. [technical] 技术的,工艺的
technical 技术的;工艺的,工业的;专业的,专门的
~ application 工业上应用〈塑料〉
~ bulletin 技术公报
~ characteristic 技术特征

~ chemistry 工业化学
~ condition 技术条件
~ data 技术数据
~ engineering 工程技术
~ exchange 技术交流
~ feature 技术特点
~ grade 工业级〈产品纯度〉,工业品位
~ - grade plastic 工业用塑料,工程塑料
~ guideline concetration [TRK] 工业指标浓度〈毒物学〉
~ information service 技术信息服务
~ inspection report 技术检验报告
~ know-how 技术诀窍,(专门)技术知识〈指不属专利保护范畴的关键性技术知识〉
~ manual 技术手册,技术规范
~ mo(u)lding 工业用成型件
~ norms 技术标准,技术规范
~ order 技术说明
~ parameter 技术参数
~ purpose 工业上应用
~ regulation 技术规程,技术规范,说明书
~ resin 工业用树脂
~ scale 工业规模
~ schedule 工艺过程,工艺规程
~ service 技术服务
~ specification 技术规格,技术规范
technique 技术,技巧;方法
~ of manufacture 制造技术
technological 工艺的,技术的
~ change 工艺更新,技术改造
~ characteristics 工艺特性
~ document 技术文件
~ process 工艺过程,工艺流程
technology 工艺学,工艺规格,工业技术
tee T字形丁字形;T形管;丁字形三通管
~ - connection T形连接〈管〉
~ joint [T - joint] T形接头,T形焊接
~ joint forming slot weld T形平头对接焊

~ joint with double fillet weld 双角焊缝的T形接头
~ joint with square butt weld T形平头对接焊
- piece T形接头;T形管,丁字管
~ pipe T形管,丁字管
~ welding T形焊接
teeth 齿;表面不平整,表面粗糙〈增强塑料断裂面〉
tefalisation 聚四氟乙烯涂层
Teflon 〈商〉特氟隆〈聚四氟乙烯〉
~ - coating 聚四氟乙烯涂层
~ coining 聚四氟乙烯铸压
~ moulding 聚四氟乙烯成型
Tego 泰戈
~ film 〈商〉泰戈膜〈一种酚醛树脂胶黏膜〉
~ glue film 〈商〉泰戈胶膜〈一种酚醛胶胶黏膜〉
telechelic polymer 远螯.聚合物,遥爪聚合物
teledynamic 伸缩料筒冲力
~ injection mo(u)lding 伸缩料筒冲力注塑成型,振动注塑成型
~ mo(u)lding 伸缩料筒冲力注塑成型
telegraphing 层合塑料或成型件内部缺陷图解
telephone 电话(机)
~ cable 电话线,通信电缆
~ case 电话机(外)壳
telescopic 套管的;伸缩的
~ baffle 伸缩挡板
~ flow 层流〈塑料熔体〉
~ joint 套管连接
~ tube 伸缩套管
telescoping rod 可伸缩棒杆
television 电视机(的)
~ case 电视机(外)壳
~ set 电视机(整件)
televisor 电视机
tellurium [Te] 碲
telogen 调聚剂
telomer 调聚物

telomeric 调聚的
~ reaction 调聚反应
telomerization 调节聚合(作用),调聚反应
temp. [temperature] 温度
temper 退火,热处理
temperature 温度
~ at extruder die 挤塑机机头物料温度
~ cabinet 调温室
~ change 温度变动
~ coefficient 温度系数
~ coefficient of viscosity 黏度温度系数
~ conditions 温度条件
~ control 温度控制,温度调节
~ controller 温度控制器,温度调节器
~ control system 温度控制系统,温度调节系统
~ control unit 温度控制装置
~ dependence (对)温度依赖性
~ dependence of velocity 速度与温度的关系
~ difference 温差
~ distortion 热变形
~ during heating time 加热时温度
~ gradient 温度梯度,温度差
~ gradient plate 温度梯度板
~ increment 温度升高〈指所升高程度〉
~ indicator 温度指示器
~ of combustion 燃烧温度
~ of deflection under load 负荷下变形温度
~ of deformation under flexural load 弯曲负荷下变形温度
~ of vulcanizing 硫化温度
~ - operated controller 温度控制器
~ profile 温度分布图
~ range 温度范围
~ recorder 温度记录仪
~ reduction 温度降低
~ - regulating device 温度调节装置,调温装置
~ regulation of the mould 模具的温度调节,模温调节
~ regulating system 温度调节系统
~ regulator 温度调节器
~ - resistant 耐热的
~ - resistant plastic 耐温塑料,耐热性塑料
~ rise 温度上升
~ sensitive adhesive 热敏黏合剂
~ sensor 温度传感器
~ sequence 温度变化过程
~ time 高温固化时间〈粘接时〉
~ - time limit 温度-时间极限
~ - time reduced law 温度-时间换算法则
~ traverse 温度变化
~ uniformity 温度均匀性
~ variation 温度变化
~ - viscosity curve 温度-黏度曲线
tempering 退火,热处理
~ channel 退火槽
template 模板,样板,型板
~ polymerization 模板聚合,母体聚合
temple 拉伸装置
~ roller 齿辊
templet 模板,样板,型板
temporary 暂时的
~ repair 临时修理
~ set 瞬时形变;弹性形变
tenacious 坚韧的;黏滞的
tenacity 韧度;韧性;耐断强度;黏性
~ test 韧性试验,黏(着)性试验
tendency roll 间接传动滑辊
tendency to… 倾向于…
~ cracking 易于龟裂
~ oxidize 氧化倾向
tensile 拉力的,张力的,拉伸的
~ adhesion 耐拉黏着力
~ bar 拉伸试条
~ compliance 拉伸柔量
~ creep 拉伸蠕变
~ creep test 拉伸蠕变试验
~ dumb - bell 哑呤形拉伸试条
~ elasticity 拉伸弹性
~ elongation 拉伸伸长

stress at offset yield point

~ energy 拉伸能
~ failure 拉断,拉伸破坏
~ fatigue test 拉伸疲劳试验
~ force 拉力,张力
~ heat distortion temperature 拉伸热变形温度
~ impact 拉伸冲击
~ impact resistance test 耐拉伸冲击试验
~ impact strength 拉伸冲击强度
~ impact strength, notched 缺口拉伸冲击强度
~ impact strength, unnotched 无缺口拉伸冲击强度
~ impact test 拉伸冲击试验
~ load 拉伸负载,张力负载
~ modulus 拉伸模量,拉伸弹性模量
~ product 张力积
~ property 拉伸性能
~ recovery 拉伸回复力
~ shear test 拉伸剪切试验
~ strain 拉伸应变
~ strength 拉伸强度
~ strength at break 断裂拉伸强度
~ strength at yield 屈服拉伸强度
~ strength limit 拉伸强度极限
~ (-strength)test 拉伸强度试验
~ (-strength)tester 拉伸强度试验机
~ stress 拉伸应力
stress at offset yield point 偏置屈服点拉伸应力
~ stress at yield 屈服拉伸应力
~ stress relaxation 拉伸应力松弛
~ test 拉伸试验
~ test specimen 拉伸试片
~ tester 拉伸(强度)试验机
~ testing machine 拉伸(强度)试验机
~ viscosity 拉伸黏度
~ yield 拉伸屈服(点);扯断伸长率
~ yield point 拉伸屈服点
~ yield strength [TYS] 拉伸屈服强度
tensility 可拉伸性
tensimeter 张力计
tensiometer 张力计;伸长计;拉伸强度

试验机
tension 张力,拉力,应力拉伸
~ draft 张力牵伸
~ element 张力件
~ fibre 受拉纤维
~ fracture 拉断
~ -free 无张力的
~ -free state 无张力状态
~ load 拉伸负载
~ modulus 张力模量
~ roll 拉紧辊,张力辊,牵引辊
~ roller 拉紧辊,张力辊,牵引辊
~ set 永久变形
~ shear strength 拉伸剪切强度
~ shear test 拉伸剪切试验
~ spring 拉力弹簧
~ strain 拉伸应变
~ stress 拉伸应力
~ test 拉伸(强度)试验
~ test specimen 拉伸试样
~ tester 拉伸(强度)试验机
~ testing machine 拉伸(强度)试验机
~ winding 张力收卷
tensonmeter 拉伸仪,伸长计
tentative 暂行的,试行的;试验性的
~ experiment [test] 初步试验,探索性实验
~ specification 暂行规范
~ standard 暂行标准
tenter 拉幅机;绷架〈薄膜〉
~ conveyor 拉幅机传送装置〈薄膜取向〉
~ dryer 框架式干燥机
tentering 拉幅,横向拉伸
~ machine 拉幅机
TEP [triethyl phosphate] 磷酸三乙酯〈增塑剂,溶剂〉
TEPA [tetraethylene pentamine] 四-1,2-亚乙基五胺,四乙撑五胺〈固化剂、硫化剂〉
terbium [Tb] 铽
terephthalate plasticizer 对苯二甲酸酯增塑剂
terephthaldehyde resin 对苯二甲醛树脂

terephthalic acid [TPA] 对苯二甲酸
terminal 末端;线端;终端
　～ board 接线盒板
　～ bond 端键
　～ box 接线盒
　～ chain 端链
　～ face 端面
　～ group 端基
　～ pressure 终始压力,末压力
　～ socket 端子插座
　～ switch 极限开关,终端开关
　～ temperature 终始温度;末端温度
terminated polymer 封端聚合物
termination 终止(作用)
　～ of chain 键的终止
　～ reaction 终止反应
ternary 三元的
　～ colo(u)rs 三元混合色
　～ dyes 三元混合染料
　～ mixed polyamide 三元共聚聚酰胺
　～ mixture 三元混合物
　～ polymerization 三元共聚(作用)
terpene resin 萜烯树脂
terpenic acid 萜烯酸
terpolymer 三元共聚物,三元聚合物
terpolymerization 三元共聚
terra 土,地
　～ alba 石膏粉
tertiary [tert-] 叔,第三
　～ - butyl hydroxyanisole 叔丁基羟基茴香醚〈抗氧剂〉
　～ creep 第三期蠕变,三重蠕变
Terylene 〈商〉涤纶,特丽纶〈聚对苯二甲酸乙二醇酯纤维〉
TES [thermal energy storage] 热能储存
test 试验,测试,检验
　～ atmosphere 试验环境
　～ bar 试(验)棒
　～ bed 试验台
　～ bench 试验台
　～ center 试验中心
　～ conditions 试验条件,试验规范
　～ cup 试(验)杯〈测定黏度用〉
　～ data 试验数据

～ desk 试验台
～ determination 试验测定
～ documentation 试验报告,试验资料
～ dye 试验用染料
～ equipement 试验设备
～ fence 试验曝晒架
～ foaming 试验发泡
～ for durability 耐久性试验,寿命试验
～ for flash point in open cup 开杯闪点试验
～ for thermal stability 热稳定性试验
～ formulation 试验配方
～ glass 试管
～ head 测试头
～ in the field 现场试验,野外试验
～ installation 试验设备
～ load 试验负荷
～ methods 试验方法
～ mixture 试验混合物
～ mo(u)ld 试验用模具
～ norm 试验标准
～ number 试验次数
～ of materials 材料试验
～ of rated performance 额定性能试验
～ parameters 试验参数
～ piece 试片,试样,试件
～ piece model 试验样品
～ plate 检验片,试板
～ portion 试量〈准确称出的剂量〉
～ report 试验报告
～ requirement specification 试验技术规范
～ results 试验结果
～ rig 试验设备
～ room 试验室
～ run 试运转,试车
～ sample 试样
～ sieve 试验用网筛
～ specifications 试验规程
～ specimen 试样,试件,试片
～ specimen for tension test 拉伸(强度)试验用试样
～ stand 试验台
～ surface 测试表面

tester

~ to failure　破坏性试验
~ tube　试管
~ tube heat ageing test　试管热老化试验
~ under load　负载试验
~ unit　试验装置
~ - use　运用试验,生产试验
~ value　测试值
~ working　试运转,试车
tester　试验机,测定器；检验器
testimonial　鉴定书
testing　试验,测试；检验
~ certificate　试验检定证书
~ equiment　试验设备
~ in conditioned atmosphere　改善环境的测试〈耐候性试验〉
~ laboratory　检验室
~ machine　试验机,测试机,检验装置
~ method　试验方法
~ of adhesives　黏合剂测试
~ of plastics　塑料测试
~ pendulum　测试用摆锤
~ press　试样压机,试片压机
~ stand　试验台
~ technique　测试方法
TETD [tetraethyl thiuram disulfide]　二硫化四乙基秋兰姆〈促进剂〉
tetra　〈词头〉四
~ - 2 - ethylhexyl pyromellitate　均苯四甲酸四(2 - 乙基己)酯〈增塑剂〉
~ pack　四面体包装
Tetrabasic lead fumarate　四碱式反丁烯二酸铅,四盐基富马酸铅〈热稳定剂〉
tetrabromo　〈词头〉四溴
3,5,3',5' - - - 4,4' - dihydroxy - diphenylsulfon　3,5,3',5' - 四溴 4,4' - 二羟基二苯砜〈阻燃剂〉
tetrabromobisphenol A　四溴双酚 A〈阻燃剂〉
~ allylether　四溴双酚 A 烯丙基醚〈阻燃剂〉
~ bis(2,3 - dibromopropyl) ether　四溴双酚 A 双(2,3 - 二溴丙基)醚〈阻燃剂〉

~ bis(2 - hydroxy ethyl) ether　四溴双酚 A 双(2 - 羟乙氧基)醚〈阻燃剂〉
tetrabromobisphenol S　四溴双酚 S〈阻燃剂〉
tetrabromobutane [TBB]　四溴丁烷〈阻燃剂〉
tetrabromoethane　四溴乙烷
tetrabromophthalic anhydride [TBPA]　四溴邻苯二甲酸酐,四溴苯酐〈阻燃剂〉
tetrabromoxylene　四溴二甲苯〈阻燃剂〉
tetrabutyl　四丁基
~ thiuram disulfide [TBTD]　二硫化四丁基秋兰姆〈促进剂〉
~ thiuram monosulfide　一硫化四丁基秋兰姆〈促进剂〉
~ titanate [TBT]　钛酸四丁酯〈催化剂〉
tetracarboxy butane [TBC]　四羧丁酯
tetrachlorobisphenol A　四氯双酚 A〈阻燃剂〉
tetrachlorophthalic anhydride　四氯邻苯二甲酸酐,四氯苯酐〈阻燃剂〉
tetradeenne dioic acid　十四烷二酸
tetraethyl thiuram disulfide [TETD]　二硫化四乙基秋兰姆〈促进剂〉
tetraethylene　四 - 1,2 - 亚乙基,四乙烯,四乙撑〈俗称〉
~ glycol　四甘醇
~ glycol di(2 - ethyl hexoate)　四甘醇二(2 - 乙基酸)酯〈增塑剂〉
~ glycol diheptanoate　四甘醇二庚二酸酯〈增塑剂〉
~ glycol dimethacrylate　三缩四乙二醇甲基丙烯酸酯
~ glycol monostearate　四甘醇单硬脂酸酯
~ pentamine [TEPA]　四乙撑五胺〈固化剂,硫化剂〉
tetrafluoro　四氟
~ - p - hydroquinone　四氟对氢醌
~ - p - phenylene diisocyanate　四氟亚苯基二异氰酸酯
tetrafluoroethylene [TFE]　四氟乙烯

~ - ethylene copolymer 四氟乙烯-乙烯共聚物
fibre 四氟乙烯纤维
~ - hexafluoroethylene copolymer 四氟乙烯-六氟乙烯共聚物
~ hexafluoropropylene 四氟乙烯-六氟丙烯共聚物
~ hexafluoropropylene copolymers aqueous dispersion 四氟乙烯-六氟丙烯共聚物水分散液
~ perfluoroalkyl vinyl ether copolymer 四氟乙烯-全氟烷基乙烯基醚共聚物
~ polymer 四氟乙烯聚合物,聚四氟乙烯
~ resin 四氟乙烯树脂
tetragonal 四方形的
~ crystal 四方晶体
~ structure 四方晶体结构
~ system 四方晶系
tetrahedron 四面体
tetrahydrofuran [THF] 四氢呋喃
~ polymer 四氢呋喃聚合物
~ - type adhesive 四氢呋喃型黏合剂
tetrahydrofurfuiyl oleate [THFO] 油酸四氢呋喃甲酯〈增塑剂〉
tetrahydrophthalic anhydride 四氢邻苯二甲酸酐,四氢苯酐〈固化剂〉
2,2',4,4' - tetrahydroxybenzophenone 2,2',4,4' - 四羟基二苯甲酮〈紫外线吸收剂〉
tetraisopropyl di (dioctylphosphito) titanate 双(亚磷酸二辛酯基)钛酸四异丙酯〈偶联剂〉
tetrakis 〈词头〉四个
~ (2 - cyanoethyl) phosphonium bromide 溴化四(2-氰乙基)鏻〈阻燃剂〉
~ (2,4 - di - tert - butyl) phenyl - 4,4' - diphenylbis phosphonate ester 四(2,4-二叔丁基苯基-4,4'-联苯基)双膦酸酯〈抗氧剂〉
~ (hydroxymethyl) phosphonium chloride 氯化四(羟甲基)鏻〈阻燃剂〉

tetramer 四聚体,四聚物
tetramcthyl ethylene diamine [TMEDA] 四甲基乙二胺〈聚氨酯泡沫塑料用催化剂〉
tetramethylsilane [TMS] 四甲基(甲)硅烷
tetramethylthiuram 四甲基秋兰姆
~ disulfide [TMTD] 二硫化四甲基秋兰姆〈促进剂〉
~ monosulfide [TMTM] 一硫化四甲基秋兰姆〈促进剂〉
tetramethylene maleic anhydride 四亚甲基马来酸酐,四亚甲基顺酐〈固化剂〉
tetraoctyl pyromellitate [TOPM] 均苯四甲酸四辛酯,均苯四甲酸四(2-乙己)酯〈增塑剂〉
tetraoctyloxytitanium 四辛氧基钛
- di(dilauryl phosphite) 四辛氧基钛双(二月桂基亚磷酸酯)〈偶联剂〉
~ di (ditridecyl phosphite) 四辛氧基钛双[双(十三烷基)亚磷酸酯]〈偶联剂〉
tetraphenylethane 四苯基乙烷
Tetron 〈商〉特特纶〈聚酯纤维〉
tex 特(克斯)〈纤维细度单位,每1000米重量克数〉
textile 织物,纺织品;纺织的
~ coating 织物涂布
~ fabric 织物,纺织品
~ fibres 纺织纤维
~ finish 织物(后)处理,织物整理
~ flocks pl 纺织纤维屑〈填料〉
~ flooring 织物铺地面
~ glass 纺织玻璃纤维〈塑料用增强材料〉
~ glass continuous filament yarn 纺织玻璃纤维连续长纱
~ glass fabric 纺织玻璃纤维织物〈塑料用增强材料〉
~ glass fibre 纺织玻璃纤维
~ glass filament products 纺织玻璃纤维长丝产品
~ glass filament yarn 纺织玻璃纤维长丝纱

textural

- ~ glass mat 纺织玻璃纤维毡〈塑料用增强材料〉
- ~ glass multifilament products 纺织玻璃纤维复丝产品,玻璃纤维复丝织品
- ~ glass spun yarn 纺织玻璃纤维细纱
- ~ glass staple fibre products 纺织玻璃(定长)短纤维制品
- ~ glass staple fibre yarn 纺织玻璃(定长)短纤维纱
- ~ industry 纺织工业
- ~ -like product 织物状产品
- ~ materials 纺织材料;纺织品
- ~ processing 纺织加工
- ~ -reinforced plastic 织物增强塑料
- ~ size 纺织浆料;织物处理剂
- ~ stress 织物应力
- ~ treatment 织物处理

textural 结构上的
- ~ measurement 结构测定

texture 构造,结构;织物结构;纹理;网纹
- ~ effect 结构效应;(织物)起绒结构效应
- ~ finishing 织物质地整理

textured 变形的
- ~ fabric 花式织物
- ~ filament yarn 变形长丝
- ~ finish 花纹整理,花纹处理
- ~ molecularly oriented fibre 分子取向的变形纤维
- ~ sheeting 皱纹薄膜片

texturing 合成纤维变形工艺〈一种特殊表面〉

texturized 卷曲变形的
- ~ mat 卷曲变形垫
- ~ yarn 卷曲变形丝;膨体纱

TF [thin film] 薄膜

TFCE [trifluorochlorethylene] 三氟氯乙烯

TFE [tetrafluoroethylene] 四氟乙烯

TFM foam mo(u)lding TFM 泡沫成型,气体反压泡沫塑料注塑法〈用专门的加热-冷却系统和气密封结构模具制造热塑性泡沫塑料〉

TFM process TFM 泡沫成型,气体反压泡沫塑料注塑法

TFOT thin film oven test 薄膜耐热试验

Tg [glass transition temperature] 玻璃态转化温度,玻璃化转变温度

TG [thermogravimetry] 热失重测量术,热重(分析)法

TGA [thermogravimetric analysis] 热解重量分析

th. [thermal] 热的

Th [thorium] 钍

thallium [Tl] 铊

thaw 熔化;融化

theory 理论;学说
- ~ of adhesion 黏合理论
- ~ of combustion 燃烧理论
- ~ of elasticity 弹性理论
- ~ of plasticity 塑性理论
- ~ of stability 稳定性理论
- ~ of strength 强度理论
- ~ of structure 结构力学

thermal 热
- ~ ageing 热老化
- ~ ageing test 热老化试验
- ~ analysis 热分析
- ~ baffle 隔热板
- ~ balance 热平衡
- ~ barrier 绝热层,保温层
- ~ black 热炭黑
- ~ bonding 热黏合
- ~ breakdown 热破坏
- ~ capacity 热容量,热功率
- ~ characteristics 热特性
- ~ checking 热(龟)裂
- ~ coefficient of cubical expansion 体积热膨胀系数
- ~ coefficient of expansion 热膨胀系数
- ~ coefficient of linear expansion 线性热膨胀系数
- ~ conductance 传热系数
- ~ conduction 热传导
- ~ conductivity 导热性;导热系数〈不受厚度影响的均质材料的〉
- ~ crack 热开裂

thermal

- ~ cracking 热裂解;热开裂
- ~ creep 热蠕变
- ~ decomposition 热分解
- ~ deformation 热变形
- ~ deformation under load 负荷下热变形
- ~ degradation 热降解
- ~ depolymerization 热解聚(作用)
- ~ diffusion 热扩散
- ~ diffusivity 热扩散系数;热扩散性
- ~ dilatation 热膨胀
- ~ dilation 热膨胀
- ~ discolorating 热褪色,热变色
- ~ dissipation 热耗散
- ~ distortion 热变形
- ~ efficiency 热效率
- ~ embrittlement 热脆裂
- ~ endurance 耐热性
- ~ endurance properties 热稳定性,热耐久性,耐热性
- ~ energy storage,[TES] 热能储存
- ~ expansion [dilatation] 热膨胀
- ~ expansion coefficient 热膨胀系数
- ~ expansion moulding process 热膨胀模成型
- ~ expansivity 热膨胀性;热膨胀系数
- ~ extension 热伸长
- ~ fatigue 热疲劳
- ~ fatigue test 热疲劳试验
- ~ foamed plastics 热发泡沫塑料
- ~ forming 热成型
- ~ impulse sealer 热脉冲热合机
- ~ impulse sealing 热脉冲热合
- ~ impulse welder 热脉冲焊接机
- ~ impulse welding 热脉冲焊接
- ~ -initiated polymerization 热引发聚合(作用)
- ~ instability 热不稳定性
- ~ insulant 绝热材料
- ~ insulating board 隔热板,绝热板
- ~ insulating material 绝热材料,保温材料
- ~ insulating property 绝热性
- ~ insulating value 绝热值
- ~ insulation 隔热,绝热,保温,热绝缘;保温层,隔热层
- ~ insulation board 隔热板,绝热板
- ~ insulation interlayer 隔热夹层
- ~ insulation material 绝热材料
- ~ insulation test 绝热试验
- ~ insulator 绝热体
- ~ laminating 热层合,热复合
- ~ molecular motion 热分子运动
- ~ motion 热运动;热振动
- ~ oxidative degradation 热氧化降解
- ~ oxidative stability 热氧化稳定性
- ~ plasticization 热塑炼
- ~ polymerization 热聚合(作用)
- ~ property 热性能
- ~ quench process 热急冷吹塑法
- ~ resilience 热回弹性〈受热后形变恢复〉
- ~ resistance 耐热性,耐热度;热阻
- ~ sealing 热合
- ~ sensibility 热敏性
- ~ sensitizer 热敏剂
- ~ shock 冷热骤变
- ~ shock resistance 耐冷热骤变性
- ~ shock test 冷热骤变试验
- ~ shrinkage 热收缩,热收缩量,热收缩率
- ~ shrinkage differential 热收缩差异
- ~ shrinkage stress 热收缩应力
- ~ spectrum analysis 热光谱分析
- ~ stability 热稳定性,耐热性
- ~ stability index 热稳定性指数
- ~ stability under load 荷载热稳定性
- ~ stability test 热稳定性试验
- ~ stabilization 热稳定(作用,化)
- ~ stabilizer 热稳定剂
- ~ state 热状态
- ~ state quantity 热状态参数
- ~ strain 热应变
- ~ stress 热应力
- ~ stress cracking [TSC] 热应力开裂
- ~ stress fatigue 热应力疲劳
- ~ stretch 热拉伸
- ~ testing 热性能试验

~ transition 热转化
~ transmissivity 热透射率；热传递系数
~ transmittance test 热传导试验
~ treatment 热处理
~ unit 热量单位
thermally 热（地）
~ foamed plastics 热发泡沫塑料
~ insulated 绝热的
~ insulating mould coating 模具隔热涂层〈注塑模具〉
~ sensitive resin 热敏性树脂
~ stable 热稳定的
~ stable polymer 热稳定性聚合物
thermionic 热离子的
~ current 热离子电流
~ emission 热离子发射
~ valve 热离子管
thermistor 热敏电阻
thermo package 热包装
thermoanalytical measurement 热分析测定
thermoanalysis 热分析
thermobalance 热天平
thermoband （电）热带
~ process 电热带焊接法
~ tape 电阻加热带〈热带焊接〉
~ welding 电热带焊接
thermochalk 示温粉笔
thennochrome rod 测温棒，热敏性棒，热变色棒
thermochromic 热变色的
thermochromism 热致变色
thermocolo(u)r 示温颜料；热敏颜料
thermocompression bonding 热压黏合
thermocompressor 热压机
thermoconductivlty 热传导率
thermocouple 热电偶；温差电偶
~ needle 热电偶针
~ thermometer 热电偶温度计
~ tip 热电偶端，热电偶接点
~ well 热电偶管
thermodiffiision 热扩散
thermodilatometry 热膨胀计测定法

thermoelastic 热弹性的
~ inversion 热弹转换
~ inversion point 热弹转换点
~ state 热弹（性）态
thermoelasticity 热弹性
thermoelectric 热电的；温差电的
~ cell 温差电池
~ couple 热电偶，温差电偶
~ effect 热电效应，温差电效应，塞贝克效应
~ material 热电材料
~ method 热电法
~ power 温差电势率
thermoelectrometry 热电测定法
thermoform 热成型制品
thermoformed part 热成型件
thermoforming 热成型
~ ability 热成型性
~ machine 热成型机
~ mo(u)ld 热成型模具
thermogram 热谱图
thermographic tranfer process 热象转印法，热压贴花法
thermograph 温度记录器
thermography 温度记录法
thermogravimetric 热失重的
~ analysis [TGA] 热失重分析
~ curve 热失重曲线
thermogravimetry [TG] 热重法，热失重测量术，热重法
thermogrip applicator 热黏涂敷机
thermohardening 热固性的
thermoindicator 温度指示计，示温纸
thermoionic emission 热离子发射
thermolabel paper （自黏）热敏纸
thermolaminating 热层合
thermolysis 热分解（作用）
thermomagnetcmietry 热磁力测定法
thermomechanical 热机械的
~ measurement 热机械测量
~ pretreatment 热机械预处理
thermometer 温度计
thermooxidation 热氧化（作用）
thermooxidative 热氧化的

~ degradation 热氧化降解
~ resistance 耐热氧化性
~ stability 热氧化稳定性
thermooxidizing 热氧化
~ degradation 热氧化降解
~ stability 热氧化稳定性
thermopaint 示温涂料,示温漆
thermopair 热电偶
thermopaper 测温纸,示温纸
thermoparticulat analysis 热粒子分析
thermopigment 示温颜料
thermopin (带有气体、液体内冷却的)模具铜销钉
thermoplast 热塑性塑料
thermoplastic [TP] 热塑性塑料;热塑性的
~ adhesive 热塑性黏合剂
~ composites 热塑性复合材料
~ elastomer [TPE] 热塑性弹性体
~ elastomer of polyvinyl chloride 聚氯乙烯热塑性弹性体
~ fibre 热塑性纤维
~ film 热塑性塑料薄膜
~ foam mo(u)lding 热塑性泡沫塑料成型
~ injection [TP-injection] 热塑性塑料注塑
~ laminate 热塑性层压塑料
~ lining 热塑性塑料衬里
~ material 热塑性材料
~ matrix 热塑性基体
~ olfinic elastomer [TPO] 烯烃热塑性弹性体
~ plastics 热塑性塑料
~ poly(1-butene) 聚(1-丁烯)热塑性塑料
~ polyester in general 热塑性聚酯的通称
~ polymer 热塑性聚合物
~ polyurethane [TPU] 热塑性聚氨酯
~ processing 热塑性塑料加工
~ range 热塑性状态,黏滞流体态
~ reinforced with short glass fibre 短玻璃纤维增强热塑性塑料

~ resin 热塑性树脂
~ rubber 热塑性橡胶
~ semifinished material 热塑性塑料半成品
~ sheet material 热塑性塑料片[板]材
~ silicone 热塑性有机硅
~ synthetic material 热塑性合成材料
~ tile 热塑性塑料砖
~ urethanes [TPU] 热塑性聚氨酯
~ urethane elastomer 热塑性聚氨酯弹性体
~ welding strip 热塑性焊条
~ with short-fibre forcement 短纤维增强热塑性塑料
thermoplasticity 热塑性
thermopolymer 热聚(合)物
thermopolymerization 热聚合
thermoregulator 调温器,温度调节器
thermorheological simplicity 热流变单一性
thermosensitive 热敏的
thermoset [TS] 热固性塑料;热固性的
~ extrusion 热固性塑料挤塑
~ injection mo(u)lding 热固性塑料注塑成型
~ laminate 热固性层合塑料
~ plastic 热固性塑料
~ processing 热固性塑料加工
~ resin 热固性树脂
~ urethane [TSUR] 热固性聚氨酯
thermosetting 热固性的
~ acrylic resin 热固性丙烯酸树脂
~ aciylics 热固性丙烯酸树脂
~ adhesive 热固性黏合剂
~ composites 热固性复合材料
~ fibre 热固性纤维
~ injection mo(u)lder 热固性(树脂)注塑成型机
~ injection [mo(u)lding machine] 热固性材料注塑成型机
~ material 热固性材料
~ matrix 热固性基体
~ mo(u)lding compound 热固性模塑料

thermosoftening

- ~ mo(u)lding material 热固性模塑料
- ~ mo(u)lding powder 热固性模塑粉
- ~ plastic 热固性塑料
- ~ plastic injection 热固性塑料注塑
- ~ plasticizer 热固性增塑剂
- ~ polyester resin 热固性聚酯树脂
- ~ polymer 热固性聚合物
- ~ powders 热固性粉末
- ~ reinforced plastics 热固性增强塑料
- ~ resin 热固性树脂
- ~ resin adhesive 热固性树脂黏合剂
- ~ yarn 热定形纱

thermosoftening 热柔性
thermosol 热溶胶
thermostability 热稳定性
thermostable 耐热的,热稳定的
- ~ polymer 耐热性聚合物,热稳定性聚合物

thermostat 恒温器;恒温箱
- ~ container 恒温箱

thermostress 热应力
thermotropic liquid crystal polymer 热致液晶聚合物
thermovaporimetric analysis 热挥发度分析
thermoviscosity 热黏度
thermoweld 热焊
thermowelding 热焊接
theta θ 〈希腊字母〉
- ~ condition θ 条件
- ~ solvent θ 溶剂〈使高分子间作用力减小的溶剂〉
- ~ state θ 状态〈聚合物〉
- ~ temperature θ 温度,弗路里温度

THF [tetrahydrofuran] 四氢呋喃
THFO [tetrahydrofurfuryl oleate] 油酸四氢呋喃甲酯〈增塑剂〉
thiazine 噻嗪
- ~ colo(u)ring matters 噻嗪染料
- ~ dyes 噻嗪染料

thiazole dyes 噻唑染料
2-(4-thiazolyl) benzmiidazol 2-(4-噻唑基)苯并咪唑〈防霉剂〉
thick 厚度;厚的;浓厚的;稠密的;粗的

- ~ article 厚壁制品
- ~ film 厚膜
- ~ line 粗线
- ~ liquor 黏稠液
- ~ mo(u)lding compound [TMC] 厚片模塑料
- ~ plank 厚板
- ~ section mo(u)lding 厚壁模塑件
- ~ sheet 厚片材
- ~ sheeting 厚膜带;厚片材
- ~ stock 厚坯
- ~ wall 厚壁
- ~ -walled 厚壁的
- ~ -walled part 厚壁部件
- ~ -walled tube 厚壁管
- ~ -walled vessel 厚壁容器

thicken 增稠,稠化;变厚
thickened printing colo(u)r 增稠颜料
thickener 增稠剂;增稠器
thickening 增稠,增厚
- ~ agent 增稠剂,增黏剂
- ~ drum 增稠转筒

thickness 厚度;稠度;浓度
- ~ change 厚度变化;厚度差,厚度不匀
- ~ control of parison 型坯厚度控制
- ~ controller 厚度控制器
- ~ ga(u)ge 厚度计,厚度规
- ~ ga(u)ging 厚度测定
- ~ indicator 测厚计,厚度显示器
- ~ measurement 厚度测定
- ~ measurement with laser 激光测厚
- ~ of glue layer [line] 黏胶层厚度
- ~ of wall 壁厚
- ~ piece 厚薄规
- ~ tester 厚度计,测厚仪

thin 薄的;稀薄的,细的;浅色
- ~ coating 薄层涂覆
- ~ film 薄膜
- ~ film evaporator 薄膜蒸发器
- ~ -ga(u)ge film 薄型薄膜
- ~ ga(u)ge sheet 薄型片材
- ~ -layer chromatography 薄层色谱法
- ~ -layer evaporator 薄膜蒸发器

~ liquor 稀薄液
~ section 薄壁；微型切片
section mo(u)lding 薄壁模塑件
~ sheeting 薄膜带
~ - shell 薄壳
~ spot 薄点，亮点
~ stock 薄坯
~ wall 薄壁
~ wall casting 薄壁铸塑件
~ walled 薄壁的
~ walled article 薄壁制品
~ walled mo(u)lding 薄壁成型件
~ - walled tube 薄壁管
~ - walled vessel 薄壁容器
thinner 稀释剂
~ [thinning]ratio 稀释比；稀释剂量
thinning 稀薄化，稀释化，淡化
thiobis 硫代双
2,2'~ [ethyl-3-(3,5-di-tert-butyl-4-hydroxyphenyl)ropionate]
2,2'-硫代双[3-(3,5-二叔丁基-4-羟基苯基)丙酸乙酯]〈抗氧剂〉
2,2'~(4-methyl-6-tert-butylphenol) 2,2'-硫代双(4-甲基-6-叔丁基苯酚)〈抗氧剂〉
2,2'~(4-methyl-6-tert-butylphenol) 2,2'-硫代双(4-甲基-6-叔丁基苯酚)〈抗氧剂2246-S〉
4,4'~(2-methyl-6-tert-butylphenol) 4,4'-硫代双(2-甲基-6-叔丁基苯酚)〈抗氧剂736〉
1,1'~(2-naphthol) 1,1'-硫代双(2-萘酚)
4,4'~(6-tert-butyl-m-cresol) 4,4'-硫代双(6-叔丁基间甲酚)〈抗氧剂300〉
4,4'~(6-tert-butyl-o-cresol) 4,4'-硫代双(6-叔丁基邻甲酚)〈抗氧剂736〉
2,2'~(p-tert-octylphenolate butylamine nickel 2,2'-硫代双(对叔辛基酚)镍-正丁胺络合物〈光稳定剂1084〉

2,2'~(p-tert-octylphenolate)nickel 2'-硫代双(对叔辛基酚)镍〈光稳定剂 AM 101〉
thiocarbamide 硫脲
thiocyano dyestuff 氰硫染料
thiodiethylene glycol bis(β-amino crotonate) 硫代二乙二醇双(β-氨基丁烯酸)酯〈热稳定剂〉
thioether 硫醚
thiofiiran 噻吩
thiogenic dyes 硫化染料
thioindigo 硫靛
~ dyes 硫靛染料
~ vat dyes 硫靛还原染料
Thiokol 〈商〉瑟奥可〈聚硫橡胶〉
~ liquid polymer 液态聚硫橡胶
thionic acid 硫羰酸；连多硫酸
thionone colo(u)rs 硫酮染料
thiophene 噻吩
thioplast 聚硫塑料
thiourea 硫脲
~ -formaldehyde resin 硫脲甲醛树脂
~ -phenol-formaldehyde chelating resin 硫脲酚醛螯合树脂
~ resin 硫脲树脂
thixotrope 触变胶体
thixotropic 触变
~ agent 触变剂
~ behaviour 触变行为
~ coefficient 触变系数
~ dispersion 触变分散体
~ filler 触变填充剂
~ flow 触变流动
~ fluid 触变性流体
~ index 触变指数
~ plastic substance 触变塑性物质
~ plasticsol 触变性增塑糊
~ resin 触变性树脂
~ thickener 触变增稠剂
~ viscoplasticity 触变黏塑性
thixotroping agent 触变剂
thixotropy 触变性
thorium [Th] 钍
thoroughly purified 完全纯净

thread 螺纹;线
- ~ accumulator 纱线存储器
- ~ angle 螺纹角
- ~ axis 螺纹轴,丝扣轴
- ~ count 螺纹数,纱线数;织物经纬密度
- ~ depth 螺槽深度,螺纹深度
- ~ fit 螺纹配合
- ~ form 螺纹断面形状
- ~ – forming insert 攻丝嵌件
- ~ – forming screw 攻丝螺钉
- ~ guidance 导纱器
- ~ guide 导纱器
- ~ insert 螺纹嵌件
- ~ nomenclature 螺纹名称
- ~ pin 丝扣销钉
- ~ pitch 螺距
- ~ Plug 螺纹插塞,螺纹模塞〈美〉
- ~ – profile angle 螺纹牙形角
- ~ representation 螺纹标准图表
- ~ – sealing paste 螺纹密封糊
- ~ spindle 螺纹轴
- ~ standard 螺纹标准
- ~ tube 导纱管
- ~ types 螺纹种类

threaded 螺纹的
- ~ connection 螺纹连接,丝扣连接
- ~ fitting 螺纹接口
- ~ nozzle adaptor 螺纹注嘴连接件
- ~ outlet 螺丝排泄口
- ~ pipe 螺纹管
- ~ rod 螺杆
- ~ sleeve 螺纹套筒,有扣套管

threading machine 攻螺纹机

threadlike 线状的
- ~ molecule 线型分子

three 三(的)
- ~ – bowl calender 三辊压延机
- ~ channel nozzles 三料道注嘴
- ~ – channel technique 三料道工艺〈共挤塑〉
- ~ dimensional 三维的,三度空间的,立体的
- ~ dimensional constitution 体型结构;三维结构
- ~ dimensional construction 三维结构
- ~ dimensional crimp 三维卷曲,立体卷曲
- ~ dimensional crosslinked network 三维交联网络,立体交联网络
- ~ dimensional effect 立体效应
- ~ dimensional movement 三维运动;空间运动
- ~ dimensional net structure 三维网状结构,立体网络结构
- ~ dimensional network 体型网构,三维网构
- ~ dimensional network polymer 体型网状聚合物
- ~ dimensional polycondensation 体型缩聚,三维缩聚
- ~ dimensional polymer 体型聚合物;三维聚合物
- ~ dimensional stress 三维应力,三度空间应力
- ~ dimensional structure 体型结构;三维结构
- ~ dimensional weave 三向织物
- ~ – element model 三元模型〈黏弹性体力学模型〉
- ~ – end twill [three harness satin] 三页斜纹
- ~ – harness satin 三页斜纹〈玻璃纤维〉
- ~ – layer 三层(的)
- ~ – layer film 三层薄膜
- ~ parameter model 三参数模型
- ~ – phase current 三相电流
- ~ – phase induction motor 三相感应电动机
- ~ – plate 三板
- ~ – plate injection mo(u)ld 三板式注塑模具
- ~ – plate mo(u)ld 三板式模具
- ~ – ply laminate 三层层合板
- ~ – ply lamination 三层合〈挤塑〉,夹心材料挤塑成型
- ~ – roll 三辊

~ – roll calender 三辊压延机
~ – roll "I" calender I 型三辊压延机
~ – roll mill 三辊磨,三辊碾磨机
~ – roll stack 三辊轧光机
~ – roll stretching assembly 三辊拉伸装置
~ – speed gear 三级(速度)齿轮
~ stage screw extruder 三段式螺杆挤塑机
~ – way ball valve 三通球阀
~ – way valve 三通阀
threshold limit value [TLV] 阈限值〈毒性极限值〉
throat 卸料斗,加料颈,进料口
~ section 喉颈段
– thickness 焊缝厚度
throttle 节流
~ quotient 节流系数〈挤塑机〉
~ valve 节流
~ zone 节流段〈双螺杆挤塑机〉
throttling calorimeter 节流量热器
through 通过,贯穿;完全,彻底
~ – bolt 贯穿螺栓,穿钉
~ bore 透孔
~ circulation 完全循环(空气)
~ crack 贯穿裂缝
~ – curing 完全固化,彻底固化
~ dielectric heating 贯穿电介质加热,贯穿高频电流加热,贯穿感应加热
~ dry 干透的
~ drying 贯流干燥
~ hole 透孔
~ impregnation 浸透
~ – type insert 贯穿嵌件
~ – type insert pin 贯穿嵌杆
~ – way valve 直通阀
~ welding 焊透
throughput 生产量[率];物料通过量
~ capacity 生产能力;物料通过量
~ rate 生产率
throwing 喷胶〈辊涂布时〉;捻丝〈纺〉
~ roll 离心辊〈用于芯材分离的辊〉
thru [through]

thrust 推力;止推力
~ ball bearing 止推滚珠轴承
~ bearing 止推[推力]轴承
~ bearing capacity 止推轴承的承载能力
~ load 推力负荷;纵向负荷,轴向负荷
~ roller 压紧辊
thulium [Tu] 铥
thumber 制动器
TI [Technical Institute] 技术研究所
TI [thallium] 铊
Ti [titanium] 钛
TI – polymer TI 聚合物〈聚酰亚胺模塑料〉
TIC [total inorganic carbon] 总无机碳
tie 连接;捆扎;条带
~ bar 连接杆,拉杆;横梁
~ bar distance 拉杆间距
~ bar space 拉杆距离
~ bolt 加固螺栓,固定螺栓;地脚螺栓
~ coat 增黏涂层,粘结层
~ – lon 可封闭式聚酰胺袋〈用于蒸汽消毒〉〈美〉
~ rod 连接杆,拉杆
tigerlily 卷丹红〈桔红色〉
tight 紧密的,坚固的
~ – sealing 密封
~ weave 密织布
~ weld 致密焊缝;密封焊接
tightener 密封件;张紧器
tightness 密封性,不透气性
tilted 倾斜的,有角度的
~ cylinder 斜筒
~ cylinder dryer with eccentric axis 偏心轴斜筒式干燥器
~ cylinder mixer 斜筒式混合机
~ mixer 倾斜式混合机
tilting 倾斜,倾倒,翻转
~ chute 斜槽
~ head press 斜头式压机
~ mixer 倾斜式混合机
~ table method 倾斜台法〈润湿性能测试〉

~ viscometer 倾斜式黏度计
timber 木材
timberflex wallpaper 带有塑料防潮层的木质壁纸
time 时间;时期,阶段,时代;定时的,计时的
  ~ adjusting device 定时装置,时限调整装置
  ~ - compression curve 时间－压缩曲线
  ~ constant 时间常数
  ~ controller 时间控制装置
  ~ - delay 时间延迟,时间滞后
  ~ - dependent 随时间而变化的,依赖于时间的
  ~ - dependent deformation 依赖时间的变形,随时间的变形
  ~ - dependent effect 时间依赖效应
  ~ - dependent properties 随时间改变的性能
  ~ efficiency 时间效率
  ~ for complete injection stroke 整个注塑行程时间
  ~ - independent elasticity [ideally elastic behavior] 非时间依赖性弹性,理想弹性特性
  ~ interval 时间间隔
  ~ limit 期限,限期
  ~ of dissolution 溶解时间
  ~ of follow - up pressure 保压时间,续压时间〈注塑〉
  ~ of gate feeze - off 浇口凝结时间
  ~ of heat 加热时间
  ~ of relaxation 松弛时间
  ~ of run 运转时间
  ~ of vulcanizing 硫化时间
  ~ on test 试验延续时间
  ~ profile 时间分布曲线
  ~ - proof 耐久的,耐用的,长寿命的
  ~ retardation 时间推迟
  ~ suitable to moulding 启用期
  ~ - temperature dependence 时间－温度依存关系
  ~ - to - break 断裂时间

~ to fracture curve 时间破坏曲线〈蠕变试验〉
~ under pressure 加压时间,受压时间
~ unit 时间单位
~ - yield 蠕变
~ yield limit 时间屈服极限〈蠕变试验〉
timer 计时器;定时器
~ controlled 受控定时器
timing 定时,计时
~ control 定时控制,时间控制
~ cycle 定期循环
~ register 定时记录
~ sprocket 定时链轮
~ unit 定时装置,计时器
tin [Sn]锡;镀锡
~ foil 锡箔
~ stabilizer (有机)锡稳定剂
tin [can] 听,白铁罐
tinct 色泽;着色的
tinctorial 着色的
~ power 着色力
~ property 着色性能
~ strength 着色力,着色强度
~ substance 着色剂
~ yield 着色量
tinge 淡色调
tinned goods 罐头食品
tint 色调;着色;浅色
~ mark 色点,色斑
tintage 上色
tinting 调色;着色
~ colo(u)r [material] 调色料,调色剂;着色染料
~ pigment 着色颜料
~ power 着色(能)力
~ strength 着色力
~ value 色度指标;光泽值
tintometer 色调计
TIOTM [triisooctyl trimellitate] 偏苯三(甲)酸三异辛酯
tip 倾斜,尖端;接点;焊嘴
~ of thermocouple 热电偶端,热电偶接点

~ - over 倾倒;翻转
~ printing 凸纹着色轧花〈压花塑料制品表面产生双色效应〉
~ - up 倒出,倾翻
tire [tyre] 轮胎
tissue 织物;薄纱;薄绢
~ paper 薄(叶)纸;纱纸
~ shearing test 织物剪切试验
tissuethene 薄型聚乙烯薄膜〈用于包装香味和防油物〉
titanate 钛酸盐[酯]
~ couping agent 钛酸酯偶联剂
~ couplers 钛酸酯偶联剂
titania [titanium dioxide] 二氧化钛
titanium [Ti] 钛
~ boride fibre 硼化钛纤维
~ butylate [tetrabutyl litanate] 钛酸(四)丁酯
~ carbide fibre 碳化钛纤维
~ di(dioctyl pyrophosphato)oxyac-etate 双(二辛基焦磷酰氧基)氧代醋酸钛〈偶联剂〉
~ dioxide 二氧化钛,钛白〈填料,光稳定剂,化纤助剂〉
~ oxide 氧化钛
~ tetrachloride 四氯化钛〈催化剂〉
δ~ trichloride 三氯化钛 δ-变体〈催化剂〉
~ white 钛白,二氧化钛
~ white powder 钛白粉
titanosiloxane polymer 铁硅氧烷聚合物
titrate 滴定
titration 滴定(法)
TLC [Tagliabue open and closed cup] 塔利亚贝氏开闭杯(闪点)试验
TLV [threshold limit value] 阈极限值〈毒性极限值〉
T. M. technical manual 技术手册
T. M. trade mark 商标
TMA [trimellitic anhydride] 偏苯三甲酸酐〈固化剂〉
TMC [thick moulding compound] 厚片模塑料
TMDI [trimethyl hexane diisocyanate] 三甲基己烷二异氰酸酯
TMEDA [tetranrcthyl ethylene diamine] 四甲基乙二胺
TMP test methods and procedure 试验方法和步骤
TMPD [trimethyl pentanediol glycol] 三甲基戊二醇
TMS [tetramethyl silane] 四甲基(甲)硅烷
TMTD [tetramethyl thiuram disulfide] 二硫化四甲基秋兰姆〈促进剂〉
TMTM [tetramethyl thiuram monosul-fide] 一硫化四甲基秋兰姆〈促进剂〉
TNP [tri(nonylphenyl) phosphite] 亚磷酸三(壬基苯)酯〈抗氧剂〉
TOC [total organic carbon] (废水中)有机化合碳总量
toe dog 小撑杆;夹持装置
TOF [trioctyl phosphate] 磷酸三辛酯
toggle 肘节
~ action 肘节作用
~ arm 肘节臂
~ clamp 肘节合模装置,连杆合模装置
~ closing system 肘节式闭模系统
~ lever 肘节连杆
~ - lever locking unit 肘节杆式锁模装置
~ lever press 肘节连杆式压机
~ link system 肘节式机构
~ locking system 肘节式锁模系统,连杆式锁模装置〈注塑机〉
~ mechanism 肘节机构
~ plate 肘板;摆板
~ press 肘节连杆式压机
~ system 肘节式系统,连杆式系统
~ type injection mo(u)lding machine 肘节式注塑成型机
~ type lock 肘节式锁模系统
~ type mo(u)ld clamping system 肘节式合模装置,连杆式合(锁)模装置
~ type tabletting machine 肘节式制锭机
tolerance 公差,容许,容限

toluene
~ limit 容许极限
~ part 精细制件
toluene 甲苯
~ diisocyanate [TDI] 甲苯二异氰酸酯
~ -formaldehyde resin 甲苯甲酸树脂
~ resin 甲苯树脂
o,p-~ **sulfonamide** 2,4-甲苯横酰胺〈增塑剂〉
~ sulfonamide formaldehyde resin 甲苯磺酰胺-甲醛树脂〈增塑剂〉
~ -p-sulfonyl hydrazide 对甲苯磺酰肼〈发泡剂〉
P-~ **sulfonylsemicarbazide** 对甲苯磺酰氨基脲〈发泡剂〉
toluol 甲苯
toluylene diisocyanate [TDI] 甲苯二异氰酸酯
tolyl 甲苯基
~ group 甲苯基
p-(p-~ **sulfonylamido**) **diphenylamine** 对(对甲苯基磺酰胺基)二苯胺〈防老剂〉
ton 吨
tone 色调
toner 调色剂〈着色力高的有机颜料〉
~ brown 色淀棕
~ yellow 色淀黄
toning 调色
tongue 舌;键,销,榫
~ and groove joint 键槽式连接
~ tear strength 舌形(试样)撕裂强度
~ tear strength test 舌形(试样)撕裂强度试验
tool 工具;模具;仪器,设备;方法
~ changing 模具更换
~ contour 模具轮廓
~ division 模具拆卸
~ element 模具零件
~ for bottle box 瓶箱用模具,瓶用周转箱模具
~ holder 模具支承架
~ kit 成套工具,工具箱,工具包
~ location spigot 模具对位板

~ made from plastics 塑料制的模具,塑料模具
~ mark 模具标志
~ maker 模具制造者
~ making 模具制造
~ opening stroke 开模行程
~ plastics 制模具用塑料
~ room 模具室;工具室
~ set 成套工具,工具箱
~ slip 模具滑移
~ steel 工具(用)钢
~ strength 模具强度
~ temperature 模具温度
~ temperature regulation 模具调温
~ test 模具试验
~ tolerance 模具公差
~ with jaw actuation 可拆模具
tooling 机械加工模具;加工塑料用模具
~ line 模接合线
~ resin 制模具用的树脂
tooth 齿(状物)
~ crown 齿顶,齿冠
~ depth 齿槽深度
~ face (齿轮)齿面
~ fillet 齿根圆角,齿根倒圆
~ flank (齿轮)齿侧面
~ pitch 齿距,齿节
~ profile (齿轮)齿(纵)断面
~ rounding machine 齿(棱)倒圆机
~ space (齿轮)齿间距
~ top 齿顶
toothed (锯)齿形的
~ attrition mill 齿形碾磨
~ blade 锯齿形叶片
~ blade impeller 锯齿形叶片搅拌叶轮
~ conveyor belt 齿形运输带
~ disk attrition mill 齿形盘式碾磨
~ disk mill 齿盘磨
~ plate 齿状板
~ roller 齿辊
~ wheel 齿轮
TOP [trioctyl phosphate, i.e. tris(2-ethylhexyl)phosphate] 磷酸三辛酯〈即磷

酸三(2-乙基己酸)酯〉
top 顶端,上面
~ bite 上辊隙
~ blow 顶吹,上吹
~ blowing 顶吹法,上吹法
~ casting 顶铸塑
~ circle (齿轮)顶圆
~ clamp 上夹紧器
~ clamp plate 上夹模板〈注塑模具〉
~ coat paste 表面涂糊料
~ coat(ing) 表面涂层;面漆
~ cutter 上部刀具
~ ejection 上模脱模
~ electrode 上部电极〈高频焊接机〉
~ -entering 从上进入
~ feed 顶部加料;上部喂料
~ flash 上溢料〈吹塑〉
~ force 上模,阳模;顶柱塞
~ force mo(u)ld 袴塑压模
~ force press 上柱塞式压机
~ knockout mo(u)ld 上脱模式模具
~ lamination 层压面层,面层板
~ layer 表面层;涂盖层〈焊接〉
~ mounting plate 上托模板
~ plate 顶板,上模板
~ ram 顶柱塞
~ ram press 顶柱塞式压机
~ roll 顶辊;上辊
~ sheet 贴面片材
~ side lock 顶侧闭锁
topaz 黄玉;黄玉黄〈浅棕黄色〉
topformer process 顶面成型法
topforming 顶面成型〈热成型〉
Topham box 托范式纺丝罐
TORM [tetraoctyl pyromellitate, i. e. tetrakis(2-ethylhexyl) pyromellitate] 均苯四甲酸四辛酯〈即均苯四酸四(2-乙基己基)酯,增塑剂〉
topochemical reaction 局部化学反应
topochemistry 局部化学
topping 贴面;贴胶;套色
~ printing 套色印花,套色印刷
torch lamp 喷灯
torpedo 分料梭,鱼雷型分梭,鱼雷头

~ detaiis 鱼雷头部件
~ extrdsioil head 鱼雷芯挤塑模头
~ head 鱼雷头;鱼雷芯机头
~ preplasticator 分料梭式预塑化器
~ screw 鱼雷头螺杆
~ -spreader 鱼雷形分流梭
~ transfer plasticizing 分料梭传递塑化
~ -type screw 鱼雷型螺杆
torque 扭矩,转矩
~ arm 转矩臂
~ breakdown test 转矩破裂试验
~ curve 扭矩曲线
~ force 扭力
~ modulus 扭转模量
~ moment 扭转矩
~ motor 转矩电动机
~ on screw 螺杆扭矩
~ rheometer 扭矩流变仪
~ test 扭转试验
~ wrench 转矩扳手
torquemeter 扭矩计
torsion 扭力,扭转
~ and tensile testing machine 扭拉试验机
~ balance 扭力天平
~ braid 扭辫,扭带
~ dynamometer 扭力计
~ modulus 扭转模量
~ moment 扭(力)矩
~ pendulum 扭摆
~ pendulum method 扭摆振动法
~ pendulum test 扭摆试验〈用于测定塑料材料剪切弹性模量和机械阻尼〉
~ spring 扭转簧
~ strength 耐扭强度
~ temperature 扭曲温度
~ test 扭力试验,扭转试验
~ vibration test 耐扭振动试验
~ viscosimeter 扭力黏度计
torsional 扭转的
~ braid analysis 扭辫分析
~ braid analyzer 扭辫分析仪
~ creep 扭转蠕变
~ deformation 扭转形变

torsionmeter
- ~ dynamometer　扭力测力计
- ~ elasticity　扭弹性
- ~ endurance limit　扭转疲劳极限
- ~ fatigue-strength　扭转疲劳强度
- ~ fracture　扭转断裂
- ~ modulus　扭曲模量,耐扭模量
- ~ moment　扭矩
- ~ pendulum　扭摆
- ~ pendulum analysis　扭摆分析
- ~ rigidity　扭转刚性;扭转刚度
- ~ rigidity modulus　扭转刚性模量
- ~ shear strength　扭剪强度
- ~ stability　扭转稳定性
- ~ stiffness　扭转刚度
- ~ strain　扭转应变
- ~ strength　扭转强度
- ~ stress　扭转应力,扭曲应力
- ~ tension　扭拉伸应力
- ~ test　扭力试验,扭力测试
- ~ tester　扭转试验机,扭力测试仪
- ~ vibration　扭转振动
- ~ vibration analysis　扭转振动分析
- ~ vibration measurement　扭转振动测定
- ~ vibration rheometer　扭转振动流变仪

torsionmeter　扭力计
tortoise shell substitute　龟壳型代用品
tortuosity　弯扭,曲折
torus　圆环面;旋转面
TOSCA [Toxic Substances Control Act]　毒物管理条例〈美国〉
Toshiba and Asahi foam mo(u)lding　东芝和旭公司泡沫塑料成型法〈气体反压注塑法〉
tot. [total]　总的,全部的
total　全部(的),总计(的)
- ~ area　总面积
- ~ carbon [TC]　(废水中)总碳含量
- ~ content　总含量
- ~ cycle time　总循环时间
- ~ deflection　总变形
- ~ draft　总牵伸;总拉伸
- ~ efficiency　总效率
- ~ foaming restraint　总发泡制约
- ~ fracture load　总断裂载荷
- ~ heat　总热量
- ~ height　总高度
- ~ inorganic carbon, [TIC]　(废水中)无机化合碳总量
- ~ installed power　总装配功率
- ~ labor　全体工作人员
- ~ I/D ratio　总长径比〈螺杆〉
- ~ length　总长度
- ~ luminous reflectance　总反光能力,全视觉反射率〈塑料表面〉
- ~ organic carbon, [TOC]　(废水中)有机化合碳总量
- ~ output　总产量,总生产率
- ~ penetration energy　总穿透能
- ~ power output　总功率输出(量)
- ~ pressure loss　总压力耗损
- ~ restraint　总防溢
- ~ strain　总应变
- ~ surface　实际表面积,有效表面积
- ~ time　总时间
- ~ torque　总扭力
- ~ transmittance　总透光率
- ~ volume shrinkage　总体积收缩
- ~ weight　总重量
- ~ width　总宽(度)
- ~ wrap　总抱辊(度)

TOTM [trioctyl trimellilate, i. e. tris(2-ethylhexyl)promellitate]　偏苯三甲酸三辛酯〈即偏苯三甲酸三(2-乙基己基)酯,增塑剂〉
touch　触摸;接触;手感〈纺织物〉
- ~ switch　触摸开关
- ~ temperature　触摸温度

tough　韧性的;黏滞的
- ~ -brittle-transition　韧脆转化
- ~ failure　韧性破坏
- ~ hardness　韧硬度
- ~ resilient plastic　韧弹性塑料
- ~ unsaturated polyester resins　韧性不饱和聚酯树脂

toughen　增韧,增黏稠
toughened polystyrene [TPS]　增韧聚苯乙烯

toughener 增韧剂
toughening agent 增韧剂
toughness 韧性;柔韧性;黏稠性
~ index 韧性指数
~ test 韧性试验,韧性测试
tow 丝束
tower packing [column packing] 填塔料
tox. toxicity 毒性
toxic (有)毒的;毒药;毒物
~ ingredient 有毒成分
~ substance 毒质
toxicity 毒性
~ quotient 毒性系数
~ test 毒性试验
toxicological evaluation 毒性测定
toxicology 毒物学
TP [thermoplastic] 热塑性塑料;热塑性的
TP elastomer [thermoplastic elastower] 热塑性弹性体
TP-injection [thermoplastics injection] 热塑性塑料注塑
TP-polyester [thermoplastics polyester] 热塑性聚酯〈美〉
TPA [terephthalic acid] 对苯二甲酸
TPE [thermoplastic elastomer] 热塑性弹性体
TPO [thermoplastic olfinic elastomer] 烯烃热塑性弹性体
TPP [triphenyl phosphate] 磷酸三苯
TPR [thermoplastic rubbers] 热塑性橡胶
TPS [toughened polystyrene] 增韧聚苯乙烯
TPSF [thermoplastic structural foam] 热塑性结构泡沫塑料
TPU [thermoplastic urethanes] 热塑性聚氨酯
TPUR [thermoplastic polyurethane] 热塑性聚氨酯
TPX [methyl pentene polymer] 甲基戊烯聚合物
TR [thio rubber] 聚硫橡胶〈一种多硫化合物〉
trace 痕量;微量〈添加塑料助剂时〉

~ analysis 痕量分析
~ component 痕量组分
~ concentration 微量浓度
~ element analysis 痕量元素分析
~ impurity 痕量杂质
~ of acid 酸痕量
~ quantity 痕量
tracer 示踪剂;曳光剂;传感器
~ -controlled 同传感器控制的
~ head 探测头
tracing film 插图用薄膜
track 径(迹)
~ of ball bearing 滚珠轴承导槽
~ resistance 耐漏电性
tracking 漏迹,电弧径迹,径迹
~ index 电弧径迹指数
~ material 炭迹爬电材料
~ resistance 耐漏电性,耐电弧径迹性
traction 牵引(力)
~ coefficient 牵引系数
tractive 牵引的
~ force 牵引力
~ output 牵引功率
~ property 牵引性能
~ speed 牵引速率[度]
trade 贸易,交易;商业
~ mark 商标
~ moulder 定制模塑厂
~ [brand] name 商品名称
traditional colo(u)r 传统色
traffic 交通,运输;交易贸易
~ engineering 交通工程
~ paint 路标漆;路线漆
tragacanth gum 黄蓍胶
trail-car [trailer] 拖车,挂车
trailing blade 曳滑刮刀
~ coater 曳滑刮刀涂布机
~ coating 曳滑刮刀涂布
trailing edge of flight 螺纹后缘
train 机组;系列
tramp(material) 混人物;外来杂质
trans 〈词头〉反(式)〈作此解时在物名中排斜体〉;〈正体〉超,跨
~ -configuration 反式构型

~ - form 反式结构
cis -- ~ isomer 顺反异构体
~ - isomerism 反式异构(现象)
cis -- ~ isomerization 顺反异构化
~ - polyisoprene 反式聚甲基丁二烯
~ - tactic 有规反式构型
trans. transparent 透明的
transannular 跨环的
~ polymerization 跨环聚合
transcalent 透热的
transcriber 转录器;复制装置
transducer 转换器;换能器;传感器
~ conditioner (能量)转换能调节器
transfer 转移;传递
~ area 接触面
~ bonding 转移黏合
~ car 搬运车
~ chamber 压铸料槽
~ chamber retainer plate 压铸料槽托模板
~ coating 转移涂布
~ coefficient 转移系数,传递系数
~ cull 压铸成型涂料
~ cylinder 压铸成型料筒
~ decorating 转印装饰法
~ equipment 传送设备
~ feed 连续自动供料
~ forming 压铸模塑
~ function 传递函数,转换函数
~ laminating 转移层合
~ line 连续流水作业生产线
~ machine 传送机;压铸机
~ mo(u)ld 压铸成型模具
~ moulded 压铸的
~ moulded part 压铸部件
~ mo(u)lding 压铸成型
~ moulding compound 压铸模塑料
~ mo(u)lding machine 压铸模塑机
~ mo(u)lding press 压铸成型机
~ mo(u)lding pressure 压铸模塑压力
~ moulding with one plunger 单压料塞压铸成型
~ moulding with two plungers 双压料塞压铸成型

~ of heat 热传递,传热
~ paper 转印纸(印刷)
~ plunger 压铸成型柱塞
~ plunger cylinder 压铸成型料筒
~ plunger retainer plate 压铸成型柱塞托板
~ pot 压铸料槽
~ pot mould 压铸料槽式模具
~ pot retainer 压铸料槽式托板
~ press 压铸成型机
~ printing 转印
~ ram 压铸成型活塞
~ roll 传料辊;传墨辊
~ slug 压铸成型涂料
~ tool 压铸成型模具
~ welder 移动式焊接机
~ well 压铸料槽
~ wheel 送料叶轮
transform 变换,改变,转换;变化
transformation 转变;变换;变化
~ coefficient 变换系数
~ constant 变换常数
~ piece 变压器
~ point 转变点,临界点
~ range 转变范围
~ rate 转变速率
~ stress 变换应力
~ zone 过渡段〈挤塑机螺杆〉
transformer 变压器
transient 瞬时的,瞬变的,过渡的,暂时的
~ condition 瞬变工况,过渡工况
~ creep 瞬时蠕变,过渡蠕变
~ interference – image 瞬变干涉图
~ process 瞬变过程
~ section [zone] 突变段,过渡段
~ state 过渡状态
~ temperature 瞬变温度
~ zone [section] 突变区,过渡区
transistor 晶体管
transit time 过渡时间
transition 转变,过渡
~ fitting 过渡配合
~ heat 转变热

~ layer 过渡层
~ point 转变点
~ region 转变段,过渡段
~ section 转变段,过渡段
~ section of screw 螺杆的过渡段
~ shrinkage 相转变收缩(性)
~ stage 过渡阶段
~ state 过渡状态
~ structure 过渡结构
~ temperature 转变温度
~ zone 过渡段
translucence 半透明性
translucency 半透明性;半透明度
translucent 半透明的
~ body 半透明体
~ coating 半透明涂层
~ coloured 半透明色的
~ plastic 半透明塑料
~ rubber 半透明橡胶
transmissibility 透过率,透过系数
~ for radiation 辐射透过率
transmission 传动,传递;透射
~ belt 传动带
~ coefficient 传递系数;透射系数
~ density 透射密度,光密度
~ dynamometer 传动式测力计
~ electron microscope 透射电子显微镜
~ factor 透射因数
~ gear 传动齿轮,变速齿轮
~ line 传输线
~ of heat 热的传递;传热
~ of light 光的透射
~ optical density 透射光密度
~ rate 变速,传动速率
~ ratio 传动比
~ -scanning electron microscope 透射-扫描电子显微镜
~ spectrum 透射光谱
~ ultrasonic welding 间接超声焊接;透射超声焊接
transmissivity 透射率;透光度
transmit 传送;透射;透光
transmittance 透射率;渗透性〈辐射、声、热〉
transmittancy 透射率,透光度
transmitted light 透射光
transmitter (声波)发送器〈超声波检验〉
~ valve 发射管〈焊接装置和干燥装置的高频发生器〉
transmute 蜕变,变化,改变
transmuted wood 改性木材;浸胶木材〈树脂浸渍木材〉
transojet technique 热固性塑料注塑法
transparence 透明性
transparency 透明性;透明度
transparent 透明的
~ ABS resin 透明 ABS 树脂,透明的苯乙烯-丙烯腈-丁二烯共聚树脂
~ body 透明体
~ coating 透明涂层
~ container 透明容器
~ coloured 透明色的
~ film 透明薄膜
~ layer 透明层
~ material 透明材料
~ package 透明包装
~ plastics 透明塑料
~ polyamide 透明聚酰胺
~ rubber 透明橡胶
transpiration cooling 蒸发冷却
transport 运输;传递
~ band 运输带
~ belt 运输带
~ zone 输送区域
transportation vessel 运输槽
transported fluid 输送液体,输送流体(管道中)
transporting chain 传输链
transversal 横向的
~ crossing 45度纵横交叉
~ cutting 横切
~ cutting saw 横切锯
~ orientation 横向取向
transverse 横(向)的
~ bending resilience 横向弯曲弹性
~ bending strength 横向弯曲强度

~ - channel flow 螺槽横流
~ compressive modulus 横向压缩模量
~ contraction 横向收缩
~ contraction ratio 横向收缩比;泊松比
~ crack 横向裂纹
~ cutting machine 横向裁剪机
~ dielectric heating 横过式高频电加热
~ direction 横向
~ elasticity 横向弹性,弯曲弹性
~ failure 横向断裂
~ flow 横流
~ flow viscometer 横流式黏度计
~ modulus 横向弹性模量
~ regularity 横向均匀度
~ relaxation 横向松弛
~ section 横剖面,横截面
~ shear 横向剪切
~ shrinkage 横向收缩
~ strain 横向应变
~ strain sensor 横向应变传感器
~ strength 横向强度,弯曲强度
~ stress 横向应力,弯曲应力
~ test‑piece 挠曲试样
~ warping 横向翘曲,横向扭曲
trap 水封;气封;气味密封;收集器
trapezoid 梯形(的)〈美〉
~ tear strength 梯形试样撕裂强度
trapezoidal 梯形的
~ - runner 梯形断面流道
trapped 截留
~ - air process 吹塑薄膜挤塑法;夹气吹塑法
~ gas 夹气〈制品缺陷〉
~ sheet forming 夹片热成型,气压热成型,夹片压气热成型
trapping 收集;截留
~ corner (模具)死角
~ of air 含空气气泡
trash 废料,废物;垃圾;杂质
~ bag 废物［废料、垃圾］袋
~ can 废物箱
~ chute 废物滑槽

~ content 含杂率
~ mark 浇口痕
~ receptacle 废料桶
travel 行程,冲程;运转
~ around the rolls 绕轴旋转
~ of piston 活塞行程
travelling 移动的
~ band conveyor 移动式运输机
~ belt 运输带,传送带
~ crane 移动式起重机
~ mixer 移动式混合器
~ paddle 活动桨叶
~ paddle mixer 活动桨叶混合器
Traver process 脱拉维法,电晕放电表面处理
travers 〈复〉横向条纹
traverse 横臂,横梁;横向移动;拉杆,连接杆
traversing 移动
~ feed arm 往复进料杆
~ gear 运输设备
~ gun 移动式喷雾枪
~ mixing head 摆动式混合头
trawl 鱼网
tray 装料盘;托盘
~ cap 塔(盘)泡罩
~ dryer 盘式干燥器
~ of a column 塔盘,塔板
treacle stage 黏液阶段;凝胶状态
treacliness 黏滞性
tread 胎面;踩踏
~ test 踩踏试验〈地毯〉
treadle (脚)踏板
treat 处理;浸渍
treated (已经)处理过的
~ roofing 铺顶油毡
~ tape 浸胶带
treater 浸渍机
~ roll 浸渍辊
treating 浸渍;处理
~ bath 浸渍浴
~ liquid 浸渍液;处理液
~ oven 干燥炉〈层合〉
~ roll 浸渍辊

treatise 论文〈科学〉
~ treatment 处理,加工;浸渍
tremolite 透闪石
trench 沟槽;管沟
trend 趋向;发展方向〈技术〉
tri 〈词头〉三
~ (2-bromo-3-chloropropyl) phosphate 磷酸三(2-溴-3-氯丙)酯〈阻燃剂〉
~ (butoxy ethyl) phosphate 磷酸三(丁氧基乙)酯〈增塑剂〉
~ -n-butyl citrate 柠檬酸三正丁酯〈增塑剂〉
2,4,6- ~ -tert-butyl phenol 2,4,6-三叔丁基苯酚,抗氧剂246
~ -n-butyl phosphate [TBP] 磷酸三正丁酯〈增塑剂〉
~ -coating method 浸渍涂布法
~ (2,3-dibromopropyl) phosphate 磷酸三(2,3-二溴丙基)酯〈阻燃剂〉
~ (2,3-dichloropropyl) phosphate 磷酸三(2,3-二氯丙基)酯〈阻燃剂〉
~ -dimethylphenyl phosphate 磷酸三(二甲苯)酯
~ (2-ethylhexyl) phosphate 磷酸三(2-乙基己)酯〈增塑剂〉
~ (2-ethylhexyl) trimellitate 偏苯三甲酸三(2-乙基己)酯〈增塑剂〉
~ -n-hexyl trimellitate 偏苯三甲酸三正己酯〈增塑剂〉
2,4,6- ~ -(2'-hydroxy-4'-n-butyl-oxyphenyl)-1,3,5-triazine 2,4,6-三(2'-羟基-4'-正丁氧基苯基)-1,3,5-三嗪〈紫外线吸收剂〉
~ -hydroxy ethylmethyl ammonium methosulfate 硫酸甲酯三羟乙基甲基铵盐,抗静电剂TM
~ (monochloropropyl) phosphate 磷酸三(一氯丙)酯〈阻燃剂〉
~ (nonylphenyl) phosphite [TNP] 亚磷酸三(壬基苯)酯〈抗氧剂〉
~ -n-octyl-n-decyl trimellitate 偏苯三甲酸三(正辛正癸)酯〈增塑剂〉
~ -n-octyl trimellitate 偏苯三甲酸

三正辛酯〈增塑剂〉
triacetate 三乙酸[醋酸]酯;甘油三乙酸[醋酸]酯;三乙酸[醋酸]纤维素
triacetin 三醋精,甘油三乙酸酯
triacetyl cellulose 三乙酰基纤维素
triage 筛余料
trial 试用,试验,试运转;试验性的
~ production 产品试制,试生产
~ quantity 试验量
~ run 试运转
~ test 探索性试验
trialkyl aluminium 三烷基铝
triallyl 三烯丙基
~ cyanurate [TAC] 三烯丙基氰尿酸酯
~ cyanurate polyester 氰尿酸三烯丙酯聚酯
~ phosphate [TAP] 磷酸三烯丙酯
triangular 三角形的
~ calender 三角形三辊压延机,A型压延机
~ filler rod 三角形焊条
~ welding spline 三角形焊条
triazine 三嗪
~ dyes 三嗪染料
~ resin 三嗪树脂〈清漆,涂料,胶黏剂〉
tribasic 三碱式,三盐基〈俗称〉
~ lead maleate 三碱式马来酸铅〈热稳定剂〉
~ lead sulphate 三碱式硫酸铅〈热稳定剂〉
tribromoneopentyl alcohol 三溴新戊醇〈阻燃剂〉
tribromophenol 三溴苯酚〈阻燃剂〉
tribromophenyl 三溴苯基
~ allyl ether 三溴苯基·烯丙基醚〈阻燃剂〉
~ dibromopropyl ether 三溴苯基·二溴丙基醚〈阻燃剂〉
tributyl 三丁基
~ borate 硼酸三丁酯
~ phosphate [TBP] 磷酸三丁酯〈增塑剂〉

~ phosphite 亚磷酸三丁酯〈抗氧剂〉
tributyltin 三丁基锡
  ~ acetate 乙酸三丁基锡〈防霉剂〉
  ~ chloride 氯化三丁基锡〈防霉剂〉
  ~ fluoride 氟化三丁基锡〈防霉剂〉
  ~ fumarate 富马酸三丁基锡〈防霉剂〉
tricapryl trimellitate 偏苯三甲酸三仲辛酯〈增塑剂〉
trichlorobutylene oxide [TCBO] 三氯代-1,4-环氧丁烷,三氯代丁撑氧
trichloroethyl phosphate [TCEP] 磷酸三氯乙酯〈阻燃剂〉
trichloroethylene [TCE] 三氯乙烯
trichlorofluoromethane 三氯氟甲烷
trichlorometfayl 三氯甲基
N-(trichloromethylthio) phthalimide N-(三氯甲基硫代)邻苯二甲酰亚胺〈防霉剂〉
triclinic system 三斜晶系
tricresyl phosphate [TCP] 磷酸三甲苯酯〈增塑剂〉
tridecyl phosphite 亚磷酸三癸酯
tridimensional 三维的;三度(空间)的
  ~ polymer 三维交联聚合物
Tridyne process 特利达因法,二段式压铸成型法
trietliane chloride 三氯乙烷
triethanolamine 三乙醇胺〈活化剂〉
triethyl 三乙基
  ~ aluminium 二乙基铝
  ~ citrate [TEC] 柠檬酸三乙酯〈增塑剂〉
  ~ phosphate [TEP] 磷酸三乙酯〈增塑剂,溶剂〉
  ~ silanol 三乙基(甲)硅醇
triethylamine 三乙胺〈催化剂〉
triethylene 三亚乙基,三乙烯
  ~ diamine 三亚乙基二胺
  ~ glycol 三甘醇,二缩三乙二醇
  ~ glycol bis-3-(3-tert-butyl-4-hyd-roxy-5-methylphenyl) propionate 三甘醇双-3-(3-叔丁基-4-羟基-5-甲基苯基)丙酸酯〈抗氧剂〉

~ glycol caprylate-caprate 三甘醇辛酸癸酸酯〈增塑剂〉
~ glycol diacetate 三甘醇二乙酸酯〈增塑剂〉
~ glycol dibenzoate 三甘醇二苯甲酸酯〈增塑剂〉
~ glycol dicaprylate 三甘醇二辛酸酯〈增塑剂〉
~ glycol di (2-ethylbutyrate) 三甘醇二(2-乙基丁酸)酯〈增塑剂〉
~ glycol di (2-ethylhexoate) 三甘醇二(2-乙基己酸)酯〈增塑剂〉
~ glycol diheptanoate 三甘醇二庚酸酯〈增塑剂〉
~ glycol dipelargonate 三甘醇二壬酸酯〈增塑剂〉
~ glycol dipropionate 三甘醇二丙酸酯
~ tetramine [TTA] 三-1,2-亚乙基四胺,〈俗称〉三乙撑四胺〈固化剂〉
trifluorochloroethylene [TFCE] 三氟氯乙烯
trifluoroethylene resin 三氟乙烯树脂
trifluoromonochloroethylene 三氟氯乙烯
  ~ resin 三氟氯乙烯树脂
triformol 三聚甲醛
triglycol 三甘醇
trihydrazinotriazine 三肼基三嗪,2,4,6-三肼基-1,3,5-三嗪〈发泡剂〉
triisobutyl aluminum 三异丁基铝〈催化剂〉
triisodecyl 三异癸基
  ~ phosphite [TIDP] 亚磷酸三异癸酯〈抗氧剂〉
  ~ trimellitate 偏苯三甲酸三异癸酯〈增塑剂〉
triisononyl trimellitate 偏苯三甲酸三异壬酯〈增塑剂〉
triisooctyl 三异辛基
  ~ phosphite [HOP] 亚磷酸三异辛酯〈抗氧剂〉
  ~ trimellitate [TIOTM] 偏苯三甲酸三异辛基酯〈增塑剂〉
triisopropyl 三异丙基

~ phosphite 亚磷酸三异丙酯〈抗氧剂〉
triisopropylphenyl phosphate 磷酸二异丙苯酯〈增塑剂〉
trilauryl 三月桂基
 ~ phosphite [TLP] 亚磷酸三月桂酯〈抗氧剂〉
 ~ trithiophosphite [TLTIP] 三硫代磷酸三月桂酯〈抗氧剂〉
trim 修整;切毛边;边角料
 ~ knife 修剪刀;切毛边刀
 ~ line 裁切线
 ~ panel 装饰衬板
 ~ size 裁切尺寸
 ~ strip 装饰条
trimellitic anhydride [TMA] 偏苯三甲酸酐〈固化剂〉
trimer 三聚体,三聚物
trimerization 三聚(作用)
trimethyl 三甲基
 ~ borate 硼酸三甲酯
 ~ hexamethylene diamine 三甲基六甲撑二胺〈固化剂〉
 ~ hexane diisocyanate [TMDI] 三甲基己烷二异氰酸酯
 1,7,7- ~ -2- oxobicyclo [2.2.1] heptane 1,7,7-三甲基-2-氧代二环 [2.2.1] 庚烷〈增塑剂〉
 2,2,4- ~ -1,3- pentanediol diisobutyrate 2,2,4-三甲基-1,3-戊二醇二异丁酸酯〈增塑剂〉
 ~ pentanediol glycol [TMPD] 三甲基戊二醇
 ~ phosphate 磷酸三甲酯〈阻燃剂〉
 1,3,5- ~ -2,4,6- tris (3,5- di- tert- butyl-4-hydroxybenzyl) benzene 1,3,5-三甲基-2,4,6-三(3,5-二叔丁基-4-羟基苄基)苯〈抗氧剂330〉
trimethylamine 三甲胺〈助剂〉
trimethylchlorosilane 三甲基氯甲硅烷
trimethylolethane tribenzoate 三羟甲基乙烷三苯甲酸酯〈增塑剂〉
trimethylolpropane 三羟甲基丙烷

~ triacrylate 三羟甲基丙烷三丙烯酸酯
~ trimethacrylate 三羟甲基丙烷三甲基丙烯酸酯
trimethylsilanol 三甲基甲硅醇
trimmed edges 修边
trimmer 切边机
trimming 修剪;切边,修边
 ~ cutter 修切机;切边机
 ~ die 切边模
 ~ knife 修边刀;剪切刀
 ~ machine 修边机;切边机
 ~ press 冲压切边机
trimmings 切屑;废边角料
trinitrotoluene 硝酸基甲苯
trioctyl 三辛基
 ~ borate 硼酸三辛酯〈抗氧剂TOB〉
 ~ mellitate 偏苯三甲酸三辛酯〈增塑剂〉
 phosphate [TOF] 磷酸三辛酯〈增塑剂〉
 ~ phosphite 亚磷酸三辛酯〈抗氧剂〉
 ~ trimellilate, [TOTM] 偏苯三酸三辛酯〈增塑剂〉
triphenyl 三苯基
 ~ phosphate [TPP] 磷酸三苯酯〈增塑剂,阻燃剂〉
 ~ phosphite 亚磷酸三苯酯〈螯合剂,抗氧剂〉
triphenylmethane dye 三苯甲烷染料
triple 三(倍)的,三(层)的
 ~ bond 三键
 ~ dyes 三重染色
 ~ laminating 三层层合〈织物/泡沫塑料/织物〉
 ~ link 三键
 ~ package 三层包装
 ~ roll (er) mill 三辊磨
 ~ thread 三线螺纹
tripolite 硅藻土
tripolymer 三聚物
tris 三(个)
 ~ (bromocresyl) phosphate 磷酸三(溴甲苯)酯〈阻燃剂〉
 1,3,5- ~ (4- tert- butyl-3-hydrox-

y)2,6 - dimethylbenzyl)1,3,5 - triazine - 2,4,6 - (1H,3H,5H) - trione 1,3,5 - 三(4 - 叔丁基 - 3 - 羟基 - 2,6 - 二甲基苄基)1,3,5 - 三嗪 - 2,4,6 - (1H,3H,5H) - 三酮〈抗氧剂 1790〉

~ (2 - chloroethyl) phosphate 磷酸三(2 - 氯乙)酯〈增塑剂〉

~ (2 - chloroethyl) phospite 亚磷酸三(2 - 氯乙)酯〈抗氧剂〉

~ (2,3 - dibromopropyl - 1) isocyanurate 三(2,3 - 二溴丙基 - 1)异氰脲酸酯〈阻燃剂〉

~ (3,5 - di - tert - butyl - 4 - hydroxyben - zyl) isocyanurate 异氰脲酸三(3,5 - 二叔丁基 - 4 - 羟基苄基)酯〈抗氧剂 3114〉

1,3,5 - ~ (3,5 - di - tert - butyl - 4 - hydr - oxybenzyl) - s - triazine - 2,4,6 - (1H,3H,5H) trione 1,3,5 - 三(3,5 - 二叔丁基 - 4 - 羟基苄基)均三嗪 - 2,4,6 - (1H,3H,5H) 三酮〈抗氧剂 3114〉

~ [β - (3,5 - di - tert - butyl - 4 - hydrox - yphenyl) propyonyloxyethyl] isocya - nurate 异氰脲酸三[β - (3,5 - 二叔丁基 - 4 - 羟基苯基)丙酰氧基乙]酯〈抗氧剂 3125〉

~ (2,4 - di - tert butylphenyl) phosphite 三(2,4 - 二叔丁基苯基)亚磷酸酯〈抗氧剂 168〉

~ (dichloropropyl) phosphate 磷酸二(二氯丙)酯〈增塑剂〉

2,4,6 - ~ (dimethylaminoethyl) phenol 2,4,6 - 三(二甲胺基乙基)苯酚〈固化剂〉

~ (2 - ethyl hexyl) phosphate 亚磷酸三(2 - 乙基己)酯〈抗氧剂〉

2,4,6 - ~ (2' - hydroxy - 4' - n - butoxy phenyl) - 1,3,5 - triadine 2,4,6 - 三(2' - 羟基 - 4' - 正丁氧苯基) - 1,3,5 - 三嗪〈紫外线吸收剂〉

1,1,3 - ~ (2 - methyl - 4 - hydroxy - 5 - terf - butyl phenyl) butane 1,1,3 - 三(2 - 甲基 - 4 - 羟基 - 5 - 叔丁基苯基)丁烷〈抗氧剂 CA〉

~ (mixed mono - and dinonylphenyl) phosphate 亚磷酸三(单壬基苯和二壬基苯混合酯)〈分解剂、抗氧剂、稳定剂〉

~ (nonylphenyl) phosphate 三(壬基)亚磷酸酯〈防老剂 TOP〉

~ (1,2,2,6,6 - pentamethyl - 4 - piperi - dine) phosphite 三(1,2,2,6,6 - 五甲基 - 4 - 哌啶基)亚磷酸酯〈光稳定剂 GW - 540〉

trisazo dye 三偶氮染料
tristearyl 三(十八烷基)酯
~ citrate 柠檬酸三(十八烷基)酯〈润滑剂〉
~ phosphite 亚磷酸三(十八烷基)酯〈抗氧剂〉
tritactic polymer 三规聚合物
tritolyl phosphate 磷酸三甲苯酯〈增塑剂〉
triturable 可粉化的
triturate 磨粉,粉除;磨碎物
trituration mill 研磨机
trixylyl phosphate 磷酸三(二甲苯)酯〈增塑剂〉
TRK [technicalguideline concentration] 工业指标浓度〈毒物学〉
trolley 手推车,空中吊运车
Trommsdorf ertect 脱洛姆斯多夫效应,凝胶效应
tropic 热带(的)
tropical 热带的
~ climate 热带气候
tropicalisation 热带化
~ test 热带性能试验
tropicalized plastic 耐热带气候的塑料
trouble - free 无故障
troublesome impurity 有害杂质
trough 槽
~ dryer 槽型干燥器
~ mixer 槽型混合器
~ steam dryer 槽型蒸汽干燥器
~ - type mixer 槽型混合器

trowel 涂抹;刮刀,刮铲〈用于黏合剂涂布〉
~ adhesive 涂抹高黏度黏合剂
trowelabllity 可涂抹性
trowelling 手工抹涂
~ compound 腻子,填平料
truck 卡车,运货车
~ dryer 栅格式干燥器
true 真实的;准确的
~ colo(u)r 真色度
~ copolymer 真共聚物
~ density 真密度
~ fault 真实故障
~ fibre 粗细均匀纤维
~ hardness 真硬度
~ helix winding 真螺旋形卷绕
~ horsepower 实际马力
~ plasticizer 真增塑剂
~ polymer 真聚合物
~ polymerization 真聚合(作用)
~ purity 真纯度
~ solvent 真溶剂
~ specific gravity 真比重,真相对密度
~ specific heat 实际比热容
~ strain 真应变
~ stress 真应力,实际应力
~ resin 天然树脂
~ tensile strength 最大拉伸强度
~ to shape 形状正确的
~ to size 尺寸准确的
~ -union ball valve 旋紧球阀
~ viscosity 真黏度
trueness 真实性,精确度
truly 真正地,正确地
~ conical screw 真正圆锥形的螺杆
~ positive mould 真阳模
trunk fiber VF 硬化纸板
trunnion 轴颈;耳轴
truss 捆,扎;桁架
trussing 捆扎,桁架系统
TS [thermoset] 热固性塑料;热固性的
TS-injection [thermosetting plastics injection] TS注塑,热固性塑料注塑
TSC [thermal stress cracking] 热应力开裂
TSO process 共注塑法〈发泡的〉
TSUR [thermost urethane] 热固性聚氨酯
TTA [triethylene tetramine] 三-1,2-亚乙基四胺,三乙撑四胺〈固化剂〉
Tu [thulium] 铥
tub 盆,桶,槽
tube 管(材);软管
~ cut off 割断管
~ cutter 切管机,管材切割机
~ expander 扩管器
~ extruding press 管材挤塑机
~ extrusion 管材挤塑
~ extrusion die 管材挤塑模头
~ fittings 管配件
~ -flange lap joint 管-法兰搭接
~ grid 管网
~ heat exchanger 管式热交换器
~ heater 管式加热器
~ mill 管式磨
~ mo(u)ld 吹管模
~ -on die 套筒式电缆挤塑口模
~ package 管材包装
~ trimming machine 管材修整机
~ winding machine 管材缠绕机,绕管机
tubing 管(材);软管;制管
~ die 挤管模头〈软管〉
~ head 挤管机头〈软管〉
~ machine 挤管机
tubular 管状的,管形的;筒形的;空心的
~ blown film 管状吹塑薄膜,吹塑管膜
~ braiding machine 管状编带机
~ die 环形模头;挤管模头
~ dryer 管状干燥器
~ exchanger 管式换热器
~ fabric 圆筒织物
~ fibre 空心纤维
~ film 管膜,管状薄膜
~ film blowing 管膜吹塑
~ film cooling 管膜冷却
~ film die 管膜挤塑模头

~ film extrusion head 管膜挤塑机头
~ film method 管膜(生产)法
~ film process 管膜成型法
~ flow 管内物料流动
~ foam extrusion 管状发泡挤塑
~ goods 管材
~ grid 管网
~ heat exchanger 管式换热器
~ heater 管式加热器
~ insert 管形嵌件
~ insulating principle 管状绝缘原理
~ knitting 圆编〈织物〉
~ lay-flat film 平折管膜
~ liner 圆筒形衬里
~ parison 管坯,管状型坯〈吹塑〉
~ plunger 环形柱塞
~ plunger accumulator head 环形柱塞式储料器
~ reactor 管式反应器
~ reflector heater 筒形反射加热器
~ resistance heater 管状电阻加热器
~ rivet 空心铆钉
tuck tape 衬里带
tufcote process 泡沫带制造法
tuft 绒束;簇绒
tufted fabric 栽绒织物;簇绒织物
tufting process 簇绒法
fumble 滚转,转鼓;抛光
~ blasting 转鼓抛光
~ finishing 转鼓(精)加工
~ mixing 转鼓混合
~ polishing 转鼓抛光,滚筒抛光
tumbler 转鼓,滚筒
~ screen 转筒筛
~ screening machine 转筒筛分机
~ sieve 转筒筛
~ test 转鼓试验
tumbling 转鼓抛光,滚光
~ action 转筒滚动
~ agitator 滚筒搅拌机
~ barrel 转鼓;滚磨筒
~ blender 滚筒混配机,转筒混合机
~ cylinder 滚筒
~ drum 转鼓

~ diyer 转筒干燥器
~ machine 转筒机;滚光机
~ mill 滚磨机
~ mixer 转鼓混合机
~ winding machine 转鼓式缠绕机
tung oil 桐油
tungsten filament 钨丝
tunnel 隧道
~ diyer 隧道式干燥器
~ gate 隧道型浇口
~ test 隧道窑式试验
tunneling 潜道脱层
turbid 混浊的,不透明的
turbidity (混)浊度,混浊(性)
~ point (混)浊度,浊度点,混浊(性)
turbidness 混浊(度,性)
turbine 涡轮,透平
~ agitator 涡轮式搅拌机
~ configurations 涡轮形状,涡轮结构
~ impeller [mixer] 涡轮式混合机
~ strirrer 涡轮式搅拌机
~ -type impeller agitator 涡轮型叶轮搅拌机
turbo 涡轮(机),透平(机)
~ compressor 涡轮压缩机
~ disperser 涡轮分散机
~ gas absorber 涡轮气体吸收器
~ sifter 涡轮筛分机
turboblower 涡轮式鼓风机
turbodryer 涡轮式干燥器
turbomill 涡轮研磨机
turbomixer 涡轮式混合机
turbulence 湍流,涡流,紊流
turbulent 扰动的,湍流的
~ flow 湍流
~ fluidized bed 湍动流化床
turbulization 湍流化,涡流化,紊流化
turmeric 郁金姜黄〈黄色〉
turn 转动,旋转;转变,变化;车削
~ button 旋钮
~ down 翻转,翻下
~ table 转台,回转台
~ yellow 泛黄,变黄
turning 转动,旋转;车削

~ bench 车床
~ blade 旋转叶片;转动(刮)刀
~ distributor 旋转分布器
~ moment 转动力矩
~ point 转折点
~ table 旋转台
~ unit 旋转装置
turnings pl 车屑
turnover 回转;翻转
~ discharge equipment of Banbury mixer 密炼机翻转式卸料装置
~ point 转折点〈曲线〉
turntable 转台,转盘
~ injection machine 转台式注塑机
~ support 转台底座
turquoise 绿松石(色);翠蓝色(的),绿光蓝色(的)
turret 塔;转塔;六角刀架,六角转头
~ lathe 六角车架,转塔车床
~ rewind 转位重绕,转位卷取,转位缠绕
~ type injection mo(u)lding machine 转位式注塑成型机
~ winder 转位式卷绕机
T.W.;t.w. [total weight] 总重;全部重量
twill [tweel] 斜纹〈纺织物〉
twin 成双的
~ - bar bumper 双杆保险杠
~ - barrel extruder 双料筒挤塑机
~ - channeled screw 双螺槽螺杆
~ - core cable 双股电缆
~ crystal 孪晶
~ - cycle hot air oven 双循环热空气烘箱
~ - cylinder mixer 双筒混合机
~ drum dryer 双鼓干燥器
~ ebulliometer 双式沸点升高仪
~ - feed proportioning spray gun 双组分按比例供料喷枪
~ fillet weld 双面角焊缝
~ head 双头
~ - headed blow mo(u)lding machine 双头式吹塑成型机

~ - headed spray gun 双喷嘴枪,双组分喷枪
~ mo(u)ld 双槽模具
~ pack adhesive 双包装黏合剂,双组分黏合剂
~ reel - up unit 双绕卷装置
~ - roll beater 双辊搅拌器
~ screw 双螺杆
~ screw continuous kneader and compounder 双螺杆连续捏合混料机
~ screw extruder 双螺杆挤塑机
~ screw extrusion machine 双螺杆挤塑机
~ - screw flow dryer 双螺杆气流干燥器
~ screw reactor 双螺杆式反应器
~ screw type compounder 双螺杆式混料机
~ - shell 双壳
~ - shell blender 双筒混合机〈美〉
~ - shell forming 双模成型〈吹塑成型〉;双片成型〈热成型〉
~ - shell thermoforming 双片热成型
~ - shell vacuum forming 双片真空成型〈热成型〉
~ spray gun 双头喷枪
~ taper screw extruder 锥形双螺杆挤塑机
~ vacuum shaping machine 双段真空成型机
~ worm 双螺杆〈美〉
~ worm extruder 双螺杆挤塑机〈美〉
~ worm mixer 双螺杆混合机
twine 麻绳;鱼网线;合股绳,捆扎绳
twist 捻(度);扭转
~ stress relaxation 扭转应力松弛
~ test 扭转试验
~ welding 麻花焊
twisted 加捻的,合股的;扭转的
~ fibre 合股纤维
~ filament 加捻的线
~ (filament) yarn 合股纱
twister 加捻器
twisting 加捻;扭转;卷曲

~ force 扭力
~ moment 扭矩；转矩
~ strain 扭应变
~ strength 扭曲强度
~ stress 扭应力
~ stress relaxation 扭曲应力松弛
~ test 扭转试验
~ tester 扭转试验器
two 二,两,双
~ – axial drawing 双轴拉伸
~ arm kneader 双臂搅拌机
~ – cavity injection mould［two – impression injection mould］ 双腔注塑模
~ – channel 双通道的；双流道的
~ – channel screw extruder 双流道螺杆挤塑机
~ – channel technique 双流道工艺
~ – colo(u)r 双色
~ – colo(u)r extrusion 双色挤塑
~ – colo(u)r injection mo(u)lder 双色注塑成型机
~ – colo(u)r injection mo(u)lding 双色注塑成型；双色注塑成型件
~ – colo(u)r injection mo(u)lding machine 双色注塑成型机
colo(u)r marble injection mo(u)lding machine 云彩型双色注塑成型机
~ – colo(u)r mo(u)lding 双色注塑成型；双色注塑成型件
~ – colo(u)r simultaneous extrusion 双色同时挤塑
~ – component 双组分的
~ – component addhesive 双组分黏合剂
~ – component foam mo(u)lding 双组分泡沫成型
~ – component lacquer 双组分漆
~ – component spray gun 双组分喷枪
~ – component spraying 双组分喷涂
~ – dimensional 二维的,二度(空间)的,平面的
~ – dimensional crimping 二维卷曲
~ – dimensional – layer molecule 二维层状分子
~ – dimensional nucleation 二向成核(作用)
~ – dimensional phase 二维相
~ – dimensional polymer 片型聚合物
~ – dimensional stress 二维应力,二轴应力
~ – dimensional stretch 纵横,拉伸
~ – dimensional structure 二维结构,片型结构
~ – direction thrust bearing 双向推力轴承
~ – flight screw 双螺纹螺杆
hand safety release 双手式安全器〈塑料加工机〉
~ – impression injection mould 双腔注塑模
~ – layer 双层的
~ – layer film 双层薄膜
~ – layer flat film 双层平膜
~ – layer laminates 双层层压制品
~ – layer plastic pipe 双层塑料管
~ – layer tubular film 双层管膜,双层管状薄膜
~ – layer tubular film centre – fed die head 双层管膜中心进料模头
~ – level 两层的
~ – level mo(u)ld 双层模具
~ – level mo(u)lding 双层模塑
~ – pack 双组分；双包装
~ – pack composition 双组分混合料
~ – pack spray equipment 双组分喷射设备
~ – part paste 双组分糊料
~ – phase 两相的
~ – phase plastic (s)两相塑料
~ – phase polycondensation 两相缩聚
~ – phase process 两相法
~ – phase structure 两相结构
~ – plate mo(u)ld 双板模具
~ – polymer 两种聚合物
~ – polymer adhesive 两种聚合物制成的黏合剂
~ – polymer formulation 两种聚合物

配料
~ - position action controller 双位调节器
~ - position controller 双位控制器
~ - pour method 二次灌料法
~ - roll 双辊
~ - roll "I" calender I型双辊压延机
~ - roll coater with fountain feed and smoothing rolls 带有供料储槽和精轧的双辊涂布机
~ roll coater with smoothing rolls 带有精轧的双辊涂布机
~ - roll crasher 双辊压碎机
~ - roll mill 双辊滚磨机
~ - roll mill system 两辊滚磨装置
~ - roll treater 双辊浸胶机
~ - sheet head 双片状机头
~ - shot foaming technique 二步发泡法〈聚氨酯〉
~ - shot injection molding 二步注塑成型
~ - shot molding 二步成型
~ - stage 二级的,二阶段
~ - stage barrier screw 双阶屏障型螺杆
~ - stage blow mo(u)lding 二段吹塑成型法
~ - stage clamping 二阶段锁模
~ - stage ejection 二步脱模
~ - stage extruder 二段式挤塑机
~ - stage injection 二步注塑
~ - stage injection mo(u)lding machine 二级注塑成型机
~ - stage mo(u)ld 双级模具
~ - stage mo(u)lding 二步法成型
~ - stage phenolic resin 二步法酚醛树脂
~ - stage plasticator 两级塑化器
~ - stage plunger injection 两级活塞注塑
~ - stage polymerization 两步聚合
~ - stage printing 两步法印花
~ - stage resin 二步法(酚醛)树脂

~ - stage screw 双级螺杆
~ - stage stretch - blow 两步拉吹
~ - stage streching 两步拉伸
~ - stage fransfer mo(u)lding 二级式压铸成型法
~ - start screw 双头螺杆
~ - station machine 双工位机
~ - step 二步的,二级的
~ - step action controller [two - position action controller] 二步调节器
~ - step cure 二步固化;二步塑化
~ - step mixing 二步混料法
~ - step process 二步法
~ - tier 双层的
~ - tone 双色调;色差〈压塑缺陷〉
~ - tone colo(u)r 双色
~ - tone finish 双色喷漆
~ - tone mo(u)lding 色差模塑品
~ - tone stripe 双色条纹
~ - way 双向的
~ - way drawing 双向拉伸
~ - way pallet 双路托盘,双向托盘
~ - way stretch 双向拉伸
~ - way switch 双向开关
TY [total yield] 总收率,总产率;总产量
tychoback 织物上的泡沫塑料层
tying 接结;捆扎
~ rope 捆扎绳
~ string 扎绞线
tylose 泰罗斯(甲基纤维素)
type 类型,式样;型号;典型
~ approval test 定型试验
~ number 型号
~ of gate [sprue] 浇口类型
~ of structure 结构类型
~ reaction 典型反应
~ sample 标准样品
~ test 典型试验;例行试验
typical 典型的
~ cross - section 标准横断面
- form 典型式(样)
~ operation 典型操作
~ properties 典型性质
typification 标准化,规范化;典型化

tyre [tire〈美〉] 轮胎
　~ builder fabric 轮胎帘子布
　~ casing ply 外胎帘布层
　~ cord 轮胎帘子线,轮胎帘布
　~ cord fabric 轮胎帘子布
　~ fabric 轮胎帘布织物,轮胎用织物
TYS [tensile yield strength] 拉伸屈服强度

U

U [uranium] 铀
U [urea] 尿素,脲
U-box U形槽
U-notch U形缺口〈试棒〉
U-PVC [U-polyvinyl chloride, unplasticized PVC] 未增塑聚氯乙烯
Ubbelodhe viscosimeter 乌别洛德黏度计
UCC foam mo(u)lding [Union Carbide Corporation foam mo(u)lding] 联合碳化物公司发泡成型法,低压发泡注塑成型法
UCC process 联合碳化物公司发泡成型法,低压发泡注塑成型法
UF [urea-formaldehyde resin] 脲甲醛树脂
UHF [ultra hight frequence] 超高频率
UHMW-PE [ultra high molecular weight polyethylene] 超高分子量聚乙烯
UL [upper limit] 上限;最高极限
ultimate 极限的;最后的
　~ bearing strength 极限承载强度
　~ bending strength 极限弯曲强度
　~ bending stress 极限弯曲应力
　~ capacity 极限功率;最大容量
　~ compression strength 压缩极限强度
　~ elongation 极限伸长率;断裂伸长
　~ extension 极限延伸
　~ fatigue strength 极限疲劳强度
　~ fibre 单纤维
　~ life 极限寿命
　~ load 极限负荷;最大载荷
　~ mechanical property 极限力学性能
　~ output 极限功率;最大功率
　~ oxidation 极限氧化
　~ pressure 极限压力
　~ production 极限生产;总产量
　~ resilience 极限回弹性
　~ strain 极限应变
　~ strength 极限强度
　~ stress 极限应力
　~ temperature 极限温度
　~ tensile strength 极限拉伸强度
　~ tensile stress 极限拉伸应力
　~ tension 极限拉力
　~ torque 极限扭矩
　~-use temperature 最终使用温度
ultra 〈词头〉超,越
　~-accelerator 超速促进剂
　~-disperse dyes 超细分散染料
　~ fine powder 超细粉
　~ fine talc pigment 超细粒滑石粉
　~ hight 超高
　~ hight draft 超大牵伸
　~ hight frequence [UHF] 超高频率
　~ hight molecular weight 超高分子量
　~ hight molecular weight polyethylene [UHMW-PE] 超高分子量聚乙烯
　~ hight speed extrusion 超高速挤塑
　~-high vacuum 超高真空
　~ low frequency 超低频率
　~-red 红外的
　~-red ray 红外线
　~-wear 超耐磨的
ultracentrifugal method 超离心法
ultracentrifuge 超速离心机
ultrafiltration 超滤
　~ membranes 超滤膜

ultramarine 群青;佛青;云青〈着色剂〉
~ blue 群青〈颜料〉
~ blue pigments 群青颜料
ultramicrocrystal 超微结晶
ultramicron 超微细粒
ultramicrostructure 超微结构
ultramicrotome 超薄切片机
ultramicrotomy 超薄切片(法)
ultraradio frequency 超射频率
ultrasonic 超声波;超声(波)的
~ acting time 超声作用时间
~ adhesive strength 超声黏合强度
~ agitation 超声搅拌
~ attenuation 超声波消声
~ bonding 超声波粘接〈用超声波使黏合剂固化〉;超声焊接
~ cleaning 超声净化
~ control 超声控制
~ converter 超声转换器
~ curing 超声固化
~ cutting 超声波切割
~ cutting gun 超声波切割枪
~ defectoscope 超声探伤仪
~ degating 超声除浇口料
~ degradation 超声(致)降解
~ detector 超声检测器
~ energy 超声波能
~ flaw detector 超声波探伤器
~ -formed mo(u)ld cavity 超声波成型模腔
~ horn [sonotrode] 超声焊头
~ impression 超声波嵌入(法)〈金属件嵌入塑料件中〉
~ impulse-echo equipment 超声波脉冲回声设备
~ inserting 超声镶嵌
~ inspection method 超声波探伤法
~ joint 超声波焊接
~ lens 超声透镜
~ mash welding 超声波滚焊;超声波压点焊
~ material testing 超声波探伤材料试验
~ micriscope 超声显微镜
~ pulse through transmission testing 超声波脉冲透射试验
~ radiation 超声辐射
~ riveting 超声波铆接
~ scan welding 超声扫描焊接
~ scanner 超声扫描器
~ sealing 超声熔合,超声封焊
~ separation 超声分离
~ sewing 超声波熔合〈塑料〉;超声缝合〈合成织物〉
~ sewing machine 超声波熔合机〈塑料〉;超声波缝合机〈合成织物〉
~ source [sonic source] 超声波源
~ stud welding 超声波轴颈焊接
~ tester 超声探伤器
~ testing 超声波检验[探伤]
~ thickness meter 超声测厚仪
~ transducer 超声波传感器
~ transmission time 超声波传播时间
~ viscometer 超声黏度计
~ wave 超声波
~ welder 超声波焊接机
~ welding 超声波焊接
~ welding gun 超声波焊枪
~ welding machine 超声波焊接机
ultrasonically welded joint 超声波焊接
ultrasonics pl 超声波学
ultrasonoscope 超声波探测仪
ultrasound 超声
ultrastructure 超微结构
ultrathin 超薄的
~ film 超薄薄膜
~ section 超薄切片
ultraviolet [UV] 紫外线(的)
~ absorber 紫外线吸收剂
~ absorption detector 紫外线吸收检测器
~ absorption spectrometry 紫外线吸收光谱法
~ absorption spectrum 紫外线吸收

— rays 光谱
~ curing 紫外线固化〈热固性塑料〉
~ degradation 紫外线降解
~ inhibitor 紫外线抑制剂
~ light 紫外光(线)
~ microscope 紫外显微镜
~ radiation 紫外线辐射(作用)
— rays 紫外线
~ screener 紫外线滤射器；紫外线屏蔽
~ — sensitive resin 紫外敏感树脂
~ sensitizer 紫外敏感剂
~ spectrophotometer 紫外分光光度计
~ spectrophotometry 紫外分光光度测定法
~ spectroscopy 紫外线光谱法
~ spectrum 紫外光谱
~ stabilizer [UV‑stabilizer] 紫外线稳定剂；UV稳定剂；耐紫外线剂
~ transmission 紫外线透射
u. m. [undermentioned] 下述的
unbranched 无支链的
~ chain 直链
~ macro‑molecule 无支链大分子；直链大分子
~ polyethylene 无支链聚乙烯
unbreakable 不易破碎的；不破裂的；耐断裂的
umbrella roof 伞形屋顶
unaided eye 肉眼
uncoated paper 未涂布纸
uncoil 退卷
uncoiler 退卷辊〈薄膜〉
uncolo(u)red 无色的
uncombined 未化合的
uncompatibility 不相容性
unconditionally identical colours 绝对等同色
unconverted monomer 未转化单体；未聚合的单体
uncured 未固化的
undeformed 未变形的

under 在…下面；在…之中；低于；少于
~ cloth 垫布，底布，衬布
~ coat 底涂层
~ coating 内涂层；上底漆
~ cutting 凹槽
~ power 功率不足；低功率
~ stress 在受力状态下
~ study 在研究中
~ test 在试验中；处于试验阶段
~ way 在进行中；在运行中
~ weight 重量不足
undercapacity 功率不足；生产率不足
undercarriage 底架
undercoat 底涂层
undercooling 过冷
undercure 欠固化；欠熟化；欠塑化
undercut 凹槽；倒拔梢
~ angle 倒拔梢角
~ feed throat 凹入式进料口
~ insert 凹槽式嵌件
~ shape 倒拔梢形；凹形槽
underdriven mixer 下传动式混合器
underfeeding 供料不足；缺料〈压模中成型件〉
underground 地下的
~ cable 地下电缆
~ installation 地下铺设〈管〉
underpackaging 欠缺包装
underlustred 不够光泽的
underpressure 欠压；负压；真空
~ tapping 受压钻孔；受压攻螺纹
underproduce 产量不足，生产不足
undershrinking 收缩不足
undersize 尺寸不足；筛下料
undertone 淡[浅]色；底彩色
underwater 水下的
~ adhesive 水下用黏合剂
~ cut 水下切割〈造粒〉
~ cut‑off device 水下切割装置〈造粒〉
~ cut pellet 水下切粒
~ die‑face cutter 水下模面切粒机

~ granulator 水下造粒机
~ pelletizer 水下切粒机
~ pelletizing 水下切粒
undirected flow 无定向流
UNDP [United Nations Devlopment Program] 联合国发展计划署〈纽约〉
undrawn 未拉伸的
~ tow 未拉伸丝束
~ yarn 未拉伸的纱线
undulated border 波浪形边缘
undulation 起皱;波纹〈织物纤维〉
undyed 未染色的
UNEP [United Nations Environment Program] 联合国环境计划署
uneven 不均匀的
~ ageing 老化不匀
~ colo(u)r 不匀染色;染色不匀
~ distribution of plasticizer 增塑剂分布不匀〈在片材中〉
~ grain 不均匀颗粒;颗粒不匀
~ heating 加热不匀
~ printing 印花不匀
~ running 运转不平稳
~ shrinkage 缩率不匀
unevenness 不均匀性
~ of size 细度不匀〈纤维〉
unexpandered beads 未发泡珠粒
unfading 不褪色的
unfilled 未填充的
~ resin 无填料树脂;纯树脂
unflighted 无螺纹线的
unfoamed material 未发泡材料
unground resin 未研细的树脂
unhomogeneity 不均一性
uniaxial 单轴的
~ crystal 一轴晶;单轴晶
~ drawing 单轴(向)拉伸
~ elongation 单轴(向)伸长
~ load 单轴负荷
~ orientation 单轴取向
~ oriented film 单轴(向)定向薄膜
~ strength 单轴强度

~ stress 单轴向应力;单轴向张力状态
~ stretched film 单轴向拉伸薄膜
~ stretching 单轴向拉伸
~ tensile deformation 单轴拉伸变形
unicellular 闭孔(的);单孔(的)
~ foam 闭孔泡沫塑料
~ plastics 闭孔泡沫塑料
~ rubber 闭孔泡沫橡胶
unicolo(u)r 单色(的)
unidirectional 单向的
~ fabric 单向织物
~ composites 单向复合材料
~ laminate 单向层合材料
~ layer 单向层
~ prepreg 单向预浸渍料
~ reinforced composite material 单向增强复合材料
~ reinforcement 单向增强材料
~ weave 单向织物;单向织法
~ weftless tape 无纬带
UNIDO [United Nations Industrial Development Organization] 联合国工业发展组织
unif. [uniformity] 均匀性;均匀度
Unified thread 统一标准螺纹〈美、英、加拿大〉
uniform 均匀的,相等的
~ combustion 均匀燃烧
~ deformation 均匀形变
~ diffuser 均匀漫射体
diffuse reflection 均匀漫反射
~ diffuse transmission 均匀散射传输,均匀漫透射
~ diffusion 均匀漫射
~ distribution 均匀分布
~ elongation 均匀伸长
~ extension 均匀伸长
~ gauge 均匀厚度
~ mixing 均匀混合
~ mixture 均匀混合物
~ pitch screw 等螺距螺杆

uniformity	

- ~ polymer 均一聚合物
- ~ pressure 等压力;均匀压力
- ~ strain 均匀应变
- ~ strength 均匀强度
- ~ stress 均匀应力
- ~ torsion 均匀扭力,均匀扭转
- ~ wear 均匀磨损

uniformity 均匀(性,度)
- ~ index 均匀指数
- ~ varying stress 等变应力

unilateral 单向的;片面的
- ~ heating 单面加热
- ~ stretching 单向拉伸

unimolecular 单分子的
- ~ film 单分子膜
- ~ layer 单分子层

uninflammability 不燃性
uninflammable 不易燃的
union 联合,组合;一致;连接
- ~ colour 一致色泽
- ~ coupling 联管节
- ~ dyes 混染染料
- ~ joint 管接头
- ~ pipe 连接管
- ~ twist yarn 混纺股线
- ~ yarns 混纺纱

union 联合;活接头〈管〉
- ~ bush 活接头内套〈管〉
- ~ end 活接头端件,活接头堵头〈管〉
- ~ nut 锁紧螺母,外套螺母〈管〉

Union Carbide accumulator process 联合碳化物公司储料器法〈熔体发泡法〉

Union Carbide Corporation foam mo(u)lding [UCC foam mo(u)lding] 联合碳化物有限公司发泡成型法,低压发泡注塑成型法

Union Carbide process 联合碳化物公司(注射发泡)法

unit 单位;单元;装置
- ~ area 单位面积
- ~ assembly drawing 部件装配图
- ~ assembly principle 部件装配原理
- ~ cell (单元)晶胞;单元格子;单位泡孔〈泡沫塑料〉
- ~ cell constant 晶胞常数
- ~ composed system 组合构成系统
- ~ - construction mo(u)ld 组合结构式模具
- ~ construction principle 组合结构原理
- ~ - construction system 组合结构式系统
- ~ cost 单位成本
- ~ damping energy 单位阻尼能
- ~ die 组合模头,可互换件模头
- ~ elongation 单位伸长
- ~ length 单位长度
- ~ mass 单位质量
- ~ mo(u)ld 组合模具,标准模具〈可互换内嵌件的模具〉
- ~ of structure 结构单元,构件
- ~ of weight 重量单位
- ~ operation 单元操作
- ~ package 单元包装
- ~ power 单位功率
- ~ pressure 单位压力
- ~ price 单价,单位价格
- ~ processes pl 单元作业
- ~ storage energy 单位储能
- ~ strain 单位应变
- ~ stress 单位应力
- ~ tensile stress 单位拉伸应力
- ~ tension 单位拉力
- ~ tensor 单位张量
- ~ - type mo(u)ld 单元式模具
- ~ volume 单位体积
- ~ weight 单位重量

United Shoe Machinery foam mo(u)lding [USM foam mo(u)lding] 联合制鞋机械公司发泡成型法,USM 发泡成型法,气体反压注塑法

unitized 成套的;组合的
- ~ construction 组合结构
- ~ injection moulding 坯料注塑成型

~ moulding　拉嵌成型件
universal　通用的,一般的;万能的
～ adhesive　通用黏合剂,万能黏合剂,万能胶
～ contract　万能接头
～ coupling　万向联轴节,万向接头
～ impact tester　万能冲击试验仪
～ joint　万向接头
～ jointing fitting　万向连接配件
～ mixer　通用混合机
～ mo(u)ld　通用模具〈模座尺寸一定,而模槽或模芯可互换〉
～ strength tester　通用强度试验仪
～ tester　万能(材料)试验仪
～ testing machine　万能(材料)试验机
～ time [UT]　世界时间
～ wear tester　通用耐磨试验仪
unload　卸荷,卸载,卸料,卸模
unloading　卸料,卸荷;卸模
～ control　溢流控制;卸压控制〈泵〉
～ fixture　卸模工具
～ valve　卸荷阀
unlocking　脱锁
unmixed　未混合的
unmodified　未改性的
～ polystyrene　未改性聚苯乙烯,纯聚苯乙烯
～ resin　未改性树脂,纯树脂
unnotched　无缺口的
～ impact resistance　无缺口耐冲击性
～ impact strength　无缺口冲击强度
～ specimen　无缺口试样
～ test bar　无缺口试棒
unoriented　未取向的
～ bulk polymer　未取向本体聚合物
～ fibre　未取向纤维
～ film　未取向薄膜
～ rigion　未取向区(域)
unoxidizable　不可氧化的
unpackaged　未包装的,散装的
unpatented　未得到专利权的
unpitched blade　无斜度桨叶〈直形桨叶〉

unplasticized　未增塑的
～ film　未增塑薄膜
～ material　未增塑材料
～ polyvinyl chloride [UPVC]　未增塑聚氯乙烯
unprocessed　未加工过的,未处理过的
unpyrolyred　未热解的
unravel　解开,拆散;阐明,解决;分裂成纤维,纤化
unregularity　不规则性
unregulated polymer　无规聚合物
unreinforced　未增强的
～ plastic　未增强塑料
～ resin　未增强树脂
～ thermo-plastic　未增强热塑性塑料
unrestricted　无限制的,无约束的,自由的
～ flow　无限制流动,自由流动
～ gate　无制约浇口;无障碍浇口
～ plastic flow　无限制塑性流动
unroll　退卷(机)
uns. [unsymmetrical]　不对称的
unsat. [unsaturated]　不饱和的
unsaturated　不饱和的
～ bond　不饱和键
～ compound　不饱和化合物
～ glass-fibre reinforced polyester　玻璃纤维增强不饱和聚酯
～ group　不饱和基团
～ hydrocarbon　不饱和烃
～ polyester [UP]　不饱和聚酯
～ polyester based sheet mo(u)lding compound　不饱和聚酯片状模塑料
～ polyester-based wet dough mo(u)lding compounds [UP mo(u)lding compounds]　不饱和聚酯湿料团模塑料
～ polyester coating　不饱和聚酯涂层 [涂料]
～ polyester mo(u)lding compound　不饱和聚酯模塑料
～ polyester resin [UP-resin]　不饱和聚酯树脂

~ polyester resin of bisphenol A types
双酚 A 型不饱和聚酯树脂
~ polyester resin of bisphenol A and p-phthalic acid types 双酚 A 与对苯二甲酸混合改型不饱和聚酯树脂
~ polyester resin of crosslinked diallyl phthalate 苯二甲酸二烯丙酯交联的不饱和聚酯树脂
~ polyester resin of methyl aciylic glycidic ester of acrylic glycidic ester types 甲基丙烯酸缩水甘油酯或丙烯酸缩水甘油酯型不饱和聚酯树脂
~ polyester resin of m-phthalic acid types 间苯二甲酸型不饱和聚酯树脂
~ polyester resin of p-phthalic acid types 对苯二甲酸型不饱和聚酯树脂
~ polyester resin of xylene types 二甲苯型不饱和聚酯树脂
~ state 不饱和性状态
unscrewing 旋松;拆卸,折开
~ drive 旋松传动〈注塑模具〉
~ mo(u)ld 旋松式模具
unshrinkable finish 防缩整理〈纺织品〉
unsintered tape 生胶带
unsized paper 无胶纸
unskilled worker 不熟练工人,没有经验工人
unskinned structural foam 无皮结构泡沫塑料
unstable equilibrium 不稳定平衡
unsteady flow 不稳定流动
unstressing 无应力
unstretched film 未拉伸薄膜
unsupported 无支撑的,无垫衬的
~ adhesive film 无衬胶膜
~ film 无衬薄膜
~ film adhesive 无衬胶膜
~ flexible film 无衬软膜
~ material 无衬材料
~ sheet 无衬片材

untight 未密封的,不紧密的
untrimmed 未切短的;未切齐的;毛边未切去的
~ size 未切边的规格
untwisting (纱)退捻
UNU [United Nations Univeisity] 联合国大学
unwind 退卷
~ with flying splice 飞接退卷
unwinder 退卷机
unwinding 退卷;放卷;解卷
~ process 退卷过程
UP [unsaturated polyester] 不饱和聚酯
~ -based sheet mo(u)lding compound 不饱和聚酯片状模塑料
~ mo(u)lding compounds 不饱和聚酯模塑料
~ -resin 不饱和聚酯树脂
up 向上(的),上面(的)
~ and down stroke (活塞)往复冲程
~ -keep 维修;维修费,保养费
~ -stroke press 上行压力机
~ -travel stop 上行止动装置
upcoiler 卷取机
uperization 超速高温杀菌
upholstery 室内装潢
upper 上面的,上限的
~ backing plate 上垫模板
~ cavity 上模槽
~ consolute temperature 上限会溶温度
~ control limit 控制上限
~ critical solution temperature 上限临界溶解温度
~ cross head 上横梁〈压机〉
~ elbow 上肘管
~ keep plate 上夹板
~ knock-out pin 上脱模销,上顶出销
~ limit 上限
~ lip 上模唇〈平片模头〉
~ part of a mo(u)ld 上模,阳模;模具上部

~ platen 上模板;上压板〈压机〉
~ plunger 上模塞
~ ram 上柱塞〈压机〉
~ - ram transfer mo(u)lding 上柱塞压铸成型法
~ traverse 上横梁
~ yield point 上屈服点
~ yield stress 上屈服应力
~ yoke 上横梁
upset 焊缝凸起处
upstroke 上行式;上行程〈活塞〉
~ press 上压式压机,上动式液压机
uPVC, UPVC [unplasticixed polyvinyl chloride] 未增塑聚氯乙烯
UR [urethan(e)] 聚氨酯;氨基甲酸乙酯
uranium, [U] 铀
uralkyd 聚氨酯改性的醇酸树脂
urea [U] 尿素,脲
  ~ alkyld resin coating 脲醛-醇酸树脂涂料
  ~ bond 尿素键
  ~ - formaldehyde (condensation) resin [UF] 脲甲醛(缩合)树脂
  ~ - formaldyhyde mo(u)lding compounds 脲甲醛模塑料
  ~ formaldehyde mo(u)lding powder 脲甲醛模塑粉
  ~ - formaldehyde plastic 脲甲醛塑料
  ~ - formaldehyde resin [UF] 脲甲醛树脂
  ~ - formaldehyde plastic - laminated sheet 脲醛树脂层合板
  ~ laminated sheet 尿素树脂层合板
  ~ melamine alkyd resincoating 脲三聚氰胺醇酸树脂涂料
  ~ melamine formaldehyde mo(u)ld-ing powder 脲三聚氰胺甲醛模塑粉
  ~ - melamine resin 脲-三聚氰胺树脂
  ~ phosphoric acid 脲磷酸
  ~ plastic 脲醛塑料;尿素塑料

~ resin 脲醛树脂;尿素树脂
~ resinadhesive 脲醛树脂黏合剂
~ resincoating 脲醛树脂涂料
~ resinmo(u)lding compound 脲醛树脂塑料
γ - ureidopropyltriethoxysilane γ - 脲基丙基三乙氧基(甲)硅烷〈偶联剂〉
urethan(e) [UR] 聚氨酯;氨基甲酸酯;氨基甲酸乙酯
~ adhesive 聚氨酯黏合剂
~ bond 氨基甲酸乙酯键
~ coating 聚氨酯涂层,聚氨酯涂料
~ elastic fibre 聚氨酯弹性纤维,氨纶〈俗称〉
~ elastomer 聚氨酯弹性体
~ foams 聚氨酯泡沫塑料,聚氨酯泡沫体
~ modified alkyd resin 聚氨酯改性的醇酸树脂
~ oil (改性)聚氨酯用油
~ plastic 聚氨酯塑料,氨基甲酸酯塑料
~ rubber 聚氨酯橡胶
urotropin 乌洛托品,六亚甲基四胺
usability 使用性能,可适用性
~ test 使用性能试验
usable 适用的,可用的
~ life 适用期,使用寿命
USASI [United States American Standards Institute] 美国标准协会
use 使用;应用;用途
~ characteristic 使用性能
~ - life 使用寿命
~ properties 使用性能
~ ratio 利用率
~ reliability 使用可靠性
~ technology 使用技术
~ testing 使用试验,应用试验
~ value 使用价值
useful life 有效寿命
USM foam mo(u)lding [United Shoe Machinery foam mo(u)lding] 联合制鞋

机械公司发泡成型法,USM 发泡成型法,气体反压注塑法
USM process USM 法,USM 发泡成型法,联合制鞋机械公司发泡成型法
U. S. Pat. [United States Patent] 美国专利
USPO [United States Patent Office] 美国专利局
USS [United States Standard] 美国标准;美国规范
U. S. St. [United States Standard] 美国标准
usual polyester 通用聚酯
UT [Universal Time] 世界时间
utilitarian parameter 应用技术参数
utility mat 碎纤维毡,废纤维毡
utilization 利用,应用
U. T. S. ;UTS [ultimate tensile strength] 极限拉伸强度
UUC foam mo(u)lding [Union Carbide Corporation foam mo(u)lding] 联合碳化物公司发泡成型法,低压发泡注塑成型法
UV [ultraviolet] 紫外线
~ absorber [ultraviolet screening agents] 紫外线吸收剂
~ – light 紫外光(线)
~ – stabilizer 紫外线稳定剂,耐紫外线剂
UW [underwater] 水下的

V

V [vanadium] 钒
V [vinyl] 乙烯基
V [volt] 伏特
V [Vee] V 形(的)
~ – bar mo(u)ld V 形镶条模具
~ – blender V 形掺混机
~ – face die V 形表面口模
~ – joint V 形焊接,V 形接头
~ – notch V 形缺口
~ – notch test bar V 形缺口试棒
V – notched specimen V 形缺口试样
V – type V 形(的)
VA – [vinyl acetate] 乙酸乙烯酯,醋酸乙烯酯
vacublast method 真空喷砂处理法
vacuity 真空度
vacuole 空泡
~ fomiation 形成空泡,形成空腔
vacuum 真空(的)
~ agitator 真空搅拌机
~ air – cushion forming 真空气垫成型
~ air slip forming 气滑成型,真空气胀
包膜成型
~ apparatus 真空设备
~ bag 真空袋
~ bag laminating 真空袋压法
~ bae method 真空袋压法
~ bag mo(u)lding 真空气袋模塑
~ bag technique 真空袋压工艺,真空袋压法
~ bonding 真空黏合
~ break 真空解除
~ breaker 真空解除器
~ calibrating process 真空定径法
~ calibration 真空定型
~ casting 真空铸塑
~ cement 真空用胶泥,真空油灰
~ chamber 真空室;真空箱
~ coating 真空涂布
~ coating by evaporation 真空蒸镀
~ connection 真空连接
~ connection pipe 真空连接管
~ control 真空控制
~ conveyor tube 真空输送管

~ corrector 真空调节器
~ deep drawing 真空深拉(法)
~ deposition 真空沉积;真空蒸镀膜
~ diecasting 真空模铸
~ drawing 真空拉伸
~ drier [diyer] 真空干燥器
~ drum dryer 真空转换干燥器
~ drying 真空干燥
~ duct 真空导管
~ embossing 真空压花,真空压纹;真空模压
~ evaporation 真空蒸发
~ evaporator 真空蒸发器
~ extruder 真空挤塑机
~ feed 真空供料
~ feed hopper 真空供料斗
~ feeder 真空供料器
~ filter 吸滤器,真空过滤器
~ formed 真空成型的
~ former 真空成型机
~ forming 真空成型
~ forming infree-space 无模真空成型
~ forming into a female mo(u)ld 阴模真空成型
~ forming machine 真空成型机
~ forming over a male mo(u)ld 阳模真空成型
~ forming with air-pressure prestretch 气压预拉伸真空成型
~ freeze dryer 真空冷冻干燥器
~ gauge 真空计
~ hopper 真空料斗
~ hopper loader 真空料斗装料器
~ impregnating 真空浸渍
~ injection mo(u)lding 真空注塑成型
~ injection process 真空注塑法
~ kneader 真空捏合机
~ laminating 真空层压,真空层合
~ lamination 真空层压,真空层合
~ line 真空管线
~ manometer 真空压力计

~ melting 真空熔化
~ metallizer 真空镀金装置
~ metallizing 真空镀金,真空金属蒸镀
~ meter 真空计
~ -mo(u)lded pre-fabricated unit 真空成型预加工装置
~ mo(u)lding 真空成型,真空模塑
~ orifice (抽)真空口
~ outer sizing unit 真空外定径装置
~ packaging 真空包装
~ packed 真空包装的
~ pan 真空容器,真空槽
~ pipe sizer 真空管材定径套
~ piston 真空活塞
~ plug-assist forming 模塞助压真空成型
~ polymerization 真空聚合
~ port (抽)真空口
~ pre-expansion 真空预发泡
~ propeller mixer 螺旋桨式真空混合器
~ pump 真空泵
~ regulator 真空调节器
~ relief valve 真空开放阀
~ rotary dryer 旋转式真空干燥器
~ seal 真空封口
~ shelf dryer 真空干燥柜
~ shell 真空罩
~ sizing 真空定型
~ snap-back forming 真空反吸成型
~ snap-back thermoforming 真空反吸热成型
~ spraying 真空喷涂
~ still 真空釜
~ tank 真空箱
~ tank sizing unit 真空箱式定径装置
~ thermoforming 真空热成型
~ -tight 真空密封的
~ -type contact calibrator 真空型定径器,真空型校准器〈用于发泡挤出物〉

~ unit 真空装置
~ valve 真空阀
~ zone 真空段
VAE［vinyl acetate－ethylene copolymer］乙酸乙烯酯－乙烯共聚物
val.［value］值
valcanization 硫化
valcanize 硫化
valcanized gum 硫化胶
valence（化合）从；原子价
~ electron concentration 价电子密度
~ forces 价力
valent 化合价的；原子价的
~ weight 当量
validity check 有效性检查
valley 沟槽，凹部
~ cable 电缆沟
~ printing 压纹印刷，凹面印刷
valuation 评估；评价
value 价值；数值；明度〈色彩〉
valve 阀，活门
~ bag 阀门袋〈包装〉
~ body 阀体
~ control 阀门控制
~ control machanism 气门控制机构
~－gated mould 阀控浇口型模具
~ gating 阀控浇口
~－gating mo(u)ld 阀控浇口式模具
~ gear 阀动齿轮
~ holder 阀座
~ housing 阀套
~ lever 阀杆
~ nut 阀用螺母
~ packing 阀填料
~ seal 阀密封
~ seat 阀座
~ spindle 阀杆
~ stem 阀杆
valved 有阀的
~ extrusion 排气挤塑
~ nozzle 开关式喷嘴，阀控喷嘴
vamp 鞋面，鞋面皮

~ lining 鞋面衬
van der Waals 范德华〈或译范德瓦耳斯〉
~ attractive forces 范德华吸引力
~ equation (of state) 范德华状态方程式
~ forces 分子间力，范德华力
vanadium［V］钒
vanadylic chloride 三氯氧钒〈催化剂〉
vane 叶片，轮叶
vaned 有叶片的
~ disk 有叶转轮，叶轮
~ rotor 有叶转子
vanilla 香草黄〈淡杏黄色〉
vap.［vapor］蒸气
vapometer 气压表，蒸气压力表
vapour［vapor〈美〉］蒸气；蒸发
~ harrier 防潮层，不透气层
~ barrier sheet 防潮层膜片，不透气层膜片
~ cure 蒸气固化
~ degreasing (method) 蒸气脱脂（法）
~ density 蒸气密度
~ deposition［metalizing］蒸气沉积；气相镀金属，高真空蒸镀金属
~ hood 通气柜
~ liquid chromatography 气液色谱法
~ metalizing 汽相镀金属
~ nozzle 通气嘴
~－phase 气相
~－phase chromatography 气相色谱法
~－phase infrared spectrophotometry 气相红外分光光度测定（法）
~－phase inhibitor［VPI］气相（氧化）抑制剂
~－phase osmometer 蒸气压渗透计
~－phase polymerization 气相聚合
~－phase reaction 气相反应
~－phase treatment 气相处理
~ plating 气相镀敷，高真空蒸镀
~ polishing 蒸气抛光

~ pressure 蒸气压(力)
~ transmission 蒸气透过
vaporization 气化(作用)·蒸发(作用)
~ efficiency 蒸发效率
vaporizer 气化器;蒸发器
vaporizing 蒸发
~ space 蒸发空间
~ surface 蒸发面
variability 可变性
variable 可变的
~ area 可变区(域)
~ compression 可变压缩
~ compression ratio 可变压缩比
~ cycle 可变周期
~ density 可变密度
~ distance along helical channel （可）变螺距〈螺杆螺纹〉
~ expansion 可变膨胀
~ gear 变速齿轮
~ gear ratio 可变传动比
~ helical winding 可变螺旋式缠绕;变缠绕角的螺旋式缠绕
~ load 变负荷
~ modulus of elasticity 可变弹性模量
~ orifice die 可变(出口)口模
~ pitch （可)变螺距〈螺杆〉
~ pitch screw 变距螺杆
~ pump 变量泵
~ - speed gear 变速齿轮
~ - speed motion 变速运动;不等速运动
~ speed motor 变速电动机
~ - speed ratio 变速比;可变传动比
~ velocity motion 变速运动
~ viscosity 可变黏度
~ - viscosity adhesive 可变黏度的黏合剂
~ wave - length phase microscope 可变波长相显微镜
variation 变化;偏差
~ in flow 流量变化
~ in rate 速率变化
~ in size 尺寸变化;尺寸偏差
~ of shape 形状变化;变形
variegated 杂色的,斑驳的
variegation 杂色,斑驳
variety 变化;多样性;变形,品种,种类
varnish 清漆;浸液
~ colour [varnish paint] 清漆颜色
~ finish 涂清漆,清漆涂层
~ formation 漆膜形成
~ paint 清漆颜色
remover 漆膜去除剂;除清漆剂
~ resin 清漆树脂;浸渍用树脂
varnished 上清漆过的;浸渍过的
~ cambric 漆布
~ fabric 漆胶织物;浸渍树脂织物〈制层压材料用〉
~ paper 树脂浸渍纸〈制层压材料用〉
~ sheet 树脂浸渍片材
~ web 树脂浸渍基料〈增强塑料〉
varnishing 上清漆;浸渍,清漆
~ machine 清漆浸渍机
~ resin 浸渍(用)树脂
vat 大桶;大盆;容器;(染料)还原物
~ dye 还原染料;瓮染料
VC [vinyl chloride] 氯乙烯
VC/AN [vinyl chloride - acrylonitrile copolymers] 聚乙烯 - 丙烯腈共聚物
VCB [vinyl chloride - butadiene copolymers] 氯乙烯 - 丁二烯共聚物
VC/E [vinyl chloride/ethylene copolymer] 氯乙烯 - 乙烯共聚物
VC/E/MA [vinyl chloride/ethylene/methylacrylate copolymer] 氯乙烯 - 乙烯 - 丙烯酸甲酯共聚物
VC/E/VAC [vinyl chloride/ethylene/vinylacetate copolymer] 氯乙烯 - 乙烯 - 乙酸乙烯酯共聚物
VCI [Verbandder Chemischen Industrie] 化学工业联合会〈德〉
VCM [vinyl chloride monomer] 氯乙烯单体
VC/MA [vinyl chloride/maleic acid copoly-

mer] 氯乙烯-马来酸酐共聚物,氯乙烯-顺酐共聚物
VC/MA [vinyl chloride/methylacrylate copolymer] 氯乙烯-丙烯酸甲酯共聚物
VC/MMA [vinyl chloride/methyl methacrylate copolymer] 氯乙烯-甲基丙烯酸甲酯共聚物
VC/OA [vinyl chloride/octyl acrylate copolymer] 氯乙烯-丙烯酸辛酯共聚物
VC/VAC [vinyl chloride/vinyl acetate copolymer] 氯乙烯-乙酸乙烯酯共聚物,氯乙烯-醋酸乙烯酯共聚物
VC/VDC [vinyl chloride/vinylidene chloride copolymer] 氯乙烯-偏(二)氯乙烯共聚物
VCP [vinyl chloride-propylene copolymere] 氯乙烯-丙烯共聚物
VCU [vinyl chloride-urethane copolymer] 氯乙烯-氨基甲酸酯共聚物
VCVA [vinyl chloride vinyl acetate copolymer] 氯乙烯-乙酸乙烯酯共聚物,氯乙烯-醋酸乙烯酯共聚物
VD [vinylidene] 亚乙烯基
VDC [vinylidene chloride] 偏二氯乙烯,偏氯乙烯
vee　V形的
　~ notch　V形缺口
　~ -trough　V形槽
　~ -type　V形的
vegetable　植物(的)
　~ adhesive　植物黏合剂
　~ dyes　植物染料
　~ fibre　植物纤维
　~ gule　植物胶
　~ mucilage　植物黏质
　~ parchment　植物羊皮纸
　~ protein　植物蛋白
　~ rubber　植物橡胶
vehicle　载色剂,载体;展色剂〈涂料〉;赋形剂〈医药〉;车辆〈以及飞行器、运载火箭等其他运输工具〉
　~ body　车体,车身〈汽车等〉
　~ structure　车辆结构
　~ weight　车辆重量
vel. [velocity] 速度
veil　覆面毡料
velocity　速度
　~ coefficient　速度系数
　~ gradient　速度梯度
　~ limit　速度极限
　~ of impact　冲击速度
　~ of propagation　传播速度
　~ profile　速度分布图
veneer　板坯;薄(片)板;层板,胶合板;镶板;镶面;饰面
　~ board　胶合板
　~ core　胶合板芯
　~ glue　胶合板用胶黏剂
　~ sheet　胶合板,层板
　~ tape　饰面膜带
　~ taping machine　镶嵌薄片板机;铺盖薄片板机
veneering press　胶合板压机
venetian blind　活动百叶窗,软百叶窗
vent　排气口,排气孔,排气通道〈模〉;通风;换气
　~ extruder　排气式挤塑机
　~ hole　通风孔;排气孔
　~ line　排气管线
　~ pipe　排气管;通风管
　~ port　排气孔
　~ section　排气段
　~ -type extruder　排气式挤塑机
　~ valve　通风阀;泄放阀
　~ zone　排气段〈挤塑机〉
vent. [ventilation] 通风
vented　排气的;通风的
　~ barrel　排气式料筒
　~ extruder　排气式挤塑机
　~ injection moulding　排气式注塑成型
　~ plasticizing unit　排气式塑化装置
　~ reciprocating-screw injection machine

排气往复螺杆式注塑机
~ screw 排气式螺杆
~ screw injection mo(u)lding 排气螺杆式注塑成型
ventilate 通风,排气,换气
ventilating 通风的,排气的
~ air 通风空气
~ duct 通风管,通风道
ventilation 通风,排气,换气
venting 排气,放气,脱气
~ seal 通风闸,排气闸
Venturi dispersion plug 文丘里式分散料塞〈注塑机〉
verdigris green 铜绿〈浅暗绿色〉
veridian 绿色颜料
verification 校准;检定;验证
vermiculite 蛭石
vermilion red 朱红
versatile 通用的;多用途的;万能的
~ additive 多用途添加剂
~ adhesive 通用黏合剂;万能黏合剂
~ test equipment 多种用途的试验设备
versatility 通用性;多面性〈用途〉;多功能性
versus [vs] …对…;反对;与…比较
Vert-o-Mix 〈商〉行星式螺杆混合机
vertamix 立式螺旋混合器
vertical 垂直的;直立的;立式的;竖直的
~ burning test 立式耐燃性试验
~ calender 直立式压延机
~ centrifugal machine 立式离心机
~ cylinder dryer 立式滚筒干燥机
~ dryer 立式干燥器
~ evaporator 立(管)式蒸发器
~ extruder 立式挤塑机
~ extrusion 立式挤塑
~ extrusion head 立式挤塑机头
~ feed 垂直加料
~ feed opening 垂直加料口
~ filter 立式滤器
~ flash mo(u)ld 立式溢料式模具

~ flash ring 立式溢料环缝
~ front end drive extruder 前端传动立式挤塑机
~ hydraulic press 立式液压机
~ injection mo(u)lder 立式注塑成型机
~ injection mo(u)lding machine 立式注塑成型机
~ injection press 立式注塑机
~ line 垂直线
~ multipass dryer 立式多程式干燥机
~ plan 俯视图
~ section 垂直剖面
~ streak 纵向条痕〈疵点〉
~ stripe 纵向条纹〈疵点〉
~ surface 垂直面
~ to grain 垂直于木纹〈胶接〉
~ -to-grainbond 垂直于木纹粘接
~ type 立式;直立型
~ two roll calender 立式双辊压延机
~ view 俯视图
~ weld 立焊
~ wet spinning 立式湿纺法
vertically-mounted 垂直装配
very 很,极
~ slight attack,[v.sl] 极微化学侵蚀
~ soluble 极易溶解的
vesicular 多孔的,蜂窝状的
~ tissue 多孔状组织
vessel 容器;船只,舰艇
vet 仔细检查;深入测试
VF,vf [vulcanized fibre] 硬化纸板,纸粕
VF2/HFP [vinylidene fluoride/hexafluoropropylene copolymer] 偏二氟乙烯-六氟丙烯共聚物
v.g.c. [viscosity-gravity constant] 黏度-比重常数
VHF [very high frequency] 超高频
V.I;VI [viscosity index] 黏度指数
vibrate 振动;摆动
vibrating 振动,摆动
~ body 振动体

vibration

~ centre 振动中心
~ chute transporter 振动斜槽输送器
~ conveyor dryer 振动输送干燥器
~ feed chute 振动式加料槽
~ feeder 振动加料器
~ friction welding 振动摩擦焊接
~ mill 振动磨
~ plate 振动板
~ roller 振动辊
~ screen 振动筛
~ stress 振动应力
~ system 振动系统
~ trough 振动槽

vibration 振动
~ conveyor 振动输送器
~ damping 振动阻尼
~ damping properties 振动阻尼性
~ deadening material 防振材料
~ frequency 振动频率
~ injection moulding 振动注塑成型
~ insulator 防振材料
~ period 振动周期
~ -proof 防振的
~ -proof material 防振材料
~ sieve 振动筛
~ slide grinding 振动磨光
~ spectra 振动光谱
~ strength 振动强度
~ stress 振动应力
~ technique 振动技术
~ test equipment 振动试验仪
~ viscometer 振动黏度计
~ welding 振动焊接

vibrational 振动;摆动
~ device 振动装置〈用于挤塑机〉
~ microlaminatioti 振动微层合

vibrator 振动器
~ chute 振动溜槽

vibratory 振动的
~ feeder 振动加料器
~ mill 振动磨
~ welding 振动焊接

Vibro – Energy mill 振动磨
vibrometer 振动计
~ method 振动法〈测动力弹性模量〉
vibromixer 振动混合器
Vicat 维卡
~ needle 维卡针〈测试〉
~ needle indentation test 维卡针压痕试验
~ penetrometer test 维卡针入度试验
~ softening point 维卡软化点
~ softening point test 维卡软化点试验
~ softening temperature [VST] 维卡软化温度

vicidity 黏稠性
vicinal double bond 连位双键
Vickers 维克斯,维氏
~ hardness 维克斯硬度,维氏硬度
~ hardness test 维克斯硬度试验,维氏硬度试验
~ hardness tester 维克斯硬度测试仪,维氏硬度测试仪

video tape 录像磁带,电视录像带
view 视图
vinyl [V] 乙烯基
~ acetate [VA,VAC] 乙酸乙烯酯,醋酸乙烯酯
~ acetate – acrylic ester copolymer 乙酸乙烯酯－丙烯酸酯共聚物
~ acetate – crotonic acid copolymer 乙酸乙烯酯－丁烯酸共聚物
~ acetate – ethylene copolymer [VAE] 乙酸乙烯酯－乙烯共聚物
~ acetate – maleic anhydride copolymer 乙酸乙烯酯－顺丁烯二酐共聚物,醋酸乙烯酯－顺酐共聚物,醋酸乙烯酯－马来酐共聚物
~ acetate plastic 乙酸乙烯酯类塑料,醋酸乙烯酯塑料
~ acetate polymer 乙酸乙烯酯类聚合物,醋酸乙烯酯聚合物
~ acetate resin 乙酸乙烯酯类树脂,醋酸乙烯酯树脂

vinyl

~ acetate resinadhesive 乙酸乙烯酯类树脂黏合剂,醋酸乙烯酯树脂胶黏剂
~ alkyl ether polymer 乙烯基烷基醚聚合物
~ - asbestos floor tile 聚氯乙烯 - 石棉地板
~ benzene 苯乙烯
~ benzoate 乙烯基苯甲酸酯
~ blend 聚氯乙烯掺混料
~ - butyral 乙烯基丁醛
~ butyrate 乙烯基丁酸酯
~ n - butyl ether 乙烯基正丁醚
~ chloride [VC] 氯乙烯
~ chloride - acetate copolymer 氯乙烯 - 乙酸[醋酸] 乙烯酯共聚物
~ chloride - acrylic ester copolymers 氯乙烯 - 丙烯酸酯共聚物
~ chloride - acrylonitrile copolymers [VC/AN] 氯乙烯 - 丙烯腈共聚物
~ chloride - alkyl vinyl ether copolymere 氯乙烯 - 烷乙烯醚共聚物
~ chloride - butadiene copolymers [VCB] 氯乙烯 - 丁二烯共聚物
~ chloride copolymer 氯乙烯共聚物
~ chloride/ethylene copolymer [VC/E] 乙烯 - 氯乙烯共聚物
~ chloride/ ethylene/ methylacrylate, copolymer [VC/E/MA] 氯乙烯 - 乙烯 - 丙烯酸甲酯共聚物
~ chloride/ethylene/vinylacetate copolymer [VC/E/VAC] 氯乙烯 - 乙烯 - 乙酸乙烯酯共聚物
~ chloride/maleic anhydride copoly - mer [VC/MA] 氯乙烯 - 马来酸酐共聚物,氯乙烯 - 顺酐共聚物
~ chloride - raaleic ester copolymer 氯乙烯 - 马来酸共聚物〈增塑糊、涂料和胶黏剂〉
~ chloride/ methylacrylate copolymer [VC/MA] 氯乙烯 - 丙烯酸甲酯共聚物
~ chloride/methyl methaciylate ccpoly - mer [VC/MMA] 氯乙烯 - 甲基丙烯酸甲酯共聚物
~ chloride monomer [VCM] 氯乙烯单体
~ chloride/octyl acrylate copolymer [VC/OA] 氯乙烯 - 丙烯酸辛酯共聚物
~ chloride plastic 氯乙烯类塑料
~ chloride - propylene copolymers [VCP] 氯乙烯 - 丙烯共聚物
~ chloride resin 氯乙烯类树脂
~ chloride rubber 氯乙烯类橡胶
~ chloride - urethane copolymers [VCU] 氯乙烯 - 氨基甲酸酯共聚物
~ chloride/vinyl acetate copolymer, [VC/VA,VC/VAC] 一氯乙烯 - 乙酸乙烯酯共聚物
~ chloride/vinylidene chloride [VC/VDC] 氯乙烯 - 偏二氯乙烯(共聚物)
~ coating 乙烯基涂料;乙烯基涂层
~ copolymer 乙烯系共聚物
~ crotonate 乙烯基丁酸酯
~ elastomer 乙烯系弹性体
~ ester 乙烯酯
~ ethere 乙烯(基)醚类
~ ethyl ether 乙烯基乙基醚
~ fibre 乙烯基系纤维
~ fluoride 氟乙烯
~ formiate 乙烯基甲酸酯
~ isobutyl ether 异丁基乙烯基醚
~ leather 聚氯乙烯人造革
~ leather cloth 聚氯乙烯人造革
~ methy formamide polymer 甲基乙烯基甲酰胺聚合物
~ methyl erher 乙烯基甲基醚
~ modified alkyd resin 乙烯基改性的醇酸树脂
~ plastics 乙烯基塑料
~ polymer 乙烯基聚合物
~ polymerization 乙烯系聚合(作用)
~ propionate 乙烯基丙酸酯
~ pyridine rubber adhesive 乙烯基吡

vinylation

啶橡胶黏合剂
~ pyrrolidone 乙烯基吡咯烷酮
~ resin 乙烯基树脂,聚氯乙烯树脂
~ resin adhesive 聚氯乙烯树脂黏合剂
~ resin coating 聚氯乙烯树脂涂料
~ stearate 乙烯基硬脂酸酯
~ sulfone reactive dyestuff 乙烯砜型活性染料
~ toluene 乙烯基甲苯,甲基苯乙烯
~ triethoxysilane 乙烯基三乙氧基(甲)硅烷
~ - tris(β - methoxyethoxy) silane 乙烯基 - 三(β - 甲氧基乙氧基)(甲)硅烷
~ tris(tert - butyperoxy) silane 乙烯基三(叔丁基过氧)(甲)硅烷

vinylation 乙烯化作用
vinylbenzene 苯乙烯
~ polymers 苯乙烯类聚合物
N' - vinylbaizyl - N - trimethoxysilys propyl ethylenediamine salt N' - 乙烯基苄基 - N - 三甲氧基甲硅烷基丙基乙二胺盐〈偶联剂〉

vinylcarbazole 乙烯基咔唑
vinylidene [VD] 亚乙烯基
vinylidene chloride [VDC] 偏二氯乙烯
~ copolymers 偏二氯乙烯共聚物
~ plastic 偏二氯乙烯塑料
~ resin 偏二氯乙烯树脂
~ - vinyl chloride copolymers 偏二氯乙烯 - 氯乙烯共聚物

vinylidene fluoride 偏二氟乙烯
~ copolymer 偏二氟乙烯共聚物
~ /hexafluoro - propylene copolymer [VF2/HFP] 偏二氟乙烯 - 六氟丙烯共聚物
~ hexafluoropropylene rubber 偏二氟乙烯 - 六氟丙烯橡胶
~ vinyl - acetate copolymer [VDFVA] 偏二氟乙烯 - 乙[醋]酸乙烯酯共聚物
~ - vinyl methyl ether copolymer 偏二氟乙烯 - 甲基乙烯基醚共聚物
~ /vinylidene chloride copolymer [VC/VDC] 偏二氟乙烯 - 偏二氯乙烯共聚物

vinylpyridine - styrene - butadine terpolymer 乙烯基吡啶 - 苯乙烯 - 丁二烯三元共聚物
vinylsiloxane rubber 乙烯基硅橡胶
vinyltriacetoxysilane 乙烯基三乙酰氧基(甲)硅烷〈偶联剂〉
vinyltrichlorosilane 乙烯基三氯(甲)硅烷〈偶联剂〉
vinyltriethoxysilane 乙烯基三乙氧基(甲)硅烷〈偶联剂〉
vinyltris(β - melhoxyethoxy) silane 乙烯基三(β - 甲氧基乙氧基)(甲)硅烷〈偶联剂〉

violet 紫(色)的,紫罗兰(色)的,青莲色的
virgin 纯洁的,新鲜的,原来的
~ material 新料,原料,纯净物料
~ plastic 纯净塑料,原始塑料,新鲜塑料
~ rubber 原胶,新胶,新鲜橡胶

virid 青绿色的
viridescent 淡绿色的
viridity 翠绿
virtual 实际上的,有效的
~ viscosity 有效黏性
vis. [viscosity] 黏度
viscid 黏的;黏滞的
viscidity 黏度;黏性
viscoelastic 黏弹性的
~ behaviour 黏弹特性
~ body 黏弹体
~ creep 黏弹蠕变
~ cross effect 黏弹性交叉效应
~ deformation 黏弹形变
~ flow 黏弹性流动
~ fluid 黏弹性流体
~ materials 黏弹性物质
~ model 黏弹性模型

~ property　黏弹性质能
viscoelasticity　黏弹性
　　~ memory　黏弹性记忆
　　~ meter　黏弹(性测定)仪
viscoelastmnoter　黏弹(性测定)仪
viscometer　黏度计
viscometric degree of polymerization　测黏聚合度
viscometry　黏度测定法
viscose　黏胶;黏胶丝
　　~ adhesive　黏胶黏合剂
　　~ fibre　黏胶纤维
　　~ film　黏胶薄膜,透明纸,玻璃纸〈即赛璐玢〉
　　~ glue　黏胶
　　~ paper　黏胶纸〈即赛璐玢〉
　　~ rayon　黏胶人造丝;黏液丝
　　~ solution　黏胶溶液
　　~ staple fibre　黏胶短纤维
viscosimeter　黏度计
viscosimetry　黏度测定法
　　~ of creep　蠕变黏度测定法
viscosity　黏度;黏性
　　~ after ageing　陈化黏度
　　~ average　黏度平均值
　　~ average molecular weight [My]　黏均分子量
　　~ coefficient　黏度系数
　　~ constant　黏度常数
　　~ control agent　黏度调节剂
　　~ conversion　黏度换算
　　~ cup　黏度杯
　　~ /density ratio　黏度/密度比,比密黏度
　　~ depressant　降黏剂
　　~ drop　黏度下降
　　~ flow　黏流
　　~ gradient　黏度梯度
　　~ index　黏度指数
　　~ method　黏度法
　　~ modifier　黏度调节剂
　　~ number　黏数,黏度值

　　~ path　黏流曲线〈塑料熔体〉
　　~ profile　黏度分布线
　　~ ratio　黏度比,相对黏度
　　~ ratio increment　黏度比增量
　　~ reductant　减黏剂
　　~ relative increment　黏度相对增量
　　~ resistance　黏性阻力
　　~ -shear-stress curve　黏度-剪切应力曲线
　　~ stability　黏度稳定性
　　~ stabilizer　黏度稳定剂
　　~ standard liquid　黏度标准液
　　~ -temperature chart　黏度-温度曲线图
　　~ temperature relation　黏度-温度关系
　　~ test　黏性试验
　　~ tube　黏度管
　　~ unit　黏度单位
viscous　黏滞的;黏性的
　　~ deformation　黏性形变
　　~ elasticity　黏弹性
　　~ flow　黏性流
　　~ fluid　黏性流体
　　~ fracture　黏性破裂
　　~ liquid　黏性液体
　　~ solution　黏性溶液
visible　可见的
　　~ absorption spectrum　可见吸收光谱
　　~ crack　可见裂纹
　　~ defect　外观疵点
　　~ fibre　可见纤维;露丝
　　~ light transmission　可见光透射率
　　~ packaging　可见包装
vision packing　透视包装
viskowaage　黏度秤
visual　视觉的,目视的
　　~ check　观察检查
　　~ examination　外表观察;表观检验
　　~ inspection　目视检查
vitreous　玻璃(状)的;透明的
　　~ silica fiber　高硅氧玻璃纤维;透明

硅石纤维
~ state 玻璃状态;透明状态
vitrification point 玻璃化温度
vitrolite 无色透明有机玻璃;瓷板,瓷砖
VKE [Verband Kunststoff erzeugende Industrie] 塑料制造工业联合会〈德〉
v/mil [volts permil] 伏特/密耳〈介电强度〉
void 空隙;孔隙;空洞〈在非微孔塑料中塑料制品缺陷〉
~ content 空洞率;空隙率
~ extrudate 无气泡挤塑物
~ -free 无孔隙〈模塑件〉;无气孔;密实的
~ laminate 无气泡层合制品
~ volume 孔隙体积;空洞率;空隙率
voided-structure fibre 空隙结构纤维
Voigt 沃伊特
~ element 沃伊特元件
~ model 沃伊特模型
VOL;vol. [volume] 体积,容积,卷,册
Volan finish 沃兰处理剂〈玻璃纤维〉
volatile 挥发(性)的
~ blowing agent 挥发性发泡剂
~ component 挥发性成分
~ constituent 挥发性组分
~ content 挥发分含量,挥发物含量
~ loss 挥发损失
~ loss layer 挥发损失层
~ loss on heating 加热挥发损失
~ matter 挥发物
~ matter loss 挥发物损失
~ solvent 挥发性溶剂
~ substance 挥发物
~ thinner 挥发性稀释剂
~ vehicle 挥发性展色剂
volatility 挥发性
volatilization 挥发(作用)
volatilize 挥发
volt 伏特〈电压单位〉
voltage 电压
~ breakdown 击穿电压

~ rating 额定电压
~ resistance 耐电压性
volume 体积;容积
~ change 体积变化
~ compressibility 体积压缩率
~ contraction 体积收缩
~ defomation 体积形变
~ density 体积密度
~ diffusion 体积扩散
~ elasticity 体积弹性
~ expansibility 体膨胀率
~ expansion 体积膨胀
~ fraction 体积分数
~ loss 体积损耗
~ mo(u)lded per shot 每次注塑成型的体积量
~ modulus of elasticity 体积弹性模量
~ of a batch 批量大小
~ of stroke 注料杆推料体积,工作容积
~ percentage of closed cells 闭孔体积百分率〈泡沫塑料〉
~ plastic 日用型塑料;大宗塑料
~ production 成批生产;大量生产
~ resistance 体积电阻
~ resistivity 体积电阻率
~ shrinkage 体积收缩
~ stability 体积稳定性
~ strain 体积应变
~ swell 体积溶胀
~ -temperature coefficient 体积-温度系数
~ viscoelasticity 体积黏弹性
~ viscosity 体积黏度
~ wear 体积磨耗
~ weight ratio 容积重量比
volumetric 容量的,容积的;体积的
~ capacity 容量
~ efficiency (螺杆)体积效率
~ expansion coefficient 体积膨胀系数
~ extrusion rate 体积挤出速率
~ feed 定容加料

~ feeder　定容加料器
~ feeding　定容加料
~ leakage flow across flights　凹螺棱体积漏流率〈挤塑〉
~ loading　定容加料
~ mo(u)ld shrinkage　体积模压收缩
~ pressure flow along channel　沿螺槽体积压力流
~ rate of discharge　体积出料速率
~ thermoanalysis　容量热分析
~ thermal analysis　容量热分析
vortex　旋涡;涡流
~ characteristic　旋涡特性
V. P. ; v. p. [vapor pressure]　蒸气压
VPE [vulcanized polyethylene]　硫化聚乙烯
VPI [vapour-phase inhibitor]　气相(氧化)抑制剂
VS; v. s [versus]　对……;反对;与……比较
v. sl. [very slight]　极微

VST; [Vicat softening temperature]　维卡软化温度
vulcanise [vulctinize]　硫化;硬化
vulcanite　硬质橡胶
vulcanizate　硫化橡胶;橡皮
vulcanization　硫化;硬化
~ accelerator　硫化促进剂
vulcanizator　硫化剂
vulcanized　硫化的;硬化的
~ elastomeric material　硫化弹性材料
~ fibre [VF]　硬化纸板
~ fibre sheet　硬化纸板材
~ latex　硫化胶乳
~ rubber　硫化橡胶;橡皮
~ sole　硫化底〈鞋〉
~ vegetable oils　硫化植物油〈软化剂〉
vulcanizing　硫化的
~ agent　硫化剂
~ machine　硫化机
~ temperature　硫化温度
~ time　硫化时间

W

W [watt]　瓦特
W [tungsten]　钨
wadding　填塞物
wafer　薄片,垫片
~ -type butterfly valve　片状蝶阀
wainscot　壁板,护墙板
waist　腰部〈瓶〉
Walke-Steel swing hardness tester　沃尔克-斯蒂尔摆杆硬度计
walkway　通道
wall　壁
~ board　墙壁板
~ cloth　壁布
~ covering　墙面涂料,墙面覆盖层
~ deposits　壁上沉积层
~ dispenser　墙壁配件
~ hook　墙壁钩,壁上固定钩

~ inlet　墙壁通入孔〈管〉
~ inlet fitting　墙壁通入孔装配件〈管〉
~ inlet plug　墙壁通入孔塞〈管配件〉
~ panel　墙板
~ panelling　护墙板
~ paper　壁纸
~ primer　墙壁底层涂料
~ stress　墙壁应力
~ thickness　壁厚
~ thickness control　壁厚控制〈吹塑〉
~ tile　墙砖
~ unevenness　壁厚不均
walnut　胡桃〈木,果〉
~ shell flour　胡桃壳粉〈填料〉
wandering spanner　游动扳手
warehouse　仓库
warm　(发)热;(加)热

warmer

- ~ colo(u)r 暖色
- ~ forging 热锻塑
- ~ -setting adhesive 中温固化黏合剂
- ~ spreading 热涂〈黏合剂〉
- ~ up 预热
- ~ up time 升温时间;预热时间

warmer 加热器,加热辊

warming 加热

- ~ up 加热;预热
- ~ up mill 加热辊;预热辊压机
- ~ up roll 加热辊
- ~ up time 预热时间

warp 翘曲,扭曲,弯曲〈塑料制品〉;经纱〈纺〉

- ~ and filling thread 经纬纱
- ~ and weft 经纬纱〈织造中〉
- ~ tape 经纱带材〈并幅布织造〉
- ~ tear resistance 经向耐撕裂性
- ~ thread 经纱〈纺〉

warpage 翘曲,扭曲,弯曲

- ~ test 翘曲试验

warped surface 翘曲面

warping 翘曲,扭曲,弯曲〈塑料制品缺陷〉;整经〈纺〉

- ~ stress 翘曲应力

wash 洗涤,冲洗

- ~ bottle 洗瓶
- ~ coating 反面涂层
- ~ -fast 耐洗的
- ~ off resistance test 耐冲洗试验
- ~ primer 洗涤底漆
- ~ temperature 洗涤温度

washablity 耐洗性

washable 可洗的,耐洗的

washer 洗涤器;衬垫,垫圈

washing tank 洗涤槽

waste 废物,废品;边角料;垃圾

- ~ box 废料箱
- ~ cotton 废棉
- ~ gas 废气
- ~ liquor 废液
- ~ materials 废料
- ~ matter 废物
- ~ mould 废模具;废模型

- ~ paper compression press 废纸压缩机
- ~ paper preparation 废纸预处理
- ~ pipe 排泄管
- ~ plastic 废塑料;再生塑料
- ~ product 废品,废物,废料
- ~ recovering 废料回收
- ~ steam 废蒸汽
- ~ stock 废料
- ~ treatment 废物处理
- ~ water 废水;废液
- ~ water purification 废水净化

water 水

- ~ absorbability 吸水性,吸湿性
- ~ absorbing 吸水的,吸潮的
- ~ absoiption 吸水性;吸水率;吸水量
- ~ bag 水袋
- ~ barrier 防水层
- ~ -based adhesive 水基黏合剂
- ~ bath 水浴,水槽
- ~ -bath flat-film method 水浴平膜法
- ~ boil test 煮沸试验〈硬度测定〉
- ~ channel 水通道,冷却水道〈模具中〉
- ~ collar 水冷套〈模具冷却〉
- ~ column [W.Co] 水柱
- ~ consumption 耗水量
- ~ containing plastic 含水塑料
- ~ containing unsaturated polyester resin 含水不饱和聚酯树脂
- ~ content 水分,含水量;含水率
- ~ cooled former 水冷定型模
- ~ cooled roll 水冷辊
- ~ cooled screw 水冷却螺杆
- ~ cooler 水冷却器
- ~ cooler circulctor 水循环冷却器,循环水冷却系统
- ~ cooling groove 冷却水(流)槽,冷却水通道
- ~ cooling tank 水冷却槽,水冷水罐
- ~ -driven hydropneumatic plant 水压传动装置
- ~ -extended polyesters, [WEP] 水扩充聚酯
- ~ -fast 耐水的;不透水的

- ~ gas 水煤气
- ~ glass 水玻璃；硅酸钠
- ~ immersion stability test 水浸稳定性试验
- ~ impermeability 不透水性
- ~ – in – oil emulsion,［WO］ 水油乳液
- ~ inlet 进水口
- ~ jacket （冷却）水夹套〈模具〉
- ~ jet cutting 喷水切割
- ~ jet pump 水喷泵
- ~ layer 水层
- ~ leakiness 透水性
- ~ line 水管线；水通道,冷却水道〈模具中〉
- ~ mark 水印〈纸〉
- ~ miscibility 水混合度〈酚醛、氨基树脂初期缩合物〉
- ~ outlet 出水口
- ~ paint 水（溶）性涂料,水性漆
- ~ permeability 透水性
- ~ pipe 水管
- ~ pollution 水污染
- ~ preheater 水预处理
- ~ quench 水骤冷,水急冷,水淬
- ~ – quenched film 水骤冷薄膜
- ~ quenching 水浴骤冷
- ~ quenching method 水浴骤冷法
- ~ – reducible［water – thinnable］水可稀释的
- ~ – repellency 憎水性；防水性
- ~ repellent 防水的
- ~ resistance 耐水性
- ~ resistance test 耐水性试验
- ~ – resistant adhesive 耐水性黏合剂
- ~ retention tester 保水性测量仪
- ~ return flow 循环回流水
- ~ – ring pelletizer 水环式切粒机
- ~ shrinkage 缩水率〈织物〉
- ~ sluice valve 排水阀
- ~ soak test 水浸试验
- ~ softening 水软化
- ~ – soluble 水溶（性）的
- ~ – soluble coating 水溶性涂料
- ~ – soluble film 水溶性薄膜
- ~ – soluble paint 水溶性涂料,水溶性漆
- ~ – soluble polymer 水溶性聚合物
- ~ – soluble resin 水溶性树脂
- ~ – soluble urea resinvarnish 水溶性脲醛树脂清漆
- ~ solubility 水溶性
- ~ sorption 吸水作用
- ~ spray 喷水〈冷却挤塑物〉
- ~ stopper 止水器；止水模板
- ~ tank 水槽,水箱,水桶
- ~ – thinnable［water – reducible］水可稀释的
- ~ – tight 不透水的,防渗的
- ~ – tight seal 不透水密封
- ~ tightness 不透水性
- ~ treatment plant 水处理装置
- ~ trough 水槽
- ~ vapour permeability 透水蒸气性
- ~ vapour transmission［water vapour pemeability,WVT］透水蒸气性
- ~ vatpo(u)r transmission rate,［WVTR］透水蒸气率
- ~ way 水通道,水路〈模具中〉
- ~ white 水白色
- **waterborne** 水生的,带水的
  - ~ coating［WB coating］水稀释涂料
  - ~ paint 水性颜料
- **waterless coating** 无水涂布
- **waterproof** 耐水的,防水的,不透水的
  - ~ adhesive 耐水黏合剂
  - ~ sheeting 防水密封带
- **waterproofing** 耐水的,防水的,不透水的
  - ~ agent 防水剂
  - ~ membrane 防水薄膜
- **watery fraction** 含水馏分
- **watt**［W］瓦特〈电功率单位〉
  - ~ density 功率密度
- **wattage** 瓦(特)数
- **wattle – based adhesive** 金合欢树脂黏合剂
- **wave** 波,波浪,波动,波纹

waved
- ~ field 波磁场
- ~ length 波长
- ~ length dispersive spectrometer 波长色散光谱仪
- ~ pattern 波纹图案
- ~ range 波范围

waved 波动的,波纹的
- ~ plate 波纹板
- ~ stress 波动应力

waviness 波纹;波痕
- ~ phenomenon 波痕波纹现象

wax 蜡
- ~ burnishing 蜡油抛光
- ~ paper 蜡纸
- ~ – polyethylene blend 蜡-聚乙烯掺混料,蜡-聚乙烯混配料
- ~ – polymer lubricant 蜡-聚合物润滑剂

waxing 涂蜡,上蜡
waxphenol 蜡酚树脂
way 路(线);手段,方法
WB coating [waterborne coating] 水稀释涂料
W. Co. [water column] 水柱
wdth [width] 宽度
weak 弱的;浅色的
- ~ acid 弱酸
- ~ and power current engineering 弱强电流工程
- ~ boundary layer 弱界面层
- ~ caustic liquor 弱碱液
- ~ colo(u)r 浅色
- ~ current 弱电流
- ~ solvent 弱溶剂
- ~ spot 脆弱点〈织疵〉;浅色斑

weakly 弱(的)
- ~ acidic ion exchange resin 弱酸性离子交换树脂
- ~ basic anion ion exchange fibre 弱碱性阴离子交换纤维
- ~ basic ion exchange resin 离子交换树脂

weakness 弱点,缺点;缺陷〈成型件上〉
wear 磨耗
- ~ hardness 耐磨硬度
- ~ layer 耐磨层〈地板〉
- ~ rate 磨耗速率
- ~ resistance 耐磨性
- ~ resistant property 耐磨性
- ~ surface 耐磨面层〈地板〉
- ~ test 磨耗试验,耐磨试验
- ~ testing machine 磨耗试验机
- ~ trace 磨损痕迹

wearability 耐磨性
wearing 磨耗的,磨损的
- ~ behaviour 磨耗性能
- ~ capacity 耐磨性,磨损量
- ~ plate 耐磨板

weather 天气,气候;风化,老化
- ~ ageing 气候老化
- ~ exposure test 户外暴露试验
- ~ – proof adhesive 耐气候老化黏合剂
- ~ resistance 耐气候老化性
- ~ resistance test 耐气候老化性试验
- ~ resistant adhesive 耐气候老化黏合剂

weatherability 耐候性
- ~ test 耐候试验

weathering 气候老化,自然老化;风蚀
- ~ ageing 气候老化
- ~ resistance 耐气候老化性
- ~ test 耐气候性试验,耐气候老化性试验;风蚀试验

Weather – O – meter 韦瑟 O 型耐候试验仪
weatherometer 耐气候老化试验仪
- ~ – test 耐气候老化试验仪试验

weave 玻璃纤维织物,玻璃布;编织,织造
- ~ of the fabrie 织物的编织
- ~ point 织造点
- ~ structure 织物结构

weaving 编织,织造
- ~ motion 〈横向〉摆动〈焊接〉
- ~ tape 织造带

web 卷材,卷筒材;料片;带条;织物
- ~ centre 料片中心〈拉幅机〉

~ coating 卷筒料涂布
~ dryer 片材状干燥器,多程式干燥器
~ gate 隔膜状浇口
~ guide 料片导向
~ impregnation 料片浸渍
~ -like material 带条状材料
~ plate 筋板
~ -shaped 带条状的
~ -shaped material 带条状材料
~ speed 带材速率[度]
~ stiffening 加强肋
~ storing unit 料片存储器
~ tension 基料张力
~ travel 料片行程
~ -type material 带条状材料
webbing 拉丝性;带条;网纹;熔折,熔塌
webcr 韦伯〈磁通量标准国际单位〉
weblike 带条状的
wedge 楔(状物)
~ gate valve 楔型闸阀
~ -type kerf 楔形劈子
~ -type mo(u)ld 楔型模具
~ -type mo(u)ld clamping system 楔型模具锁模装置系统
~ -type scarfed joint 楔型斜接接头
weft 纬纱
~ thread 纬纱线
~ yarn 纬纱
weigh 称量,定量
~ batcher 称量配料斗
~ batching 称量配料
~ bridge 桥式称,台称,地称
~ bucket 称料斗
~ feeding 定量加料,计重加料
weighed portion 称(出的)量;试量
weigher 称量器
~ self-adjusting 自调式称量器
weighing 称重,称量
~ and closing machine 称量封装机
~ batcher 称量配料斗
~ capacity 称重能力,称重范围
~ machine 称量器
weight 重量
~ accumulator 重负荷蓄力器

~ average degree of polymerization 重均聚合度
~ average molecular weight [Mw] 重均分子
~ batcher 计重供料器
~ batching 计重供料
~ change 重量变化
~ content 重量百分率
~ distribution curve 重量分布曲线
~ distribution function 重量分布函数
~ feed 定重加料,定量加料,计重加料
~ feeder 定量加料器,计重加料器〈美〉
~ feeding 定量加料,计重加料
~ feeding device 定量加料装置,计重加料装置
~ feeding equipment 定量加料设备,计重加料设备
~ fraction crystallinity 重量分数法晶度
~ gain 重量增加
~ -inquantity 称量
~ loading 计重装料
~ loss 重量损失
~ mo(u)lded per shot 每次注塑料重量
~ per cent 重量百分率
~ per unit length 单位长度的重量
~ per unit volume 单位体积的重量
~ saving 减轻重量;节省重量
~ -starved feeding 欠重加料
~ unevenness 重量不匀率
~ uniformity 重量均匀度
~ variation 重量变化
weighting 加重
~ agent 增量剂
~ material 填充物,填料
weightometer 自动秤,自动称重
weir plate 溢流堰板
Weissenberg 韦森贝格
~ extuder 韦森贝格挤塑机
~ effect 韦森贝格效应
~ rheogoniometer [WRG] 韦森贝格流

变仪,锥 - 板旋转流变仪
weld 焊接,熔焊;熔接,熔合;焊缝;焊接点
- ~ area 焊接面,焊接区分界面;熔合面
- ~ bead 焊道,焊接缝凸起处,焊珠,焊蚕
- ~ camber 焊缝蚕
- ~ cracking 焊区开裂(性)
- ~ defect 焊接缺陷
- ~ edge 焊接熔合口
- ~ edgewise 沿边焊接
- ~ face 焊缝(表)面
- ~ factor 焊接效率
- ~ hem 焊(接)边
- ~ joint 焊接接头;焊缝;焊接
- ~ joint factor 焊接因数;焊接相对强度
- ~ line 熔合纹,熔合线,汇流纹,接缝线,焊接线
- ~ manipulator 焊接操纵器
- ~ mark 熔合纹,熔接痕
- ~ opening 焊接坡口
- ~ positioner [weld manipulator] 焊接转换装置
- ~ quality 焊接质量,焊缝质量
- ~ reinforcement 焊缝补强
- ~ riveting 铆接成型,超声波铆接
- ~ seam 焊缝
- ~ shape 焊接形状
- ~ stress 焊接应力
- ~ strength 焊缝强度
- ~ structure 焊接结构
- ~ surface 焊缝(表)面
- ~ test 焊接试验
- ~ throat 焊缝喉部;焊缝高度
- ~ width 焊缝宽度
- ~ zone 焊接区(域);熔合区

weldability 可焊性
- ~ test 可焊性试验

weldable 可焊的
- ~ material 可焊材料
- ~ plastic 可焊塑料
- ~ primer 可焊底漆

weldbonding [weld - bonding] 焊接粘接〈焊接与粘接相结合〉
welded 焊接了的
- ~ blank 焊接坯件
- ~ compensator 焊接补偿器
- ~ construction 焊接结构
- ~ expansion joint [piece] 焊接膨胀补偿器
- ~ joint 焊接接合,焊接
- ~ joint in plastics 塑料焊接(接头)
- ~ junction 焊接(连接)
- ~ seam 焊缝
- ~ sleeve 焊接套筒
- ~ steel frame 焊接钢架

welder 焊(接)机;焊工
- ~ screen 焊工遮光罩
- ~ shield 焊工防护罩

welding 焊接,熔接
- ~ additive 焊接添加料
- ~ and separating edge [welding and cutting edge] 焊割边
- ~ and separating electrode 焊割电极〈高频焊接〉
- ~ anvil 焊砧
- ~ apparatus 焊接器具,焊接机
- ~ area 焊接面,焊接区分界面
- ~ assembly 焊接构件
- ~ base material 焊接基材,焊接用材料
- ~ bead 焊蚕
- ~ bench 焊台
- ~ by both side 双面焊
- ~ condition 焊接条件,焊接参数
- ~ construction 焊接结构
- ~ defect 焊接缺陷
- ~ details 焊接详图
- ~ die 焊模,焊接工具
- ~ direction 焊接方向
- ~ distortion 焊接变形
- ~ drum 焊接转筒
- ~ edge 焊接熔合口
- ~ edge of the die 电极焊边
- ~ electrode 电焊条
- ~ equipment 焊接设备

~ filler 焊条
~ fixture 焊接夹具
~ flash 不理想的焊瘤;焊接凸起外,焊骨
~ frequency 焊接频率
~ groove 焊接坡口
~ gun 焊枪
~ gun tip 焊枪嘴
~ head 焊头
~ head tip 焊头嘴,焊接机嘴
~ jig 焊接夹具
~ joint 焊接(接合)
~ joint strength 焊接接合强度
~ junction [joint] 焊接(接合)
~ layer 焊层
~ line 熔接线,熔合线
~ machine 焊接机
~ material 焊接(填充)材料
~ method 焊接方法
~ nozzle 焊嘴
~ of plastics 塑料焊接
~ output 焊接效率
~ parameter 焊接参数
~ parent material 焊接基材
~ part 焊接件
~ pliers pl 焊接夹钳
~ press 焊接压机
~ pressure 焊接压力
~ primer 焊接底料
~ process 焊接方法
~ rate 焊接速率[度]
~ region 焊接区
~ rod 焊条
~ rollers 焊接滚轮
~ run 焊层;焊缝
~ schedule 焊接规范
~ seam 焊缝
~ sequence 焊接工序
~ set 焊接设备
~ shackle 焊接能量聚集器
~ spline 三角焊条
~ stress 焊接应力
~ table 焊接工作台
~ technique 焊接技术

~ temperature 焊接温度
~ test 可焊接性试验
~ time 焊接时间
~ tip 焊嘴
~ tool 焊接工具
~ torch 气焊嘴,焊(接)炬,焊熔接炬
~ upset 焊接凸起处
~ width 焊接宽度
~ wire 焊丝,焊条
~ with resistance tapes 电阻加热带焊接
~ zone 焊接区(域),熔合区
weldment 焊(接)件
~ damage 焊件损伤
well 井
~ type nozzle 井型注料嘴
welt(ing) 边,缘,贴边;滚边带
Werner-Pfleiderer mixer 维尔纳-普弗莱德勒型混合器
wet 潮湿(的)
~ adhesive bonding 湿黏合
~ ageing 湿法老化,湿(态)老化
~ applicator 喷雾器
~ bag method 湿袋模压法
~ blending 湿渗合
~ bonding 湿法黏合
~ bonding strength 湿态粘接强度
~ colo(u)r 湿(颜)色料
~ density 湿密度
~ electrical property 湿态电性能
~ extrusion 湿态挤塑
~ fastness 耐湿性;湿牢度
~ feed 湿(材)料
~ film 湿膜
~ film thickness 湿膜厚度
~ flex 湿态挠曲
~ flexural strength 湿态弯曲强度
~ grinding 湿磨法
~ heat test 湿热试验
~ hiding power 湿态遮盖力
~ lamainate 湿层合;层合物
~ laminate process 湿层合法
~ laminating 湿态层合
~ lamination 湿态层合

~ laying 湿法成网
~ lay – up 湿铺叠
~ lay – up lamination 湿铺叠层压
~ lay – up mo(u)lding 湿铺叠成型
~ lay – up process 湿铺叠法
~ milling 湿态研磨
~ – mix glue 水混胶料;湿法混合胶料
~ mixing 湿式混合
~ modulus 湿模量
~ moulding 湿法成型
~ – on – wet coating 湿–湿涂布
~ – on – wet painting 湿–湿–涂料
~ – out 打湿,浸透,浸润
~ – out rate 浸润速率
~ – out spraying 喷雾(打湿)
~ process 湿法成型
~ processing 湿处理
~ – proof 防潮的
~ sanding 湿砂磨
~ spinning 湿纺,湿法纺丝
~ stock 湿(材)料
– strength 湿强度〈粘接〉
~ strength of adhesion 湿态黏合强度
~ strength resin 增湿强用树脂
~ surface 湿润(表)面
~ – tack – adhesive 湿性黏合剂
~ tenacity 湿强度
~ tensile strength 湿拉伸强度
~ test 湿试法
~ treatment 湿处理
~ tumbling 湿态滚光
~ waxing 湿涂 [打,上] 蜡
~ winding 湿法缠绕
~ weight 湿重
wetness 潮湿,湿度
wettability 湿润性;润湿性;吸湿性
~ power 润湿率,吸湿能力
~ test 吸湿性试验
wettable polar surface 可湿润极性表面〈粘接件〉
wetted surface 湿润(表)面
wetter 润湿剂
wetting 润湿

~ agent 润湿剂;铺展剂,涂铺剂,展涂剂
~ colo(u)re 润湿性色料
~ force 润湿力
~ heat 润湿热
~ power 润湿力
wetting property 润湿性
wheel 车轮;旋转
~ abrator 喷丸清理机,砂轮清理机
~ mill 轮碾机,碾磨机
~ roll 抛 [磨] 光辊
whirl 旋转;涡流
~ point 旋涡点
~ sinter bath 旋涡烧结浴
~ sinter process 旋涡烧结法
~ sintering 旋涡烧结
whisker 晶须,须晶〈增强剂〉
~ crystal 须状晶体
~ reinforced 须晶增强的
~ reinforced composite 须晶增强复合材料
white 白,白色(的)
~ agent 增白剂
~ black 白炭黑〈二氧化硅,填料〉
~ carbon 白炭黑〈二氧化硅,填料〉
~ deposit 白色沉积物
~ finish 增白处理
~ lac 白色漆
~ lacquer 白色漆
~ liquor 白色液体
~ lead 铅白;白铅粉
~ mica 白云母
~ oil 白油〈溶剂,润滑剂〉
~ point temperature 白化温度点
~ shellac 白虫胶
~ speck 白点
~ sphere 白球
~ spirit 漆用石油溶剂,200 号溶剂油,松香水〈俗称〉
whiteness 白(色);洁白度
whitening 增白
whitewash 白涂料;白垩灰浆;刷白
whiting 白垩粉
Whitworth(screw) thread 惠氏螺纹

whizzer 离心(干燥)机
whole 全(部)的,整(个)的
　~ depth 全齿深(齿轮)
　~ depth circle [root circle] 齿根圆
　~ disc 整圆盘
　~ part impregnation technique 整件浸渍法
wide 宽的,宽阔的
　~ - angle radiator - type multiple heater 广角辐射型多级加热器
　~ film 宽膜
　~ slit die 宽缝模头
　~ strip 宽带材
width 宽度
　~ of flat section 平坦部分宽度〈试样〉
　~ of flight [screw] land 螺纹顶宽〈螺杆〉
　~ of thread 螺纹宽度
　~ overall 总宽度,整体宽度
Willert temperature control system 威勒特温度控制系统
Williams parallel plate plastometer 威廉斯平行板塑度计
wind 卷取,卷绕;风,通风
　~ loading 收卷载荷;风载,风压载荷
　~ nut 叶轮式螺母
　~ off 退卷;放卷
　~ on 缠绕;卷取
　~ screen 挡风板,挡风玻璃〈汽车〉
　~ separator 风分离器
　~ shield 挡风板,挡风玻璃〈汽车〉
　~ tunnel 风道
　~ - up degrees 扭转角的度数
　~ - up device 收卷装置,卷取装置
　~ - up ratio 卷取比;缠绕系数
　~ - up roll 卷取辊
　~ - up station 卷取工位
winder 缠绕机;收卷机,卷取机
　~ block 收卷[卷取]机卷筒
winding 收卷,卷绕,缠绕
　~ angle 缠绕角
　~ biaxial 双轴向缠绕
　~ central angle 缠绕中心角
　~ device 收卷装置,缠绕装置

~ down frame 卷取机构
~ drum 卷取滚筒
~ equipment 收卷装置,缠绕装置
~ machine 收卷机,缠绕机
~ mandrel 收卷芯轴
~ - off device 退卷装置
~ - off speed 退卷速率[度]
~ pattern 缠绕线型
~ rate 缠绕速度
~ rate ratio 缠绕速比
~ reel 卷绕筒
~ regulator 卷绕调节器
~ roller 卷取辊
~ shaft 缠绕轴;卷轴
~ speed 缠绕速率[度],收卷速率[度]
~ spindle 收卷芯轴;卷轴
~ technique 卷取技术
~ tension 缠绕张力
~ - up 收卷,卷取,缠绕
~ - up equipment 卷取设备
~ - up reel 卷取轴筒
~ - up roller 卷辊
~ - up speed ratio 卷取速比
~ - up tension 卷绕张力
~ - up unit 卷取装置
window 亮点,白点〈薄膜〉;窗
~ bag 窗式袋〈包装〉
~ board 窗台
~ fitting 窗配件
~ frame 窗框
~ ledge capping 窗(台)遮光板
~ package 窗包装
~ screen 窗纱
~ spew 亮点
~ strip 窗玻璃密封条
~ visor 遮阳板
~ wiper 车窗刮水器
windshield 挡风板,挡风玻璃〈汽车〉
~ visor 挡风玻璃遮阳板〈汽车〉
~ wiper 挡风玻璃刮水器〈汽车〉
wing 叶片,翼,叶轮;挡泥板〈汽车〉
~ - beater mill 叶片 - 搅拌式研磨机
wipe 擦净

wiper
~ off 擦去,揩净
~ - roll machine 擦辊机
wiper 脱模钳;脱模器;刮水器〈汽车挡风玻璃〉
~ blade 刮板;导向片;刮水器刮片〈汽车挡风玻璃〉
wiping 擦净
~ paint 擦拭涂料
~ solvent 擦拭用溶剂
wire 电线,电缆;金属丝;金属丝网
~ and cable 电线与电缆
~ cloth 金属丝布
~ cloth sieve 金属丝布筛
~ coater 电线包覆机,电线包覆装置
~ coating 电缆包覆
~ - coating compound 电缆包覆料
~ coating enamel 电线涂覆磁漆
~ coating resin 电线包覆树脂
~ conduit 电线导管
~ covering 线缆涂覆
~ - covering compound 线缆涂覆料
~ diameter 电线直径
~ duct 电线套管
~ - gauze 金属丝网
~ helix [wire spiral coil] 金属丝螺旋线圈
~ insulating compound 电线绝缘料
~ insulating ribbon 电线绝缘带
~ mesh 金属网,线网
~ mesh screen 金属网筛
~ resistance strainga(u)ge 线式电阻应变仪
~ rope 金属丝绳、钢丝绳
~ sheathing 导线护套
~ spiral coil 金属丝螺旋线圈
~ strainga(u)ge 线式应变仪
~ train 电线包覆机组
~ tube 电线导管
~ winder 电线收卷机
~ - winding machine 电线收卷机
~ - wound coating rod 线绕涂布杆,梅依尔杆
~ - wound doctor 线绕刮涂器
~ - wound doctor kiss coater 线绕刮刀舔涂机

wiring diagram 布线 [配线,接线,线路,装配] 图
with the machine direction 纵向,与机器同方向
withdraw 退绕
withdrawal 退回,撤回;退绕;旋松;排出
~ roller 退绕辊
withdrawing 放回,退回;退绕
~ device 牵引退绕装置
~ from the mo(u)ld 脱模
withstand 抵抗,耐得住
withstanding 经得起,顶得住
~ fire 耐火的
~ voltage 耐电压的
~ voltage test 耐电压试验
Wittier expander roll 韦脱勒展幅辊
WO [water-in-oil emulsion] 水油乳液
Wöhler stress-cycle diagram 威勒应力循环曲线
wollastonite 硅灰石〈一种硅酸钙矿,填料〉
womtest 耐气候老化试验仪试验
wood 木材
~ adhesive 木材黏合剂
alcohol 甲醇,木醇
~ - base laminate 木基层合板;木基层合制品
~ bonding 木材粘接,木材黏合
~ chipboard 刨花板
~ chips pl 木屑
~ dust 木屑
~ fibre 木纤维
~ filling 木粉填充
~ flour 木粉〈填料〉
~ flour filler 木粉填料
~ flour phenolics 木粉填充酚醛塑料
~ gluing 木材胶黏
~ grainfinishing 木纹装饰
~ graining 木纹装饰
~ meal 木粉〈填料〉
~ oil 木油
~ plastic 木材塑料
~ - plastics composite 木材复合塑料
~ pulp 木(纸)浆

~ pulp paper 木浆纸
~ rosin 木松香(天然树脂)
~ shaving 刨花
~ - to - metal bonding 木材与金属粘接
~ - to - wood adhesive 木材与木材黏合剂,木材黏合剂
~ - to - wood glue 木材与木材胶黏剂,木材胶黏剂〈美〉
~ veneer 木胶合板
~ waste panel 废木板
~ wool 刨花,锯屑
wooden 木(质)的
~ form [pattern] 木型〈制层压塑料〉
~ grid 木质格框,木质木栅
~ jig 木质夹具;木质模型
~ pattern shop 模型车间
work 功;工作,操作,加工
~ bench 工作台
~ capacity 生产能力,工作能力
~ dia 工作直径
~ hardening 加工硬化
~ in - process 在加工中
~ of adhesion 黏附功
~ of rupture 断裂功
~ piece 工件,加工件
~ table 工作台
~ - to - break 断裂功
workbability 可加工性,可塑性
~ of coatings 涂料的加工性能
worked material 回收料;再生料
workers and staff 全体工作人员
working 工作,操作,运转;加工
~ ability 工作能力,加工能力
~ accuracy 加工精度
~ area 工作面积;加工面积
~ atmosphere 工作环境
~ capacity 生产能力,工作量
~ condition 工作条件,使用条件;生产条件,加工条件
~ cycle 工作循环
~ drawing 施工图,生产图,加工图
~ efficiency 工作效率
~ hour 工作小时,工时

~ hours (一个人)工作时间
~ instruction 操作规程
~ life 适用期,使用期,储存期;工作寿命
~ load 工作荷载
~ method 操作方法;加工方法
~ order 操作工序;加工单
~ place 工作地点,现场
~ platen area 模板工作面积
~ pressure 工作压力,操作压力
~ range 工作范围
~ scheme 工作计划,作业进度表
~ specification 操作规程
~ stress 工作应力,允许应用力
~ temperature 工作温度
~ time 运转时间;贴合时间
~ width 加工宽度;工作宽度
works 工厂
~ engineer 工厂工程师
workshop drawing 车间用加工图,车间生产图
worm 螺[蜗]杆,螺纹;旋管,蛇管
~ conveyor 螺旋式输送机
~ die 螺杆式模头
~ extruder 螺杆挤塑机,蜗杆挤塑机
~ feeder 螺旋供料器
~ gear stud 涡轮轴
~ gear (ing) 涡轮(传动)
~ - like chain 螺旋状链
~ - like polymer 螺旋状聚合物
~ of constant pitch 等螺距螺杆
~ pipe 蛇形管
~ piston 螺杆式柱塞;螺杆式注料杆
~ pitch 螺杆螺距
~ shaft 蜗轮轴
~ thread 蜗杆螺纹
worn - out 磨损了的
~ parts 磨损的零件
~ surface 磨损表面
worsted yarn 精纺毛纱
wound 缠绕的
~ article 缠绕制品
~ body 缠绕物件
~ product 缠绕制品

woven 编织的;纺织的
- backing fabric （地毯）编织背衬带
- belt 编织带
- cloth 编织布;玻璃纤维织物〈美〉
- fabric 织物,纺织布
- filaments 编织纤维;(玻璃)纤维织物
- glass fabric 玻璃纤维布
- – glass filament fabric 玻璃长丝织物〈增强材料〉
- – glass roving fabric 玻纤粗纱布
- – glass staple fibre fabric 玻璃短纤维织物〈增强材料〉
- goods 织物,纺织品
- narrow fabric 机织带子,机织狭幅织物
- packaging material 编织包装材料
- plastics filaments 塑料长丝织物
- roving （无捻）粗纱布,(无捻)粗纱织物
- roving fibric （无捻）粗纱布
- scrim 平纹织物
- staple fibres 短纤维织物
- tape 编织带

WPI [World Patents Index] 世界专利索引

wracking 破损;扭变〈包装〉

wrap 包扎,包装;包装纸;抢辊(度)
- angle 包角
- – around transition section （螺杆）等深过渡段
- – up 收卷

wrapped 包装的;缠绕的

- cable 包覆电缆
- tube 缠绕软管

wrapper 包装;包装纸

wrapping 包扎,包装;包装材料
- film 包装薄膜
- machine 包装机
- material 包装材料
- pipe machine 卷管机
- plate 包装板

WRG [Weissenberg rheogoniometer] 韦森伯格流变仪

wringer roll 脱水滚;挤压辊

wrinkle 起皱,皱纹,皱褶,褶皱;折叠
- formation 形成皱纹〈片材〉
- – free 无皱纹的
- mark 皱痕
- – proof 耐皱褶
- recovery 褶皱回复
- – recovery tester 褶皱回复试验仪
- resistance 耐皱性
- – resistant 防皱的〈纺织品〉
- varnish 皱纹清漆

wrinkling 皱纹;起皱

wirite – out meter 自记仪器

wt [watt] 瓦特

wt. [weight] 重量

WVP [water vapor permeability] 水蒸气渗透性

WVT [water vapo(u)r transmission] 透水蒸气性

WVTR [water – vapo(u)r transmission rate] 水蒸气透过率

X

X;x. mol fraction 摩尔分数

X [xenon] 氙

x – alloy X合金〈挤塑机机筒衬里的特硬合金〉

X – ray X – 射线
- absorption X – 射线吸收
- analysis X – 射线分析
- crystallographic analysis X – 射线结晶(结构)分析
- crystallography X – 射线结晶学

~ diagram　X-射线图；伦琴射线图
~ diffraction　X-射线衍射
~ diffaction angle　X-射线衍射角
~ diffraction line　X-射线衍射线条
~ diffraction microscope　X-射线衍射显微镜
~ diffraction pattern　X-射线衍射图像
~ difeactometer　X-射线衍射仪
~ inspection　X-射线检查
~ low-angle scattering　X-射线小角散射
~ microscopy　X-射线显微法
~ pattern　X-射线图像
~ photoelectron spectroscopic study　X-射线光电子谱分析
~ photograph　X-光照片
~ scattering　X-射线散射
~ small-angle scattering　X-射线小角散射
~ spectrometer　X-射线分光计
XABS [elastomeric copolymer from acrylonitrile, butadiene, styrene and a carboxylic group containing monomer] 丙烯腈-丁二烯-苯乙烯和一个含羧基单体的共聚弹性体
xanthene dye　𠮟吨染料
xanthophyllite　绿脆云母
xenon [X]　氙
~ are light ageing　氙弧光老化
~ fademeter　氙光耐晒牢度试验仪
~ lamp　氙灯
~ test　耐氙光照试验
~ test apparatus　耐氙光照试验装置
XLPE [crosslinked polyethylene] 交联聚乙烯
Xn. [number average degree of poly-merization] 数(量平)均聚合度

XNBR [elastomeric copolymer from acrylonitrile, butadiene and a car-boxylic group containing monomer] 丙烯腈-丁二烯和一个含羧基单体的共聚弹性体
XPS [expandable (expanded) polystyrene] 可发性聚苯乙烯
XPS [expanded polystyrene] 发泡聚苯乙烯
XSBR [elastomeric copolymer from styrene, butadiene and a carboxylic group containing monomer] 苯乙烯-丁二烯和一个含羧基单体的共聚弹性体
Xw. [weight average degree of polymer-ization] 重(量平)均聚合度
XWt [experimental weight] 实验重量
p-xylene　对二甲苯
xylene　二甲苯
~ -formaldehyde resin　二甲苯甲醛树脂
~ resin　二甲苯(甲醛)树脂
~ resin adhesive　二甲苯(甲醇)树脂黏合剂
~ resin modified phenolic mo(u)ld-ing powder　二甲苯(甲醛)树脂改性酚醛模塑粉
m- - -diamine　间二甲苯二胺〈固化剂〉
xylenol　二甲酚
~ -formaldehyde resin　二甲酚甲醛树脂
~ resin　二甲酚(甲醛)树脂
xylol　二甲苯
~ -formaldehyde resin　二甲苯甲醛树脂
xylon　木质, 木纤维
xylyl diphenyl phosphate　磷酸二甲苯·二苯酯〈增塑剂〉

Y

Y Y – branch 叉形件;Y形支管,45°支管
Y – check valve [angle – seat check valve] Y形止回阀,斜座止回阀
Y – globe valve. [angle scat valve] Y形球阀,斜座阀
Y – piece Y形件
Y pipe Y形管
Y – type Y形
Yankee dryer 扬基干燥器,筒形干燥器
yarn 纱,丝束,线;细股(绳)
　~ cop 纱管
　~ count 纱线支数
　~ number 纱线支数
　~ oiling unit 线浸油装置
　~ package 纱筒;纱线卷装
　~ strength tester 纱线强力试验仪
　~ tension 纱线张力
Yb [yearbook] 年鉴;年刊
yd [yard] 码
yearly production 年产量
yeast 酵母;泡沫
yel. [yellow] 黄色.
yellow 黄(色);黄(色)的;变黄,泛黄
　~ dyes 黄色染料
　~ index 黄色指数
　~ lithopone 黄立德粉
　~ pigments 黄色颜料
　~ spot 黄斑
yellowing 变黄,泛黄
　~ factor 泛黄系数;黄变度
　~ on ageing 老化黄变
　~ [yellow]resistance 耐泛黄性
yellowish 带黄色的
yellowness 泛黄色
　~ coefficient 黄色系数
　~ index 黄色指数
yield 屈服;产量,收率
　~ behavior 屈服性能
　~ bending moment 屈服弯曲矩
　~ condition 屈服条件
　~ force 屈服力
　~ limit 屈服极限,屈服点
　~ limit curve 屈服极限曲线〈蠕变试验〉
　~ point 屈服点;流动点;击穿点
　~ point elongation 屈服点伸长
　~ point test 屈服点试验
　~ pressure 屈服压力;流压
　~ properties 流动性
　~ strain 屈服应变
　~ strength 屈服强度
　~ stress 屈服应力;流动应力
　~ temperature 屈服温度;流动温度
　~ value 屈服值;流动度值
yielability 可屈服性
yielding 易弯曲的;屈服(性的);流动(性的)
　~ condition 屈服条件
　~ point 屈服点
　~ strain 永久应变;塑性应变
yld [yield] 产率,产量
YIL [yellow indicator lamp] 黄色指示灯
YO [yearly output] 年产量
yoke 模框;横梁
Young's modulus 杨氏模量,弹性模量
　~ inelasticity 杨氏弹性模量
　~ inflexure 杨氏弯曲模量
　~ intension 杨氏拉伸模量
　~ intraction 杨氏牵引模量
　~ of elasticity 杨氏弹性模量
YF [yield point] 屈服点
YP [yield pressure] 屈服压力
YR;yr [year] 年
YS [YS][yield strength] 屈服强度
YSBR [thermoplastics block copolymer from styrene and butadiene] 苯乙烯 – 丁二烯热塑性嵌段共聚物

YXSBR [thermoplastic block copolymer from styrene and butadiene containing carboxylic group] 含有羧基的苯乙烯和丁二烯热塑性嵌段共聚物

Z

Z；z [zero] 零
Z-averagemolecular weight [Mz] Z-均分子量
Z-blade mixer Z形桨叶式混合机；曲拐式搅拌机
Z-calender Z形压延机
Z-shaped blade kneader [sygma type kneader] Z形桨叶捏和机
Z-shaped blade mixer Z形桨叶式混合机
Z-shaped kneader Z形（桨叶式）捏合机
Z-twist （纱）Z形捻，反手捻
Z-type calender Z形（四辊）压延机
zein 玉米蛋白
~ fiber 玉米纤维
~ plastics 玉米塑料〈一种蛋白质塑料，已很少使用〉
zero 零，零点；零度
~ adjuster 零位调节器
~ adjustment 调零，零位调整
~ adjustment control 调零控制
~ balance 零点平衡；零位调整
~ creep 无蠕变
~ damage 无损伤
~ defect 无缺陷
~ deflection 无偏差
~ elongation 无零伸长
~ error 无误差
~ line 基线
~ meter screw 无计量段螺杆
~ point 零点
~ pressure mo(u)lding 无压成型
~ pressure resin 无压树脂
~ setting 调零，零位调整
~ shear viscosity 零剪切黏度
~ strength temperature 零强度温度

~ strength time 零强度时间
~ -strength-time test 零强度-时间试验
~ temperature 零度，基准点温度
~ -twist yam 无捻纱
~ wind 零缠绕，纵向缠绕
Z. F. [zero frequency] 零频率
Ziegler 齐格勒
~ (-Natta) catalyst 齐格勒（-纳塔）催化剂
~ polymerization catalyst 齐格勒聚合催化剂
~ process 齐格勒（聚合）法
zigzag-shaped polymer 锯齿型聚合物
zinc [Zn] 锌
~ acetate 乙酸锌，醋酸锌〈催化剂〉
~ acetate activated carbon catalyst 乙酸锌-活性炭催化剂〈催化剂〉
~ borate 硼酸锌〈催燃剂〉
~ n-butyl xanthate 正丁基黄原酸锌〈促进剂 ZBX〉
~ calcium catalyst 锌钙催化剂〈催化剂〉
~ chloride 氯化锌〈催化剂〉
~ chromate 铬酸锌
~ chrome yellow 梓铬黄〈着色剂〉
~ dibutyldithiocarbamate 二丁基二硫代氨基甲酸锌〈促进剂 BZ〉
~ diethyldithiocarbamate 二乙基二硫代氨基甲酸锌〈促进剂 ZDC〉
~ dimethyldithiocarbamate 二甲基二硫代氨基甲酸锌〈促进剂 PZ〉
~ dust primer 锌粉底涂料，锌粉漆
~ 2-ethylhexoate 2-乙基己酸锌〈热稳定剂〉
~ ethyl phenyl dithiocarbamate 乙基苯基二硫代氨基甲酸锌〈促进剂 PX〉

~ isopropyl xanthate 异丙基黄原酸锌〈促进剂 ZIP〉
~ laurate 月桂酸锌〈热稳定剂〉
~ -2- mercaptobenzimidazole 2-巯基苯并咪唑锌盐〈防老剂 MBZ〉
~ -2- mercaptobenzothiazole 2-巯基苯并噻唑锌盐〈促进剂 MZ〉
~ octate 辛酸锌
~ oxide 氧化锌〈光稳定剂,填料,硫化活性剂〉
~ pentachlorothiophenolate 五氯硫酚锌
~ N-pentamethylene dithiocabamate N-五亚甲基二硫代氨基甲酸锌〈促进剂 ZPD〉
~ salt of 2- mercaptobenzimidazole 2-巯基苯并咪唑锌盐〈抗氧剂〉
~ stearate 硬脂酸锌〈热稳定剂〉
~ white 锌白;氧化锌
zirconium oxide 氧化锆
ZL [zero line] 基准线
zone 区域,区段;地带;层
~ heating 区段加热
~ of conversion 转化段〈螺杆〉
~ of deformation 形变区
~ of dwell 静止区
~ of fracture 断裂区
~ of oxidation 氧化层
~ temperature 区段温度
ZST test [zero-strength-time test] 零强度-时间试验
ZT [zero time] 开始时间

附录一　聚合物常用缩写词

AAS	丙烯腈－丙烯酸酯－苯乙烯三元共聚物
ABR	丙烯酸酯－丁二烯橡胶
ABS	丙烯腈－丁二烯－苯乙烯三元共聚物
ACS	丙烯腈－氯化聚乙烯－苯乙烯(共聚物)
ACS	丙烯腈－苯乙烯共聚物与氯化聚乙烯的热塑性共混物
AES	乙烯－丙烯－苯乙烯－丙烯腈共聚物
AES	丙烯腈－乙烯－丙烯－苯乙烯的热塑性四元共聚物
AS	丙烯腈－苯乙烯共聚物
AVCS	丙烯腈－乙烯咔唑－苯乙烯共聚物
BCM	酚醛(树脂)共聚物
BR	顺丁橡胶
BS	丁二烯－苯乙烯共聚物
BT	聚(1－丁烯)热塑性塑料
CF	甲酚－甲醛树脂
CFRP	碳纤维增强塑料
CLPVC	交联聚氯乙烯
CTFC	三氟氯乙烯
DAIP	间苯二甲酸二烯丙酯树脂
DAP	邻苯二甲酸二烯丙酯树脂
DD	双氰胺甲醛树脂
E－PVC	乳液法聚氯乙烯
EAA	乙烯－丙烯酸共聚物
ECTFE	乙烯－三氟氯乙烯共聚物
EHMWPE	超高分子量聚乙烯
EMA	乙烯－马来酐共聚物

续表

EMAA	乙烯-丙烯酸(甲酯)共聚物
EP	环氧树脂
EP	乙烯-丙烯共聚物
EPD	乙烯-丙烯-二烯三元共聚物
EPR	乙丙橡胶
EPS	可发性聚苯乙烯
EPSAN	乙烯-丙烯-苯乙烯-丙烯腈共聚物
ETFE	乙烯-四氟乙烯共聚物
EVA	乙烯-乙酸乙烯酯共聚物
FEP	氟化乙丙树脂
FRP	(玻璃)纤维增强塑料
FRPP	玻璃纤维增强聚丙烯
FRPVC	玻璃纤维增强聚氯乙烯
FRTP	玻璃纤维增强热塑性塑料
GFP	玻璃纤维增强环氧树脂
GFPA	玻璃纤维增强聚酰胺
GFPP	玻璃纤维增强聚丙烯
GFRP	玻璃纤维增强塑料
GFRTP	玻璃纤维增强热塑性塑料
HDPE	高密度聚乙烯
HIPS	耐高冲击聚苯乙烯
IIR	异丁烯-异戊二烯橡胶,异丁橡胶
LDPE	低密度聚乙烯
LLDPE	线型低密度聚乙烯
LMPE	低相对分子质量聚乙烯
LPPE	低压聚乙烯
MABS	甲基丙烯酸甲酯-丙烯腈-丁二烯-苯乙烯共聚物
MF	三聚氰胺-甲醛树脂

续表

NBR	丁腈橡胶
NCR	腈基氯丁橡胶
NR	天然橡胶
OPP	定向聚丙烯(薄膜)
OPS	定向聚苯乙烯(薄膜)
OPVC	定向聚氯乙烯(薄膜)
PA	聚酰胺,尼龙
PB	聚丁烯
PBAN	聚丁二烯-丙烯腈,丁腈橡胶
PBD	聚丁二烯
PBS	聚丁二烯-苯乙烯,丁苯橡胶
PC	聚碳酸酯
PE	聚乙烯
PEC	氯化聚乙烯
PO	聚烯烃
PP	聚丙烯
PS	聚苯乙烯
PTFE	聚四氟乙烯
PUR	聚氨酯
PVC	聚氯乙烯
PVCA	氯乙烯-乙酸乙烯酯共聚物
PVDC	聚偏(二)氯乙烯
PVDF	聚偏(二)氟乙烯
RP	增强塑料
SAN	苯乙烯-丙烯腈共聚物
SB	苯乙烯-丁二烯共聚物
SBR	丁苯橡胶
SBS	苯乙烯-丁二烯-苯乙烯嵌段共聚物

续表

S·PVC	悬浮法聚氯乙烯
TFCE	三氟氯乙烯
TFE	四氟乙烯
TP	热塑性塑料
FPE	热塑性弹性体
TPO	烯烃热塑性弹性体
TPUR	热塑性聚氨酯
U–PVC	未增塑聚氯乙烯
UF	脲甲醛树脂
UHMW–PE	超高分子量聚乙烯
UP	不饱和聚酯
UR	聚氨酯
VCB	氯乙烯–丁二烯共聚物
VCP	氯乙烯–丙烯共聚物
VCVA	氯乙烯–乙酸乙烯共聚物
XLPE	交联聚乙烯
XPS	可发性聚苯乙烯
YSBR	苯乙烯–丁二烯热塑性嵌段共聚物

附录二　塑料专业术语

国标	标准术语	单位	非标准术语
1033	密度	g/cm^3	比重,比容
1033	相对密度	g/cm^3	比重
1636	表观密度	g/cm^3	松密度,堆积密度,貌视密度
3682	熔体流动速率	g/10min	熔体指数,熔融指数
T2412	等规指数	%	等规度
1040	拉伸强度	MPa	抗张强度
1040	拉伸屈服强度	MPa	抗张强度
1040	拉伸断裂强度	MPa	抗张强度
1040	断裂伸长率	%	伸长率
3941	弯曲强度	MPa	挠曲强度,抗弯强度
9341	弯曲弹性模量	MPa	弹性模量,弹性模数
1843	悬臂梁冲击强度	J/m	抗冲强度
1043	简支梁冲击强度	kJ/m^2	抗冲强度
9639	落镖冲击破损质量	g	落镖冲击强度
11548	落镖中值破坏能量	J	落镖冲击强度
1041	压缩强度	MPa	抗压强度
1130	直角撕裂强度	kN/m	撕裂强度
8808	剥离力	N	剥离强度
1634	热变形温度	℃	
1633	维卡软化点	℃	软化点
5470	冲击脆化温度	℃	脆化温度
T1842	环境应力开裂	h	耐环境应力开裂
1034	吸水性		吸水率
9342	洛氏硬度		硬度

续表

国标	标准术语	单位	非标准术语
2411	邵氏硬度		硬度
2410	透光率	%	透明度
2410	雾度	%	浊度
1408	击穿电压	kV	
1408	击穿强度	kV/mm	介电强度
1409	相对介电系数		介电常数,介质常数
1409	介质损耗角正切		介电损耗角正切
1410	体积电阻系数(体积电阻率)	$\Omega \cdot cm$	体积电阻
1410	表面电阻系数(表面电阻率)	Ω	表面电阻
T6595	鱼眼	个$/1520cm^2$	晶点
3399	导热系数	$W(m \cdot K)$	导热率

附录三 聚合物的容度参数表

聚合物	δ
聚四氟乙烯	6.2
聚二甲基硅氧烷	7.3
丁基橡胶	7.7
聚乙烯	7.9
聚丙烯	8.1
聚异丁烯	8.1
天然橡胶	7.9~8.3
丁苯橡胶	8.1~8.5
无规聚丙烯	8.5
聚苯乙烯	8.6~9.7
聚丁二烯	8.6
聚硫橡胶	9.0~9.4
氯丁橡胶	9.2
丁腈橡胶	9.4~9.5
聚乙酸乙烯酯	9.4
聚甲基丙烯酸甲酯	9.0~9.5
聚氯乙烯	9.5~9.7
聚碳酸酯	9.8
聚偏二氯乙烯	9.8
脲三聚氰胺树脂	9.6~10.1
环氧树脂	9.7~10.9
乙基纤维素	10.3
聚对苯二甲酸乙二醇酯	10.7
硝化纤维素	10.7

续表

聚合物	δ
乙酸纤维素	10.9
酚醛树脂	11.5
聚酰胺	12.7~13.6
尼龙 66	13.6
聚丙烯腈	15.4

附录四 塑料专业常用法定计量单位换算表

项目	法定单位 符号	法定单位 名称或中文符号	原用单位 符号	原用单位 名称或中文符号	换算关系
密度	kg/m^3	千克/米^3	lb/in^3 lb/ft^3	磅/英寸^3 磅/英尺^3	$1lb/in^3 = 27679.9kg/m^3$ $1lb/ft^3 = 16.0185kg/m^3$
拉伸强度			kg/cm^2	千克/厘米^2	$1kg/cm^2 = 9.80665 \times 10^4 Pa$ $= 0.0980665MPa$ $\approx 0.1MPa$
弯曲强度			kg/mm^2	千克/毫米^2	$1kg/mm^2 = 9.80665 \times 10^6 MPa$ $= 9.80665MPa$
压缩强度			kg/m^2	千克/米^2	$1kg/m^2 = 9.80665Pa$
弹性模量	Pa	帕斯卡	$1bf/in^2$ (Psi)	磅力/英寸^2	$1bf/in^2 = 6894.76Pa$ $\approx 6.895kPa$ $\approx 0.006895MPa$
剪切强度			$klbf/in^2$ (ksi)	千磅力/英寸^2	$1klbf/in^2 = 6894760Pa$ $= 6894.76kPa$
布氏硬度			lbf/ft^2	磅力/英尺^2	$1lbf/ft^2 = 47.8803Pa$ $= 0.04788kPa$
冲击强度	kJ/m^2	千焦耳/米^2	$kg \cdot cm/cm^2$	千克·厘米/厘米^2	$1kg \cdot cm/cm^2 = 0.098J/cm^2$ $= 0.1J/cm^2$ $\approx 1kJ/m^2$
冲击强度	kJ/m	千焦耳/米	$kg \cdot cm/cm$	千克·厘米/厘米	$1kg \cdot cm/cm = 9.8J/m$ $= 0.0098kJ/m$ $\approx 0.01kJ/m$
冲击强度	J/m	焦耳/米	$ft \cdot lb/in$	英尺·磅/英寸	$1ft \cdot lb/in = 0.5334J/cm$ $= 53.34J/m$
冲击强度	kJ/m^2	千焦耳/米^2	$ft \cdot lb/in^2$	英尺·磅/英寸^2	$1ft \cdot lb/in^2 = 0.21J/cm^2$ $= 2.1kJ/m^2$

续表

项目	法定单位		原用单位		换算关系
	符号	名称或中文符号	符号	名称或中文符号	
撕裂强度 剥离强度 抗劈强度	N/m	牛顿/米	kgf/cm lbf/in	千克力/厘米 磅力/英寸	1kgf/cm =9.80665N/cm ≈10N/cm =980.665N/m 1lbf/in =175.12677N/m
温度	℃ K	摄氏度 开尔文	℉	华氏度	1℃ =(℉ -32)÷1.8 1K =1℃(当表示温度差和温度间隔时)